CBV500
CARBON DISULFIDE
CAS: 75-15-0 NIOSH: FF 6650000
DOT: 1131
mf: CS_2 mw: 76.13

HR: 3

PROP: Clear, colorless liquid; nearly odorless when pure. Mp: $-110.8°$, bp: $46.5°$, lel: 1.3%, uel: 50%, flash p: $-22°F$ (CC), d: 1.261 @ $20°/20°$, autoign temp: $257°F$, vap press: 400 mm @ $28°$, vap d: 2.64.

SYNS:
CARBON BISULFIDE (DOT) CARBON SULFIDE
CARBON BISULPHIDE CARBON SULPHIDE (DOT)

TOXICITY DATA: CODEN:
mmo-sat 100 μL/plate   NIOSH* 5AUG77
sce-hmn:lym 10200 μg/L   BCTKAG 14,115,81
ihl-man TCLo:40 mg/m^3 (91W   MELAAD 60,566,69
  male):REP
orl-rat TDLo:2 g/kg (6-15D   TOXID9 4,86,84
  preg):TER
ihl-rat TCLo:200 mg/m^3/24H   KHZDAN 21,257,78
  (1-21D preg):REP
orl-hmn LDLo:14 mg/kg   32ZWAA 8,225,74
ihl-hmn LCLo:4000 ppm/30M   29ZWAE -,118,68
ihl-hmn LCLo:2000 ppm/5M   TABIA2 3,231,33
unr-man LDLo:186 mg/kg   85DCAI 2,73,70
orl-rat LD50:3188 mg/kg   GISAAA 31(1),13,66
orl-mus LD50:2780 mg/kg   GSIAAA 31(1),13,66
orl-rbt LD50:2550 mg/kg   GISAAA 31(1),13,66
orl-gpg LD50:2125 mg/kg   GISAAA 31(1),13,66
ipr-gpg LDLo:400 mg/kg   AIHAAP 35,21,74
ihl-mam LCLo:2000 ppm/5M   AEPPAE 138,65,28

Reported in EPA TSCA Inventory. EPA Genetic Toxicology Program. Community Right To Know List. EPA Extremely Hazardous Substances List.

OSHA PEL: TWA 20 ppm; CL 30 ppm; Pk 100 ppm/30M/8H
ACGIH TLV: TWA 10 ppm (skin)
DFG MAK: 10 ppm (30 mg/m^3)
NIOSH REL: TWA 1 ppm; CL 10 ppm/15M

DOT Classification: Flammable Liquid, Label: Flammable Liquid; IMO: Flammable Liquid; Label: Flammable Liquid, Poison

THR: A human poison by ingestion and possibly other routes. Mildly toxic to humans by inhalation. An experimental poison by intraperitoneal route. Human reproductive effects on spermatogenesis by inhalation. An experimental teratogen. Other experimental reproductive effects. Human mutagenic data. The main toxic effect is on the central nervous system, acting as a narcotic and anesthetic in acute poisoning with death following from respiratory failure.

Aluminum powder ignites in CS_2 vapor. The vapor ignites on contact with fluorine. Reacts violently with azides; CsN_3; ClO; ethylamine diamine; ethylene imine; $Pb(N_3)_2$; LiN_3; (H_2SO_4 + permanganates); KN_3; RbN_3; NaN_3; phenylcopper-triphenylphosphine complexes. Incompatible with air; metals; oxidants. To fight fire, use water, CO_2, dry chemical, fog, mist. When heated to decomposition it emits highly toxic fumes of SO_x. For further information, see Vol. 3, No. 5 of *DPIM Report*.

HR: – Hazard rating indicating relative hazard for toxicity, fire, and reactivity with 3 denoting the worst hazard level.
See Introduction: paragraph 3, p. xix

NIOSH: – The National Institute for Occupational Safety and Health's Registry of Toxic Effects of Chemical Substances accession number.
See Introduction: paragraph 5, p. xix

CODEN: – A code which represents the cited reference for the toxicity data. Complete bibliographic citations in CODEN order are listed Section 6, Vol. I.
See Introduction: paragraph 17, p. xxix

Reviews and Status – Here are listed IARC carcinogenic evaluations, NTP carcinogenic test status, EPA Extremely Hazardous Substances List, EPA Community Right To-Know-List, EPA Genetic Toxicology Program. (GENE-TOX), and EPA TSCA status lines.
See Introduction: paragraph 18, p. xxx

Standards and Recommendations – Here are listed the OSHA PEL, ACGIH TLV, DFG MAK, and NIOSH REL workplace air levels. U.S. DOT Classification and labels are also listed.
See Introduction: paragraph 19, p. xxx

Note: For complete entry on Carbon Disulfide see pp. 711–712.

1993 UPDATE

SAX'S DANGEROUS PROPERTIES
—— *of* ——
INDUSTRIAL MATERIALS

Eighth Edition

1993 UPDATE

SAX'S DANGEROUS PROPERTIES —— *of* —— INDUSTRIAL MATERIALS

Eighth Edition

RICHARD J. LEWIS, SR.

VNR VAN NOSTRAND REINHOLD
New York

Copyright © 1994 by Van Nostrand Reinhold

Library of Congress Catalog Card Number 93-39039
ISBN 0-442-01675-1

All rights reserved. No part of this work covered by the copyright hereon may be reproduced or used in any form or by any means— graphic, electronic, or mechanical, including photocopying, recording, taping, or information storage and retrieval systems—without the written permission of the publisher.

I(T)P   Van Nostrand Reinhold is an International Thomson Publishing company. ITP logo is a trademark under license.

Printed in the United States of America

Van Nostrand Reinhold
115 Fifth Avenue
New York, NY 10003

ITP Germany
Königswinterer Str. 418
53227 Bonn
Germany

International Thomson Publishing
Berkshire House, 168–173
High Holborn, London WC1V 7AA
England

International Thomson Publishing Asia
38 Kim Tian Rd., #0105
Kim Tian Plaza
Singapore 0316

Thomas Nelson Australia
102 Dodds Street
South Melbourne 3205
Victoria, Australia

International Thomson Publishing Japan
Kyowa Building, 3F
2-2-1 Hirakawacho
Chiyada-Ku, Tokyo 102
Japan

Nelson Canada
1120 Birchmount Road
Scarborough, Ontario
M1K 5G4, Canada

16 15 14 13 12 11 10 9 8 7 6 5 4 3 2

Library of Congress Cataloging in Publication Data

Lewis, Richard J., Sr.
   1993 update to Sax's dangerous properties of industrial materials
   / Richard J. Lewis, Sr.
      p. cm.
   Includes bibliographical references and index.
   ISBN 0-442-01675-1
   1. Hazardous substances—Handbooks, manuals, etc.   I. Lewis,
Richard J., Sr. Sax's dangerous properties of industrial materials.
II. Title.
T55.3.H3L15 1994
604.7—dc20   93-39039
                                            CIP

Dedicated to GRL, for 1,001 reasons

Contents

Preface	**ix**
Introduction	**xi**
Key to Abbreviations	**xxvii**
Section **1 General Chemical Entries**	**1**
Section **2 CAS Registry Number Cross-Index**	**201**
Section **3 Synonym Cross-Index**	**205**
Section **4 References**	**251**

Preface

This update to the Eighth Edition of *Sax's Dangerous Properties of Industrial Materials* contains over 900 new entries and 200 revised entries.

Most of the new entries were selected because they are on the EPA TSCA Inventory. These are reported to be used in commerce in the United States.

Revised entries were included because they contain revised or new information of all kinds and, specifically, updates to OSHA and ACGIH air standards and recommendations.

A court order has vacated the OSHA Air Standards set in 1989 and contained in 29CFR 1910.1000. OSHA has decided to enforce only pre-1989 air standards. I have elected to include both the Transitional Limits scheduled to be effective on December 31, 1992, and the Final Rule limits, that were scheduled to be effective on September 1, 1989. These represent the current best judgment as to appropriate workplace air levels. While they may not be enforceable by OSHA, they are better guides than the OSHA air standards adopted in 1969.

Each entry concludes with a safety profile, a textual summary of the hazards presented by the entry. The discussion of human exposures lists target organs and specific effects reported.

The book consists of the substance entries and indices to aid in locating a specific entry.

Section 1 contains the CAS Number cross-index for CAS numbers for the listed materials.

Section 2 contains the prime name and synonym cross-index for the listed materials.

Section 3 contains the complete bibliographic references.

Please refer to the Introduction for an explanation of the sources of data and codes used.

Every effort has been made to include the most current and complete information. The author welcomes comments or corrections to the data presented.

Richard J. Lewis, Sr.

Introduction

The list of potentially hazardous materials includes drugs, food additives, preservatives, ores, pesticides, dyes, detergents, lubricants, soaps, plastics, extracts from plant and animal sources, plants and animals that are toxic by contact or consumption, and industrial intermediates and waste products from production processes. Some of the information refers to materials of undefined composition. The chemicals included are assumed to exhibit the reported toxic effect in their pure state unless otherwise noted. However, even in the case of a supposedly "pure" chemical, there is usually some degree of uncertainty as to its exact composition and the impurities that may be present. This possibility must be considered in attempting to interpret the data presented because the toxic effects observed could in some cases be caused by a contaminant. Some radioactive materials are included but the effect reported is the chemically produced effect rather than the radiation effect.

For each entry the following data are provided when available: the DPIM code, hazard rating, entry name; CAS number; DOT number; molecular formula; molecular weight; line structural formula; a description of the material and physical properties; and synonyms. Following this are listed toxicity data with references for reports of primary skin and eye irritation, mutation, reproductive, carcinogenic, and acute toxic dose data. The Consensus Reports section contains, where available, NTP Fifth Annual Report on Carcinogens notation, IARC reviews, NTP Carcinogenesis Testing Program results, EPA Extremely Hazardous Substances List, the EPA Genetic Toxicology Program, and the Community Right-to-Know List. We also indicate the presence of the material on the update of the EPA TSCA inventory of chemicals in use in the United States. The next grouping consists of the U.S. Occupational Safety and Health Administration's (OSHA) permissible exposure levels, the American Conference of Governmental Industrial Hygienists' (ACGIH) Threshold Limit Values (TLVs), German Research Society's (MAK) values, National Institute for Occupational Safety and Health (NIOSH) recommended exposure levels, and U.S. Department of Transportation (DOT) classifications. Each entry concludes with a safety profile that discusses the toxic and other hazards of the entry.

1. *DPIM Entry Code* identifies each entry by a unique code consisting of three letters and three numbers, for example, AAA123. The first letter of the entry code indicates the alphabetical position of the entry. Codes beginning with "A" are assigned to entries indexed with the A's. Each listing in the cross-indexes is referenced to its appropriate entry by the DPIM entry code.

2. *Entry Name* is the name of each material, selected, where possible, to be a commonly used designation.

3. *Hazard Rating* (*HR:*) is assigned to each material in the form of a number (1, 2, or 3) that briefly identifies the level of the toxicity or hazard. The letter "D" is used where the data available are insufficient to indicate a relative rating. In most cases a "D" rating is assigned when only in-vitro mutagenic or experimental reproductive data are available. Ratings are assigned on the basis of low (1), medium (2), or high (3) toxic, fire, explosive, or reactivity hazard.

The number "3" indicates an LD50 below 400 mg/kg or an LC50 below 100 ppm; or that the material is explosive, highly flammable, or highly reactive.

The number "2" indicates an LD50 of 400–4000 mg/kg or an LC50 of 100–500 ppm; or that the material is flammable or reactive.

The number "1" indicates an LD50 of 4000–40,000 mg/kg or an LC50 of 500–5000 ppm; or that the material is combustible or has some reactivity hazard.

4. *Chemical Abstracts Service Registry Number* (*CAS*) is a numeric designation assigned by the American Chemical Society's Chemical Abstracts Service and uniquely identifies a specific chemical compound. This entry allows one to conclusively identify a material regardless of the name or naming system used.

5. *DOT:* indicates a four-digit hazard code assigned by the U.S. Department of Transportation. This code is recognized internationally and is in agreement with the United Nations coding system. The code is used on transport documents, labels, and placards. It is also used to determine the regulations for shipping the material.

INTRODUCTION

6. *Molecular Formula* (*mf:*) or atomic formula (*af:*) designates the elemental composition of the material and is structured according to the Hill System (see *Journal of the American Chemical Society*, 22(8): 478–494, 1900), in which carbon and hydrogen (if present) are listed first, followed by the other elemental symbols in alphabetical order. The formulas for compounds that do not contain carbon are ordered strictly alphabetically by element symbol. Compounds such as salts or those containing waters of hydration have molecular formulas incorporating the CAS dot-disconnect convention. In this convention, the components are listed individually and separated by a period. The individual components of the formula are given in order of decreasing carbon atom count, and the component ratios given. A lower case "x" indicates that the ratio is unknown. A lower case "n" indicates a repeating, polymer-like structure. The formula is obtained from one of the cited references or a chemical reference text, or derived from the name of the material.

7. *Molecular Weight* (*mw:*) or atomic weight (*aw:*) is calculated from the molecular formula, using standard elemental molecular weights (carbon = 12.01).

8. *Structural formula* is a line formula indicating the structure of a given material.

9. *Properties* (*PROP:*) are selected to be useful in evaluating the hazard of a material and designing its proper storage and use procedures. A definition of the material is included where necessary. The physical description of the material may refer to the form, color, and odor to aid in positive identification. When available, the boiling point, melting point, density, vapor pressure, vapor density, and refractive index are given. The flash point, autoignition temperature, and lower and upper explosive limits are included to aid in fire protection and control. An indication is given of the solubility or miscibility of the material in water and common solvents. Unless otherwise indicated, temperature is given in degrees Celsius, and pressure in millimeters of mercury.

10. *Synonyms* for the entry name are listed alphabetically. Synonyms include other chemical names, common or generic names, foreign names (with the language in parentheses), or codes. Some synonyms consist in whole or in part of registered trademarks. These trademarks are not identified as such. The reader is cautioned that some synonyms, particularly common names, may be ambiguous and refer to more than one material.

11. *Skin and Eye Irritation Data* lines include, in sequence, the tissue tested (skin or eye); the species of animal tested; the total dose, and, where applicable, the duration of exposure; for skin tests only, whether open or occlusive; an interpretation of the irritation response severity when noted by the author; and the reference from which the information was extracted. Only positive irritation test results are included.

Materials that are applied topically to the skin or to the mucous membranes can elicit either (a) systemic effects of an acute or chronic nature or (b) local effects, more properly termed "primary irritation." A primary irritant is a material that, if present in sufficient quantity for a sufficient period of time, will produce a nonallergic, inflammatory reaction of the skin or of the mucous membrane at the site of contact. Primary irritants are further limited to those materials that are not corrosive. Hence, concentrated sulfuric acid is not classified as a primary irritant.

a. Primary Skin Irritation. In experimental animals, a primary skin irritant is defined as a chemical that produces an irritant response on first exposure in a majority of the test subjects. However, in some instances compounds act more subtly and require either repeated contact or special environmental conditions (humidity, temperature, occlusion, etc.) to produce a response.

The most standard animal irritation test is the Draize procedure (*Journal of Pharmacology and Experimental Therapeutics*, 82: 377–419, 1944). This procedure has been modified and adopted as a regulatory test by the Consumer Product Safety Commission (CPSC) in 16 CFR 1500.41 (formerly 21 CFR 191.11). In this test a known amount (0.5 mL of a liquid or 0.5 gram of a solid or semisolid) of the test material is introduced under a one-square-inch gauze patch. The patch is applied to the skin (clipped free of hair) of twelve albino rabbits. Six rabbits are tested with intact skin and six with abraded skin. The abrasions are minor incisions made through the stratum corneum, but are not sufficiently deep to disturb the dermis or to produce bleeding. The patch is secured in place with adhesive tape, and the entire trunk of the animal is wrapped with an impervious material, such as rubberized cloth, for a 24-hour period. The animal is immobilized during exposure. After 24 hours the patches are removed and the resulting reaction evaluated for erythema, eschar, and edema formation. The reaction is again scored at the end of 72 hours (48 hours after the initial reading), and the two readings are averaged. A material producing any degree of positive reaction is cited as an irritant.

As the modified Draize procedure described above has become the standard test specified by the U.S. government, nearly all of the primary skin irritation data either strictly adheres to the test protocol or involves only simple modifications to it. When test procedures other than those described above are reported in the literature, appropriate codes are included in the data line to indicate those deviations.

The most common modification is the lack of occlusion of the test patch, so that the treated area is left open to the atmosphere. In such cases the notation "open" appears in the irritation data line. Another frequent modification involves immersion of the whole arm or whole body in the test material or, more commonly, in a dilute aqueous solution of the test material. This type of test is often conducted on soap and detergent solutions. Immersion data are identified by the abbreviation "imm" in the data line.

The dose reported is based first on the lowest dose producing an irritant effect and second on the latest study published. The dose is expressed as follows:

(1) Single application by the modified Draize procedure is indicated by only a dose amount. If no exposure time is given, then the data are for the standard 72-hour test. For test times other than 72 hours, the dose data are given in mg (or another appropriate unit)/duration of exposure, e.g., 10 mg/24H.

(2) Multiple applications involve administration of the dose in divided portions applied periodically. The total dose of test material is expressed in mg (or another appropriate unit)/duration of exposure, with the symbol "I" indicating intermittent exposure, e.g., 5 mg/6D-I.

The method of testing materials for primary skin irritation given in the Code of Federal Regulations does not include an interpretation of the response. However, some authors do include a subjective rating of the irritation observed. If such a severity rating is given, it is included in the data line as mild ("MLD"), moderate ("MOD"), or severe ("SEV"). The Draize procedure employs a rating scheme that is included here for informational purposes only, because other researchers may not categorize irritation response in this manner.

Category	Code	Skin Reaction (Draize)
Slight (Mild)	MLD	Well-defined erythema and slight edema (edges of area well defined by definite raising)
Moderate	MOD	Moderate-to-severe erythema and moderate edema (area raised approximately 1 mm)
Severe	SEV	Severe erythema (beet redness) to slight eschar formation (injuries in depth) and severe edema (raised more than 1 mm and extending beyond area of exposure)

b. Primary Eye Irritation. In experimental animals, a primary eye irritant is defined as a chemical that produces an irritant response in the test subject on first exposure. Eye irritation study procedures developed by Draize have been modified and adopted as a regulatory test by CPSC in 16 CFR 1500.42. In this procedure, a known amount of the test material (0.1 mL of a liquid or 100 mg of a solid or paste) is placed in one eye of each of six albino rabbits; the other eye remains untreated, serving as a control. The eyes are not washed after instillation and are examined at 24, 48, and 72 hours for ocular reaction. After the recording of ocular reaction at 24 hours, the eyes may be further examined following the application of fluorescein. The eyes may also be washed with a sodium chloride solution (U.S.P. or equivalent) after the 24-hour reaction has been recorded.

A test is scored positive if any of the following effects are observed: (1) ulceration (besides fine stippling); (2) opacity of the cornea (other than slight dulling of normal luster); (3) inflammation of the iris (other than a slight deepening of the rugae or circumcorneal injection of the blood vessel); (4) swelling of the conjunctiva (excluding the cornea and iris) with eversion of the eyelid; or (5) a diffuse crimson-red color with individual vessels not clearly identifiable. A material is an eye irritant if four of six rabbits score positive. It is considered a nonirritant if none or only one of six animals exhibits irritation. If intermediate results are obtained, the test is performed again. Materials producing any degree of irritation in the eye are identified as irritants. When an author has designated a substance as either a mild, moderate, or severe eye irritant, this designation is also reported.

The dose reported is based first on the lowest dose producing an irritant effect and second on the latest study published. Single and multiple applications are indicated as described above under "Primary Skin Irritation." Test times other than 72 hours are noted in the dose. All eye irritant test exposures are assumed to be continuous, unless the reference states that the eyes were washed after instillation. In this case, the notation "rns" (rinsed) is included in the data line.

Species Exposed. Since Draize procedures for determining both skin and eye irritation specify rabbits as the test species, most of the animal irritation data are for rabbits, although any of the species listed in Table 2 may be used. We have endeavored to include as much human data as possible, since this information is directly applicable to occupational exposure, much of which comes from studies conducted on volunteers (for example, for cosmetic or soap ingredients) or from persons accidentally exposed. When accidental exposure, such as a spill, is cited, the line includes the abbreviation "nse" (nonstandard exposure). In these cases it is often very difficult to determine the precise amount of the material to which the individual was exposed. Therefore, for accidental exposures an estimate of the concentration or strength of the material, rather than a total dose amount, is generally provided.

12. *Mutation Data* lines include, in sequence, the mutation test system utilized, the species of the tested organism (and, where applicable, the route of administration or cell type), the exposure concentration or dose, and the reference from which the information was extracted.

A mutation is defined as any heritable change in genetic material. Unlike irritation, reproductive, tumorigenic, and toxic dose data, which report the results of whole-animal studies, mutation data also include studies on lower organisms such as bacteria, molds, yeasts, and insects, as well as in-vitro mammalian cell cultures. Studies of plant mutagenesis are not included. No attempt is made to evaluate the significance of the data or to rate the relative potency of the compound as a mutagenic risk to man.

Each element of the mutation line is discussed below.

a. Mutation Test System. A number of test systems are used to detect genetic alterations caused by chemicals. Additional ones may be added as they are reported in the literature. Each test system is identified by the 3-letter code shown in parentheses. For additional information about mutation tests, the reader may wish to consult the *Handbook of Mutagenicity Test Procedures*, edited by B. J. Kilbey, M. Legator, W. Nichols, and C. Ramel (Amsterdam: Elsevier Scientific Publishing Company/North-Holland Biomedical Press, 1977).

(1) The Mutation in Microorganisms (mmo) System utilizes the detection of heritable genetic alterations in microorganisms that have been exposed directly to the chemical.

(2) The Microsomal Mutagenicity Assay (mma) System utilizes an in-vitro technique that allows enzymatic activation of promutagens in the presence of an indicator organism in which induced mutation frequencies are determined.

(3) The Micronucleus Test (mnt) System utilizes the fact that chromosomes or chromosome fragments may not be incorporated into one or the other of the daughter nuclei during cell division.

(4) The Specific Locus Test (slt) System utilizes a method for detecting and measuring rates of mutation at any or all of several recessive loci.

(5) The DNA Damage (dnd) System detects the damage to DNA strands, including strand breaks, crosslinks, and other abnormalities.

(6) The DNA Repair (dnr) System utilizes methods of monitoring DNA repair as a function of induced genetic damage.

(7) The Unscheduled DNA Synthesis (dns) System detects the synthesis of DNA during usually nonsynthetic phases.

(8) The DNA Inhibition (dni) System detects damage that inhibits the synthesis of DNA.

(9) The Gene Conversion and Mitotic Recombination (mrc) System utilizes unequal recovery of genetic markers in the region of the exchange during genetic recombination.

(10) The Cytogenetic Analysis (cyt) System utilizes cultured cells or cell lines to assay for chromosomal aberrations following the administration of the chemical.

(11) The Sister Chromatid Exchange (sce) System detects the interchange of DNA in cytological preparations of metaphase chromosomes between replication products at apparently homologous loci.

(12) The Sex Chromosome Loss and Nondisjunction (sln) System measures the nonseparation of homologous chromosomes at meiosis and mitosis.

(13) The Dominant Lethal Test (dlt). A dominant lethal is a genetic change in a gamete that kills the zygote produced by that gamete. In mammals, the dominant lethal test measures the reduction of litter size by examining the uterus and noting the number of surviving and dead implants.

(14) The Mutation in Mammalian Somatic Cells (msc) System utilizes the induction and isolation of mutants in cultured mammalian cells by identification of the gene change.

(15) The Host-Mediated Assay (hma) System uses two separate species, generally mammalian and bacterial, to detect heritable genetic alteration caused by metabolic conversion of chemical substances administered to host mammalian species in the bacterial indicator species.

(16) The Sperm Morphology (spm) System measures the departure from normal in the appearance of sperm.

(17) The Heritable Translocation Test (trn) measures the transmissibility of induced translocations to subsequent generations. In mammals, the test uses sterility and reduced fertility in the progeny of the treated parent. In addition, cytological analysis of the F1 progeny or subsequent progeny of the treated parent is carried out to prove the existence of the induced translocation. In *Drosophila*, heritable translocations are detected genetically using easily distinguishable phenotypic markers, and these translocations can be verified with cytogenetic techniques.

(18) The Oncogenic Transformation (otr) System utilizes morphological criteria to detect cytological differences between normal and transformed tumorigenic cells.

(19) The Phage Inhibition Capacity (pic) System utilizes a lysogenic virus to detect a change in the genetic characteristics by the transformation of the virus from noninfectious to infectious.

(20) The Body Fluid Assay (bfa) System uses two separate species, usually mammalian and bacterial. The test substance is first administered to the host, from whom body fluid (e.g., urine, blood) is subsequently taken. This body fluid is then tested in-vitro, and mutations are measured in the bacterial species.

b. Species. Those test species that are peculiar to mutation data are designated by the 3-letter codes shown below.

	Code	Species
Bacteria	bcs	*Bacillus subtilis*
	esc	*Escherichia coli*
	hmi	*Haemophilus influenzae*
	klp	*Klebsiella pneumoniae*
	sat	*Salmonella typhimurium*
	srm	*Serratia marcescens*
Molds	asn	*Aspergillus nidulans*
	nsc	*Neurospora crassa*
Yeasts	smc	*Saccharomyces cerevisiae*
	ssp	*Schizosaccharomyces pombe*
Protozoa	clr	*Chlamydomonas reinhardi*
	eug	*Euglena gracilis*
	omi	other microorganisms
Insects	dmg	*Drosophila melanogaster*
	dpo	*Drosophila pseudo-obscura*
	grh	grasshopper
	slw	silkworm
	oin	other insects
Fish	sal	salmon
	ofs	other fish

If the test organism is a cell type from a mammalian species, the parent mammalian species is reported, followed by a colon and the cell type designation. For example, human leukocytes are coded "hmn:leu." The various cell types currently cited in this edition are as follows:

Designation	Cell Type
ast	Ascites tumor
bmr	bone marrow
emb	embryo
fbr	fibroblast
hla	HeLa cell
kdy	kidney
leu	leukocyte
lng	lung
lvr	liver
lym	lymphocyte
mmr	mammary gland
ovr	ovary
spr	sperm
tes	testis
oth	other cell types not listed above

In the case of host-mediated and body-fluid assays, both the host organism and the indicator organism are given as follows: host organism/indicator organism, e.g., "ham/sat" for a test in which hamsters were exposed to the test chemical and *S. typhimurium* was used as the indicator organism.

For in-vivo mutagenic studies, the route of administration is specified following the species designation, e.g., "mus-orl" for oral administration to mice. See Table 1 for a complete list of routes cited. The route of administration is not specified for in-vitro data.

c. Units of Exposure. The lowest dose producing a positive effect is cited. The author's calculations are used to determine the lowest dose at which a positive effect was observed. If the author fails to state the lowest effective dose, two times the control dose will be used. Ideally, the dose should be reported in universally accepted toxicological units such as milligrams of test chemical per kilogram of test animal body weight. Although this is possible in cases where the actual intake of the chemical by an organism of known weight is reported, it is not possible in many systems using insect and bacterial species. In cases where a dose is reported or where the amount can be converted to a dose unit, it is normally listed as milligrams per kilogram (mg/kg). However, micrograms (μg), nanograms (ng), or picograms (pg) per kilogram may also be used for convenience of presentation. Concentrations of gaseous materials in air are listed as parts per hundred (pph), per million (ppm), per billion (ppb), or per trillion (ppt).

Test systems using microbial organisms traditionally report exposure data as an amount of chemical per liter (L) or amount per plate, well, disc, or tube. The amount may be on the basis of weight (g, mg, μg, ng, or pg) or moles (millimoles (mmol), micromoles (μmol), nanomoles (nmol), or picomoles (pmol)). These units describe the exposure concentration rather than the dose actually taken up by the test species. Insufficient data currently exist to permit the development of dose amounts from this information. In such cases, therefore, the material concentration units used by the author are reported.

Because the exposure values reported in host-mediated and body fluid assays are doses delivered to the host organism, no attempt is made to estimate the exposure concentration to the indicator organism. The exposure values cited for host-mediated assay data are in units of milligrams (or other appropriate units of weight) of material administered per kilogram of host body weight or in parts of vapor or gas per million (ppm) parts of air (or other appropriate concentrations) by volume.

13. *Toxicity Dose Data* lines include, in sequence, the route of exposure; the species of animal studied; the toxicity measure; the amount of material per body weight or concentration per unit of air volume and, where applicable, the duration of exposure; a descriptive notation of the type of effect reported; and the reference from which the information was extracted. Only positive toxicity test results are cited in this section.

All toxic-dose data appearing in the book are derived from reports of the toxic effects produced by individual materials. For human data, a toxic effect is defined as any reversible or irreversible noxious effect on the body, any benign or malignant tumor, any teratogenic effect, or any death that has been reported to have resulted from exposure to a material via any route. For humans, a toxic effect is any effect that was reported in the source reference. There is no qualifying limitation on the duration of exposure or for the quantity or concentration of the material, or is there a qualifying limitation on the circumstances that resulted from the exposure. Regardless of the absurdity of the circumstances that were involved in a toxic exposure, it is assumed that the same circumstances could recur. For animal data, toxic effects are limited to the production of tumors, benign (neoplastigenesis) or malignant (carcinogenesis); the production of changes in the offspring resulting from action on the fetus directly (teratogenesis); and death. There is no limitation on either the duration of exposure or on the quantity or concentration of the dose of the material reported to have caused these effects.

The report of the lowest total dose administered over the shortest time to produce the toxic effect was given preference, although some editorial liberty was taken so that additional references might be cited. No restrictions were placed on the amount of a material producing death in an experimental animal nor on the time period over which the dose was given.

Each element of the toxic dose line is discussed below.

a. Route of Exposure or Administration. Although many exposures to materials in the industrial community occur via the respiratory tract or skin, most studies in the published literature report exposures of experimental animals in which the test materials were introduced primarily through the mouth by pills, in food, in drinking water, or by intubation directly into the stomach. The abbreviations and definitions of the various routes of exposure reported are given in Table 1.

INTRODUCTION

Table 1. Routes of Administration to, or Exposure of,
Animal Species to Toxic Substances

Route	Abbreviation	Definition
Eyes	eye	Administration directly onto the surface of the eye. Used exclusively for primary irritation data. See *Ocular*.
Intraaural	ial	Administration into the ear.
Intraarterial	iat	Administration into the artery.
Intracerebral	ice	Administration into the cerebrum.
Intracervical	icv	Administration into the cervix.
Intradermal	idr	Administration within the dermis by hypodermic needle
Intraduodenal	idu	Administration into the duodenum.
Inhalation	ihl	Inhalation in chamber, by cannulation, or through mask.
Implant	imp	Placed surgically within the body location described in reference.
Intramuscular	ims	Administration into the muscle by hypodermic needle.
Intraplacental	ipc	Administration into the placenta.
Intrapleural	ipl	Administration into the pleural cavity by hypodermic needle.
Intraperitoneal	ipr	Administration into the peritoneal cavity.
Intrarenal	irn	Administration into the kidney.
Intraspinal	isp	Administration into the spinal canal.
Intratracheal	itr	Administration into the trachea.
Intratesticular	itt	Administration into the testes.
Intrauterine	iut	Administration into the uterus.
Intravaginal	ivg	Administration into the vagina.
Intravenous	ivn	Administration directly into the vein by hypodermic needle.
Multiple	mul	Administration into a single animal by more than one route.
Ocular	ocu	Administration directly onto the surface of the eye or into the conjunctival sac. Used exclusively for systemic toxicity data.
Oral	orl	Per os, intragastric, feeding, or introduction with drinking water.
Parenteral	par	Administration into the body through the skin. Reference cited is not specific about the route used. Could be ipr, scu, ivn, ipl, ims, irn, or ice.
Rectal	rec	Administration into the rectum or colon in the form of enema or suppository.
Subcutaneous	scu	Administration under the skin.
Skin	skn	Application directly onto the skin, either intact or abraded. Used for both systemic toxicity and primary irritant effects.
Unreported	unr	Dose, but not route, is specified in the reference.

b. Species Exposed. Because the effects of exposure of humans are of primary concern, we have indicated, when available, whether the results were observed in man, woman, child, or infant. If no such distinction was made in the reference, the abbreviation "hmn" (human) is used. However, the results of studies on rats or mice are the most frequently reported and hence provide the most useful data for comparative purposes. The species and abbreviations used in reporting toxic dose data are listed alphabetically in Table 2.

c. Description of Exposure. In order to describe better the administered dose reported in the literature, six abbreviations are used. These terms indicate whether the dose caused death (LD) or other toxic effects (TD) and whether it was administered as a lethal concentration (LC) or toxic concentration (TC) in the inhaled air. In general, the term "Lo" is used where the number of subjects studied was not a significant number from the population or the calculated percentage of subjects showing an effect was listed as 100. The definition of terms is as follows:

*TDLo—Toxic Dose Low—*the lowest dose of a material introduced by any route, other than inhalation, over any given period of time and reported to produce any toxic effect in humans or to produce carcinogenic, neoplastigenic, or teratogenic effects in animals or humans.

*TCLo—Toxic Concentration Low—*the lowest concentration of a material in air to which humans or animals have been exposed for any given period of time that has produced any toxic effect in humans or produced a carcinogenic, neoplastigenic, or teratogenic effect in animals or humans.

*LDLo—Lethal Dose Low—*the lowest dose (other than LD50) of a material introduced by any route, other than inhalation, over any given period of time in one or more divided portions and reported to have caused death in humans or animals.

TABLE 2. Species
(with assumptions for toxic dose calculation from non-specific data*)

Species	Abbrev.	Age	Weight	Consumption (Approx.) Food (g/day)	Consumption (Approx.) Water (mL/day)	1 ppm in Food Equals, in (mg/kg/day)	Approximate Gestation Period (days)
Bird–type not specified	brd		1 kg				
Bird–wild bird species	bwd		40 gm				
Cat, adult	cat		2 kg	100	100	0.05	64 (59–68)
Child	chd	1–13 Y	20 kg				
Chicken, adult	ckn	8 W	800 gm	140	200	0.175	
Cattle	ctl		500 kg	10,000		0.02	284 (279–290)
Duck, adult (domestic)	dck	8 W	2.5 kg	250	500	0.1	
Dog, adult	dog	52 W	10 kg	250	500	0.025	62 (56–68)
Domestic animals (Goat, Sheep)	dom		60 kg	2,400		0.04	G: 152 (148–156) S: 146 (144–147)
Frog, adult	frg		33 gm				
Guinea Pig, adult	gpg		500 gm	30	85	0.06	68
Gerbil	grb		100 gm	5	5	0.05	25 (24–26)
Hamster	ham	14 W	125 gm	15	10	0.12	16 (16–17)
Human	hmn	Adult	70 kg				
Horse, Donkey	hor		500 kg	10,000		0.02	H: 339 (333–345) D: 365
Infant	inf	0–1 Y	5 kg				
Mammal (species unspecified in reference)	mam		200 gm				
Man	man	Adult	70 kg				
Monkey	mky	2.5 Y	5 kg	250	500	0.05	165
Mouse	mus	8 W	25 gm	3	5	0.12	21
Non-mammalian species	nml						
Pigeon	pgn	8 W	500 gm				
Pig	pig		60 kg	2,400		0.041	114 (112–115)
Quail (laboratory)	qal		100 gm				
Rat, adult female	rat	14 W	200 gm	10	20	0.05	22
Rat, adult male	rat	14 W	250 gm	15	25	0.06	
Rat, adult	rat	14 W	200 gm	15	25		
Rat, weanling	rat	3 W	50 gm	15	25	0.3	
Rabbit, adult	rbt	12 W	2 kg	60	330	0.03	31
Squirrel	sql		500 gm				44
Toad	tod		100 gm				
Turkey	trk	18 W	5 kg				
Woman	wmn	Adult	50 kg				270

Values given in Table 2 are within reasonable limits usually found in the published literature and are selected to facilitate calculations for data from publications in which toxic dose information has not been presented for an individual animal of the study. See, for example, *Association of Food and Drug Officials, Quarterly Bulletin,* volume 18, page 66, 1954; Guyton, *American Journal of Physiology,* volume 150, page 75, 1947; *The Merck Veterinary Manual* 5th Edition, Merck & Co., Inc., Rahway, N.J., 1979; and The UFAW *Handbook on the Care and Management of Laboratory Animals,* 4th Edition, Churchill Livingston, London, 1972. Data for lifetime exposure are calculated from the assumptions for adult animals for the entire period of exposure. For definitive dose data, the reader must review the referenced publication.

LD50—Lethal Dose Fifty—a calculated dose of a material that is expected to cause the death of 50% of an entire defined experimental animal population. It is determined from the exposure to the material, by any route other than inhalation, of a significant number from that population. Other lethal dose percentages, such as LD1, LD10, LD30, and LD99, may be published in the scientific literature for the specific purposes of the author. Such data would be published if these figures, in the absence of a calculated lethal dose (LD50), were the lowest found in the literature.

LCLo—Lethal Concentration Low—the lowest concentration of a material in air, other than LC50, that has been reported to have caused death in humans or animals. The reported concentrations may be entered for periods of exposure that are less than 24 hours (acute) or greater than 24 hours (subacute and chronic).

LC50—Lethal Concentration Fifty—a calculated concentration of a material in air, exposure to which for a specified length of time is expected to cause the death of 50% of an entire defined experimental animal population. It is determined from the exposure to the material of a significant number from that population.

The following table summarizes the above information.

d. Units of Dose Measurement. As in almost all experi-

INTRODUCTION

xviii

Category	Exposure Time	Route of Exposure	TOXIC EFFECTS Human	Animal
TDLo	Acute or Chronic	All except Inhalation	Any Non-Lethal	CAR, NEO, NEO, ETA, TER, REP
TCLo	Acute or Chronic	Inhalation	Any Non-Lethal	CAR, NEO, ETA, TER, REP
LDLo	Acute or Chronic	All except Inhalation	Death	Death
LD50	Acute	All except Inhalation	Not Applicable	Death (Statistically Determined)
LCLo	Acute or Chronic	Inhalation	Death	Death
LC50	Acute	Inhalation	Not Applicable	Death (Statistically Determined)

mental toxicology, the doses given are expressed in terms of the quantity administered per unit body weight, or quantity per skin surface area, or quantity per unit volume of the respired air. In addition, the duration of time over which the dose was administered is also listed, as needed. Dose amounts are generally expressed as milligrams (thousandths of a gram) per kilogram (mg/kg). In some cases, because of dose size and its practical presentation in the file, grams per kilogram (g/kg), micrograms (millionths of a gram) per kilogram (µg/kg), or nanograms (billionths of a gram) per kilogram (ng/kg) are used. Volume measurements of dose were converted to weight units by appropriate calculations. Densities were obtained from standard reference texts. Where densities were not readily available, all liquids were assumed to have a density of one gram per milliliter. Twenty drops of liquid are assumed to be equal in volume to one milliliter.

All body weights have been converted to kilograms (kg) for uniformity. For those references in which the dose was reported to have been administered to an animal of unspecified weight or a given number of animals in a group (e.g., feeding studies) without weight data, the weights of the respective animal species were assumed to be those listed in Table 2 and the dose is listed on a per-kilogram body-weight basis. Assumptions for daily food and water intake are found in Table 2 to allow approximation of doses for humans and species of experimental animals in cases in which the dose was originally reported as a concentration in food or water. The values presented are selections that are reasonable for the species and convenient for dose calculations.

Concentrations of a gaseous material in air are generally listed as parts of vapor or gas per million parts of air by volume (ppm). However, parts per hundred (pph or percent), parts per billion (ppb), or parts per trillion (ppt) may be used for convenience of presentation. If the material is a solid or a liquid, the concentrations are listed preferably as milligrams per cubic meter (mg/m^3) but may, as applicable, be listed as micrograms per cubic meter ($\mu g/m^3$), nanograms per cubic meter (ng/m^3), or picograms (trillionths of a gram) per cubic meter (pg/m^3) of air. For those cases in which other measurements of contaminants are used, such as the number of fibers or particles, the measurement is spelled out.

Where the duration of exposure is available, time is presented as minutes (M), hours (H), days (D), weeks (W), or years (Y). Additionally, continuous (C) indicates that the exposure was continuous over the time administered, such as ad-libitum feeding studies or 24-hour, 7-day-per-week inhalation exposures. Intermittent (I) indicates that the dose was administered during discrete periods, such as daily or twice weekly. In all cases, the total duration of exposure appears first after the kilogram body weight and a slash, and is followed by descriptive data; e.g., 10 mg/kg/3W-I indicates ten milligrams per kilogram body weight administered over a period of three weeks, intermittently in a number of separate, discrete doses. This description is intended to provide the reader with enough information for an approximation of the experimental conditions, which can be further clarified by studying the reference cited.

e. Frequency of Exposure. Frequency of exposure to the test material depends on the nature of the experiment. Frequency of exposure is given in the case of an inhalation experiment, for human exposures (where applicable), or where CAR, NEO, ETA, REP, or TER is specified as the toxic effect.

f. Duration of Exposure. For assessment of tumorigenic effect, the testing period should be the life span of the animal, or should extend until statistically valid calculations can be obtained regarding tumor incidence. In the toxic dose line, the total dose causing the tumorigenic effect is given. The duration of exposure is included to give an indication of the testing period during which the animal was exposed to this total dose. For multigenerational studies, the time during gestation when the material was administered to the mother is also provided.

g. Notations Descriptive of the Toxicology. The toxic dose line thus far has indicated the route of entry, the species involved, the description of the dose, and the amount of the dose. The next entry found on this line when a toxic exposure (TD or TC) has been listed is the toxic effect. Following a colon will be one of the notations found in Table 3. These notations indicate the organ system affected or special effects that the material produced, e.g., TER = teratogenic effect. No attempt was made to be definitive in reporting these effects because such definition requires detailed qualification that is beyond the scope of this book. The selection of the dose was based first on the lowest dose producing an effect and second on the latest study published.

14. *Reproductive Effects Data* lines include, in sequence, the reproductive effect reported, the route of exposure, the species of animal tested, the type of dose, the total dose

TABLE 3. Notations Descriptive of the Toxicology

Notation	Effects (not limited to effects listed)
ALR	Allergic systemic reaction such as might be experienced by individuals sensitized to penicillin.
BCM	Blood clotting mechanism effects—any effect that increases or decreases clotting time.
BLD	Blood effects—effect on all blood elements, electrolytes, pH, proteins, oxygen carrying or releasing capacity.
BPR	Blood pressure effects—any effect that increases or decreases any aspect of blood pressure.
CAR	Carcinogenic effects—see paragraph 9(g) in text.
CNS	Central nervous system effects—includes effects such as headaches, tremor, drowsiness, convulsions, hypnosis, anesthesia.
COR	Corrosive effects—burns, desquamation.
CUM	Cumulative effects—where material is retained by the body in greater quantities than is excreted, or the effect is increased in severity by repeated body insult.
CVS	Cardiovascular effects—such as an increase or decrease in the heart activity through effect on ventricle or auricle; fibrillation; constriction or dilation of the arterial or venous system.
DDP	Drug dependence effects—any indication of addiction or dependence.
ETA	Equivocal tumorigenic agent—see text.
EYE	Eye effect—sirritation, diploplia, cataracts, eye ground, blindness by effects to the eye or the optic nerve.
GIT	Gastrointestinal tract effects—diarrhea, constipation, ulceration.
GLN	Glandular effects—any effect on the endocrine glandular system.
IRR	Irritant effects—any irritant effect on the skin, eye, or mucous membrane.
MLD	Mild irritation effects—used exclusively for primary irritation data.
MMI	Mucous membrane effects—irritation, hyperplasia, changes in ciliary activity.
MOD	Moderate irritation effects—used exclusively for primary irritation data.
MSK	Musculo-skeletal effects—such as osteoporosis, muscular degeneration.
NEO	Neoplastic effects—see text.
PNS	Peripheral nervous system effects.
PSY	Psychotropic effects—exerting an effect upon the mind.
PUL	Pulmonary system effects—effects on respiration and respiratory pathology.
RBC	Red blood cell effects—includes the several anemias.
REP	Reproductive effects—see text.
SEV	Severe irritation effects—used exclusively for primary irritation data.
SKN	Skin effects—such as erythema, rash, sensitization of skin, petechial hemorrhage.
SYS	Systemic effects—effects on the metabolic and excretory function of the liver or kidneys.
TER	Teratogenic effects—non-transmissible changes produced in the offspring.
UNS	Unspecified effects—the toxic effects were unspecific in the reference.
WBC	White blood cell effects—effects on any of the cellular units other than erythrocytes, including any change in number or form.

amount administered, the time and duration of administration, and the reference from which the information was extracted. Only positive reproductive effects data for mammalian species are cited. Because of differences in the reproductive systems among species and the systems' varying responses to chemical exposures, no attempt is made to extrapolate animal data or to evaluate the significance of a substance as a reproductive risk to humans.

Each element of the reproductive effects data line is discussed below.

a. Reproductive Effect. For human exposure, the effects are included in the safety profile. The effects include those reported to affect the male or female reproductive systems, mating and conception success, fetal effects (including abortion), transplacental carcinogenesis, and post-birth effects on parents and offspring.

b. Route of Exposure or Administration. See Table 1 for a complete list of abbreviations and definitions of the various routes of exposure reported. For reproductive effects data, the specific route is listed either when the substance was administered to only one of the parents or when the substance was administered to both parents by the same route. However, if the substance was administered to each parent by a different route, the route is indicated as "mul" (multiple).

c. Species Exposed. Reproductive effects data are cited for mammalian species only. Species abbreviations are shown in Table 2. Also shown in Table 2 are approximate gestation periods.

d. Type of Exposure. Only two types of exposure, TDLo and TCLo, are used to describe the dose amounts reported for reproductive effects data.

e. Dose Amounts and Units. The total dose amount that

INTRODUCTION

was administered to the exposed parent is given. If the substance was administered to both parents, the individual amounts to each parent have been added together and the total amount shown. Where necessary, appropriate conversion of dose units has been made. The dose amounts listed are those for which the reported effects are statistically significant. However, human case reports are cited even when no statistical tests can be performed. The statistical test is that used by the author. If no statistic is reported, a Fisher's Exact Test is applied with significance at the 0.05 level, unless the author makes a strong case for significance at some other level.

Dose units are usually given as an amount administered per unit body weight or as parts of vapor or gas per million parts of air by volume. There is no limitation on either the quantity or concentration of the dose or the duration of exposure reported to have caused the reproductive effect(s).

f. Time and Duration of Treatment. The time when a substance is administered to either or both parents may significantly affect the results of a reproductive study, because there are differing critical periods during the reproductive cycles of each species. Therefore, to provide some indication of when the substance was administered, which should facilitate selection of specific data for analysis by the user, a series of up to four terms follows the dose amount. These terms indicate to which parent(s) and at what time the substance was administered. The terms take the general form:

(uD male/vD pre/w-xD preg/yD post)

where u = total number of days of administration to male prior to mating

v = total number of days of administration to female prior to mating

w = first day of administration to pregnant female during gestation

x = last day of administration to pregnant female during gestation

y = total number of days of administration to lactating mother after birth of offspring

If administration is to the male only, then only the first of the above four terms is shown following the total dose to the male, e.g., 10 mg/kg (5D male). If administration is to the female only, then only the second, third, or fourth term, or any combination thereof, is shown following the total dose to the female, for example:

10 mg/kg (3D pre)
10 mg/kg (3D pre/4-7D preg)
10 mg/kg (3D pre/4-7D preg/5D post)
10 mg/kg (3D pre/5D post)
10 mg/kg (4-7D preg)
10 mg/kg (4-7D preg/5D post)
10 mg/kg (5D post) (NOTE: This example indicates administration was only to the lactating mother, and only after birth of the offspring.)

If administration is to both parents, then the first term and any combination of the last three terms are listed, e.g., 10 mg/kg (5D male/3D pre/4-7D preg). If administration is continuous through two or more of the above periods, the above format is abbreviated by replacing the slash (/) with a dash (-). For example, 10 mg/kg (3D pre-5D post) indicates a total of 10 mg/kg administered to the female for three days prior to mating, on each day during gestation, and for five days following birth. Approximate gestation periods for various species are shown in Table 2.

g. Multigeneration Studies. Some reproductive studies entail administration of a substance to several consecutive generations, with the reproductive effects measured in the final generation. The protocols for such studies vary widely. Therefore, because of the inherent complexity and variability of these studies, they are cited in a simplified format as follows. The specific route of administration is reported if it was the same for all parents of all generations; otherwise the abbreviation "mul" is used. The total dose amount shown is that administered to the F0 generation only; doses to the Fn (where n = 1, 2, 3, etc.) generations are not reported. The time and duration of treatment for multigeneration studies are not included in the data line. Instead, the dose amount is followed by the abbreviation "(MGN)", e.g., 10 mg/kg (MGN). This code indicates a multigeneration study, and the reader must consult the cited reference for complete details of the study protocol.

15. *Carcinogenic Study Result.* Tumorigenic citations are classified according to the reported results of the study to aid the reader in selecting appropriate references for in-depth review and evaluation. The classification ETA (equivocal tumorigenic agent) denotes those studies reporting uncertain, but seemingly positive, results. The criteria for the three classifications are listed below. These criteria are used to abstract the data in individual reports on a consistent basis and do not represent a comprehensive evaluation of a material's tumorigenic potential to humans.

The following nine technical criteria are used to abstract the toxicological literature and classify studies that report positive tumorigenic responses. No attempts are made either to evaluate the various test procedures or to correlate results from different experiments.

(1) A citation is coded "CAR" (carcinogenic) when review of an article reveals that all the following criteria are satisfied:

(a) There is a statistically significant increase in the incidence of tumors in the test animals. The statistical test is that used by the author. If no statistic is reported, a Fisher's Exact Test is applied with significance at the 0.05 level, unless the author makes a strong case for significance at some other level.

(b) A control group of animals is used and the treated and control animals are maintained under identical conditions.

(c) The sole experimental variable between the groups is the administration or nonadministration of the test material (see (10) below).

(d) The tumors consist of autonomous populations of cells of abnormal cytology capable of invading and destroying normal tissues, or the tumors metastasize as confirmed by histopathology.

(2) A citation is coded "NEO" (neoplastic) when review of an article reveals that all the following criteria are satisfied:

(a) There is a statistically significant increase in the incidence of tumors in the test animals. The statistical test is that used by the author. If no statistic is reported, a Fisher's Exact Test is applied with significance at the 0.05 level, unless the author makes a strong case for significance at some other level.

(b) A control group of animals is used and the treated and control animals are maintained under identical conditions.

(c) The sole experimental variable between the groups is the administration or nonadministration of the test material (see (10) below).

(d) The tumors consist of autonomous populations of cells of abnormal cytology capable of invading and destroying normal tissues, or the tumors metastasize as confirmed by histopathology.

(3) A citation is coded "NEO" (neoplastic) when review of an article reveals that all the following criteria are satisfied:

(a) A statistically significant increase in the incidence of tumors in the test animals. The statistical test is that used by the author. If no statistic is reported, a Fisher's Exact Test is applied with significance at the 0.05 level, unless the author makes a strong case for significance at some other level.

(b) A control group of animals is used, and the treated and control animals are maintained under identical conditions.

(c) The sole experimental variable between the groups is the administration or nonadministration of the test material (see (10) below).

(d) The tumors consist of cells that closely resemble the tissue of origin, that are not grossly abnormal cytologically, and that may compress surrounding tissues, but that neither invade tissues nor metastasize; or

(e) The tumors produced cannot be classified as either benign or malignant.

(4) A citation is coded "ETA" (equivocal tumorigenic agent) when some evidence of tumorigenic activity is presented, but one or more of the criteria listed in (1) or (2) above are lacking. Thus, a report with positive pathological findings, but with no mention of control animals, is coded "ETA."

(5) Because an author may make statements or draw conclusions based on a larger context than that of the particular data reported, papers in which the author's conclusions differ substantially from the evidence presented in the paper are subject to review.

(6) All doses except those for transplacental carcinogenesis are reported in one of the following formats.

(a) For all routes of administration other than inhalation: cumulative dose is reported in mg (or another appropriate unit)/kg/duration of administration.

Whenever the dose reported in the reference is not in the units discussed herein, conversion to this format is made.

The total cumulative dose is derived from the lowest dose level that produces tumors in the test group.

(b) For inhalation experiments: concentration is reported in ppm (or mg/m^3)/total duration of exposure.

The concentration refers to the lowest concentration that produces tumors.

(7) Transplacental carcinogenic doses are reported in one of the following formats.

(a) For all routes of administration other than inhalation: cumulative dose is reported in mg/kg/(time of administration during pregnancy).

The cumulative dose is derived from the lowest single dose that produces tumors in the offspring. The test chemical is administered to the mother.

(b) For inhalation experiments: concentration is reported in ppm (or mg/m^3)/(time of exposure during pregnancy).

The concentration refers to the lowest concentration that produces tumors in the offspring. The mother is exposed to the test chemical.

(8) For the purposes of this listing, all test chemicals are reported as pure, unless stated to be otherwise by the author. This does not rule out the possibility that unknown impurities may have been present.

(9) A mixture of compounds whose test results satisfy the criteria in (1), (2), or (3) above is included if the composition of the mixture can be clearly defined.

(10) For tests involving promoters or initiators, a study is included if the following conditions are satisfied (in addition to the criteria in (1), (2), or (3) above):

(a) The test chemical is applied first, followed by an application of a standard promoter. A positive control group in which the test animals are subjected to the same standard promoter under identical conditions is maintained throughout the duration of the experiment. The data are only used if positive and negative control groups are mentioned in the reference.

(b) A known carcinogen is first applied as an initiator, followed by application of the test chemical as a promoter. A positive control group in which the test animals are subjected to the same initiator under identical conditions is maintained throughout the duration of the experiment. The data are used only if positive and negative control groups are mentioned in the reference.

16. *Cited Reference* is the final entry of the irritation, mutation, reproductive, tumorigenic, and toxic dose data lines. This is the source from which the information was extracted. All references cited are publicly available. No governmental classified documents have been used for source information. All references have been given a unique six-letter CODEN character code (derived from the American Society for Testing and Materials *CODEN for Periodical Titles* and the CAS *Source Index*), which identifies periodicals, serial publications, and individual published works. For those references for which no CODEN was found, the corresponding six-letter code includes asterisks (*) in the last one or two posi-

tions following the first four or five letters of an acronym for the publication title. Following the CODEN designation (for most entries) are: the number of the volume, followed by a comma; the page number of the first page of the article, followed by a comma; and a two-digit number, indicating the year of publication in this century. When the cited reference is a report, the report number is listed. Where contributors have provided information on their unpublished studies, the CODEN consists of the first three letters of the last name, the initials of the first and middle names, and a number sign (#). The date of the letter supplying the information is listed. All CODEN acronyms are listed in alphabetical order and defined in the CODEN Section.

17. *Consensus Reports* lines supply additional information to enable the reader to make knowledgeable evaluations of potential chemical hazards. Two types of reviews are listed: (a) International Agency for Research on Cancer (IARC) monograph reviews, which are published by the United Nations World Health Organization (WHO), and (b) the National Toxicology Program (NTP).

a. Cancer Reviews. In the U.N. International Agency for Research on Cancer (IARC) monographs, information on suspected environmental carcinogens is examined, and summaries of available data with appropriate references are presented. Included in these reviews are synonyms, physical and chemical properties, uses and occurrence, and biological data relevant to the evaluation of carcinogenic risk to humans. The monographs in the series contain an evaluation of approximately 1100 materials. Single copies of the individual monographs (specify volume number) can be ordered from WHO Publications Centre USA, 49 Sheridan Avenue, Albany, New York 12210, telephone (518) 436-9686.

The format of the IARC data line is as follows. The entry "IARC Cancer Review:" indicates that the carcinogenicity data pertaining to a compound has been reviewed by the IARC committee. The committee's conclusions are summarized in three words. The first word indicates whether the data pertains to humans or to animals. The next two words indicate the degree of carcinogenic risk as defined by IARC.

For experimental animals the evidence of carcinogenicity is assessed by IARC and judged to fall into one of four groups defined as follows:

(1) Sufficient Evidence of carcinogenicity is provided when there is an increased incidence of malignant tumors: (a) in multiple species or strains; (b) in multiple experiments (preferably with different routes of administration or using different dose levels); or (c) to an unusual degree with regard to the incidence, site, or type of tumor, or age at onset. Additional evidence may be provided by data on dose-response effects.

(2) Limited Evidence of carcinogenicity is available when the data suggest a carcinogenic effect but are limited because: (a) the studies involve a single species, strain, or experiment; (b) the experiments are restricted by inadequate dosage levels, inadequate duration of exposure to the agent, inadequate period of follow-up, poor survival, the use of too few animals, or inadequate reporting; or (c) the neoplasms produced often occur spontaneously and, in the past, have been difficult to classify as malignant by histological criteria alone (e.g., lung adenomas and adenocarcinomas, and liver tumors in certain strains of mice).

(3) Inadequate Evidence is available when, because of major qualitative or quantitative limitations, the studies cannot be interpreted as showing either the presence or absence of a carcinogenic effect.

(4) No Evidence applies when several adequate studies are available that show that within the limitations of the tests used, the chemical is not carcinogenic.

It should be noted that the categories *Sufficient Evidence* and *Limited Evidence* refer only to the strength of the experimental evidence that these chemicals are carcinogenic and not to the extent of their carcinogenic activity nor to the mechanism involved. The classification of any chemical may change as new information becomes available.

The evidence for carcinogenicity from studies in humans is assessed by the IARC committees and judged to fall into one of four groups defined as follows:

(1) Sufficient Evidence of carcinogenicity indicates that there is a causal relationship between the exposure and human cancer.

(2) Limited Evidence of carcinogenicity indicates that a causal relationship is credible, but that alternative explanations, such as chance, bias, or confounding, could not adequately be excluded.

(3) Inadequate Evidence, which applies to both positive and negative evidence, indicates that one of two conditions prevailed: (a) there are few pertinent data; or (b) the available studies, while showing evidence of association, do not exclude chance, bias, or confounding.

(4) No Evidence applies when several adequate studies are available which do not show evidence of carcinogenicity.

This cancer review reflects only the conclusion of the IARC committee based on the data available for the committee's evaluation. Hence, for some substances there may be a disparity between the IARC determination and the information on the tumorigenic data lines (see paragraph 15). Also, some substances previously reviewed by IARC may be reexamined as additional data become available. These substances will contain multiple IARC review lines, each of which is referenced to the applicable IARC monograph volume.

An IARC entry indicates that some carcinogenicity data pertaining to a compound have been reviewed by the IARC committee. It indicates whether the data pertain to humans or to animals and whether the results of the determination are positive, suspected, indefinite, or negative, or whether there are no data.

This cancer review reflects only the conclusion of the IARC Committee, based on the data available at the time of the Committee's evaluation. Hence, for some materials there may be disagreement between the IARC determination and the tumorigenicity information in the toxicity data lines.

b. NTP Status. The notation "NTP Fifth Annual Report on Carcinogens" indicates that the entry is listed in the fifth report made to the U.S. Congress by the National Toxicology Program (NTP) as required by law. This listing implies that the entry is assumed to be a human carcinogen.

Another NTP notation indicates that the material has been tested by the NTP under its Carcinogenesis Testing Program. These entries are also identified as National Cancer Institute (NCI), which reported the studies before the NCI Carcinogenesis Testing Program was absorbed by NTP. To obtain additional information about NTP, the Carcinogenesis Testing Program, or the status of a particular material under test, contact the Toxicology Information and Scientific Evaluation Group, NTP/TRTP/NIEHS, Mail Drop 18-01, P.O. Box 12233, Research Triangle Park, NC 27709.

c. EPA Extremely Hazardous Substances List. This list was developed by the U.S. Environmental Protection Agency (EPA) as required by the Superfund Amendments and Reauthorization Act of 1986 (SARA). Title III Section 304 requires notification by facilities of a release of certain extremely hazardous substances. These 402 substances were listed by EPA in the *Federal Register* of November 17, 1986.

d. Community Right-to-Know List. This list was developed by the EPA as required by the Superfund Amendments and Reauthorization Act of 1986 (SARA). Title III, Sections 311-312 require manufacturing facilities to prepare Material Safety Data Sheets and notify local authorities of the presence of listed chemicals. Both specific chemicals and classes of chemicals are covered by these sections.

e. EPA Genetic Toxicology Program (GENE-TOX). This status line indicates that the material has had genetic effects reported in the literature during the period 1969-1979. The test protocol in the literature is evaluated by an EPA expert panel on mutations, and the positive or negative genetic effect of the substance is reported. To obtain additional information about this program, contact GENE-TOX program, USEPA, 401 M Street, SW, TS796, Washington, DC 20460, telephone (202) 260-1513.

f. EPA TSCA Status Line. This line indicates that the material appears on the chemical inventory prepared by the Environmental Protection Agency in accordance with provisions of the Toxic Substances Control Act (TSCA). Materials reported in the inventory include those that are produced commercially in or are imported into this country. The reader should note, however, that materials already regulated by EPA under FIFRA and by the Food and Drug Administration under the Food, Drug, and Cosmetic Act, as amended, are not included in the TSCA inventory. Similarly, alcohol, tobacco, and explosive materials are not regulated under TSCA. TSCA regulations should be consulted for an exact definition of reporting requirements. For additional information about TSCA, contact EPA, Office of Toxic Substances, Washington, D.C. 20402. Specific questions about the inventory can be directed to the EPA Office of Industry Assistance, telephone (800) 424-9065.

18. *Standards and Recommendations* section contains regulations by agencies of the United States Government or recommendations by expert groups. "OSHA" refers to standards promulgated under Section 6 of the Occupational Safety and Health Act of 1970. "DOT" refers to materials regulated for shipment by the Department of Transportation. Because of frequent changes to and litigation of Federal regulations, it is recommended that the reader contact the applicable agency for information about the current standards for a particular material. Omission of a material or regulatory notation from this edition does not imply any relief from regulatory responsibility.

a. OSHA Air Contaminant Standards. The values given are for the revised standards that were published in January 13, 1989 and were scheduled to take effect from September 1, 1989 through December 31, 1992. These are noted with the entry "OSHA PEL:" followed by "TWA" or "CL," meaning either time-weighted average or ceiling value, respectively, to which workers can be exposed for a normal 8-hour day, 40-hour work week without ill effects. For some materials, TWA, CL, and Pk (peak) values are given in the standard. In those cases, all three are listed. Finally, some entries may be followed by the designation "(skin)." This designation indicates that the compound may be absorbed by the skin and that, even though the air concentration may be below the standard, significant additional exposure through the skin may be possible.

b. ACGIH Threshold Limit Values. The American Conference of Governmental Industrial Hygienists (ACGIH) Threshold Limit Values are noted with the entry "ACGIH TLV:" followed by "TWA" or "CL," meaning either time-weighted average or ceiling value, respectively, to which workers can be exposed for a normal 8-hour day, 40-hour work week without ill effects. The notation "CL" indicates a ceiling limit which must not be exceeded. The notation "skin" indicates that the material penetrates intact skin, and skin contact should be avoided even though the TLV concentration is not exceeded. STEL indicates a short-term exposure limit, usually a 15-minute time-weighted average, which should not be exceeded. Biological Exposure Indices *(BEI:)* are, according to the ACGIH, set to provide a warning level "...of biological response to the chemical, or warning levels of that chemical or its metabolic product(s) in tissues, fluids, or exhaled air of exposed workers...."

The latest annual TLV list is contained in the publication *Threshold Limit Values and Biological Exposure Indices.* This publication should be consulted for future trends in recommendations. The ACGIH TLVs are adopted in whole or in part by many countries and local administrative agencies throughout the world. As a result, these recommendations

INTRODUCTION

have a major effect on the control of workplace contaminant concentrations. The ACGIH may be contacted for additional information at 6500 Glenway Ave., Cincinnati, Ohio 45211.

c. DFG MAK. These lines contain the German Research Society's Maximum Allowable Concentration values. Those materials that are classified as to workplace hazard potential by the German Research Society are noted on this line. The MAK values are also revised annually and discussions of materials under consideration for MAK assignment are included in the annual publication together with the current values. *BAT:* indicates Biological Tolerance Value for a Working Material which is defined as, ". . . the maximum permissible quantity of a chemical compound, its metabolites, or any deviation from the norm of biological parameters induced by these substances in exposed humans." *TRK:* values are Technical Guiding Concentrations for workplace control of carcinogens. For additional information, write to Deutsche Forschungsgemeinschaft (German Research Society), Kennedyallee 40, D-5300 Bonn 2, Federal Republic of Germany. The publication *Maximum Concentrations at the Workplace and Biological Tolerance Values for Working Materials* can be obtained from Verlag Chemie GmbH, Buchauslieferung, P.O. Box 1260/1280, D-6940 Weinheim, Federal Republic of Germany.

d. NIOSH REL. This line indicates that a NIOSH criteria document recommending a certain occupational exposure has been published for this compound or for a class of compounds to which this material belongs. These documents contain extensive data, analysis, and references. The more recent publications can be obtained from the National Institute for Occupational Safety and Health, U.S. Department of Health and Human Services, 4676 Columbia Pkwy., Cincinnati, Ohio 45226.

e. DOT Classification. This is the hazard classification according to the U.S. Department of Transportation (DOT) or the International Maritime Organization (IMO). This classification gives an indication of the hazards expected in transportation, and serves as a guide to the development of proper labels, placards, and shipping instructions. The basic hazard classes include compressed gases, flammables, oxidizers, corrosives, explosives, radioactive materials, and poisons. Although a material may be designated by only one hazard class, additional hazards may be indicated by adding labels or by using other means as directed by DOT. Many materials are regulated under general headings such as "pesticides" or "combustible liquids" as defined in the regulations. These are not noted here, as their specific concentration or properties must be known for proper classification. Special regulations may govern shipment by air. This information should serve *only as a guide*, because the regulation of transported materials is carefully controlled in most countries by federal and local agencies. Because there are frequent changes to regulations, it is recommended that the reader contact the applicable agency for information about the current standards for a particular material. United States transportation regulations are found in 40 CFR, Parts 100 to 189. Contact the U.S. Department of Transportation, Materials Transportation Bureau, Washington, D.C. 20590.

19. *Safety Profiles* are text summaries of the reported hazards of the entry. The word "experimental" indicates that the reported effects resulted from a controlled exposure of laboratory animals to the substance. Toxic effects reported include carcinogenic, reproductive, acute lethal, and human nonlethal effects, skin and eye irritation, and positive mutation study results.

Human effects are identified either by *human* or more specifically by *man, woman, child,* or *infant.* Specific symptoms or organ systems effects are reported when available.

Carcinogenicity potential is denoted by the words "confirmed," "suspected," or "questionable." The substance entries are grouped into three classes based on experimental evidence and the opinion of expert review groups. The OSHA, IARC, ACGIH, and DFG MAK decision schedules are not related or synchronized. Thus, an entry may have had a recent review by only one group. The most stringent classification of any regulation or expert group is taken as governing.

Class I—Confirmed Carcinogens

These substances are capable of causing cancer in exposed humans. An entry was assigned to this class if it had one or more of the following data items present.

a. an OSHA regulated carcinogen

b. an ACGIH assignment as a human or animal carcinogen

c. a DFG MAK assignment as a confirmed human or animal carcinogen

d. an IARC assignment of human or animal sufficient evidence of carcinogenicity, or higher

e. NTP Fifth Annual Report on Carcinogens

Class II—Suspected Carcinogens

These substances may be capable of causing cancer in exposed humans. The evidence is suggestive, but not sufficient to convince expert review committees. Some entries have not yet had expert review, but contain experimental reports of carcinogenic activity. In particular, an entry is included if it has positive reports of carcinogenic endpoint in two species. As more studies are published, many Class II carcinogens will have their carcinogenicity confirmed. On the other hand, some will be judged noncarcinogenic in the future. An entry was assigned to this class if it had one or more of the following data items present.

a. an ACGIH assignment as a suspected carcinogen

b. an DFG MAK assignment as a suspected carcinogen

c. an IARC assignment of human or animal limited evidence

d. two animal studies reporting positive carcinogenic endpoint in different species

Class III—Questionable Carcinogens

For these entries there is minimal published evidence of possible carcinogenic activity. The reported endpoint is often neoplastic growth with no spread or invasion characteristic of carcinogenic pathology. An even weaker endpoint is that of *equivocal tumorigenic agent* (ETA). Reports are assigned this designation when the study was defective. The study may have lacked control animals, may have used very small sample size, often lack complete pathology reporting, or suffer many other study design defects. Many of these studies were designed for other than carcinogenic evaluation, and the reported carcinogenic effect is a by-product of the study, not the goal. The data are presented because some of the substances studied may be carcinogens. There is insufficient data to affirm or deny the possibility. An entry was assigned to this class if it had one or more of the following data items present.

a. an IARC assignment of inadequate or no evidence

b. a single human report of carcinogenicity

c. a single experimental carcinogenic report, or duplicate reports in the same species

d. one or more experimental neoplastic or equivocal tumorigenic agent reports

Fire and explosion hazards are briefly summarized in terms of conditions of flammable or reactive hazard. Materials that are incompatible with the entry are listed here. Fire and explosion hazards are briefly summarized in terms of conditions of flammable or reactive hazard. Fire-fighting materials and methods are discussed where feasible. A material with a flash point of 100°F or less is considered dangerous; if the flash point is from 100 to 200°, the flammability is considered moderate; if it is above 200°F, the flammability is considered low (the material is considered combustible).

Also included in the safety profile are disaster hazards comments, which serve to alert users of materials, safety professionals, researchers, supervisors, and fire-fighters to the dangers that may be encountered on entering storage premises during a fire or other emergency. Although the presence of water, steam, acid fumes, or powerful vibrations can cause many materials to decompose into dangerous compounds, we are particularly concerned with high temperatures (such as those resulting from a fire) because these can cause many otherwise inert chemicals to emit highly toxic gases or vapors such as NO_x, SO_x, acids, and so forth, or evolve vapors of antimony, arsenic, mercury, and the like.

Key to Abbreviations

abs — absolute
ACGIH — American Conference of Governmental Industrial Hygienists
af — atomic formula
alc — alcohol
alk — alkaline
amorph — amorphous
anhyd — anhydrous
approx — approximately
aq — aqueous
atm — atmosphere
autoign — autoignition
aw — atomic weight
BEI — ACGIH Biological Exposure Indexes
bp — boiling point
b range — boiling range
CAS — Chemical Abstracts Service
cc — cubic centimeter
CC — closed cup
CL — ceiling concentration
COC — Cleveland open cup
compd(s) — compound(s)
conc — concentration, concentrated
contg — containing
cryst, crys — crystal(s), crystalline
d — density
D — day(s)
decomp, dec — decomposition
deliq — deliquescent
dil — dilute
DOT — U.S. Department of Transportation
EPA — U.S. Environmental Protection Agency
eth — ether
(F) — Fahrenheit
FCC — Food Chemical Codex
FDA — U.S. Food and Drug Administration
flam — flammable
flash p — flash point
fp — freezing point
g, gm — gram
glac — glacial
gran — granular, granules
H, hr — hour(s)
HR: — hazard rating

htd — heated
htg — heating
hygr — hygroscopic
IARC — International Agency for Research on Cancer
immisc — immiscible
incomp — incompatible
insol — insoluble
IU — International Unit
kg — kilogram (one thousand grams)
L, l — liter
lel — lower explosive limit
liq — liquid
M — minute(s)
m^3 — cubic meter
mf — molecular formula
mg — milligram
misc — miscible
mL, ml — milliliter
mm — millimeter
mod — moderately
mp — melting point
mppcf — million particles per cubic foot
mw — molecular weight
μ — micron
ng — nanogram
NIOSH — National Institute for Occupational Safety and Health
nonflam — nonflammable
NTP — National Toxicology Program
OBS. — obsolete
OC — open cup
org — organic
ORM — other regulated material (DOT)
OSHA — Occupational Safety and Health Administration
Pa — Pascals
PEL — permissible exposure level
petr — petroleum
pg — picogram (one trillionth of a gram)
Pk — peak concentration
pmole — picomole
powd — powder
ppb — parts per billion (v/v)
pph — parts per hundred (v/v)(percent)
ppm — parts per million (v/v)
ppt — parts per trillion (v/v)

KEY TO ABBREVIATIONS

prep — preparation
PROP — properties
refr — refractive
rhomb — rhombic
S,sec — second(s)
sl, slt, sltly — slightly
sol — soluble
soln — solution
solv(s) — solvent(s)
spont — spontaneous(ly)
STEL — short-term exposure limit
subl — sublimes
TCC — Tag closed cup
tech — technical
temp — temperature
TLV — Threshold Limit Value
TOC — Tag open cup
TWA — time weighted average
U, unk — unknown, unreported

uel — upper explosive limit
μg, ug — microgram
ULC, ulc — Underwriters Laboratory Classification
USDA — U.S. Department of Agriculture
vac — vacuum
vap — vapor
vap d — vapor density
vap press — vapor pressure
visc — viscosity
vol — volume
W — week(s)
Y — year(s)
% — percent(age)
> — greater than
< — less than
≤ — less than or equal to
≥ — greater than or equal to
° — degrees of temperature in Celsius (Centigrade)
(F),°F — temperature in Fahrenheit

A

AAF275   CAS:83-32-9   **HR: 2**
ACENAPHTHENE
mf: $C_{12}H_{10}$ mw: 154.22

SYNS: ACENAPHTHYLENE, 1,2-DIHYDRO-
◊ 1,8-ETHYLENENAPHTHALENE ◊ NAPHTHYLENEETHYLENE
◊ PERIETHYLENENAPHTHALENE

TOXICITY DATA with REFERENCE
mmo-omi 3 mg   MIKBA5 54,360,85
ipr-rat LD50:600 mg/kg   GTPZAB 14(6),46,70

CONSENSUS REPORTS: Reported in EPA TSCA Inventory.

SAFETY PROFILE: Moderately toxic by intraperitoneal route. Mutation data reported. When heated to decomposition it emits acrid smoke and irritating vapors.

AAF300   CAS:82-86-0   **HR: 2**
ACENAPHTHENEDIONE
mf: $C_{12}H_6O_2$ mw: 182.18

SYN: 1,2-ACENAPHTHYLENEDIONE

TOXICITY DATA with REFERENCE
unr-rat LD50:728 mg/kg   RPTOAN 41,146,78

CONSENSUS REPORTS: Reported in EPA TSCA Inventory.

SAFETY PROFILE: Moderately toxic by an unspecified route. When heated to decomposition it emits acrid smoke and irritating vapors.

AAG250   CAS:75-07-0   **HR: 3**
ACETALDEHYDE
DOT: UN 1089
mf: C_2H_4O mw: 44.06

PROP: Colorless, fuming liquid; pungent, fruity odor. Mp: −123.5°, bp: 20.8°, lel: 4.0%, uel: 57%, flash p: −36°F (CC), d: 0.804 @ 0°/20°, autoign temp: 347°F, vap d: 1.52. Misc in water, alc, and ether.

SYNS: ACETALDEHYD (GERMAN) ◊ ACETIC ALDEHYDE ◊ ALDEHYDE ACETIQUE (FRENCH) ◊ ALDEIDE ACETICA (ITALIAN) ◊ ETHANAL ◊ ETHYL ALDEHYDE ◊ FEMA No. 2003 ◊ NCI-C56326 ◊ OCTOWY ALDEHYD (POLISH) ◊ RCRA WASTE NUMBER U001

TOXICITY DATA with REFERENCE
eye-hmn 50 ppm/15M   JIHTAB 28,262,46
skn-rbt 500 mg open MLD   UCDS** 12/13/63
eye-rbt 40 mg SEV   UCDS** 12/13/63
mma-sat 10 µL/plate   EVHPAZ 21,79,77
dnr-esc 10 µL/plate   EVHPAZ 21,79,77
sce-hmn:lym 20 ppm/48H   MUREAV 58,115,78
ipr-rat TDLo:300 mg/kg (female 8–13D post):REP   TJADAB 36,31A,87
ipr-rat TDLo:50 mg/kg (12D preg):TER   DADEDV 9,339,82
ipr-rat TDLo:50 mg/kg (12D preg):TER   DADEDV 9,339,82
ipr-rat TDLo:50 mg/kg (12D preg):REP   DADEDV 9,339,82
ivn-mus TDLo:4 g/kg (female 6D post):TER   JOANAY 132,107,81
ipr-rat TDLo:600 mg/kg (8–15D preg):TER   SEIJBO 23,13,83
ipr-rat TDLo:400 mg/kg (8–15D preg):TER   SEIJBO 23,13,83
ipr-rat TDLo:100 mg/kg (12D preg):TER   DADEDV 9,339,82
ipr-rat TDLo:400 mg/kg (8–15D preg):TER   SEIJBO 23,13,83
ihl-rat TCLo:735 ppm/6H/2Y-I:CAR   TXCYAC 41,213,86
ihl-ham TCLo:2040 ppm/7H/52W-I:ETA   EJCAAH 18,13,82
ihl-rat TCLo:1410 ppm/6H/65W-I:ETA   TXCYAC 31,123,84
ihl-hmn TCLo:134 ppm/30M:PUL   JAMAAP 165,1908,57
orl-rat LD50:661 mg/kg   AGACBH 4,125,74
ihl-rat LC50:37 g/m^3/30M   APTOA6 6,299,50
ipr-rat LDLo:500 mg/kg   JBCHA3 152,41,44
ihl-mus LC50:1500 ppm/4H   DTLVS* 4,3,80
scu-rat LD50:640 mg/kg   APTOA6 6,299,50
scu-mus LD50:560 mg/kg   APTOA6 6,299,50
ivn-mus LD50:212 mg/kg   JOANAY 128,65,79
ihl-ham LC50:17,000 ppm/4H   PEXTAR 24,162,79
itr-ham LD50:96 mg/kg   PEXTAR 24,162,79

CONSENSUS REPORTS: IARC Cancer Review: Group 2B IMEMDT 7,77,87; Animal Sufficient Evidence IMEMDT 36,101,85; Human Inadequate Evidence IMEMDT 36,101,85. On Community Right-To-Know List. Reported in EPA TSCA Inventory. EPA Genetic Toxicology Program.

OSHA PEL: (Transitional: TWA 200 ppm) TWA 100 ppm; STEL 150 ppm
ACGIH TLV: TWA 100 ppm; STEL 150 ppm (Proposed: CL 25, Animal Carcinogen)
DFG MAK: 50 ppm (90 mg/m^3), Suspected Carcinogen

DOT Classification: Flammable Liquid; Label: Flammable Liquid

SAFETY PROFILE: Suspected carcinogen with experimental carcinogenic and tumorigenic data. Poison by intratracheal and intravenous routes. A human systemic irritant by inhalation. An experimental teratogen. Other experimental reproductive effects. A skin and severe eye irritant. A narcotic. Human mutation data reported. A common air contaminant. Highly flammable liquid. Mixtures of 30–60 percent of the vapor in air ignite above 100°. It can react violently with acid anhydrides, alcohols, ketones, phenols, NH_3, HCN, H_2S, halogens, P, isocyanates, strong alkalies, and amines. Reactions with cobalt chloride, mercury(II) chlorate, or mercury(II) perchlorate form sensitive, explosive products. Polymerizes violently in the presence of traces of metals or acids. Reaction with oxygen may lead to detonation. When heated to decomposition it emits acrid smoke and fumes.

AAJ150   CAS:89-52-1   **HR: 2**
2-ACETAMIDOBENZOIC ACID
mf: $C_9H_9NO_3$ mw: 179.19

SYNS: o-ACETAMIDOBENZOIC ACID ◊ o-ACETOAMINOBENZOIC ACID ◊ N-ACETYLAMINOBENZOIC ACID ◊ 2-(ACETYLAMINO)BENZOIC ACID ◊ ACETYLANTHRANILIC ACID ◊ N-ACETYLANTHRANILIC ACID ◊ ANTHRANILIC ACID, N-ACETYL- ◊ BENZOIC ACID, 2-(ACETYLAMINO)-(9CI) ◊ 2-CARBOXYACETANILIDE

TOXICITY DATA with REFERENCE
orl-mus LD50:1114 mg/kg   FRPSAX 38,847,83

CONSENSUS REPORTS: Reported in EPA TSCA Inventory.

SAFETY PROFILE: Moderately toxic by ingestion. When heated to decomposition it emits toxic vapors of NO_x.

AAX500   CAS:108-24-7   **HR: 2**
ACETIC ANHYDRIDE
DOT: UN 1715
mf: $C_4H_6O_3$ mw: 102.10

PROP: Colorless, very mobile, strongly refractive liquid; very strong acetic odor. Mp: −73.1°, bp: 140°, flash p: 129°F (CC), d: 1.082 @ 20°/4°, lel: 2.9%, uel: 10.3%, autoign temp: 734°F, vap press: 10 mm @ 36.0°, vap d: 3.52. Somewhat sol in cold water; decomp in hot water and hot alc; misc in alc and ether.

SYNS: ACETIC ACID, ANHYDRIDE ◊ ACETIC OXIDE ◊ ACETYL ANHYDRIDE ◊ ACETYL ETHER ◊ ACETYL OXIDE ◊ ANHYDRIDE ACETIQUE (FRENCH) ◊ ANIDRIDE ACETICA (ITALIAN) ◊ AZIJNZUURANHYDRIDE (DUTCH) ◊ ESSIGSAEUREANHYDRID (GERMAN) ◊ ETHANOIC ANHYDRATE ◊ OCTOWY BEZWODNIK (POLISH)

TOXICITY DATA with REFERENCE
skn-rbt 10 mg/24H open MLD   AMIHBC 4,119,51
skn-rbt 540 mg open MLD   UCDS** 8/7/63
eye-rbt 250 μg open SEV   AMIHBC 4,119,51
orl-rat LD50:1780 mg/kg   AMIHBC 4,119,51
ihl-rat LC50:1000 ppm/4H   34ZIAG -,607,69
skn-rbt LD50:4000 mg/kg   UCDS** 8/7/63

CONSENSUS REPORTS: Reported in EPA TSCA Inventory.

OSHA PEL: CL 5 ppm
ACGIH TLV: CL 5 ppm (Proposed: TWA 5 ppm)
DFG MAK: 5 ppm (20 mg/m^3)

DOT Classification: IMO: Corrosive Material; Label: Corrosive, Flammable Liquid.

SAFETY PROFILE: Moderately toxic by inhalation, ingestion, and skin contact. A skin and severe eye irritant. Moderate fire and explosion hazard when exposed to heat or flame. Potentially explosive reactions with barium peroxide, boric acid, chromium trioxide, 1,3-diphenyltriazene, hydrochloric acid + water, hypochlorous acid, nitric acid, perchloric acid + water, peroxyacetic acid, potassium permanganate, tetrafluoroboric acid, 4-toluenesulfonic acid + water, and acetic acid + water. Reactions with ethanol + sodium hydrogen sulfate, and hydrogen peroxide form explosive products. Reactions with ammonium nitrate + hexamethylenetetrammonium acetate + nitric acid form as products the military explosives RDX and HMX. Reacts violently with N-tert-butylphthalimic acid + tetrafluoroboric acid, chromic acid, glycerol + phosphoryl chloride, and metal nitrates (e.g., copper or sodium nitrates). Incompatible with 2-aminoethanol, aniline, chlorosulfonic acid, (CrO_3 + acetic acid), ethylenediamine, ethyleneimine, glycerol, oleum, HF, permanganates, NaOH, Na_2O_2, H_2SO_4, water, N_2O_2, (glycerol + phosphoryl chloride). When heated to decomposition it emits toxic fumes; can react vigorously with oxidizing materials, will react violently on contact with water or steam. Used in production of drugs of abuse. To fight fire, use CO_2, dry chemical, water mist, alcohol foam. See also ANHYDRIDES.

AAY600   CAS:93-70-9   **HR: 1**
o-ACETOACETOCHLORANILIDE
mf: $C_{10}H_{10}ClNO_2$ mw: 211.66

SYNS: AAoC ◊ ACETOACETANILIDE, o-CHLORO- ◊ ACETOACETANILIDE, 2′-CHLORO- ◊ ACETOACET-o-CHLORANILIDE ◊ ACETOACET-o-CHLOROANILIDE ◊ ACETOACETYL-2-CHLOROANILIDE ◊ BUTANEAMIDE, N-(2-CHLOROPHENYL)-3-OXO- ◊ o-CHLOROACETOACETANILIDE ◊ 2′-CHLOROACETOACETANILIDE ◊ N-(2-CHLOROPHENYL)ACETOACETAMIDE ◊ 3-OXO-N-(2-CHLOROPHENYLBUTANAMIDE)

TOXICITY DATA with REFERENCE
orl-rat LD50:11600 mg/kg   LONZA# 10JUL81

CONSENSUS REPORTS: Reported in EPA TSCA Inventory.

SAFETY PROFILE: Slightly toxic by ingestion. When heated to decomposition it emits toxic vapors of NO_x and Cl^-.

ABG350   CAS:581-08-8   **HR: 2**
o-ACETOPHENETIDIDE
mf: $C_{10}H_{13}NO_2$ mw: 179.24

SYNS: ACETAMIDE, N-(2-ETHOXYPHENYL)-(9CI) ◊ ACETANILIDE, 2′-ETHOXY- ◊ 2-ETHOXYACETANILIDE ◊ 2′-ETHOXYACETANILIDE ◊ N-(2-ETHOXYPHENYL)ACETAMIDE

TOXICITY DATA with REFERENCE
orl-mus LD50:680 mg/kg   TXAPA9 19,20,71

CONSENSUS REPORTS: Reported in EPA TSCA Inventory.

SAFETY PROFILE: Moderately toxic by ingestion. When heated to decomposition it emits toxic vapors of NO_x.

ABH000   CAS:98-86-2   *HR: 3*
ACETOPHENONE
mf: C_8H_8O mw: 120.16

PROP: Colorless liquid or plates; sweet, pungent odor. Mp: 19.7°, bp: 202.3°, flash p: 180°F (OC), d: 1.026 @ 20°/4°, vap d: 4.14, vap press: 1 mm @ 15°, autoign temp: 1060°F. Very sol in propylene glycol and fixed oils; sol in alc, chloroform, and ether; sltly sol in water; insol in glycerin.

SYNS: ACETYLBENZENE ◊ BENZOYL METHIDE ◊ DYMEX ◊ FEMA No. 2009 ◊ HYPNONE ◊ KETONE METHYL PHENYL ◊ METHYL PHENYL KETONE ◊ 1-PHENYLETHANONE ◊ PHENYL METHYL KETONE ◊ USAF EK-496

TOXICITY DATA with REFERENCE
skn-rbt 10 mg/24H open   JIHTAB 26,269,44
skn-rbt 515 mg open MLD   UCDS** 12/27/71
eye-rbt 771 µg SEV   AJOPAA 29,1363,46
cyt-smc 10 mmol/tube   HEREAY 33,457,47
orl-rat LD50:815 mg/kg   GTPZAB 26(8),53,82
orl-mus LD50:740 mg/kg   GTPZAB 26(8),53,82
scu-mus LDLo:330 mg/kg   HDTU** -,-,33
ipr-mus LD50:200 mg/kg   NTIS** AD277-689

CONSENSUS REPORTS: Reported in EPA TSCA Inventory.
ACGIH TLV: (Proposed: 10 ppm)

SAFETY PROFILE: Poison by intraperitoneal and subcutaneous routes. Moderately toxic by ingestion. A skin and severe eye irritant. Mutation data reported. Narcotic in high concentration. A hypnotic. Combustible liquid. To fight fire use foam, CO_2, dry chemical. When heated to decomposition it emits acrid smoke and fumes. See also KETONES.

ABH150   CAS:613-91-2   *HR: 3*
ACETOPHENONE, OXIME
mf: C_8H_9NO mw: 135.18

SYN: ETHANONE, 1-PHENYL-, OXIME

TOXICITY DATA with REFERENCE
orl-mus LD50:2 g/kg   MEXPAG 11,137,64
unr-mus LD50:450 mg/kg   PCJOAU 12,227,78

CONSENSUS REPORTS: Reported in EPA TSCA Inventory.

SAFETY PROFILE: Poison by an unspecified route. Slightly toxic by ingestion. When heated to decomposition it emits toxic vapors of NO_x.

ABW550   CAS:2628-16-2   *HR: 2*
4-ACETOXYSTYRENE
mf: $C_{10}H_{10}O_2$ mw: 162.20

SYNS: p-ACETOXYSTYRENE ◊ C-908 ◊ 4-ETHENYLPHENOL ACETATE ◊ PHENOL, 4-ETHENYL-, ACETATE ◊ PHENOL, p-VINYL-, ACETATE (6CI,7CI,8CI) ◊ p-VINYLPHENOL ACETATE ◊ 4-VINYLPHENYL ACETATE

TOXICITY DATA with REFERENCE
eye-rbt 100 mg MLD   EPASR* 8EHQ-1190-1082
orl-rat LD50:1503 mg/kg   EPASR* 8EHQ-1190-1082
skn-rat LD50:>2 g/kg   EPASR* 8EHQ-1190-1082

SAFETY PROFILE: Moderately toxic by ingestion. Slightly toxic by skin contact. An eye irritant. When heated to decomposition it emits acrid smoke and irritating vapors.

ACC100   CAS:1646-26-0   *HR: 2*
2-ACETYLBENZOFURAN
mf: $C_{10}H_8O_2$ mw: 160.18

SYNS: 2-ACETYLCOUMARONE ◊ 1-(2-BENZOFURANYL)ETHANONE ◊ BENZO(b)FURAN-2-YL METHYL KETONE ◊ 2-BENZOFURANYL METHYL KETONE ◊ ETHANONE, 1-(2-BENZOFURANYL)-(9CI) ◊ KETONE, 2-BENZOFURANYL METHYL

TOXICITY DATA with REFERENCE
ipr-mus LD50:1200 mg/kg   EJMCA5 12,383,77

CONSENSUS REPORTS: Reported in EPA TSCA Inventory.

SAFETY PROFILE: Moderately toxic by intraperitoneal route. When heated to decomposition it emits acrid smoke and irritating vapors.

ACF100   CAS:1888-91-1   *HR: 2*
ACETYLCAPROLACTAM
mf: $C_8H_{13}NO_2$ mw: 155.22

SYNS: N-ACETYLCAPROLACTAM ◊ ACETYLKAPROLAKTAM ◊ 2H-AZEPIN-2-ONE, 1-ACETYLHEXAHYDRO-

TOXICITY DATA with REFERENCE
orl-uns LD50:1300 mg/kg   85JCAE-,884,86

CONSENSUS REPORTS: Reported in EPA TSCA Inventory.

SAFETY PROFILE: Moderately toxic by ingestion. When heated to decomposition it emits toxic vapors of NO_x.

ACM200   CAS:1192-62-7   *HR: D*
2-ACETYLFURAN
mf: $C_6H_6O_2$ mw: 110.12

SYN: ACETYLFURAN ◊ ETHANONE, 1-(2-FURANYL)-(9CI) ◊ FURAN, 2-ACETYL- ◊ 1-(2-FURANYL)ETHANONE ◊ 2-FURYL METHYL KETONE ◊ KETONE, 2-FURYL METHYL ◊ METHYL 2-FURYL KETONE

TOXICITY DATA with REFERENCE
mma-sat 165 nmol/plate   DFSCDX 13,353,86

dnr-bcs 5500 µg/disc   DFSCDX 13,353,86
cyt-ham:ovr 4500 FE2mol/L   CALEDQ 13,89,81

CONSENSUS REPORTS: Reported in EPA TSCA Inventory.

SAFETY PROFILE: Mutation data reported. When heated to decomposition it emits acrid smoke and irritating vapors.

ACQ270   CAS:1115-47-5   HR: 1
N-ACETYL-dl-METHIONINE
mf: $C_7H_{13}NO_3S$ mw: 191.27

SYNS: ACETYL-dl-METHIONINE ◇ dl-N-ACETYLMETHIONINE ◇ METHIONINE, N-ACETYL-, dl- ◇ dl-METHIONINE, N-ACETYL-(9CI)

TOXICITY DATA with REFERENCE
ipr-mus LD50:6700 mg/kg   AIPTAK 91,163,52

CONSENSUS REPORTS: Reported in EPA TSCA Inventory.

SAFETY PROFILE: Slightly toxic by intraperitoneal route. When heated to decomposition it emits toxic vapors of NO_x and SO_x.

ADA365   CAS:1122-54-9   HR: 2
4-ACETYLPYRIDINE
mf: C_7H_7NO mw: 121.15

SYNS: KETONE, METHYL 4-PYRIDYL ◇ METHYL 4-PYRIDYL KETONE ◇ PYRIDINE, 4-ACETYL-

TOXICITY DATA with REFERENCE
mrc-smc 9900 ppm   MUREAV 163,23,86
sln-smc 6200 ppm   MUREAV 163,23,86
ipr-mus LD50:1400 mg/kg   JMCMAR 14,551,71

CONSENSUS REPORTS: Reported in EPA TSCA Inventory.

SAFETY PROFILE: Moderately toxic by intraperitoneal route. Mutation data reported. When heated to decomposition it emits toxic vapors of NO_x.

ADC300   CAS:1866-15-5   HR: 2
ACETYLTHIOCHOLINE IODIDE
mf: $C_7H_{16}NOS•I$ mw: 289.20

SYNS: ACETYLTHIOCHOLINE DIIODIDE ◇ S-ACETYLTHIOCHOLINE IODIDE ◇ 2-(ACETYLTHIO)-N,N,N-TRIMETHYLETHANAMINIUM IODIDE ◇ AMMONIUM, (2-MERCAPTOETHYL)TRIMETHYL-, IODIDE ACETATE ◇ CHOLINE, S-ACETYLTHIO-, IODIDE ◇ ETHANAMINIUM, 2-(ACETYLTHIO)-N,N,N-TRIMETHYL-, IODIDE (9CI) ◇ ((2-MERCAPTOETHYL)TRIMETHYLAMMONIUM IODIDE ACETATE)

TOXICITY DATA with REFERENCE
ivn-mus LD50:1800 µg/kg   CSLNX* NX#02898

CONSENSUS REPORTS: Reported in EPA TSCA Inventory.

SAFETY PROFILE: Moderately toxic by intraperitoneal route. When heated to decomposition it emits toxic vapors of NO_x, SO_x, and I^-.

ADI775   CAS:92-81-9   HR: 2
ACRIDAN
mf: $C_{13}H_{11}N$ mw: 181.25

SYNS: ACRIDANE ◇ ACRIDINE, 9,10-DIHYDRO-(9CI) ◇ CARBAZINE ◇ 9,10-DIHYDROACRIDINE

TOXICITY DATA with REFERENCE
orl-rat LD50:2140 mg/kg   JPMSAE 63,1068,74
scu-mus LD50:3630 µg/kg   PSEBAA 78,392,51

SAFETY PROFILE: Poison by subcutaneous route. Moderately toxic by ingestion. When heated to decomposition it emits toxic vapors of NO_x.

ADI825   CAS:578-95-0   HR: D
9-ACRIDANONE
mf: $C_{13}H_9NO$ mw: 195.23

SYNS: ACRIDANONE ◇ 9(10H)-ACRIDINONE (9CI) ◇ ACRIDONE ◇ 9-ACRIDONE ◇ 9(10H)-ACRIDONE

TOXICITY DATA with REFERENCE
dnd-uns:lyms 10 pph   BIPMAA 11,2537,72

CONSENSUS REPORTS: Reported in EPA TSCA Inventory.

SAFETY PROFILE: Mutation data reported. When heated to decomposition it emits toxic vapors of NO_x.

AEE100   CAS:880-52-4   HR: 2
N-(1-ADAMANTYL)ACETAMIDE
mf: $C_{12}H_{19}NO$ mw: 193.32

SYN: ACETAMIDE, N-(1-ADAMANTYL)-

TOXICITY DATA with REFERENCE
ipr-mus LD50:520 mg/kg   PCJOAU 14,185,80

CONSENSUS REPORTS: Reported in EPA TSCA Inventory.

SAFETY PROFILE: Moderately toxic by intraperitoneal route. When heated to decomposition it emits toxic vapors of NO_x.

AEH500   CAS:321-30-2   HR: 3
ADENINE SULFATE
mf: $C_5H_5N_5•1/2H_2O_4S$

SYNS: ADENINSULFAT ◇ 1H-PURIN-6-AMINE, SULFATE

TOXICITY DATA with REFERENCE
ipr-rat LD50:200 mg/kg   AIPTAK 232,302,78
ipr-mus LD50:750 mg/kg   TXAPA9 47,229,79

CONSENSUS REPORTS: Reported in EPA TSCA Inventory.

SAFETY PROFILE: Poison by intraperitoneal route. When heated to decomposition it emits toxic vapors of NO_x and SO_x.

AEK100   CAS:58-64-0   HR: 2
ADENOSINE DIPHOSPHATE
mf: $C_{10}H_{15}N_5O_{10}P_2$ mw: 427.24

SYNS: ADENOSINE 5′-DIPHOSPHATE ◇ ADENOSINE DIPHOSPHORIC ACID ◇ ADENOSINE 5′-DIPHOSPHORIC ACID ◇ ADENOSINE PYROPHOSPHATE ◇ ADENOSINE 5′-PYROPHOSPHATE ◇ ADENOSINE 5′-PYROPHOSPHORIC ACID ◇ ADENOSINE, 5′-(TRIHYDROGEN DIPHOSPHATE) (9CI) ◇ ADENOSINE, 5′-(TRIHYDROGEN PYROPHOSPHATE) ◇ 5′-ADENYLPHOSPHORIC ACID ◇ ADP ◇ 5′-ADP ◇ ADP (NUCLEOTIDE)

TOXICITY DATA with REFERENCE
oth-hmn:oth 100 FE2mol/L   JIDEAE 65,52,75
ipr-mus LD50:3333 mg/kg   PCJOAU 20,160,86

CONSENSUS REPORTS: Reported in EPA TSCA Inventory.

SAFETY PROFILE: Moderately toxic by intraperitoneal route. Human mutation data reported. When heated to decomposition it emits toxic vapors of NO_x and PO_x.

AEM100   CAS:987-65-5   HR: 3
ADENOSINE 5′-(TETRAHYDROGENTRIPHOSPHATE), DISODIUM SALT
mf: $C_{10}H_{14}N_5O_{13}P_3 \cdot 2Na$ mw: 551.18

SYNS: ADENOSINE TRIPHOSPHATE DISODIUM ◇ ADETPHOS ◇ ATP DISODIUM ◇ ATP DISODIUM SALT ◇ DISODIUM ADENOSINE TRIPHOSPHATE ◇ DISODIUM ADENOSINE 5′-TRIPHOSPHATE ◇ DISODIUM ATP ◇ DISODIUM DIHYDROGEN ATP ◇ SODIUM ATP

TOXICITY DATA with REFERENCE
orl-rat LD50:>2 g/kg   DRUGAY 6,20,82
scu-rat LD50:>2 g/kg   DRUGAY 6,20,82
ivn-rat LD50:380 mg/kg   DRUGAY 6,20,82
orl-mus LD50:>2 g/kg   DRUGAY 6,20,82
scu-mus LD50:>2 g/kg   DRUGAY 6,20,82
ivn-mus LD50:266 mg/kg   DRUGAY 6,20,82

CONSENSUS REPORTS: Reported in EPA TSCA Inventory.

SAFETY PROFILE: Poison by intravenous route. Slightly toxic by ingestion and subcutaneous route. When heated to decomposition it emits toxic vapors of NO_x and SO_x.

AEN250   CAS:124-04-9   HR: 3
ADIPIC ACID
DOT: NA 9077
mf: $C_6H_{10}O_4$ mw: 146.16

PROP: White monoclinic prisms. Mp: 152°, flash p: 385°F (CC), d: 1.360 @ 25°/4°, vap press: 1 mm @ 159.5°, vap d: 5.04, autoign temp: 788°F, bp: 337.5°. Very sol in alc. Sol in acetone, water = 1.4% @15°; 0.6% @ 15° in ether.

SYNS: ACIFLOCTIN ◇ ACINETTEN ◇ ADILACTETTEN ◇ ADIPINIC ACID ◇ 1,4-BUTANEDICARBOXYLIC ACID ◇ FEMA No. 2011 ◇ 1,6-HEXANEDIOIC ACID ◇ KYSELINA ADIPOVA (CZECH) ◇ MOLTEN ADIPIC ACID

TOXICITY DATA with REFERENCE
eye-rbt 20 mg/24H SEV   28ZPAK -,51,72
orl-rat LDLo:3600 mg/kg   28ZPAK -,51,72
ipr-rat LD50:275 mg/kg   JAFCAU 5,759,57
orl-mus LD50:1900 mg/kg   JAFCAU 5,759,57
ipr-mus LD50:275 mg/kg   TXAPA9 32,566,75
ivn-mus LD50:680 mg/kg   JAFCAU 5,759,57

CONSENSUS REPORTS: Reported in EPA TSCA Inventory.
ACGIH TLV: (Proposed: TWA 5 mg/^{3})

SAFETY PROFILE: Poison by intraperitoneal route. Moderately toxic by other routes. A severe eye irritant. Combustible when exposed to heat or flame; can react with oxidizing materials. When heated to decomposition it emits acrid smoke and fumes.

AER250   CAS:111-69-3   HR: 3
ADIPONITRILE
DOT: UN 2205
mf: $C_6H_8N_2$ mw: 108.16

PROP: Water-white liquid, practically odorless. Mp: 2.3°, bp: 295°, flash p: 199.4°F (OC), d: 0.965 @ 20°/4°, vap d: 3.73.

SYNS: ADIPIC ACID DINITRILE ◇ ADIPIC ACID NITRILE ◇ ADIPODINITRILE ◇ 1,4-DICYANOBUTANE ◇ HEXANEDINITRILE ◇ HEXANEDIOIC ACID DINITRILE ◇ NITRILE ADIPICO (ITALIAN) ◇ TETRAMETHYLENE CYANIDE

TOXICITY DATA with REFERENCE
orl-rat LD50:155 mg/kg   GISAAA 49(12),40,84
ihl-rat LC50:1710 mg/m^3/4H   TOXID9 1,76,81
orl-mus LD50:172 mg/kg   ARTODN 57,88,85
ipr-mus LD50:40 mg/kg   NTIS** AD691-490
orl-rbt LD50:22 mg/kg   GISAAA 49(12),40,84
scu-gpg LD50:50 mg/kg   MELAAD 46,221,55

CONSENSUS REPORTS: EPA Extremely Hazardous Substances List. Reported in EPA TSCA Inventory. Cyanide and its compounds are on the Community Right-To-Know List.
ACGIH TLV: (Proposed: TWA 2 ppm (skin))
NIOSH REL: TWA 18 mg/m^3

DOT Classification: IMO: Poison B; Label: St. Andrews Cross

SAFETY PROFILE: Poison by inhalation, ingestion, subcutaneous, and intraperitoneal routes. The nitrile group will behave as a cyanide when ingested or absorbed in the body. It produces disturbances of the respiration and circulation, irritation of the stomach and intestines, and loss of weight. Its low vapor pressure at room temperature makes exposure to harmful concentrations of its vapors unlikely if handled with reasonable care in well-ventilated areas. Flammable when exposed to heat or flame. When heated to decomposition it emits toxic fumes of CN^-. Can react with oxidizing materials. To fight fire, use foam, CO_2, dry chemical. See also HYDROCYANIC ACID and NITRILES.

AGF300   CAS:2179-57-9   HR: 3
ALLYL DISULFIDE
mf: $C_6H_{10}S_2$ mw: 146.28

SYNS: ALLYL DISULPHIDE ◇ DIALLYL DISULFIDE ◇ DIALLYL DISULPHIDE ◇ DISULFIDE, DI-2-PROPENYL (9CI) ◇ 4,5-DITHIA-1,7-OCTADIENE ◇ 2-PROPENYL DISULPHIDE

TOXICITY DATA with REFERENCE
orl-rat LD50:260 mg/kg   FCTOD7 26,297,88
skn-rbt LD50:3600 mg/kg   FCTOD7 26,297,88

CONSENSUS REPORTS: Reported in EPA TSCA Inventory.

SAFETY PROFILE: Poison by ingestion. Moderately toxic by skin contact. When heated to decomposition it emits toxic vapors of SO_x.

AHA135   CAS:1317-25-5   HR: 1
ALUMINUM CHLOROHYDROXYALLANTOINATE
mf: $C_4H_9Al_2ClN_4O_7$ mw: 314.58

SYNS: ALCA ◇ ALCLOXA ◇ ALUMINIUM CHLOROHYDROXYALLANTOINATE ◇ ALUMINUM, CHLORO((2,5-DIOXO-4-IMIDAZOLIDINYL)UREATO)TETRAHYDROXYDI- ◇ ALUMINUM, CHLOROTETRAHYDROXY((2-HYDROXY-5-OXO-2-IMIDAZOLIN-4-YL)UREATO)DI- ◇ CHLORHYDROXYALUMINUM ALLANTOINATE ◇ CHLOROTETRAHYDROXY((2-HYDROXY-5-OXO-2-IMIDAZOLIN-4-YL)UREATO)DIALUMINUM

TOXICITY DATA with REFERENCE
eye-rbt 24% MLD   YKKZAJ 102,89,82
orl-rat LD50:>8 g/kg   DRUGAY -,68,90
scu-rat LD50:>8 g/kg   DRUGAY -,68,90
orl-mus LD50:>8 g/kg   DRUGAY -,68,90
scu-mus LD50:>8 g/kg   DRUGAY -,68,90
unr-mus LD50:15,800 mg/kg   AMPMAR 40,63,79
ACGIH TLV: TWA 2 mg(Al)/m^3

CONSENSUS REPORTS: Reported in EPA TSCA Inventory.

SAFETY PROFILE: Slightly toxic by several routes. An eye irritant. When heated to decomposition it emits toxic vapors of NO_x and Cl^-.

AHA275   CAS:300-92-5   HR: 1
ALUMINUM DISTEARATE
mf: $C_{36}H_{71}AlO_5$ mw: 611.05

SYNS: ALUMINUM DISTEARATE (ACGIH) ◇ ALUMINUM HYDROXIDE DISTEARATE ◇ ALUMINUM, HYDROXYBIS(OCTADECANOATO-O)-(9CI) ◇ ALUMINUM, HYDROXYBIS(STEARATO)- ◇ ALUMINUM HYDROXYDISTEARATE ◇ SPECIAL M
ACGIH TLV: TWA 10 mg/m^3

CONSENSUS REPORTS: Reported in EPA TSCA Inventory.

SAFETY PROFILE: A nuisance dust. When heated to decomposition it emits acrid smoke and irritating vapors.

AHC600   CAS:555-31-7   HR: 1
ALUMINUM(II) ISOPROPYLATE
mf: $C_9H_{21}O_3 \cdot Al$ mw: 204.28

SYNS: ALUMINUM ISOPROPOXIDE ◇ TRIISOPROPOXYALUMINUM

TOXICITY DATA with REFERENCE
orl-rat LD50:11300 mg/kg   AIHAAP 30,470,69
ACGIH TLV: TWA 2 mg(Al)/m^3

CONSENSUS REPORTS: Reported in EPA TSCA Inventory.

SAFETY PROFILE: Slightly toxic by ingestion. When heated to decomposition it emits acrid smoke and irritating vapors.

AHH825   CAS:637-12-7   HR: 1
ALUMINUM TRISTEARATE
mf: $C_{54}H_{105}O_6 \cdot Al$ mw: 877.57

SYNS: ALUGEL 34TN ◇ ALUMINIUM STEARATE ◇ ALUMINUM STEARATE ◇ METASAP XX ◇ OCTADECANOIC ACID, ALUMINUM SALT ◇ ROFOB 3 ◇ SA 1500 ◇ STEARIC ACID, ALUMINUM SALT ◇ TRIBASIC ALUMINUM STEARATE
ACGIH TLV: TWA 10 mg/m^3

CONSENSUS REPORTS: Reported in EPA TSCA Inventory.

SAFETY PROFILE: A nuisance dust. When heated to decomposition it emits acrid smoke and irritating vapors.

AHQ300   CAS:88-64-2   HR: 3
3-AMINOACETANILIDE-4-SULFONIC ACID
mf: $C_8H_{10}N_2O_4S$ mw: 230.26

SYNS: ACETANILIDE-4-SULFONIC ACID, 3-AMINO- ◇ BENZENESULFONIC ACID, 4-ACETAMIDO-2-AMINO-

TOXICITY DATA with REFERENCE
ivn-mus LD50:320 mg/kg   CSLNX* NX#00379

CONSENSUS REPORTS: Reported in EPA TSCA Inventory.

SAFETY PROFILE: Poison by intravenous route. When heated to decomposition it emits toxic vapors of NO_x and SO_x.

AIB300   CAS:83-07-8   HR: 3
4-AMINOANTIPYRINE
mf: $C_{11}H_{13}N_3O$ mw: 203.27

SYNS: AAP ◇ 4-AMINOANTIPYRENE ◇ AMINOANTIPYRIN
◇ AMINOANTIPYRINE ◇ AMINOAZOPHENAZONE ◇ AMINOPHENAZONE
◇ 4-AMINOPHENAZONE ◇ 4-AMMINOANTIPIRINA ◇ AMPYRONE
◇ 1,5-DIMETHYL-2-PHENYL-4-AMINOPYRAZOLINE ◇ METAPIRAZONE
◇ 3-PYRAZOLIN-5-ONE, 4-AMINO-2,3-DIMETHYL-1-PHENYL-
◇ 3H-PYRAZOL-3-ONE, 4-AMINO-1,2-DIHYDRO-1,5-DIMETHYL-2-PHENYL-
◇ SOLNAPYRIN-A

TOXICITY DATA with REFERENCE
mmo-sat 5 FE2mol/plate   MUREAV 206,317,88
dnr-esc 312 µg/well   MUREAV 133,161,84
orl-rat LD50:1700 mg/kg   BCFAAI 117,638,78
ipr-rat LD50:1200 mg/kg   JPETAB 99,171,50
orl-mus LD50:800 mg/kg   CCCCAK 47,636,82
ipr-mus LD50:270 mg/kg   ARZNAD 10,820,60

CONSENSUS REPORTS: Reported in EPA TSCA Inventory.

SAFETY PROFILE: Poison by intraperitoneal route. Moderately toxic by ingestion. Mutation data reported. When heated to decomposition it emits acrid smoke and irritating vapors.

AIB340   CAS:1123-54-2   *HR: 3*
6-AMINO-8-AZAPURINE
mf: $C_4H_4N_6$ mw: 136.14

SYNS: 7-AMINO-1H-v-TRIAZOLO(4,5-d)PYRIMIDINE ◇ 8-AZAADENINE
◇ 8-AZAPURINE, 6-AMINO- ◇ 1H-v-TRIAZOLO(4,5-d)PYRIMIDIN-7-AMINE

TOXICITY DATA with REFERENCE
dnd-esc 20 2mol/L   MUREAV 89,95,81
ipr-mus LD50:315 mg/kg   PCJOAU 11,224,77

CONSENSUS REPORTS: Reported in EPA TSCA Inventory.

SAFETY PROFILE: Poison by intraperitoneal route. Mutation data reported. When heated to decomposition it emits toxic vapors of NO_x.

AIB350   CAS:104-23-4   *HR: D*
4'-AMINOAZOBENZENE-4-SULFONIC ACID
mf: $C_{12}H_{11}N_3O_3S$ mw: 277.32

SYNS: 4-AMINOAZOBENZENE-4'-SULPHONIC ACID ◇ p-((p-AMINO-PHENYL)AZO)BENZENESULFONIC ACID ◇ BENZENESULFONIC ACID, p-((p-AMINOPHENYL)AZO)- ◇ C.I. FOOD YELLOW 6

TOXICITY DATA with REFERENCE
otr-ham:kdy 25 mg/L   BJCAAI 38,34,78

CONSENSUS REPORTS: Reported in EPA TSCA Inventory.

SAFETY PROFILE: Mutation data reported. When heated to decomposition it emits toxic vapors of NO_x and SO_x.

AIC825   CAS:556-18-3   *HR: 2*
4-AMINOBENZALDEHYDE
mf: C_7H_7NO mw: 121.15

SYNS: p-AMINOBENZALDEHYDE ◇ BENZALDEHYDE, 4-AMINO-

TOXICITY DATA with REFERENCE
ipr-mus LD50:912 mg/kg   FEPRA7 6,348,47

CONSENSUS REPORTS: Reported in EPA TSCA Inventory.

SAFETY PROFILE: Moderately toxic by intraperitoneal route. When heated to decomposition it emits toxic vapors of NO_x.

AID620   CAS:88-68-6   *HR: 1*
2-AMINOBENZAMIDE
mf: $C_7H_8N_2O$ mw: 136.17

SYNS: o-AMINOBENZAMIDE ◇ ANTHRANILAMIDE ◇ ANTHRANILIMIDIC ACID ◇ BENZAMIDE, o-AMINO- ◇ BENZAMIDE, 2-AMINO-(9CI) ◇ 2-CARBAMOYLANILINE

TOXICITY DATA with REFERENCE
orl-brd LD50:1 g/kg   AECTCV 12,355,83

CONSENSUS REPORTS: Reported in EPA TSCA Inventory.

SAFETY PROFILE: Slightly toxic by ingestion. When heated to decomposition it emits toxic vapors of NO_x.

AID700   CAS:1197-55-3   *HR: 2*
4-AMINOBENZENEACETIC ACID
mf: $C_8H_9NO_2$ mw: 151.18

SYNS: ACETIC ACID, (p-AMINOPHENYL)- ◇ (p-AMINOPHENYL)ACETIC ACID ◇ 4-AMINOPHENYLACETIC ACID ◇ p-AMINO-α-TOLUIC ACID ◇ BENZENEACETIC ACID, 4-AMINO-(9CI) ◇ 4-CARBOXYMETHYLANILINE

TOXICITY DATA with REFERENCE
mma-sat 2500 µg/plate   DEMAEP 2,163,86
ipr-mus LD50:3500 mg/kg   FRPSAX 13,286,58

CONSENSUS REPORTS: Reported in EPA TSCA Inventory.

SAFETY PROFILE: Moderately toxic by intraperitoneal route. Mutation data reported. When heated to decomposition it emits toxic vapors of NO_x.

AIN150   CAS:619-45-4   *HR: 3*
p-AMINOBENZOIC ACID METHYL ESTER
mf: $C_8H_9NO_2$ mw: 151.18

SYNS: BENZOIC ACID, p-AMINO-, METHYL ESTER ◇ METHYL p-AMINOBENZOATE

TOXICITY DATA with REFERENCE
ipr-mus LD50:237 mg/kg   JMCMAR 17,900,74

CONSENSUS REPORTS: Reported in EPA TSCA Inventory.

SAFETY PROFILE: Poison by intraperitoneal route. When heated to decomposition it emits toxic vapors of NO_x.

AJA550　　　　　　CAS:104-13-2　　　　　HR: 3
p-AMINOBUTYLBENZENE
mf: $C_{10}H_{15}N$　　mw: 149.26

SYNS: 1-AMINO-4-BUTYLBENZENE ◇ ANILINE, 4-BUTYL- ◇ BENZENAMINE, 4-BUTYL-(9CI) ◇ p-n-BUTYLANILINE ◇ 4-BUTYLBENZENAMINE

TOXICITY DATA with REFERENCE
ipr-mus LD50:81 mg/kg　　JMCMAR 17,900,74

CONSENSUS REPORTS: Reported in EPA TSCA Inventory.

SAFETY PROFILE: Poison by intraperitoneal route. When heated to decomposition it emits toxic vapors of NO_x.

AJC750　　　　　　CAS:1688-71-7　　　　HR: 3
4'-AMINOBUTYROPHENONE
mf: $C_{10}H_{13}NO$　　mw: 163.24

SYNS: p-AMINOBUTYROPHENONE ◇ 1-(4-AMINOPHENYL)-1-BUTANONE ◇ 1-BUTANONE, 1-(4-AMINOPHENYL)-(9CI) ◇ BUTYROPHENONE, 4'-AMINO-

TOXICITY DATA with REFERENCE
orl-rat LD50:84 mg/kg　　GEPHDP 14,465,83
orl-mus LD50:133 mg/kg　　GEPHDP 14,465,83
ipr-mus LD50:183 mg/kg　　GEPHDP 14,465,83
orl-qal LD50:178 mg/kg　　AECTCV 12,355,83
orl-brd LD50:42200 µg/kg　　AECTCV 12,355,83

CONSENSUS REPORTS: Reported in EPA TSCA Inventory.

SAFETY PROFILE: Poison by ingestion and intraperitoneal routes. When heated to decomposition it emits toxic vapors of NO_x.

AJE300　　　　　　CAS:327-57-1　　　　　HR: D
2-AMINOCAPROIC ACID
mf: $C_6H_{13}NO_2$　　mw: 131.20

SYNS: α-AMINOCAPROIC ACID ◇ 2-AMINOHEXANOIC ACID ◇ (S)-2-AMINOHEXANOIC ACID ◇ CAPRINE ◇ GLYCOLEUCINE ◇ NORLEUCINE ◇ NORLEUCINE, L- ◇ L-NORLEUCINE (9CI) ◇ L-(+)-NORLEUCINE

TOXICITY DATA with REFERENCE
mrc-esc 300 mg/L　　JGMIAN 8,45,53
mmo-omi 100 mg/L　　MUREAV 12,349,71

CONSENSUS REPORTS: Reported in EPA TSCA Inventory.

SAFETY PROFILE: Mutation data reported. When heated to decomposition it emits toxic vapors of NO_x.

AJE400　　　　　　CAS:719-59-5　　　　　HR: 2
2-AMINO-5-CHLOROBENZOPHENONE
mf: $C_{13}H_{10}ClNO$　　mw: 231.69

SYNS: (2-AMINO-5-CHLOROPHENYL)PHENYLMETHANONE ◇ METHANONE, (2-AMINO-5-CHLOROPHENYL)PHENYL-

TOXICITY DATA with REFERENCE
ipr-mus LD50:681 mg/kg　　IJSIDW 44,1,82

CONSENSUS REPORTS: Reported in EPA TSCA Inventory.

SAFETY PROFILE: Moderately toxic by intraperitoneal route. When heated to decomposition it emits toxic vapors of NO_x and Cl^-.

AKC800　　　　　　CAS:1937-19-5　　　　HR: 2
AMINOGUANIDINE HYDROCHLORIDE
mf: $CH_6N_4 \cdot ClH$　　mw: 110.57

SYNS: GUANIDINE, AMINO-, HYDROCHLORIDE ◇ GUANYLHYDRAZINE HYDROCHLORIDE ◇ HYDRAZINECARBOXIMIDAMIDE HYDROCHLORIDE

TOXICITY DATA with REFERENCE
scu-rat LDLo:2984 mg/kg　　JPETAB 28,251,26

CONSENSUS REPORTS: Reported in EPA TSCA Inventory.

SAFETY PROFILE: Moderately toxic by subcutaneous route. When heated to decomposition it emits toxic vapors of NO_x, HCl, and Cl^-.

AKD925　　　　　　CAS:102-56-7　　　　　HR: 3
AMINOHYDROQUINONE DIMETHYL ETHER
mf: $C_8H_{11}NO_2$　　mw: 153.20

SYNS: ANILINE, 2,5-DIMETHOXY- ◇ BENZENAMINE, 2,5-DIMETHOXY-(9CI) ◇ C.I. 35811 ◇ 2,5-DIMETHOXYANILINE ◇ 2,5-DIMETHOXYBENZENAMINE

TOXICITY DATA with REFERENCE
orl-mus LD50:120 mg/kg　　GTPZAB 4(2),30,60
orl-brd LD50:100 mg/kg　　TXAPA9 21,315,72

CONSENSUS REPORTS: Reported in EPA TSCA Inventory.

SAFETY PROFILE: Poison by ingestion. When heated to decomposition it emits toxic vapors of NO_x.

AKM500　　　　　　CAS:120-35-4　　　　　HR: 3
3-AMINO-4-METHOXY BENZANILIDE
mf: $C_{14}H_{14}N_2O_2$　　mw: 242.30

SYN: BENZANILIDE, 3-AMINO-4-METHOXY-

TOXICITY DATA with REFERENCE
ivn-mus LD50:320 mg/kg　　CSLNX* NX#01183

CONSENSUS REPORTS: Reported in EPA TSCA Inventory.

SAFETY PROFILE: Poison by intravenous route. When heated to decomposition it emits toxic vapors of NO_x.

AKO350   CAS:2379-90-0   HR: 2
1-AMINO-2-METHOXY-4-OXYANTHRAQUINONE
mf: $C_{15}H_{11}NO_4$ mw: 269.27

SYNS: ACETOQUINONE LIGHT PINK RLZ ◇ 1-AMINO-4-HYDROXY-2-METHOXY-9,10-ANTHRACENEDIONE ◇ 1-AMINO-4-HYDROXY-2-METHOXYANTHRAQUINONE ◇ 1A-2MO-4OA ◇ 9,10-ANTHRACENEDIONE, 1-AMINO-4-HYDROXY-2-METHOXY-(9CI) ◇ ANTHRAQUINONE, 1-AMINO-4-HYDROXY-2-METHOXY- ◇ ARTISIL BRILLIANT PINK RFS ◇ CELLITON FAST PINK RF ◇ CELLITON FAST PINK RFA-CF ◇ CERVEN DISPERZNI 4 ◇ C.I. 60755 ◇ C.I. DISPERSE RED 4 ◇ CILLA FAST PINK RF ◇ DIANIX FAST PINK R ◇ DISPERSE PINK Zh ◇ DISPERSE RED-4 ◇ DISPERSE ROSE Zh ◇ ESTEROQUINONE LIGHT PINK RLL ◇ FENACET FAST PINK RF ◇ INTERCHEM ACETATE FAST PINK DNA ◇ MIKETON FAST PINK RL ◇ MIKETON POLYESTER PINK RL ◇ NYLOQUINONE PINK B ◇ PALANIL PINK RF ◇ PERILITON BRILLIANT PINK R ◇ SAMARON PINK RFL ◇ SUPRACET FAST PINK 2R

TOXICITY DATA with REFERENCE
ipr-rat LD50:2600 mg/kg   GTPZAB 21(12),27,77

CONSENSUS REPORTS: Reported in EPA TSCA Inventory.

SAFETY PROFILE: Moderately toxic by intraperitoneal route. When heated to decomposition it emits toxic vapors of NO_x.

AKY950   CAS:95-84-1   HR: D
2-AMINO-4-METHYLPHENOL
mf: C_7H_9NO mw: 123.17

TOXICITY DATA with REFERENCE
mmo-sat 333 µg/plate   EMMUEG 11(Suppl 12),1,88
mma-sat 333 µg/plate   EMMUEG 11(Suppl 12),1,88

CONSENSUS REPORTS: Reported in EPA TSCA Inventory.

SAFETY PROFILE: Mutation data reported. When heated to decomposition it emits toxic vapors of NO_x.

ALI240   CAS:84-89-9   HR: 2
5-AMINO-1-NAPHTHALENESULFONIC ACID
mf: $C_{10}H_9NO_3S$ mw: 223.26

SYN: 1-NAPHTHALENESULFONIC ACID, 5-AMINO-

TOXICITY DATA with REFERENCE
orl-rat LD50:>5 g/kg   GISAAA 51(1),87,86
ipr-rat LD50:2880 mg/kg   GISAAA 51(1),87,86
orl-mus LD50:>5 g/kg   GISAAA 51(1),87,86
ipr-mus LD50:2990 mg/kg   GISAAA 51(1),87,86

CONSENSUS REPORTS: Reported in EPA TSCA Inventory.

SAFETY PROFILE: Moderately toxic by intraperitoneal route. Low toxicity by ingestion. When heated to decomposition it emits toxic vapors of NO_x and SO_x.

ALJ500   CAS:86-97-5   HR: 3
5-AMINO-2-NAPHTHOL
mf: $C_{10}H_9NO$ mw: 159.20

SYN: 2-NAPHTHOL, 5-AMINO-

TOXICITY DATA with REFERENCE
ivn-mus LD50:180 mg/kg   CSLNX* NX#04025

CONSENSUS REPORTS: Reported in EPA TSCA Inventory.

SAFETY PROFILE: Poison by intravenous route. When heated to decomposition it emits toxic vapors of NO_x.

ALW100   CAS:722-27-0   HR: D
4-AMINOPHENYL DISULFIDE
mf: $C_{12}H_{12}N_2S_2$ mw: 248.38

SYNS: ANILINE, p,p'-DITHIODI- ◇ BENZENAMINE, 4,4'-DITHIOBIS-(9CI) ◇ 4,4'-DITHIOBISBENZENAMINE ◇ p,p'-DITHIODIANILINE ◇ 4,4'-DITHIODIANILINE

TOXICITY DATA with REFERENCE
mma-sat 100 µg/plate   MUREAV 67,123,79

CONSENSUS REPORTS: Reported in EPA TSCA Inventory.

SAFETY PROFILE: Mutation data reported. When heated to decomposition it emits toxic vapors of NO_x and SO_x.

ALX100   CAS:1783-81-9   HR: 2
m-AMINOPHENYL METHYL SULFIDE
mf: C_7H_9NS mw: 139.23

SYNS: m-AMINOTHIOANISOLE ◇ 3-AMINOTHIOANISOLE ◇ ANILINE, m-(METHYLTHIO)- ◇ BENZENAMINE, 3-(METHYLTHIO)-(9CI) ◇ 3-METHYLMERCAPTOANILINE ◇ m-(METHYLTHIO)ANILINE ◇ 3-(METHYLTHIO)BENZENAMINE

TOXICITY DATA with REFERENCE
orl-qal LD50:750 mg/kg   AECTCV 12,355,83
orl-brd LD50:750 mg/kg   AECTCV 12,355,83

CONSENSUS REPORTS: Reported in EPA TSCA Inventory.

SAFETY PROFILE: Moderately toxic by ingestion. When heated to decomposition it emits toxic vapors of NO_x and SO_x.

AMA010   CAS:1603-41-4   HR: 3
6-AMINO-3-PICOLINE
mf: $C_6H_8N_2$ mw: 108.16

SYNS: 2-AMINO-5-METHYLPYRIDINE ◇ 3-PICOLINE, 6-AMINO- ◇ 2-PYRIDINAMINE, 5-METHYL-

TOXICITY DATA with REFERENCE
orl-rat LD50:200 mg/kg   85JCAE-,841,86
scu-mus LD50:110 mg/kg   AJEBAK 36,491,58
skn-gpg LD50:400 mg/kg   85JCAE-,841,86

CONSENSUS REPORTS: Reported in EPA TSCA Inventory.

SAFETY PROFILE: Poison by ingestion, subcutaneous, and skin contact routes. When heated to decomposition it emits toxic vapors of NO_x.

AMK700   CAS:580-22-3   **HR: D**
2-AMINOQUINOLINE
mf: $C_9H_8N_2$ mw: 144.19

SYNS: 2-QUINOLINAMINE (9CI) ◇ QUINOLINE, 2-AMINO-

TOXICITY DATA with REFERENCE
mma-sat 500 nmol/plate   ABCHA6 42,861,78

CONSENSUS REPORTS: Reported in EPA TSCA Inventory.

SAFETY PROFILE: Mutation data reported. When heated to decomposition it emits toxic vapors of NO_x.

AMK725   CAS:580-17-6   **HR: 3**
3-AMINOQUINOLINE
mf: $C_9H_8N_2$ mw: 144.19

SYNS: 3-QUINOLINEAMINE ◇ QUINOLINE, 3-AMINO-

TOXICITY DATA with REFERENCE
mma-sat 5 FE2mol/plate   MUREAV 187,191,87
ipr-mus LD50:150 mg/kg   FATOAO 41,708,78
ivn-mus LD50:180 mg/kg   CSLNX* NX#03890

CONSENSUS REPORTS: Reported in EPA TSCA Inventory.

SAFETY PROFILE: Poison by intraperitoneal and intravenous route. Mutation data reported. When heated to decomposition it emits toxic vapors of NO_x.

AML600   CAS:634-60-6   **HR: 3**
2-AMINORESORCINOL HYDROCHLORIDE
mf: $C_6H_7NO_2 \cdot ClH$ mw: 161.60

SYN: RESORCINOL, 2-AMINO-, HYDROCHLORIDE

TOXICITY DATA with REFERENCE
orl-rat LDLo:500 mg/kg   JPETAB 90,260,47
ipr-rat LD50:30 mg/kg   PHBUA9 3,337,55

CONSENSUS REPORTS: Reported in EPA TSCA Inventory.

SAFETY PROFILE: Poison by intraperitoneal route. Moderately toxic by ingestion. When heated to decomposition it emits toxic vapors of NO_x.

AMS675   CAS:104-96-1   **HR: 2**
4-AMINOTHIOANISOLE
mf: C_7H_9NS mw: 139.23

SYNS: p-AMINOPHENYL METHYL SULFIDE ◇ p-AMINOTHIOANISOLE ◇ ANILINE, p-(METHYLTHIO)- ◇ BENZENAMINE, 4-(METHYLTHIO)-(9CI) ◇ 4-(METHYLTHIO)ANILINE ◇ 4-(METHYLTHIO)BENZENAMINE ◇ p-THIOANISIDINE ◇ p-THIOMETHOXYANILINE

TOXICITY DATA with REFERENCE
orl-qal LD50:562 mg/kg   AECTCV 12,355,83

CONSENSUS REPORTS: Reported in EPA TSCA Inventory.

SAFETY PROFILE: Moderately toxic by ingestion. When heated to decomposition it emits toxic vapors of NO_x and SO_x.

AMS750   CAS:1004-40-6   **HR: 3**
6-AMINO-2-THIOURACIL
mf: $C_4H_5N_3OS$ mw: 143.18

SYN: URACIL, 6-AMINO-2-THIO-

TOXICITY DATA with REFERENCE
ipr-mus LD50:370 mg/kg   ARZNAD 31,1713,81

CONSENSUS REPORTS: Reported in EPA TSCA Inventory.

SAFETY PROFILE: Poison by intraperitoneal route. When heated to decomposition it emits toxic vapors of NO_x and SO_x.

AMU550   CAS:344-72-9   **HR: 3**
2-AMINO-4-(TRIFLUOROMETHYL)-5-THIAZOLECARBOXYLIC ACID ETHYL ESTER
mf: $C_7H_7F_3N_2O_2S$ mw: 240.22

SYN: 5-THIAZOLECARBOXYLIC ACID, 2-AMINO-4-(TRIFLUOROMETHYL)-, ETHYL ESTER

TOXICITY DATA with REFERENCE
ivn-mus LDLo:75 mg/kg   CBCCT* 6,142,54

CONSENSUS REPORTS: Reported in EPA TSCA Inventory.

SAFETY PROFILE: Poison by intravenous route. When heated to decomposition it emits toxic vapors of NO_x and SO_x.

AMW100   CAS:932-52-5   **HR: D**
5-AMINOURACIL
mf: $C_4H_5N_3O_2$ mw: 127.12

SYNS: 2,4(1H,3H)-PYRIMIDINEDIONE, 5-AMINO- ◇ URACIL, 5-AMINO-

TOXICITY DATA with REFERENCE
mmo-esc 300 mg/L   JGMIAN 18,543,58
dni-hmn:hlas 5 mmol/L   RAREAE 37,334,69

CONSENSUS REPORTS: Reported in EPA TSCA Inventory.

SAFETY PROFILE: Human mutation data reported. When heated to decomposition it emits toxic vapors of NO_x.

ANB100   CAS:1863-63-4   HR: 3
AMMONIUM BENZOATE
mf: $C_7H_5O_2 \cdot H_4N$ mw: 139.17

SYNS: BENZOIC ACID, AMMONIUM SALT ◇ VULNOC AB

TOXICITY DATA with REFERENCE
orl-rat LD50:825 mg/kg   GISAAA 51(1),75,86
orl-mus LD50:235 mg/kg   GISAAA 51(1),75,86
ivn-rbt LDLo:400 mg/kg   JPETAB 44,81,32

CONSENSUS REPORTS: Reported in EPA TSCA Inventory.

SAFETY PROFILE: Poison by ingestion and intravenous routes. When heated to decomposition it emits toxic vapors of NO_x and NH_3.

AND250   HR: 3
AMMONIUM CADMIUM CHLORIDE
mf: $4NH_4Cl \cdot CdCl_2$ mw: 397.3

PROP: Colorless, rhombic crystals. D: 2.01; sol in water.

CONSENSUS REPORTS: Cadmium and its compounds are on the Community Right-To-Know List.

OSHA PEL: TWA 5 μg(Cd)/m^3
ACGIH TLV: TWA 0.05 mg(Cd)/m^3 (Proposed: TWA 0.01 mg(Cd)/m^3 (dust), Suspected Human Carcinogen; 0.002 mg(Cd)/m^3 (respirable dust), Suspected Human Carcinogen); BEI: 10 μg/g creatinine in urine; 10 μg/L in blood
DFG BAT: Blood: 1.5 μg/dL; Urine: 15 μg/dL; Suspected Carcinogen
NIOSH REL: (Cadmium) Reduce to lowest feasible level.

SAFETY PROFILE: Confirmed human carcinogen. A poison. When heated to decomposition it emits toxic fumes of NH_3, NO_x, and Cl$^-$. See also CADMIUM COMPOUNDS.

ANO750   CAS:1113-38-8   HR: 3
AMMONIUM OXALATE
DOT: NA 2449
mf: $C_2H_2O_4 \cdot 2H_3N$ mw: 124.12

PROP: Colorless crystals. Mp: decomp; d: 1.50. Sltly sol in water.

SYNS: ETHANEDIOIC ACID DIAMMONIUM SALT ◇ OXALIC ACID, DIAMMONIUM SALT

DOT Classification: ORM-A; Label: None

CONSENSUS REPORTS: Reported in EPA TSCA Inventory.

SAFETY PROFILE: A poison. Can react violently with (NaOCl + ammonium acetate). When heated to decomposition it can emit toxic fumes of NH_3 and NO_x. See also OXALATES.

ANT600   CAS:528-94-9   HR: 3
AMMONIUM SALICYLATE
mf: $C_7H_5O_3 \cdot H_4N$ mw: 155.17

SYNS: 2-HYDROXYBENZOIC ACID MONOAMMONIUM SALT ◇ SALICYLIC ACID, MONOAMMONIUM SALT ◇ SALICYL-VASOGEN

TOXICITY DATA with REFERENCE
orl-hmn TDLo:57 mg/kg:GIT   JPETAB 36,319,29
par-rat LDLo:600 mg/kg   JPETAB 36,319,29
par-mus LDLo:550 mg/kg   JPETAB 36,319,29
ivn-dog LDLo:467 mg/kg   AIPTAK 51,398,35

CONSENSUS REPORTS: Reported in EPA TSCA Inventory.

SAFETY PROFILE: Poison by intravenous route. Moderately toxic by parenteral route. Human systemic effects by ingestion: nausea or vomiting. When heated to decomposition it emits toxic vapors of NH_3.

ANU200   CAS:1002-89-7   HR: 1
AMMONIUM STEARATE
mf: $C_{18}H_{35}O_2 \cdot H_4N$ mw: 301.58

SYNS: AMMONIUM STEARATE ◇ OCTADECANOIC ACID, AMMONIUM SALT ◇ STEARIC ACID, AMMONIUM SALT

ACGIH TLV: TWA 10 mg/m^3

CONSENSUS REPORTS: Reported in EPA TSCA Inventory.

SAFETY PROFILE: A nuisance dust. When heated to decomposition it emits toxic vapors of NH_3.

ANY750   HR: 3
AMMONIUM VANADO-ARSENATE
mf: $H_{40}N_{10}O_5 \cdot 3As_2O_5 \cdot 4O_4V_2$ mw: 1228.78

TOXICITY DATA with REFERENCE
scu-rat LDLo:246 mg/kg   AJSNAO 1,347,17
ivn-rbt LDLo:75 mg/kg   AJSNAO 1,347,17

CONSENSUS REPORTS: Arsenic and its compounds are on the Community Right-To-Know List.
ACGIH TLV: TWA 0.2 mg(As)/m^3 (Proposed: 0.01 mg(As)/m^3; Human Carcinogen)
NIOSH REL: (Vanadium Compounds) CL 0.05 mg(V)/m^3/15M; (Arsenic, Inorganic) CL 2 μg(As)/m^3/15M

SAFETY PROFILE: Poison by subcutaneous and intravenous routes. See ARSENIC and VANADIUM COMPOUNDS. When heated to decomposition it emits very toxic NO_x, NH_3, and As.

AOE200   CAS:598-74-3   HR: 3
iso-AMYLAMINE
mf: $C_5H_{13}N$ mw: 87.19

SYNS: 2-BUTANAMINE, 3-METHYL-(9CI) ◇ 1,2-DIMETHYLPROPAN-

AMINE ◇ 1,2-DIMETHYLPROPYLAMINE ◇ 3-METHYL-2-BUTANAMINE ◇ PROPYLAMINE, 1,2-DIMETHYL-

TOXICITY DATA with REFERENCE
ipr-mus LD50:279 mg/kg   JJPAAZ 17,475,67

CONSENSUS REPORTS: Reported in EPA TSCA Inventory.

SAFETY PROFILE: Poison by intraperitoneal route. When heated to decomposition it emits toxic vapors of NO_x.

AOF800   CAS:110-53-2   HR: 2
n-AMYL BROMIDE
mf: $C_5H_{11}Br$ mw: 151.07

SYNS: AMYL BROMIDE ◇ 1-BROMOPENTANE ◇ PENTANE, 1-BROMO- ◇ PENTYL BROMIDE ◇ n-PENTYL BROMIDE ◇ 1-PENTYL BROMIDE

TOXICITY DATA with REFERENCE
ipr-mus LD50:1250 mg/kg   GTPZAB 20(12),52,76
ihl-uns LC50:26800 mg/m³   GTPZAB 18(4),55,74

CONSENSUS REPORTS: Reported in EPA TSCA Inventory.

SAFETY PROFILE: Moderately toxic by intraperitoneal route. Slightly toxic by inhalation. When heated to decomposition it emits toxic vapors of Br^-.

AOM200   CAS:686-31-7   HR: 3
tert-AMYL PEROXY-2-ETHYLHEXANOATE, technically pure (DOT)
DOT: UN 2898
mf: $C_{13}H_{26}O_3$ mw: 230.39

SYNS: 2-ETHYLPEROXYHEXANOIC ACID tert-PENTYL ESTER ◇ PEROXYHEXANOIC ACID, 2-ETHYL-, tert-PENTYL ESTER

DOT Classification: Organic Peroxide; Label: Organic Peroxide; Flammable Liquid; Label: Flammable Liquid, Organic Peroxide

CONSENSUS REPORTS: Reported in EPA TSCA Inventory.

SAFETY PROFILE: Flammable liquid and peroxide. When heated to decomposition it emits acrid smoke and irritating vapors.

AOM325   CAS:136-81-2   HR: 2
o-AMYLPHENOL
mf: $C_{11}H_{16}O$ mw: 164.27

SYNS: o-PENTYLPHENOL ◇ 2-PENTYLPHENOL ◇ PHENOL, o-PENTYL- ◇ PHENOL, 2-PENTYL-(9CI)

TOXICITY DATA with REFERENCE
orl-rat LD50:700 mg/kg   JPETAB 53,218,35

CONSENSUS REPORTS: Reported in EPA TSCA Inventory.

SAFETY PROFILE: Moderately toxic by ingestion. When heated to decomposition it emits acrid smoke and irritating vapors.

AOT100   CAS:1075-76-9   HR: 1
3-ANILINOPROPIONITRILE
mf: $C_9H_{10}N_2$ mw: 146.21

SYNS: β-ANILINOPROPIONITRILE ◇ N-(CYANOETHYL)ANILINE ◇ N-(β-CYANOETHYL)ANILINE ◇ N-(2-CYANOETHYL)ANILINE ◇ 2-PHENYL-AMINOPROPIONITRILE ◇ PROPANENITRILE, 3-ANILINO- ◇ PROPANENITRILE, 3-(PHENYLAMINO)- ◇ PROPIONITRILE, 3-ANILINO-

TOXICITY DATA with REFERENCE
orl-uns LD50:4500 mg/kg   PJPPAA 32,223,80

CONSENSUS REPORTS: Reported in EPA TSCA Inventory.

SAFETY PROFILE: Slightly toxic by ingestion. When heated to decomposition it emits toxic vapors of NO_x.

AOT300   CAS:591-31-1   HR: D
m-ANISALDEHYDE
mf: $C_8H_8O_2$ mw: 136.16

SYNS: BENZALDEHYDE, 3-METHOXY-(9CI) ◇ m-METHOXY-BENZALDEHYDE ◇ 3-METHOXYBENZALDEHYDE

TOXICITY DATA with REFERENCE
sce-hmn:lyms 1 mmol/L   MUREAV 206,17,88

CONSENSUS REPORTS: Reported in EPA TSCA Inventory.

SAFETY PROFILE: Mutation data reported. When heated to decomposition it emits acrid smoke and irritating vapors.

AOU600   CAS:100-09-4   HR: 3
p-ANISIC ACID
mf: $C_8H_8O_3$ mw: 152.16

SYNS: 4-ANISIC ACID ◇ DRACONIC ACID ◇ KYSELINA 4-METHOXY-BENZOOVA ◇ p-METHOXYBENZOIC ACID ◇ 4-METHOXYBENZOIC ACID

TOXICITY DATA with REFERENCE
scu-mus LD50:400 mg/kg   CKFRAY 31,236,82

CONSENSUS REPORTS: Reported in EPA TSCA Inventory.

SAFETY PROFILE: Poison by subcutaneous route. When heated to decomposition it emits acrid smoke and irritating vapors.

AVO890   CAS:536-90-3   HR: 2
m-ANISIDINE
mf: C_7H_9NO mw: 123.17

SYNS: m-AMINOANISOLE ◇ 3-AMINOANISOLE ◇ m-ANISYLAMINE ◇ BENZENAMINE, 3-METHOXY-(9CI) ◇ 3-METHOXYANILINE ◇ 3-METHOXYBENZENAMINE

TOXICITY DATA with REFERENCE
cyt-ham:ovr 160 mg/L   EMMUEG 10(Suppl 10),1,87
sce-ham:ovr 50 mg/L   EMMUEG 10(Suppl 10),1,87
orl-qal LD50:562 mg/kg   AECTCV 12,355,83
orl-brd LD50:562 mg/kg   AECTCV 12,355,83

CONSENSUS REPORTS: Reported in EPA TSCA Inventory.

SAFETY PROFILE: Moderately toxic by ingestion. Mutation data reported. When heated to decomposition it emits toxic vapors of NO_x.

AOY450   CAS:104-92-7   *HR: 2*
ANISYL BROMIDE
mf: C_7H_7BrO mw: 187.05

SYNS: ANISOLE, p-BROMO- ◇ BENZENE, 1-BROMO-4-METHOXY-(9CI) ◇ p-BROMANISOLE ◇ p-BROMOANISOLE ◇ 4-BROMOANISOLE ◇ 1-BROMO-4-METHOXYBENZENE ◇ p-BROMOPHENYL METHYL ETHER ◇ p-METHOXYBROMOBENZENE ◇ 4-METHOXYBROMOBENZENE ◇ p-METHOXYPHENYL BROMIDE ◇ 4-METHOXYPHENYL BROMIDE

TOXICITY DATA with REFERENCE
orl-mus LD50:2200 mg/kg   GISAAA 44(12),19,79
ipr-mus LD50:1186 mg/kg   GISAAA 44(12),19,79

CONSENSUS REPORTS: Reported in EPA TSCA Inventory.

SAFETY PROFILE: Moderately toxic by ingestion and intraperitoneal routes. When heated to decomposition it emits toxic vapors of Br^-.

APG550   CAS:723-62-6   *HR: 2*
ANTHRACENE-9-CARBOXYLIC ACID
mf: $C_{15}H_{10}O_2$ mw: 222.25

SYNS: 9-ANTHROIC ACID ◇ 9-CARBOXYANTHRACENE

TOXICITY DATA with REFERENCE
pic-esc 127 µg/well   MUREAV 260,349,91
ipr-mus LD50:750 mg/kg   JMCMAR 11,1020,68

CONSENSUS REPORTS: Reported in EPA TSCA Inventory.

SAFETY PROFILE: Moderately toxic by intraperitoneal route. Mutation data reported. When heated to decomposition it emits acrid smoke and irritating vapors.

APG700   CAS:2150-60-9   *HR: D*
2-ANTHRACENESULFONIC ACID, 4,8-DIAMINO-9,10-DIHYDRO-1,5-DIHYDROXY-9,10-DIOXO-, MONOSODIUM SALT
mf: $C_{14}H_9N_2O_7S \bullet Na$ mw: 372.30

SYNS: ACID ALIZARINE SAPPHIRE SE ◇ ACID ANTHRAQUINONE BLUE ◇ ACID BLUE 43 ◇ ACID LEATHER BLUE G ◇ ACILAN SAPPHIROL SE ◇ ALIZARIN BRILLANT BLUE BS ◇ ALIZARINE AZUROL SE ◇ ALIZARINE BLUE SE ◇ ALIZARINE BRILLANT BLUE BS ◇ ALIZARINE LIGHT BLUE SE ◇ ALIZARINE MSE ◇ ALIZARINE SAPPHIRE SE ◇ ALIZARINE SAPPHIROL SE ◇ ALIZARIN LIGHT BLUE SE ◇ ANTHRAQUINONE ACID BLUE ◇ BUCACID ALIZARINE LIGHT BLUE SE ◇ CALCOCID ALIZARINE BLUE SE ◇ C.I. 63000 ◇ C.I. ACID BLUE 43 ◇ ERIO FAST CYANINE SE ◇ FENAZO LIGHT BLUE AE ◇ KITON FAST BLUE G ◇ MITSUI ALIZARINE SAPHIROL SE ◇ OXANAL FAST BLUE G ◇ TERTRACID LIGHT BLUE SE ◇ VONDACID BLUE SE

TOXICITY DATA with REFERENCE
mma-sat 100 µg/plate   MUREAV 68,307,79

CONSENSUS REPORTS: Reported in EPA TSCA Inventory.

SAFETY PROFILE: Mutation data reported. When heated to decomposition it emits toxic vapors of NO_x and SO_x.

API800   CAS:609-86-9   *HR: 3*
ANTHRANILIC ACID, 3,5-DIIODO-
mf: $C_7H_5I_2NO_2$ mw: 388.93

SYNS: 2-AMINO-3,5-DIIODOBENZOIC ACID ◇ BENZOIC ACID, 2-AMINO-3,5-DIIODO-(9CI) ◇ 3,5-DIIODOANTHRANILIC ACID

TOXICITY DATA with REFERENCE
unr-rat LD50:180 mg/kg   JAPMA8 42,721,53
orl-mus LD50:900 mg/kg   QJPPAL 19,483,46

CONSENSUS REPORTS: Reported in EPA TSCA Inventory.

SAFETY PROFILE: Poison by an unspecified route. Moderately toxic by ingestion. When heated to decomposition it emits toxic vapors of NO_x.

APJ500   CAS:133-18-6   *HR: 1*
ANTHRANILIC ACID, PHENETHYL ESTER
mf: $C_{15}H_{15}NO_2$ mw: 241.31

PROP: White to yellow crystals; grape odor.

SYNS: BENZOIC ACID, 2-AMINO-, 2-PHENYLETHYL ESTER ◇ BENZYLCARBINYL ANTHRANILATE ◇ β-PHENETHYL-o-AMINOBENZOATE ◇ PHENETHYL ANTHRANILATE ◇ 2-PHENYLETHYL-o-AMINOBENZOATE ◇ PHENYLETHYL ANTHRANILATE ◇ 2-PHENYL-ETHYL ANTHRANILATE

TOXICITY DATA with REFERENCE
skn-rbt 500 mg/24H MOD   FCTXAV 14,659,76

CONSENSUS REPORTS: Reported in EPA TSCA Inventory.

SAFETY PROFILE: A skin irritant. See also ESTERS. When heated to decomposition it emits toxic fumes of NO_x.

APM500   CAS:90-44-8   *HR: D*
ANTHRONE
mf: $C_{14}H_{10}O$ mw: 194.24

SYNS: 9(10H)-ANTHRACENONE ◇ CARBOTHRONE ◇ 9,10-DIHYDRO-9-OXOANTHRACENE

TOXICITY DATA with REFERENCE
mmo-sat 60 µg/plate   MUREAV 157,149,85

CONSENSUS REPORTS: Reported in EPA TSCA Inventory.

SAFETY PROFILE: Mutation data reported. When heated to decomposition it emits acrid smoke and irritating vapors.

AQV500   CAS:627-75-8   HR: 2
d-ARGININE HYDROCHLORIDE
mf: $C_6H_{14}N_4O_2 \cdot ClH$ mw: 210.70

SYNS: ARGININE, MONOHYDROCHLORIDE, d- ◇ d-ARGININE, MONOHYDROCHLORIDE (9CI)

TOXICITY DATA with REFERENCE
ipr-rat LD50:3582 mg/kg   ABBIA4 64,319,56

CONSENSUS REPORTS: Reported in EPA TSCA Inventory.

SAFETY PROFILE: Moderately toxic by intraperitoneal route. When heated to decomposition it emits toxic vapors of NO_x, HCl, and Cl^-.

ARA250   CAS:98-50-0   HR: 3
ARSANILIC ACID
mf: $C_6H_8AsNO_3$ mw: 217.06

PROP: Needles from aq solns. Mp: 232°, bp: decomp, $-H_2O$ @ 15°. Very sol in hot water and alc; insol in ether and benzene.

SYNS: p-AMINOBENZENEARSONIC ACID ◇ 4-AMINOBENZENEARSONIC ACID ◇ AMINOPHENYLARSINE ACID ◇ p-AMINOPHENYLARSINE ACID ◇ p-AMINOPHENYLARSINIC ACID ◇ 4-AMINOPHENYLARSONIC ACID ◇ p-ANILINEARSONIC ACID ◇ ANTOXYLIC ACID ◇ p-ARSANILIC ACID ◇ 4-ARSANILIC ACID ◇ ATOXYLIC ACID

TOXICITY DATA with REFERENCE
orl-rat LD50:>1000 mg/kg   TXAPA9 18,185,71
ipr-rat LDLo:400 mg/kg   JPETAB 80,393,44
ipr-mus LD50:291 mg/kg   JMCMAR 9,221,66
ivn-mus LD50:100 mg/kg   CSLNX* NX#06774

CONSENSUS REPORTS: IARC Cancer Review: Animal Inadequate Evidence IMEMDT 23,39,80. Reported in EPA TSCA Inventory. Arsenic and its compounds are on the Community Right-To-Know List.

OSHA PEL: TWA 0.5 mg(As)/m^3
ACGIH TLV: TWA 0.2 mg(As)/m^3

SAFETY PROFILE: Poison by intravenous, and intraperitoneal routes. Moderately toxic by ingestion. Flammable, decomposes with heat to yield flammable vapors. When heated to decomposition or on contact with acid or acid fumes it emits highly toxic fumes of As and NO_x. See also ARSENIC COMPOUNDS and ANILINE.

ARA750   CAS:7440-38-2   HR: 3
ARSENIC
DOT: UN 1558

af: As   aw: 74.92

PROP: Silvery to black, brittle, crystalline, or amorphous metalloid. Mp: 814° @ 36 atm, bp: subl @ 612°, d: black crystals 5.724 @ 14°; black amorphous 4.7, vap press: 1 mm @ 372° (subl). Insol in water; sol in HNO_3.

SYNS: ARSEN (GERMAN, POLISH) ◇ ARSENIC, metallic (DOT) ◇ ARSENICALS ◇ ARSENIC-75 ◇ ARSENIC BLACK ◇ COLLOIDAL ARSENIC ◇ GREY ARSENIC ◇ METALLIC ARSENIC

TOXICITY DATA with REFERENCE
cyt-mus-ipr 4 mg/kg/48H-I   EXPEAM 37,129,81
orl-rat TDLo:605 µg/kg (35 W preg):REP   GISAAA (8)30,77
orl-rat TDLo:605 µg/kg (35 W preg):REP   GISAAA (8)30,77
orl-rat TDLo:580 µg/kg (female 30W pre):TER   FATOAO 41,620,78
orl-man TDLo:76 mg/kg/12Y-I:CAR   RMCHAW 99,664,71
imp-rbt TDLo:75 mg/kg:ETA   ZEKBAI 52,425,42
orl-man TDLo:7857 mg/kg/55Y:SKN   CMAJAX 120,168,79
orl-man TDLo:7857 mg/kg/55Y:GIT   CMAJAX 120,168,79
orl-rat LD50:763 mg/kg   GTPZAB 31(12),53,87
orl-mus LD50:145 mg/kg   GTPZAB 31(12),53,87
ipr-mus LD50:46,200 µg/kg   GTPZAB 31(12),53,87
scu-rbt LDLo:300 mg/kg   ASBIAL 24,442,38
ipr-gpg LDLo:10 mg/kg   CRSBAW 81,164,18
scu-gpg LDLo:300 mg/kg   ASBIAL 24,442,38

CONSENSUS REPORTS: NTP Fifth Annual Report on Carcinogens. IARC Cancer Review: Group 1 IMEMDT 7,100,87; Human Sufficient Evidence IMEMDT 23,39,80; Human Inadequate Evidence IMEMDT 2,48,73. Reported in EPA TSCA Inventory. Arsenic and its compounds are on the Community Right-To-Know List.

OSHA PEL: TWA 0.01 mg(As)/m^3; Cancer Hazard
ACGIH TLV: TWA 0.2 mg(As)/m^3 (Proposed: 0.01 mg(As)/m^3; Human Carcinogen)
DFG TRK: 0.2 mg/m^3 calculated as arsenic in that portion of dust that can possibly be inhaled.
NIOSH REL: CL 2 µg(As)/m^3

DOT Classification: Poison B; Label: Poison

SAFETY PROFILE: Confirmed human carcinogen producing liver tumors. Poison by subcutaneous, intramuscular, and intraperitoneal routes. Human systemic skin and gastrointestinal effects by ingestion. An experimental teratogen. Other experimental reproductive effects. Mutation data reported. Flammable in the form of dust when exposed to heat or flame or by chemical reaction with powerful oxidizers such as bromates, chlorates, iodates, peroxides, lithium, NCl_3, KNO_3, $KMnO_4$, Rb_2C_2, $AgNO_4$, NOCl, IF_5, CrO_3, ClF_3, ClO, BrF_3, BrF_5, BrN_3, RbC_3BCH, CsC_3BCH. Slightly explosive in the form of dust when exposed to flame. When heated or on contact with acid or acid fumes, it emits highly toxic fumes; can react vigorously on contact with oxidizing materials. Incompatible with bromine azide, dirubidium acetylide, halogens, palladium, zinc, platinum, NCl_3, $AgNO_3$, CrO_3, Na_2O_2, hexafluoroisopropylideneamino lithium.

ARB000   CAS:10102-53-1   **HR: 3**
m-ARSENIC ACID
mf: $AsHO_3$ mw: 123.93

SYN: METAARSENIC ACID

CONSENSUS REPORTS: Reported in EPA TSCA Inventory. Arsenic and its compounds are on the Community Right-To-Know List.

OSHA PEL: TWA 0.01 mg(As)/m^3; Cancer Hazard
ACGIH TLV: TWA 0.2 mg(As)/m^3 (Proposed: 0.01 mg(As)/m^3; Human Carcinogen)
DFG MAK: Human Carcinogen
NIOSH REL: (Arsenic, Inorganic) CL 2 µg(As)/m^3/15M

SAFETY PROFILE: Confirmed human carcinogen. When heated to decomposition it emits toxic fumes of arsenic. See also ARSENIC COMPOUNDS.

ARB250   CAS:7778-39-4   **HR: 3**
o-ARSENIC ACID
DOT: UN 1553/UN 1554
mf: AsH_3O_4 mw: 141.95

SYNS: ACIDE ARSENIQUE LIQUIDE (FRENCH) ◇ ARSENATE ◇ ARSENIC ACID, liquid (DOT) ◇ ARSENIC ACID, solid (DOT) ◇ DESICCANT L-10 ◇ HI-YIELD DESICCANT H-10 ◇ ORTHOARSENIC ACID ◇ RCRA WASTE NUMBER P010 ◇ ZOTOX ◇ ZOTOX CRAB GRASS KILLER

TOXICITY DATA with REFERENCE
cyt-hmn:leu 7200 nmol/L   MUREAV 88,73,81
cyt-hmn:fbr 100 ppb   MUREAV 88,73,81
ipr-rat TDLo:30 mg/kg (9D preg):TER   JTSCDR 4,405,79
orl-mus TDLo:120 mg/kg (7–15D preg):TER   TJADAB 15,301,77
orl-mus TDLo:120 mg/kg (7–15D preg):TER   TJADAB 15,301,77
ipr-rat TDLo:30 mg/kg (9D preg):TER   JTSCDR 4,405,79
orl-rat LD50:48 mg/kg   FMCHA2 -,C18,83
orl-dog LDLo:10 mg/kg   FDWU** -,-,31
orl-rbt LDLo:5 mg/kg   FDWU** -,-,31
orl-pgn LDLo:100 mg/kg   FDWU** -,-,31
orl-ckn LDLo:125 mg/kg   FDWU** -,-,31

CONSENSUS REPORTS: Reported in EPA TSCA Inventory. Arsenic and its compounds are on the Community Right-To-Know List.

OSHA PEL: TWA 0.01 mg(As)/m^3; Cancer Hazard
ACGIH TLV: TWA 0.2 mg(As)/m^3 (Proposed: 0.01 mg(As)/m^3; Human Carcinogen)
DFG MAK: Human Carcinogen
NIOSH REL: (Arsenic, Inorganic) CL 2 µg(As)/m^3/15M

DOT Classification: Poison B; Label: Poison

SAFETY PROFILE: Confirmed human carcinogen. Poison by ingestion. An experimental teratogen. Human mutation data reported. When heated to decomposition it emits toxic fumes of arsenic. See also ARSENIC COMPOUNDS.

ARB750   CAS:7778-44-1   **HR: 3**
ARSENIC ACID, CALCIUM SALT (2583)
DOT: UN 1573
mf: As_2O_8•3Ca   mw: 398.08

PROP: Colorless, amorphous powder. D: 3.620. Solubility in water: 0.013/100 @ 25°.

SYNS: ARSENIATE de CALCIUM (FRENCH) ◇ CALCIUMARSENAT ◇ CALCIUM ARSENATE (MAK) ◇ CALCIUM ORTHOARSENATE ◇ KALZIUMARSENIAT (GERMAN) ◇ TRICALCIUMARSENAT (GERMAN) ◇ TRICALCIUM ARSENATE

TOXICITY DATA with REFERENCE
itr-rat TDLo:1600 µg/kg:ETA   IJCNAW 24,786,79
itr-ham TDLo:120 mg/kg/15W-C:NEO   CALEDQ 27,99,85
itr-ham TD:214 mg/kg/15W-I:NEO   IJCNAW 40,220,87
orl-rat LD50:20 mg/kg   AFDOAQ 15,122,51
skn-rat LD50:2400 mg/kg   28ZEAL 5,35,76
orl-mus LD50:794 mg/kg   AMRL** TR-72-62,72
orl-dog LD50:38 mg/kg   85DPAN -,-,71/76
orl-rbt LDLo:50 mg/kg   JPETAB 39,246,30
orl-mam LD50:35 mg/kg   PCOC** -,170,66

CONSENSUS REPORTS: NTP Fifth Annual Report on Carcinogens. IARC Cancer Review: Group 1 IMEMDT 7,100,87; Human Sufficient Evidence IMEMDT 23,39,80; Animal No Evidence IMEMDT 2,48,73; Animal Inadequate Evidence IMEMDT 23,39,80. Reported in EPA TSCA Inventory. Arsenic and its compounds are on the Community Right-To-Know List. EPA Extremely Hazardous Substances List.

OSHA PEL: TWA 0.01 mg(As)/m^3; Cancer Hazard
ACGIH TLV: TWA 0.2 mg(As)/m^3 (Proposed: 0.01 mg(As)/m^3; Human Carcinogen)
DFG MAK: Human Carcinogen
NIOSH REL: CL 2 µg(As)/m^3/15M

DOT Classification: Poison B; Label: Poison

SAFETY PROFILE: Confirmed human carcinogen. Poison by ingestion. Moderately toxic by skin contact. When heated to decomposition it emits toxic fumes of arsenic.

ARC000   CAS:7778-43-0   **HR: 3**
ARSENIC ACID, DISODIUM SALT
mf: Na_2HAsO_4•$7H_2O$ mw: 312.01

PROP: Colorless powder, effloresces. D: 1.88, mp: − $7H_2O$ @ 130°, bp: decomp @ 150°. Solubility in water: 61/100 @ 15°, sol in glycerin.

SYNS: DISODIUM ARSENATE ◇ DISODIUM ARSENIC ACID ◇ DISODIUM HYDROGEN ARSENATE ◇ DISODIUM HYDROGEN ORTHOARSE-

NATE ◇ DISODIUM MONOHYDROGEN ARSENATE ◇ SODIUM ACID ARSENATE ◇ SODIUM ARSENATE ◇ SODIUM ARSENATE DIBASIC, anhydrous

TOXICITY DATA with REFERENCE
cyt-hmn:leu 7200 μmol/L   MUREAV 88,73,81
mrc-bcs 100 mmol/L   MUREAV 77,109,80
ipr-rat LDLo:30 mg/kg   JPETAB 58,454,36

CONSENSUS REPORTS: Reported in EPA TSCA Inventory. Arsenic and its compounds are on the Community Right-To-Know List.

OSHA PEL: TWA 0.5 mg(As)/m^3: Cancer Hazard
ACGIH TLV: TWA 0.2 mg(As)/m^3 (Proposed: 0.01 mg(As)/m^3; Human Carcinogen)
NIOSH REL: (Arsenic, Inorganic) CL 2 μg(As)/m^3/15M
DFG MAK: Human Carcinogen

SAFETY PROFILE: Confirmed human carcinogen. Poison by intraperitoneal route. Human mutation data reported. When heated to decomposition it emits toxic fumes of arsenic. See ARSENIC COMPOUNDS.

ARC250　　　CAS:10048-95-0　　　**HR: 3**
ARSENIC ACID, DISODIUM SALT, HEPTAHYDRATE
mf: AsHO$_4$•2Na•7H$_2$O   mw: 427.05

SYNS: DISODIUM ARSENATE, HEPTAHYDRATE ◇ SODIUM ACID ARSENATE, HEPTAHYDRATE ◇ SODIUM ARSENATE, DIBASIC, HEPTAHYDRATE ◇ SODIUM ARSENATE HEPTAHYDRATE

TOXICITY DATA with REFERENCE
orl-mus TDLo:120 mg/kg (female 10D post):TER   EVHPAZ 19,219,77
ipr-rat TDLo:30 mg/kg (female 9D post):TER   TJADAB 10,153,74
ipr-mus TDLo:45 mg/kg (female 9D post):REP   AEHLAU 24,62,72
orl-mus TDLo:120 mg/kg (female 10D post):REP   EVHPAZ 19,219,77
ipr-mus TDLo:40 mg/kg (female 9D post):TER   NTIS** CONF-771017
ipr-rat TDLo:20 mg/kg (female 9D post):TER   TJADAB 10,153,74
ipr-mus TDLo:40 mg/kg (female 9D post):TER   NTIS** CONF-771017
ipr-rat TDLo:20 mg/kg (female 9D post):TER   TJADAB 10,153,74
ipr-rat TDLo:20 mg/kg (female 10D post):TER   TJADAB 10,153,74
ipr-mus TDLo:40 mg/kg (female 9D post):TER   TJADAB 6,235,72
ipr-rat TDLo:20 mg/kg (female 10D post):TER   TJADAB 10,153,74
ims-mus LD50:87,360 μg/kg   EXMDA4 (440),312,78
scu-gpg LDLo:50 mg/kg   BMJOAE 2,217,13

CONSENSUS REPORTS: NTP Fifth Annual Report on Carcinogens. Arsenic and its compounds are on the Community Right-To-Know List.

OSHA PEL: TWA 0.01 mg(As)/m^3; Cancer Hazard
ACGIH TLV: TWA 0.2 mg(As)/m^3 (Proposed: 0.01 mg(As)/m^3; Human Carcinogen)
NIOSH REL: (Arsenic, Inorganic) CL 2 μg(As)/m^3/15M
DFG MAK: Human Carcinogen

SAFETY PROFILE: Confirmed human carcinogen. Poison by subcutaneous route. An experimental teratogen. Other experimental reproductive effects. See also ARSENIC COMPOUNDS. When heated to decomposition it emits toxic fumes of arsenic.

ARC500　　　CAS:7774-41-6　　　**HR: 3**
ARSENIC ACID, HEMIHYDRATE
mf: AsH$_3$O$_4$•1/2H$_2$O   mw: 150.96

PROP: White, translucent crystals. Mp: 35.5°, bp: – H$_2$O @ 160°, d: 2.0–**2.5.**

SYNS: ARSENIC ACID, solid (DOT) ◇ ORTHOARSENIC ACID HEMIHYDRATE

TOXICITY DATA with REFERENCE
ivn-rbt LD50:6 mg/kg   MEIEDD 11,126,89

CONSENSUS REPORTS: Arsenic and its compounds are on the Community Right-To-Know List.

OSHA PEL: TWA 0.01 mg(As); Cancer Hazard
ACGIH TLV: TWA 0.2 mg(As)/m^3 (Proposed: 0.01 mg(As)/m^3; Human Carcinogen)
NIOSH REL: (Arsenic, Inorganic) CL 2 μg(As)/m^3/15M
DFG MAK: Human Carcinogen.

DOT Classification: Poison B; Label: Poison

SAFETY PROFILE: Confirmed human carcinogen. Poison by intravenous route. When heated to decomposition it emits toxic fumes of arsenic. See also ARSENIC COMPOUNDS.

ARD000　　　CAS:10103-50-1　　　**HR: 3**
ARSENIC ACID, MAGNESIUM SALT
DOT: UN 1622
mf: AsH$_3$O$_4$•7Mg   mw: 312.12

PROP: Monoclinic, white crystals. D: 2.60–2.61.

SYNS: ARSENIATE de MAGNESIUM (FRENCH) ◇ MAGNESIUM ARSENATE ◇ MAGNESIUM ARSENATE PHOSPHOR

TOXICITY DATA with REFERENCE
orl-rat LDLo:280 mg/kg   TXAPA9 1,156,59
orl-mus LD50:315 mg/kg   IRGGAJ 20,21,63
orl-rbt LDLo:80 mg/kg   AIHAAP 19,504,58

CONSENSUS REPORTS: Reported in EPA TSCA Inventory. Arsenic and its compounds are on the Community Right-To-Know List.

OSHA PEL: TWA 0.01 mg(As)/m^3; Cancer Hazard
ACGIH TLV: TWA 0.2 mg(As)/m^3 (Proposed: 0.01 mg(As)/m^3; Human Carcinogen)
DFG MAK: Human Carcinogen
NIOSH REL: (Arsenic, Inorganic) CL 2 µg(As)/m^3/15M

DOT Classification: Poison B; Label: Poison

SAFETY PROFILE: Confirmed human carcinogen. Poison by ingestion. When heated to decomposition it emits toxic fumes of arsenic. See also ARSENIC COMPOUNDS.

ARD250   CAS:7784-41-0   HR: 3
ARSENIC ACID, MONOPOTASSIUM SALT
DOT: UN 1677
mf: AsH$_2$O$_4$•K   mw: 180.04

SYNS: MACQUER'S SALT ◇ MONOPOTASSIUM ARSENATE ◇ MONOPOTASSIUM DIHYDROGEN ARSENATE ◇ POTASSIUM ACID ARSENATE ◇ POTASSIUM ARSENATE ◇ POTASSIUM DIHYDROGEN ARSENATE ◇ POTASSIUM HYDROGEN ARSENATE

TOXICITY DATA with REFERENCE
cyt-hmn:leu 1 µmol/L   CNREA8 25,980,65

CONSENSUS REPORTS: NTP Fifth Annual Report on Carcinogens. IARC Cancer Review: Human Sufficient Evidence IMEMDT 23,39,80. Reported in EPA TSCA Inventory. Arsenic and its compounds are on the Community Right-To-Know List.

OSHA PEL: TWA 0.01 mg(As)/m^3, Cancer Hazard
ACGIH TLV: TWA 0.2 mg(As)/m^3 (Proposed: 0.01 mg(As)/m^3; Human Carcinogen)
NIOSH REL: (Arsenic, Inorganic) CL 2 µg(As)/m^3/15M

DOT Classification: Poison B; Label: Poison.

SAFETY PROFILE: Confirmed human carcinogen. Mutation data reported. When heated to decomposition it emits toxic fumes of arsenic. See also ARSENIC COMPOUNDS.

ARD500   CAS:15120-17-9   HR: 3
ARSENIC ACID, MONOSODIUM SALT
mf: AsO$_3$•Na   mw: 145.91

SYNS: ARSENIC ACID, SODIUM SALT (9CI) ◇ SODIUM ARSENATE ◇ SODIUM METAARSENATE ◇ SODIUM MONOHYDROGEN ARSENATE

TOXICITY DATA with REFERENCE
sln-dmg-orl 2 µmol/L   CNJGA8 11,677,69
slt-dmg-orl 100 µmol   CNJGA8 17,55,75

CONSENSUS REPORTS: Arsenic and its compounds are on the Community Right-To-Know List.

OSHA PEL: TWA 0.5 mg(As)/m^3: Cancer Hazard
ACGIH TLV: TWA 0.2 mg(As)/m^3 (Proposed: 0.01 mg(As)/m^3; Human Carcinogen)
DFG MAK: Human Carcinogen
NIOSH REL: (Arsenic, Inorganic) CL 2 µg(As)/m^3/15M

SAFETY PROFILE: Confirmed human carcinogen. A poison. Mutation data reported. See also ARSENIC COMPOUNDS. When heated to decomposition it emits toxic fumes of arsenic.

ARD600   CAS:10103-60-3   HR: 3
ARSENIC ACID, MONOSODIUM SALT
mf: AsH$_2$O$_4$•Na   mw: 163.93

SYNS: MONOSODIUM ARSENATE ◇ SODIUM ARSENATE ◇ SODIUM DIHYDROGEN ARSENATE ◇ SODIUM DIHYDROGEN ORTHOARSENATE

TOXICITY DATA with REFERENCE
ivn-rbt LDLo:45 mg/kg   JPETAB 23,107,24

OSHA PEL: TWA 0.5 mg(As)/m^3: Cancer Hazard
ACGIH TLV: TWA 0.2 mg(As)/m^3 (Proposed: 0.01 mg(As)/m^3; Human Carcinogen)
NIOSH REL: CL 2 µg(As)/m^3/15M
DFG MAK: Human Carcinogen

CONSENSUS REPORTS: Arsenic and its compounds are on the Community Right-To-Know List.

SAFETY PROFILE: Confirmed human carcinogen. Poison by intravenous route. When heated to decomposition it emits toxic fumes of arsenic.

ARD750   CAS:7631-89-2   HR: 3
ARSENIC ACID, SODIUM SALT
DOT: UN 1685
mf: AsH$_3$O$_4$•7Na   mw: 202.94

SYNS: FATSCO ANT POISON ◇ SODIUM ARSENATE (DOT) ◇ SODIUM ORTHOARSENATE ◇ SWEENEY'S ANT-GO

TOXICITY DATA with REFERENCE
cyt-hmn:lym 2 µmol/L   ADREDL 267,91,80
orl-mus TDLo:475 mg/kg (female 8–12D post):REP
   TCMUD8 6,361,86
orl-mus TDLo:475 mg/kg (female 8–12D post):REP
   TCMUD8 6,361,86
orl-mus TDLo:120 mg/kg (female 11D post):TER   JEPTDQ
   1(6),857,78
orl-mus TDLo:120 mg/kg (female 9–11D post):TER
   TJADAB 8,98,73
orl-mus TDLo:120 mg/kg (female 9–11D post):TER
   TJADAB 8,98,73
orl-mus TDLo:120 mg/kg (female 11D post):REP   JEPTDQ
   1(6),857,78
ipr-mus TDLo:40 mg/kg (female 9D post):TER   JEPTDQ
   1(6),857,78
ipr-mus TDLo:40 mg/kg (female 9D post):TER   JEPTDQ
   1(6),857,78
ipr-mus TDLo:40 mg/kg (female 9D post):TER   JEPTDQ
   1(6),857,78
orl-mus TDLo:120 mg/kg (female 9D post):TER   JEPTDQ
   1(6),857,78
scu-mus TDLo:10 mg(As)/kg/20D-C:ETA   VDGPAN 55,289,71

scu-mus TDLo:10 mg(As)/kg/1–20D:ETA   VDGPAN 55,289,71
ivn-mus TDLo:10 mg(As)/kg/20W-I:ETA   VDGPAN 55,289,71
ipr-rat LDLo:49 mg/kg   JPETAB 58,454,36
ivn-rat LDLo:85 mg/kg   JPETAB 33,270,28
orl-rbt LDLo:51 mg/kg   JPETAB 33,270,28
ivn-rbt LDLo:28 mg/kg   JPETAB 33,270,28

CONSENSUS REPORTS: NTP Fifth Annual Report on Carcinogens. IARC Cancer Review: Human Sufficient Evidence IMEMDT 23,39,80; Animal Inadequate Evidence IMEMDT 2,48,73; IMEMDT 23,39,80. Reported in EPA TSCA Inventory. Arsenic and its compounds are on the Community Right-To-Know List.

OSHA PEL: TWA 0.01 mg(As)/m^3; Cancer Hazard
ACGIH TLV: TWA 0.2 mg(As)/m^3 (Proposed: 0.01 mg(As)/m^3; Human Carcinogen)
NIOSH REL: CL 2 μg(As)/m^3/15M

DOT Classification: Poison B; Label: Poison

SAFETY PROFILE: Confirmed human carcinogen with experimental tumorigenic data. Poison by ingestion, intravenous, and intraperitoneal routes. An experimental teratogen. Other experimental reproductive effects. Mutation data reported. When heated to decomposition it emits toxic fumes of As and Na$_2$O. See also ARSENIC COMPOUNDS.

ARE500   CAS:8028-73-7   *HR: 3*
ARSENICAL DUST
DOT: UN 1562

SYNS: ARSENICAL FLUE DUST ◇ FLUE DUST, ARSENIC CONTAINING

TOXICITY DATA with REFERENCE
itr-rat TDLo:120 mg/kg/15W-I:ETA   EVHPAZ 19,191,77

CONSENSUS REPORTS: Reported in EPA TSCA Inventory. Arsenic and its compounds are on the Community Right-To-Know List.

OSHA PEL: TWA 0.5 mg(As)/m^3
ACGIH TLV: TWA 0.2 mg(As)/m^3 (Proposed: 0.01 mg(As)/m^3; Human Carcinogen)
NIOSH REL: CL 2 μg(As)/m^3/15M

DOT Classification: Poison B; Label: Poison

SAFETY PROFILE: A poison. Questionable carcinogen with experimental tumorigenic data. See also ARSENIC COMPOUNDS.

ARF250   CAS:7784-33-0   *HR: 3*
ARSENIC(III) BROMIDE
DOT: UN 1555
mf: AsBr$_3$ mw: 314.65

PROP: Colorless, rhombic crystals. Mp: 32.8°, bp: 220.0°, vap press: 1 mm @ 41.8°, d: 3.3972 @ 25°, (liq) 3.3282.

SYNS: ARSENIC TRIBROMIDE ◇ ARSENOUS BROMIDE ◇ ARSENOUS TRIBROMIDE ◇ TRIBROMOARSINE

CONSENSUS REPORTS: Reported in EPA TSCA Inventory. Arsenic and its compounds are on the Community Right-To-Know List.

OSHA PEL: TWA 0.01 mg(As)/m^3; Cancer Hazard
ACGIH TLV: TWA 0.2 mg(As)/m^3 (Proposed: 0.01 mg(As)/m^3; Human Carcinogen)
NIOSH REL: (Arsenic, Inorganic) CL 2 μg(As)/m^3/15M

DOT Classification: Poison B; Label: Poison

SAFETY PROFILE: Confirmed carcinogen. A poison. See also ARSENIC COMPOUNDS and BROMIDES. When heated to decomposition it emits very toxic fumes of As and Br$^-$.

ARF500   CAS:7784-34-1   *HR: 3*
ARSENIC CHLORIDE
DOT: UN 1560
mf: AsCl$_3$ mw: 181.28

PROP: Colorless, oily liquid. D: 2.15 @ 25°, mp: −16°, bp: 130°. Decomp in water and by UV light; misc in chloroform, CCl$_4$, ether, iodine, P, S, alkali iodides, oils and fats. Vap d: 6.25, vap press: 10 mm @ 23.5°.

SYNS: ARSENIC BUTTER ◇ ARSENIC(III) CHLORIDE ◇ ARSENIOUS CHLORIDE ◇ ARSENOUS CHLORIDE ◇ ARSENOUS TRICHLORIDE (9CI) ◇ CHLORURE d'ARSENIC (FRENCH) ◇ CHLORURE ARSENIEUX (FRENCH) ◇ FUMING LIQUID ARSENIC ◇ TRICHLOROARSINE ◇ TRICHLORURE d'ARSENIC (FRENCH)

TOXICITY DATA with REFERENCE
cyt-hmn:leu 600 nmol/L   MUREAV 88,73,81
mrc-bcs 30 μL/disc   MUREAV 77,109,80
otr-ham:emb 3 μmol/L   CNREA8 39,193,79
ihl-mus LCLo:338 ppm/10M   HBTXAC 1,324,56
ihl-cat LCLo:100 mg/m^3/1H   ZGEMAZ 13,523,21

CONSENSUS REPORTS: Reported in EPA TSCA Inventory. Arsenic and its compounds are on the Community Right-To-Know List. EPA Extremely Hazardous Substances List.

OSHA: Cancer Hazard
ACGIH TLV: TWA 0.2 mg(As)/m^3 (Proposed: 0.01 mg(As)/m^3; Human Carcinogen)
NIOSH REL: (Arsenic, Inorganic) CL 2 μg(As)/m^3/15M

DOT Classification: Poison B; Label: Poison

SAFETY PROFILE: A poison via inhalation. See also ARSENIC COMPOUNDS and CHLORIDES. Very poisonous; fumes in air. Mutation data reported. When heated to decomposition it emits very toxic fumes of As and Cl$^-$. Highly reactive. Explodes with Na, K, and Al on impact.

ARF750 HR: 3
ARSENIC COMPOUNDS

SYN: ARSENICALS

CONSENSUS REPORTS: Arsenic and its compounds are on the Community Right-To-Know List.

OSHA PEL: Inorganic: TWA 0.01 mg(As)/m^3; Cancer Hazard; Organic: TWA 0.5 mg(As)/m^3
ACGIH TLV: TWA 0.2 mg(As)/m^3 (Proposed: (inorganic compounds) 0.01 mg(As)/m^3; Human Carcinogen)
NIOSH REL: CL 2 µg(As)/m^3/15M

SAFETY PROFILE: Inorganic compounds are confirmed human carcinogens producing tumors of the mouth, esophagus, larynx, bladder, and paranasal sinus. A recognized carcinogen of the skin, lungs, and liver. Used as insecticides, herbicides, silvicides, defoliants, desiccants, and rodenticides. Poisoning from arsenic compounds may be acute or chronic. Acute poisoning usually results from swallowing arsenic compounds; chronic poisoning from either swallowing or inhaling. Acute allergic reactions to arsenic compounds used in medical therapy have been fairly common, the type and severity of reaction depending upon the compound. Inorganic arsenicals are more toxic than organics. Trivalent is more toxic than pentavalent. Acute arsenic poisoning (from ingestion) results in marked irritation of the stomach and intestines with nausea, vomiting, and diarrhea. In severe cases, the vomitus and stools are bloody and the patient goes into collapse and shock with weak, rapid pulse, cold sweats, coma, and death. Chronic arsenic poisoning, whether through ingestion or inhalation, may manifest itself in many different ways. There may be disturbances of the digestive system such as loss of appetite, cramps, nausea, constipation, or diarrhea. Liver damage may occur, resulting in jaundice. Disturbances of the blood, kidneys, and nervous system are not infrequent. Arsenic can cause a variety of skin abnormalities including itching, pigmentation, and even cancerous changes. A characteristic of arsenic poisoning is the great variety of symptoms that can be produced. Dangerous; when heated to decomposition, or when metallic arsenic contacts acids or acid fumes, or when water solutions of arsenicals are in contact with active metals such as Fe, Al, Zn, highly toxic fumes of arsenic are emitted.

In treating acute poisoning from ingestion BAL (dimercaptol) is of questionable effectiveness for acute and chronic poisoning with trivalent arsenicals, such as arsenic trioxide, arsine, and arsenites. It is of no value for pentavalent arsenicals, such as cacodylic acid, methanearsonic acid, sodium cacodylate, MSMA, DSMA, arsanilic acid, arsenic acid, and arsenates. Vomiting and gastric lavage are the preferred emergency treatments for acute arsenical poisoning. More recent medical treatment of arsenical poisoning uses exchange transfusion and dialysis. Note: Arsenic compounds are common air contaminants.

ARG750 CAS:7784-45-4 HR: 3
ARSENIC IODIDE
DOT: NA 1557
mf: AsI$_3$ mw: 455.62

PROP: Red hexagonal crystals. Mp: 141.8°; bp: 403°; d: 4.38 @ 13°. Solubility: in water: 6/100 @ 25°, in CS$_2$: 5.2/100.

SYNS: ARSENIC TRIIODIDE ◊ ARSENOUS IODIDE ◊ ARSENOUS TRIIODIDE (9CI) ◊ TRIIODOARSINE

CONSENSUS REPORTS: Reported in EPA TSCA Inventory. Arsenic and its compounds are on the Community Right-To-Know List.

OSHA PEL: TWA 0.01 mg(As)/m^3; Cancer Hazard
ACGIH TLV: TWA 0.2 mg(As)/m^3 (Proposed: 0.01 mg(As)/m^3; Human Carcinogen)
NIOSH REL: (Arsenic, Inorganic) CL 2 µg(As)/m^3/15M

DOT Classification: Poison B; Label: Poison

SAFETY PROFILE: A poison. See also ARSENIC COMPOUNDS and IODIDES. Can form a shock-sensitive compound with sodium or potassium. When heated to decomposition it emits very toxic fumes of I$^-$ and arsenic.

ARH250 HR: 3
ARSENIC PENTASULFIDE
mf: As$_2$S$_5$ mw: 310.2

PROP: Brownish-yellow, glassy, amorphous, highly refractive mass. Mp: 500° (subl).

CONSENSUS REPORTS: Arsenic and its compounds are on the Community Right-To-Know List.

OSHA PEL: TWA 0.01 mg(As)/m^3
ACGIH TLV: TWA 0.2 mg(As)/m^3 (Proposed: 0.01 mg(As)/m^3; Human Carcinogen)
NIOSH REL: CL 2 µg(As)/m^3/15M

SAFETY PROFILE: See also ARSENIC COMPOUNDS and SULFIDES. Flammable in the form of dust when exposed to heat or flame. Explosive when intimately mixed with powerful oxidizers, such as Cl$_2$, KNO$_3$, or chlorates. Will react with water and steam to produce toxic and flammable vapors. Incompatible with water, steam, and strong oxidizers.

ARH500 CAS:1303-28-2 HR: 3
ARSENIC PENTOXIDE
DOT: UN 1559
mf: As$_2$O$_5$ mw: 229.84

PROP: White, amorphous, deliquescent solid. Mp: decomp @ 800°, d: 4.32. Sol in alc. Solubility in water: 65.8/100 @ 20°.

SYNS: ANHYDRIDE ARSENIQUE (FRENCH) ◊ ARSENIC ACID ◊ ARSE-

NIC ACID ANHYDRIDE ◇ ARSENIC ANHYDRIDE ◇ ARSENIC OXIDE ◇ ARSENIC(V) OXIDE ◇ DIARSENIC PENTOXIDE ◇ RCRA WASTE NUMBER P011 ◇ ZOTOX

TOXICITY DATA with REFERENCE
cyt-hmn:leu 1200 nmol/L   MUREAV 88,73,81
mrc-bcs 50 mmol/L   MUREAV 77,109,80
itt-rat TDLo:4597 µg/kg (male 1D pre):REP   JRPFA4 7,21,64
itt-rat TDLo:4597 µg/kg (male 1D pre):REP   JRPFA4 7,21,64
orl-rat LD50:8 mg/kg   28ZEAL 4,50,69
orl-mus LD50:55 mg/kg   IRGGAJ 20,21,63
ivn-rbt LDLo:6 mg/kg   NTIS** PB214-270

CONSENSUS REPORTS: NTP Fifth Annual Report on Carcinogens. IARC Cancer Review: Human Sufficient Evidence IMEMDT 23,39,80. Reported in EPA TSCA Inventory. Arsenic and its compounds are on the Community Right-To-Know List. EPA Extremely Hazardous Substances List.

OSHA PEL: Cancer Hazard
ACGIH TLV: TWA 0.2 mg(As)/m^3 (Proposed: 0.01 mg(As)/m^3; Human Carcinogen)
DFG MAK: Human Carcinogen
NIOSH REL: CL 2 µg(As)/m^3/15M

DOT Classification: Poison B; Label: Poison

SAFETY PROFILE: Confirmed human carcinogen. Poison by ingestion and intravenous routes. Experimental reproductive effects. Mutation data reported. Reacts vigorously with Rb_2C_2. When heated to decomposition it emits toxic fumes of arsenic. See also ARSENIC COMPOUNDS.

ARI000   CAS:1303-33-9   *HR: 3*
ARSENIC SULFIDE
DOT: NA 1557
mf: As_2S_3 mw: 246.04

PROP: Yellow or red crystals. Bp: 707°, d: 3.43; mp: 312°. Insol in water; sol in alkalies.

SYNS: ARSENIC SESQUISULFIDE ◇ ARSENIC SULFIDE YELLOW ◇ ARSENIC SULPHIDE ◇ ARSENIC TRISULFIDE ◇ ARSENIC YELLOW ◇ ARSENIOUS SULPHIDE ◇ ARSENOUS SULFIDE ◇ C.I. 77086 ◇ DIARSENIC TRISULFIDE ◇ KING'S YELLOW ◇ ORPIMENT

TOXICITY DATA with REFERENCE
scu-rat TDLo:125 mg/kg:ETA   BJCAAI 20,190,66
itr-mus TDLo:73885 µg/kg/15W-C:ETA   CALEDQ 27,99,85

CONSENSUS REPORTS: IARC Cancer Review: Human Sufficient Evidence IMEMDT 23,39,80. Reported in EPA TSCA Inventory. Arsenic and its compounds are on the Community Right-To-Know List.

OSHA PEL: Cancer Hazard
ACGIH TLV: TWA 0.2 mg(As)/m^3 (Proposed: 0.01 mg(As)/m^3; Human Carcinogen)
NIOSH REL: (Arsenic, Inorganic) CL 2 µg(As)/m^3/15M

DOT Classification: Poison B; Label: Poison

SAFETY PROFILE: Confirmed human carcinogen with experimental tumorigenic data. A poison. Reacts violently with H_2O_2, (KNO_3 + S). When heated to decomposition or on contact with acid or acid fumes it emits highly toxic fumes of SO_2, H_2S, and As. Reacts with water or steam to emit toxic and flammable vapors.

ARI250   CAS:7784-35-2   *HR: 3*
ARSENIC TRIFLUORIDE
mf: AsF_3 mw: 131.92

PROP: Colorless liquid. D: 3.01, mp: −5.95, bp: 51°, vap press: 100 mm @ 13.2°, 400 mm @ 41.5°. Insol in water; sol in alc, benzene, and mercury.

SYNS: ARSENIC FLUORIDE ◇ ARSENOUS FLUORIDE ◇ TRIFLUOROARSINE

TOXICITY DATA with REFERENCE
ihl-mus LCLo:2000 mg/m^3/10M   NDRC** NDCrc-132,Aug,42

CONSENSUS REPORTS: Reported in EPA TSCA Inventory. Arsenic and its compounds are on the Community Right-To-Know List.

OSHA PEL: TWA 0.01 mg(As)/m^3; Cancer Hazard
ACGIH TLV: TWA 0.2 mg(As)/m^3 (Proposed: 0.01 mg(As)/m^3; Human Carcinogen)
NIOSH REL: (Arsenic, Inorganic) CL 2 µg(As)/m^3/15M

SAFETY PROFILE: Confirmed human carcinogen. A poison by inhalation. Strong reaction with P_2O_3. When heated to decomposition it emits very toxic fumes of As and F$^-$. See also FLUORIDES and ARSENIC COMPOUNDS.

ARJ500   CAS:14060-38-9   *HR: 3*
ARSENIOUS ACID, SODIUM SALT
mf: AsH_3O_3•7Na   mw: 286.88

PROP: Colorless or grayish-white powder. D: 1.87.

SYNS: ARSONIC ACID, SODIUM SALT (9CI) ◇ ARSENIOUS ACID, SODIUM SALT POLYMERS ◇ NATRIUMARSENIT (GERMAN) ◇ SODIUM ORTHOARSENITE

TOXICITY DATA with REFERENCE
ipr-rat LDLo:9 mg/kg   JPETAB 58,454,36
orl-frg LDLo:600 mg/kg   HBAMAK 4,1289,35
scu-frg LDLo:200 mg/kg   HBAMAK 4,1289,35

CONSENSUS REPORTS: Arsenic and its compounds are on the Community Right-To-Know List.

OSHA PEL: TWA 0.01 mg(As)/m^3; Cancer Hazard
ACGIH TLV: TWA 0.2 mg(As)/m^3 (Proposed: 0.01 mg(As)/m^3; Human Carcinogen)
NIOSH REL: (Arsenic, Inorganic) CL 2 µg(As)/m^3/15M

SAFETY PROFILE: Confirmed human carcinogen. Poison by intraperitoneal and subcutaneous routes. Moderately toxic by ingestion. When heated to decomposition it emits toxic fumes of arsenic.

ARJ750   CAS:1303-18-0   HR: 3
ARSENOPYRITE
mf: AsFeS   mw: 162.83

SYNS: ARSENOMARCASITE ◇ MISPICKEL

TOXICITY DATA with REFERENCE
ivn-mus LDLo:200 mg/kg   JNCIAM 1,241,40

CONSENSUS REPORTS: Arsenic and its compounds are on the Community Right-To-Know List.

OSHA PEL: TWA 0.01 mg(As)/m^3
ACGIH TLV: TWA 0.2 mg(As)/m^3 (Proposed: 0.01 mg(As)/m^3; Human Carcinogen)
NIOSH REL: (Arsenic, Inorganic) CL 2 µg(As)/m^3/15M

SAFETY PROFILE: Poison by intravenous route. When heated to decomposition it emits very toxic fumes of As and SO$_x$.

ARJ900   CAS:63951-03-1   HR: 3
ARSENOXIDE SODIUM
mf: C$_6$H$_5$AsNO$_2$•Na   mw: 221.03

SYN: PHENOL, 2-AMINO-4-ARSENOSO-, SODIUM SALT

TOXICITY DATA with REFERENCE
ivn-rat LDLo:20 mg/kg   MADCAJ 6,195,37

OSHA PEL: TWA 0.5 mg(As)/m^3
ACGIH TLV: TWA 0.2 mg(As)/m^3 (Proposed: 0.01 mg(As)/m^3; Human Carcinogen)

SAFETY PROFILE: Poison by intravenous route. When heated to decomposition it emits toxic fumes of NO$_x$ and As.

ARX300   CAS:283-24-9   HR: 3
3-AZABICYCLO(3.2.2)NONANE
mf: C$_8$H$_{15}$N   mw: 125.24

TOXICITY DATA with REFERENCE
ivn-mus LD50:56 mg/kg   CSLNX* NX#01357

CONSENSUS REPORTS: Reported in EPA TSCA Inventory.

SAFETY PROFILE: Poison by intravenous route. When heated to decomposition it emits toxic vapors of NO$_x$.

ASK925   CAS:538-41-0   HR: 3
p-AZOANILINE
mf: C$_{12}$H$_{12}$N$_4$ mw: 212.28

SYNS: ANILINE, 4,4′-AZODI- ◇ 4,4′-AZOBISBENZENAMINE ◇ 4,4′-AZODIANILINE ◇ BENZENAMINE, 4,4′-AZOBIS-(9CI) ◇ p-DIAMINOAZOBENZENE ◇ 4,4′-DIAMINOAZOBENZENE

TOXICITY DATA with REFERENCE
mma-sat 100 µg/plate   MUTAEX 4,115,89
ivn-mus LD50:180 mg/kg   CSLNX* NX#01451

CONSENSUS REPORTS: Reported in EPA TSCA Inventory.

SAFETY PROFILE: Poison by intravenous route. Mutation data reported. When heated to decomposition it emits toxic vapors of NO$_x$.

ASL500   CAS:2638-94-0   HR: 2
4,4′-AZOBIS(4-CYANOPENTANOIC ACID)
mf: C$_{12}$H$_{16}$N$_4$O$_4$ mw: 280.32

SYNS: AZOBIS(CYANOVALERIC ACID) ◇ 4,4′-AZOBIS(4-CYANOVALERIC ACID) ◇ KYSELINA 4,4′-AZO-BIS-(4-KYANVALEROVA) ◇ PENTANOIC ACID, 4,4′-AZOBIS(4-CYANO)- (9CI) ◇ VALERIC ACID, 4,4′-AZOBIS(4-CYANO)-

TOXICITY DATA with REFERENCE
ipr-mus LD50:666 mg/kg   85JCAE-,922,86

CONSENSUS REPORTS: Reported in EPA TSCA Inventory.

SAFETY PROFILE: Moderately toxic by intraperitoneal route. When heated to decomposition it emits toxic vapors of NO$_x$.

ASP600   CAS:275-51-4   HR: 3
AZULENE
mf: C$_{10}$H$_8$ mw: 128.18

SYNS: BICYCLO(5.3.0)DECAPENTAENE ◇ BICYCLO(0.3.5)DECA-1,3,5,7,9-PENTAENE ◇ BICYCLO(5.3.0)-DECA-2,4,6,8,10-PENTAENE ◇ CYCLOPENTACYCLOHEPTENE

TOXICITY DATA with REFERENCE
orl-rat LD50:>4 g/kg   DRUGAY 6,13,82
ipr-rat LD50:180 mg/kg   DRUGAY 6,13,82
scu-rat LD50:520 mg/kg   DRUGAY 6,13,82
orl-mus LD50:>3 g/kg   DRUGAY 6,13,82
ipr-mus LD50:108 mg/kg   DRUGAY 6,13,82
scu-mus LD50:145 mg/kg   DRUGAY 6,13,82
ivn-mus LD50:56 mg/kg   CSLNX* NX#07952

CONSENSUS REPORTS: Reported in EPA TSCA Inventory.

SAFETY PROFILE: Poison by intraperitoneal, intravenous, and subcutaneous routes. When heated to decomposition it emits acrid smoke and irritating vapors.

B

BAI800 CAS:1191-79-3 **HR: 3**
BARIUM CADMIUM STEARATE
mf: $C_{72}H_{140}O_8 \cdot Ba \cdot Cd$ mw: 1383.86

SYNS: CADMIUM BARIUM STEARATE ◊ OCTADECANOIC ACID, BARIUM CADMIUM SALT (4:1:1) (9CI) ◊ STEARIC ACID, BARIUM CADMIUM SALT (4:1:1)

TOXICITY DATA with REFERENCE
orl-rat LD50:3171 mg/kg   GISAAA 35(2),98,70
orl-mus LD50:1381 mg/kg   41HTAH -,14,78

OSHA PEL: TWA 5 µg(Cd)/m^3
ACGIH TLV: TWA 0.01 mg(Cd)/m^3; Suspected Carcinogen
NIOSH REL: TWA reduce to lowest feasible level

CONSENSUS REPORTS: Reported in EPA TSCA Inventory.

SAFETY PROFILE: Confirmed human carcinogen. Moderately toxic by ingestion. When heated to decomposition it emits toxic fumes of Ba and Cd.

BAW250 CAS:205-99-2 **HR: 3**
BENZ(e)ACEPHENANTHRYLENE
mf: $C_{20}H_{12}$ mw: 252.32

PROP: Mp: 168°.

SYNS: 3,4-BENZ(e)ACEPHENANTHRYLENE
◊ 2,3-BENZFLUORANTHENE ◊ 3,4-BENZFLUORANTHENE
◊ BENZO(b)FLUORANTHENE ◊ BENZO(e)FLUORANTHENE
◊ 2,3-BENZOFLUORANTHENE ◊ 3,4-BENZOFLUORANTHENE
◊ 2,3-BENZOFLUORANTHRENE ◊ B(b)F

TOXICITY DATA with REFERENCE
mma-sat 31 nmol/plate   CRNGDP 6,1023,85
otr-ham:lng 100 µg/L   TXCYAC 17,149,80
sce-ham-ipr 900 mg/kg/24H   MUREAV 66,65,79
imp-rat TDLo:5 mg/kg:ETA   JJIND8 71,539,83
skn-mus TDLo:88 ng/kg/120W-I:CAR   ARGEAR 50,266,80
ipr-mus TDLo:5046 µg/kg/15D-I:NEO   CALEDQ 34,15,87
scu-mus TDLo:72 mg/kg/9W-I:ETA   AICCA6 19,490,63
skn-mus TD:72 mg/kg/60W-I:ETA   CANCAR 12,1194,59
imp-rat TD:5 mg/kg:ETA   50NNAZ 7,571,83
skn-mus TD:4037 µg/kg/20D-I:ETA   CRNGDP 6,1023,85

CONSENSUS REPORTS: NTP Fifth Annual Report on Carcinogens. IARC Cancer Review: Group 2B IMEMDT 7,56,87, Animal Sufficient Evidence IMEMDT 32,147,83; IMEMDT 3,69,73. EPA Genetic Toxicology Program.
ACGIH TLV: Suspected Carcinogen

SAFETY PROFILE: Confirmed carcinogen with experimental carcinogenic and tumorigenic data. Mutation data reported. When heated to decomposition it emits acrid smoke and irritating fumes.

BBC250 CAS:56-55-3 **HR: 3**
BENZ(a)ANTHRACENE
mf: $C_{18}H_{12}$ mw: 228.30

PROP: Colorless leaflets or plates. Bp: 400°, mp: 160°.

SYNS: BA ◊ BENZANTHRACENE ◊ 1,2-BENZANTHRACENE
◊ 1,2-BENZ(a)ANTHRACENE ◊ 1,2-BENZANTHRAZEN (GERMAN)
◊ BENZANTHRENE ◊ 1,2-BENZANTHRENE ◊ BENZOANTHRACENE
◊ BENZO(a)ANTHRACENE ◊ 1,2-BENZOANTHRACENE ◊ BENZO(a)PHENANTHRENE ◊ BENZO(b)PHENANTHRENE ◊ 2,3-BENZOPHENANTHRENE
◊ 2,3-BENZPHENANTHRENE ◊ NAPHTHANTHRACENE ◊ RCRA WASTE NUMBER U018 ◊ TETRAPHENE

TOXICITY DATA with REFERENCE
mma-sat 4 µg/plate   CRNGDP 5,747,84
msc-hmn:lym 9 µmol/L   DTESD7 10,277,82
dni-hmn:oth 10 µmol/L   CNREA8 42,3676,82
dnd-mus-skn 192 µmol/kg   CRNGDP 5,231,84
skn-mus TDLo:18 mg/kg:NEO   CNREA8 38,1699,78
scu-mus TDLo:2 mg/kg:ETA   CNREA8 15,632,55
imp-mus TDLo:80 mg/kg:CAR   BJCAAI 22,825,68
skn-mus TD:18 mg/kg:ETA   CNREA8 38,1705,78
skn-mus TD:360 mg/kg/56W-I:ETA   CNREA8 11,892,51
skn-mus TD:240 mg/kg/1W-I:NEO   BJCAAI 9,177,55
ivn-mus LDLo:10 mg/kg   JNCIAM 1,225,40

CONSENSUS REPORTS: NTP Fifth Annual Report on Carcinogens. IARC Cancer Review: Group 2A IMEMDT 7,56,87, Animal Sufficient Evidence IMEMDT 32,135,83; IMEMDT 3,45,73. EPA Genetic Toxicology Program. Reported in EPA TSCA Inventory.
ACGIH TLV: (Proposed: Suspected Human Carcinogen)

SAFETY PROFILE: Confirmed carcinogen with experimental carcinogenic, neoplastigenic, tumorigenic data by skin contact and other routes. Poison by intravenous route. Human mutation data reported. It is found in oils, waxes, smoke, food, drugs. When heated to decomposition it emits acrid smoke and irritating fumes.

BBM750 CAS:1670-14-0 **HR: 2**
BENZENECARBOXIMIDAMIDE HYDROCHLORIDE
mf: $C_7H_8N_2 \cdot ClH$ mw: 156.63

SYN: BENZAMIDINE, HYDROCHLORIDE

TOXICITY DATA with REFERENCE
ipr-mus LD50:580 mg/kg   BIREBV 20,1045,79

CONSENSUS REPORTS: Reported in EPA TSCA Inventory.

SAFETY PROFILE: Moderately toxic by intraperitoneal route. When heated to decomposition it emits toxic vapors of NO_x, HCl, and Cl^-.

BBO500   CAS:88-96-0   *HR: 1*
1,2-BENZENEDICARBOXAMIDE
mf: $C_8H_8N_2O_2$ mw: 164.18

SYNS: NCI-C03612 ◇ P-D ◇ PHTHALAMIDE ◇ o-PHTHALIC ACID DIAMIDE

TOXICITY DATA with REFERENCE
ipr-rat LD50:4004 mg/kg   APFRAD 48,23,90
ipr-mus LD50:4104 mg/kg   APFRAD 48,23,90

CONSENSUS REPORTS: Reported in EPA TSCA Inventory.

CONSENSUS REPORTS: NTP Carcinogenesis Bioassay (feed): No Evidence: mouse, rat NCITR* NCI-TR-161,79

SAFETY PROFILE: Mildly toxic by intraperitoneal route. When heated to decomposition it emits toxic vapors of NO_x.

BBS300   CAS:80-17-1   *HR: 3*
BENZENESULFONIC HYDRAZIDE
mf: $C_6H_8N_2O_2S$ mw: 172.22

SYNS: BENZENESULFOHYDRAZIDE ◇ BENZENESULFONIC ACID, HYDRAZIDE ◇ BENZENESULFONOHYDRAZIDE ◇ BENZENESULFONYL HYDRAZIDE ◇ BENZENESULFONYL HYDRAZINE ◇ BENZENE SULPHONOHYDRAZIDE ◇ CELOGEN BSH ◇ ChKhZ 9 ◇ GENITRON BSH ◇ HYDRAZIDE BSG ◇ NITROPORE OBSH ◇ PHENYLSULFOHYDRAZIDE ◇ PHENYLSULFONYL HYDRAZIDE ◇ PHENYLSULFONYLHYDRAZINE ◇ POROFOR BSH ◇ POROFOR-BSH-PULVER ◇ POROFOR ChKhZ 9

TOXICITY DATA with REFERENCE
orl-rat LDLo:50 mg/kg   IPSTB3 3,93,76

CONSENSUS REPORTS: Reported in EPA TSCA Inventory.

SAFETY PROFILE: Poison by ingestion. When heated to decomposition it emits toxic vapors of NO_x and SO_x

BBY990   CAS:76-93-7   *HR: 2*
BENZILIC ACID
mf: $C_{14}H_{12}O_3$ mw: 228.26

SYNS: ACIDE DIPHENYLHYDROXYACETIQUE ◇ BENZENEACETIC ACID, α-HYDROXY-α-PHENYL-(9CI) ◇ DIPHENYLGLYCOLIC ACID ◇ α-α-DIPHENYLGLYCOLIC ACID ◇ DIPHENYLHYDROXYACETIC ACID ◇ HYDROXYDIPHENYLACETIC ACID ◇ α-HYDROXY-α-PHENYLBENZENEACETIC ACID

TOXICITY DATA with REFERENCE
orl-mus LD50:2 g/kg   AIPTAK 116,154,58
scu-mus LD50:1300 mg/kg   AIPTAK 116,154,58

CONSENSUS REPORTS: Reported in EPA TSCA Inventory.

SAFETY PROFILE: Moderately toxic by subcutaneous route. Slightly toxic by ingestion. When heated to decomposition it emits acrid smoke and irritating vapors.

BCE475   CAS:2634-33-5   *HR: 2*
1,2-BENZISOTHIAZOL-3(2H)-ONE
mf: C_7H_5NOS mw: 151.19

SYNS: 1,2-BENZISOTHIAZOLIN-3-ONE ◇ PROXEL PL

TOXICITY DATA with REFERENCE
orl-rat LD50:1020 mg/kg   PLRCAT 3,385,71
orl-mus LD50:1150 mg/kg   PLRCAT 3,385,71

CONSENSUS REPORTS: Reported in EPA TSCA Inventory.

SAFETY PROFILE: Moderately toxic by ingestion. When heated to decomposition it emits toxic vapors of NO_x and SO_x.

BCJ200   CAS:203-12-3   *HR: D*
2,13-BENZOFLUORANTHENE
mf: $C_{18}H_{10}$ mw: 226.28

SYNS: BENZO(ghi)FLUORANTHENE ◇ BENZO(mno)FLUORANTHENE ◇ 7,10-BENZOFLUORANTHENE

TOXICITY DATA with REFERENCE
mma-sat 10 µg/plate   MUREAV 174,247,86

CONSENSUS REPORTS: IARC Cancer Review: Group 3 IMEMDT 7,56,87; Animal Inadequate Evidence IMEMDT 32,171,83; Human No Adequate Data IMEMDT 32,171,83. Reported in Reported in EPA TSCA Inventory.

SAFETY PROFILE: Mutation data reported. When heated to decomposition it emits acrid smoke and irritating vapors.

BCJ800   CAS:243-17-4   *HR: D*
2,3-BENZOFLUORENE
mf: $C_{17}H_{12}$ mw: 216.29

SYN: 11H-BENZO(b)FLUORENE

TOXICITY DATA with REFERENCE
mma-sat 15 µg/plate   MUREAV 174,247,86

CONSENSUS REPORTS: IARC Cancer Review: Group 3 IMEMDT 7,56,87; Animal Inadequate Evidence IMEMDT 32,183,83; Human No Adequate Data IMEMDT 32,183,83. Reported in EPA TSCA Inventory.

SAFETY PROFILE: Mutation data reported. When heated to decomposition it emits acrid smoke and irritating vapors.

BCS400 CAS:574-66-3 HR: 2
BENZOPHENONE, OXIME
mf: $C_{13}H_{11}NO$ mw: 197.25

SYNS: BENZOPHENOXIME ◇ DIPHENYL KETOXIME ◇ DIPHENYLMETHANONE OXIME ◇ (DIPHENYLMETHYLENE)HYDROXYLAMINE ◇ METHANONE, DIPHENYL-, OXIME (9CI)

TOXICITY DATA with REFERENCE
unr-mus LD50:560 mg/kg   PCJOAU 12,227,78

CONSENSUS REPORTS: Reported in EPA TSCA Inventory.

SAFETY PROFILE: Moderately toxic by unspecified route. When heated to decomposition it emits toxic vapors of NO_x.

BDJ800 CAS:93-91-4 HR: 2
BENZOYLACETONE
mf: $C_{10}H_{10}O_2$ mw: 162.20

SYNS: ACETOACETOPHENONE ◇ α-ACETYLACETOPHENONE ◇ 2-ACETYLACETOPHENONE ◇ ACETYLBENZOYLMETHANE ◇ BENZOYL-ACETON ◇ 1,3-BUTANEDIONE, 1-PHENYL-

TOXICITY DATA with REFERENCE
unr-rat LDLo:600 mg/kg   BCPCA6 14,1325,65

CONSENSUS REPORTS: Reported in EPA TSCA Inventory.

SAFETY PROFILE: Moderately toxic by an unspecified route. When heated to decomposition it emits acrid smoke and irritating vapors.

BDK800 CAS:135-57-9 HR: 1
o-(BENZOYLAMINO)PHENYL DISULFIDE
mf: $C_{26}H_{20}N_2O_2S_2$ mw: 456.60

SYNS: BENZANILIDE, 2′,2′′′-DITHIOBIS- ◇ BENZAMIDE, N,N′-(DITHIODI-2,1-PHENYLENE)BIS- ◇ BIS(o-BENZAMIDOPHENYL) DISULFIDE ◇ BIS(2-BENZAMIDOPHENYL) DISULFIDE ◇ BIS-o-BENZOYL-AMINOFENYL-DISULFID ◇ o,o′-DIBENZAMIDODIPHENYL DISULFIDE ◇ DI-o-BENZAMIDOPHENYL DISULPHIDE ◇ 2,2′-DIBENZOYLAMINO-DIPHENYL DISULFIDE ◇ 2′,2′′′-DITHIOBISBENZANILIDE ◇ 2′,2′′′-DITHIODIBENZANILIDE ◇ N,N′-(DITHIODI-2,1-PHENYLENE)BISBENZAMIDE ◇ PEPTAZIN BAFD ◇ PEPTISANT 1O ◇ PEPTON 22

TOXICITY DATA with REFERENCE
eye-rbt 500 mg/24H MLD   85JCAE-,1007,86

CONSENSUS REPORTS: Reported in EPA TSCA Inventory.

SAFETY PROFILE: An eye irritant. When heated to decomposition it emits toxic vapors of NO_x and SO_x.

BDL850 CAS:85-52-9 HR: 2
2-BENZOYLBENZOIC ACID
mf: $C_{14}H_{10}O_3$ mw: 226.24

SYNS: BENZOIC ACID, o-BENZOYL- ◇ BENZOIC ACID, 2-BENZOYL- ◇ BENZOPHENONE-2-CARBOXYLIC ACID

TOXICITY DATA with REFERENCE
orl-rat LD50:4600 mg/kg   GTPZAB 15(11),52,71
orl-mus LD50:880 mg/kg   GTPZAB 15(11),52,71

CONSENSUS REPORTS: Reported in EPA TSCA Inventory.

SAFETY PROFILE: Moderately toxic by ingestion. When heated to decomposition it emits acrid smoke and irritating vapors.

BDW100 CAS:599-71-3 HR: 1
BENZSULFOHYDROXAMIC ACID
mf: $C_6H_7NO_3S$ mw: 173.20

SYN: HYDROXAMIC ACID, BENZSULFO-

TOXICITY DATA with REFERENCE
scu-mus LDLo:1 g/kg   AIPTAK 12,447,04

CONSENSUS REPORTS: Reported in EPA TSCA Inventory.

SAFETY PROFILE: Slightly toxic by subcutaneous route. When heated to decomposition it emits toxic vapors of NO_x and SO_x.

BDX000 CAS:140-11-4 HR: 3
BENZYL ACETATE
mf: $C_9H_{10}O_2$ mw: 150.19

PROP: Colorless liquid; sweet, floral fruity odor. Mp: −51.5°, bp: 213.5°, flash p: 216°F (CC), d: 1.06, autoign temp: 862°F, vap press: 1 mm @ 45°, vap d: 5.1, refr index: 1.501. Sol in alc, most fixed oils, propylene glycol; insol in glycerin and water @ 214°.

SYNS: ACETIC ACID BENZYL ESTER ◇ ACETIC ACID PHENYLMETHYL ESTER ◇ α-ACETOXYTOLUENE ◇ BENZYL ETHANOATE ◇ FEMA No. 2135 ◇ NCI-C06508

TOXICITY DATA with REFERENCE
skn-rbt 100 mg/24H MOD   CTOIDG 94(8),41,79
dnr-bcs 21 mg/disc   OIGZDE 34,267,85
mma-hmn:lyms 1500 mg/L   MUREAV 196,61,88
mma-mus:lyms 500 mg/L   MUREAV 196,61,88
msc-mus:lyms 700 mg/L   SCIEAS 236,933,87
orl-rat TDLo:258 g/kg/2Y-I:NEO   NTPTR* NTP-TR-250,86
orl-mus TDLo:258 g/kg/2Y-I:NEO   NTPTR* NTP-TR-250,86
ihl-hmn TCLo:50 ppm:PSY,PUL,GLN   TGNCDL 2,31,61
orl-rat LD50:2490 mg/kg   FCTXAV 2,327,64
orl-mus LD50:830 mg/kg   GISAAA 50(7),17,85
ihl-mus LCLo:1300 mg/m³/22H   AGGHAR 5,1,33
ihl-cat LC50:245 ppm/8H   AMIHAB 21,28,60
skn-cat LDLo:10 g/kg   JPETAB 84,358,45
orl-rbt LD50:2200 mg/kg   GISAAA 50(7),17,85
scu-rbt LDLo:3000 mg/kg   AGGHAR 5,1,33
orl-gpg LD50:2200 mg/kg   GISAAA 50(7),17,85
scu-gpg LDLo:3000 mg/kg   AGGHAR 5,1,33

CONSENSUS REPORTS: IARC Cancer Review: Group 3 IMEMDT 7,56,87; Animal Limited Evidence IMEMDT 40,109,86. NTP Carcinogenesis Studies (gavage); Some Evidence: mouse, rat NTPTR* NTP-TR-250,86. Reported in EPA TSCA Inventory.
ACGIH TLV: (Proposed: TWA 10 ppm, Animal Carcinogen)

SAFETY PROFILE: A poison by inhalation. Moderately toxic by ingestion and subcutaneous routes. Human systemic effects by inhalation: an antipsychotic, unspecified respiratory and urinary system effects. Questionable carcinogen with experimental tumorigenic data. Combustible liquid. To fight fire, use alcohol foam, CO_2. When heated to decomposition it emits irritating fumes. See also ESTERS.

BDX090   CAS:1214-39-7   **HR: 2**
6-BENZYLADENINE
mf: $C_{12}H_{11}N_5$ mw: 225.28

SYNS: ABG 3034 ◇ ADENINE, N-BENZYL- ◇ BA ◇ 6-BA ◇ BA (GROWTH STIMULANT) ◇ BAP ◇ 6-BAP ◇ BAP (GROWTH STIMULANT) ◇ BENZYLADENINE ◇ N-BENZYLADENINE ◇ N^6-BENZYLADENINE ◇ BENZYLAMINOPURINE ◇ N^6-(BENZYLAMINO)PURINE ◇ 6-(BENZYLAMINO)PURINE ◇ 6-(N-BENZYLAMINO)PURINE ◇ N-(PHENYLMETHYL)-1H-PURIN-6-AMINE ◇ 1H-PURIN-6-AMINE, N-(PHENYLMETHYL)-(9CI) ◇ SD 4901 ◇ SQ 4609

TOXICITY DATA with REFERENCE
oth-hmn:leu 100 nmol/L   EXPEAM 32,29,76
oth-hmn:leu 10 µmol/L   EXPEAM 32,29,76
orl-rat LD50:2125 mg/kg   TOIZAG 19,336,72
orl-mus LD50:1300 mg/kg   TOIZAG 19,336,72
skn-mus LD50:>5 g/kg   TOIZAG 19,336,72
scu-mus LD50:>2300 mg/kg   TOIZAG 19,336,72

CONSENSUS REPORTS: Reported in EPA TSCA Inventory.

SAFETY PROFILE: Moderately toxic by ingestion and skin contact. Human mutation data reported. When heated to decomposition it emits toxic vapors of NO_x.

BDY750   CAS:103-14-0   **HR: 2**
4-(BENZYLAMINO)PHENOL
mf: $C_{13}H_{13}NO$ mw: 199.27

SYN: PHENOL, p-(BENZYLAMINO)-

TOXICITY DATA with REFERENCE
orl-rat LDLo:500 mg/kg   JPETAB 90,260,47

CONSENSUS REPORTS: Reported in EPA TSCA Inventory.

SAFETY PROFILE: Moderately toxic by ingestion. When heated to decomposition it emits toxic vapors of NO_x.

BEE800   CAS:1833-31-4   **HR: 3**
BENZYLCHLORODIMETHYLSILANE
mf: $C_9H_{13}ClSi$ mw: 184.76

SYN: SILANE, BENZYLCHLORODIMETHYL-

TOXICITY DATA with REFERENCE
ivn-mus LD50:56 mg/kg   CSLNX* NX#04165

CONSENSUS REPORTS: Reported in EPA TSCA Inventory.

SAFETY PROFILE: Poison by intravenous route. When heated to decomposition it emits toxic vapors of Cl^-.

BEL900   CAS:122-18-9   **HR: 1**
BENZYLDIMETHYLCETYLAMMONIUM CHLORIDE
mf: $C_{25}H_{46}N \cdot Cl$ mw: 396.17

SYNS: AMMONIUM, BENZYLDIMETHYLHEXADECYL-, CHLORIDE ◇ BENZENEMETHANAMINIUM, N-HEXADECYL-N,N-DIMETHYL-, CHLORIDE ◇ BENZYLDIMETHYLHEXADECYLAMMONIUM CHLORIDE ◇ WINZER SOLUTION

TOXICITY DATA with REFERENCE
eye-rbt 150 mg MLD   ARZNAD 9,349,59
eye-gpg 500 mg MOD   ARZNAD 9,349,59

CONSENSUS REPORTS: Reported in EPA TSCA Inventory.

SAFETY PROFILE: An eye irritant. When heated to decomposition it emits toxic vapors of NO_x and Cl^-.

BFG600   CAS:101-82-6   **HR: 2**
2-BENZYLPYRIDINE
mf: $C_{12}H_{11}N$ mw: 169.24

SYNS: 2-(PHENYLMETHYL)PYRIDINE ◇ PYRIDINE, 2-BENZYL- ◇ PYRIDINE, 2-(PHENYLMETHYL)-(9CI)

TOXICITY DATA with REFERENCE
scu-mus LD50:1500 mg/kg   AEPPAE 227,129,55

CONSENSUS REPORTS: Reported in EPA TSCA Inventory.

SAFETY PROFILE: Moderately toxic by subcutaneous route. When heated to decomposition it emits toxic vapors of NO_x.

BFI400   CAS:2284-30-2   **HR: 3**
4-BENZYL RESORCINOL
mf: $C_{13}H_{12}O_2$ mw: 200.25

SYN: RESORCINOL, 4-BENZYL-

TOXICITY DATA with REFERENCE
ivn-mus LD50:73 mg/kg   BJPCAL 22,221,64

CONSENSUS REPORTS: Reported in EPA TSCA Inventory.

SAFETY PROFILE: Poison by intravenous route. When

BFL300 CAS:56-37-1 HR: 3
BENZYLTRIETHYLAMMONIUM CHLORIDE
mf: $C_{13}H_{22}N \bullet Cl$ mw: 227.81

SYNS: AMMONIUM, BENZYLTRIETHYL-, CHLORIDE ◊ BENZENEMETHANAMINIUM, N,N,N-TRIETHYL-, CHLORIDE (9CI) ◊ TEBAC ◊ N,N,N-TRIETHYLBENZENEMETHANAMINIUM CHLORIDE ◊ TRIETHYLBENZYLAMMONIUM CHLORIDE

TOXICITY DATA with REFERENCE
ivn-mus LD50:18 mg/kg   CSLNX* NX#01867

CONSENSUS REPORTS: Reported in EPA TSCA Inventory.

SAFETY PROFILE: Poison by intravenous route. When heated to decomposition it emits toxic vapors of NO_x and Cl^-.

BFN600 CAS:633-65-8 HR: 3
BERBERINE HYDROCHLORIDE
mf: $C_{20}H_{18}NO_4 \bullet Cl$ mw: 371.84

SYNS: BERBERINE CHLORIDE ◊ BERBINIUM, 7,8,13,13a-TETRADEHYDRO-9,10-DIMETHOXY-2,3-(METHYLENEDIOXY)-,CHLORIDE ◊ BENZO(g)(1,3)BENZODIOXOLO(5,6-a)QUINOLIZINIUM,5,6-DIHYDRO-9,10-DIMETHOXY-, CHLORIDE (9CI) ◊ BERBERINIUM CHLORIDE

TOXICITY DATA with REFERENCE
dnd-uns:lyms 22 μmol/L   IJBBBQ 18,245,81
orl-rat LD50:>15 g/kg   KSRNAM 8,654,74
orl-mus LD50:>29,586 mg/kg   KSRNAM 8,654,74
ipr-mus LD50:37 mg/kg   JPETAB 104,253,52

CONSENSUS REPORTS: Reported in EPA TSCA Inventory.

SAFETY PROFILE: Poison by intraperitoneal route. Slightly toxic by ingestion. Mutation data reported. When heated to decomposition it emits toxic vapors of NO_x and Cl^-.

BFV760 CAS:22298-29-9 HR: 3
BETAMETHASONE BENZOATE
mf: $C_{29}H_{33}FO_6$ mw: 496.62

SYNS: BETAMETHASONE 17-BENZOATE ◊ BETHAMETHASONE 17-BENZOATE ◊ MS-1112

TOXICITY DATA with REFERENCE
scu-rat TDLo:80 μg/kg (female 9–18D post):REP   OYYAA2 10,661,75
skn-rbt TDLo:7500 μg/kg (female 7–18D post):REP   OYYAA2 10,685,75
skn-rbt TDLo:7500 μg/kg (female 7–18D post):TER   OYYAA2 10,685,75
scu-rat TDLo:10 mg/kg (female 9–18D post):TER   OYYAA2 10,661,75
scu-rat TDLo:1 mg/kg (female 9–18D post):TER   OYYAA2 10,661,75
skn-rbt TDLo:7500 μg/kg (female 7–18D post):REP   OYYAA2 10,685,75
scu-rat TDLo:80 μg/kg (1–8D preg):REP   OYYAA2 10,661,75
scu-rat LD50:194 mg/kg   TXAPA9 8,250,66

SAFETY PROFILE: Poison by subcutaneous route. An experimental teratogen. Other experimental reproductive effects. When heated to decomposition it emits toxic fumes of F^-.

BGB275 CAS:482-89-3 HR: D
($\Delta^{2,2'}$)-BIINDOLINE-3,3'-DIONE
mf: $C_{16}H_{10}N_2O_2$ mw: 262.28

SYNS: ($\Delta^{2,2'}$)-BIPSEUDOINDOXYL ◊ 11669 BLUE ◊ BLUE NO. 201 ◊ C.I. 73000 ◊ C.I. VAT BLUE 1 ◊ CYSTOCEVA ◊ D&C BLUE NO. 6 ◊ D and C BLUE NO. 6 ◊ DIINDOGEN ◊ (2,2'-BIINDOLINE)-3,3'-DIONE ◊ INDIGO ◊ INDIGO BLUE ◊ INDIGO CIBA ◊ INDIGO CIBA SL ◊ INDIGO J ◊ INDIGO N ◊ INDIGO NAC ◊ INDIGO NACCO ◊ INDIGO P ◊ INDIGO PLN ◊ INDIGO POWDER W ◊ INDIGO PURE BASF ◊ INDIGO PURE BASF POWDER K ◊ INDIGO SYNTHETIC ◊ INDIGOTIN ◊ INDIGO VS ◊ 3H-INDOL-3-ONE,2(1,3-DIHYDRO-3-OXO-2H-INDOL-2-YLIDENE)-1,2-DIHYDRO-(9CI) ◊ LITHOSOL DEEP BLUE V ◊ MITSUI INDIGO PASTE ◊ MITSUI INDIGO PURE ◊ MODR KYPOVA 1 ◊ MONOLITE FAST NAVY BLUE BV ◊ NCI-C61392 ◊ SYNTHETIC INDIGO ◊ SYNTHETIC INDIGO TS ◊ VAT BLUE 1 ◊ VULCAFIX BLUE R ◊ VULCAFOR BLUE A ◊ VULCANOSINE DARK BLUE L ◊ VYNAMON BLUE A

TOXICITY DATA with REFERENCE
mma-sat 500 nmol/plate   CRNGDP 3,1321,82

CONSENSUS REPORTS: Reported in EPA TSCA Inventory.

SAFETY PROFILE: Mutation data reported. When heated to decomposition it emits toxic vapors of NO_x.

BGC100 CAS:602-09-5 HR: 3
(1,1'-BINAPHTHALENE)-2,2'-DIOL
mf: $C_{20}H_{14}O_2$ mw: 286.34

SYNS: β-BINAPHTHOL ◊ 1,1'-BI-2-NAPHTHOL ◊ BIS-β-NAPHTHOL ◊ 2,2'-DIHYDROXYBINAPHTHALENE ◊ 2,2'-DIHYDROXYDINAPHTHYL ◊ 2,2'-DINAPHTHOL

TOXICITY DATA with REFERENCE
orl-mus LDLo:42 mg/kg   AECTCV 14,111,85

CONSENSUS REPORTS: Reported in EPA TSCA Inventory.

SAFETY PROFILE: Poison by ingestion. When heated to decomposition it emits acrid smoke and irritating vapors.

BGO600 CAS:553-26-4 HR: 3
4,4'-BIPYRIDINE
mf: $C_{10}H_8N_2$ mw: 156.20

SYNS: γ,γ-BIPYRIDYL ◇ 4,4-BIPYRIDYL ◇ 4,4′-BIPYRIDYL ◇ 4,4′-DIPYRIDINE ◇ γ,γ-DIPYRIDYL ◇ 4,4-DIPYRIDYL ◇ 4,4′-DIPYRIDYL ◇ 4-(4-PYRIDYL)PYRIDINE

TOXICITY DATA with REFERENCE
orl-rat LD50:172 mg/kg   JTEHD6 10,363,82

CONSENSUS REPORTS: Reported in EPA TSCA Inventory.

SAFETY PROFILE: Poison by ingestion. When heated to decomposition it emits toxic vapors of NO_x.

BGO775   CAS:515-69-5   **HR: 2**
BISABOLOL
mf: $C_{15}H_{26}O$ mw: 222.41

SYNS: (–)-α-BISABOLOL ◇ α-4-DIMETHYL-α-(4-METHYL-3-PENTENYL)-3-CYCLOHEXENE-1-METHANOL ◇ 5-HEPTEN-2-OL, 6-METHYL-2-(4-METHYL-3-CYCLOHEXEN-1-YL)- ◇ 6-METHYL-2-(4-METHYL-3-CYCLOHEXEN-1-YL)-5-HEPTEN-2-OL

TOXICITY DATA with REFERENCE
orl-rat LD50:14,850 mg/kg   ARZNAD 19,615,69
orl-mus LD50:11,350 mg/kg   ARZNAD 19,615,69

CONSENSUS REPORTS: Reported in EPA TSCA Inventory.

SAFETY PROFILE: Moderately toxic by ingestion. When heated to decomposition it emits acrid smoke and irritating vapors.

BHB100   CAS:94-01-9   **HR: 1**
1,3-BIS(BENZOYLOXY)BENZENE
mf: $C_{20}H_{14}O_4$ mw: 318.34

SYNS: 1,3-BENZENEDIOL, DIBENZOATE ◇ RESORCINOL, DIBENZOATE

TOXICITY DATA with REFERENCE
ipr-mus LD50:8000 mg/kg   JAPMA8 46,185,57

CONSENSUS REPORTS: Reported in EPA TSCA Inventory.

SAFETY PROFILE: Slightly toxic by intraperitoneal route. When heated to decomposition it emits acrid smoke and irritating vapors.

BHB300   CAS:140-28-3   **HR: 3**
1,2-BIS(BENZYLAMINO)ETHANE
mf: $C_{16}H_{20}N_2$ mw: 240.38

SYNS: BENZATHINE ◇ BENZATIN ◇ DBED ◇ N,N′-DIBENZYL-ETHYLENEDIAMINE ◇ ETHYLENEDIAMINE, N,N′-DIBENZYL- ◇ USAF DO-53

TOXICITY DATA with REFERENCE
orl-mus LD50:388 mg/kg   CNCRA6 52,579,68
ipr-mus LD50:50 mg/kg   NTIS** AD277-689
par-mus LD50:80 mg/kg   ANTCAO 4,633,54

CONSENSUS REPORTS: Reported in EPA TSCA Inventory.

SAFETY PROFILE: Poison by ingestion, intraperitoneal, and parenteral routes. When heated to decomposition it emits toxic vapors of NO_x.

BHL200   CAS:995-33-5   **HR: 3**
4,4-BIS(tert-BUTYLPEROXY)VALERIC ACID BUTYL ESTER
DOT: UN 2140/UN 2141
mf: $C_{17}H_{34}O_6$ mw: 334.51

SYNS: n-BUTYL-4,4-DI-(tert-BUTYLPEROXY)VALERATE, not >52% with inert solid (UN2141) (DOT) ◇ n-BUTYL-4,4-DI-(tert-BUTYLPEROXY)VALERATE, technically pure (UN2140)(DOT) ◇ PENTANOIC ACID, 4,4-BIS((1,1-DIMETHYLETHYL)DIOXY)-, BUTYL ESTER ◇ TRIGONOX 17/40 ◇ VALERIC ACID, 4,4-BIS(tert-BUTYLPEROXY)-, BUTYL ESTER

DOT Classification: Organic Peroxide; Label: Organic Peroxide (UN2141); Flammable Liquid; Label: Flammable Liquid, Organic Peroxide

CONSENSUS REPORTS: Reported in EPA TSCA Inventory.

SAFETY PROFILE: Flammable liquid and peroxide. When heated to decomposition it emits acrid smoke and irritating vapors.

BHM300   CAS:119-80-2   **HR: 3**
BIS(2-CARBOXYPHENYL) DISULFIDE
mf: $C_{14}H_{10}O_4S_2$ mw: 306.36

SYNS: BENZOIC ACID, 2,2′-DITHIOBIS-(9CI) ◇ BENZOIC ACID, 2,2′-DITHIODI- ◇ BIS(o-CARBOXYPHENYL) DISULFIDE ◇ 2,2′-DITHIOBIS(BENZOIC ACID) ◇ 2,2′-DITHIODIBENZOESAEURE ◇ 2,2′-DITHIODIBENZOIC ACID

TOXICITY DATA with REFERENCE
ipr-mus LD50:367 mg/kg   ARZNAD 21,284,71

CONSENSUS REPORTS: Reported in EPA TSCA Inventory.

SAFETY PROFILE: Poison by intraperitoneal route. When heated to decomposition it emits toxic vapors of SO_x.

BIM800   CAS:101-76-8   **HR: 2**
BIS(p-CHLOROPHENYL)METHANE
mf: $C_{13}H_{10}Cl_2$ mw: 237.13

SYNS: DI-(p-CHLOROPHENYL)METHANE ◇ DI-(4-CHLOROPHENYL)METHANE ◇ METHANE, BIS(4-CHLOROPHENYL)-

TOXICITY DATA with REFERENCE
orl-rat LD50:1 g/kg   JPETAB 88,359,46
orl-mus LDLo:1500 mg/kg   JPETAB 88,400,46

CONSENSUS REPORTS: Reported in EPA TSCA Inventory.

BIY600   CAS:1518-15-6   **HR: 3**
1,4-BIS(DICYANOMETHYLENE)CYCLOHEXANE
mf: $C_{12}H_8N_4$ mw: 208.24

SYN: $\Delta^{1-\alpha}$-4-α'-CYCLOHEXANEDIMALONONITRILE

TOXICITY DATA with REFERENCE
ivn-mus LD50:56 mg/kg   CSLNX* NX#05268

CONSENSUS REPORTS: Reported in EPA TSCA Inventory.

SAFETY PROFILE: Poison by intravenous route. When heated to decomposition it emits toxic vapors of NO_x.

BJA250   CAS:105-18-0   **HR: 3**
1,4-BIS(DIETHYLAMINO)-2-BUTYNE
mf: $C_{12}H_{24}N_2$ mw: 196.38

SYNS: 2-BUTYNYLENEDIAMINE, N,N,N′N′-TETRAETHYL- ◇ N,N,N′,N′-TETRAETHYL-2-BUTYNYLENEDIAMINE

TOXICITY DATA with REFERENCE
ivn-mus LD50:56 mg/kg   CSLNX* NX#04930

CONSENSUS REPORTS: Reported in EPA TSCA Inventory.

SAFETY PROFILE: Poison by intravenous route. When heated to decomposition it emits toxic vapors of NO_x.

BJB500   CAS:14239-68-0   **HR: 3**
BIS(DIETHYLDITHIOCARBAMATO)CADMIUM
mf: $C_{10}H_{20}CdN_2S_4$ mw: 408.96

SYNS: CADMIUM DIETHYL DITHIOCARBAMATE ◇ ETHYL CADMATE ◇ ETHYL TUADS

TOXICITY DATA with REFERENCE
mmo-sat 10 μg/plate   MUREAV 68,313,79
dnd-esc 1 μmol/L   ARTODN 46,277,80
orl-mus TDLo:7100 mg/kg/78W-I:ETA   NTIS** PB223-159
scu-mus TDLo:1000 mg/kg:ETA   NTIS** PB223-159

CONSENSUS REPORTS: Reported in EPA TSCA Inventory. Cadmium and its compounds are on the Community Right-To-Know List.
OSHA PEL: TWA 5 μg(Cd)/m³
ACGIH TLV: TWA 0.05 mg(Cd)/m³ (Proposed: TWA 0.01 mg(Cd)/m³ (dust), Suspected Human Carcinogen; 0.002 mg(Cd)/m³ (respirable dust), Suspected Human Carcinogen); BEI: 10 μg/g creatinine in urine; 10 μg/L in blood
DFG BAT: Blood 1.5 μg/dL; Urine 15 μg/dL, Suspected Carcinogen
NIOSH REL: (Cadmium) Reduce to lowest feasible level.

SAFETY PROFILE: Confirmed human carcinogen with experimental tumorigenic data. Mutation data reported. When heated to decomposition it emits very toxic fumes of NO_x and SO_x. See also CADMIUM COMPOUNDS and CARBAMATES.

BJF600   CAS:366-29-0   **HR: D**
4,4′-BIS(N,N-DIMETHYLAMINO)BIPHENYL
mf: $C_{16}H_{20}N_2$ mw: 240.38

SYNS: BENZIDINE, N,N,N′,N′-TETRAMETHYL- ◇ (1,1′-BIPHENYL)-4,4′-DIAMINE, N,N,N′,N′-TETRAMETHYL- ◇ N,N,N′,N′-TETRAMETHYLBENZIDINE ◇ N,N,N′,N′-TETRAMETHYL-p,p′-BENZIDINE

TOXICITY DATA with REFERENCE
mma-sat 10 nmol/plate   EMMUEG 10,263,87

CONSENSUS REPORTS: Reported in EPA TSCA Inventory.

SAFETY PROFILE: Mutation data reported. When heated to decomposition it emits toxic vapors of NO_x.

BKB100   CAS:2158-76-1   **HR: D**
2-(N,N-BIS(2-HYDROXYETHYL)AMINO)-1,4-BENZOQUINONE
mf: $C_{10}H_{13}NO_4$ mw: 211.24

SYNS: 1,4-BENZOQUINONE,2-(N,N-BIS(2-HYDROXYETHYL)AMINO)- ◇ 2,5-CYCLOHEXADIENE-1,4-DIONE, 2-(BIS(2-HYDROXYETHYL)AMINO)- ◇ DI(2′-HYDROXYETHYL)AMINO-1,4-BENZOQUINONE

TOXICITY DATA with REFERENCE
dnd-mus:lyms 2 mmol/L   CNREA8 48,1727,88
dnd-mus:lyms 1 mmol/L   CNREA8 44,78,84

CONSENSUS REPORTS: Reported in EPA TSCA Inventory.

SAFETY PROFILE: Mutation data reported. When heated to decomposition it emits toxic vapors of NO_x.

BKD800   CAS:120-86-5   **HR: 2**
N,N′-BIS(2-HYDROXYETHYL)-DITHIOOXAMIDE
mf: $C_6H_{12}N_2O_2S_2$ mw: 208.32

SYNS: OXAMIDE, N,N′-BIS(2-HYDROXYETHYL)DITHIO- ◇ USAF MK-5

TOXICITY DATA with REFERENCE
ipr-mus LD50:750 mg/kg   NTIS** AD277-689

CONSENSUS REPORTS: Reported in EPA TSCA Inventory.

SAFETY PROFILE: Moderately toxic by intraperitoneal route. When heated to decomposition it emits toxic vapors of NO_x and SO_x.

BKH200   CAS:131-54-4   **HR: D**
BIS(2-HYDROXY-4-METHOXYPHENYL)METHANONE
mf: $C_{15}H_{14}O_5$ mw: 274.29

SYNS: BENZOPHENONE-6 ◇ BENZOPHENONE, 2,2′-DIHYDROXY-4,4′-

DIMETHOXY- ◇ CYASORB UV 12 ◇ METHANONE, BIS(2-HYDROXY-4-METHOXYPHENYL)-(9CI) ◇ UVINUL D 49

TOXICITY DATA with REFERENCE
mma-sat 10 µg/plate   JACTDZ 2(5),35,83

CONSENSUS REPORTS: Reported in EPA TSCA Inventory.

SAFETY PROFILE: Mutation data reported. When heated to decomposition it emits acrid smoke and irritating vapors.

BKO600   CAS:101-70-2   HR: 2
BIS(4-METHOXYPHENYL)AMINE
mf: $C_{14}H_{15}NO_2$ mw: 229.30

SYNS: BENZENAMINE, 4-METHOXY-N-(4-METHOXYPHENYL)- ◇ 4-BIPHENYLAMINE, 4,4′-DIMETHOXY- ◇ BIS(p-ANISYLAMINE) ◇ BIS(p-METHOXYPHENYL)AMINE ◇ DI-p-ANISYLAMINE ◇ p,p′-DIMETHOXYDIPHENYLAMINE ◇ 4,4′-DIMETHOXYDIPHENYL-AMINE ◇ DI-p-METHOXYPHENYLAMINE ◇ TERMOFLEKS A

TOXICITY DATA with REFERENCE
cyt-ham:lng 30 mg/L   MUREAV 241,175,90
orl-rat LD50:2470 mg/kg   KCRZAE 26(9),28,67
orl-mus LD50:2500 mg/kg   KCRZAE 26(9),28,67

CONSENSUS REPORTS: Reported in EPA TSCA Inventory.

SAFETY PROFILE: Moderately toxic by ingestion. Mutation data reported. When heated to decomposition it emits toxic vapors of NO_x.

BKP500   CAS:2475-44-7   HR: 3
1,4-BIS(METHYLAMINO)-9,10-ANTHRACENEDIONE
mf: $C_{16}H_{14}N_2O_2$ mw: 266.32

SYNS: ANTHRAQUINONE, 1,4-BIS(METHYLAMINO)- ◇ C.I. DISPERSE BLUE 78 ◇ C.I. SOLVENT BLUE 78 ◇ C.I. SOLVENT BLUE 93 ◇ DIARESIN BLUE K ◇ DISPERSE BLUE 78 ◇ DISPERSE BLUE 110 ◇ MACROLEX BLUE FR ◇ SOLVENT BLUE 78 ◇ SOLVENT BLUE 93

TOXICITY DATA with REFERENCE
mma-sat 5 mg/plate   EMMUEG 19(Suppl 20),8,92
ivn-mus LD50:180 mg/kg   CSLNX* NX#01356

CONSENSUS REPORTS: Reported in EPA TSCA Inventory.

SAFETY PROFILE: Poison by intravenous route. Mutation data reported. When heated to decomposition it emits toxic vapors of NO_x.

BKW100   CAS:1304-85-4   HR: 1
BISMUTH HYDROXIDE NITRATE OXIDE
mf: $Bi_5H_9N_4O_{22}$ mw: 1462.03

SYNS: BASIC BISMUTH NITRATE ◇ BISMUTH MAGISTERY ◇ BISMUTH SUBNITRATE ◇ BISMUTH SUBNITRICUM ◇ BISMUTH WHITE ◇ BISMUTHYL NITRATE ◇ BLANC DE FARD ◇ C.I. 77169 ◇ C.I. PIGMENT WHITE 17 ◇ COSMETIC WHITE ◇ FLAKE WHITE ◇ MAGISTERY OF BISMUTH ◇ NOVISMUTH ◇ PAINT WHITE ◇ SNOWCAL 5SW ◇ SPANISH WHITE ◇ VICALIN

TOXICITY DATA with REFERENCE
orl-inf TDLo:259 mg/kg:BLD   JAMAAP 133,1280,47
orl-inf LDLo:1 g/kg   34ZIAG -,134,69

CONSENSUS REPORTS: Reported in EPA TSCA Inventory.

SAFETY PROFILE: Human systemic effects by ingestion: methemoglobinemia and carboxyhemoglobin. When heated to decomposition it emits toxic vapors of NO_x and Bi.

BKW600   CAS:1304-76-3   HR: 1
BISMUTH OXIDE
mf: Bi_2O_3 mw: 465.96

SYNS: BISMUTHOUS OXIDE ◇ BISMUTH(3+) OXIDE ◇ BISMUTH SESQUIOXIDE ◇ BISMUTH TRIOXIDE ◇ BISMUTH YELLOW ◇ C.I. 77160 ◇ DIBISMUTH TRIOXIDE

TOXICITY DATA with REFERENCE
orl-rat LD50:5 g/kg   GTPZAB 30(6),16,86
orl-mus LD50:10 g/kg   GTPZAB 30(6),16,86

CONSENSUS REPORTS: Reported in EPA TSCA Inventory.

SAFETY PROFILE: Slightly toxic by ingestion. When heated to decomposition it emits toxic vapors of Bi.

BLA600   CAS:645-15-8   HR: 2
BIS(4-NITROPHENYL) PHOSPHATE
mf: $C_{12}H_9N_2O_8P$ mw: 340.20

SYNS: BIS(p-NITROPHENYL) PHOSPHATE ◇ BNPP ◇ DI-p-NITROPHENYL PHOSPHATE ◇ PHENOL, p-NITRO-, HYDROGEN PHOSPHATE ◇ PHOSPHORIC ACID, BIS(p-NITROPHENYL) ESTER ◇ PHOSPHORSAEURE-BIS-(p-NITRO-PHENYLESTER)

TOXICITY DATA with REFERENCE
ipr-mus LD50:410 mg/kg   HSZPAZ 348,609,67

CONSENSUS REPORTS: Reported in EPA TSCA Inventory.

SAFETY PROFILE: Moderately toxic by intraperitoneal route. When heated to decomposition it emits toxic vapors of NO_x and PO_x.

BLA800   CAS:860-39-9   HR: 3
BIS(2-NITRO-4-TRIFLUOROMETHYLPHENYL) DISULFIDE
mf: $C_{14}H_6F_6N_2O_4S_2$ mw: 444.34

SYNS: DISULFIDE, BIS(2-NITRO-α-α-α-TRIFLUORO-p-TOLYL) ◇ USAF MA-9

TOXICITY DATA with REFERENCE
ipr-mus LD50:50 mg/kg   NTIS** AD277-689

CONSENSUS REPORTS: Reported in EPA TSCA Inventory.

SAFETY PROFILE: Poison by intraperitoneal route. When heated to decomposition it emits toxic vapors of NO_x, SO_x, and F^-.

BMR100   CAS:103-88-8   HR: 3
4-BROMOACETANILIDE
mf: C_8H_8BrNO mw: 214.08

SYNS: ACETAMIDE, N-(4-BROMOPHENYL)- ◊ ACETANILIDE, p-BROMO- ◊ ACETANILIDE, 4'-BROMO- ◊ ANTISEPSIN ◊ ASEPSIN ◊ p-BROMOACETANILIDE ◊ 4'-BROMOACETANILIDE ◊ p-BROMO-N-ACETANILIDE ◊ BROMOANILIDE ◊ BROMOANTIFEBRIN ◊ USAF DO-40

TOXICITY DATA with REFERENCE
ipr-mus LD50:250 mg/kg   NTIS** AD277-689

CONSENSUS REPORTS: Reported in EPA TSCA Inventory.

SAFETY PROFILE: Poison by intraperitoneal route. When heated to decomposition it emits toxic vapors of NO_x and Br^-.

BMT400   CAS:578-57-4   HR: 2
2-BROMOANISOLE
mf: C_7H_7BrO mw: 187.05

SYNS: ANISOLE, o-BROMO- ◊ ANISYL BROMIDE ◊ BENZENE, 1-BROMO-2-METHOXY-(9CI) ◊ o-BROMOANISOLE ◊ 1-BROMO-2-METHOXYBENZENE ◊ o-BROMOPHENYL METHYL ETHER ◊ o-METHOXYBROMOBENZENE ◊ 2-METHOXYBROMOBENZENE ◊ o-METHOXYPHENYL BROMIDE ◊ 2-METHOXYPHENYL BROMIDE

TOXICITY DATA with REFERENCE
orl-mus LD50:2466 mg/kg   GISAAA 44(12),19,79
ipr-mus LD50:1544 mg/kg   GISAAA 44(12),19,79

CONSENSUS REPORTS: Reported in EPA TSCA Inventory.

SAFETY PROFILE: Moderately toxic by ingestion and intraperitoneal routes. When heated to decomposition it emits toxic vapors of Br^-.

BMT700   CAS:1122-91-4   HR: 3
4-BROMOBENZALDEHYDE
mf: C_7H_5BrO mw: 185.03

SYNS: BENZALDEHYDE, p-BROMO- ◊ BENZALDEHYDE, 4-BROMO-(9CI) ◊ p-BROMOBENZALDEHYDE

TOXICITY DATA with REFERENCE
orl-mus LD50:1230 mg/kg   GISAAA 44(12),19,79
ipr-mus LD50:389 mg/kg   GISAAA 44(12),19,79

CONSENSUS REPORTS: Reported in EPA TSCA Inventory.

SAFETY PROFILE: Poison by intraperitoneal route. Moderately toxic by ingestion. When heated to decomposition it emits toxic vapors of Br^-.

BMU100   CAS:586-76-5   HR: 2
4-BROMOBENZOIC ACID
mf: $C_7H_5BrO_2$ mw: 201.03

SYNS: BENZOIC ACID, p-BROMO- ◊ BENZOIC ACID, 4-BROMO-(9CI) ◊ p-BROMOBENZOIC ACID ◊ p-CARBOXYBROMOBENZENE

TOXICITY DATA with REFERENCE
orl-mus LD50:1059 mg/kg   GISAAA 44(12),19,79
ipr-mus LD50:536 mg/kg   GISAAA 44(12),19,79

CONSENSUS REPORTS: Reported in EPA TSCA Inventory.

SAFETY PROFILE: Moderately toxic by ingestion and intraperitoneal routes. When heated to decomposition it emits toxic vapors of Br^-.

BMW290   CAS:2113-57-7   HR: 2
3-BROMOBIPHENYL
mf: $C_{12}H_9Br$ mw: 233.12

SYN: BIPHENYL, 3-BROMO-

TOXICITY DATA with REFERENCE
ipr-mus LDLo:500 mg/kg   CBCCT* 6,217,54

CONSENSUS REPORTS: Reported in EPA TSCA Inventory.

SAFETY PROFILE: Moderately toxic by intraperitoneal route. When heated to decomposition it emits toxic vapors of Br^-.

BMW300   CAS:92-66-0   HR: D
4-BROMOBIPHENYL
mf: $C_{12}H_9Br$ mw: 233.12

TOXICITY DATA with REFERENCE
mma-sat 50 µg/plate   BCPCA6 27,1245,78

CONSENSUS REPORTS: Reported in EPA TSCA Inventory.

SAFETY PROFILE: Mutation data reported. When heated to decomposition it emits toxic vapors of Br^-.

BNB800   CAS:112-29-8   HR: 2
1-BROMODECANE
mf: $C_{10}H_{21}Br$ mw: 221.22

SYNS: DECANE, 1-BROMO- ◊ DECYL BROMIDE ◊ n-DECYL BROMIDE ◊ 1-DECYL BROMIDE

TOXICITY DATA with REFERENCE
ipr-mus LD50:4070 mg/kg   GTPZAB 20(12),52,76
ihl-uns LC50:4200 mg/m^3 GTPZAB 18(4),55,74

CONSENSUS REPORTS: Reported in EPA TSCA Inventory.

SAFETY PROFILE: Moderately toxic by inhalation route. Mildly toxic by intraperitoneal routes. When heated to decomposition it emits toxic vapors of Br$^-$.

BNH100　　　CAS:728-84-7　　　**HR: 3**
2-BROMO-1,3-DIPHENYL-1,3-PROPANEDIONE
mf: $C_{15}H_{11}BrO_2$　　mw: 303.17

SYN: 1,3-PROPANEDIONE, 2-BROMO-1,3-DIPHENYL-

TOXICITY DATA with REFERENCE
ipr-mus LDLo:31200 µg/kg　　CBCCT* 4,232,52

CONSENSUS REPORTS: Reported in EPA TSCA Inventory.

SAFETY PROFILE: Poison by intraperitoneal route. When heated to decomposition it emits toxic vapors of Br$^-$.

BNI600　　　CAS:927-68-4　　　**HR: D**
BROMOETHYL ACETATE
mf: $C_4H_7BrO_2$　　mw: 167.02

SYNS: 2-BROMOETHYLACETATE ◇ ETHANOL, 2-BROMO-, ACETATE

TOXICITY DATA with REFERENCE
mmo-sat 10 mg/plate　　EVHPAZ 21,79,77
dnr-esc 10 mg/plate　　EVHPAZ 21,79,77

CONSENSUS REPORTS: Reported in EPA TSCA Inventory.

SAFETY PROFILE: Mutation data reported. When heated to decomposition it emits toxic vapors of Br$^-$.

BNI650　　　CAS:107-09-5　　　**HR: D**
2-BROMOETHYLAMINE
mf: C_2H_6BrN　　mw: 124.00

SYNS: 2-AMINOETHYL BROMIDE ◇ 2-BROMOETHANAMINE ◇ β-BROMOETHYLAMINE ◇ ETHANAMINE, 2-BROMO-(9CI) ◇ ETHYLAMINE, 2-BROMO-

TOXICITY DATA with REFERENCE
mmo-sat 500 µmol/L　　MUREAV 118,229,83

CONSENSUS REPORTS: Reported in EPA TSCA Inventory.

SAFETY PROFILE: Mutation data reported. When heated to decomposition it emits toxic vapors of NO_x and Br$^-$.

BNM100　　　CAS:565-74-2　　　**HR: 2**
2-BROMOISOVALERIC ACID
mf: $C_5H_9BrO_2$　　mw: 181.05

SYNS: α-BROMOISOVALERIC ACID ◇ 2-BROMO-3-METHYLBUTANOIC ACID ◇ 2-BROMO-3-METHYLBUTYRIC ACID ◇ BUTANOIC ACID, 2-BROMO-3-METHYL-(9CI) ◇ BUTYRIC ACID, 2-BROMO-3-METHYL-

TOXICITY DATA with REFERENCE
orl-rat LD50:769 mg/kg　　EPASR* 8EHQ-0188-0714
skn-rat LD50:1410 mg/kg　　EPASR* 8EHQ-0188-0714

CONSENSUS REPORTS: Reported in EPA TSCA Inventory.

SAFETY PROFILE: Moderately toxic by ingestion and skin contact. When heated to decomposition it emits toxic vapors of Br$^-$.

BNS200　　　CAS:90-11-9　　　**HR: 2**
1-BROMONAPHTHALENE
mf: $C_{10}H_7Br$　　mw: 207.08

SYNS: α-BROMONAPHTHALENE ◇ NAPHTHALENE, 1-BROMO-

TOXICITY DATA with REFERENCE
ipr-mus LD50:810 mg/kg　　GTPZAB 20(12),52,76

CONSENSUS REPORTS: Reported in EPA TSCA Inventory.

SAFETY PROFILE: Moderately toxic by intraperitoneal route. When heated to decomposition it emits toxic vapors of Br$^-$.

BNV775　　　CAS:106-37-6　　　**HR: 2**
p-BROMOPHENYL BROMIDE
mf: $C_6H_4Br_2$　　mw: 235.92

SYNS: BENZENE, p-DIBROMO- ◇ BENZENE, 1,4-DIBROMO-(9CI) ◇ p-DIBROMOBENZENE ◇ 1,4-DIBROMOBENZENE

TOXICITY DATA with REFERENCE
orl-mus LD50:3120 mg/kg　　GISAAA 44(12),19,79
ipr-mus LD50:1891 mg/kg　　GISAAA 44(12),19,79

CONSENSUS REPORTS: Reported in EPA TSCA Inventory.

SAFETY PROFILE: Moderately toxic by ingestion and intraperitoneal routes. When heated to decomposition it emits toxic vapors of Br$^-$.

BNW550　　　CAS:622-88-8　　　**HR: 2**
4-BROMOPHENYL HYDRAZINE HYDROCHLORIDE
mf: $C_6H_7BrN_2 \cdot ClH$　　mw: 223.52

SYN: HYDRAZINE, 1-(p-BROMOPHENYL)-, HYDROCHLORIDE

TOXICITY DATA with REFERENCE
orl-rat LDLo:500 mg/kg　　JPETAB 90,260,47

CONSENSUS REPORTS: Reported in EPA TSCA Inventory.

SAFETY PROFILE: Moderately toxic by ingestion. When heated to decomposition it emits toxic vapors of NO_x, HCl, Cl$^-$, and Br$^-$.

BOB550 CAS:2114-00-3 HR: 3
2-BROMOPROPIOPHENONE
mf: C_9H_9BrO mw: 213.09

SYNS: α-BROMOPROPIOPHENONE ◊ PROPIOPHENONE, 2-BROMO- ◊ TL 336

TOXICITY DATA with REFERENCE
ihl-mus LCLo:1600 mg/m^3/10M   NDRC** NDCrc-132,AUG42
ivn-mus LD50:56 mg/kg   CSLNX* NX#02729

CONSENSUS REPORTS: Reported in EPA TSCA Inventory.

SAFETY PROFILE: Poison by intravenous route. Moderately toxic by inhalation. When heated to decomposition it emits toxic vapors of Br$^-$.

BOB600 CAS:109-04-6 HR: 3
2-BROMOPYRIDINE
mf: C_5H_4BrN mw: 158.01

SYN: PYRIDINE, 2-BROMO-

TOXICITY DATA with REFERENCE
ipr-mus LDLo:31,300 μg/kg   CBCCT* 4,322,52

CONSENSUS REPORTS: Reported in EPA TSCA Inventory.

SAFETY PROFILE: Poison by intraperitoneal route. When heated to decomposition it emits toxic vapors of NO$_x$ and Br$^-$.

BOD550 CAS:1113-59-3 HR: 3
3-BROMOPYRUVIC ACID
mf: $C_3H_3BrO_3$ mw: 166.97

SYNS: 3-BROMO-2-OXOPROPANOIC ACID ◊ 3-BROMOPYRUVATE ◊ BROMOPYRUVIC ACID ◊ β-BROMOPYRUVIC ACID ◊ PROPANOIC ACID, 3-BROMO-2-OXO-(9CI) ◊ PYRUVIC ACID, BROMO-

TOXICITY DATA with REFERENCE
ipr-mus LD50:72 mg/kg   JPETAB 123,48,58

CONSENSUS REPORTS: Reported in EPA TSCA Inventory.

SAFETY PROFILE: Poison by intraperitoneal route. When heated to decomposition it emits toxic vapors of NO$_x$ and Br$^-$.

BOD600 CAS:87-12-7 HR: 2
5-BROMOSALICYL-4-BROMOANILIDE
mf: $C_{13}H_9Br_2NO_2$ mw: 371.05

SYNS: BENZAMIDE, 5-BROMO-N-(4-BROMOPHENYL)-2-HYDROXY- ◊ p-BROMANILID KYSELINY 5-BROMSALICYLOVE ◊ 3-BROMO-6-HYDROXYBENZ-p-BROMANILIDE ◊ 4′,5-DIBROMOSALICYLANILIDE ◊ DIBROMSALAN ◊ NSC-20527 ◊ SALICYLANILIDE, 4′,5-DIBROMO- ◊ TEMASEPT

TOXICITY DATA with REFERENCE
orl-rat LD50:410 mg/kg   IMSUAI 39,56,70

CONSENSUS REPORTS: Reported in EPA TSCA Inventory.

SAFETY PROFILE: Moderately toxic by ingestion. When heated to decomposition it emits toxic vapors of NO$_x$ and Br$^-$.

BOG255 CAS:106-38-7 HR: 2
p-BROMOTOLUENE
mf: C_7H_7Br mw: 171.05

SYNS: PARABROMOTOLUENE ◊ TOLUENE, p-BROMO-

TOXICITY DATA with REFERENCE
ipr-mus LD50:1741 mg/kg   GTPZAB 20(12),52,76
ihl-uns LC50:1300 mg/m^3 GTPZAB 18(4),55,74

CONSENSUS REPORTS: Reported in EPA TSCA Inventory.

SAFETY PROFILE: Moderately toxic by inhalation and intraperitoneal routes. When heated to decomposition it emits toxic vapors of Br$^-$.

BOG260 CAS:95-46-5 HR: 2
2-BROMOTOLUENE
mf: C_7H_7Br mw: 171.05

SYNS: BENZENE, 1-BROMO-2-METHYL-(9CI) ◊ 1-BROMO-2-METHYLBENZENE ◊ o-BROMOTOLUENE ◊ 2-METHYLBROMOBENZENE ◊ o-METHYLPHENYL BROMIDE ◊ TOLUENE, o-BROMO- ◊ o-TOLYL BROMIDE ◊ 2-TOLYL BROMIDE

TOXICITY DATA with REFERENCE
orl-mus LD50:1864 mg/kg   GISAAA 44(12),19,79
ipr-mus LD50:1358 mg/kg   GISAAA 44(12),19,79

CONSENSUS REPORTS: Reported in EPA TSCA Inventory.

SAFETY PROFILE: Moderately toxic by ingestion and intraperitoneal route. When heated to decomposition it emits toxic vapors of Br$^-$.

BOG300 CAS:591-17-3 HR: 2
3-BROMOTOLUENE
mf: C_7H_7Br mw: 171.05

SYNS: BENZENE, 1-BROMO-3-METHYL- ◊ m-BROMOTOLUENE ◊ 5-BROMOTOLUENE ◊ m-METHYLBROMOBENZENE ◊ 3-METHYLBROMOBENZENE ◊ TOLUENE, m-BROMO- ◊ m-TOLYL BROMIDE

TOXICITY DATA with REFERENCE
orl-mus LD50:1436 mg/kg   GISAAA 44(12),19,79
ipr-mus LD50:1215 mg/kg   GISAAA 44(12),19,79

CONSENSUS REPORTS: Reported in EPA TSCA Inventory.

SAFETY PROFILE: Moderately toxic by ingestion and in-

BOG255   CAS:583-68-6   HR: D
2-BROMOTOLUIDINE

SYNS: 4-METHYL-2-BROMOANILINE ◊ p-TOLUIDINE, 2-BROMO-

TOXICITY DATA with REFERENCE
mma-sat 1 μmol/plate   MUREAV 77,317,80

CONSENSUS REPORTS: Reported in EPA TSCA Inventory.

SAFETY PROFILE: Mutation data reported. When heated to decomposition it emits toxic vapors of Br⁻.

BOI500   CAS:765-09-3   HR: 3
1-BROMOTRIDECANE
mf: $C_{13}H_{27}Br$ mw: 263.31

SYN: TRIDECANE, 1-BROMO-

TOXICITY DATA with REFERENCE
ivn-mus LD50:180 mg/kg   CSLNX* NX#03504

CONSENSUS REPORTS: Reported in EPA TSCA Inventory.

SAFETY PROFILE: Poison by intravenous route. When heated to decomposition it emits toxic vapors of Br⁻.

BOL303   CAS:2374-05-2   HR: 2
4-BROMO-2,6-XYLENOL
mf: C_8H_9BrO mw: 201.08

SYNS: 4-BROMO-2,6-DIMETHYLPHENOL ◊ 2,6-XYLENOL, 4-BROMO-

TOXICITY DATA with REFERENCE
ipr-mus LD50:650 mg/kg   JMPCAS 2,201,60

CONSENSUS REPORTS: Reported in EPA TSCA Inventory.

SAFETY PROFILE: Moderately toxic by intraperitoneal route. When heated to decomposition it emits toxic vapors of Br⁻.

BOW255   CAS:107-01-7   HR: 2
2-BUTENE
mf: C_4H_8 mw: 56.12

SYNS: β-BUTYLENE ◊ PSEUDOBUTYLENE

TOXICITY DATA with REFERENCE
ihl-mus LC50:425 ppm   85JCAE -,12,86

CONSENSUS REPORTS: Reported in EPA TSCA Inventory.

SAFETY PROFILE: Moderately toxic by inhalation. When heated to decomposition it emits acrid smoke and irritating vapors.

BOX300   CAS:110-64-5   HR: 3
2-BUTENE-1,4-DIOL
mf: $C_4H_8O_2$ mw: 88.12

SYNS: AGRISYNTH B2D ◊ 2-BUTENE, 1,4-DIHYDROXY- ◊ 1,4-DIHYDROXY-2-BUTENE

TOXICITY DATA with REFERENCE
orl-rat LD50:1250 mg/kg   GAFCC*
ipr-rat LD50:327 mg/kg   JPPMAB 26,597,74

CONSENSUS REPORTS: Reported in EPA TSCA Inventory.

SAFETY PROFILE: Poison by intraperitoneal route. Moderately toxic by ingestion. When heated to decomposition it emits acrid smoke and irritating vapors.

BPM660   CAS:1852-16-0   HR: 2
N-(BUTOXYMETHYL)-2-PROPENAMIDE
mf: $C_8H_{15}NO_2$ mw: 157.24

SYNS: ACRYLAMIDE, N-BUTOXYMETHYL- ◊ N-(BUTOXYMETHYL)ACRYLAMIDE ◊ N-BUTOXYMETHYLAKRYLAMID ◊ 2-PROPENAMIDE, N-(BUTOXYMETHYL)-(9CI)

TOXICITY DATA with REFERENCE
orl-rat LD50:1030 mg/kg   85JCAE -,706,86

CONSENSUS REPORTS: Reported in EPA TSCA Inventory.

SAFETY PROFILE: Moderately toxic by ingestion. When heated to decomposition it emits toxic vapors of NO_x.

BPU750   CAS:123-86-4   HR: 3
n-BUTYL ACETATE
DOT: UN 1123
mf: $C_6H_{12}O_2$ mw: 116.18

PROP: Colorless liquid; strong fruity odor. Bp: 126°, fp: −73.5°, ULC: 50–60, lel: 1.4%, uel: 7.5%, flash p: 72°F, d: 0.88 @ 20°/20°, refr index: 1.393–1.396, autoign temp: 797°F, vap press: 15 mm @ 25°. Misc with alc, ether, and propylene glycol; sltly sol in water.

SYNS: ACETATE de BUTYLE (FRENCH) ◊ ACETIC ACID n-BUTYL ESTER ◊ BUTILE (ACETATI di) (ITALIAN) ◊ BUTYLACETAT (GERMAN) ◊ BUTYL ACETATE ◊ 1-BUTYL ACETATE ◊ BUTYLACETATEN (DUTCH) ◊ BUTYLE (ACETATE de) (FRENCH) ◊ BUTYL ETHANOATE ◊ FEMA No. 2174 ◊ OCTAN n-BUTYLU (POLISH)

TOXICITY DATA with REFERENCE
eye-hmn 300 ppm   JIHTAB 25,282,43
skn-rbt 500 mg/24H MOD   FCTXAV 17,509,79
skn-rbt 500 mg/24H MLD   85JCAE -,355,86
eye-rbt 20 mg SEV   AMIHBC 10,61,54
ihl-rat TCLo:1500 ppm/7H (female 7–16D post):TER   NTIS** PB83-258038
ihl-hmn TCLo:200 ppm:NOSE,EYE,PUL   JIHTAB 25,282,43
orl-rat LD50:13100 mg/kg   85GMAT -,28,82
ihl-rat LC50:2000 ppm/4H   NPIRI* 1,7,74

orl-mus LD50:7060 mg/kg　YKYUA6 32,1241,81
ihl-mus LC50:6 g/m^3/2H　YKYUA6 32,1241,81
ipr-mus LD50:1230 mg/kg　SCCUR* -,2,61
ihl-cat LCLo:68 g/m^3/72M　AGGHAR 5,1,33
orl-rbt LD50:3200 mg/kg　85GMAT -,28,82
orl-gpg LDLo:4700 mg/kg　FCTXAV 17,509,79
ihl-gpg LCLo:67 g/m^3/4H　FCTXAV 17,515,79
ipr-gpg LDLo:1500 mg/kg　AIHAAP 35,21,74

CONSENSUS REPORTS: Reported in EPA TSCA Inventory.

OSHA PEL: (Transitional: TWA 150 ppm) TWA 150 ppm; STEL 200 ppm
ACGIH TLV: TWA 150 ppm; STEL 200 ppm (Proposed: TWA 20 ppm)
DFG MAK: 200 ppm (950 mg/m^3)

DOT Classification: Flammable Liquid; Label: Flammable Liquid

SAFETY PROFILE: Moderately toxic by intraperitoneal route. Mildly toxic by inhalation and ingestion. An experimental teratogen. A skin and severe eye irritant. Human systemic effects by inhalation: conjunctiva irritation, unspecified nasal and respiratory system effects. A mild allergen. High concentrations are irritating to eyes and respiratory tract and cause narcosis. Evidence of chronic systemic toxicity is inconclusive. Flammable liquid. Moderately explosive when exposed to flame. Ignites on contact with potassium tert-butoxide. To fight fire, use alcohol foam, CO_2, dry chemical. When heated to decomposition it emits acrid and irritating fumes. See also ESTERS.

BPW050　　　CAS:107-58-4　　　***HR: 2***
N-tert-BUTYLACRYLAMIDE
mf: $C_7H_{13}NO$　　mw: 127.21

SYNS: ACRYLAMIDE, N-tert-BUTYL- ◊ N-(1,1-DIMETHYLETHYL)-2-PROPENAMIDE ◊ 2-PROPENAMIDE, N-(1,1-DIMETHYLETHYL)-(9CI)

TOXICITY DATA with REFERENCE
orl-mus LD50:941 mg/kg　ARTODN 47,179,81

CONSENSUS REPORTS: Reported in EPA TSCA Inventory.

SAFETY PROFILE: Moderately toxic by ingestion. When heated to decomposition it emits toxic vapors of NO_x.

BPX000　　　CAS:75-65-0　　　***HR: 3***
tert-BUTYL ALCOHOL
DOT: UN 1120
mf: $C_4H_{10}O$　　mw: 74.14

PROP: Colorless liquid or rhombic prisms or plates. Mp: 25.3°, bp: 82.8°, flash p: 50°F (CC), d: 0.7887 @ 20°/4°, autoign temp: 896°F, vap press: 40 mm @ 24.5°, vap d: 2.55, lel: 2.4%, uel: 8.0%.

SYNS: ALCOOL BUTYLIQUE TERTIAIRE (FRENCH) ◊ tert-BUTANOL ◊ BUTANOL TERTIAIRE (FRENCH) ◊ tert-BUTYL HYDROXIDE ◊ 1,1-DIMETHYLETHANOL ◊ 2-METHYL-2-PROPANOL ◊ NCI-C55367 ◊ TRIMETHYLCARBINOL

TOXICITY DATA with REFERENCE
orl-mus TDLo:103 g/kg (female 6–20D post):REP　JPETAB 222,294,82
orl-mus TDLo:135 g/kg (female 6–20D post):REP　JPETAB 222,294,82
orl-mus TDLo:103 g/kg (female 6–20D post):REP　JPETAB 222,294,82
ihl-rat TCLo:5000 ppm/7H (female 1–22D post):TER　TJADAB 35,56A,87
orl-rat LD50:3500 mg/kg　SCIEAS 116,663,52
ipr-mus LD50:933 mg/kg　SCCUR* -,2,61
ivn-mus LD50:1538 mg/kg　AIPTAK 135,330,62
orl-rbt LD50:3559 mg/kg　IMSUAI 41,31,72
par-frg LDLo:12 g/kg　AIPTAK 50,296,35

CONSENSUS REPORTS: Community Right-To-Know List. Reported in EPA TSCA Inventory. EPA Genetic Toxicology Program.

OSHA PEL: (Transitional: TWA 100 ppm) TWA 100 ppm; STEL 150 ppm
ACGIH TLV: TWA 100 ppm; STEL 150 ppm (Proposed: TWA 100 ppm)
DFG MAK: 100 ppm (300 mg/m^3)

DOT Classification: Flammable Liquid; Label: Flammable Liquid

SAFETY PROFILE: Moderately toxic by ingestion, intravenous, and intraperitoneal routes. An experimental teratogen. Other experimental reproductive effects. Dangerous fire hazard when exposed to heat or flame. Moderately explosive in the form of vapor when exposed to flame. Ignites on contact with potassium-sodium alloys. To fight fire, use alcohol foam, CO_2, dry chemical. Incompatible with oxidizing materials, H_2O_2. See also n-BUTYL ALCOHOL and ALCOHOLS.

BPY100　　　CAS:513-49-5　　　***HR: 3***
sec-BUTYLAMINE, (S)-
mf: $C_4H_{11}N$　　mw: 73.16

SYNS: (+)-2-BUTYLAMINE ◊ S-2-BUTYLAMINE

TOXICITY DATA with REFERENCE
orl-rat LD50:380 mg/kg　28ZEAL 5,33,76
orl-dog LD50:225 mg/kg　28ZEAL 5,33,76
skn-rbt LD50:2500 mg/kg　28ZEAL 5,33,76

CONSENSUS REPORTS: Reported in EPA TSCA Inventory.

SAFETY PROFILE: Poison by ingestion. Moderately toxic by skin contact. When heated to decomposition it emits toxic vapors of NO_x.

BQV600   CAS:88-60-8   *HR: 2*
6-tert-BUTYL-m-CRESOL
mf: $C_{11}H_{16}O$ mw: 164.27

SYNS: 2-(1,1-DIMETHYLETHYL)-5-METHYLPHENOL ◇ PHENOL, 2-tert-BUTYL-5-METHYL-

TOXICITY DATA with REFERENCE
orl-mus LD50:1080 mg/kg   JAPMA8 38,366,49

SAFETY PROFILE: Moderately toxic by ingestion. When heated to decomposition it emits acrid smoke and irritating vapors.

BRQ050   CAS:541-33-3   *HR: 1*
BUTYLIDENE CHLORIDE
mf: $C_4H_8Cl_2$ mw: 127.02

SYNS: BUTANE, 1,1-DICHLORO- ◇ 1,1-DICHLOROBUTANE

TOXICITY DATA with REFERENCE
orl-mus LD50:4859 mg/kg   JPPMAB 3,169,51

CONSENSUS REPORTS: Reported in EPA TSCA Inventory.

SAFETY PROFILE: Slightly toxic by ingestion. When heated to decomposition it emits toxic vapors of Cl⁻.

BRR700   CAS:2052-15-5   *HR: 1*
n-BUTYL LEVULINATE
mf: $C_9H_{16}O_3$ mw: 172.25

SYNS: BUTYL LAEVULINATE ◇ n-BUTYL LAEVULINATE ◇ BUTYL LEVULINATE ◇ BUTYL 4-OXOPENTANOATE ◇ 4-KETOPENTANOIC ACID BUTYL ESTER ◇ LEVULINIC ACID, BUTYL ESTER ◇ PENTANOIC ACID, 4-OXO-, BUTYL ESTER (9CI)

TOXICITY DATA with REFERENCE
skn-rbt 500 mg/24H MLD   FCTOD7 21,655,83

CONSENSUS REPORTS: Reported in EPA TSCA Inventory.

SAFETY PROFILE: A skin irritant. When heated to decomposition it emits acrid smoke and irritating vapors.

BRU300   CAS:464-07-3   *HR: 1*
tert-BUTYL METHYL CARBINOL
mf: $C_6H_{14}O$ mw: 102.20

SYNS: 2-BUTANOL, 3,3-DIMETHYL- ◇ 3,3-DIMETHYL-2-BUTANOL ◇ PINACOLYL ALCOHOL (6CI)

TOXICITY DATA with REFERENCE
ihl-rat LCLo:3600 ppm/2.3H   JJATDK 7,307,87

CONSENSUS REPORTS: Reported in EPA TSCA Inventory.

SAFETY PROFILE: Slightly toxic by inhalation. When heated to decomposition it emits acrid smoke and irritating vapors.

BRU790   CAS:2219-82-1   *HR: 3*
2-tert-BUTYL-6-METHYLPHENOL
mf: $C_{11}H_{16}O$ mw: 164.27

SYN: PHENOL, 2-tert-BUTYL-6-METHYL-

TOXICITY DATA with REFERENCE
ivn-mus LD50:120 mg/kg   JMCMAR 23,1350,80

CONSENSUS REPORTS: Reported in EPA TSCA Inventory.

SAFETY PROFILE: Poison by intravenous route. When heated to decomposition it emits acrid smoke and irritating vapors.

BRU800   CAS:98-27-1   *HR: 3*
4-tert-BUTYL-2-METHYLPHENOL
mf: $C_{11}H_{16}O$ mw: 164.27

SYN: PHENOL, 4-tert-BUTYL-2-METHYL-

TOXICITY DATA with REFERENCE
ipr-mus LD50:81 mg/kg   JMCMAR 18,868,75
ivn-mus LD50:180 mg/kg   JMCMAR 23,1350,80

CONSENSUS REPORTS: Reported in EPA TSCA Inventory.

SAFETY PROFILE: Poison by intravenous and intraperitoneal routes. When heated to decomposition it emits acrid smoke and irritating vapors.

BSB000   CAS:142-77-8   *HR: 1*
BUTYL OLEATE
mf: $C_{22}H_{42}O_2$ mw: 338.64

PROP: Liquid. Bp: 173°, flash p: 356°F(OC), d: 0.873, vap d: 11.3.

SYN: (Z)-9-OCTADECENOIC ACID BUTYL ESTER ◇ OLEIC ACID, BUTYL ESTER ◇ PLASTHALL 503 ◇ UNIFLEX BYO

TOXICITY DATA with REFERENCE
skn-rbt 500 mg/24H MOD   FCTXAV 17,241,79

CONSENSUS REPORTS: Reported in EPA TSCA Inventory.

SAFETY PROFILE: A skin irritant. Combustible when exposed to heat or flame. To fight fire, use CO_2, dry chemical. Incompatible with oxidizing materials. When heated to decomposition it emits acrid smoke and irritating fumes. See also ESTERS; BUTYL ALCOHOL; and OLEIC ACID.

BSC600   CAS:109-13-7   *HR: 3*
tert-BUTYL PERISOBUTYRATE
DOT: UN 2142/UN 2562
mf: $C_8H_{16}O_3$ mw: 160.24

SYNS: tert-BUTYL PEROXYISOBUTYRATE ◇ tert-BUTYL PEROXYISOBUTYRATE, >52% but not >77% in solution (UN2142)(DOT) ◇ tert-BUTYL PEROXYISOBUTYRATE, >77% in solution (DOT) ◇ tert-BUTYL PER-

OXYISOBUTYRATE, not >52% in solution (UN2562) (DOT) ◇ ESPEROX 24M ◇ LUPERSOL 8 ◇ PEROXYISOBUTYRIC ACID, tert-BUTYL ESTER ◇ PROPANEPEROXOIC ACID, 2-METHYL-, 1,1-DIMETHYLETHYL ESTER

DOT Classification: Forbidden (>77% in solution); Organic Peroxide; Label: Organic Peroxide; Flammable Liquid; Label: Flammable Liquid, Organic Peroxide

CONSENSUS REPORTS: Reported in EPA TSCA Inventory.

SAFETY PROFILE: Explosive and flammable peroxide. Handle very carefully in concentrated solutions. When heated to decomposition it emits acrid smoke and irritating vapors.

BSD150   CAS:2372-21-6   **HR: 3**
tert-BUTYL PEROXYISOPROPYL CARBONATE, technically pure (DOT)
DOT: UN 2103
mf: $C_8H_{16}O_4$ mw: 176.24

SYNS: KAYACARBON BIC ◇ LUPERSOL TBIC ◇ LUPERSOL TBIC-M75 ◇ PEROXYCARBONIC ACID, OO-tert-BUTYL O-ISOPROPYL ESTER

DOT Classification: Organic Peroxide; Label: Organic Peroxide; Flammable Liquid; Label: Flammable Liquid, Organic Peroxide

CONSENSUS REPORTS: Reported in EPA TSCA Inventory.

SAFETY PROFILE: Flammable liquid and peroxide. Handle carefully. When heated to decomposition it emits acrid smoke and irritating vapors.

BSE460   CAS:88-18-6   **HR: 3**
2-t-BUTYLPHENOL
mf: $C_{10}H_{14}O$ mw: 150.24

SYN: PHENOL, o-(tert-BUTYL)-

TOXICITY DATA with REFERENCE
ipr-mus LD50:82 mg/kg   JMCMAR 18,868,75

CONSENSUS REPORTS: Reported in EPA TSCA Inventory.

SAFETY PROFILE: Poison by intraperitoneal route. When heated to decomposition it emits acrid smoke and irritating vapors.

BSG300   CAS:780-11-0   **HR: 2**
3-tert-BUTYLPHENYL N-METHYLCARBAMATE
mf: $C_{12}H_{17}NO_2$ mw: 207.30

SYNS: CARBAMIC ACID, METHYL-, 3-tert-BUTYLPHENYL ESTER ◇ H-22 ◇ KNOCKBAL ◇ PHENOL, 3-(1,1-DIMETHYLETHYL)-, METHYLCARBAMATE (9CI) ◇ RE 5030 ◇ TBPMC ◇ TERBAM

TOXICITY DATA with REFERENCE
orl-mus LD50:470 mg/kg   85AREA 1,44,77
skn-mus LD50:2660 mg/kg   JPIFAN (4),28,70
orl-rbt LD50:505 mg/kg   JPIFAN (4),28,70

CONSENSUS REPORTS: Reported in EPA TSCA Inventory.

SAFETY PROFILE: Moderately toxic by ingestion and skin contact. When heated to decomposition it emits toxic vapors of NO_x.

BSJ550   CAS:539-32-2   **HR: 3**
3-BUTYLPYRIDINE
mf: $C_9H_{13}N$ mw: 135.23

SYNS: 3-n-BUTYLPYRIDINE ◇ PYRIDINE, 3-BUTYL- ◇ 1-(3-PYRIDYL)BUTANE

TOXICITY DATA with REFERENCE
ipr-mus LD50:270 mg/kg   JPETAB 88,82,46
ivn-mus LD50:59 mg/kg   JPETAB 88,82,46

CONSENSUS REPORTS: Reported in EPA TSCA Inventory.

SAFETY PROFILE: Poison by intravenous and intraperitoneal routes. When heated to decomposition it emits toxic vapors of NO_x.

BSP500   CAS:98-51-1   **HR: 2**
p-tert-BUTYLTOLUENE
mf: $C_{11}H_{16}$ mw: 148.27

PROP: Colorless liquid.

SYNS: p-METHYL-tert-BUTYLBENZENE ◇ 1-METHYL-4-tert-BUTYLBENZENE ◇ TBT

TOXICITY DATA with REFERENCE
eye-hmn 5 ppm/2H   AMIHBC 9,227,54
skn-rbt 500 mg/24H MLD   AMIHBC 9,227,54
eye-rbt 100 mg   AMIHBC 9,227,54
ihl-hmn TCLo:10 ppm/3M:GIT   AMIHBC 9,227,54
ihl-hmn TCLo:20 ppm/5M:EYE,IRR,GIT   28ZRAQ -,156,60
orl-rat LD50:1500 mg/kg   AMIHBC 9,227,54
ihl-rat LC50:1500 mg/m^3/4H   AMIHBC 9,227,54
orl-mus LD50:900 mg/kg   AMIHBC 9,227,54
ihl-mus LC50:248 ppm/2H   AMIHBC 9,227,54
orl-rbt LD50:2000 mg/kg   AMIHBC 9,227,54

CONSENSUS REPORTS: Reported in EPA TSCA Inventory.

OSHA PEL: (Transitional: TWA 10 ppm) TWA 10 ppm; STEL 20 ppm
ACGIH TLV: TWA 10 ppm; STEL 20 ppm. (Proposed: TWA 1 ppm)
DFG MAK: 10 ppm (60 mg/m^3)

SAFETY PROFILE: Moderately toxic by inhalation and ingestion. A skin and human eye irritant. Human systemic effects by inhalation: nausea or vomiting, conjunctiva irritation, unspecified effects on the sense of taste. Inhalation of vapors causes irritation of lungs and depression of central nervous system. Prolonged exposure may result in damage to liver and kidneys. Flammable when exposed to heat or flame. Incompatible with oxidizing materials. When heated to decomposition it emits acrid smoke and fumes.

BSR900 CAS:1779-51-7 **HR: 3**
n-BUTYLTRIPHENYLPHOSPHONIUM BROMIDE
mf: $C_{22}H_{24}P \cdot Br$ mw: 399.34

SYN: PHOSPHONIUM, BUTYLTRIPHENYL-, BROMIDE

TOXICITY DATA with REFERENCE
ivn-mus LD50:56 mg/kg   CSLNX* NX#06771

CONSENSUS REPORTS: Reported in EPA TSCA Inventory.

SAFETY PROFILE: Poison by intravenous route. When heated to decomposition it emits toxic vapors of PO_x and Br^-.

BSS300 CAS:689-11-2 **HR: 2**
sec-BUTYLUREA
mf: $C_5H_{12}N_2O$ mw: 116.19

SYN: UREA, sec-BUTYL-

TOXICITY DATA with REFERENCE
par-mus LDLo:2789 mg/kg   JPETAB 52,216,34

CONSENSUS REPORTS: Reported in EPA TSCA Inventory.

SAFETY PROFILE: Moderately toxic by parenteral. When heated to decomposition it emits toxic vapors of NO_x.

BSS310 CAS:1118-12-3 **HR: 2**
tert-BUTYLUREA
mf: $C_5H_{12}N_2O$ mw: 116.19

SYNS: (1,1-DIMETHYLETHYL)UREA ◇ UREA, tert-BUTYL- ◇ UREA, (1,1-DIMETHYLETHYL)-(9CI)

TOXICITY DATA with REFERENCE
orl-mus LD50:3050 mg/kg   AIPTAK 219,103,76

CONSENSUS REPORTS: Reported in EPA TSCA Inventory.

SAFETY PROFILE: Moderately toxic by ingestion. When heated to decomposition it emits toxic vapors of NO_x.

BSS550 CAS:105-77-1 **HR: 2**
BUTYLXANTHIC DISULFIDE
mf: $C_{10}H_{18}O_2S_4$ mw: 298.52

SYNS: BIS-BUTYLXANTHOGEN ◇ CPB ◇ DI(BUTOXYTHIOCARBONYL) DISULFIDE ◇ DIBUTYL DIXANTHOGEN ◇ DIBUTYLDIXANTOGENATE ◇ DIBUTYL XANTHOGEN DISULFIDE ◇ DITHIOBIS(THIOFORMIC ACID) O,O-DIBUTYL ESTER ◇ DXG ◇ FORMIC ACID, DITHIOBIS(THIO-, O,O-DIBUTYL ESTER ◇ THIOPEROXYDICARBONIC ACID, DIBUTYL ESTER

TOXICITY DATA with REFERENCE
orl-mus LD50:2700 mg/kg   GISAAA 47(3),88,82

CONSENSUS REPORTS: Reported in EPA TSCA Inventory.

SAFETY PROFILE: Moderately toxic by ingestion. When heated to decomposition it emits toxic vapors of SO_x.

BSW550 CAS:106-31-0 **HR: 1**
BUTYRIC ANHYDRIDE
mf: $C_8H_{14}O_3$ mw: 158.22

SYNS: ANHYDRID KYSELINY MASELNE ◇ BUTANOIC ACID, ANHYDRIDE (9CI) ◇ BUTANOIC ANHYDRIDE ◇ BUTYRANHYDRID ◇ BUTYRIC ACID ANHYDRIDE ◇ n-BUTYRIC ACID ANHYDRIDE ◇ n-BUTYRIC ANHYDRIDE ◇ BUTYRYL OXIDE

TOXICITY DATA with REFERENCE
orl-rat LD50:8790 mg/kg   85JCAE -,321,86
orl-mus LD30: 2 g/kg   85GMAT -,31,82

CONSENSUS REPORTS: Reported in EPA TSCA Inventory.

SAFETY PROFILE: Mildly toxic by ingestion. When heated to decomposition it emits acrid smoke and irritating vapors.

BSY300 CAS:2494-56-6 **HR: 3**
BUTYRYLCHOLINE IODIDE
mf: $C_9H_{20}NO_2 \cdot I$ mw: 301.20

SYN: AMMONIUM, (2-BUTYRYLOXYETHYL)TRIMETHYL-, IODIDE

TOXICITY DATA with REFERENCE
ivn-mus LD50:56 mg/kg   CSLNX* NX#01428

CONSENSUS REPORTS: Reported in EPA TSCA Inventory.

SAFETY PROFILE: Poison by intravenous route. When heated to decomposition it emits toxic vapors of NO_x.

C

CAD000 CAS:7440-43-9 **HR: 3**
CADMIUM
af: Cd aw: 112.40

PROP: Hexagonal crystals, silver-white, malleable metal. Mp: 320.9°, bp: 767 ± 2°, d: 8.642, vap press: 1 mm @ 394°.

SYNS: C.I. 77180 ◊ COLLOIDAL CADMIUM ◊ KADMIUM (GERMAN)

TOXICITY DATA with REFERENCE

cyt-ham:ovr 1 μmol/L CGCGBR 26,251,80
orl-rat TDLo:155 mg/kg (male 13W pre):REP BECTA6 20,96,78
scu-rat TDLo:250 μg/kg (female 19D post):REP APTOD9 19,A122,80
orl-rat TDLo:21500 μg/kg (multi):TER ENVRAL 22,466,80
orl-rat TDLo:155 mg/kg (male 13W pre):REP BECTA6 20,96,78
orl-mus TDLo:1700 mg/kg (female 8–12D post):REP TCMUD8 6,361,86
orl-mus TDLo:448 mg/kg (multi):TER AEHLAU 23,102,71
ivn-rat TDLo:8 mg/kg (female 8–15D post):TER JJATDK 1,264,81
orl-rat TDLo:220 mg/kg (female 1–22D post):TER TOLED5 11,233,82
orl-rat TDLo:21500 μg/kg (multi):REP ENVRAL 22,466,80
ipr-rat TDLo:1124 μg/kg (male 1D pre):REP TXAPA9 41,194,77
orl-rat TDLo:23 mg/kg (female 1–22D post):TER PSEBAA 158,614,78
ivn-rat TDLo:1250 μg/kg (female 14D post):TER JJATDK 1,264,81
ivn-rat TDLo:1250 μg/kg (female 9D post):TER JJATDK 1,264,81
par-ham TDLo:2 mg/kg (female 8D post):TER TJADAB 31,52A,85
ivn-rat TDLo:1250 μg/kg (female 9D post):TER JJATDK 1,264,81
ivn-rat TDLo:1250 μg/kg (female 14D post):TER JJATDK 1,264,81
ihl-wmn TCLo:129 μg/m^3/20Y-C:CAR AJIMD8 10,153,86
ims-rat TDLo:40 mg/kg/4W-I:CAR JEPTDQ 1(1),51,77 ZHYGAM 31,224,85
ims-rat TD:70 mg/kg:ETA BJCAAI 18,124,64
ims-rat TD:63 mg/kg:ETA NATUAS 193,592,62
ims-rat TD:45 mg/kg/4W-I:NEO NCIUS* PH-43-64-886,SEPT,71
ihl-man TCLo:88 μg/m^3/8.6Y:KID AEHLAU 28,147,74
ihl-hmn LCLo:39 mg/m^3/20M AIHAAP 31,180,70
unk-man LDLo:15 mg/kg 85DCAI 2,73,70
orl-rat LD50:225 mg/kg TXAPA9 41,667,77
ihl-rat LC50:25 mg/m^3/30M SAIGBL 16,212,74
ipr-rat LD50:4 mg/kg TXAPA9 41,667,77
scu-rat LD50:9 mg/kg TXAPA9 41,667,77
ivn-rat LD50:1800 μg/kg JJATDK 1,264,81
orl-mus LD50:890 mg/kg 41HTAH -,14,78
ihl-mus LCLo:170 mg/m^3 NTIS** PB158-508
orl-rbt LDLo:70 mg/kg AMPMAR 34,127,73
scu-rbt LDLo:6 mg/kg PROTA* -,-,55
ivn-rbt LDLo:5 mg/kg JOGBAS 35,693,28
ims-ham LDLo:25 mg/kg NCIUS* PH-43-64-886

CONSENSUS REPORTS: NTP Fifth Annual Report on Carcinogens. IARC Cancer Review: Group 2A IMEMDT 7,139,87; Animal Sufficient Evidence IMEMDT 11,39,76; IMEMDT 2,74,73. Cadmium and its compounds are on the Community Right-To-Know List. Reported in EPA TSCA Inventory. EPA Genetic Toxicology Program.

OSHA PEL: TWA 5 μg(Cd)/m^3
ACGIH TLV: Dust and Salts: TWA 0.05 mg(Cd)/m^3 (Proposed: TWA 0.01 mg(Cd)/m^3 (dust), Suspected Human Carcinogen; 0.002 mg(Cd)/m^3 (respirable dust), Suspected Human Carcinogen); BEI: 10 μg/g creatinine in urine; 10 μg/L in blood. (Proposed: 5 μg/g creatinine in urine; 5 μg/L in blood.)
DFG BAT: Blood 1.5 μg/dL; Urine 15 μg/dL. MAK: Suspected Carcinogen.
NIOSH REL: (Cadmium) Reduce to lowest feasible level

SAFETY PROFILE: Confirmed human carcinogen with experimental carcinogenic, tumorigenic, and neoplastigenic data. A human poison by inhalation and possibly other routes. Poison experimentally by ingestion, inhalation, intraperitoneal, subcutaneous, intramuscular, and intravenous routes. In humans inhalation causes an excess of protein in the urine. Experimental teratogenic and reproductive effects. Mutation data reported. The dust ignites spontaneously in air and is flammable and explosive when exposed to heat, flame, or by chemical reaction with oxidizing agents, metals, HN$_3$, Zn, Se, and Te. Explodes on contact with hydrazoic acid. Violent or explosive reaction when heated with ammonium nitrate. Vigorous reaction when heated with nitryl fluoride. When heated to a high temperature it emits toxic fumes of Cd. See also CADMIUM COMPOUNDS.

CAD250 CAS:543-90-8 **HR: 3**
CADMIUM(II) ACETATE
mf: C$_2$H$_4$O$_2$•1/2Cd mw: 116.25

PROP: Monoclinic, colorless crystals; odor of acetic acid. Mp: 256°, bp: decomp, d: 2.341.

SYNS: ACETIC ACID, CADMIUM SALT ◇ BIS(ACETOXY)CADMIUM ◇ CADMIUM ACETATE (DOT) ◇ CADMIUM DIACETATE ◇ C.I. 77185

TOXICITY DATA with REFERENCE
cyt-hmn:lym 10 nmol/L　MUREAV 85,236,81
otr-ham:emb 1 μmol/L　CNREA8 39,193,79
dnd-ham:emb 1 μmol/L　CNREA8 39,193,79
ipr-rat TDLo:2 mg/kg (female 20D post):REP　BECTA6 23,25,79
ipr-rat TDLo:1 mg/kg (14D preg):TER　BECTA6 23,25,79
ipr-rat TDLo:2371 μg/kg (12D preg):REP　BECTA6 20,206,78
scu-rat TDLo:2325 ng/kg (female 1D pre):REP　JRPFA4 17,559,68
ipr-rat TDLo:2371 μg/kg (12D preg):TER　BECTA6 20,206,78
ipr-rat TDLo:2371 μg/kg (12D preg):TER　BECTA6 20,206,78
ipr-mus LD50:14 mg/kg　TXAPA9 49,41,79

CONSENSUS REPORTS: Reported in EPA TSCA Inventory. EPA Genetic Toxicology Program. Cadmium and its compounds are on the Community Right-To-Know List.

OSHA PEL: TWA 5 μg(Cd)/m^3
ACGIH TLV: TWA 0.05 mg(Cd)/m^3 (Proposed: TWA 0.01 mg(Cd)/m^3 (dust), Suspected Human Carcinogen; 0.002 mg(Cd)/m^3 (respirable dust), Suspected Human Carcinogen); BEI: 10 μg/g creatinine in urine; 10 μg/L in blood
NIOSH REL: (Cadmium) Reduce to lowest feasible level.

SAFETY PROFILE: Confirmed human carcinogen. Poison by intraperitoneal route. An experimental teratogen. Other experimental reproductive effects. Human mutation data reported. When heated to decomposition it emits toxic fumes of Cd. See also CADMIUM COMPOUNDS.

CAD275　　CAS:5743-04-4　　*HR: 3*
CADMIUM ACETATE DIHYDRATE
mf: $C_4H_6O_4$•Cd•$2H_2O$　mw: 266.54

PROP: Crystals, becoming anhydrous at 130°; slt acetic acid odor. D: 2.01, 2.341 (anhydrous), mp: 255° (anhydrous). Sol in water and alc.

SYNS: ACETIC ACID, CADMIUM SALT, DIHYDRATE ◇ CADMIUM DIACETATE DIHYDRATE

TOXICITY DATA with REFERENCE
cyt-hmn:lyms 1 mg/L　CYGEDX 12(3),46,78

OSHA PEL: TWA 5 μg(Cd)/m^3
ACGIH TLV: TWA 0.01 mg(Cd)/m^3; Suspected Carcinogen.

SAFETY PROFILE: Confirmed human carcinogen. Mutation data reported. When heated to decomposition it emits toxic fumes of Cd.

CAD325　　CAS:22750-53-4　　*HR: 3*
CADMIUM AMIDE
mf: CdH_4N_2　mw: 144.46

SYN: CADMIUM DIAMIDE

CONSENSUS REPORTS: Cadmium compounds are on the Community Right-To-Know List.

OSHA PEL: TWA 5 μg(Cd)/m^3
ACGIH TLV: TWA 0.05 mg(Cd)/m^3 (Proposed: TWA 0.01 mg(Cd)/m^3 (dust), Suspected Human Carcinogen; 0.002 mg(Cd)/m^3 (respirable dust), Suspected Human Carcinogen); BEI: 10 μg/g creatinine in urine; 10 μg/L in blood
NIOSH REL: (Cadmium) Reduce to lowest feasible level.

SAFETY PROFILE: Confirmed human carcinogen. May explode if heated. Reacts violently with water. When heated to decomposition it emits toxic fumes of Cd and NO_x. See also CADMIUM COMPOUNDS and AMIDES.

CAD350　　CAS:14215-29-3　　*HR: 3*
CADMIUM AZIDE
mf: CdN_6　mw: 196.45

$Cd(N_3)_2$

SYN: CADMIUM DIAZIDE

CONSENSUS REPORTS: Cadmium compounds are on the Community Right-To-Know List.

OSHA PEL: TWA 5 μg(Cd)/m^3
ACGIH TLV: TWA 0.05 mg(Cd)/m^3 (Proposed: TWA 0.01 mg(Cd)/m^3 (dust), Suspected Human Carcinogen; 0.002 mg(Cd)/m^3 (respirable dust), Suspected Human Carcinogen); BEI: 10 μg/g creatinine in urine; 10 μg/L in blood
NIOSH REL: (Cadmium) Reduce to lowest feasible level.

SAFETY PROFILE: Confirmed human carcinogen. The dry solid is an unstable heat- and friction-sensitive explosive. When heated to decomposition it emits toxic fumes of NO_x and Cd. See also CADMIUM COMPOUNDS and AZIDES.

CAD500　　CAS:7495-93-4　　*HR: 3*
CADMIUM BIS(2-ETHYLHEXYL) PHOSPHITE
mf: $C_{32}H_{68}O_6P_2$•Cd　mw: 723.34

SYN: BIS(2-ETHYLHEXYL) ESTER PHOSPHOROUS ACID CADMIUM SALT

TOXICITY DATA with REFERENCE
ipr-mus LDLo:250 mg/kg　CBCCT* 7,790,55

CONSENSUS REPORTS: Cadmium and its compounds are on the Community Right-To-Know List.

OSHA PEL: TWA 5 μg(Cd)/m³
ACGIH TLV: TWA 0.05 mg(Cd)/m³ (Proposed: TWA 0.01 mg(Cd)/m³ (dust), Suspected Human Carcinogen; 0.002 mg(Cd)/m³ (respirable dust), Suspected Human Carcinogen); BEI: 10 μg/g creatinine in urine; 10 μg/L in blood
NIOSH REL: (Cadmium) Reduce to lowest feasible level.

SAFETY PROFILE: Confirmed human carcinogen. Poison by intraperitoneal route. When heated to decomposition it emits toxic fumes of PO_x and Cd. See also CADMIUM COMPOUNDS.

CAD600   CAS:7789-42-6   *HR: 3*
CADMIUM BROMIDE
mf: Br_2Cd mw: 272.22

PROP: Pearly hexagonal crystals, hygroscopic. Mp: 566°, bp: 963°, d: 5.192. Sol in water and alc; moderately sol in acetone; sltly sol in ethanol.

SYN: CADMIUM DIBROMIDE

CONSENSUS REPORTS: Reported in EPA TSCA Inventory.

OSHA PEL: TWA 5 μg(Cd)/m³
ACGIH TLV: TWA 0.01 mg(Cd)/m³; Suspected Carcinogen.

SAFETY PROFILE: Confirmed human carcinogen. When heated to decomposition it emits toxic fumes of Cd and Br^-.

CAD750   CAS:2191-10-8   *HR: 3*
CADMIUM CAPRYLATE
mf: $C_{16}H_{30}O_4 \cdot Cd$ mw: 398.86

SYN: OCTANOIC ACID, CADMIUM SALT (2:1)

TOXICITY DATA with REFERENCE
orl-rat LD50:950 mg/kg   JHEMA2 18,144,74
itr-rat LDLo:10 mg/kg   JHEMA2 18,144,74
orl-mus LD50:300 mg/kg   JHEMA2 18,144,74

CONSENSUS REPORTS: Reported in EPA TSCA Inventory. Cadmium and its compounds are on the Community Right-To-Know List.

OSHA PEL: TWA 5 μg(Cd)/m³
ACGIH TLV: TWA 0.05 mg(Cd)/m³ (Proposed: TWA 0.01 mg(Cd)/m³ (dust), Suspected Human Carcinogen; 0.002 mg(Cd)/m³ (respirable dust), Suspected Human Carcinogen); BEI: 10 μg/g creatinine in urine; 10 μg/L in blood
NIOSH REL: (Cadmium) Reduce to lowest feasible level.

SAFETY PROFILE: Confirmed human carcinogen. Poison by ingestion and intratracheal routes. When heated to decomposition it emits toxic fumes of Cd. See also CADMIUM COMPOUNDS.

CAD800   CAS:513-78-0   *HR: 3*
CADMIUM CARBONATE
mf: $CO_3 \cdot Cd$ mw: 172.41

SYNS: CADMIUM MONOCARBONATE ◇ CARBONIC ACID, CADMIUM SALT ◇ CHEMCARB

TOXICITY DATA with REFERENCE
sce-ham:ovr 870 nmol/L   ENMUDM 7,381,85
orl-mus LD50:310 mg/kg   GTPZAB 25(2),42,81

OSHA PEL: TWA 5 2g(Cd)/m³
ACGIH TLV: TWA 0.01 mg(Cd)/m³; Suspected Carcinogen
NIOSH REL: (Cadmium, dust and fume): lowest feasible concentration.

CONSENSUS REPORTS: Reported in EPA TSCA Inventory.

SAFETY PROFILE: Confirmed human carcinogen. Poison by ingestion. Mutation data reported. When heated to decomposition it emits toxic fumes of cadmium.

CAE000   *HR: 3*
CADMIUM CHLORATE
mf: $CdCl_2O_6$ mw: 279.31

$Cd(ClP_3)_2$

PROP: Colorless, deliquescent prisms. Mp: 80°, d: 2.28 @ 18°.

CONSENSUS REPORTS: Cadmium and its compounds are on the Community Right-To-Know List.

OSHA PEL: TWA 5 μg(Cd)/m³
ACGIH TLV: TWA 0.05 mg(Cd)/m³ (Proposed: TWA 0.01 mg(Cd)/m³ (dust), Suspected Human Carcinogen; 0.002 mg(Cd)/m³ (respirable dust), Suspected Human Carcinogen); BEI: 10 μg/g creatinine in urine; 10 μg/L in blood
NIOSH REL: (Cadmium) Reduce to lowest feasible level.

SAFETY PROFILE: Confirmed human carcinogen. A powerful oxidizing agent. Flammable by chemical reaction with reducing agents. Moderate explosion hazard when shocked or exposed to heat. Violent or explosive reaction with sulfides (e.g., copper(II) sulfide (explodes); antimony(II) sulfide; arsenic(III) sulfide; tin(II) sulfide; tin(IV) sulfide). When heated to decomposition it emits toxic fumes of Cd and Cl^-. See also CHLORATES.

CAE250   CAS:10108-64-2   *HR: 3*
CADMIUM CHLORIDE
mf: $CdCl_2$ mw: 183.30

PROP: Hexagonal, colorless crystals. Mp: 568°, d: 4.047 @ 25°, vap press: 10 mm @ 656°, bp: 960°.

SYNS: CADDY ◇ CADMIUM DICHLORIDE ◇ KADMIUMCHLORID (GERMAN) ◇ VI-CAD

CADMIUM CHLORIDE CAE250

TOXICITY DATA with REFERENCE
dni-hmn:hla 250 µmol/L   MUREAV 92,427,82
cyt-ofs-mul 630 µg/L/4W-C   BECTA6 36,199,86
orl-mus TDLo:6 mg/kg (female 15–19D post):REP
    JEPTDQ 1(3),187,78
orl-mus TDLo:6 mg/kg (female 15–19D post):REP
    JEPTDQ 1(3),187,78
orl-rat TDLo:48 mg/kg (female 90D pre):REP   NTIS** CONF-771017
ihl-rat TCLo:130 µg/m^3 (female 1–19D post):TER
    GTPZAB 31(8),25,87
orl-rat TDLo:14677 µg/kg (female 6–14D post):TER
    TXAPA9 23,222,72
ipr-rat TDLo:2933 µg/kg (female 10D post):TER   TJADAB 7,237,73
orl-rat TDLo:130 mg/kg (female 7–16D post):TER
    JJATDK 2,255,82
scu-rat TDLo:7 mg/kg (female 12D post):TER   TJADAB 35,74A,87
ihl-rat TCLo:200 µg/m^3/24H (female 1–21D post):TER
    TXCYAC 10,297,78
scu-ham TDLo:5 mg/kg (female 1D pre):REP   BIREBV 29,249,83
orl-rat TDLo:652 mg/kg (female 7–16D post):REP   JJATDK 2,255,82
ipr-mus TDLo:1 mg/kg (male 1D pre):REP   JPETAB 187,641,73
orl-rbt TDLo:990 mg/kg (female 1–6D post):REP   JRPFA4 70,323,84
orl-rat TDLo:652 mg/kg (female 7–16D post):REP   JJATDK 2,255,82
scu-rbt TDLo:5 mg/kg (female 1D pre):REP   CCPTAY 26,181,82
scu-mus TDLo:3 mg/kg (female 1D pre):REP   TJADAB 16,127,77
ihl-rat TCLo:200 µg/m^3/24H (female 1–21D post):REP
    TXCYAC 10,297,78
orl-rat TDLo:652 mg/kg (female 7–16D post):REP   JJATDK 2,255,82
orl-rat TDLo:4890 ng/kg (male 30D pre):REP   EVHPAZ 13,59,76
scu-rat TDLo:1250 µg/kg (male 1D pre):REP   ABMGAJ 5,153,60
scu-rat TDLo:1250 µg/kg (male 1D pre):REP   ABMGAJ 5,153,60
orl-mus TDLo:248 g/kg (female 1–19D post):TER   JONUAI 109,1640,79
ipr-rat TDLo:2933 µg/kg (female 10D post):TER   TJADAB 7,237,73
orl-rat TDLo:280 mg/kg (female 6–19D post):TER
    NATUAS 239,231,72
ipr-rat TDLo:2 mg/kg (female 10D post):TER   ARTODN 59,443,87
ipr-mus TDLo:1833 µg/kg (female 7D post):TER   NTIS** CONF-771017
orl-mus TDLo:6 mg/kg (female 15–19D post):TER
    JEPTDQ 1(3),187,78
ipr-mus TDLo:1833 µg/kg (female 7D post):TER   NTIS** CONF-771017
ipr-rat TDLo:2933 µg/kg (female 10D post):TER   TJADAB 7,237,73
scu-rat TDLo:4892 µg/kg (female 19D post):TER   TXCYAC 18,103,80
orl-rat TDLo:17 mg/kg (male 6W pre):TER   EESADV 4,51,80
scu-rat TDLo:32 mg/kg (female 12–15D post):TER
    TJADAB 23,75,81
orl-rat TDLo:14,677 µg/kg (female 6–14D post):TER
    TXAPA9 23,222,72
ihl-rat TCLo:20 µg/m^3/23H/78W-C:CAR   JJIND8 70,367,83
scu-rat TDLo:3666 µg/kg:CAR   PAACA3 24,84,83
ims-rat TDLo:4500 µg/kg:ETA   ARPAAQ 83,493,67
par-rat TDLo:1700 µg/kg:ETA   ARPAAQ 83,493,67
scu-mus TDLo:5499 µg/kg:ETA   JNCIAM 31,745,63
scu-rat TD:4500 µg/kg:ETA   ARPAAQ 83,493,67
ihl-rat TC:41 µg/m^3/23H/78-W-C:CAR   JJIND8 70,367,83
ihl-rat TC:82 µg/m^3/23H/78W-C:CAR   JJIND8 70,367,83
scu-rat TD:7332 µg/kg:CAR   PAACA3 24,84,83
scu-rat TD:5499 µg/kg:NEO   PSEBAA 115,653,64
scu-rat TD:3666 mg/kg:NEO   CNREA8 43,4575,83
scu-rat TD:7332 mg/kg:NEO   CNREA8 43,4575,83
scu-rat TD:40,770 ng/kg/2D-I:CAR   CNREA8 48,4656,88
orl-wmn LDLo:3 g/kg:BPR,GIT   BMJOAE 292,1559,86
orl-rat LD50:88 mg/kg   AFDOAQ 15,122,51
ipr-rat LD50:1800 µg/kg   EVHPAZ 28,89,79
orl-mus LD50:60 mg/kg   APTOA6 48,108,81
ihl-mus LC50:2300 mg/m^3 NTIS** PB158-508
ipr-mus LD50:9300 µg/kg   NEZAAQ 32,472,77
scu-mus LD50:3200 µg/kg   APTOA6 48,108,81
ivn-mus LD50:3500 µg/kg   TXAPA9 53,510,80
ihl-dog LC90:420 mg/m^3/30M   JIHTAB 29,302,47
ivn-dog LDLo:5 mg/kg   EQSSDX 1,1,75
scu-cat LDLo:25 mg/kg   EQSSDX 1,1,75
ivn-cat LDLo:5 mg/kg   HBAMAK 4,1289,35
orl-rbt LDLo:70 mg/kg   EQSSDX 1,1,75

CONSENSUS REPORTS: Animal Sufficient Evidence IMEMDT 11,39,76; IMEMDT 2,74,73. EPA Genetic Toxicology Program. Cadmium and its compounds are on the Community Right-To-Know List. Reported in EPA TSCA Inventory.

OSHA PEL: TWA 5 µg(Cd)/m^3
ACGIH TLV: TWA 0.05 mg(Cd)/m^3 (Proposed: TWA 0.01 mg(Cd)/m^3 (dust), Suspected Human Carcinogen; 0.002 mg(Cd)/m^3 (respirable dust), Suspected Human Carcinogen); BEI: 10 µg/g creatinine in urine; 10 µg/L in blood
DFG MAK: Animal Carcinogen, Suspected Human Carcinogen
NIOSH REL: (Cadmium) Reduce to lowest feasible level.

SAFETY PROFILE: Confirmed human carcinogen with experimental carcinogenic and tumorigenic data. Poison by ingestion, inhalation, skin contact, intraperitoneal, subcutaneous, intravenous, and possibly other routes. Human systemic effects by ingestion: blood pressure, acute pulmonary edema, hypermotility, diarrhea. Experimental teratogenic and reproductive effects. Human mutation data reported. Reacts violently with BrF_3 and K. When heated to decomposition it emits very toxic fumes of Cd and Cl^-. See also CADMIUM COMPOUNDS and CHLORIDES.

CAE375 CAS:72589-96-9 *HR: 3*
CADMIUM CHLORIDE, DIHYDRATE
mf: $CdCl_2 \cdot 2H_2O$ mw: 219.34

TOXICITY DATA with REFERENCE
orl-mus TDLo:15 mg/kg (5D male):TER AXVMAW 34,399,80
itt-mam TDLo:29 µg/kg (1D male):REP BECTA6 26,233,81
scu-rat TDLo:6580 µg/kg:ETA ARGEAR 36,119,70

CONSENSUS REPORTS: Cadmium and its compounds are on the Community Right-To-Know List.

OSHA PEL: TWA 5 µg(Cd)/m^3
ACGIH TLV: TWA 0.05 mg(Cd)/m^3 (Proposed: TWA 0.01 mg(Cd)/m^3 (dust), Suspected Human Carcinogen; 0.002 mg(Cd)/m^3 (respirable dust), Suspected Human Carcinogen); BEI: 10 µg/g creatinine in urine; 10 µg/L in blood
DFG MAK: Animal Carcinogen, Suspected Human Carcinogen
NIOSH REL: (Cadmium) Reduce to lowest feasible level.

SAFETY PROFILE: Confirmed human carcinogen with experimental tumorigenic data. An experimental teratogen. Other experimental reproductive effects. When heated to decomposition it emits toxic fumes of Cl^- and Cd. See also CADMIUM CHLORIDE, CADMIUM COMPOUNDS, and CHLORIDES.

CAE425 CAS:7790-78-5 *HR: 3*
CADMIUM CHLORIDE, HYDRATE (2585)
mf: $CdCl_2 \cdot 5/2H_2O$ mw: 228.35

TOXICITY DATA with REFERENCE
dni-hmn:lym 28 µmol/L IAAAAM 79,83,86
scu-rat TDLo:2 mg/kg (male 1D pre):REP CCPTAY 27,521,83
scu-rat TDLo:7 mg/kg (male 1D pre):REP JOENAK 34,329,66
scu-rat TDLo:2 mg/kg (male 1D pre):REP CCPTAY 27,521,83
ipr-rat TDLo:6237 µg/kg:REP ESKGA2 29,P-46,83
orl-mus LD50:194 mg/kg JTEHD6 22,35,87
ipr-mus LD50:4567 µg/kg TXAPA9 63,461,82

CONSENSUS REPORTS: Cadmium and its compounds are on the Community Right-To-Know List.

OSHA PEL: TWA 5 µg(Cd)/m^3
ACGIH TLV: TWA 0.05 mg(Cd)/m^3 (Proposed: TWA 0.01 mg(Cd)/m^3 (dust), Suspected Human Carcinogen; 0.002 mg(Cd)/m^3 (respirable dust), Suspected Human Carcinogen); BEI: 10 µg/g creatinine in urine; 10 µg/L in blood
DFG MAK: Animal Carcinogen, Suspected Human Carcinogen
NIOSH REL: (Cadmium) Reduce to lowest feasible level.

SAFETY PROFILE: Confirmed human carcinogen. Poison by ingestion and intraperitoneal routes. Experimental reproductive effects. Human mutation data reported. When heated to decomposition it emits toxic fumes of Cl^- and Cd. See also CADMIUM CHLORIDE, CADMIUM COMPOUNDS, and CHLORIDES.

CAE500 CAS:35658-65-2 *HR: 3*
CADMIUM CHLORIDE, MONOHYDRATE
mf: $CdCl_2 \cdot H_2O$ mw: 201.32

TOXICITY DATA with REFERENCE
orl-rat TDLo:179 mg/kg (male 1D pre):REP CALEDQ 36,307,87
orl-rat TDLo:65 mg/kg/2Y-C:ETA CALEDQ 9,191,80
scu-rat TDLo:4478 µg/kg:CAR CALEDQ 36,307,87

CONSENSUS REPORTS: Cadmium and its compounds are on the Community Right-To-Know List.

OSHA PEL: TWA 5 µg(Cd)/m^3
ACGIH TLV: TWA 0.05 mg(Cd)/m^3 (Proposed: TWA 0.01 mg(Cd)/m^3 (dust), Suspected Human Carcinogen; 0.002 mg(Cd)/m^3 (respirable dust), Suspected Human Carcinogen); BEI: 10 µg/g creatinine in urine; 10 µg/L in blood
DFG MAK: Animal Carcinogen, Suspected Human Carcinogen
NIOSH REL: (Cadmium) Reduce to lowest feasible level.

SAFETY PROFILE: Confirmed human carcinogen with experimental carcinogenic and tumorigenic data. Experimental reproductive effects. When heated to decomposition it emits very toxic fumes of Cd and Cl^-. See also CADMIUM CHLORIDE, CADMIUM COMPOUNDS, and CHLORIDES.

CAE750 *HR: 3*
CADMIUM COMPOUNDS

TOXICITY DATA with REFERENCE
ihl-hmn TCLo:1500 µg/m^3/14Y-I:CAR,PUL ANYAA9 271,273,76

CONSENSUS REPORTS: Cadmium and its compounds are on the Community Right-To-Know List.

OSHA PEL: TWA 5 µg(Cd)/m³
ACGIH TLV: Dust and Salts: TWA 0.05 mg(Cd)/m³ (Proposed: TWA 0.01 mg(Cd)/m³ (dust), Suspected Human Carcinogen; 0.002 mg(Cd)/m³ (respirable dust), Suspected Human Carcinogen); BEI: 10 µg/g creatinine in urine; 10 µg/L in blood
DFG BAT: Blood 1.5 µg/dL; Urine 15 µg/dL. MAK: Suspected Carcinogen.
NIOSH REL: (Cadmium) Reduce to lowest feasible level.

SAFETY PROFILE: Confirmed human carcinogens producing lung tumors. Poison by ingestion. The irritating and emetic action is so violent, however, that little of the cadmium has time to be absorbed and fatal poisoning rarely ensues. Experimental carcinogens and teratogens. Cases of human poisoning have been reported from ingestion of food or beverages prepared or stored in cadmium-plated containers. Inhalation of fumes or dusts affects the respiratory tract and the kidneys. Brief exposure to high concentrations may result in pulmonary edema and death. Fatal concentrations may be breathed without sufficient discomfort to warn a worker to leave the exposure site. Cadmium oxide fumes can cause metal fume fever resembling that caused by zinc oxide fumes. When heated to decomposition cadmium compounds emit toxic fumes of Cd.

CAF500   HR: 3
CADMIUM DICYANIDE
mf: C_2CdN_2 mw: 164.44

CONSENSUS REPORTS: Cadmium and its compounds and Cyanide and its compounds are on the Community Right-To-Know List.

OSHA PEL: TWA 5 µg(Cd)/m³
ACGIH TLV: TWA 0.05 mg(Cd)/m³ (Proposed: TWA 0.01 mg(Cd)/m³ (dust), Suspected Human Carcinogen; 0.002 mg(Cd)/m³ (respirable dust), Suspected Human Carcinogen); BEI: 10 µg/g creatinine in urine; 10 µg/L in blood
NIOSH REL: (Cadmium) Reduce to lowest feasible level.

SAFETY PROFILE: Confirmed human carcinogen. A poison. Incompatible with magnesium. When heated to decomposition it emits toxic fumes of Cd and CN⁻. See also CADMIUM COMPOUNDS and CYANIDE.

CAF750   CAS:15954-91-3   HR: 3
CADMIUM(II) EDTA COMPLEX

SYN: (ETHYLENEDINITRILO)TETRAACETIC ACID CADMIUM(II) COMPLEX

TOXICITY DATA with REFERENCE
ipr-mus LD50:7800 µg(Cd)/kg   PABIAQ 11,853,63

CONSENSUS REPORTS: Cadmium and its compounds are on the Community Right-To-Know List.

OSHA PEL: TWA 5 µg(Cd)/m³
ACGIH TLV: TWA 0.05 mg(Cd)/m³ (Proposed: TWA 0.01 mg(Cd)/m³ (dust), Suspected Human Carcinogen; 0.002 mg(Cd)/m³ (respirable dust), Suspected Human Carcinogen); BEI: 10 µg/g creatinine in urine; 10 µg/L in blood
NIOSH REL: (Cadmium) Reduce to lowest feasible level.

SAFETY PROFILE: Confirmed human carcinogen. Poison by intraperitoneal route. When heated to decomposition it emits toxic fumes of NO_x and Cd.

CAG000   CAS:14486-19-2   HR: 3
CADMIUM FLUOBORATE
mf: B_2CdF_8 mw: 286.02

SYNS: CADMIUM FLUOROBORATE ◊ TL 1026

TOXICITY DATA with REFERENCE
orl-rat LDLo:250 mg/kg   NCNSA6 5,27,53
ihl-mus LCLo:650 mg/m³/10M   NDRC** No.9-4-1-19,44

CONSENSUS REPORTS: Reported in EPA TSCA Inventory. Cadmium and its compounds are on the Community Right-To-Know List.

OSHA PEL: TWA 5 µg(Cd)/m³
ACGIH TLV: TWA 0.05 mg(Cd)/m³ (Proposed: TWA 0.01 mg(Cd)/m³ (dust), Suspected Human Carcinogen; 0.002 mg(Cd)/m³ (respirable dust), Suspected Human Carcinogen); BEI: 10 µg/g creatinine in urine; 10 µg/L in blood
NIOSH REL: (Cadmium) Reduce to lowest feasible level.

SAFETY PROFILE: Confirmed human carcinogen. Poison by ingestion and inhalation. When heated to decomposition it emits very toxic fumes of Cd and F⁻. See TETRAFLUOROBORATE.

CAG250   CAS:7790-79-6   HR: 3
CADMIUM FLUORIDE
mf: CdF_2 mw: 150.40

PROP: Cubic, white crystals. Mp: 1100°, bp: 1758°, d: 6.64, vap press: 1 mm @ 1112°.

SYN: CADMIUM FLUORURE (FRENCH)

TOXICITY DATA with REFERENCE
scu-frg LDLo:280 mg/kg   CRSBAW 124,133,37

CONSENSUS REPORTS: Reported in EPA TSCA Inventory. Cadmium and its compounds are on the Community Right-To-Know List.

OSHA PEL: TWA 5 µg(Cd)/m³
ACGIH TLV: TWA 0.05 mg(Cd)/m³ (Proposed: TWA 0.01 mg(Cd)/m³ (dust), Suspected Human Carcinogen; 0.002 mg(Cd)/m³ (respirable dust), Suspected Human Carcinogen); BEI: 10 µg/g creatinine in urine; 10 µg/L in blood.
NIOSH REL: (Cadmium) Reduce to lowest feasible level.

SAFETY PROFILE: Confirmed human carcinogen. Poison

by subcutaneous route. Violent reaction with K. When heated to decomposition it emits very toxic fumes of Cd and F⁻. See also FLUORIDES and CADMIUM COMPOUNDS.

CAG500 CAS:17010-21-8 **HR: 3**
CADMIUM FLUOSILICATE
mf: CdF_6Si mw: 254.49

PROP: Hexagonal, colorless crystals.

SYN: TL 1070

TOXICITY DATA with REFERENCE
orl-rat LDLo:100 mg/kg   NCNSA6 5,27,53
ihl-mus LCLo:670 mg/m³/10M   NDRC** No.9-4-1-19,44

CONSENSUS REPORTS: Cadmium and its compounds are on the Community Right-To-Know List.

OSHA PEL: TWA 5 µg(Cd)/m³
ACGIH TLV: TWA 0.05 mg(Cd)/m³ (Proposed: TWA 0.01 mg(Cd)/m³ (dust), Suspected Human Carcinogen; 0.002 mg(Cd)/m³ (respirable dust), Suspected Human Carcinogen); BEI: 10 µg/g creatinine in urine; 10 µg/L in blood
NIOSH REL: (Cadmium) Reduce to lowest feasible level.

SAFETY PROFILE: Confirmed human carcinogen. Poison by ingestion and inhalation. When heated to decomposition it emits very toxic fumes of Cd and F⁻.

CAG750 CAS:16039-55-7 **HR: 3**
CADMIUM LACTATE
mf: $C_6H_{10}O_6 \cdot Cd$ mw: 290.56

PROP: Needles.

SYN: LACTIC ACID, CADMIUM SALT

CONSENSUS REPORTS: Cadmium and its compounds are on the Community Right-To-Know List.

OSHA PEL: TWA 5 µg(Cd)/m³
ACGIH TLV: TWA 0.05 mg(Cd)/m³ (Proposed: TWA 0.01 mg(Cd)/m³ (dust), Suspected Human Carcinogen; 0.002 mg(Cd)/m³ (respirable dust), Suspected Human Carcinogen); BEI: 10 µg/g creatinine in urine; 10 µg/L in blood
NIOSH REL: (Cadmium) Reduce to lowest feasible level.

SAFETY PROFILE: Confirmed human carcinogen. A poison. When heated to decomposition it emits toxic fumes of Cd. See also CADMIUM COMPOUNDS.

CAG775 CAS:2605-44-9 **HR: 3**
CADMIUM LAURATE
mf: $C_{24}H_{46}O_4 \cdot Cd$ mw: 511.10

SYNS: CADMIUM DILAURATE ◇ CADMIUM DODECANOATE ◇ DODECANOIC ACID, CADMIUM SALT (9CI) ◇ LAURIC ACID, CADMIUM SALT (2:1)

TOXICITY DATA with REFERENCE
orl-rat LD50:2370 mg/kg   41HTAH -,14,78
orl-mus LD50:1060 mg/kg   41HTAH -,14,78

CONSENSUS REPORTS: Reported in EPA TSCA Inventory.

OSHA PEL: TWA 5 µg(Cd)/m³
ACGIH TLV: TWA 0.01 mg(Cd)/m³; Suspected Carcinogen

SAFETY PROFILE: Confirmed human carcinogen. Moderately toxic by ingestion. When heated to decomposition it emits toxic fumes of Cd.

CAH000 CAS:10325-94-7 **HR: 3**
CADMIUM NITRATE
mf: CdN_2O_6 mw: 236.42

PROP: White, prismatic needles; hygroscopic. Mp: 350°.

SYNS: CADMIUM DINITRATE ◇ NITRIC ACID, CADMIUM SALT

TOXICITY DATA with REFERENCE
mrc-bcs 5 mmol/L   MUREAV 77,109,80
unr-rat LD50:200 mg/kg   GISAAA 50(3),57,85
orl-mus LD50:100 mg/kg   41HTAH -,14,78
ihl-mus LC50:3850 mg/m³   NTIS** PB1580508

CONSENSUS REPORTS: Reported in EPA TSCA Inventory. EPA Genetic Toxicology Program. Cadmium and its compounds are on the Community Right-To-Know List.

OSHA PEL: TWA 5 µg(Cd)/m³
ACGIH TLV: TWA 0.05 mg(Cd)/m³ (Proposed: TWA 0.01 mg(Cd)/m³ (dust), Suspected Human Carcinogen; 0.002 mg(Cd)/m³ (respirable dust), Suspected Human Carcinogen); BEI: 10 µg/g creatinine in urine; 10 µg/L in blood
NIOSH REL: (Cadmium) Reduce to lowest feasible level.

SAFETY PROFILE: Confirmed human carcinogen. Poison by ingestion and possibly other routes. Moderately toxic by inhalation. Mutation data reported. When heated to decomposition it emits very toxic fumes of Cd and NO_x. See also CADMIUM COMPOUNDS and NITRATES.

CAH250 CAS:10022-68-1 **HR: 3**
CADMIUM(II) NITRATE TETRAHYDRATE (1:2:4)
mf: $N_2O_6 \cdot Cd \cdot 4H_2O$ mw: 308.50

SYNS: DUSICNAN KADEMNATY (CZECH) ◇ NITRIC ACID, CADMIUM SALT, TETRAHYDRATE

TOXICITY DATA with REFERENCE
skn-rbt 500 mg/24H SEV   28ZPAK -,12,72
eye-rbt 20 mg/24H MOD   28ZPAK -,12,72
mmo-esc 6 µmol/L   ENVRAL 26,279,85
orl-rat LD50:300 mg/kg   28ZPAK -,12,72

CONSENSUS REPORTS: Cadmium and its compounds are on the Community Right-To-Know List.

OSHA PEL: TWA 5 μg(Cd)/m³
ACGIH TLV: TWA 0.05 mg(Cd)/m³ (Proposed: TWA 0.01 mg(Cd)/m³ (dust), Suspected Human Carcinogen; 0.002 mg(Cd)/m³ (respirable dust), Suspected Human Carcinogen); BEI: 10 μg/g creatinine in urine; 10 μg/L in blood
NIOSH REL: (Cadmium) Reduce to lowest feasible level.

SAFETY PROFILE: Confirmed human carcinogen. Poison by ingestion. A severe skin and moderate eye irritant. Mutation data reported. See also CADMIUM COMPOUNDS, CADMIUM NITRATE, and NITRATES. When heated to decomposition it emits very toxic fumes of Cd and NO_x.

CAH500   CAS:1306-19-0   **HR: 3**
CADMIUM OXIDE
mf: CdO   mw: 128.40

PROP: (1) Amorphous, brown crystals; (2) cubic, brown crystals. Mp (1): <1426°, mp (2): decomp @ 950°, bp: 1559°, d (1): 6.95, d (2): 8.15, vap press: 1 mm @ 1000°.

SYNS: KADMU TLENEK (POLISH) ◇ NCI-C02551

TOXICITY DATA with REFERENCE
ihl-rat TCLo:23 μg/m³/5H (15W pre/1–20D preg):REP
    TOLED5 22,53,84
ihl-rat TCLo:183 μg/m³/5H (15W pre/1–20D preg):REP
    TOLED5 22,53,84
ihl-rat TCLo:183 μg/m³/5H (15W pre/1–20D preg):REP
    TOLED5 22,53,84
ihl-rat TCLo:91 μg/m³ (female 1–19D post):TER   GTPZAB 31(8),25,87
scu-rat TDLo:90 mg/kg:NEO   BJCAAI 20,190,66
ihl-hmn TCLo:8630 μg/m³/5H   YAKUD5 22,455,80
ihl-man TCLo:500 μg/m³/5Y-I:NOSE,KID   QJMEA7 38,425,69
ihl-man TCLo:40 μg/m³:CVS,KID   GISAAA 45(10)22,80
orl-rat LD50:72 mg/kg   YAKUD5 22,455,80
ihl-rat LC50:780 mg/m³/10M   NTIS** PB158-508
ipr-rat LD50:12 mg/kg   ZDKAA8 38(9),18,78
orl-mus LD50:72 mg/kg   41HTAH -,14,78
ihl-mus LC50:340 mg/m³/10M   NTIS** PB158-508
ihl-dog LC50:400 mg/m³/10M   YAKUD5 22,455,80
ihl-mky LC50:15 g/m³/10M   NTIS** PB158-508
ihl-rbt LC50:3 g/m³/15M   NTIS** PB158-508
ihl-gpg LC50:3 g/m³/15M   NTIS** PB158-508

CONSENSUS REPORTS: NTP Fifth Annual Report on Carcinogens. IARC Cancer Review: Group 2A IMEMDT 7,139,87; Human Inadequate Evidence IMEMDT 2,74,73; IMEMDT 11,39,76; Animal Sufficient Evidence IMEMDT 11,39,76; IMEMDT 2,74,73. Reported in EPA TSCA Inventory. EPA Extremely Hazardous Substances List. Cadmium and its compounds are on the Community Right-To-Know List.

OSHA PEL: TWA 5 μg(Cd)/m³
ACGIH TLV: TWA 0.05 mg(Cd)/m³ (Proposed: TWA 0.01 mg(Cd)/m³ (dust), Suspected Human Carcinogen; 0.002 mg(Cd)/m³ (respirable dust), Suspected Human Carcinogen); BEI: 10 μg/g creatinine in urine; 10 μg/L in blood
DFG MAK: Suspected Carcinogen
NIOSH REL: (Cadmium) Reduce to lowest feasible level.

SAFETY PROFILE: Confirmed human carcinogen with experimental neoplastigenic data. Poison by ingestion, inhalation, and intraperitoneal routes. An experimental teratogen. Other experimental reproductive effects. Human systemic effects by inhalation include: change in the sense of smell, change in heart rate, blood pressure increase, an excess of protein in the urine, and other kidney or bladder changes. Mixtures with magnesium explode when heated. When heated to decomposition it emits toxic fumes of Cd. See also CADMIUM COMPOUNDS.

CAH750   **HR: 3**
CADMIUM OXIDE FUME
mf: CdO   mw: 128.40

SYN: CADMIUM FUME

TOXICITY DATA with REFERENCE
ihl-hmn LCLo:2500 mg/m³   JIHTAB 29,279,47
ihl-man TCLo:8630 μg/m³/5H:PUL   BJIMAG 23,292,66
ihl-rat LC50:500 mg/m³/10M   JIHTAB 29,279,47
ihl-mus LCLo:700 mg/m³/10M   JIHTAB 29,279,47
ihl-dog LC50:4000 mg/m³/10M   JIHTAB 29,279,47
ihl-mky LC50:15000 mg/m³/10M   JIHTAB 29,279,47
ihl-rbt LC50:2500 mg/m³/10M   JIHTAB 29,279,47
ihl-gpg LC50:3500 mg/m³/10M   JIHTAB 29,279,47

CONSENSUS REPORTS: Reported in EPA TSCA Inventory. Cadmium and its compounds are on the Community Right-To-Know List.

OSHA PEL: TWA 5 μg(Cd)/m³
ACGIH TLV: TWA 0.05 mg(Cd)/m³ (Proposed: TWA 0.01 mg(Cd)/m³ (dust), Suspected Human Carcinogen; 0.002 mg(Cd)/m³ (respirable dust), Suspected Human Carcinogen); BEI: 10 μg/g creatinine in urine; 10 μg/L in blood
NIOSH REL: (Cadmium) Reduce to lowest feasible level.

SAFETY PROFILE: Confirmed human carcinogen. Poison by inhalation. Moderately toxic to humans by inhalation. Human pulmonary system effects by inhalation, including: coughing, difficult breathing, and cyanosis. A strong irritant via inhalation. When heated to decomposition it emits toxic fumes of Cd. See also CADMIUM OXIDE and CADMIUM COMPOUNDS.

CAI000   CAS:13477-17-3   **HR: 3**
CADMIUM PHOSPHATE
mf: $Cd_3O_8P_2 \cdot 4H_2O$ mw: 599.22

PROP: Amorphous or colorless crystals. Mp: 1500°.

SYN: TL 1182

TOXICITY DATA with REFERENCE
ihl-mus LCLo:650 mg/m^3/10M   NDRC** No.9-4-1-19,44

CONSENSUS REPORTS: Reported in EPA TSCA Inventory. Cadmium and its compounds are on the Community Right-To-Know List.

OSHA PEL: TWA 5 μg(Cd)/m^3
ACGIH TLV: TWA 0.05 mg(Cd)/m^3 (Proposed: TWA 0.01 mg(Cd)/m^3 (dust), Suspected Human Carcinogen; 0.002 mg(Cd)/m^3 (respirable dust), Suspected Human Carcinogen); BEI: 10 μg/g creatinine in urine; 10 μg/L in blood
NIOSH REL: (Cadmium) Reduce to lowest feasible level.

SAFETY PROFILE: Confirmed human carcinogen. Poison by inhalation. When heated to decomposition it emits toxic fumes of Cd and PO$_x$. See CADMIUM COMPOUNDS and PHOSPHATES.

CAI125   CAS:12014-28-7   **HR: 3**
CADMIUM PHOSPHIDE
mf: Cd$_3$P$_2$ mw: 399.18

CONSENSUS REPORTS: Cadmium compounds are on the Community Right-To-Know List.

OSHA PEL: TWA 5 μg(Cd)/m^3
ACGIH TLV: Dust and Salts: TWA 0.05 mg(Cd)/m^3 (Proposed: TWA 0.01 mg(Cd)/m^3 (dust), Suspected Human Carcinogen; 0.002 mg(Cd)/m^3 (respirable dust), Suspected Human Carcinogen); BEI: 10 μg/g creatinine in urine; 10 μg/L in blood
DFG BAT: Blood 1.5 μg/dL; Urine 15 μg/dL. MAK: Suspected Carcinogen.
NIOSH REL: (Cadmium) Reduce to lowest feasible level.

SAFETY PROFILE: Confirmed human carcinogen. Explosive reaction with concentrated nitric acid. When heated to decomposition it emits toxic fumes of PO$_x$ and Cd. See also CADMIUM COMPOUNDS and PHOSPHIDES.

CAI250   **HR: 3**
CADMIUM PROPIONATE
mf: C$_6$H$_{10}$CdO$_5$ mw: 258.55

CONSENSUS REPORTS: Cadmium and its compounds are on the Community Right-To-Know List.

OSHA PEL: TWA 5 μg(Cd)/m^3
ACGIH TLV: Dust and Salts: TWA 0.05 mg(Cd)/m^3 (Proposed: TWA 0.01 mg(Cd)/m^3 (dust), Suspected Human Carcinogen; 0.002 mg(Cd)/m^3 (respirable dust), Suspected Human Carcinogen); BEI: 10 μg/g creatinine in urine; 10 μg/L in blood
DFG BAT: Blood 1.5 μg/dL; Urine 15 μg/dL. MAK: Suspected Carcinogen.

NIOSH REL: (Cadmium) Reduce to lowest feasible level.

SAFETY PROFILE: Confirmed human carcinogen. The salt has exploded. Incompatible with 3-pentanone vapor. When heated to decomposition it emits toxic fumes of Cd. See also CADMIUM COMPOUNDS.

CAI350   CAS:18897-36-4   **HR: 3**
CADMIUM 2-PYRIDINETHIONE
mf: C$_{10}$H$_8$CdN$_2$O$_2$S$_2$ mw: 364.72

SYNS: CADMIUM,BIS(1-HYDROXY-2(1H)-PYRIDINETHIONATO)- ◇ CADMIUM PT ◇ CdPT

TOXICITY DATA with REFERENCE
orl-rat LD50:240 mg/kg   TOANDB 3,1,79
ivn-rbt LD50:1340 μg/kg   TOANDB 3,1,79

OSHA PEL: TWA 5 μg(Cd)/m^3
ACGIH TLV: TWA 0.01 mg(Cd)/m^3; Suspected Carcinogen
NIOSH REL: (Cadmium): TWA reduce to lowest feasible level.

SAFETY PROFILE: Confirmed human carcinogen. Poison by ingestion and intravenous routes. When heated to decomposition it emits toxic fumes of NO$_x$, SO$_x$, and Cd.

CAI400   CAS:19010-79-8   **HR: 3**
CADMIUM SALICYLATE
mf: C$_{14}$H$_{10}$CdO$_6$ mw: 386.64

PROP: Monohydrate small needles or plates. Mp: 242°. Sltly sol in cold water, methanol, eth; Very sol in boiling water.

SYNS: BIS(2-HYDROXYBENZOATO-O^1,O^2-), (T-4)-CADMIUM (9CI) ◇ CADMIUM, BIS(SALICYLATO)-

TOXICITY DATA with REFERENCE
orl-rat LD50:1200 mg/kg   41HTAH -,14,78
orl-mus LD50:164 mg/kg   41HTAH -,14,78

OSHA PEL: TWA 5 μg(Cd)/m^3
ACGIH TLV: TWA 0.01 mg(Cd)/m^3; Suspected Carcinogen

SAFETY PROFILE: Confirmed human carcinogen. Poison by ingestion. When heated to decomposition it emits toxic fumes of Cd.

CAI500   **HR: 3**
CADMIUM SELENIDE
mf: CdSe   mw: 191.36

PROP: Preparative hazard.

CONSENSUS REPORTS: Cadmium and its compounds as well as selenium and its compounds are on the Community Right-To-Know List.

OSHA PEL: TWA 5 µg(Cd)/m^3
ACGIH TLV: Dust and Salts: TWA 0.05 mg(Cd)/m^3 (Proposed: TWA 0.01 mg(Cd)/m^3 (dust), Suspected Human Carcinogen; 0.002 mg(Cd)/m^3 (respirable dust), Suspected Human Carcinogen); BEI: 10 µg/g creatinine in urine; 10 µg/L in blood
DFG BAT: Blood 1.5 µg/dL; Urine 15 µg/dL. MAK: Suspected Carcinogen; 0.1 mg(Se)/m^3.
NIOSH REL: (Cadmium) Reduce to lowest feasible level.

SAFETY PROFILE: Confirmed human carcinogen. Selenium compounds are considered to be poisons. When heated to decomposition it emits toxic fumes of Cd and Se. See also CADMIUM COMPOUNDS and SELENIUM COMPOUNDS.

CAI750   CAS:141-00-4   **HR: 3**
CADMIUM SUCCINATE
mf: C$_4$H$_4$O$_4$•Cd   mw: 228.48

SYNS: CADMINATE ◇ SUCCINIC ACID, CADMIUM SALT (1:1)

TOXICITY DATA with REFERENCE
orl-rat LD50:660 mg/kg   FMCHA2 -,D53,80
orl-mus LD50:312 mg/kg   28ZEAL 5,35,76
ipr-mus LD50:270 mg/kg   AIPTAK 128,391,60

CONSENSUS REPORTS: Reported in EPA TSCA Inventory. Cadmium and its compounds are on the Community Right-To-Know List.

OSHA PEL: TWA 5 µg(Cd)/m^3
ACGIH TLV: Dust and Salts: TWA 0.05 mg(Cd)/m^3 (Proposed: TWA 0.01 mg(Cd)/m^3 (dust), Suspected Human Carcinogen; 0.002 mg(Cd)/m^3 (respirable dust), Suspected Human Carcinogen); BEI: 10 µg/g creatinine in urine; 10 µg/L in blood
DFG BAT: Blood 1.5 µg/dL; Urine 15 µg/dL. MAK: Suspected Carcinogen.
NIOSH REL: (Cadmium) Reduce to lowest feasible level.

SAFETY PROFILE: Confirmed human carcinogen. Poison by ingestion and intraperitoneal routes. Moderately toxic by ingestion. When heated to decomposition it emits toxic fumes of Cd. See also CADMIUM COMPOUNDS.

CAJ000   CAS:10124-36-4   **HR: 3**
CADMIUM SULFATE (1581)
mf: O$_4$S•Cd   mw: 208.46

PROP: Rhombic, white crystals. Mp: 1000°, d: 4.691.

SYNS: CADMIUM SULFATE ◇ CADMIUM SULPHATE ◇ SULFURIC ACID, CADMIUM(2+) SALT ◇ SULPHURIC ACID, CADMIUM SALT (1:1)

TOXICITY DATA with REFERENCE
mrc-bcs 5 mmol/L   MUREAV 77,109,80
dnd-rat:lvr 30 µmol/L   MUREAV 113,357,83
msc-mus:lym 150 µg/L   JTEHD6 9,367,82
par-mus TDLo:28 µg/kg (female 12D post):TER   IGSBAL 92,65,76
ipr-mus TDLo:1030 µg/kg (female 9D post):TER   TJADAB 29,427,84
ipr-mus TDLo:5150 µg/kg (female 9D post):TER   TJADAB 32,407,85
ipr-mus TDLo:5150 µg/kg (female 9D post):REP   TJADAB 29,427,84
ipr-mus TDLo:3 mg/kg (female 9D post):TER   TJADAB 27,54A,83
par-mus TDLo:28 µg/kg (female 12D post):TER   IGSBAL 92,65,76
ivn-ham TDLo:2 mg/kg (female 8D post):TER   BECTA6 22,175,79
ivn-ham TDLo:2 mg/kg (female 8D post):TER   BECTA6 22,175,79
ipr-mus TDLo:2570 µg/kg (female 9D post):TER   TJADAB 29,427,84
ipr-ham TDLo:2800 µg/kg (female 8D post):TER   TJADAB 29(2),30A,84
orl-rat LD50:280 mg/kg   41HTAH -,14,78
orl-mus LD50:88 mg/kg   41HTAH -,14,78
ipr-mus LD50:12,760 µg/kg   COREAF 256,1043,63
orl-dog LDLo:105 mg/kg   EQSSDX 1,1,75
scu-dog LDLo:27 mg/kg   EQSSDX 1,1,75
scu-frg LDLo:105 mg/kg   HBAMAK 4,1317,35

CONSENSUS REPORTS: NTP Fifth Annual Report on Carcinogens. IARC Cancer Review: Group 2A IMEMDT 7,139,87; Animal Sufficient Evidence IMEMDT 11,39,76; IMEMDT 2,74,73. Reported in EPA TSCA Inventory. EPA Genetic Toxicology Program. Cadmium and its compounds are on the Community Right-To-Know List.

OSHA PEL: TWA 5 µg(Cd)/m^3
ACGIH TLV: TWA 0.05 mg(Cd)/m^3 (Proposed: TWA 0.01 mg(Cd)/m^3 (dust), Suspected Human Carcinogen; 0.002 mg(Cd)/m^3 (respirable dust), Suspected Human Carcinogen); BEI: 10 µg/g creatinine in urine; 10 µg/L in blood
DFG MAK: Suspected Carcinogen
NIOSH REL: (Cadmium) Reduce to lowest feasible level.

SAFETY PROFILE: Confirmed human carcinogen with experimental carcinogenic data. Poison by ingestion, subcutaneous, and intraperitoneal routes. Experimental teratogenic and reproductive effects. Mutation data reported. See also CADMIUM COMPOUNDS and SULFATES. When heated to decomposition it emits very toxic fumes of Cd and SO$_x$.

CAJ250   CAS:7790-84-3   **HR: 3**
CADMIUM SULFATE (1:1) HYDRATE (3:8)
mf: O$_4$S•Cd•3H$_2$O   mw: 256.51

SYNS: CADMIUM SULFATE OCTAHYDRATE ◇ SULFURIC ACID, CADMIUM SALT, HYDRATE

CAJ500 CADMIUM SULFATE TETRAHYDRATE

TOXICITY DATA with REFERENCE
dnd-esc 3 μmol/L JOBAAY 133,75,78
cyt-ham:fbr 10 μmol/L/1H MUREAV 40,125,76
ivn-ham TDLo:2 mg/kg (8D preg):REP TJADAB 21,181,80
ivn-ham TDLo:2 mg/kg (8D preg):TER TJADAB 21,181,80
ivn-ham TDLo:2 mg/kg (female 8D post):TER TJADAB 21,181,80
ivn-ham TDLo:2 mg/kg (8D preg):TER TJADAB 21,181,80
ivn-ham TDLo:2 mg/kg (female 8D post):TER TJADAB 21,181,80
ivn-ham TDLo:2 mg/kg (female 8D post):TER TJADAB 21,181,80
scu-rat TDLo:60 mg/kg/2Y-I:NEO AOHYA3 16,111,73
scu-rat TD:15 mg/kg/2Y-I:ETA AOHYA3 16,111,73

CONSENSUS REPORTS: IARC Cancer Review: Animal Sufficient Evidence IMEMDT 2,74,73. Cadmium and its compounds are on the Community Right-To-Know List.

OSHA PEL: TWA 5 μg(Cd)/m^3
ACGIH TLV: TWA 0.05 mg(Cd)/m^3 (Proposed: TWA 0.01 mg(Cd)/m^3 (dust), Suspected Human Carcinogen; 0.002 mg(Cd)/m^3 (respirable dust), Suspected Human Carcinogen); BEI: 10 μg/g creatinine in urine; 10 μg/L in blood
NIOSH REL: (Cadmium) Reduce to lowest feasible level.

SAFETY PROFILE: Confirmed human carcinogen with experimental tumorigenic and neoplastigenic data. Experimental teratogenic and reproductive effects. Mutation data reported. When heated to decomposition it emits very toxic fumes of Cd and SO$_x$. See also CADMIUM SULFATE, CADMIUM COMPOUNDS, and SULFATES.

CAJ500 CAS:13477-21-9 HR: 3
CADMIUM SULFATE TETRAHYDRATE
mf: O$_4$S•Cd•4H$_2$O   mw: 280.54

SYN: SULFURIC ACID, CADMIUM SALT, TETRAHYDRATE

TOXICITY DATA with REFERENCE
scu-rat TDLo:20 mg/kg/10W-I:NEO BJCAAI 18,667,64

CONSENSUS REPORTS: Cadmium and its compounds are on the Community Right-To-Know List.

OSHA PEL: TWA 5 μg(Cd)/m^3
ACGIH TLV: TWA 0.05 mg(Cd)/m^3 (Proposed: TWA 0.01 mg(Cd)/m^3 (dust), Suspected Human Carcinogen; 0.002 mg(Cd)/m^3 (respirable dust), Suspected Human Carcinogen); BEI: 10 μg/g creatinine in urine; 10 μg/L in blood
NIOSH REL: (Cadmium) Reduce to lowest feasible level.

SAFETY PROFILE: Confirmed human carcinogen with experimental neoplastigenic data. When heated to decomposition it emits very toxic fumes of Cd and SO$_x$. See also CADMIUM COMPOUNDS and CADMIUM SULFATE OCTAHYDRATE.

CAJ750 CAS:1306-23-6 HR: 3
CADMIUM SULFIDE
mf: CdS   mw: 144.46

PROP: Hexagonal, yellow-orange crystals. Mp: 1750 @ 100 atm, bp: subl in N$_2$, d: 4.82.

SYNS: AURORA YELLOW ◇ CADMIUM GOLDEN 366 ◇ CADMIUM LEMON YELLOW 527 ◇ CADMIUM ORANGE ◇ CADMIUM PRIMROSE 819 ◇ CADMIUM SULPHIDE ◇ CADMIUM YELLOW ◇ CADMOPUR YELLOW ◇ CAPSEBON ◇ C.I. 77199 ◇ C.I. PIGMENT ORANGE 20 ◇ C.I. PIGMENT YELLOW 37 ◇ FERRO YELLOW ◇ GREENOCKITE ◇ NCI-C02711

TOXICITY DATA with REFERENCE
cyt-hmn:leu 62 μg/L PJACAW 48,133,72
otr-ham:emb 1 mg/L CNREA8 42,2757,82
dnd-ham:ovr 10 mg/L CRNGDP 3,657,82
scu-rat TDLo:90 mg/kg:CAR BJCAAI 20,190,66
ims-rat TDLo:120 mg/kg:ETA BJCAAI 20,190,66
scu-rat TD:135 mg/kg:ETA PBPHAW 14,47,78
scu-rat TD:250 mg/kg:ETA NATUAS 198,1213,63
orl-rat LD50:7080 mg/kg 41HTAH -,14,78
orl-mus LD50:1166 mg/kg 41HTAH -,14,78
ihl-mus LCLo:1350 mg/m^3 NTIS** PB158-508

CONSENSUS REPORTS: NTP Fifth Annual Report on Carcinogens. IARC Cancer Review: Group 2A IMEMDT 7,139,87; Animal Sufficient Evidence IMEMDT 11,39,76; IMEMDT 2,74,73. EPA Genetic Toxicology Program. Cadmium and its compounds are on the Community Right-To-Know List. Reported in EPA TSCA Inventory.

OSHA PEL: TWA 5 μg(Cd)/m^3
ACGIH TLV: TWA 0.05 mg(Cd)/m^3 (Proposed: TWA 0.01 mg(Cd)/m^3 (dust), Suspected Human Carcinogen; 0.002 mg(Cd)/m^3 (respirable dust), Suspected Human Carcinogen); BEI: 10 μg/g creatinine in urine; 10 μg/L in blood
DFG MAK: Suspected Carcinogen
NIOSH REL: (Cadmium) Reduce to lowest feasible level.

SAFETY PROFILE: Confirmed human carcinogen with experimental carcinogenic and tumorigenic data. Moderately toxic by ingestion and inhalation. Human mutation data reported. When heated to decomposition it emits very toxic fumes of Cd and SO$_x$. See also CADMIUM COMPOUNDS and SULFIDES.

CAK000 HR: 3
CADMIUM THERMOVACUUM AEROSOL
mf: Cd   mw: 112.40

SYN: AEROSOL of THERMOVACUUM CADMIUM

TOXICITY DATA with REFERENCE
unr-rat LD50:1365 mg/kg GTPZAB 22(5),6,78
unr-mus LD50:815 mg/kg GTPZAB 22(5),6,78

CONSENSUS REPORTS: Cadmium and its compounds are on the Community Right-To-Know List.

OSHA PEL: TWA 5 µg(Cd)/m^3
ACGIH TLV: TWA 0.05 mg(Cd)/m^3 (Proposed: TWA 0.01 mg(Cd)/m^3 (dust), Suspected Human Carcinogen; 0.002 mg(Cd)/m^3 (respirable dust), Suspected Human Carcinogen); BEI: 10 µg/g creatinine in urine; 10 µg/L in blood
NIOSH REL: (Cadmium) Reduce to lowest feasible level.

SAFETY PROFILE: Confirmed human carcinogen. Moderately toxic by an unspecified route. When heated to decomposition it emits very toxic fumes of Cd. See also CADMIUM and CADMIUM COMPOUNDS.

CAK250   CAS:73419-42-8   *HR: 3*
CADMIUM-THIONEINE
mf: $C_{18}H_{30}N_6O_4S_2$•Cd   mw: 571.06

PROP: Cadmium(II) is bound to the protein thioneine from rat or rabbit liver (BCPCA6 26,25,77).

TOXICITY DATA with REFERENCE
ivn-rat LD50:280 µg/kg   BCPCA6 26,25,77

CONSENSUS REPORTS: Cadmium and its compounds are on the Community Right-To-Know List.

OSHA PEL: TWA 5 µg(Cd)/m^3
ACGIH TLV: TWA 0.05 mg(Cd)/m^3 (Proposed: TWA 0.01 mg(Cd)/m^3 (dust), Suspected Human Carcinogen; 0.002 mg(Cd)/m^3 (respirable dust), Suspected Human Carcinogen); BEI: 10 µg/g creatinine in urine; 10 µg/L in blood
NIOSH REL: (Cadmium) Reduce to lowest feasible level.

SAFETY PROFILE: Confirmed human carcinogen. Deadly poison by intravenous route. When heated to decomposition it emits very toxic fumes of NO_x, SO_x, and Cd. See also CADMIUM COMPOUNDS.

CAM000   CAS:5902-95-4   *HR: 3*
CALCIUM ACID METHYL ARSONATE
mf: $C_2H_8As_2O_6$•Ca   mw: 318.02

SYNS: CALAR ◊ CALCIUM ACID METHANEARSONATE ◊ CALCIUM HYDROGEN METHANEARSONATE ◊ CALCIUM METHANEARSONATE ◊ CAMA ◊ SUPER CRAB-E-RAD-CALAR ◊ SUPER DAL-E-RAD ◊ SUPER DAL-E-RAD-CALAR ◊ USAF AN-11

TOXICITY DATA with REFERENCE
ipr-mus LD50:500 mg/kg   NTIS** AD414-344
unr-mam LD50:4000 mg/kg   FMCHA2 -,C241,83

CONSENSUS REPORTS: Arsenic and its compounds are on the Community Right-To-Know List.

OSHA PEL: TWA 0.5 mg(As)/m^3
ACGIH TLV: TWA 0.2 mg(As)/m^3 (Proposed: 0.01 mg(As)/m^3; Human Carcinogen)

SAFETY PROFILE: Moderately toxic by intraperitoneal and possibly other routes. Arsenic compounds are considered to be poisons. An herbicide. When heated to decomposition it emits toxic fumes of As. See also ARSENIC COMPOUNDS and CALCIUM COMPOUNDS.

CAM500   CAS:27152-57-4   *HR: 3*
CALCIUM ARSENITE
DOT: NA 1574
mf: As_2O_6•3Ca   mw: 366.08

PROP: White, granular powder.

SYNS: ARSENIOUS ACID, CALCIUM SALT ◊ CALCIUM ARSENITE, solid (DOT) ◊ MONOCALCIUM ARSENITE

CONSENSUS REPORTS: NTP Fifth Annual Report on Carcinogens. Arsenic and its compounds are on the Community Right-To-Know List.

OSHA PEL: Cancer Hazard
ACGIH TLV: TWA 0.2 mg(As)/m^3 (Proposed: 0.01 mg(As)/m^3; Human Carcinogen)
NIOSH REL: (Inorganic Arsenic) CL 0.002 mg(As)/m^3/15M

DOT Classification: Poison B; Label: Poison

SAFETY PROFILE: Confirmed carcinogen. A poison by inhalation and ingestion. When heated to decomposition it emits toxic fumes of As. See also ARSENIC COMPOUNDS and CALCIUM COMPOUNDS.

CAR780   CAS:62-33-9   *HR: 2*
CALCIUM DISODIUM ETHYLENEDIAMINETETRAACETATE
mf: $C_{10}H_{12}CaN_2O_8$•2Na   mw: 374.30

SYNS: ACETIC ACID, (ETHYLENEDINITRILO)TETRA-, CALCIUM DISODIUM SALT ◊ ADSORBONAC ◊ ANTALLIN ◊ CALCIATE(2–), ((ETHYLENEDINITRILO)TETRAACETATO)-, DISODIUM ◊ CALCITETRACEMATE DISODIUM ◊ CALCIUM DISODIUM EDATHAMIL ◊ CALCIUM DISODIUM EDETATE ◊ CALCIUM DISODIUM EDTA ◊ CALCIUM DISODIUM (ETHYLENEDINITRILO)TETRAACETATE ◊ CALCIUM DISODIUM VERSENATE ◊ CALCIUM EDTA ◊ CALCIUM TITRIPLEX ◊ DI-SODIUM CALCIUM EDTA ◊ DISODIUM CALCIUM ETHYLENEDIAMINETETRAACETATE ◊ EDATHAMIL CALCIUM DISODIUM ◊ EDETAMIN ◊ EDETAMINE ◊ EDETATE CALCIUM ◊ EDETIC ACID CALCIUM DISODIUM SALT ◊ EDTACAL ◊ EDTA CALCIUM DISODIUM SALT ◊ ETHYLENEDIAMINETETRAACETIC ACID, CALCIUM DISODIUM CHELATE ◊ LEDCLAIR ◊ MONOCALCIUM DISODIUM EDTA ◊ MOSATIL ◊ RIKELATE CALCIUM ◊ SODIUM CALCIUM EDETATE ◊ SORMETAL ◊ TETACIN ◊ TETACIN-CALCIUM ◊ TETAZINE ◊ VERSENE CA

TOXICITY DATA with REFERENCE
orl-rat LD50:10 g/kg   TXAPA9 5,142,63
ipr-rat LD50:3850 mg/kg   JPETAB 117,20,56
ivn-rat LD50:3 g/kg   CLDND*-,188,90
ipr-mus LD50:4500 mg/kg   CLDND* 5,142,63
orl-rbt LD50:7 g/kg   TXAPA9 5,142,63
ipr-rbt LD50:6 g/kg   DRUGAY-,188,90
ivn-rbt LDLo:4 g/kg   FEPRA7 11,321,52

CONSENSUS REPORTS: Reported in EPA TSCA Inventory.

SAFETY PROFILE: Moderately toxic by intraperitoneal route. Mildly toxic by ingestion and intravenous routes. When heated to decomposition it emits toxic vapors of NO_x.

CAT775   CAS:471-34-1   **HR: 1**
CALCIUM MONOCARBONATE
mf: $CO_3 \cdot Ca$ mw: 100.09

SYNS: AEROMATT ◇ AKADAMA ◇ ALBACAR ◇ ALBACAR 5970 ◇ ALBAFIL ◇ ALBAGLOS ◇ ALBAGLOS SF ◇ ALLIED WHITING ◇ ATOMIT ◇ ATOMITE ◇ AX 363 ◇ BF 200 ◇ BRILLIANT 15 ◇ BRITOMYA M ◇ CARBONIC ACID, CALCIUM SALT (1:1) ◇ CALCENE CO ◇ CALCICOLL ◇ CALCIDAR 40 ◇ CALCILIT 8 ◇ CALCIUM CARBONATE (1:1) ◇ CALIBRITE ◇ CAL-LIGHT SA ◇ CALMOS ◇ CALMOTE ◇ CALOFIL A 4 ◇ CALOFORT S ◇ CALOFORT U ◇ CALOFOR U 50 ◇ CALOPAKE F ◇ CALOPAKE HIGH OPACITY ◇ CALSEEDS ◇ CALTEC ◇ CAMEL-CARB ◇ CAMEL-TEX ◇ CAMEL-WITE ◇ CARBITAL 90 ◇ CARBIUM ◇ CARBIUM MM ◇ CARBOREX 2 ◇ CARUSIS P ◇ CCC G-WHITE ◇ CCC No.AA OO-LITIC ◇ CCR ◇ CCW ◇ CHEMCARB ◇ C.I. PIGMENT WHITE 18 ◇ CLEFNON ◇ CRYSTIC PREFIL S ◇ DACOTE ◇ DOMAR ◇ DURAMITE ◇ DURCAL 10 ◇ EGRI M 5 ◇ ESKALON 100 ◇ FILTEX WHITE BASE ◇ FINNCARB 6002 ◇ GAROLITE SA ◇ GILDER'S WHITING ◇ HAKUENKA CC ◇ HAKUENKA R 06 ◇ HOMOCAL D ◇ HYDROCARB 60 ◇ K 250 ◇ KOTAMITE ◇ KREDAFIL 150 EXTRA ◇ KREDAFIL RM 5 ◇ KS 1300 ◇ KULU 40 ◇ LEVIGATED CHALK ◇ MARBLEWHITE 325 ◇ MARFIL ◇ MC-T ◇ MICROCARB ◇ MICROMIC CR 16 ◇ MICROMYA ◇ MICROWHITE 25 ◇ MONOCALCIUM CARBONATE ◇ MSK-C ◇ MULTIFLEX MM ◇ N 34 ◇ NCC 45 ◇ NEOANTICID ◇ NEOLITE F ◇ NON-FER-AL ◇ NS (carbonate) ◇ NS 100 (carbonate) ◇ NS 200 (filler) ◇ NZ ◇ OA-A 1102 ◇ OMYA ◇ OMYA BLH ◇ OMYACARB F ◇ OMYALENE G 200 ◇ OMYALITE 90 ◇ OS-CAL ◇ PIGMENT WHITE 18 ◇ P-LITE 500 ◇ POLCARB ◇ PREPARED CHALK ◇ PS 100 (carbonate) ◇ PURECAL ◇ PURECALO ◇ PZ ◇ QUEENSGATE WHITING ◇ RED BALL ◇ R JUTAN ◇ ROYAL WHITE LIGHT ◇ RX 2557 ◇ SHIPRON A ◇ SILVER W ◇ SL 700 ◇ SMITHKO KALKARB WHITING ◇ SNOWCAL ◇ SNOWFLAKE WHITE ◇ SNOW TOP ◇ SOCAL ◇ SOCAL E 2 ◇ SOFTON 1000 ◇ SS 30 (carbonate) ◇ SS 50 (carbonate) ◇ SSB 100 ◇ STANWHITE 500 ◇ STURCAL D ◇ SUNLIGHT 700 ◇ SUPER 1500 ◇ SUPERCOAT ◇ SUPERMITE ◇ SUPER MULTIFEX ◇ SUPER-PFLEX ◇ SUPER 3S ◇ SUPER SSS ◇ SURFEX MM ◇ SURFIL S ◇ SUSPENSO ◇ SYLACAUGA 88B ◇ T 130-2500 ◇ TAMA PEARL TP 121 ◇ TANCAL 100 ◇ TM 1 (filler) ◇ TONASO ◇ TOYOFINE TF-X ◇ TP 121 (filler) ◇ TP 222 ◇ ULTRA-PFLEX ◇ UNIBUR 70 ◇ VEVETONE ◇ VICRON ◇ VICRON 31-6 ◇ VIENNA WHITE ◇ VIGOT 15 ◇ WHICA BA ◇ WHITCARB W ◇ WHITE-POWDER ◇ WHITING ◇ WHITON 450 ◇ WINNOFIL S ◇ WITCARB ◇ WITCARB P ◇ WITCARB REGULAR ◇ YORK WHITE ◇ ZG 301

TOXICITY DATA with REFERENCE
skn-rbt 500 mg/24H MOD   28ZPAK -,267,72
eye-rbt 750 μg/24H SEV   28ZPAK -,267,72
orl-rat LD50:6450 mg/kg   28ZPAK -,267,72

CONSENSUS REPORTS: Reported in EPA TSCA Inventory.

SAFETY PROFILE: Mildly toxic by ingestion. A skin and severe eye irritant. When heated to decomposition it emits acrid smoke and irritating vapors.

CBF700   CAS:105-60-2   **HR: 3**
CAPROLACTAM
mf: $C_6H_{11}NO$ mw: 113.18

PROP: White crystals. Mp: 69°, vap press: 6 mm @ 120°.

SYNS: AMINOCAPROIC LACTAM ◇ 6-AMINOHEXANOIC ACID CYCLIC LACTAM ◇ 2-AZACYCLOHEPTANONE ◇ 6-CAPROLACTAM ◇ ω-CAPROLACTAM (MAK) ◇ CAPROLATTAME (FRENCH) ◇ CYCLOHEXANONE ISOOXIME ◇ EPSILON KAPROLAKTAM (POLISH) ◇ HEXAHYDRO-2-AZEPINONE ◇ HEXAHYDRO-2H-AZEPIN-2-ONE ◇ 6-HEXANELACTAM ◇ HEXANONE ISOXIME ◇ HEXANONISOXIM (GERMAN) ◇ 1,6-HEXOLACTAM ◇ e-KAPROLAKTAM (CZECH) ◇ 2-KETOHEXAMETHYLENIMINE ◇ NCI-C50646 ◇ 2-OXOHEXAMETHYLENIMINE ◇ 2-PERHYDROAZEPINONE

TOXICITY DATA with REFERENCE
skn-rbt 500 mg/24H MLD   28ZPAK -,149,72
eye-rbt 20 mg/24H MOD   28ZPAK -,149,72
slt-dmg-orl 5 mmol/L   PMRSDJ 5,313,85
mmo-smc 100 mg/L   PMRSDJ 5,271,85
cyt-hmn:lym 270 mg/L   PMRSDJ 5,457,85
orl-rbt TDLo:3450 mg/kg (female 6–28D post):TER   JJATDK 7,317,87
orl-rat TDLo:10 g/kg (female 6–15D post):REP   JJATDK 7,317,87
ihl-rat TCLo:125 mg/m^3/24H (76D male):REP   GTPZAB 19(10),40,75
ihl-hmn TCLo:100 ppm:PUL   AIHAAP 34,384,73
orl-rat LD50:1210 mg/kg   NTPTR* NTP-TR-214,82
ihl-rat LC50:300 mg/m^3/2H   85GMAT -,32,82
ihl-rat LC50:300 mg/m^3/2H   85GMAT -,32,82
orl-mus LD50:930 mg/kg   GTPZAB 10(10),54,66
ihl-mus LC50:450 mg/m^3 GTPZAB 10(10),54,66
ipr-mus LD50:650 mg/kg   JPMSAE 60,1058,71
scu-mus LDLo:750 mg/kg   AEXPBL 50,199,1903
skn-rbt LDLo:1438 mg/kg   AIHAAP 30,470,69
scu-frg LDLo:2800 mg/kg   AEXPBL 50,199,1903

CONSENSUS REPORTS: IARC Cancer Review: Group 4 IMEMDT 7,56,87; Animal No Evidence IMEMDT 39,247,86. Reported in EPA TSCA Inventory.

OSHA PEL: Dust: 1 mg/m^3; STEL 3 mg/m^3; Vapor: 5 ppm; STEL 10 ppm
ACGIH TLV: TWA Dust: 1 mg/m^3; Vapor: 5 ppm; STEL 10 ppm
DFG MAK: 25 mg/m^3

SAFETY PROFILE: Moderately toxic by ingestion, skin contact, intraperitoneal, and subcutaneous routes. Human systemic effects by inhalation: cough. An experimental teratogen. Other experimental reproductive effects. Human mutation data reported. A skin and eye irritant. Potentially explosive reaction with acetic acid + dinitrogen trioxide. When heated to decomposition it emits toxic fumes of NO_x.

CBF705   CAS:762-16-3   **HR: 2**
CAPROLYL PEROXIDE
DOT: NA 2129/UN 2129
mf: $C_{16}H_{30}O_4$ mw: 286.46

SYNS: CAPRYL PEROXIDE ◇ CAPRYLYL PEROXIDE ◇ CAPRYLYL PEROXIDE (DOT) ◇ CAPRYLYL PEROXIDE SOLUTION (DOT) ◇ DICAPRYLYL PEROXIDE ◇ DIOCTANOYL PEROXIDE ◇ n-OCTANOYL PEROXIDE (DOT)

◇ PERKADOX SE 8 ◇ PEROXIDE, BIS(1-OXOOCTYL) (9CI) ◇ PEROXIDE, OCTANOYL

DOT Classification: Organic Peroxide; Label: Organic Peroxide

CONSENSUS REPORTS: Reported in EPA TSCA Inventory.

SAFETY PROFILE: A peroxide. Handle carefully. When heated to decomposition it emits acrid smoke and irritating vapors.

CBW750   CAS:630-08-0   *HR: 3*
CARBON MONOXIDE
DOT: UN 1016/NA 9202
mf: CO   mw: 28.01

PROP: Colorless, odorless gas. Mp: −207°, bp: −191.3°, lel: 12.5%, uel: 74.2%, d: (gas) 1.250 g/L @ 0°, (liquid) 0.793, autoign temp: 1128°F.

SYNS: CARBONE (OXYDE de) (FRENCH) ◇ CARBONIC OXIDE ◇ CARBONIO (OSSIDO di) (ITALIAN) ◇ CARBON MONOXIDE, CRYOGENIC liquid (DOT) ◇ CARBON OXIDE (CO) ◇ EXHAUST GAS ◇ FLUE GAS ◇ KOHLENMONOXID (GERMAN) ◇ KOHLENOXYD (GERMAN) ◇ KOOLMONOXYDE (DUTCH) ◇ OXYDE de CARBONE (FRENCH) ◇ WEGLA TLENEK (POLISH)

TOXICITY DATA with REFERENCE

ihl-mus TCLo:65 ppm/24H (female 7–18D post):REP
    TJADAB 29(2),8B,84
ihl-rat TCLo:1 mg/m^3/24H (female 72D pre):REP   HYSAAV 35(4-6),277,70
ihl-rat TCLo:150 ppm/24H (1–22D preg):REP   NETOD7 2,7,80
ihl-rbt TCLo:180 ppm/24H (female 1–30D post):REP
    LANCAO 2,1220,72
ihl-rbt TCLo:180 ppm/24H (female 1–30D post):REP
    LANCAO 2,1220,72
ihl-mus TCLo:8 pph/1H (female 8D post):TER   FPNJAG 11,301,58
ihl-mus TCLo:125 ppm/24H (female 7–18D post):TER
    TJADAB 30,253,84
ihl-rat TCLo:1 mg/m^3/24H (72D pre):REP   HYSAAV 35(4-6),277,70
ihl-mus TCLo:8 pph/1H (female 8D post):REP   FPNJAG 11,301,58
ihl-mus TCLo:250 ppm/7H (female 6–15D post):REP
    TJADAB 19,385,79
ihl-rat TCLo:1 mg/m^3/24H (72D pre):REP   HYSAAV 35(4-6),277,70
ihl-rat TCLo:1 mg/m^3/24H (72D pre):REP   HYSAAV 35(4-6),277,70
ihl-rat TDLo:150 ppm/24H (1–22D preg):TER   TXAPA9 56,370,80
ihl-mus TCLo:8 pph/1H (female 8D post):TER   FPNJAG 11,301,58
ihl-mus TCLo:250 ppm/7H (female 6–15D post):TER
    TJADAB 19,385,79
ihl-hmn TCLo:600 mg/m^3/10M   GTPZAB 31(4),34,87
ihl-man LCLo:4000 ppm/30M   29ZWAE -,207,68
ihl-man TCLo:650 ppm/45M:CNS,BLD   AIHAAP 34,212,73
ihl-hmn LCLo:5000 ppm/5M   TABIA2 3,231,33
ihl-rat LC50:1807 ppm/4H   TXAPA9 17,752,70
ihl-mus LC50:2444 ppm/4H   TXAPA9 17,752,70
ihl-dog LCLo:4000 ppm/46M   HBAMAK 4,1360,35
ihl-rbt LCLo:4000 ppm   HBAMAK 4,1360,35
ihl-gpg LC50:5718 ppm/4H   TXAPA9 17,752,70
ihl-mam LCLo:5000 ppm/5M   AEPPAE 138,65,28
ihl-bwd LD50:1334 ppm   AECTCV 12,355,83

CONSENSUS REPORTS: Reported in EPA TSCA Inventory.

OSHA PEL: (Transitional: TWA 50 ppm) TWA 35; CL 200 ppm

ACGIH TLV: 25 ppm; BEI: less than 8% carboxyhemoglobin in blood at end of shift; less than 40 ppm CO in end-exhaled air at end of shift. (Proposed: less than 3.5% carboxyhemoglobin in blood at end of shift; less than 20 ppm CO in end-exhaled air at end of shift.)

DFG MAK: 30 ppm (33 mg/m^3); BAT: 5% in blood at end of shift

NIOSH REL: (Carbon Monoxide) TWA 35 ppm; CL 200 ppm

DOT Classification: Flammable Gas; Label: Flammable Gas and Poison Gas

SAFETY PROFILE: Mildly toxic by inhalation in humans but has caused many fatalities. Experimental teratogenic and reproductive effects. Human systemic effects by inhalation: changes in psychophysiological tests and methemoglobinemia-carboxhemoglobinemia. Can cause asphyxiation by preventing hemoglobin from binding oxygen. After removal from exposure, the half-life of elimination from the blood is one hour. Chronic exposure effects can occur at lower concentrations. A common air contaminant. Acute cases of poisoning resulting from brief exposures to high concentrations seldom result in any permanent disability if recovery takes place. Chronic effects as the result of repeated exposure to lower concentrations have been described, particularly in the Scandinavian literature. Auditory disturbances and contraction of the visual fields have been demonstrated. Glycosuria does occur, and heart irregularities have been reported. Other workers have found that where the poisoning has been relatively long and severe, cerebral congestion and edema may occur, resulting in long-lasting mental or nervous damage. Repeated exposure to low concentration of the gas, up to 100 ppm in air, is generally believed to cause no signs of poisoning or permanent damage. Industrially, sequelae are rare, as exposure, though often severe, is usually brief. It is a common air contaminant.

     A dangerous fire hazard when exposed to flame. Severe

explosion hazard when exposed to heat or flame. Violent or explosive reaction on contact with bromine trifluoride, bromine pentafluoride, chlorine dioxide, or peroxodisulfuryl difluoride. Mixture of liquid CO with liquid O_2 is explosive. Reacts with sodium or potassium to form explosive products sensitive to shock, heat, or contact with water. Mixture with copper powder + copper(II) perchlorate + water forms an explosive complex. Mixture of liquid CO with liquid dinitrogen oxide is a rocket propellant combination. Ignites on warming with iodine heptafluoride. Ignites on contact with cesium oxide + water. Potentially explosive reaction with iron(III) oxide between 0° and 150°C. Exothermic reaction with ClF_3, $(Li + H_2O)$, NF_3, OF_2, $(K + O_2)$, Ag_2O, $(Na + NH_3)$. To fight fire, stop flow of gas.

CBY000   CAS:56-23-5   HR: 3
CARBON TETRACHLORIDE
DOT: UN 1846
mf: CCl_4 mw: 153.81

PROP: Colorless liquid; heavy, ethereal odor. Mp: −22.6°, bp: 76.8°, fp: −22.9°, flash p: none, d: 1.597 @ 20°, vap press: 100 mm @ 23.0°.

SYNS: BENZINOFORM ◊ CARBONA ◊ CARBON CHLORIDE ◊ CARBON TET ◊ CZTEROCHLOREK WEGLA (POLISH) ◊ ENT 4,705 ◊ FASCIOLIN ◊ FLUKOIDS ◊ METHANE TETRACHLORIDE ◊ NECATORINA ◊ NECATORINE ◊ PERCHLOROMETHANE ◊ R 10 ◊ RCRA WASTE - NUMBER U211 ◊ TETRACHLOORKOOLSTOF (DUTCH) ◊ TETRACHLOORMETAAN ◊ TETRACHLORKOHLENSTOFF, (GERMAN) ◊ TETRACHLORMETHAN (GERMAN) ◊ TETRACHLOROCARBON ◊ TETRACHLOROMETHANE ◊ TETRACHLORURE de CARBONE (FRENCH) ◊ TETRACLOROMETANO (ITALIAN) ◊ TETRACLORURO di CARBONIO (ITALIAN) ◊ TETRAFINOL ◊ TETRAFORM ◊ TETRASOL ◊ UNIVERM ◊ VERMOESTRICID

TOXICITY DATA with REFERENCE
skn-rbt 4 mg MLD   XEURAQ MDDC-1715
skn-rbt 500 mg/24H MLD   85JCAE-,91,86
eye-rbt 2200 μg/30S MLD   XEURAQ MDDC-1715
eye-rbt 500 mg/24H MLD   85JCAE -,91,86
mmo-sat 20 uL/L   EJMBA2 18,213,83
mmo-asn 5000 ppm   MUREAV 147,288,85
ihl-rat TCLo:250 ppm/8H (female 10–15D post):REP
   DABBBA 32,2021,71
ihl-rat TCLo:250 ppm/8H (female 10–15D post):REP
   DABBBA 32,2021,71
orl-rat TDLo:3 g/kg (14D preg):TER   BEXBAN 82,1262,76
ihl-rat TCLo:300 ppm/7H (female 6–15D post):TER
   TXAPA9 28,452,74
orl-rat TDLo:2 g/kg (7–8D preg):REP   85DJA5 -,95,71
ipr-rat TDLo:5 g/kg (male 1D pre):REP   TXCYAC 10,39,78
ipr-rat TDLo:71,500 mg/kg (male 15D pre):REP   EXPEAM 22,395,66
orl-rat TDLo:7691 mg/kg (male 10D pre):REP   ESKHA5 (99),156,81
par-rat TDLo:2384 mg/kg (female 18D post):TER
   BEXBAN 76,1467,73
ihl-rat TCLo:300 ppm/7H (female 6–15D post):TER
   TXAPA9 28,452,74
ihl-rat TCLo:300 ppm/7H (female 6–15D post):TER
   TXAPA9 28,452,74
scu-rat TDLo:15600 mg/kg/12W-I:ETA   JJIND8 38,891,67
orl-mus TDLo:4400 mg/kg/19W-I:NEO   JJIND8 20,431,58
par-mus TDLo:305 g/kg/30W-I:ETA   BEXBAN 89,845,80
orl-ham TDLo:9250 mg/kg/30W-I:ETA   JJIND8 26,855,61
orl-mus TD:12 g/kg/88D-I:NEO   JJIND8 4,385,44
scu-rat TD:100 g/kg/25W-I:ETA   KRMJAC 12,37,65
scu-rat TD:31 g/kg/12W-I:ETA   JJIND8 45,1237,70
scu-rat TD:182 g/kg/70W-I:CAR   JJIND8 44,419,70
orl-mus TD:8580 mg/kg/9W-I:NEO   JJIND8 4,385,44
orl-mus TD:57600 mg/kg/12W-I:NEO   JJIND8 2,197,41
ihl-hmn TCLo:20 ppm:GIT   85CYAB 2,136,59
orl-wmn TDLo:1800 mg/kg:EYE,CNS   TXMDAX 69,86,73
orl-man TDLo:1700 mg/kg:CNS,PUL,GIT   SAMJAF 49,635,75
ihl-hmn LCLo:1000 ppm   PCOC** -,198,66
ihl-hmn TCLo:45 ppm/3D:CNS,GIT   LANCAO 1,360,60
ihl-hmn TCLo:317 ppm/30M:GIT   JAMAAP 103,962,34
ihl-hmn LCLo:5 pph/5M   TABIA2 3,231,33
unk-man LDLo:93 mg/kg   85DCAI 2,73,70
orl-rat LD50:2350 mg/kg   ARTODN 54,275,83
ihl-rat LC50:8000 ppm/4H   NPIRI* 1,16,74
skn-rat LD50:5070 mg/kg   SPEADM 78-1,16,78
ipr-rat LD50:1500 mg/kg   XEURAQ MDDC-1715
orl-mus LD50:8263 mg/kg   JPPMAB 3,169,51
ihl-mus LC50:9526 ppm/8H   JIDHAN 29,382,47
ipr-mus LD50:572 mg/kg   PHMCAA 10,172,68
orl-dog LDLo:1000 mg/kg   QJPPAL 7,205,34
ihl-dog LCLo:14,620 ppm/8H   NIHBAZ 191,1,49
ipr-dog LD50:1500 mg/kg   TXAPA9 10,119,67
ivn-dog LDLo:125 mg/kg   QJPPAL 7,205,34
ihl-cat LCLo:38,110 ppm/2H   HBAMAK 4,1405,35
scu-cat LDLo:300 mg/kg   JPETAB 63,153,38

CONSENSUS REPORTS: NTP Fifth Annual Report on Carcinogens. IARC Cancer Review: Group 2B IMEMDT 7,143,87; Animal Sufficient Evidence IMEMDT 20,371,79; IMEMDT 1,53,72; Human Inadequate Evidence IMEMDT 1,53,72; Human Limited Evidence IMEMDT 20,371,79. Community Right-To-Know List. EPA Genetic Toxicology Program. Reported in EPA TSCA Inventory.

OSHA PEL: (Transitional: TWA 10 ppm; CL 25 ppm; PK 200 ppm/5 min) TWA 2 ppm
ACGIH TLV: TWA 5 ppm; STEL 30 (skin); Suspected Human Carcinogen (Proposed: TWA 5; STEL 10 (skin); Animal Carcinogen)
DFG MAK: 10 ppm (65 mg/m^3); BEI: 1.6 mL/m^3 in alveolar air 1 hour after exposure; Suspected Carcinogen
NIOSH REL: (Carbon Tetrachloride) CL 2 ppm/60M

DOT Classification: ORM-A; Label: None; Poison B; Label: Poison

SAFETY PROFILE: Confirmed carcinogen with experimental carcinogenic, neoplastigenic, and tumorigenic data. A human poison by ingestion and possibly other routes. Poison by subcutaneous and intravenous routes. Mildly toxic by inhalation. Human systemic effects by inhalation and ingestion: nausea or vomiting, pupillary constriction, coma, antipsychotic effects, tremors, somnolence, anorexia, unspecified respiratory system and gastrointestinal system effects. Experimental teratogenic and reproductive effects. An eye and skin irritant. Damages liver, kidneys, and lungs. Mutation data reported. A narcotic. Individual susceptibility varies widely. Contact dermatitis can result from skin contact.

Carbon tetrachloride has a narcotic action resembling that of chloroform, though not as strong. Following exposure to high concentrations, the victim may become unconscious, and, if exposure is not terminated, death can follow from respiratory failure. The aftereffects following recovery from narcosis are more serious than those of delayed chloroform poisoning, usually taking the form of damage to the kidneys, liver, and lungs. Exposure to lower concentrations, insufficient to produce unconsciousness, usually results in severe gastrointestinal upset and may progress to serious kidney and hepatic damage. The kidney lesion is an acute nephrosis; the liver involvement consists of an acute degeneration of the central portions of the lobules. When recovery takes place, there may be no permanent disability. Marked variation in individual susceptibility to carbon tetrachloride exists; some persons appear to be unaffected by exposures that seriously poison their fellow workers. Alcoholism and previous liver and kidney damage seem to render the individual more susceptible. Concentrations on the order of 1000 to 1500 ppm are sufficient to cause symptoms if exposure continues for several hours. Repeated daily exposure to such concentration may result in poisoning.

Though the common form of poisoning following industrial exposure is usually one of gastrointestinal upset, which may be followed by renal damage, other cases have been reported in which the central nervous system has been affected, resulting in the production of polyneuritis, narrowing of the visual fields, and other neurological changes. Prolonged exposure to small amounts of carbon tetrachloride has also been reported as causing cirrhosis of the liver.

Locally, a dermatitis may be produced following long or repeated contact with the liquid. The skin oils are removed and the skin becomes red, cracked, and dry. The effect of carbon tetrachloride on the eyes, either as a vapor or as a liquid, is one of irritation with lacrimation and burning.

Industrial poisoning is usually acute with malaise, headache, nausea, dizziness, and confusion, which may be followed by stupor and sometimes loss of consciousness. Symptoms of liver and kidney damage may follow later with development of dark urine, sometimes jaundice and liver enlargement, followed by scanty urine, albuminuria, and renal casts; uremia may develop and cause death. Where exposure has been less acute, the symptoms are usually headache, dizziness, nausea, vomiting, epigastric distress, loss of appetite, and fatigue. Visual disturbances (blind spots, spots before the eyes, a visual "haze," and restriction of the visual fields), secondary anemia, and, occasionally a slight jaundice may occur. Dermatitis may be noticed on the exposed parts.

Forms impact-sensitive explosive mixtures with particulates of many metals, e.g., aluminum (when ball milled or heated to 152° in a closed container); barium (bulk metal also reacts violently); beryllium; potassium (200 times more shock-sensitive than mercury fulminate); potassium-sodium alloy (more sensitive than potassium); lithium; sodium; zinc (burns readily). Also forms explosive mixtures with chlorine trifluoride; calcium hypochlorite (heat sensitive); calcium disilicide (friction and pressure sensitive); triethyldialuminum trichloride (heat sensitive); decaborane(14) (impact sensitive); dinitrogen tetraoxide. Violent or explosive reaction on contact with fluorine. Forms explosive mixtures with ethylene between 25° and 105° and between 30 and 80 bar. Potentially explosive reaction on contact with boranes. 9:1 mixtures of methanol and CCl_4 react exothermically with aluminum, magnesium, or zinc. Potentially dangerous reaction with dimethyl formamide, 1,2,3,4,5,6-hexachlorocyclohexane, or dimethylacetamide when iron is present as a catalyst. CCl_4 has caused explosions when used as a fire extinguisher on wax and uranium fires. Incompatible with aluminum trichloride, dibenzoyl peroxide, potassium-tert-butoxide. Vigorous exothermic reaction with allyl alcohol, $Al(C_2H_5)_3$, (benzoyl peroxide + C_2H_4), BrF_3, diborane, disilane, liquid O_2, Pu, ($AgClO_4$ + HCl), potassium-tert-butoxide, tetraethylenepentamine, tetrasilane, trisilane, Zr. When heated to decomposition it emits toxic fumes of Cl^- and phosgene. It has been banned from household use by the FDA. See also CHLORINATED HYDROCARBONS, ALIPHATIC.

CCH850 CAS:590-46-5 HR: 1
(CARBOXYMETHYL)TRIMETHYLAMMONIUM CHLORIDE
mf: $C_5H_{12}NO_2 \cdot Cl$ mw: 153.63

SYNS: AMMONIUM, (CARBOXYMETHYL)TRIMETHYL-, CHLORIDE ◇ GLYKOKOLLBETAIN-CHLORID ◇ METHANAMINIUM, 1-CARBOXY-N,N,N-TRIMETHYL-, CHLORIDE

TOXICITY DATA with REFERENCE
scu-mus LD50:8 g/kg   ABMGAJ 3,28,59

CONSENSUS REPORTS: Reported in EPA TSCA Inventory.

SAFETY PROFILE: Slightly toxic by subcutaneous route. When heated to decomposition it emits toxic vapors of NO_x and Cl^-.

CCK655 CAS:461-05-2 HR: 1
dl-CARNITINE CHLORIDE
mf: $C_7H_{15}NO_3 \cdot Cl$ mw: 196.68

CCP850 CATECHOL

SYNS: AMMONIUM,(3-CARBOXY-2-HYDROXYPROPYL)TRIMETHYL-, CHLORIDE, (+ −)- ◊ BICARNESINE ◊ (+ −)-(3-CARBOXY-2-HYDROXYPROPYL)TRIMETHYLAMMONIUM CHLORIDE ◊ (+ −)-CARNITINE CHLORIDE ◊ (+ −)-CARNITINE HYDROCHLORIDE ◊ d,l-CARNITINE HYDROCHLORIDE ◊ dl-CARNITINE HYDROCHLORIDE ◊ 1-PROPANAMINIUM, 3-CARBOXY-2-HYDROXY-N,N,N-TRIMETHYL-, CHLORIDE, (+ −)-(9CI)

TOXICITY DATA with REFERENCE
scu-rat LD50:10 g/kg   ABMGAJ 3,28,59
scu-mus LD50:6 g/kg   ABMGAJ 3,28,59

CONSENSUS REPORTS: Reported in EPA TSCA Inventory.

SAFETY PROFILE: Mildly toxic by subcutaneous route. When heated to decomposition it emits toxic vapors of NO_x and Cl^-.

CCP850   CAS:120-80-9   **HR: 3**
CATECHOL
mf: $C_6H_6O_2$ mw: 110.12

PROP: Colorless crystals. Mp: 105°, bp: 246°, flash p: 261°F (CC), d: 1.341 @ 15°, vap press: 10 mm @ 118.3°, vap d: 3.79. Sol in water, chloroform, and benzene; very sol in alc and ether.

SYNS: o-BENZENEDIOL ◊ 1,2-BENZENEDIOL ◊ CATECHIN ◊ C.I. 76500 ◊ C.I. OXIDATION BASE 26 ◊ o-DIHYDROXYBENZENE ◊ 1,2-DIHYDROXYBENZENE ◊ o-DIOXYBENZENE ◊ o-DIPHENOL ◊ DURAFUR DEVELOPER C ◊ FOURAMINE PCH ◊ FOURRINE 68 ◊ o-HYDROQUINONE ◊ o-HYDROXYPHENOL ◊ 2-HYDROXYPHENOL ◊ NCI-C55856 ◊ OXYPHENIC ACID ◊ PELAGOL GREY C ◊ o-PHENYLENEDIOL ◊ PYROCATECHIN ◊ PYROCATECHINIC ACID ◊ PYROCATECHOL ◊ PYROCATECHUIC ACID

TOXICITY DATA with REFERENCE
mrc-smc 300 mg/L   MUREAV 135,109,84
dni-hmn:hla 200 µmol/L   MUREAV 92,427,82
dns-rat-orl 1 g/kg   JJIND8 74,1283,85
scu-rat TDLo:5 mg/kg (1D pre):REP   ENDOAO 57,466,55
orl-rat LD50:260 mg/kg   AFREAW 3,197,51
scu-rat LDLo:110 mg/kg   AIPTAK 176,193,68
orl-mus LD50:260 mg/kg   AFREAW 3,197,51
ipr-mus LD50:175 mg/kg   BEXBAN 61,291,66
scu-mus LD50:247 mg/kg   INHEAO 5,143,67
ivn-dog LDLo:40 mg/kg   HBTXAC 1,62,56
skn-rbt LD50:800 mg/kg   AIHAAP 37,596,76
ipr-gpg LDLo:150 mg/kg   HBTXAC 1,62,55
par-frg LDLo:160 mg/kg   AEPPAE 166,437,32

CONSENSUS REPORTS: IARC Cancer Review: Group 3 IMEMDT 7,56,87; Animal Inadequate Evidence IMEMDT 15,155,77. EPA Extremely Hazardous Substances List. Reported in EPA TSCA Inventory. EPA Genetic Toxicology Program.

OSHA PEL: TWA 5 ppm (skin)
ACGIH TLV: TWA 5 ppm (skin)

SAFETY PROFILE: Poison by ingestion, subcutaneous, intraperitoneal, intravenous, and parenteral routes. Moderately toxic by skin contact. Experimental reproductive effects. Can cause dermatitis on skin contact. An allergen. Human mutation data reported. Questionable carcinogen. Systemic effects similar to those of phenol. Combustible when exposed to heat or flame; can react vigorously with oxidizing materials. Hypergolic reaction with concentrated nitric acid. To fight fire, use water, CO_2, dry chemical. When heated to decomposition it emits acrid smoke and irritating fumes. See also PHENOL.

CDN550   CAS:75-87-6   **HR: 2**
CHLORAL
mf: C_2HCl_3O mw: 147.38

SYNS: ACETALDEHYDE, TRICHLORO-(9CI) ◊ ANHYDROUS CHLORAL ◊ CLORALIO ◊ GRASEX ◊ RCRA WASTE NUMBER U034 ◊ TRICHLOROACETALDEHYDE ◊ 2,2,2-TRICHLOROACETALDEHYDE ◊ TRICHLOROETHANAL

TOXICITY DATA with REFERENCE
mmo-sat 1 mg/plate   STEVA8 46,229,85
mma-sat 10 mg/plate   STEVA8 46,229,85
mma-smc 1 g/L   MUREAV 155,53,85
ipr-mus LD50:600 mg/kg   BCFAAI 111,293,72

CONSENSUS REPORTS: Reported in EPA TSCA Inventory.

SAFETY PROFILE: Moderately toxic by intraperitoneal route. Mutation data reported.

CDS050   CAS:115-27-5   **HR: 3**
CHLORENDIC ANHYDRIDE
mf: $C_9H_2Cl_6O_3$ mw: 370.81

SYN: 5-NORBORNENE-2,3-DICARBOXYLIC ANHYDRIDE, 1,4,5,6,7,7-HEXACHLORO-

TOXICITY DATA with REFERENCE
orl-rat LD50:2300 mg/kg   GTPZAB 26(4),49,82
ihl-rat LC50:>1 g/m^3 GTPZAB 26(4),49,82
orl-mus LD50:2400 mg/kg   GTPZAB 26(4),49,82
ihl-cat LC50:>1 g/m^3 GTPZAB 26(4),49,82
ihl-rbt LC50:>1 g/m^3 GTPZAB 26(4),49,82

SAFETY PROFILE: Poison by inhalation. Moderately toxic by ingestion. When heated to decomposition it emits toxic vapors of Cl^-.

CDZ050   CAS:588-07-8   **HR: 3**
3′-CHLOROACETANILIDE
mf: C_8H_8ClNO mw: 169.62

SYN: m-CHLOROACETANILIDE

TOXICITY DATA with REFERENCE
ipr-rat LD50:350 mg/kg   JAPMA8 48,204,59
ipr-mus LD50:610 mg/kg   JAPMA8 48,204,59

CONSENSUS REPORTS: Reported in EPA TSCA Inventory.

SAFETY PROFILE: Poison by intraperitoneal route. When heated to decomposition it emits toxic vapors of NO_x and Cl^-.

CEG800   CAS:121-50-6   *HR: 3*
4-CHLORO-3-AMINOBENZOTRIFLUORIDE
mf: $C_7H_5ClF_3N$ mw: 195.58

SYNS: 3-AMINO-4-CHLOROBENZOTRIFLUORIDE ◇ 3-AMINO-4-CHLORO-α-α-α-TRIFLUOROTOLUENE ◇ AZOENE FAST ORANGE RD SALT ◇ 2-CHLORO-5-(TRIFLUOROMETHYL)ANILINE ◇ 6-CHLORO-α-α-α-TRIFLUORO-m-TOLUIDINE ◇ C.I. 37050 ◇ DAITO ORANGE SALT RD ◇ DIAZO FAST ORANGE RD ◇ FAST ORANGE RD OIL ◇ FAST ORANGE RD SALT ◇ FAST ORANGE SALT RD ◇ FAST ORANGE SALT RDA ◇ FAST ORANGE SALT RDN ◇ HILTOSAL FAST ORANGE RD SALT ◇ ORANGE SALT NRD ◇ SANYO FAST ORANGE SALT RD ◇ m-TOLUIDINE, 6-CHLORO-α-α-α-TRIFLUORO- ◇ USAF MA-13

TOXICITY DATA with REFERENCE
ipr-mus LD50:100 mg/kg   NTIS** AD277-689

CONSENSUS REPORTS: Reported in EPA TSCA Inventory.

SAFETY PROFILE: Poison by intraperitoneal route. When heated to decomposition it emits toxic vapors of NO_x, F^- and Cl^-.

CEI100   CAS:131-09-9   *HR: 1*
2-CHLOROANTHRAQUINONE
mf: $C_{14}H_7ClO_2$ mw: 242.66

SYNS: 9,10-ANTHRACENEDIONE, 2-CHLORO- ◇ ANTHRAQUINONE, 2-CHLORO- ◇ 2-CHLORO-9,10-ANTHRACENEDIONE

TOXICITY DATA with REFERENCE
ipr-rat LD50:4310 mg/kg   GISAAA 49(4),90,84

CONSENSUS REPORTS: Reported in EPA TSCA Inventory.

SAFETY PROFILE: Mildly toxic by intraperitoneal route. When heated to decomposition it emits toxic vapors of Cl^-.

CEI600   CAS:104-88-1   *HR: 2*
p-CHLOROBENZALDEHYDE
mf: C_7H_5ClO mw: 140.57

SYNS: BENZALDEHYDE, p-CHLORO- ◇ 4-CHLOROBENZALDEHYDE ◇ p-CHLOROBENZENECARBOXALDEHYDE

TOXICITY DATA with REFERENCE
orl-rat LD50:1575 mg/kg   GTPZAB 31(2),50,87
orl-mus LD50:1400 mg/kg   GTPZAB 31(2),50,87

CONSENSUS REPORTS: Reported in EPA TSCA Inventory.

SAFETY PROFILE: Moderately toxic by ingestion. When heated to decomposition it emits toxic vapors of Cl^-.

CEJ600   CAS:100-03-8   *HR: 3*
p-CHLOROBENZENESULFINIC ACID
mf: $C_6H_5ClO_2S$ mw: 176.62

SYN: BENZENESULFINIC ACID, p-CHLORO-

TOXICITY DATA with REFERENCE
orl-rat LD50:>500 mg/kg   JAFCAU 25,501,77
ivn-mus LD50:56 mg/kg   CSLNX* NX#04978

CONSENSUS REPORTS: Reported in EPA TSCA Inventory.

SAFETY PROFILE: Poison by intravenous route. Moderately toxic by ingestion. When heated to decomposition it emits toxic vapors of SO_x and Cl^-.

CEL290   CAS:535-80-8   *HR: 2*
3-CHLOROBENZOIC ACID
mf: $C_7H_5ClO_2$ mw: 156.57

SYNS: ACIDO m-CLOROBENZOICO ◇ BENZOIC ACID, 3-CHLORO- ◇ m-CHLOROBENZOIC ACID ◇ BENZOIC ACID, m-CHLORO-

TOXICITY DATA with REFERENCE
ipr-rat LD50:750 mg/kg   BCFAAI 112,53,73

CONSENSUS REPORTS: Reported in EPA TSCA Inventory.

SAFETY PROFILE: Moderately toxic by intraperitoneal route. When heated to decomposition it emits toxic vapors of Cl^-.

CEL300   CAS:74-11-3   *HR: 2*
4-CHLOROBENZOIC ACID
mf: $C_7H_5ClO_2$ mw: 156.57

SYNS: ACIDO p-CLOROBENZOICO ◇ BENZOIC ACID, p-CHLORO- ◇ BENZOIC ACID, 4-CHLORO-(9CI) ◇ p-CARBOXYCHLOROBENZENE ◇ p-CHLORBENZOIC ACID ◇ p-CHLOROBENZOIC ACID ◇ CHLORODRACYLIC ACID

TOXICITY DATA with REFERENCE
orl-rat LD50:1170 mg/kg   JAFCAU 21,794,73
ipr-rat LD50:1000 mg/kg   BCFAAI 112,53,73
orl-mus LD50:1170 mg/kg   JAFCAU 23,724,75

CONSENSUS REPORTS: Reported in EPA TSCA Inventory.

SAFETY PROFILE: Moderately toxic by ingestion and intraperitoneal routes. When heated to decomposition it emits toxic vapors of Cl^-.

CEM850   CAS:615-18-9   *HR: 1*
2-CHLOROBENZOXAZOLE
mf: C_7H_4ClNO mw: 153.57

SYN: BENZOXAZOLE, 2-CHLORO-

TOXICITY DATA with REFERENCE
orl-mus LD50:2 g/kg   MDCHAG 4(1),337,64

CONSENSUS REPORTS: Reported in EPA TSCA Inventory.

SAFETY PROFILE: Slightly toxic by ingestion. When heated to decomposition it emits toxic vapors of NO_x and Cl^-.

CEP300   CAS:140-53-4   *HR: 3*
4-CHLOROBENZYL CYANIDE
mf: C_8H_6ClN mw: 151.60

SYNS: ACETONITRILE, (p-CHLOROPHENYL)- ◇ BENZENEACETONITRILE, 4-CHLORO- ◇ 4-CHLOR-BENZYL-CYANID ◇ 4-CHLOROBENZENEACETONITRILE ◇ p-CHLOROBENZYL CYANIDE ◇ p-CHLOROPHENYLACETONITRILE ◇ (4-CHLOROPHENYL)ACETONITRILE ◇ 2-(4-CHLOROPHENYL)ACETONITRILE

TOXICITY DATA with REFERENCE
ipr-mus LD50:27 mg/kg   PCBPBS 2,95,72
ivn-mus LD50:56 mg/kg   CSLNX* NX#07883

CONSENSUS REPORTS: Reported in EPA TSCA Inventory.

SAFETY PROFILE: Poison by intravenous and intraperitoneal routes. When heated to decomposition it emits toxic vapors of NO_x and Cl^-.

CEQ800   CAS:544-47-8   *HR: 2*
p-CHLOROBENZYLPSEUDOTHIURONIUM CHLORIDE
mf: $C_8H_9ClN_2S \cdot ClH$ mw: 237.16

SYNS: 2-(4-CHLOROBENZYL)-2-THIOPSEUDOUREAHYDROCHLORIDE ◇ PSEUDOUREA, 2-(p-CHLOROBENZYL)-2-THIO-, MONOHYDROCHLORIDE

TOXICITY DATA with REFERENCE
orl-mus LDLo:1600 mg/kg   AECTCV 14,111,85

CONSENSUS REPORTS: Reported in EPA TSCA Inventory.

SAFETY PROFILE: Moderately toxic by ingestion. When heated to decomposition it emits toxic vapors of NO_x, SO_x, HCl, and Cl^-.

CEU300   CAS:628-20-6   *HR: 3*
4-CHLOROBUTANENITRILE
mf: C_4H_6ClN mw: 103.56

SYNS: BUTANENITRILE, 4-CHLORO-(9CI) ◇ BUTYRONITRILE, 4-CHLORO- ◇ γ-CHLOROBUTYRONITRILE ◇ 4-CHLOROBUTYRONITRILE

TOXICITY DATA with REFERENCE
orl-mus LD50:53,380 μg/kg   ARTODN 55,47,84

CONSENSUS REPORTS: Reported in EPA TSCA Inventory.

SAFETY PROFILE: Poison by ingestion. When heated to decomposition it emits toxic vapors of NO_x and Cl^-.

CFB600   CAS:104-83-6   *HR: 2*
1-CHLORO-4-CHLOROMETHYLBENZENE
mf: $C_7H_6Cl_2$ mw: 161.03

SYN: TOLUENE, p,α-DICHLORO-

TOXICITY DATA with REFERENCE
unr-rat LD50:1075 mg/kg   GISAAA 55(5),13,90
unr-mus LD50:1156 mg/kg   GISAAA 55(5),13,90
unr-gpg LD50:5625 mg/kg   GISAAA 55(5),13,90

CONSENSUS REPORTS: Reported in EPA TSCA Inventory.

SAFETY PROFILE: Moderately toxic by unspecified routes. When heated to decomposition it emits toxic vapors of Cl^-.

CFE500   CAS:615-74-7   *HR: 2*
6-CHLORO-m-CRESOL
mf: C_7H_7ClO mw: 142.59

SYNS: PHENOL, 2-CHLORO-5-METHYL- ◇ 2-CHLORO-5-METHYLPHENOL

TOXICITY DATA with REFERENCE
orl-qal LD50:562 mg/kg   AECTCV 12,355,83
orl-brd LD50:562 mg/kg   AECTCV 12,355,83

CONSENSUS REPORTS: Reported in EPA TSCA Inventory.

SAFETY PROFILE: Moderately toxic by ingestion. When heated to decomposition it emits toxic vapors of Cl^-.

CFF600   CAS:766-66-5   *HR: 2*
2-CHLOROCYCLOHEPTANONE
mf: $C_7H_{11}ClO$ mw: 146.63

SYNS: α-CHLOROCYCLOHEPTANONE ◇ CYCLOHEPTANONE, 2-CHLORO-

TOXICITY DATA with REFERENCE
ipr-mus LD50:980 mg/kg   COREAF 254,2683,62

CONSENSUS REPORTS: Reported in EPA TSCA Inventory.

SAFETY PROFILE: Moderately toxic by intraperitoneal route. When heated to decomposition it emits toxic vapors of Cl^-.

CFJ100   CAS:447-31-4   *HR: 2*
α-CHLORODEOXYBENZOIN
mf: $C_{14}H_{11}ClO$ mw: 230.70

SYNS: ACETOPHENONE, α-CHLORO-α-PHENYL- ◇ ACETOPHENONE, 2-CHLORO-2-PHENYL-(8CI) ◇ α-CHLOROBENZYL PHENYL KETONE ◇ 2-CHLORO-1,2-DIPHENYLETHANONE ◇ 2-CHLORO-2-PHENYLACETOPHENONE ◇ DESYL CHLORIDE ◇ ETHONE, 2-CHLORO-1,2-DIPHENYL-(9CI)

TOXICITY DATA with REFERENCE
ipr-mus LDLo:500 mg/kg   CBCCT* 5,284,53

CONSENSUS REPORTS: Reported in EPA TSCA Inventory.

SAFETY PROFILE: Moderately toxic by intraperitoneal route. When heated to decomposition it emits toxic vapors of Cl$^-$.

CGC100   CAS:119-21-1   **HR: 3**
1-CHLORO-2,4-DIMETHOXY-5-NITROBENZENE
mf: $C_8H_8ClNO_4$ mw: 217.62

SYN: BENZENE, 5-CHLORO-1-NITRO-2,4-DIMETHOXY-

TOXICITY DATA with REFERENCE
orl-brd LD50:100 mg/kg   TXAPA9 21,315,72

CONSENSUS REPORTS: Reported in EPA TSCA Inventory.

SAFETY PROFILE: Poison by ingestion. When heated to decomposition it emits toxic vapors of NO_x and Cl$^-$.

CGM225   CAS:393-75-9   **HR: 2**
4-CHLORO-3,5-DINITRO-α-α-α-TRIFLUOROTOLUENE
mf: $C_7H_2ClF_3N_2O_4$ mw: 270.56

SYNS: BENZENE,2-CHLORO-1,3-DINITRO-5-(TRIFLUOROMETHYL)- ◇ BENZOTRIFLUORIDE, 4-CHLORO-3,5-DINITRO- ◇ 4-CHLORO-3,5-DINITROBENZOTRIFLUORIDE ◇ 3,5-DINITRO-4-CHLORO-α-α-α-TRIFLUOROTOLUENE ◇ TOLUENE, 4-CHLORO-3,5-DINITRO-α-α-α-TRIFLUORO-

TOXICITY DATA with REFERENCE
orl-rat LD50:930 mg/kg   AISSAW 19,351,83

CONSENSUS REPORTS: Reported in EPA TSCA Inventory.

SAFETY PROFILE: Moderately toxic by ingestion. When heated to decomposition it emits toxic vapors of NO_x, Cl$^-$, and F$^-$.

CGO600   CAS:542-58-5   **HR: 3**
2-CHLOROETHYL ACETATE
mf: $C_4H_7ClO_2$ mw: 122.56

SYNS: ACETOXYETHYL CHLORIDE ◇ 2-CHLORETHYLACETAT ◇ 2-CHLOROETHANOL ACETATE ◇ β-CHLOROETHYL ACETATE ◇ ETHANOL, 2-CHLORO-, ACETATE

TOXICITY DATA with REFERENCE
orl-mus LD50:250 mg/kg   ZHYGAM 26,17,80

CONSENSUS REPORTS: Reported in EPA TSCA Inventory.

SAFETY PROFILE: Poison by ingestion. When heated to decomposition it emits toxic vapors of Cl$^-$.

CHI950   CAS:350-30-1   **HR: D**
2-CHLORO-1-FLUORO-4-NITROBENZENE
mf: $C_6H_3ClFNO_2$ mw: 175.55

SYNS: BENZENE, 2-CHLORO-1-FLUORO-4-NITRO- ◇ 3-CHLORO-4-FLUORONITROBENZENE

TOXICITY DATA with REFERENCE
mmo-sat 102 µg/plate   MUREAV 116,217,83

SAFETY PROFILE: Mutation data reported. When heated to decomposition it emits toxic vapors of NO_x, F$^-$, and Cl$^-$.

CHU100   CAS:615-48-5   **HR: 3**
5-CHLORO-2-MERCAPTOANILINE HYDROCHLORIDE
mf: C_6H_6ClNS•ClH   mw: 196.10

SYN: ANILINE, 5-CHLORO-2-MERCAPTO-, HYDROCHLORIDE

TOXICITY DATA with REFERENCE
ivn-mus LD50:56 mg/kg   CSLNX* NX#00461

CONSENSUS REPORTS: Reported in EPA TSCA Inventory.

SAFETY PROFILE: Poison by intravenous route. When heated to decomposition it emits toxic vapors of NO_x, SO_x, HCl, and Cl$^-$.

CIH100   CAS:1006-99-1   **HR: 2**
5-CHLORO-2-METHYLBENZOTHIAZOLE
mf: C_8H_6ClNS mw: 183.66

SYNS: BENZOTHIAZOLE, 5-CHLORO-2-METHYL- ◇ USAF EK-P-4382

TOXICITY DATA with REFERENCE
ipr-mus LD50:500 mg/kg   NTIS** AD277-689

CONSENSUS REPORTS: Reported in EPA TSCA Inventory.

SAFETY PROFILE: Moderately toxic by intraperitoneal route. When heated to decomposition it emits toxic vapors of NO_x, SO_x, and Cl$^-$.

CJG825   CAS:83-42-1   **HR: 3**
2-CHLORO-6-NITROTOLUENE
mf: $C_7H_6ClNO_2$ mw: 171.59

SYNS: BENZENE, 1-CHLORO-2-METHYL-3-NITRO- ◇ 6-CHLORO-2-NITROTOLUENE ◇ TOLUENE, 2-CHLORO-6-NITRO-

TOXICITY DATA with REFERENCE
ivn-mus LD50:56 mg/kg   CSLNX* NX#03505

CONSENSUS REPORTS: Reported in EPA TSCA Inventory.

SAFETY PROFILE: Poison by intravenous route. When heated to decomposition it emits toxic vapors of NO_x and Cl$^-$.

CJL100 CAS:92-39-7 HR: 3
2-CHLOROPHENOTHIAZINE
mf: $C_{12}H_8ClNS$ mw: 233.72

SYNS: PHENOTHIAZINE, 2-CHLORO- ◊ 10H-PHENOTHIAZINE, 2-CHLORO-

TOXICITY DATA with REFERENCE
dni-bac-esc 160 mg/L   BCPCA6 26,1205,77
oth-esc 160 mg/L   BCPCA6 26,1205,77
ivn-mus LD50:18 mg/kg   CSLNX* NX#00590

CONSENSUS REPORTS: Reported in EPA TSCA Inventory.

SAFETY PROFILE: Poison by intravenous route. Mutation data reported. When heated to decomposition it emits toxic vapors of NO_x, SO_x, and Cl^-.

CJU200 CAS:92-00-2 HR: 1
N-(3-CHLOROPHENYL)DIETHANOLAMINE
mf: $C_{10}H_{14}ClNO_2$ mw: 215.70

SYNS: N,N-BIS(2-HYDROXYETHYL)CHLOROANILIDE ◊ N-(m-CHLOROPHENYL)DIETHANOLAMINE ◊ DIETHANOLAMINOCHLOROBENZENE ◊ N,N-DIETHANOLANILIDE, 3-CHLORO- ◊ DIETHANOLCHLOROANILIDE ◊ N,N-DIHYDROXYETHYL-m-CHLOROANILINE ◊ N,N-DIHYDROXYETHYL-3-CHLOROANILINE ◊ EMERY 5715 ◊ EMERY 5717 ◊ ETHANOL, 2,2'-((3-CHLOROPHENYL) IMINO)BIS-

TOXICITY DATA with REFERENCE
orl-rat LD50:5150 mg/kg   LONZA# 13JUL81

CONSENSUS REPORTS: Reported in EPA TSCA Inventory.

SAFETY PROFILE: Mildly toxic by ingestion. When heated to decomposition it emits toxic vapors of NO_x and Cl^-.

CKA550 CAS:555-60-2 HR: 3
((3-CHLOROPHENYL)HYDRAZONO)PROPANEDINITRILE
mf: $C_9H_5ClN_4$ mw: 204.63

SYNS: CARBONYL CYANIDE, 3-CHLOROPHENYLHYDRAZONE ◊ CCCP ◊ CCP ◊ MESOXALONITRILE, (m-CHLOROPHENYL)HYDRAZONE (8CI) ◊ PROPANEDINITRILE, (3-CHLOROPHENYL)HYDRAZONO)-(9CI)

TOXICITY DATA with REFERENCE
ipr-mus LD50:8 mg/kg   BCPCA6 18,1389,69

CONSENSUS REPORTS: Reported in EPA TSCA Inventory.

SAFETY PROFILE: Poison by intraperitoneal route. When heated to decomposition it emits toxic vapors of NO_x and Cl^-.

CKP725 CAS:627-30-5 HR: 2
3-CHLOROPROPANOL
mf: C_3H_7ClO mw: 94.55

SYNS: 3-CHLORPROPAN-1-OL ◊ 1-PROPANOL, 3-CHLORO- ◊ TRIMETHYLENE CHLOROHYDRIN

TOXICITY DATA with REFERENCE
mma-sat 33 μg/plate   ENMUDM 9(Suppl 9),1,87
orl-mus LD50:2300 mg/kg   ZHYGAM 26,17,80

CONSENSUS REPORTS: Reported in EPA TSCA Inventory.

SAFETY PROFILE: Moderately toxic by ingestion. Mutation data reported.

CLD600 CAS:130-16-5 HR: 2
5-CHLORO-8-QUINOLINOL
mf: C_9H_6ClNO mw: 179.61

SYNS: 5-CHLORO-8-HYDROXYQUINOLINE ◊ CHLOROXYQUINOLINE ◊ 8-QUINOLINOL, 5-CHLORO-

TOXICITY DATA with REFERENCE
mma-sat 50 nmol/plate   MUREAV 42,335,77
orl-gpg LDLo:1200 mg/kg   PSEBAA 28,484,31

CONSENSUS REPORTS: Reported in EPA TSCA Inventory.

SAFETY PROFILE: Moderately toxic by ingestion. Mutation data reported. When heated to decomposition it emits toxic vapors of NO_x and Cl^-.

CLD800 CAS:635-93-8 HR: 3
5-CHLOROSALICYLALDEHYDE
mf: $C_7H_5ClO_2$ mw: 156.57

SYN: SALICYLALDEHYDE, 5-CHLORO-

TOXICITY DATA with REFERENCE
ivn-mus LD50:56 mg/kg   CSLNX* NX#05073

CONSENSUS REPORTS: Reported in EPA TSCA Inventory.

SAFETY PROFILE: Poison by intravenous route. When heated to decomposition it emits toxic vapors of NO_x and Cl^-.

CLD825 CAS:321-14-2 HR: 3
5-CHLOROSALICYLIC ACID
mf: $C_7H_5ClO_3$ mw: 172.57

SYNS: BENZOIC ACID, 5-CHLORO-2-HYDROXY- ◊ SALICYLIC ACID, 5-CHLORO-

TOXICITY DATA with REFERENCE
orl-rat LD50:250 mg/kg   PHMGBN 9,164,73
ipr-mus LDLo:250 mg/kg   CBCCT* 7,792,55

CONSENSUS REPORTS: Reported in EPA TSCA Inventory.

SAFETY PROFILE: Poison by ingestion and intraperitoneal

CLJ800   CAS:89-68-9   HR: 2
6-CHLOROTHYMOL
mf: $C_{10}H_{13}ClO$ mw: 184.68

SYNS: 4-CHLORO-5-METHYL-2-(1-METHYLETHYL)PHENOL ◇ CHLOROTHYMOL ◇ CHLORTHYMOL ◇ PHENOL, 4-CHLORO-5-METHYL-2-(1-METHYLETHYL)-(9CI) ◇ THYMOL, 6-CHLORO-

TOXICITY DATA with REFERENCE
scu-mus LD50:2460 mg/kg   SIZSAR 3,73,52

CONSENSUS REPORTS: Reported in EPA TSCA Inventory.

SAFETY PROFILE: Moderately toxic by subcutaneous route. When heated to decomposition it emits toxic vapors of Cl⁻.

CLK227   CAS:87-63-8   HR: 3
6-CHLORO-o-TOLUIDINE
mf: C_7H_8ClN mw: 141.61

SYNS: 2-AMINO-3-CHLOROTOLUENE ◇ 3-CHLORO-2-AMINOTOLUENE ◇ 6-CHLORO-2-METHYLANILINE ◇ 6-CHLORO-2-TOLUIDINE ◇ o-TOLUIDINE, 6-CHLORO-

TOXICITY DATA with REFERENCE
dni-mus-orl 200 mg/kg   MUREAV 46,305,77
scu-cat LDLo:200 mg/kg   AHBAAM 110,12,33

CONSENSUS REPORTS: Reported in EPA TSCA Inventory.

SAFETY PROFILE: Poison by subcutaneous route. Mutation data reported. When heated to decomposition it emits toxic vapors of NO_x and Cl⁻.

CLO600   CAS:94-76-8   HR: 3
CHLOROTOLYLTHIOGLYCOLIC ACID
mf: $C_9H_9ClO_2S$ mw: 216.69

SYNS: ACETIC ACID, ((4-CHLORO-2-METHYL)PHENYL)THIO- ◇ 4-CHLORO-2-METHYLPHENYLTHIOGLYCOLIC ACID ◇ RED 3B ACID

TOXICITY DATA with REFERENCE
ipr-mus LD50:150 mg/kg   NTIS** AD691-490

CONSENSUS REPORTS: Reported in EPA TSCA Inventory.

SAFETY PROFILE: Poison by intraperitoneal route. When heated to decomposition it emits toxic vapors of SO_x and Cl⁻.

CLX300   CAS:132-22-9   HR: 3
CHLORPHENIRAMINE
mf: $C_{16}H_{19}ClN_2$ mw: 274.82

SYNS: ALLERGICAN ◇ ALLERGISAN ◇ 2-(p-CHLORO-α-(2-(DIMETHYLAMINO)ETHYL)BENZYL)PYRIDINE ◇ CHLOROPHENYL-PYRIDAMINE ◇ 1-(p-CHLOROPHENYL)-1-(2-PYRIDYL)-3-DIMETHYLAMINOPROPANE ◇ 1-(p-CHLOROPHENYL)-1-(2-PYRIDYL)-3-N,N-DIMETHYLPROPYLAMINE ◇ 4-CHLOROPHENIRAMINE ◇ CHLOROPIRIL ◇ CHLOROPROPHENPYRIDAMINE ◇ CHLORPHENAMINE ◇ CHLOR-PROPHENPYRIDAMINE ◇ CHLOR-TRIMETON ◇ CHLOR-TRIPOLON ◇ CLORFENIRAMINA ◇ CLOROPIRIL ◇ HAYNON ◇ HISTADUR ◇ PIRITON ◇ POLARONIL ◇ PYRIDINE, 2-(p-CHLORO-α-(2-(DIMETHYLAMINO)ETHYL)BENZYL)- ◇ 2-PYRIDINEPROPANAMINE, γ-(4-CHLOROPHENYL)-N,N-DIMETHYL-(9CI)

TOXICITY DATA with REFERENCE
orl-rat LD50:118 mg/kg   MEWEAC 17,2791,66
orl-mus LD50:121 mg/kg   MEWEAC 17,2791,66
ipr-mus LD50:125 mg/kg   YKKZAJ 92,1339,72
scu-mus LD50:160 mg/kg   BCFAAI 111,293,72
ivn-mus LD50:20 mg/kg   MEWEAC 17,2791,66
ivn-rbt LD50:22 mg/kg   MEWEAC 17,2791,66

CONSENSUS REPORTS: Reported in EPA TSCA Inventory.

SAFETY PROFILE: Poison by ingestion, intraperitoneal, subcutaneous, and intravenous routes. When heated to decomposition it emits toxic vapors of NO_x and Cl⁻.

CMA100   CAS:2921-88-2   HR: 3
CHLORPYRIFOS
DOT: NA 2783
mf: $C_9H_{11}Cl_3NO_3PS$ mw: 350.59

SYNS: BRODAN ◇ O,O-DIAETHYL-O-3,5,6-TRICHLOR-2-PYRIDYLMONOTHIOPHOSPHAT (GERMAN) ◇ O,O-DIETHYL-O-3,5,6-TRICHLORO-2-PYRIDYL PHOSPHOROTHIOATE ◇ DOWCO 179 ◇ DURSBAN ◇ DURSBAN F ◇ ENT 27,311 ◇ ERADEX ◇ LORSBAN ◇ OMS-0971 ◇ PYRINEX ◇ 3,5,6-TRICHLORO-2-PYRIDINOL-O-ESTER with O,O-DIETHYL PHOSPHOROTHIOATE

TOXICITY DATA with REFERENCE
cyt-dmg-orl 50 ppb/3S   ENMUDM 5,835,83
orl-mus TDLo:250 mg/kg (female 6–15D post):REP   TXAPA9 54,31,80
orl-man TDLo:300 mg/kg:PNS   ARTODN 59,176,86
orl-rat LD50:82 mg/kg   TXAPA9 14,515,69
ihl-rat LD50:78 mg/kg   BECTA6 19,113,78
skn-rat LD50:202 mg/kg   TXAPA9 14,515,69
orl-mus LD50:60 mg/kg   JESEDU 13,11,78
ihl-mus LD50:94 mg/kg   BECTA6 19,113,78
ipr-mus LD50:192 mg/kg   TXAPA9 65,144,82
orl-rbt LD50:1000 mg/kg   SPEADM 78-1,45,78
skn-rbt LD50:2000 mg/kg   GUCHAZ 6,203,73

CONSENSUS REPORTS: EPA Genetic Toxicology Program.

OSHA PEL: TWA 0.2 mg/m³ (skin)
ACGIH TLV: TWA 0.2 mg/m³ (skin)

DOT Classification: ORM-A; Label: None

SAFETY PROFILE: Poison by ingestion, intraperitoneal, skin contact, and inhalation routes. Human systemic effects by ingestion: paresthesia, muscle weakness, coma. Experimental reproductive effects: developmental toxicity. Mutation data reported. When heated to decomposition it emits very toxic fumes of Cl⁻, NO_x, PO_x, and SO_x.

CMF260   CAS:66-23-9   **HR: 3**
CHOLINE ACETATE (ESTER), BROMIDE
mf: $C_7H_{16}NO_2$•Br   mw: 226.15

SYNS: ACETOXYETHYL-TRIMETHYLAMMONIUM BROMIDE ◊ ACETYLCHOLINE BROMHYDRATE ◊ ACETYLCHOLINE BROMIDE ◊ ACETYLCHOLINE HYDROBROMIDE ◊ 2-(ACETYLOXY)-N,N,N-TRIMETHYLETHANAMINIUM BROMIDE ◊ CHOLINE, ACETYL-, BROMIDE ◊ ETHANAMINIUM, 2-(ACETYLOXY)-N,N,N-TRIMETHYL-, BROMIDE (9CI) ◊ PRAGMOLINE ◊ TONOCHOLIN B

TOXICITY DATA with REFERENCE
scu-mus LD50:170 mg/kg   JPETAB 103,62,51

CONSENSUS REPORTS: Reported in EPA TSCA Inventory.

SAFETY PROFILE: Poison by subcutaneous route. When heated to decomposition it emits toxic vapors of NO_x and Br⁻.

CMF800   CAS:123-41-1   **HR: 3**
CHOLINE HYDROXIDE
mf: $C_5H_{14}NO$•HO   mw: 121.21

SYNS: BURSINE ◊ FAGINE ◊ GOSSYPINE ◊ LURIDINE ◊ SINCALINE ◊ SINKALIN ◊ SINKALINE ◊ VIDINE

TOXICITY DATA with REFERENCE
ivn-mus LD50:21400 μg/kg   THERAP 23,1357,68

CONSENSUS REPORTS: Reported in EPA TSCA Inventory.

SAFETY PROFILE: Poison by intravenous route. When heated to decomposition it emits toxic vapors of NO_x.

CMO500   CAS:2429-74-5   **HR: D**
C.I. DIRECT BLUE 15, TETRASODIUM SALT
mf: $C_{34}H_{28}N_6O_{16}S_4$•4Na   mw: 996.88

SYNS: AIREDALE BLUE D ◊ AIZEN DIRECT SKY BLUE 5BH ◊ AMANIL SKY BLUE ◊ ATLANTIC SKY BLUE A ◊ ATUL DIRECT SKY BLUE ◊ AZINE SKY BLUE 5B ◊ BELAMINE SKY BLUE A ◊ BENZANIL SKY BLUE ◊ BENZO SKY BLUE S ◊ BENZO SKY BLUE A-CF ◊ CHLORAMINE SKY BLUE A ◊ CHLORAMINE SKY BLUE 4B ◊ CHROME LEATHER PURE BLUE ◊ C.I. 24400 ◊ C.I. DIRECT BLUE 15 ◊ CRESOTINE PURE BLUE ◊ DIACOTTON SKY BLUE 5B ◊ DIAMINE SKY BLUE CI ◊ DIAPHTAMINE PURE BLUE ◊ DIAZOL PURE BLUE 4B ◊ DIPHENYL BRILLIANT BLUE ◊ DIPHENYL SKY BLUE 6B ◊ DIRECT BLUE 10G ◊ DIRECT BLUE 15 ◊ DIRECT BLUE HH ◊ DIRECT PURE BLUE ◊ DIRECT PURE BLUE M ◊ DIRECT SKY BLUE A ◊ ENIANIL PURE BLUE AN ◊ FENAMIN SKY BLUE ◊ HISPAMIN SKY BLUE 3B ◊ KAYAKU DIRECT SKY BLUE 5B ◊ MITSUI DIRECT SKY BLUE 5B ◊ MODR PRIMA 15 ◊ NAPHTAMINE BLUE 10G ◊ NCI C61290 ◊ NIAGARA BLUE 4B ◊ NIAGARA SKY BLUE ◊ NIPPON DIRECT SKY BLUE ◊ NITTO DIRECT SKY BLUE 5B ◊ PHENAMINE SKY BLUE A ◊ PONTAMINE SKY BLUE 5BX ◊ PONTACYL SKY BLUE 4BX ◊ SHIKISO DIRECT SKY BLUE 5B ◊ SKY BLUE 4B ◊ SKY BLUE 5B ◊ TERTRODIRECT BLUE F ◊ VONDACEL BLUE HH

TOXICITY DATA with REFERENCE
mmo-sat 500 μg/plate   VHTODE 22,413,80
mma-sat 300 nmol/plate   MUREAV 116,305,83
ipr-rat TDLo:140 mg/kg (female 8D post):TER   TJADAB 2,85,69
ipr-rat TDLo:70 mg/kg (female 8D post):REP   PSEBAA 127,215,68
ipr-rat TDLo:200 mg/kg (female 8D post):REP   PSEBAA 127,215,68
ipr-rat TDLo:200 mg/kg (female 8D post):REP   PSEBAA 127,215,68   TJADAB 2,85,69

CONSENSUS REPORTS: Reported in EPA TSCA Inventory.

SAFETY PROFILE: An experimental teratogen. Other experimental reproductive effects. Mutation data reported. When heated to decomposition it emits very toxic fumes of NO_x, Na_2O, and SO_x.

CMP910   CAS:485-71-2   **HR: 3**
CINCHONIDINE
mf: $C_{19}H_{22}N_2O$ mw: 294.43

SYNS: CINCHONAN-9-OL, (8-α-9R)-(9CI) ◊ (−)-CINCHONIDINE ◊ (8S,9R)-CINCHONIDINE ◊ CINCHOVATINE ◊ α-QUINIDINE ◊ 2-QUINUCLIDINEMETHANOL, α-4-QUINOLYL-5-VINYL-

TOXICITY DATA with REFERENCE
ipr-rat LD50:206 mg/kg   APTOA6 4,265,48

CONSENSUS REPORTS: Reported in EPA TSCA Inventory.

SAFETY PROFILE: Poison by intraperitoneal route. When heated to decomposition it emits toxic vapors of NO_x.

CMP970   CAS:621-79-4   **HR: 3**
CINNAMAMIDE
mf: C_9H_9NO mw: 147.19

SYNS: 2-BENZYLIDENEACETAMIDE ◊ 3-PHENYLACRYLAMIDE ◊ 3-PHENYLPROPENAMIDE ◊ 3-PHENYL-2-PROPENAMIDE ◊ 2-PROPENAMIDE, 3-PHENYL-(9CI)

TOXICITY DATA with REFERENCE
orl-mus LD50:1600 mg/kg   CHTPBA 8,202,73
ivn-rbt LDLo:62 mg/kg   COREAF 153,895,11

CONSENSUS REPORTS: Reported in EPA TSCA Inventory.

SAFETY PROFILE: Poison by intravenous route. Moderately toxic by ingestion. When heated to decomposition it emits toxic vapors of NO_x.

CMP980   CAS:140-10-3   **HR: 3**
trans-CINNAMIC ACID
mf: $C_9H_8O_2$ mw: 148.17

SYNS: trans-β-CARBOXYSTYRENE ◊ CINNAMIC ACID, (E)- ◊ (E)-CIN-

NAMIC ACID ◇ trans-3-PHENYLACRYLIC ACID ◇ (E)-3-PHENYL-2-PROPENOIC ACID ◇ 2-PROPENOIC ACID, 3-PHENYL-, (E)-(9CI)

TOXICITY DATA with REFERENCE
orl-brd LD50:100 mg/kg   AECTCV 12,355,83

CONSENSUS REPORTS: Reported in EPA TSCA Inventory.

SAFETY PROFILE: Poison by ingestion. When heated to decomposition it emits acrid smoke and irritating vapors.

CMQ100   CAS:90-50-6   ***HR: 3***
CINNAMIC ACID, 3,4,5-TRIMETHOXY-
mf: $C_{12}H_{14}O_5$ mw: 238.26

SYNS: O-METHYLSINAPIC ACID ◇ 2-PROPENOIC ACID, 3-(3,4,5-TRIMETHOXYPHENYL)-(9CI) ◇ 3,4,5-TRIMETHOXYCINNAMIC ACID ◇ 3,4,5-TRIMETHOXYPHENYLACRYLIC ACID ◇ 3-(3,4,5-TRIMETHOXYPHENYL)-2-PROPENOIC ACID

TOXICITY DATA with REFERENCE
orl-brd LD50:422 mg/kg   AECTCV 12,355,83

CONSENSUS REPORTS: Reported in EPA TSCA Inventory.

SAFETY PROFILE: Moderately toxic by ingestion. When heated to decomposition it emits acrid smoke and irritating vapors.

CMQ850   CAS:122-69-0   ***HR: 1***
CINNAMYL CINNAMATE
mf: $C_{18}H_{16}O_2$ mw: 264.34

SYNS: CINNAMIC ACID, CINNAMYL ESTER ◇ CINNAMYL ALCOHOL, CINNAMATE ◇ CINNAMYLESTER KYSELINY SKORICOVE ◇ PHENYLALLYL CINNAMATE ◇ 3-PHENYL-2-PROPEN-1-YL CINNAMATE ◇ 3-PHENYL-2-PROPENYL 3-PHENYL-2-PROPENOATE ◇ 2-PROPENOIC ACID, 3-PHENYL-, 3-PHENYL-2-PROPENYL ESTER (9CI) ◇ STYRACIN

TOXICITY DATA with REFERENCE
orl-rat LD50:5000 mg/kg   FCTXAV 13,753,75

CONSENSUS REPORTS: Reported in EPA TSCA Inventory.

SAFETY PROFILE: Mildly toxic by ingestion. When heated to decomposition it emits acrid smoke and irritating vapors.

CMS205   CAS:62865-26-3   ***HR: 3***
C.I. PIGMENT YELLOW 35

SYNS: B-3-Zh ◇ CADMIUM GOLDEN ◇ CADMIUM LEMON ◇ CADMIUM PRIMROSE ◇ CADMIUM SULFIDE mixed with ZINC SULFIDE (1:1) ◇ C.I. 77205

OSHA PEL: TWA 5 μg(Cd)/m^3
ACGIH TLV: TWA 0.01 mg(Cd)/m^3; Suspected Carcinogen
NIOSH REL: (Cadmium): TWA reduce to lowest feasible level.

SAFETY PROFILE: Confirmed human carcinogen. When heated to decomposition it emits toxic fumes of Cd.

CNA250   CAS:7440-48-4   ***HR: 3***
COBALT
af: Co   aw: 58.93

PROP: Gray, hard, magnetic, ductile, somewhat malleable metal. Exists in two allotropic forms. At room temperature, the hexagonal form is more stable than the cubic form; both forms can exist at room temperature. Stable in air or toward water at ordinary temperatures. D 8.92, mp 1493°, bp about 3100°, Brinell hardness: 125, latent heat of fusion 62 cal/g, latent heat of vaporization 1500 cal/g, specific heat (15–100°): 0.1056 cal/g/°C. Readily sol in dil HNO_3; very slowly attacked by HCl or cold H_2SO_4. The hydrated salts of cobalt are red, and the sol salts form red solns that become blue on adding conc HCl.

SYNS: AQUACAT ◇ C.I. 77320 ◇ COBALT-59 ◇ KOBALT (GERMAN, POLISH) ◇ NCI-C60311 ◇ SUPER COBALT

TOXICITY DATA with REFERENCE
ims-rat TDLo:126 mg/kg:NEO   NATUAS 173,822,54
imp-rbt TDLo:75 mg/kg:ETA   ZEKBAI 52,425,42
ims-rat TD:126 mg/kg:NEO   BJCAAI 10,668,56
orl-rat LDLo:1500 mg/kg   15CYAT 2,1026,63
ipr-rat LDLo:250 mg/kg   EQSSDX 1,1,75
ivn-rat LDLo:100 mg/kg   EQSSDX 1,1,75
itr-rat LDLo:25 mg/kg   NTIS** AEC-TR-6710
ipr-mus LDLo:100 mg/kg   EQSSDX 1,1,75
orl-rbt LDLo:750 mg/kg   AIPTAK 62,347,39
ivn-rbt LDLo:100 mg/kg   EQSSDX 1,1,75

CONSENSUS REPORTS: Reported in EPA TSCA Inventory. Cobalt and its compounds are on the Community Right-To-Know List.

OSHA PEL: (Transitional: TWA 0.1 mg/m^3) TWA 0.05 mg/m^3
ACGIH TLV: (metal, dust, and fume) TWA 0.05 mg(Co)/m^3 (Proposed: TWA 0.02 mg(Co)/m^3; Animal Carcinogen)
DFG TRK: 0.5 mg/m^3 calculated as cobalt in that portion of dust that can possibly be inhaled in the production of cobalt powder and catalysts; hard metal (tungsten carbide) and magnet production (processing of powder, machine pressing, and mechanical processing of unsintered articles); other cobalt alloys and compounds: 0.1 mg/m^3 calculated as cobalt in that portion of dust that can possibly be inhaled. Animal Carcinogen, Suspected Human Carcinogen.
NIOSH REL: (Cobalt): Insufficient evidence for recommending limit.

SAFETY PROFILE: Confirmed carcinogen with experimental neoplastigenic and tumorigenic data. Poison by intravenous, intratracheal, and intraperitoneal routes. Moderately toxic by ingestion. Inhalation of the dust may cause pulmonary damage. The powder may cause dermatitis. Ingestion of soluble salts produces nausea and vomiting by local irritation. Powdered cobalt ignites spontaneously in air. Flammable when exposed to heat or flame. Explosive reaction with hydrazinium nitrate, ammonium nitrate + heat, and 1,3,4,7-tetramethylisoindole (at 390°C). Ignites on contact with bromine pentafluoride. Incandescent reaction with acetylene or nitryl fluoride. See also COBALT COMPOUNDS.

CNB475 CAS:513-79-1 **HR: 2**
COBALT(2+) CARBONATE
mf: $CO_3 \cdot Co$ mw: 118.94

SYNS: CARBONIC ACID, COBALT(2+) SALT (1:1) ◊ C.I. 77353 ◊ COBALT CARBONATE ◊ COBALT CARBONATE (1:1) ◊ COBALT MONOCARBONATE ◊ COBALTOUS CARBONATE

TOXICITY DATA with REFERENCE
orl-rat LD50:640 mg/kg   SchF## 16MAY86

CONSENSUS REPORTS: IARC Cancer Review: Group 2B; Human Inadequate Evidence IMEMDT 52,363,91

CONSENSUS REPORTS: Reported in EPA TSCA Inventory.

SAFETY PROFILE: Moderately toxic by ingestion. When heated to decomposition it emits toxic fumes of Co.

CNF390 CAS:53-84-9 **HR: 1**
CODEHYDROGENASE I
mf: $C_{21}H_{27}N_7O_{14}P_2$ mw: 663.49

SYNS: ADENINE-NICOTINAMIDE DINUCLEOTIDE ◊ CODEHYDRASE I ◊ COENZYME I ◊ COZYMASE ◊ COZYMASE I ◊ DIPHOSPHOPYRIDINE NUCLEOTIDE ◊ DPN ◊ ENZOPRIDE ◊ NAD ◊ NAD+ ◊ β-NAD ◊ NADIDE ◊ NICOTINAMIDE-ADENINE DINUCLEOTIDE ◊ NICOTINEAMIDE ADENINE DINUCLEOTIDE ◊ PYRIDINIUM, 3-CARBAMOYL-1-β-D-RIBOFURANOSYL-, HYDROXIDE, 5′-ESTER with ADENOSINE 5′-5′-(TRIHYDROGEN PYROPHOSPHATE), inner salt

TOXICITY DATA with REFERENCE
mmo-sat 100 μg/plate   ABCHA6 45,327,81
mma-sat 100 μg/plate   ABCHA6 45,327,81
ipr-mus LD50:4333 mg/kg   PCJOAU 20,160,86

CONSENSUS REPORTS: Reported in EPA TSCA Inventory.

SAFETY PROFILE: Mildly toxic by intraperitoneal route. Mutation data reported. When heated to decomposition it emits toxic vapors of NO_x and PO_x.

CNF400 CAS:53-59-8 **HR: 2**
CODEHYDROGENASE II
mf: $C_{21}H_{28}N_7O_{17}P_3$ mw: 743.47

SYNS: CODEHYDRASE II ◊ COENZYME II ◊ COZYMASE II ◊ NADP ◊ β-NADP ◊ NAD PHOSPHATE ◊ PYRIDINIUM, 3-CARBAMOYL-1-β-D-RIBOFURANOSYL-, HYDROXIDE, 5′,5′-ESTER with ADENOSINE 2′-(DIHYDROGEN PHOSPHATE) 5′-(TRIHYDROGEN PYROPHOSPHATE), inner salt ◊ TPN ◊ β-TPN ◊ TPN (NUCLEOTIDE) ◊ TRIPHOSPHOPYRIDINE NUCLEOTIDE

TOXICITY DATA with REFERENCE
ipr-mus LD50:3166 mg/kg   PCJOAU 20,160,86

CONSENSUS REPORTS: Reported in EPA TSCA Inventory.

SAFETY PROFILE: Moderately toxic by intraperitoneal route. When heated to decomposition it emits toxic vapors of NO_x and PO_x.

CNN500 CAS:10290-12-7 **HR: 3**
COPPER ORTHOARSENITE
DOT: UN 1586
mf: $AsCuHO_3$ mw: 187.47

PROP: Yellowish-green powder. Mp: decomp.

SYNS: ACID COPPER ARSENITE ◊ AIR-FLO GREEN ◊ ARSONIC ACID, COPPER(2+) SALT (1:1) (9CI) ◊ COPPER ARSENITE, solid (DOT) ◊ CUPRIC ARSENITE ◊ CUPRIC GREEN ◊ SCHEELES GREEN ◊ SCHEELE'S MINERAL ◊ SWEDISH GREEN

CONSENSUS REPORTS: Arsenic and its compounds, as well as copper and its compounds, are on the Community Right-To-Know List.

OSHA PEL: Cancer Hazard
ACGIH TLV: TWA 0.2 mg(As)/m^3 (Proposed: 0.01 mg(As)/m^3; Human Carcinogen)
NIOSH REL: (Arsenic, Inorganic) CL 0.002 mg(As)/m^3/15M

DOT Classification: Poison B; Label: Poison

SAFETY PROFILE: Poison. When heated to decomposition it emits toxic fumes of As. See also ARSENIC COMPOUNDS and COPPER COMPOUNDS.

CNU850 CAS:614-60-8 **HR: 3**
trans-o-COUMARIC ACID
mf: $C_9H_8O_3$ mw: 164.17

SYNS: CINNAMIC ACID, o-HYDROXY-, (E)- ◊ trans-o-HYDROXYCINNAMIC ACID ◊ trans-2-HYDROXYCINNAMIC ACID ◊ (E)-3-(2-HYDROXYPHENYL)-2-PROPENOIC ACID ◊ 2-PROPENOIC ACID, 3-(2-HYDROXYPHENYL)-, (E)-(9CI)

TOXICITY DATA with REFERENCE
ivn-mus LD50:180 mg/kg   CSLNX* NX#02587

CONSENSUS REPORTS: Reported in EPA TSCA Inventory.

SAFETY PROFILE: Poison by intravenous route. When heated to decomposition it emits acrid smoke and irritating vapors.

CNU875 CAS:496-41-3 **HR: 3**
COUMARILIC ACID
mf: $C_9H_6O_3$ mw: 162.15

SYN: 2-BENZOFURANCARBOXYLIC ACID

TOXICITY DATA with REFERENCE
ivn-mus LD50:320 mg/kg   CSLNX* NX#02495

CONSENSUS REPORTS: Reported in EPA TSCA Inventory.

SAFETY PROFILE: Poison by intravenous route. When heated to decomposition it emits acrid smoke and irritating vapors.

CNX400   CAS:596-27-0   HR: 3
o-CRESOLPHTHALEIN
mf: $C_{22}H_{18}O_4$ mw: 346.40

SYNS: 3′,3″-DIMETHYLPHENOLPHTHALEIN ◇ PHENOLPHTHALEIN, 3′,3″-DIMETHYL-

TOXICITY DATA with REFERENCE
ivn-mus LD50:320 mg/kg   CSLNX* NX#02167

CONSENSUS REPORTS: Reported in EPA TSCA Inventory.

SAFETY PROFILE: Poison by intravenous route. When heated to decomposition it emits acrid smoke and irritating vapors.

COF400   CAS:599-64-4   HR: 3
p-CUMYLPHENOL
mf: $C_{15}H_{16}O$ mw: 212.31

SYNS: p-(α-CUMYL)PHENOL ◇ p-(α-α-DIMETHYLBENZYL)PHENOL ◇ 4-(DIMETHYLPHENYLMETHYL)PHENOL ◇ 4-HYDROXYDIPHENYL-DIMETHYLMETHANE ◇ 4-(1-METHYL-1-PHENETHYL)PHENOL ◇ PHENOL, p-(α-α-DIMETHYLBENZYL)- ◇ PHENOL, 4-(1-METHYL-1-PHENETHYL)-(9CI)

TOXICITY DATA with REFERENCE
orl-frg LD50:335 mg/kg   GTPZAB 12(12),44,68

CONSENSUS REPORTS: Reported in EPA TSCA Inventory.

SAFETY PROFILE: Poison by ingestion. When heated to decomposition it emits acrid smoke and irritating vapors.

COR550   CAS:1071-27-8   HR: 2
(3-CYANOPROPYL) TRICHLOROSILANE
mf: $C_4H_6Cl_3NSi$ mw: 202.55

SYNS: BUTYRONITRILE, 4-(TRICHLOROSILYL)- ◇ SILANE, (3-CYANOPROPYL)TRICHLORO- ◇ SILANE, TRICHLORO(3-CYANOPROPYL)- ◇ TRICHLOR-3-KYANPROPYLSILAN

TOXICITY DATA with REFERENCE
orl-rat LD50:2830 mg/kg   AIHAAP 23,95,62

CONSENSUS REPORTS: Reported in EPA TSCA Inventory.

SAFETY PROFILE: Moderately toxic by ingestion. When heated to decomposition it emits toxic vapors of NO_x and Cl^-.

COW930   CAS:947-04-6   HR: 2
CYCLODODECALACTAM
mf: $C_{12}H_{23}NO$ mw: 197.36

SYNS: AZACYCLOTRIDECAN-2-ONE ◇ 2-AZACYCLOTRIDECANONE

TOXICITY DATA with REFERENCE
ipr-mus LD50:500 mg/kg   JPMSAE 60,1058,71

CONSENSUS REPORTS: Reported in EPA TSCA Inventory.

SAFETY PROFILE: Moderately toxic by intraperitoneal route. When heated to decomposition it emits toxic vapors of NO_x.

CPB120   CAS:482-54-2   HR: 2
1,2-CYCLOHEXANEDIAMINETETRAACETIC ACID
mf: $C_{14}H_{22}N_2O_8$ mw: 346.38

SYNS: ACETIC ACID, (1,2-CYCLOHEXYLENEDINITRILO)TETRA- ◇ CDTA ◇ CGTA ◇ CHEL 600 ◇ COMPLEXON IV ◇ 1,2-CYCLOHEXANEDIAMINE-N,N,N′,N′-TETRAACETIC ACID ◇ 1,2-CYCLOHEXYLENEDIAMINETETRAACETIC ACID ◇ (1,2-CYCLOHEXYLENEDINITRILO)TETRAACETIC ACID ◇ CYDTA ◇ DCTA ◇ 1,2-DIAMINOCYCLOHEXANETETRAACETIC ACID ◇ 1,2-DIAMINO-CYCLOHEXANE-N,N′-TETRAACETIC ACID ◇ GLYCINE, N,N^1-1,2-CYCLOHEXANEDIYLBIS(N-(CARBOXYMETHYL))- (9CI) ◇ KOMPLEXON IV ◇ KYSELINA 1,2-CYKLOHEXYLENDIAMINTETRAOCTOVA ◇ OCTA

TOXICITY DATA with REFERENCE
ipr-rat LD50:413 mg/kg   TOLED5 32,37,86
ipr-mus LD50:150 mg/kg   NTIS** AD691-490

CONSENSUS REPORTS: Reported in EPA TSCA Inventory.

SAFETY PROFILE: Moderately toxic by intraperitoneal route. When heated to decomposition it emits toxic vapors of NO_x.

CPB200   CAS:765-87-7   HR: D
1,2-CYCLOHEXANEDIONE
mf: $C_6H_8O_2$ mw: 112.14

SYN: 1,2-DIOXOCYCLOHEXANE

TOXICITY DATA with REFERENCE
mmo-sat 1 mg/plate   MUREAV 67,367,79

CONSENSUS REPORTS: Reported in EPA TSCA Inventory.

SAFETY PROFILE: Mutation data reported. When heated to decomposition it emits acrid smoke and irritating vapors.

CPC300   CAS:78-18-2   HR: 2
CYCLOHEXANONE PEROXIDE
mf: $C_{12}H_{22}O_5$ mw: 246.34

SYNS: CYCLOHEXANOL,1-((1-HYDROPEROXYCYCLOHEXYL)DIOXY)- ◇ 1-HYDROPEROXYCYCLOHEXYL-1-HYDROXYCYCLOHEXYL PEROXIDE ◇ 1-HYDROXY-1-HYDROPEROXYDICYCLOHEXYL PEROXIDE ◇ 1-

HYDROXY-1'-HYDROPEROXYDICYCLOHEXYL PEROXIDE ◊ PEROXIDE, 1-HYDROPEROXYCYCLOHEXYL 1-HYDROXYCYCLOHEXYL

TOXICITY DATA with REFERENCE
eye-rbt 80 mg/1M RNS SEV   ZAARAM 8,25,58
par-mus LD50:2 g/kg   NCPBBY Jan/Feb,69

CONSENSUS REPORTS: Reported in EPA TSCA Inventory.

SAFETY PROFILE: Slightly toxic by parenteral route. Severe eye irritant. When heated to decomposition it emits acrid smoke and irritating vapors.

CPD750   CAS:100-40-3   **HR: 3**
CYCLOHEXENYLETHYLENE
mf: C_8H_{12} mw: 108.20

PROP: Liquid. Bp: 128°, fp: −109°, flash p: 60°F (TOC), d: 0.832 @ 20°/4°, autoign temp: 517°F, vap press: 25.8 mm @ 38°, vap d: 3.76.

SYNS: BUTADIENE DIMER ◊ 4-ETHENYL-1-CYCLOHEXENE ◊ NCI-C54999 ◊ 1,2,3,4-TETRAHYDROSTYRENE ◊ 1-VINYLCYCLOHEXENE-3 ◊ 1-VINYLCYCLOHEX-3-ENE ◊ 4-VINYLCYCLOHEXENE ◊ 4-VINYL-CYCLOHEXENE-1 ◊ 4-VINYL-1-CYCLOHEXENE

TOXICITY DATA with REFERENCE
orl-mus TDLo:103 g/kg/2Y-I:CAR,REP   JTEHD6 21,507,87
skn-mus TDLo:16 g/kg/54W-I:ETA   JNCIAM 31,41,63
orl-mus TD:103 g/kg/2Y-I:NEO,REP   NTPTR* NTP-TR-303,86
orl-mus TD:206 g/kg/2Y-I:CAR   NTPTR* NTP-TR-303,86
orl-rat LD50:2563 mg/kg   AIHAAP 30,470,69
ihl-rat LCLo:8000 ppm/4H   AIHAAP 30,470,69
ihl-mus LC50:27,000 mg/m^3 IARC** 11,-,76
skn-rbt LD50:16,640 mg/kg   AIHAAP 30,470,69

CONSENSUS REPORTS: IARC Cancer Review: Group 3 IMEMDT 7,56,87; Animal Inadequate Evidence IMEMDT 11,277,76; Animal Limited Evidence IMEMDT 39,181,86. NTP Carcinogenesis Studies (gavage); Clear Evidence: mouse NTPTR* NTP-TR-303,86; Inadequate Studies: rat NTPTR* NTP-TR-303,86. Reported in EPA TSCA Inventory.
ACGIH TLV: TWA 0.1 ppm; Suspected Human Carcinogen

SAFETY PROFILE: Moderately toxic by ingestion and inhalation. Mildly toxic by skin contact. Questionable carcinogen with experimental carcinogenic, neoplastigenic, and tumorigenic data. Experimental reproductive effects. Dangerous fire hazard when exposed to heat, flame, or oxidizers. Can react with oxidizers. To fight fire, use foam, CO_2, dry chemical.

CPG700   CAS:103-00-4   **HR: 3**
1-CYCLOHEXYLAMINO-2-PROPANOL
mf: $C_9H_{19}NO$ mw: 157.29

SYNS: 2-PROPANOL, 1-(CYCLOHEXYLAMINO)- ◊ USAF DO-19

TOXICITY DATA with REFERENCE
ipr-mus LD50:100 mg/kg   NTIS** AD277-689

CONSENSUS REPORTS: Reported in EPA TSCA Inventory.

SAFETY PROFILE: Poison by intraperitoneal route. When heated to decomposition it emits toxic vapors of NO_x.

CPI400   CAS:542-18-7   **HR: 1**
CYCLOHEXYL CHLORIDE
mf: $C_6H_{11}Cl$ mw: 118.62

SYNS: CHLOROCYCLOHEXANE ◊ CYCLOHEXANE, CHLORO- ◊ MONOCHLOROCYCLOHEXANE

TOXICITY DATA with REFERENCE
ihl-rat LCLo:31 g/m^3 GTPZAB 10(1),49,66
ihl-mus LC50:31 g/m^3/2H   85GMAT-,36,82

CONSENSUS REPORTS: Reported in EPA TSCA Inventory.

SAFETY PROFILE: Slightly toxic by inhalation. When heated to decomposition it emits toxic vapors of Cl^-.

CPY800   CAS:936-52-7   **HR: 3**
N-(1-CYCLOPENTEN-1-YL)-MORPHOLINE
mf: $C_9H_{15}NO$ mw: 153.25

SYNS: MORPHOLINE, 4-(1-CYCLOPENTEN-1-YL)- ◊ (1-MORPHOLINOCYCLOPENTENE)

TOXICITY DATA with REFERENCE
ivn-mus LD50:320 mg/kg   CSLNX* NX#02169

CONSENSUS REPORTS: Reported in EPA TSCA Inventory.

SAFETY PROFILE: Poison by intravenous route. When heated to decomposition it emits toxic vapors of NO_x.

CQL300   CAS:63-37-6   **HR: D**
CYTIDINE MONOPHOSPHATE
mf: $C_9H_{14}N_3O_8P$ mw: 323.23

SYNS: CMP ◊ 5'-CMP ◊ CMP (nucleotide) ◊ CYTIDINE 5'-(DIHYDROGENPHOSPHATE) ◊ CYTIDINE 5'-MONOPHOSPHATE ◊ CYTIDINE 5'-MONOPHOSPHORIC ACID ◊ CYTIDINE 5'-PHOSPHATE ◊ CYTIDINE 5'-PHOSPHORIC ACID ◊ CYTIDYLIC ACID ◊ 5'-CYTIDYLIC ACID

TOXICITY DATA with REFERENCE
oth-hmn:oth 1 mmol/L   JIDEAE 65,52,75

CONSENSUS REPORTS: Reported in EPA TSCA Inventory.

SAFETY PROFILE: Human mutation data reported. When heated to decomposition it emits toxic vapors of NO_x and PO_x.

CQL400   CAS:65-47-4   HR: D
CYTIDINE-5′-TRIPHOSPHATE
mf: $C_9H_{16}N_3O_{14}P_3$ mw: 483.19

SYNS: CTP ◊ 5′-CTP ◊ CYTIDINE, 5′-(TETRAHYDROGEN TRIPHOSPHATE) ◊ CYTIDINE 5′-TRIPHOSPHORIC ACID

TOXICITY DATA with REFERENCE
oth-hmn:oth 1 mmol/L   JIDEAE 65,52,75

CONSENSUS REPORTS: Reported in EPA TSCA Inventory.

SAFETY PROFILE: Human mutation data reported. When heated to decomposition it emits toxic vapors of NO_x and PO_x.

CQM600   CAS:71-30-7   HR: 2
CYTOSINIMINE
mf: $C_4H_5N_3O$ mw: 111.12

SYNS: 4-AMINO-2-HYDROXYPYRIMIDINE ◊ 4-AMINO-2(1H)-PYRIMIDINONE ◊ CYTOSINE (8CI) ◊ 2(1H)-PYRIMIDINONE, 4-AMINO-

TOXICITY DATA with REFERENCE
ipr-mus LD50:>2222 mg/kg   JPETAB 207,504,78

CONSENSUS REPORTS: Reported in EPA TSCA Inventory.

SAFETY PROFILE: Moderately toxic by intraperitoneal route. When heated to decomposition it emits toxic vapors of NO_x.

D

DAA825 CAS:1928-38-7 **HR: D**
2,4-D METHYL ESTER
mf: $C_9H_8Cl_2O_3$ mw: 235.07

SYN: ACETIC ACID, (2,4-DICHLOROPHENOXY)-, METHYL ESTER ◇ METHYL 2,4-D ESTER

TOXICITY DATA with REFERENCE
scu-mus TDLo:945 mg/kg (female 6–14D post):REP   NTIS** PB223-160

SAFETY PROFILE: An experimental teratogen. When heated to decomposition it emits toxic vapors of Cl^-.

DAF350 CAS:541-02-6 **HR: 1**
DECAMETHYLCYCLOPENTASILOXANE
mf: $C_{10}H_{30}O_5Si_5$ mw: 370.85

SYNS: CYCLOPENTASILOXANE, DECAMETHYL- ◇ DEKAMETHYLCYKLOPENTASILOXAN

TOXICITY DATA with REFERENCE
skn-rbt 500 mg/24H MLD   85JCAE -,1234,86
eye-rbt 500 mg/24H MLD   85JCAE -,1234,86

CONSENSUS REPORTS: Reported in EPA TSCA Inventory.

SAFETY PROFILE: A skin and eye irritant. When heated to decomposition it emits acrid smoke and irritating vapors.

DAG650 CAS:646-25-3 **HR: 3**
1,10-DECANEDIAMINE
mf: $C_{10}H_{24}N_2$ mw: 172.36

TOXICITY DATA with REFERENCE
orl-rat LDLo:500 mg/kg   JPETAB 90,260,47
ipr-mus LDLo:125 mg/kg   CBCCT* 4,377,52

CONSENSUS REPORTS: Reported in EPA TSCA Inventory.

SAFETY PROFILE: Poison by intraperitoneal route. Moderately toxic by ingestion. When heated to decomposition it emits toxic vapors of NO_x.

DAH475 CAS:762-12-9 **HR: 2**
DECANOYL PEROXIDE
DOT: UN 2120
mf: $C_{20}H_{38}O_4$ mw: 342.58

SYNS: DECANOX ◇ DECANOYL PEROXIDE, technically pure (DOT) ◇ DIDECANOYL PEROXIDE ◇ PERKADOX SE 10 ◇ PEROXIDE, BIS(1-OXODECYL)

DOT Classification: Organic Peroxide; Label: Organic Peroxide   CFRGBR 49,172.101,90

CONSENSUS REPORTS: Reported in EPA TSCA Inventory.

SAFETY PROFILE: An organic peroxide. Handle carefully. When heated to decomposition it emits acrid smoke and irritating vapors.

DAJ500 CAS:2082-84-0 **HR: 3**
DECYLTRIMETHYLAMMONIUM BROMIDE
mf: $C_{13}H_{30}N \cdot Br$ mw: 280.35

SYN: AMMONIUM, DECYLTRIMETHYL-, BROMIDE

TOXICITY DATA with REFERENCE
ivn-rat LD50:5500 µg/kg   APTOA6 47,17,80
ivn-mus LD50:2800 µg/kg   APTOA6 47,17,80

CONSENSUS REPORTS: Reported in EPA TSCA Inventory.

SAFETY PROFILE: Poison by intravenous route. When heated to decomposition it emits toxic vapors of NO_x and Br^-.

DAZ050 CAS:951-78-0 **HR: D**
2′-DEOXYURIDINE
mf: $C_9H_{12}N_2O_5$ mw: 228.23

SYNS: DEOXYRIBOSE URACIL ◇ DEOXYURIDINE ◇ 2′-DESOXYURIDINE ◇ dU ◇ URACIL DESOXYURIDINE ◇ dURD ◇ URIDINE, 2′-DEOXY-

TOXICITY DATA with REFERENCE
pic-esc 2 mmol/L   MUREAV 178,177,87
dni-ham:lng 1 mmol/L   BICMBE 64,809,82
msc-ham:lng 1 mmol/L   BICMBE 64,809,82

CONSENSUS REPORTS: Reported in EPA TSCA Inventory.

SAFETY PROFILE: Mutation data reported. When heated to decomposition it emits toxic vapors of NO_x.

DBO100 CAS:614-94-8 **HR: D**
2,4-DIAMINOANISOLE DIHYDROCHLORIDE
mf: $C_7H_{10}N_2O \cdot 2ClH$ mw: 211.11

SYNS: 1,3-BENZENEDIAMINE, 4-METHOXY-, DIHYDROCHLORIDE ◇ m-PHENYLENEDIAMINE, 4-METHOXY-, DIHYDROCHLORIDE

TOXICITY DATA with REFERENCE
mma-sat 10 µg/plate   MUREAV 79,289,80
bfa-rat:sat 100 mg/kg   MUREAV 79,289,80

CONSENSUS REPORTS: Reported in EPA TSCA Inventory.

SAFETY PROFILE: Mutation data reported. When heated to decomposition it emits toxic vapors of NO_x, HCl, and Cl^-.

DBQ190   CAS:535-87-5   **HR: D**
3,5-DIAMINOBENZOIC ACID
mf: $C_7H_8N_2O_2$ mw: 152.17

SYN: BENZOIC ACID, 3,5-DIAMINO-

TOXICITY DATA with REFERENCE
dns-rat:lvr 10 µmol/L   MUREAV 206,183,88

CONSENSUS REPORTS: Reported in EPA TSCA Inventory.

SAFETY PROFILE: Mutation data reported. When heated to decomposition it emits toxic vapors of NO_x.

DBQ200   CAS:618-56-4   **HR: 3**
3,5-DIAMINOBENZOIC ACID DIHYDROCHLORIDE
mf: $C_7H_8N_2O_2 \cdot 2ClH$ mw: 225.09

SYN: BENZOIC ACID, 3,5-DIAMINO-, DIHYDROCHLORIDE

TOXICITY DATA with REFERENCE
ivn-mus LD50:180 mg/kg   CSLNX* NX#00464

CONSENSUS REPORTS: Reported in EPA TSCA Inventory.

SAFETY PROFILE: Poison by intravenous route. When heated to decomposition it emits toxic vapors of NO_x, HCl, and Cl^-.

DBQ250   CAS:128-94-9   **HR: 1**
4,5-DIAMINOCHRYSAZIN
mf: $C_{14}H_{10}N_2O_4$ mw: 270.26

SYNS: 9,10-ANTHRACENEDIONE, 1,8-DIAMINO-4,5-DIHYDROXY- ◇ ANTHRAQUINONE, 1,8-DIAMINO-4,5-DIHYDROXY- ◇ 1,8-DIAMINOCHRYSAZINE ◇ 1,8-DIAMINO-4,5-DIHYDROXYANTHRACHINON ◇ 1,8-DIAMINO-4,5-DIHYDROXYANTHRAQUINONE ◇ 1,8-DIAMINO-4,5-DIHYDROXY-9,10-ANTHRAQUINONE ◇ 4,5-DIAMINO-1,8-DIHYDROXYANTHRAQUINONE ◇ 1,8-DIHYDROXY-4,5-DIAMINOANTHRACHINON

TOXICITY DATA with REFERENCE
eye-rbt 100 mg/24H MOD   85JCAE -,653,86

CONSENSUS REPORTS: Reported in EPA TSCA Inventory.

SAFETY PROFILE: An eye irritant. When heated to decomposition it emits toxic vapors of NO_x.

DBR450   CAS:81-63-0   **HR: D**
1,4-DIAMINO-2,3-DIHYDROANTHRAQUINONE
mf: $C_{14}H_{12}N_2O_2$ mw: 240.28

SYNS: 9,10-ANTHRACENEDIONE, 1,4-DIAMINO-2,3-DIHYDRO- ◇ ANTHRAQUINONE, 2,3-DIHYDRO-1,4-DIAMINO-

TOXICITY DATA with REFERENCE
mmo-sat 100 µg/plate   MUREAV 40,203,76
mma-sat 500 µg/plate   MUREAV 40,203,76

CONSENSUS REPORTS: Reported in EPA TSCA Inventory.

SAFETY PROFILE: Mutation data reported. When heated to decomposition it emits toxic vapors of NO_x.

DCG800   CAS:7784-44-3   **HR: 3**
DIAMMONIUM HYDROGEN ARSENATE
DOT: UN 1546
mf: $AsH_3O_4 \cdot 2H_3N$ mw: 176.03

PROP: White powder or crystals. Mp: decomp to yield NH_3.

SYNS: AMMONIUM ACID ARSENATE ◇ AMMONIUM ARSENATE, solid (DOT) ◇ DIAMMONIUM ARSENATE ◇ DIAMMONIUM MONOHYDROGEN ARSENATE ◇ DIBASIC AMMONIUM ARSENATE ◇ SECONDARY AMMONIUM ARSENATE

CONSENSUS REPORTS: Arsenic and its compounds are on the Community Right-To-Know List.

OSHA PEL: Cancer Hazard
ACGIH TLV: TWA 0.2 mg(As)/m^3 (Proposed: 0.01 mg(As)/m^3; Human Carcinogen)
NIOSH REL: (Inorganic Arsenic) CL 0.002 mg(As)/m^3/15M

DOT Classification: Poison B; Label: Poison.

SAFETY PROFILE: A poison. When heated to decomposition it emits very toxic fumes of As, NO_x, and NH_3. See also ARSENIC.

DDB500   CAS:132-64-9   **HR: D**
DIBENZOFURAN
mf: $C_{12}H_8O$ mw: 168.20

SYNS: 2,2'-BIPHENYLENE OXIDE ◇ DIBENZO(b,d)FURAN ◇ DIPHENYLENE OXIDE

TOXICITY DATA with REFERENCE
sce-ham:ovr 10 mg/L   EMMUEG 10(Suppl 10),1,87

CONSENSUS REPORTS: Reported in EPA TSCA Inventory.

SAFETY PROFILE: Mutation data reported. When heated to decomposition it emits acrid smoke and irritating vapors.

DDE250   CAS:787-84-8   **HR: D**
1,2-DIBENZOYLHYDRAZINE
mf: $C_{14}H_{12}N_2O_2$ mw: 240.28

SYN: HYDRAZINE, 1,2-DIBENZOYL-

TOXICITY DATA with REFERENCE
ivn-mus LD50:56 mg/kg   CSLNX* NX#01359

CONSENSUS REPORTS: Reported in EPA TSCA Inventory.

SAFETY PROFILE: Poison by intravenous route. When heated to decomposition it emits toxic vapors of NO_x.

DDF700   CAS:122-65-6   *HR: 3*
N,N′-DIBENZYLDITHIOOXAMIDE
mf: $C_{16}H_{16}N_2S_2$ mw: 300.46

SYNS: OXAMIDE, N,N′-DIBENZYLDITHIO- ◇ USAF MK-1

TOXICITY DATA with REFERENCE
ipr-mus LD50:50 mg/kg   NTIS** AD277-689

CONSENSUS REPORTS: Reported in EPA TSCA Inventory.

SAFETY PROFILE: Poison by intraperitoneal route. When heated to decomposition it emits toxic vapors of NO_x and SO_x.

DDF800   CAS:122-75-8   *HR: 3*
N,N′-DIBENZYLETHYLENEDIAMINE DIACETATE
mf: $C_{16}H_{20}N_2 \cdot 2C_2H_4O_2$ mw: 360.50

SYNS: DBED DIACETATE ◇ ETHYLENEDIAMINE, N,N′-DIBENZYL-, DIACETATE

TOXICITY DATA with REFERENCE
ims-mus LD50:138 mg/kg   ANTCAO 4,633,54

CONSENSUS REPORTS: Reported in EPA TSCA Inventory.

SAFETY PROFILE: Poison by intramuscular route. When heated to decomposition it emits toxic vapors of NO_x.

DDI900   CAS:77-48-5   *HR: 3*
DIBROMANTINE
mf: $C_5H_6Br_2N_2O_2$ mw: 285.95

SYNS: DIBROMANTIN ◇ N,N′-DIBROMODIMETHYLHYDANTOIN ◇ 1,3-DIBROMO-5,5-DIMETHYL-2,4-IMIDAZOLIDINEDIONE ◇ HYDANTOIN, 1,3-DIBROMO-5,5-DIMETHYL- ◇ 2,4-IMIDAZOLIDINEDIONE, 1,3-DIBROMO-5,5-DIMETHYL-(9CI)

TOXICITY DATA with REFERENCE
orl-rat LD50:760 mg/kg   EPASR* 8EHQ-0281-0382
ihl-rat LCLo:29 g/m³/1H   EPASR* 8EHQ-0281-0382
skn-rbt LDLo:20 g/kg   EPASR* 8EHQ-0581-0382

CONSENSUS REPORTS: Reported in EPA TSCA Inventory.

SAFETY PROFILE: Poison by inhalation route. Moderately toxic by ingestion. Slightly toxic by skin contact. When heated to decomposition it emits toxic vapors of NO_x and Br^-.

DDK050   CAS:108-36-1   *HR: 2*
1,3-DIBROMOBENZENE
mf: $C_6H_4Br_2$ mw: 235.92

SYNS: BENZENE, m-DIBROMO- ◇ BENZENE, 1,3-DIBROMO-(9CI) ◇ m-DIBROMOBENZENE

TOXICITY DATA with REFERENCE
orl-mus LD50:2250 mg/kg   GISAAA 44(12),19,79
ipr-mus LD50:900 mg/kg   GISAAA 44(12),19,79

CONSENSUS REPORTS: Reported in EPA TSCA Inventory.

SAFETY PROFILE: Moderately toxic by ingestion and intraperitoneal routes. When heated to decomposition it emits toxic vapors of Br^-.

DDN900   CAS:93-52-7   *HR: D*
(1,2-DIBROMOETHYL)BENZENE
mf: $C_8H_8Br_2$ mw: 263.98

SYNS: BENZENE, (1,2-DIBROMOETHYL)- ◇ α-β-DIBROMOETHYLBENZENE ◇ 1,2-DIBROMO-1-PHENYLETHANE ◇ DOWSPRAY 9

TOXICITY DATA with REFERENCE
mmo-sat 20 µmol/L   CRNGDP 2,499,81

CONSENSUS REPORTS: Reported in EPA TSCA Inventory.

SAFETY PROFILE: Mutation data reported. When heated to decomposition it emits toxic vapors of Br^-.

DDN950   CAS:540-49-8   *HR: D*
1,2-DIBROMOETHYLENE
mf: $C_2H_2Br_2$ mw: 185.86

SYN: ETHYLENE, 1,2-DIBROMO-

TOXICITY DATA with REFERENCE
mmo-sat 100 µg/plate   TECSDY 15,101,87
mma-sat 10 µg/plate   TECSDY 15,101,87

CONSENSUS REPORTS: Reported in EPA TSCA Inventory.

SAFETY PROFILE: Mutation data reported. When heated to decomposition it emits toxic vapors of Br^-.

DDO450   CAS:661-95-0   *HR: 1*
1,2-DIBROMOHEXAFLUOROPROPANE
mf: $C_3Br_2F_6$ mw: 309.85

SYNS: 1,2-DIBROMO-1,1,2,3,3,3-HEXAFLUOROPROPANE ◇ PROPANE, 1,2-DIBROMO-1,1,2,3,3,3-HEXAFLUORO-

TOXICITY DATA with REFERENCE
orl-rat LDLo:8624 mg/kg   RADLAX 105,323,72

CONSENSUS REPORTS: Reported in EPA TSCA Inventory.

SAFETY PROFILE: Mildly toxic by ingestion. When heated to decomposition it emits toxic vapors of F⁻ and Br⁻.

DDQ150   CAS:594-34-3   *HR: D*
1,2-DIBROMO-2-METHYLPROPANE
mf: $C_4H_8Br_2$ mw: 215.94

SYN: PROPANE, 1,2-DIBROMO-2-METHYL-

TOXICITY DATA with REFERENCE
mmo-sat 10 μmol/plate   CNREA8 34,2576,74
dnr-esc 10 mg/plate   CNREA8 34,2576,74

CONSENSUS REPORTS: Reported in EPA TSCA Inventory.

SAFETY PROFILE: Mutation data reported. When heated to decomposition it emits toxic vapors of Br⁻.

DDQ500   CAS:99-28-5   *HR: 3*
2,6-DIBROMO-4-NITROPHENOL
mf: $C_6H_3Br_2NO_3$ mw: 296.92

SYN: PHENOL, 2,6-DIBROMO-4-NITRO-

TOXICITY DATA with REFERENCE
ivn-mus LD50:56 mg/kg   CSLNX* NX#03498

CONSENSUS REPORTS: Reported in EPA TSCA Inventory.

SAFETY PROFILE: Poison by intravenous route. When heated to decomposition it emits toxic vapors of NO_x and Br⁻.

DDR150   CAS:615-58-7   *HR: 3*
2,4-DIBROMOPHENOL
mf: $C_6H_4Br_2O$ mw: 251.92

SYN: PHENOL, 2,4-DIBROMO-

TOXICITY DATA with REFERENCE
orl-mus LD50:282 mg/kg   GISAAA 44(12),19,79
ipr-mus LD50:160 mg/kg   GISAAA 44(12),19,79

CONSENSUS REPORTS: Reported in EPA TSCA Inventory.

SAFETY PROFILE: Poison by ingestion and intraperitoneal routes. When heated to decomposition it emits toxic vapors of Br⁻.

DDS500   CAS:600-05-5   *HR: D*
2,3-DIBROMOPROPIONIC ACID
mf: $C_3H_4Br_2O_2$ mw: 231.89

SYNS: 2,3-DIBROMOPROPANOIC ACID ◇ α-β-DIBROMOPROPIONIC ACID ◇ PROPIONIC ACID, 2,3-DIBROMO- ◇ PROPANOIC ACID, 2,3-DIBROMO-(9CI)

TOXICITY DATA with REFERENCE
mmo-sat 10 μg/plate   APTOA6 45,112,79
mma-sat 10 μg/plate   APTOA6 45,112,79

CONSENSUS REPORTS: Reported in EPA TSCA Inventory.

SAFETY PROFILE: Mutation data reported. When heated to decomposition it emits toxic vapors of Br⁻.

DDT500   CAS:141-17-3   *HR: 1*
DIBUTOXYETHOXYETHYL ADIPATE
mf: $C_{22}H_{42}O_8$ mw: 434.64

SYNS: ADIPIC ACID, BIS(2-(2-BUTOXYETHOXY)ETHYL) ESTER ◇ HEXANEDIOIC ACID, BIS(2-(2-BUTOXYETHOXY)ETHYL) ESTER (9CI) ◇ TP-95 ◇ WAREFLEX

TOXICITY DATA with REFERENCE
orl-rat LD50:6 g/kg   NPIRI* 2,16,75

CONSENSUS REPORTS: Reported in EPA TSCA Inventory.

SAFETY PROFILE: Slightly toxic by ingestion. When heated to decomposition it emits acrid smoke and irritating vapors.

DDV500   CAS:719-22-2   *HR: 1*
2,6-DI-tert-BUTYL-p-BENZOQUINONE
mf: $C_{14}H_{20}O_2$ mw: 220.34

SYNS: p-BENZOQUINONE, 2,6-DI-tert-BUTYL- ◇ DBQ

TOXICITY DATA with REFERENCE
ipr-mus LD50:2270 mg/kg   TOLED5 6,173,80

CONSENSUS REPORTS: Reported in EPA TSCA Inventory.

SAFETY PROFILE: Moderately toxic by intraperitoneal route. When heated to decomposition it emits acrid smoke and irritating vapors.

DEA100   CAS:88-27-7   *HR: 2*
2,6-DI-tert-BUTYL-α-(DIMETHYLAMINO)-p-CRESOL
mf: $C_{17}H_{29}NO$ mw: 263.47

SYNS: AGIDOL 3 ◇ p-CRESOL, 2,6-DI-tert-BUTYL-α-(DIMETHYLAMINO)- ◇ ETHYL 703 ◇ ETHYL ANTIOXIDANT 703 ◇ F 1 ◇ F 1 (ANTIOXIDANT) ◇ OMI ◇ PHENOL, 4-((DIMETHYLAMINO)METHYL)-2,6-BIS(1,1-DIMETHYLETHYL)-(9CI)

TOXICITY DATA with REFERENCE
orl-rat LD50:1030 mg/kg   IPSTB3 3,93,76

CONSENSUS REPORTS: Reported in EPA TSCA Inventory.

SAFETY PROFILE: Moderately toxic by ingestion. When heated to decomposition it emits toxic vapors of NO_x.

DEC800   CAS:88-58-4   *HR: 2*
2,5-DI-t-BUTYLHYDROQUINONE
mf: $C_{14}H_{22}O_2$ mw: 222.36

SYNS: 2,5-DI-tert-BUTYLBENZENE-1,4-DIOL ◇ HYDROQUINONE, 2,5-DI-tert-BUTYL-

TOXICITY DATA with REFERENCE
orl-ham TDLo:134 g/kg/24W-C:NEO   CRNGDP 12,1341,91
orl-rat LDLo:800 mg/kg   KODAK* 21MAY71

CONSENSUS REPORTS: Reported in EPA TSCA Inventory.

SAFETY PROFILE: Moderately toxic by ingestion. Questionable carcinogen with experimental neoplastigenic data. When heated to decomposition it emits acrid smoke and irritating vapors.

DEC900   CAS:1421-49-4   **HR: D**
3,5-DI-tert-BUTYL-4-HYDROXYBENZOIC ACID
mf: $C_{15}H_{22}O_3$ mw: 250.37

SYN: BENZOIC ACID, 3,5-DI-tert-BUTYL-4-HYDROXY-

TOXICITY DATA with REFERENCE
dni-hmn:lyms 25 µmol/L   BBRCA9 80,963,78

CONSENSUS REPORTS: Reported in EPA TSCA Inventory.

SAFETY PROFILE: Human mutation data reported. When heated to decomposition it emits acrid smoke and irritating vapors.

DEF090   CAS:2406-25-9   **HR: 3**
DI-tert-BUTYL NITROXIDE
mf: $C_8H_{18}NO$ mw: 144.27

SYN: DTBN ◇ NITROXIDE, BIS(1,1-DIMETHYLETHYL) (9CI) ◇ NITROXIDE, DI-tert-BUTYL

TOXICITY DATA with REFERENCE
orl-mus LD50:505 mg/kg   JPETAB 141,349,63
ivn-mus LD50:53800 µg/kg   JPETAB 141,349,63

CONSENSUS REPORTS: Reported in EPA TSCA Inventory.

SAFETY PROFILE: Poison by intravenous route. Moderately toxic by ingestion. When heated to decomposition it emits toxic vapors of NO_x.

DEG100   CAS:128-39-2   **HR: 3**
2,6-DI-tert-BUTYLPHENOL
mf: $C_{14}H_{22}O$ mw: 206.36

SYNS: 2,6-BIS(tert-BUTYL)PHENOL ◇ ETHANOX 701 ◇ PHENOL, 2,6-DI-tert-BUTYL-

TOXICITY DATA with REFERENCE
ivn-mus LD50:120 mg/kg   JMCMAR 23,1350,80

CONSENSUS REPORTS: Reported in EPA TSCA Inventory.

SAFETY PROFILE: Poison by intravenous route. When heated to decomposition it emits acrid smoke and irritating vapors.

DEH300   CAS:1187-33-3   **HR: 3**
N,N-DIBUTYLPROPIONAMIDE
mf: $C_{11}H_{23}NO$ mw: 185.35

SYN: PROPIONAMIDE, N,N-DIBUTYL-

TOXICITY DATA with REFERENCE
ipr-mus LDLo:125 mg/kg   CBCCT* 5,288,53

CONSENSUS REPORTS: Reported in EPA TSCA Inventory.

SAFETY PROFILE: Poison by intraperitoneal route. When heated to decomposition it emits toxic vapors of NO_x.

DEH700   CAS:1962-75-0   **HR: 2**
DIBUTYL TEREPHTHALATE
mf: $C_{16}H_{22}O_4$ mw: 278.38

SYN: 1,4-BENZENEDICARBOXYLIC ACID, DIBUTYL ESTER ◇ TEREPHTHALIC ACID, DIBUTYL ESTER

TOXICITY DATA with REFERENCE
ipr-mus LDLo:1392 mg/kg   JPMSAE 56,1446,67

SAFETY PROFILE: Moderately toxic by intraperitoneal route. When heated to decomposition it emits acrid smoke and irritating vapors.

DEJ100   CAS:78-04-6   **HR: 2**
DIBUTYLTIN MALEATE
mf: $C_{12}H_{20}O_4Sn$ mw: 347.01

SYNS: ADVASTAB DBTM ◇ ADVASTAB T290 ◇ ADVASTAB T340 ◇ BT 31 ◇ 2,2-DIBUTYL-1,3,2-DIOXASTANNEPIN-4,7-DIONE ◇ DIBUTYL(MALEOYLDIOXY)TIN ◇ DIBUTYLSTANNYLENE MALEATE ◇ 1,3,2-DIOXASTANNEPIN-4,7-DIONE, 2,2-DIBUTYL- ◇ IRGasTAB T 4 ◇ IRGasTAB T 150 ◇ IRGasTAB T 290 ◇ KS 4B ◇ MA300A ◇ MARKURE UL2 ◇ NUODEX V 1525 ◇ STANCLERET 157 ◇ STANN RC 40F ◇ STAVINOR 1300SN ◇ STAVINOR SN 1300 ◇ TN 3J ◇ TVS-MA 300 ◇ TVS-N 2000E

TOXICITY DATA with REFERENCE
orl-mus LDLo:470 mg/kg   AECTCV 14,111,85

CONSENSUS REPORTS: Reported in EPA TSCA Inventory.

SAFETY PROFILE: Moderately toxic by ingestion. When heated to decomposition it emits toxic fumes of Sn.

DEL300   CAS:773-76-2   **HR: 1**
5,7-DICHLOR-8-HYDROXYCHINOLIN
mf: $C_9H_5Cl_2NO$ mw: 214.05

SYNS: CHLOFUCID ◇ CHLOROXINE ◇ CHLOROXYQUINOLINE ◇ CHLORQUINOL ◇ CHQ ◇ CLOFUZID ◇ DICHLOROHYDROXYQUINOLINE ◇ 5,7-DICHLORO-8-HYDROXYQUINOLINE ◇ 5,7-DICHLOROOXINE ◇ DICHLOROQUINOLINOL ◇ DICHLOROXIN ◇ 5,7-DICHLOROXINE

◇ ENDIARON ◇ QUESYL ◇ 8-QUINOLINOL, 5,7-DICHLORO- ◇ QUINOLOR ◇ QUIXALIN

TOXICITY DATA with REFERENCE
orl-cat LDLo:2 g/kg   ARZNAD 22,1307,72

CONSENSUS REPORTS: Reported in EPA TSCA Inventory.

SAFETY PROFILE: Slightly toxic by ingestion. When heated to decomposition it emits toxic vapors of NO_x and Cl^-.

DEM300   CAS:683-72-7   HR: 2
2,2-DICHLOROACETAMIDE
mf: $C_2H_3Cl_2NO$ mw: 127.96

SYNS: ACETAMIDE, DICHLORO- ◇ ACETAMIDE, 2,2-DICHLORO- (8CI,9CI) ◇ DICHLOROACETAMIDE

TOXICITY DATA with REFERENCE
ipr-mus LDLo:1750 mg/kg   JACSAT 63,1437,41

CONSENSUS REPORTS: Reported in EPA TSCA Inventory.

SAFETY PROFILE: Moderately toxic by intraperitoneal route. When heated to decomposition it emits toxic vapors of NO_x and Cl^-.

DEO290   CAS:554-00-7   HR: 3
2,4-DICHLOROANILINE
mf: $C_6H_5Cl_2N$ mw: 162.02

SYNS: ANILINE, 2,4-DICHLORO- ◇ BENZENAMINE, 2,4-DICHLORO-(9CI) ◇ 2,4-DICHLORANILIN ◇ 2,4-DICHLOROBENZENAMINE

TOXICITY DATA with REFERENCE
orl-rat LD50:1600 mg/kg   TSCAT* OTS 206512
ipr-rat LD50:400 mg/kg   TSCAT* OTS 206512
orl-mus LD50:400 mg/kg   TSCAT* OTS 206512
ipr-mus LD50:400 mg/kg   TSCAT* OTS 206512
orl-cat LDLo:113 mg/kg   AEXPBL 72,241,13

CONSENSUS REPORTS: Reported in EPA TSCA Inventory.

SAFETY PROFILE: Poison by ingestion and intraperitoneal routes. When heated to decomposition it emits toxic vapors of NO_x and Cl^-.

DEP600   CAS:95-50-1   HR: 3
o-DICHLOROBENZENE
DOT: UN 1591
mf: $C_6H_4Cl_2$ mw: 147.00

PROP: Clear liquid. Mp: −17.5°, bp: 180–183°, fp: −22°, flash p: 151°F, d: 1.307 @ 20°/20°, vap d: 5.05, autoign temp: 1198°F, lel: 2.2%, uel: 9.2%.

SYNS: CHLOROBEN ◇ CHLORODEN ◇ CLOROBEN ◇ DCB ◇ o-DICHLORBENZENE ◇ o-DICHLOR BENZOL ◇ 1,2-DICHLORO-BENZENE (MAK) ◇ DICHLOROBENZENE, ORTHO, liquid (DOT) ◇ DILANTIN DB ◇ DILATIN DB ◇ DIZENE ◇ DOWTHERM E ◇ NCI-C54944 ◇ ODB ◇ ODCB ◇ ORTHODICHLOROBENZENE ◇ ORTHODICHLOROBENZOL ◇ RCRA WASTE NUMBER U070 ◇ SPECIAL TERMITE FLUID ◇ TERMITKIL

TOXICITY DATA with REFERENCE
eye-rbt 100 mg/30S rns MLD   AMIHAB 17,180,58
spm-rat-ipr 250 mg/kg   JACTDZ 4(2),224,85
ipr-rat TDLo:50 mg/kg (1D male):REP   JACTDZ 4(1),224,85
ihl-rat TCLo:200 ppm/6H (6–15D preg):TER   FAATDF 5,190,85
orl-rat LD50:500 mg/kg   WRPCA2 7,135,68
ihl-rat LCLo:821 ppm/7H   AMIHAB 17,180,58
ipr-rat LD50:840 mg/kg   MEPAAX 20,519,69
orl-mus LD50:4386 g/kg   YKYUA6 32,471,81
ivn-mus LDLo:400 mg/kg   JPBAA7 44,281,37
orl-rbt LD50:500 mg/kg   85ARAE 3,32,76/77
ivn-rbt LDLo:250 mg/kg   JPBAA7 44,281,37
orl-gpg LDLo:2000 mg/kg   14CYAT 2,1336,63
ihl-gpg LCLo:800 ppm/24H   JPBAA7 44,281,37

CONSENSUS REPORTS: IARC Cancer Review: Group 3 IMEMDT 7,192,87; Animal Inadequate Evidence IMEMDT 7,231,74, IMEMDT 29,213,82; Human Inadequate Evidence IMEMDT 7,231,74, IMEMDT 29,213,82. Reported in EPA TSCA Inventory. Community Right-To-Know List.

OSHA PEL: CL 50 ppm
ACGIH TLV: TWA 25 ppm, STEL 50 ppm
DFG MAK: 50 ppm (300 mg/m^3)

DOT Classification: ORM-A; Label: None; IMO: Poison B; Label: St. Andrews Cross

SAFETY PROFILE: Poison by ingestion and intravenous routes. Moderately toxic by inhalation and intraperitoneal routes. An experimental teratogen. Other experimental reproductive effects. An eye, skin, and mucous membrane irritant. Causes liver and kidney injury. Questionable carcinogen. Mutation data reported. A pesticide. Flammable when exposed to heat or flame. Can react vigorously with oxidizing materials. To fight fire, use water, foam, CO_2, or dry chemical. Slow reaction with aluminum may lead to explosion during storage in a sealed aluminum container. When heated to decomposition it emits toxic fumes of Cl^-. See also BENZENE CHLORIDE and CHLORINATED HYDROCARBONS, AROMATIC.

DEP800   CAS:106-46-7   HR: 3
p-DICHLOROBENZENE
DOT: UN 1592
mf: $C_6H_4Cl_2$ mw: 147.00

PROP: White crystals, penetrating odor. Mp: 53°, bp: 173.4°, flash p: 150°F (CC), d: 1.4581 @ 20.5°/4°, vap press: 10 mm @ 54.8°, vap d: 5.08.

SYNS: p-CHLOROPHENYL CHLORIDE ◇ p-DICHLOORBENZEEN (DUTCH) ◇ 1,4-DICHLOORBENZEEN (DUTCH) ◇ p-DICHLORBENZOL (GERMAN) ◇ 1,4-DICHLOR-BENZOL (GERMAN) ◇ DI-CHLORICIDE ◇ 1,4-DICHLOROBENZENE (MAK) ◇ DICHLOROBENZENE, PARA, solid

(DOT) ◊ p-DICHLOROBENZOL ◊ p-DICLOROBENZENE (ITALIAN) ◊ 1,4-DICLOROBENZENE (ITALIAN) ◊ EVOLA ◊ NCI-C54955 ◊ PARACIDE ◊ PARA CRYSTALS ◊ PARADI ◊ PARADICHLORBENZOL (GERMAN) ◊ PARADICHLOROBENZENE ◊ PARADICHLOROBENZOL ◊ PARADOW ◊ PARAMOTH ◊ PARANUGGETS ◊ PARAZENE ◊ PDB ◊ PDCB ◊ PERSIA-PERAZOL ◊ RCRA WASTE NUMBER U070 ◊ RCRA WASTE NUMBER U071 ◊ RCRA WASTE NUMBER U072 ◊ SANTOCHLOR

TOXICITY DATA with REFERENCE
eye-hmn 80 ppm   AMIHAB 14,138,56
mmo-asn 200 mg/L   CJMIAZ 16,369,70
orl-rat TDLo:10 g/kg (female 6–15D post):TER   BECTA6 37,164,86
ihl-rbt TCLo:800 ppm/6H (female 6–18D post):TER   FAATDF 5,190,85
orl-rat TDLo:7500 mg/kg (female 6–15D post):TER   BECTA6 37,164,86
orl-rat TDLo:155 g/kg/2Y-I:CAR   NTPTR* NTP-TR-319,87
orl-mus TDLo:155 g/kg/2Y-I:CAR   NTPTR* NTP-TR-319,87
orl-hmn TDLo:300 mg/kg:EYE,PUL,GIT   PCOC** -,851,66
orl-hmn LDLo:857 mg/kg   34ZIAG-,210,69
unr-hmn LDLo:357 mg/kg   YKYUA6 31,1499,80
unr-man LDLo:221 mg/kg   85DCAI 2,73,70
orl-rat LD50:500 mg/kg   WRPCA2 9,119,70
ipr-rat LD50:2562 mg/kg   JAPMA8 38,124,49
orl-mus LD50:2950 mg/kg   GUCHAZ 6,183,73
ipr-mus LD50:2 g/kg   MUTAEX 2,111,87
scu-mus LD50:5145 mg/kg   TOIZAG 20,772,73
orl-rbt LD50:2830 mg/kg   YKYUA6 29,453,78
orl-gpg LDLo:2800 mg/kg   14CYAT 2,1338,63

CONSENSUS REPORTS: NTP Fifth Annual Report on Carcinogens. IARC Cancer Review: Group 2B IMEMDT 7,192,87; Animal Inadequate Evidence IMEMDT 7,231,74; IMEMDT 29,213,82. Human Inadequate Evidence IMEMDT 7,231,74; Reported in EPA TSCA Inventory. EPA Genetic Toxicology Program. Community Right-To-Know List.

OSHA PEL: (Transitional: TWA 75 ppm) TWA 75 ppm; STEL 110 ppm
ACGIH TLV: TWA 75 ppm; STEL 110 ppm; (Proposed: 10 ppm; Animal Carcinogen)
DFG MAK: 75 ppm (450 mg/m^3)

DOT Classification: ORM-A; Label: None; IMO: Poison B; Label: St. Andrews Cross

SAFETY PROFILE: Confirmed carcinogen with experimental carcinogenic data. An experimental teratogen. A human poison by an unspecified route. Moderately toxic to humans by ingestion. Moderately toxic experimentally by ingestion, subcutaneous, and intraperitoneal routes. Mildly toxic by subcutaneous route. Other experimental reproductive effects. Human systemic effects by ingestion: unspecified changes in the eyes, lungs, thorax and respiration, and decreased motility or constipation. Can cause liver injury in humans. A human eye irritant. Mutation data reported. A fumigant. Flammable when exposed to heat, flame, or oxidizers. Dangerous; can react vigorously with oxidizing materials. To fight fire, use water, foam, CO_2, dry chemical. When heated to decomposition it emits toxic fumes of Cl^-. See also CHLORINATED HYDROCARBONS, AROMATIC.

DER100   CAS:50-84-0   HR: 2
2,4-DICHLOROBENZOIC ACID
mf: $C_7H_4Cl_2O_2$ mw: 191.01

SYN: BENZOIC ACID, 2,4-DICHLORO-

TOXICITY DATA with REFERENCE
scu-mus LD50:1200 mg/kg   BCPCA6 13,1538,64
orl-mus LD50:830 mg/kg   SKEZAP 20,332,79

CONSENSUS REPORTS: Reported in EPA TSCA Inventory.

SAFETY PROFILE: Moderately toxic by ingestion and subcutaneous route. When heated to decomposition it emits toxic vapors of Cl^-.

DEU650   CAS:926-57-8   HR: 2
1,3-DICHLORO-2-BUTENE
mf: $C_4H_6Cl_2$ mw: 125.00

SYN: 2-BUTENE, 1,3-DICHLORO-

TOXICITY DATA with REFERENCE
ihl-rat LC50:3930 mg/m^3 ZKMAAX (6),66,69
ihl-mus LC50:4400 mg/m^3 ZKMAAX (6),66,69

CONSENSUS REPORTS: Reported in EPA TSCA Inventory.

SAFETY PROFILE: Moderately toxic by inhalation. When heated to decomposition it emits toxic vapors of Cl^-.

DEV000   CAS:764-41-0   HR: 3
1,4-DICHLORO-2-BUTENE
mf: $C_4H_6Cl_2$ mw: 125.00

PROP: Colorless liquid. Mp: 1–3°; bp: 156°; d: 1.183 @ 25°/4°.

SYNS: DCB ◊ 1,4-DCB ◊ 1,4-DICHLOROBUTENE-2 (MAK) ◊ RCRA WASTE NUMBER U074

TOXICITY DATA with REFERENCE
skn-rbt 10 mg/24H open SEV   AMIHBC 4,119,51
eye-rbt 20 mg open SEV   AMIHBC 4,119,51
mmo-sat 1 mmol/L   ARTODN 41,249,79
mma-sat 1 mmol/L   ARTODN 41,249,79
sln-dmg-orl 2 mmol/L/3D-I   35WYAM -,63,76
cyt-rat-ihl 1700 µg/m^3/30D-I   ZKMAAX 25,335,85
orl-rat TDLo:750 µg/m^3(75D male pre):REP   GISAAA 51(7),77,86
ihl-rat TCLo:5 ppm/6H (6-15D preg):TER   TXAPA9 64,125,82
ihl-rat TCLo:1 ppm/6H/82W-I:CAR   EPASR* 8EHQ-0985-0567

ihl-rat TC:100 ppb/6H/82W-I:NEO   EPASR* 8EHQ-0985-0567
orl-rat LD50:89 mg/kg   AMIHBC 4,119,51
ihl-rat LCLo:62 ppm/4H   AMIHBC 4,119,51
orl-mus LD50:190 mg/kg   GTPZAB 29(4),49,85
ihl-mus LC50:920 mg/m^3 GTPZAB 29(4),49,85
ivn-mus LD50:56 mg/kg   CSLNX* NX#01103
skn-rbt LD50:620 mg/kg   AMIHBC 4,119,51

CONSENSUS REPORTS: Reported in EPA TSCA Inventory. EPA Genetic Toxicology Program.
ACGIH TLV: (Proposed: TWA 0.005 ppm; Suspected Human Carcinogen)
DFG MAK: Animal Carcinogen, Suspected Human Carcinogen

SAFETY PROFILE: Confirmed carcinogen with experimental carcinogenic and neoplastigenic data. Poison by ingestion, inhalation, and intravenous routes. Moderately toxic by skin contact. An experimental teratogen. Other experimental reproductive effects. Mutation data reported. A severe skin and eye irritant. When heated to decomposition it emits toxic fumes of Cl$^-$. See also CHLORINATED HYDROCARBONS, ALIPHATIC.

DFF809   CAS:75-34-3   *HR: 3*
1,1-DICHLOROETHANE
DOT: UN 2362
mf: $C_2H_4Cl_2$ mw: 98.96

PROP: Colorless liquid; aromatic, ethereal odor; hot, saccharine taste. Mp: −97.7°, lel: 5.6%, bp: 57.3°, flash p: 22°F (TOC), d: 1.174 @ 20°/4°, vap press: 230 mm @ 25°, vap d: 3.44, autoign temp: 856°F.

SYNS: AETHYLIDENCHLORID (GERMAN) ◇ CHLORINATED HYDROCHLORIC ETHER ◇ CHLORURE d'ETHYLIDENE (FRENCH) ◇ CLORURO di ETILIDENE (ITALIAN) ◇ 1,1-DICHLOORETHAAN (DUTCH) ◇ 1,1-DICHLORAETHAN (GERMAN) ◇ 1,1-DICLOROETANO (ITALIAN) ◇ ETHYLIDENE CHLORIDE ◇ ETHYLIDENE DICHLORIDE ◇ NCI-C04535 ◇ RCRA WASTE NUMBER U076

TOXICITY DATA with REFERENCE
ihl-rat TCLo:6000 ppm/7H (6–15D preg):TER   TXAPA9 28,452,74
orl-mus TDLo:185 g/kg/78W-I:ETA,TER   NCITR* NCI-CG-TR-66,78
orl-mus TD:1300 g/kg/78W-I:ETA,TER   NCITR* NCI-CG-TR-66,78
orl-rat LD50:725 mg/kg   HYSAAV 32,349,67
ihl-rat LCLo:16,000 ppm/4H   JIDHAN 31,343,49

CONSENSUS REPORTS: NCI Carcinogenesis Bioassay (gavage); Inadequate Studies: mouse, rat NCITR* NCI-CG-TR-66,78. Reported in EPA TSCA Inventory.

OSHA PEL: TWA 100 ppm
ACGIH TLV: TWA 100 ppm
DFG MAK: 100 ppm (400 mg/m^3)
NIOSH REL: (1,1-Dichloroethane): Handle with caution.

DOT Classification: Flammable Liquid; Label: Flammable Liquid

SAFETY PROFILE: Moderately toxic by ingestion. Experimental teratogenic effects. Questionable carcinogen with experimental tumorigenic data. Liver damage reported in experimental animals. A very dangerous fire hazard and moderate explosion hazard when exposed to heat or flame; can react vigorously with oxidizing materials. To fight fire, use alcohol foam, water, foam, CO_2, dry chemical. When heated to decomposition it emits highly toxic fumes of phosgene and Cl$^-$.

DFL100   CAS:498-67-9   *HR: 3*
(DICHLOROFLUOROMETHYL)BENZENE
mf: $C_7H_5Cl_2F$ mw: 179.02

SYNS: BENZENE, (DICHLOROFLUOROMETHYL)- ◇ α-α-DICHLORO-α-FLUOROTOLUENE ◇ TOLUENE, α-α-DICHLORO-α-FLUORO-

TOXICITY DATA with REFERENCE
orl-rat LD50:450 mg/kg   GTPZAB 31(4),46,87
ihl-rat LC50:2 g/m^3 GTPZAB 31(4),46,87
orl-mus LD50:1875 mg/kg   GTPZAB 31(4),46,87
ihl-mus LC50:3100 mg/m^3 GTPZAB 31(4),46,87
ihl-uns LC50:3200 mg/m^3 GTPZAB 30(3),6,86

CONSENSUS REPORTS: Reported in EPA TSCA Inventory.

SAFETY PROFILE: Poison by inhalation route. Moderately toxic by ingestion. When heated to decomposition it emits toxic vapors of F$^-$ and Cl$^-$.

DFL720   CAS:2460-49-3   *HR: D*
4,5-DICHLOROGUAIACOL
mf: $C_7H_6Cl_2O_2$ mw: 193.03

SYNS: 4,5-DICHLORO-2-METHOXYPHENOL ◇ PHENOL, 4,5-DICHLORO-2-METHOXY-

TOXICITY DATA with REFERENCE
mmo-smc 50 mg/L   MUREAV 119,273,83

SAFETY PROFILE: Mutation data reported. When heated to decomposition it emits toxic vapors of Cl$^-$.

DFM025   CAS:356-18-3   *HR: 3*
1,2-DICHLORO-1,2,3,3,4,4-HEXAFLUOROCYCLOBUTANE
mf: $C_4Cl_2F_6$ mw: 232.94

SYNS: CYCLOBUTANE, 1,2-DICHLOROHEXAFLUORO-
◇ CYCLOBUTANE, 1,2-DICHLORO-1,2,3,3,4,4-HEXAFLUORO-(9CI)
◇ 1,2-DICHLOROHEXAFLUOROCYCLOBUTANE
◇ 1,2-DICHLOROPERFLUOROCYCLOBUTANE

TOXICITY DATA with REFERENCE
ihl-mus LCLo:21 pph/30S   ANASAB 16,3,61

SAFETY PROFILE: Poison by inhalation. When heated to decomposition it emits toxic vapors of NO_x and Cl^-.

DFM050   CAS:706-79-6   HR: 3
1,2-DICHLOROHEXAFLUOROCYCLOPENTENE
mf: $C_5Cl_2F_6$ mw: 244.95

SYNS: CYCLOPENTENE, 1,2-DICHLOROHEXAFLUORO- ◊ CYCLOPENTENE, 1,2-DICHLORO-3,3,4,4,5,5-HEXAFLUORO-(9CI) ◊ 1,2-DICHLORO-3,3,4,4,5,5-HEXAFLUOROCYCLOPENTENE ◊ 1,2-DICHLOROPERFLUOROCYCLOPENTENE

TOXICITY DATA with REFERENCE
orl-rat LD50:280 mg/kg   GISAAA 30(11),6,65
orl-mus LD50:276 mg/kg   GISAAA 30(11),6,65
ihl-mus LC50:2100 mg/m^3/2H   85JCAE-,163,86
orl-rbt LD50:280 mg/kg   GISAAA 30(11),6,65
orl-gpg LD50:280 mg/kg   GISAAA 30(11),6,65

CONSENSUS REPORTS: Reported in EPA TSCA Inventory.

SAFETY PROFILE: Poison by ingestion. Moderately toxic by inhalation. When heated to decomposition it emits toxic vapors of F^- and Cl^-.

DFN425   CAS:130-20-1   HR: 2
3,3′-DICHLOROINDANTHRONE
mf: $C_{28}H_{12}Cl_2N_2O_4$ mw: 511.32

SYN: AHCOVAT BLUE BCF ◊ ALIZANTHRENE BLUE RC ◊ AMANTHRENE BLUE BCL ◊ 5,9,14,18-ANTHRAZINETETRONE, 7,16-DICHLORO-6,15-DIHYDRO- ◊ ATIC VAT BLUE BC ◊ BENZADONE BLUE RC ◊ BLUE K ◊ CALCOLOID BLUE BLC ◊ CALCOLOID BLUE BLD ◊ CALCOLOID BLUE BLFD ◊ CALCOLOID BLUE BLR ◊ CALEDON BLUE XRC ◊ CARBANTHRENE BLUE BCF ◊ CARBANTHRENE BLUE BCS ◊ CARBANTHRENE BLUE RBCF ◊ CARBANTHRENE BLUE RCS ◊ C.I. 69825 ◊ CIBANONE BLUE FG ◊ CIBANONE BLUE FGF ◊ CIBANONE BLUE FGL ◊ CIBANONE BLUE GF ◊ C.I. VAT BLUE 6 ◊ D and C BLUE No. 9 ◊ DICHLOROINDANTHRONE ◊ 7,16-DICHLOROINDANTHRONE ◊ FENAN BLUE BCS ◊ FENANTHREN BLUE BC ◊ FENANTHREN BLUE BD ◊ HARMONE B 79 ◊ HELANTHRENE BLUE BC ◊ INDANTHREN BLUE BC ◊ INDANTHREN BLUE BCA ◊ INDANTHREN BLUE BCS ◊ INDANTHRENE BLUE BC ◊ INDANTHRENE BLUE BCF ◊ INDO BLUE B-I ◊ INDO BLUE WD 279 ◊ INDOTONER BLUE B 79 ◊ INTRAVAT BLUE GF ◊ MIKETHRENE BLUE BC ◊ MIKETHRENE BLUE BCS ◊ MONOLITE FAST BLUE 2RV ◊ MONOLITE FAST BLUE 2RVSA ◊ NAVINON BLUE BC ◊ NAVINON BRILLIANT BLUE RCL ◊ NIHONTHRENE BLUE BC ◊ NIHONTHRENE BRILLIANT BLUE RCL ◊ NYANTHRENE BLUE BFP ◊ OSTANTHREN BLUE BCL ◊ OSTANTHREN BLUE BCS ◊ PALANTHRENE BLUE BC ◊ PALANTHRENE BLUE BCA ◊ PARADONE BLUE RC ◊ PERNITHRENE BLUE BC ◊ PONSOL BLUE BCS ◊ PONSOL BLUE BF ◊ PONSOL BLUE BFD ◊ PONSOL BLUE BFDP ◊ PONSOL BLUE BFN ◊ PONSOL BLUE BFND ◊ PONSOL BLUE BFP ◊ RESINATED INDO BLUE B 85 ◊ ROMANTRENE BLUE FBC ◊ SANDOTHRENE BLUE NG ◊ SANDOTHRENE BLUE NGR ◊ SANDOTHRENE BLUE NGW ◊ SOLANTHRENE BLUE B ◊ SOLANTHRENE BLUE F-SBA ◊ SOLANTHRENE BLUE SB ◊ TINON BLUE GF ◊ TINON BLUE GL ◊ VAT BLUE 6 ◊ VAT BLUE KD ◊ VAT FAST BLUE BCS ◊ VAT GREEN B ◊ VAT SKY BLUE K ◊ VAT SKY BLUE KD ◊ VAT SKY BLUE KP 2F

TOXICITY DATA with REFERENCE
orl-mus LD50:1800 mg/kg   GNAMAP 14,152,75
skn-mus LD50:25 g/kg   GNAMAP 14,152,75

SAFETY PROFILE: Moderately toxic by ingestion. Mildly toxic by skin contact. When heated to decomposition it emits toxic vapors of NO_x and Cl^-.

DFN450   CAS:1324-55-6   HR: 1
DICHLOROISOVIOLANTHRONE
mf: $C_{34}H_{14}Cl_2O_2$ mw: 525.38

SYNS: AHCOVAT BRILLIANT VIOLET 2R ◊ AHCOVAT BRILLIANT VIOLET 4R ◊ AMANTHRENE BRILLIANT VIOLET RR ◊ ARLANTHRENE VIOLET 4R ◊ ATIC VAT BRILLIANT PURPLE 4R ◊ BENZADONE BRILLIANT PURPLE 2R ◊ BENZADONE BRILLIANT PURPLE 4R ◊ BENZO(rst)PHENANTHRO(10,1,2-cde)PENTAPHENE-9,18-DIONE, DICHLORO- ◊ BRILLIANT VIOLET K ◊ CALCOLOID VIOLET 4RD ◊ CALCOLOID VIOLET 4RP ◊ CALEDON BRILLIANT PURPLE 4R ◊ CALEDON BRILLIANT PURPLE 4RP ◊ CALEDON PRINTING PURPLE 4R ◊ CARBANTHRENE BRILLIANT VIOLET 4R ◊ CARBANTHRENE VIOLET 2R ◊ CARBANTHRENE VIOLET 2RP ◊ C.I. 60010 ◊ CIBANONE VIOLET F 4R ◊ CIBANONE VIOLET F 2RB ◊ CIBANONE VIOLET 2R ◊ CIBANONE VIOLET 4R ◊ C.I. PIGMENT VIOLET 31 ◊ C.I. VAT VIOLET 1 (8CI) ◊ FENANTHREN BRILLIANT VIOLET 2R ◊ FENANTHREN BRILLIANT VIOLET 4R ◊ INDANTHREN BRILLIANT VIOLET 4R ◊ INDANTHREN BRILLIANT VIOLET RR ◊ INDANTHRENE BRILLIANT VIOLET 4R ◊ INDANTHRENE BRILLIANT VIOLET RR ◊ INDANTHREN PRINTING VIOLET F 4R ◊ INDOFAST VIOLET LAKE ◊ NIHONTHRENE BRILLIANT VIOLET 4R ◊ NIHONTHRENE BRILLIANT VIOLET RR ◊ NYANTHRENE BRILLIANT VIOLET 4R ◊ PONOLITH FAST VIOLET 4RN ◊ SANDOTHRENE VIOLET N 4R ◊ SANDOTHRENE VIOLET N 2RB ◊ SANDOTHRENE VIOLET 4R ◊ SOLANTHRENE BRILLIANT VIOLET F 2R ◊ SYMULER FAST VIOLET R ◊ TINON VIOLET B 4RP ◊ TINON VIOLET 4R ◊ TINON VIOLET 2RB ◊ VAT BRIGHT VIOLET K ◊ VAT BRILLIANT VIOLET K ◊ VAT BRILLIANT VIOLET KD ◊ VAT BRILLIANT VIOLET KP ◊ VIOLET KYPOVA 1 ◊ VIOLET PIGMENT 31

TOXICITY DATA with REFERENCE
orl-mus LD50:6700 mg/kg   85JCAE -,1329,86

CONSENSUS REPORTS: Reported in EPA TSCA Inventory.

SAFETY PROFILE: Mildly toxic by ingestion. When heated to decomposition it emits toxic vapors of Cl^-.

DFS700   CAS:675-62-7   HR: 3
DICHLOROMETHYL-3,3,3-TRIFLUOROPROPYLSILANE
mf: $C_4H_7Cl_2F_3Si$ mw: 211.10

SYN: SILANE, DICHLOROMETHYL(3,3,3-TRIFLUOROPROPYL)-

TOXICITY DATA with REFERENCE
ihl-mus LC50:300 mg/m^3/2H   85JCAE -,1223,86

CONSENSUS REPORTS: Reported in EPA TSCA Inventory.

SAFETY PROFILE: Poison by inhalation. When heated to decomposition it emits toxic vapors of F^- and Cl^-.

DFX500 CAS:576-24-9 **HR: 2**
2,3-DICHLOROPHENOL
mf: $C_6H_4Cl_2O$ mw: 163.00

TOXICITY DATA with REFERENCE
orl-mus LD50:2376 mg/kg   TOLED5 29,39,85

CONSENSUS REPORTS: Reported in EPA TSCA Inventory.

SAFETY PROFILE: Moderately toxic by ingestion. When heated to decomposition it emits toxic vapors of Cl^-.

DFX850 CAS:583-78-8 **HR: 2**
2,5-DICHLOROPHENOL
mf: $C_6H_4Cl_2O$ mw: 163.00

SYN: PHENOL, 2,5-DICHLORO-

TOXICITY DATA with REFERENCE
sce-mus-ipr 210 mg/kg   JACTDZ 2(2),249,83
orl-rat LD50:580 mg/kg   NTIS** PB85-143766
orl-mus LD50:946 mg/kg   TOLED5 29,39,85

CONSENSUS REPORTS: Reported in EPA TSCA Inventory.

SAFETY PROFILE: Moderately toxic by ingestion. Mutation data reported. When heated to decomposition it emits toxic vapors of Cl^-.

DFY425 CAS:95-77-2 **HR: 2**
3,4-DICHLOROPHENOL
mf: $C_6H_4Cl_2O$ mw: 163.00

SYN: PHENOL, 3,4-DICHLORO-

TOXICITY DATA with REFERENCE
orl-mus LD50:1685 mg/kg   TOLED5 29,39,85

CONSENSUS REPORTS: Reported in EPA TSCA Inventory.

SAFETY PROFILE: Moderately toxic by ingestion. When heated to decomposition it emits toxic vapors of Cl^-.

DFY450 CAS:591-35-5 **HR: 2**
3,5-DICHLOROPHENOL
mf: $C_6H_4Cl_2O$ mw: 163.00

SYN: PHENOL, 3,5-DICHLORO-

TOXICITY DATA with REFERENCE
orl-mus LD50:2389 mg/kg   TOLED5 29,39,85

CONSENSUS REPORTS: Reported in EPA TSCA Inventory.

SAFETY PROFILE: Moderately toxic by ingestion. When heated to decomposition it emits toxic vapors of Cl^-.

DGB480 CAS:2621-62-7 **HR: 1**
N-(2,5-DICHLOROPHENYL)ACETAMIDE
mf: $C_8H_7Cl_2NO$ mw: 204.06

SYNS: ACETAMIDE, N-(2,5-DICHLOROPHENYL)-(9CI) ◇ ACETANILIDE, 2′,5′-DICHLORO- ◇ 2,5-DICHLORACETANILID ◇ 2′,5′-DICHLORO-ACETANILIDE

TOXICITY DATA with REFERENCE
orl-rat LD50:4100 mg/kg   85JCAE -,579,86

CONSENSUS REPORTS: Reported in EPA TSCA Inventory.

SAFETY PROFILE: Mildly toxic by ingestion. When heated to decomposition it emits toxic vapors of NO_x and Cl^-

DGC850 CAS:305-15-7 **HR: 2**
(2,5-DICHLOROPHENYL)HYDRAZINE
mf: $C_6H_6Cl_2N_2$ mw: 177.04

SYN: HYDRAZINE, (2,5-DICHLOROPHENYL)-

TOXICITY DATA with REFERENCE
orl-rat LDLo:500 mg/kg   NCNSA6 5,20,53

CONSENSUS REPORTS: Reported in EPA TSCA Inventory.

SAFETY PROFILE: Moderately toxic by ingestion. When heated to decomposition it emits toxic vapors of NO_x and Cl^-.

DGJ250 CAS:86-98-6 **HR: 3**
4,7-DICHLOROQUINOLINE
mf: $C_9H_5Cl_2N$ mw: 198.05

SYNS: QUINOLINE, 4,7-DICHLORO- ◇ TL 1473

TOXICITY DATA with REFERENCE
mma-sat 500 nmol/plate   MUREAV 42,335,77
scu-mus LDLo:80 mg/kg   NDRC** 30101,8,45

CONSENSUS REPORTS: Reported in EPA TSCA Inventory.

SAFETY PROFILE: Poison by subcutaneous route. Mutation data reported. When heated to decomposition it emits toxic vapors of NO_x and Cl^-.

DGJ950 CAS:2213-63-0 **HR: 3**
2,3-DICHLOROQUINOXALINE
mf: $C_8H_4Cl_2N_2$ mw: 199.04

SYN: QUINOXALINE, 2,3-DICHLORO-

TOXICITY DATA with REFERENCE
ivn-mus LD50:5600 µg/kg   CSLNX* NX#03256

CONSENSUS REPORTS: Reported in EPA TSCA Inventory.

SAFETY PROFILE: Poison by intravenous route. When heated to decomposition it emits toxic vapors of NO_x and Cl^-.

DGM700   CAS:95-73-8   *HR: 2*
2,4-DICHLOROTOLUENE
mf: $C_7H_6Cl_2$ mw: 161.03

SYNS: BENZENE, 2,4-DICHLORO-1-METHYL-(9CI) ◊ 2,4-DICHLORO-1-METHYLBENZENE ◊ TOLUENE, 2,4-DICHLORO-

TOXICITY DATA with REFERENCE
orl-rat LD50:4600 mg/kg   GISAAA 53(2),80,88
orl-mus LD50:2900 mg/kg   GISAAA 53(2),80,88
orl-gpg LD50:5 g/kg   GISAAA 53(2),80,88

CONSENSUS REPORTS: Reported in EPA TSCA Inventory.

SAFETY PROFILE: Moderately toxic by ingestion. When heated to decomposition it emits toxic vapors of Cl^-.

DGV650   CAS:1561-49-5   *HR: 2*
DICYCLOHEXYL PEROXIDE CARBONATE
DOT: UN 2152/UN 2153
mf: $C_{14}H_{22}O_6$ mw: 286.36

SYNS: DICYCLOHEXYL PEROXYDICARBONATE, not >91% with water (UN2153) (DOT) ◊ DICYCLOHEXYL PEROXYDICARBONATE, technically pure (UN2152) (DOT) ◊ PEROXYDICARBONIC ACID, DICYCLOHEXYL ESTER

DOT Classification: Organic Peroxide; Label: Organic Peroxide

CONSENSUS REPORTS: Reported in EPA TSCA Inventory.

SAFETY PROFILE: A peroxide. Handle carefully. When heated to decomposition it emits acrid smoke and irritating vapors.

DGV700   CAS:84-61-7   *HR: 1*
DICYCLOHEXYL PHTHALATE
mf: $C_{20}H_{26}O_4$ mw: 330.46

SYNS: 1,2-BENZENEDICARBOXYLIC ACID, DICYCLOHEXYL ESTER ◊ ERGOPLAST.FDC ◊ HF 191 ◊ KP 201 ◊ PHTHALIC ACID, DICYCLOHEXYL ESTER ◊ UNIMOLL 66

TOXICITY DATA with REFERENCE
orl-rat LD50:30 g/kg   EVHPAZ 3,61,73
orl-rat TDLo:10,500 mg/kg/7D-C   APTOA6 51,217,82

CONSENSUS REPORTS: Reported in EPA TSCA Inventory.

SAFETY PROFILE: Mildly toxic by ingestion. When heated to decomposition it emits acrid smoke and irritating vapors.

DHJ200   CAS:109-89-7   *HR: 3*
DIETHYLAMINE
DOT: UN 1154
mf: $C_4H_{11}N$ mw: 73.16

PROP: Colorless liquid, ammoniacal odor. Mp: −38.9°, bp: 55.5°, flash p: −0.4°F, d: 0.7108 @ 20°/20°, autoign temp: 594°F, vap press: 400 mm @ 38.0°, vap d: 2.53, lel: 1.8%, uel: 10.1%.

SYNS: 2-AMINOPENTANE ◊ DIAETHYLAMIN (GERMAN) ◊ N,N-DIETHYLAMINE ◊ DIETILAMINA (ITALIAN) ◊ DWUETYLOAMINA (POLISH) ◊ N-ETHYL-ETHANAMINE

TOXICITY DATA with REFERENCE
skn-rbt 10 mg/24H MLD   AMIHBC 4,119,51
skn-rbt 500 mg open MLD   UCDS** 5/21/71
eye-rbt 50 µg open SEV   AMIHBC 4,119,51
ihl-mam LC50:5000 mg/m³   TPKVAL 14,80,75
orl-rat LD50:540 mg/kg   AEHLAU 1,343,60
ihl-rat LC50:4000 ppm/4H   AEHLAU 1,343,60
skn-rbt LD50:820 mg/kg   UCDS** 5/21/71

CONSENSUS REPORTS: Reported in EPA TSCA Inventory.

OSHA PEL: (Transitional: TWA 25 ppm) TWA 10 ppm; STEL 25 ppm
ACGIH TLV: TWA 10 ppm; STEL 25 ppm (Proposed: TWA 5 ppm)
DFG MAK: 10 ppm (30 mg/m³)

DOT Classification: Flammable Liquid; Label: Flammable Liquid.

SAFETY PROFILE: Moderately toxic by ingestion, inhalation, and skin contact. A skin and severe eye irritant. Exposure to strong vapor can cause severe cough and chest pains. Contact with liquid can damage eyes, possibly permanently; contact with skin causes necrosis and vesication. A very dangerous fire hazard when exposed to heat, flame, or oxidizers. To fight fire, use alcohol foam, CO_2, dry chemical. Explodes on contact with dicyanofurazan. Violent reaction with sulfuric acid. Ignites on contact with cellulose nitrate of sufficiently high surface area. When heated to decomposition it emits toxic fumes of NO_x. See also AMINES.

DIB300   CAS:105-16-8   *HR: 1*
2-(DIETHYLAMINO)ETHYL METHACRYLATE
mf: $C_{10}H_{19}NO_2$ mw: 185.30

SYNS: DAKTOSE B ◊ 2-DIETHYLAMINOETHYLESTER KYSELINY METHAKRYLOVE ◊ DIETHYLAMINOETHYL METHACRYLATE ◊ β-(DIETHYLAMINO)ETHYL METHACRYLATE ◊ 2-(N,N-DIETHYLAMINO) ETHYL METHACRYLATE ◊ METHACRYLIC ACID, 2-(DIETHYLAMINO) ETHYL ESTER ◊ 2-PROPENOIC ACID, 2-METHYL-, 2-(DIETHYLAMINO) ETHYL ESTER (9CI)

TOXICITY DATA with REFERENCE
orl-rat LD50:4696 mg/kg   85GMAT -,51,82

ihl-rat LC50:11 g/m³/4H 85GMAT -,51,82
ihl-mus LC50:12,100 mg/m³/2H 85GMAT -,51,82

CONSENSUS REPORTS: Reported in EPA TSCA Inventory.

SAFETY PROFILE: Mildly toxic by ingestion and inhalation. When heated to decomposition it emits toxic vapors of NO_x.

DIJ300 CAS:1563-01-5 HR: 3
3-(4-DIETHYLAMINO-2-HYDROXYPHENYLAZO)-4-HYDROXYBENZENESULFONIC ACID
mf: $C_{16}H_{19}N_3O_5S$ mw: 365.44

SYN: BENZENESULFONIC ACID, 3-(4-DIETHYLAMINO-2-HYDROXYPHENYLAZO)-4-HYDROXY-

TOXICITY DATA with REFERENCE
ivn-mus LD50:320 mg/kg CSLNX* NX#03409

CONSENSUS REPORTS: Reported in EPA TSCA Inventory.

SAFETY PROFILE: Poison by intravenous route. When heated to decomposition it emits toxic vapors of NO_x and SO_x.

DIO400 CAS:91-68-9 HR: D
3-(DIETHYLAMINO)PHENOL
mf: $C_{10}H_{15}NO$ mw: 165.26

SYNS: m-(DIETHYLAMINO)PHENOL ◇ PHENOL, m-(DIETHYLAMINO)- ◇ PHENOL, 3-(DIETHYLAMINO)-(9CI)

TOXICITY DATA with REFERENCE
mma-sat 1 µg/plate EMMUEG 11(Suppl 12),1,88

CONSENSUS REPORTS: Reported in EPA TSCA Inventory.

SAFETY PROFILE: Mutation data reported. When heated to decomposition it emits toxic vapors of NO_x.

DIS650 CAS:579-66-8 HR: 2
2,6-DIETHYLANILINE
mf: $C_{10}H_{15}N$ mw: 149.26

SYNS: ANILINE, 2,6-DIETHYL- ◇ BENZENAMINE, 2,6-DIETHYL-(9CI) ◇ 2,6-DIETHYLBENZENAMINE

TOXICITY DATA with REFERENCE
mma-sat 250 ng/plate BECTA6 35,696,85
orl-rat LD50:1800 mg/kg FAATDF 3,285,83
orl-rat TDLo:10080 mg/kg/20D-C FAATDF 3,285,83

CONSENSUS REPORTS: Reported in EPA TSCA Inventory.

SAFETY PROFILE: Moderately toxic by ingestion. Mutation data reported. When heated to decomposition it emits toxic vapors of NO_x.

DIV800 HR: 3
DIETHYLCADMIUM
mf: $C_4H_{10}Cd$ mw: 170.5

PROP: An oil; decomp by moisture. D: 1.6562, mp: −21°, bp: 64°.

CONSENSUS REPORTS: Cadmium and its compounds are on the Community Right-To-Know List.

OSHA PEL: TWA 5 µg(Cd)/m³
ACGIH TLV: TWA 0.05 mg(Cd)/m³ (Proposed: TWA 0.01 mg(Cd)/m³ (dust), Suspected Human Carcinogen; 0.002 mg(Cd)/m³ (respirable dust), Suspected Human Carcinogen); BEI: 10 µg/g creatinine in urine; 10 µg/L in blood
DFG BAT: Blood 1.5 µg/dL; Urine 15 µg/dL, Suspected Carcinogen
NIOSH REL: (Cadmium) Reduce to lowest feasible level.

SAFETY PROFILE: Confirmed human carcinogen. A poison. A dangerous fire and explosion hazard. Explodes when heated rapidly to 130°C. On exposure to air it forms white fumes that turn brown and explode. The vapor explodes when heated to 180°C. When heated to decomposition it emits highly toxic fumes of cadmium. See also CADMIUM COMPOUNDS.

DJQ850 CAS:140-80-7 HR: 3
N,N-DIETHYL-4-METHYLTETRAMETHYLENEDIAMINE
mf: $C_9H_{22}N_2$ mw: 158.33

SYN: TETRAMETHYLENEDIAMINE, N,N-DIETHYL-4-METHYL-

TOXICITY DATA with REFERENCE
ivn-mus LD50:180 mg/kg CSLNX* NX#05228

CONSENSUS REPORTS: Reported in EPA TSCA Inventory.

SAFETY PROFILE: Poison by intravenous route. When heated to decomposition it emits toxic vapors of NO_x.

DJU700 CAS:1006-59-3 HR: 3
2,6-DIETHYLPHENOL
mf: $C_{10}H_{14}O$ mw: 150.24

SYN: PHENOL, 2,6-DIETHYL-

TOXICITY DATA with REFERENCE
ipr-mus LD50:230 mg/kg JMPCAS 2,201,60
ivn-mus LD50:100 mg/kg JMCMAR 23,1350,80

CONSENSUS REPORTS: Reported in EPA TSCA Inventory.

SAFETY PROFILE: Poison by intravenous and intraperitoneal route. When heated to decomposition it emits acrid smoke and irritating vapors.

DJV300   CAS:1665-59-4   **HR: 3**
N,N-DIETHYL-N′-PHENYLETHYLENEDIAMINE
mf: $C_{12}H_{20}N_2$ mw: 192.34

SYNS: ETHYLENEDIAMINE, N,N-DIETHYL-N′-PHENYL- ◊ 1167 F

TOXICITY DATA with REFERENCE
ivn-rbt LDLo:50 mg/kg   AIPAAV 63,400,39

CONSENSUS REPORTS: Reported in EPA TSCA Inventory.

SAFETY PROFILE: Poison by intravenous route. When heated to decomposition it emits toxic vapors of NO_x.

DJX250   CAS:1114-51-8   **HR: 2**
N,N-DIETHYLPROPANAMIDE
mf: $C_7H_{15}NO$ mw: 129.23

SYNS: DIETHYLAMIDE of PROPIONIC ACID ◊ N,N-DIETHYL-PROPIONAMIDE ◊ PROPANAMIDE, N,N-DIETHYL-(9CI) ◊ PROPIONAMIDE, N,N-DIETHYL-

TOXICITY DATA with REFERENCE
ipr-mus LD50:770 mg/kg   DIPHAH 18,245,66

CONSENSUS REPORTS: Reported in EPA TSCA Inventory.

SAFETY PROFILE: Moderately toxic by intraperitoneal route. When heated to decomposition it emits toxic vapors of NO_x.

DKB160   CAS:636-09-9   **HR: 2**
DIETHYL TEREPHTHALATE
mf: $C_{12}H_{14}O_4$ mw: 222.26

SYNS: 1,4-BENZENEDICARBOXYLIC ACID, DIETHYL ESTER ◊ TEREPHTHALIC ACID, DIETHYL ESTER

TOXICITY DATA with REFERENCE
ipr-mus LDLo:1111 mg/kg   JPMSAE 56,1446,67

CONSENSUS REPORTS: Reported in EPA TSCA Inventory.

SAFETY PROFILE: Moderately toxic by intraperitoneal route. When heated to decomposition it emits acrid smoke and irritating vapors.

DKD500   CAS:757-44-8   **HR: 1**
DIETHYL (2-(TRIETHOXYSILYL)ETHYL)PHOSPHONIC ACID
mf: $C_{12}H_{29}O_6PSi$ mw: 328.47

SYNS: 2-(DIETHOXYPHOSPHINYL)ETHYLTRIETHOXYSILANE ◊ PHOSPHONIC ACID, (2-(TRIETHOXYSILYL)ETHYL)-, DIETHYL ESTER

TOXICITY DATA with REFERENCE
orl-rat LD50:17200 mg/kg   TXAPA9 28,313,74

CONSENSUS REPORTS: Reported in EPA TSCA Inventory.

SAFETY PROFILE: Mildly toxic by ingestion. When heated to decomposition it emits toxic vapors of PO_x.

DKD650   CAS:634-95-7   **HR: 2**
1,1-DIETHYLUREA
mf: $C_5H_{12}N_2O$ mw: 116.19

SYNS: asym-DIETHYLUREA ◊ N,N-DIETHYLUREA ◊ UREA, 1,1-DIETHYL- ◊ UREA, N,N-DIETHYL-(9CI)

TOXICITY DATA with REFERENCE
ipr-mus LDLo:2905 mg/kg   JPETAB 54,188,35

CONSENSUS REPORTS: Reported in EPA TSCA Inventory.

SAFETY PROFILE: Moderately toxic by intraperitoneal route. When heated to decomposition it emits toxic vapors of NO_x.

DKF700   CAS:367-25-9   **HR: D**
2,4-DIFLUOROANILINE
mf: $C_6H_5F_2N$ mw: 129.12

SYNS: ANILINE, 2,4-DIFLUORO- ◊ BENZENAMINE, 2,4-DIFLUORO-(9CI) ◊ 2,4-DIFLUOROBENZENAMINE

TOXICITY DATA with REFERENCE
mma-sat 1 mg/plate   EMMUEG 11(Suppl 12),1,88

CONSENSUS REPORTS: Reported in EPA TSCA Inventory.

SAFETY PROFILE: Mutation data reported. When heated to decomposition it emits toxic vapors of NO_x and F^-.

DKH250   CAS:312-30-1   **HR: 3**
4,4′-DIFLUORO-3,3-DINITRODIPHENYL SULFONE
mf: $C_{12}H_6F_2N_2O_6S$ mw: 344.26

SYNS: BENZENE, 1,1′-SULFONYLBIS(4-FLUORO-3-NITRO)- ◊ BIS(4-FLUORO-3-NITROPHENYL)SULFONE ◊ p,p′-DIFLUORO-m,m′-DINITRODIPHENYL SULFONE ◊ 3,3′-DINITRO-4,4′-DIFLUORODIPHENYL SULFONE ◊ NDS ◊ SULFONE, BIS(4-FLUORO-3-NITROPHENYL)

TOXICITY DATA with REFERENCE
ivn-mus LD50:100 mg/kg   CSLNX* NX#03495

CONSENSUS REPORTS: Reported in EPA TSCA Inventory.

SAFETY PROFILE: Poison by intravenous route. When heated to decomposition it emits toxic vapors of NO_x, SO_x, and F^-.

DKH300   CAS:624-72-6   **HR: D**
1,2-DIFLUOROETHANE
mf: $C_2H_4F_2$ mw: 66.06

SYNS: ETHANE, 1,2-DIFLUORO- ◊ FC143 ◊ FLUOROCARBON FC143 ◊ FREON 152

TOXICITY DATA with REFERENCE
mma-sat 50 pph/72H-C   TXAPA9 72,15,84

SAFETY PROFILE: Mutation data reported. When heated to decomposition it emits toxic vapors of F⁻.

DKH900   CAS:446-35-5   **HR: D**
2,4-DIFLUORONITROBENZENE
mf: $C_6H_3F_2NO_2$ mw: 159.10

SYN: BENZENE, 2,4-DIFLUORO-1-NITRO-

TOXICITY DATA with REFERENCE
mmo-sat 80 nL/plate   MUREAV 116,217,83

CONSENSUS REPORTS: Reported in EPA TSCA Inventory.

SAFETY PROFILE: Mutation data reported. When heated to decomposition it emits toxic vapors of NO_x and F⁻.

DLV900   CAS:2346-00-1   **HR: 2**
4,5-DIHYDRO-2-METHYLTHIAZOLE
mf: C_4H_7NS mw: 101.18

SYNS: METHYL-2 Δ-2 THIAZOLINE ◊ THIAZOLE, 4,5-DIHYDRO-2-METHYL- ◊ 2-THIAZOLINE, 2-METHYL-

TOXICITY DATA with REFERENCE
ipr-mus LD50:600 mg/kg   EJMCA5 20,16,85

CONSENSUS REPORTS: Reported in EPA TSCA Inventory.

SAFETY PROFILE: Moderately toxic by intraperitoneal route. When heated to decomposition it emits toxic vapors of NO_x and SO_x.

DMN450   CAS:947-42-2   **HR: 3**
DIHYDROXYDIPHENYLSILANE
mf: $C_{12}H_{12}O_2Si$ mw: 216.33

SYNS: DIFENYL-DIHYDROXYSILAN ◊ SILANE, DIHYDROXYDIPHENYL-

TOXICITY DATA with REFERENCE
orl-mus LD50:2150 mg/kg   85JCAE -,1237,86
ivn-mus LD50:180 mg/kg   CSLNX* NX#04052

CONSENSUS REPORTS: Reported in EPA TSCA Inventory.

SAFETY PROFILE: Poison by intravenous route. Moderately toxic by ingestion. When heated to decomposition it emits acrid smoke and irritating vapors.

DMT500   CAS:150-25-4   **HR: 2**
N,N-DIHYDROXYETHYL GLYCINE
mf: $C_6H_{13}NO_4$ mw: 163.20

SYN: GLYCINE, N,N-DIHYDROXYETHYL-

TOXICITY DATA with REFERENCE
ipr-mus LD50:1540 mg/kg   REPMBN 10,391,62

CONSENSUS REPORTS: Reported in EPA TSCA Inventory.

SAFETY PROFILE: Moderately toxic by intraperitoneal route. When heated to decomposition it emits toxic vapors of NO_x.

DMW250   CAS:131-53-3   **HR: D**
2,2′-DIHYDROXY-4-METHOXYBENZOPHENONE
mf: $C_{14}H_{12}O_4$ mw: 244.26

SYNS: ADVASTAB 47 ◊ BENZOPHENONE-8 ◊ BENZOPHENONE, 2,2′-DIHYDROXY-4-METHOXY- ◊ CYASORB UV 24 ◊ CYASORB UV 24 LIGHT ABSORBER ◊ DIOXYBENZON ◊ DIOXYBENZONE ◊ METHANONE, (2-HYDROXY-4-METHOXYPHENYL)(2-HYDROXYPHENYL)-(9CI) ◊ SPECTRA-SORB UV 24 ◊ UF 2 ◊ UV 24

TOXICITY DATA with REFERENCE
mmo-sat 12500 μg/L   ENMUDM 4,340,82
mma-sat 3 μg/plate   ENMUDM 8(Suppl 7),1,86
mma-mus:lyms 32 μg/plate   JACTDZ 2(5),35,83

CONSENSUS REPORTS: Reported in EPA TSCA Inventory.

SAFETY PROFILE: Mutation data reported. When heated to decomposition it emits acrid smoke and irritating vapors.

DNF300   CAS:618-76-8   **HR: 2**
3,5-DIIODO-4-HYDROXYBENZOIC ACID
mf: $C_7H_4I_2O_3$ mw: 389.91

SYN: BENZOIC ACID, 3,5-DIIODO-4-HYDROXY-

TOXICITY DATA with REFERENCE
orl-mus LD50:4000 mg/kg   JAPMA8 43,495,54
ipr-mus LD50:1000 mg/kg   JMPCAS 2,213,60

CONSENSUS REPORTS: Reported in EPA TSCA Inventory.

SAFETY PROFILE: Moderately toxic by ingestion and intraperitoneal routes. When heated to decomposition it emits toxic vapors of I⁻.

DNN830   CAS:100-18-5   **HR: 2**
1,4-DIISOPROPYLBENZENE
mf: $C_{12}H_{18}$ mw: 162.30

SYNS: BENZENE, 1,4-BIS(1-METHYLETHYL)-(9CI) ◊ BENZENE, p-DIISOPROPYL- ◊ 1,4-BIS(1-METHYLETHYL)BENZENE ◊ p-DIISOPROPYLBENZENE ◊ p-DIISOPROPYLBENZOL

TOXICITY DATA with REFERENCE
orl-mus LD50:3400 mg/kg   85GMAT -,54,82
ipr-mus LD50:1650 mg/kg   GTPZAB 14(2),41,70

CONSENSUS REPORTS: Reported in EPA TSCA Inventory.

SAFETY PROFILE: Moderately toxic by ingestion and intraperitoneal route. When heated to decomposition it emits acrid smoke and irritating vapors.

DNN900   CAS: 95-29-4   **HR: 2**
N,N-DIISOPROPYL-2-BENZOTHIAZOLESULFENAMIDE
mf: $C_{13}H_{18}N_2S_2$ mw: 266.45

SYNS: 2-BENZOTHIAZOLESULFENAMIDE, N,N-DIISOPROPYL- ◊ DIPAC

TOXICITY DATA with REFERENCE
orl-mus LD50:3892 mg/kg   GTPZAB 8(7),39,64

CONSENSUS REPORTS: Reported in EPA TSCA Inventory.

SAFETY PROFILE: Moderately toxic by ingestion. When heated to decomposition it emits toxic vapors of NO_x and SO_x.

DNP700   CAS: 121-05-1   **HR: 3**
N,N-DIISOPROPYL ETHYLENEDIAMINE
mf: $C_8H_{20}N_2$ mw: 144.30

SYNS: ETHYLENEDIAMINE, N,N-DIISOPROPYL- ◊ USAF AM-2

TOXICITY DATA with REFERENCE
ipr-mus LD50:200 mg/kg   NTIS** AD277-689

CONSENSUS REPORTS: Reported in EPA TSCA Inventory.

SAFETY PROFILE: Poison by intraperitoneal route. When heated to decomposition it emits toxic vapors of NO_x.

DNT300   CAS: 522-75-8   **HR: 1**
4,4'-DIISOTHIOINDIGO
mf: $C_{16}H_8O_2S_2$ mw: 296.36

SYNS: ANTINOLO RED B ◊ ($\Delta^{2,2'}$(3H,3'H)-BIBENZO(b)THIOPHENE)-3,3'-DIONE ◊ C.I. 73300 ◊ CIBA PINK B ◊ C.I. VAT RED 41 ◊ DURINDONE PRINTING RED B ◊ DURINDONE RED B ◊ DURINDONE RED BP ◊ HELIANE RED 5B ◊ HELINDON RED BB ◊ ISOTHIOINDIGO ◊ TETRA PINK B ◊ THIOINDIGO ◊ THIOINDIGO RED B ◊ THIOINDIGO RED S ◊ TINA PINK B ◊ TYRIAN RED A-5B ◊ VAT RED 5B

TOXICITY DATA with REFERENCE
ipr-rat LD50:4170 mg/kg   GISAAA 50(8),91,85

CONSENSUS REPORTS: Reported in EPA TSCA Inventory.

SAFETY PROFILE: Mildly toxic by intraperitoneal route. When heated to decomposition it emits toxic vapors of SO_x.

DNV610   CAS: 2418-14-6   **HR: 2**
2,3-DIMERCAPTOSUCCINIC ACID
mf: $C_4H_6O_4S_2$ mw: 182.22

SYNS: BUTANEDIOIC ACID, 2,3-DIMERCAPTO-(9CI) ◊ 2,3-DIMERCAPTOBUTANEDIOIC ACID ◊ DIMERCAPTOSUCCINIC ACID ◊ α-β-DIMERCAPTOSUCCINIC ACID ◊ SUCCINIC ACID, 2,3-DIMERCAPTO- ◊ SUXIMER

TOXICITY DATA with REFERENCE
orl-rat LD50:4 g/kg   YHHPAL 15,335,80
orl-mus LD50:6 g/kg   YHHPAL 15,335,80
ipr-mus LD50:2478 mg/kg   ARTODN 61,321,88

CONSENSUS REPORTS: Reported in EPA TSCA Inventory.

SAFETY PROFILE: Moderately toxic by intraperitoneal route. Slightly toxic by ingestion. When heated to decomposition it emits toxic vapors of SO_x.

DOQ800   CAS: 124-40-3   **HR: 3**
DIMETHYLAMINE
DOT: UN 1032/UN 1160
mf: C_2H_7N mw: 45.10

SYNS: DIMETHYLAMINE, anhydrous (DOT) ◊ DIMETHYLAMINE, aqueous solution (DOT) ◊ DIMETHYLAMINE, solution (DOT) ◊ DMA ◊ N-METHYLMETHANAMINE ◊ RCRA WASTE NUMBER U092

TOXICITY DATA with REFERENCE
eye-rbt 50 mg/5M   BJIMAG 23,153,66
cyt-rat-ihl 50 μg/m^3 GISAAA 36(11),9,71
orl-rat LD50:698 mg/kg   HYSAAV 32,329,67
ihl-rat LC50:4540 ppm/6H   AIHAAP 43,411,82
orl-mus LD50:316 mg/kg   HYSAAV 32,329,67
ihl-mus LC50:7650 ppm/2H   AIHAAP 43,411 82
orl-rbt LD50:240 mg/kg   HYSAAV 32,329,67
ivn-rbt LD50:4 g/kg   MEIEDD 10,470,83
orl-gpg LD50:240 mg/kg   HYSAAV 32,329,67
ihl-mam LC50:3700 mg/m^3 TPKVAL 14,80,75

CONSENSUS REPORTS: EPA Genetic Toxicology Program. Reported in EPA TSCA Inventory.

OSHA PEL: TWA 10 ppm
ACGIH TLV: TWA 5 ppm, STEL 15 ppm
DFG MAK: 10 ppm (18 mg/m^3)

DOT Classification: Flammable Gas; Label: Flammable Gas (UN1032); Flammable Liquid; Label: Flammable Liquid (UN1160)

SAFETY PROFILE: Poison by ingestion. Moderately toxic by inhalation and intravenous routes. Mutation data reported. An eye irritant. Corrosive to the eyes, skin, and mucous membranes. A flammable gas. When heated to decomposition it emits toxic fumes of NO_x. Incompatible with acrylaldehyde, fluorine, and maleic anhydride

DOU650   CAS: 619-84-1   **HR: 3**
p-DIMETHYLAMINO BENZOIC ACID
mf: $C_9H_{11}NO_2$ mw: 165.21

SYN: BENZOIC ACID, p-(DIMETHYLAMINO)-

TOXICITY DATA with REFERENCE
ivn-mus LD50:180 mg/kg   CSLNX* NX#04362

CONSENSUS REPORTS: Reported in EPA TSCA Inventory.

SAFETY PROFILE: Poison by intravenous route. When heated to decomposition it emits toxic vapors of NO_x.

DOZ100   CAS:1421-89-2   **HR: 3**
DIMETHYLAMINOETHANOL ACETATE
mf: $C_6H_{13}NO_2$ mw: 131.20

SYNS: ACETIC ACID, 2-(DIMETHYLAMINO)ETHYL ESTER ◊ 2-DIMETHYLAMINOETHANOL ACETATE ◊ DIMETHYLAMINOETHYL ACETATE ◊ 2-(DIMETHYLAMINO)ETHYL ACETATE

TOXICITY DATA with REFERENCE
ipr-mus LD50:104 mg/kg   IJNEAQ 8,131,69

CONSENSUS REPORTS: Reported in EPA TSCA Inventory.

SAFETY PROFILE: Poison by intraperitoneal route. When heated to decomposition it emits toxic vapors of NO_x.

DPB300   CAS:2439-35-2   **HR: 3**
DIMETHYLAMINOETHYL ACRYLATE
mf: $C_7H_{13}NO_2$ mw: 143.21

SYNS: ACRYLIC ACID, 2-(DIMETHYLAMINO)ETHYL ESTER ◊ ADAME ◊ 2-PROPENOIC ACID, 2-(DIMETHYLAMINO)ETHYL ESTER (9CI)

TOXICITY DATA with REFERENCE
orl-rat LD50:455 mg/kg   EPASR* 8EHQ-1190-1119
ihl-rat LC50:66 mg/m^3/4H   EPASR* 8EHQ-0391-1119

SAFETY PROFILE: Poison by inhalation route. Moderately toxic by ingestion. When heated to decomposition it emits toxic vapors of NO_x.

DQM100   CAS:582-60-5   **HR: 3**
5,6-DIMETHYLBENZIMIDAZOLE
mf: $C_9H_{10}N_2$ mw: 146.21

SYN: BENZIMIDAZOLE, 5,6-DIMETHYL-

TOXICITY DATA with REFERENCE
ipr-mus LD50:400 mg/kg   RPTOAN 41,249,78

CONSENSUS REPORTS: Reported in EPA TSCA Inventory.

SAFETY PROFILE: Poison by intraperitoneal route. When heated to decomposition it emits toxic vapors of NO_x.

DQM850   CAS:619-04-5   **HR: 3**
3,4-DIMETHYLBENZOIC ACID
mf: $C_9H_{10}O_2$ mw: 150.19

SYN: BENZOIC ACID, 3,4-DIMETHYL-

TOXICITY DATA with REFERENCE
ipr-mus LD50:316 mg/kg   JMCMAR 11,1020,68

CONSENSUS REPORTS: Reported in EPA TSCA Inventory.

SAFETY PROFILE: Poison by intraperitoneal route. When heated to decomposition it emits acrid smoke and irritating vapors.

DQR350   CAS:81-26-5   **HR: 3**
2,2'-DIMETHYL-1,1'-BIANTHRAQUINONE
mf: $C_{30}H_{18}O_4$ mw: 442.48

SYNS: 1,1'-BIANTHRACENE-9,9',10,10'-TETRAONE, 2,2'-DIMETHYL- ◊ 2,2'-DIMETHYL-1,1'-BIANTHRACENE-9,9',10,10'-TETRONE

TOXICITY DATA with REFERENCE
ivn-mus LD50:180 mg/kg   CSLNX* NX#04855

CONSENSUS REPORTS: Reported in EPA TSCA Inventory.

SAFETY PROFILE: Poison by intravenous route. When heated to decomposition it emits acrid smoke and irritating vapors.

DQS100   CAS:1134-35-6   **HR: 3**
4,4'-DIMETHYL-2,2'-BIPYRIDINE
mf: $C_{12}H_{12}N_2$ mw: 184.26

SYN: 2,2'-BIPYRIDINE, 4,4'-DIMETHYL-

TOXICITY DATA with REFERENCE
ipr-mus LD50:78700 μg/kg   TOXIA6 23,815,85

CONSENSUS REPORTS: Reported in EPA TSCA Inventory.

SAFETY PROFILE: Poison by intraperitoneal route. When heated to decomposition it emits toxic vapors of NO_x.

DQV300   CAS:760-79-2   **HR: 2**
N,N-DIMETHYLBUTYRAMIDE
mf: $C_6H_{13}NO$ mw: 115.20

SYN: BUTYRAMIDE, N,N-DIMETHYL-

TOXICITY DATA with REFERENCE
ipr-mus LD50:2110 mg/kg   AIHAAP 32,539,71
ivn-mus LD50:1620 mg/kg   AIHAAP 32,539,71
ivn-rbt LD50:790 mg/kg   AIHAAP 32,539,71

CONSENSUS REPORTS: Reported in EPA TSCA Inventory.

SAFETY PROFILE: Moderately toxic by intraperitoneal and intravenous routes. When heated to decomposition it emits toxic vapors of NO_x.

DQW800   CAS:506-82-1   HR: 3
DIMETHYLCADMIUM
mf: C_2H_6Cd mw: 142.47

PROP: Oil, decomp by water, foul odor. D: 1.984; mp: −4.5°; bp: 106°.

CONSENSUS REPORTS: Cadmium and its compounds are on the Community Right-To-Know List.

OSHA PEL: TWA 5 µg(Cd)/m^3
ACGIH TLV: TWA 0.05 mg(Cd)/m^3 (Proposed: TWA 0.01 mg(Cd)/m^3 (dust), Suspected Human Carcinogen; 0.002 mg(Cd)/m^3 (respirable dust), Suspected Human Carcinogen); BEI: 10 µg/g creatinine in urine; 10 µg/L in blood
DFG BAT: Blood 1.5 µg/dL; Urine 15 µg/dL, Suspected Carcinogen
NIOSH REL: (Cadmium) Reduce to lowest feasible level.

SAFETY PROFILE: Confirmed human carcinogen. Contact with air produces the friction-sensitive explosive dimethyl cadmium peroxide. Explodes when heated above 150°C. Ignition may occur on contact with air if the surface area is large. See also CADMIUM COMPOUNDS.

DRI600   CAS:110-70-3   HR: 3
N,N′-DIMETHYLDIAMINOETHANE
mf: $C_4H_{12}N_2$ mw: 88.18

SYNS: 1,2-BIS(METHYLAMINO)ETHANE ◇ 2,5-DIAZAHEXANE ◇ N,N′-DIMETHYLETHANEDIAMINE ◇ N,N′-DIMETHYL-1,2-ETHANEDIAMINE ◇ N,N′-DIMETHYLETHYLENEDIAMINE ◇ sym-DIMETHYLETHYLENEDIAMINE ◇ ETHYLENEDIAMINE, N,N′-DIMETHYL- ◇ 1,2-ETHANEDIAMINE, N,N′-DIMETHYL-(9CI)

TOXICITY DATA with REFERENCE
ipr-mus LD50:200 mg/kg   EJMCA5 17,235,82

CONSENSUS REPORTS: Reported in EPA TSCA Inventory.

SAFETY PROFILE: Poison by intravenous route. When heated to decomposition it emits toxic vapors of NO_x.

DRJ100   CAS:2618-77-1   HR: 2
2,5-DIMETHYL-2,5-DI-(BENZOYLPEROXY)HEXANE
DOT: UN 2172/UN 2173
mf: $C_{22}H_{26}O_6$ mw: 386.48

SYNS: BENZENECARBOPEROXOIC ACID, 1,1,4,4-TETRAMETHYL-1,4-BUTANEDIYL ESTER ◇ 2,5-DIMETHYL-2,5-DI-(BENZOYLPEROXY)HEXANE, not >82% with inert solid (UN2173) (DOT) ◇ 2,5-DIMETHYL-2,5-DI-(BENZOYLPEROXY)HEXANE, technically pure (UN2172) (DOT) ◇ PEROXYBENZOIC ACID, 1,1,4,4-TETRAMETHYLTETRAMETHYLENE ESTER

DOT Classification: Organic Peroxide; Label: Organic Peroxide

CONSENSUS REPORTS: Reported in EPA TSCA Inventory.

SAFETY PROFILE: A peroxide. Handle carefully. When heated to decomposition it emits acrid smoke and irritating vapors.

DRJ825   CAS:1068-27-5   HR: 2
2,5-DIMETHYL-2,5-DI(tert-BUTYLPEROXY)HEXYNE-3
DOT: UN 2158/UN 2159
mf: $C_{16}H_{30}O_4$ mw: 286.46

SYNS: 2,5-DIMETHYL-2,5-DI-(tert-BUTYLPEROXY)HEXYNE-3, not >52% with inert solid (UN2159) (DOT) ◇ 2,5-DIMETHYL-2,5-DI-(tert-BUTYLPEROXY)HEXYNE-3, technically pure (UN2158)(DOT) ◇ 3-HEXYNE, 2,5-DIMETHYL-2,5-DI(t-BUTYLPEROXY)-

TOXICITY DATA with REFERENCE
ipr-mus LD50:1850 mg/kg   BSPII* 1/75-19B

DOT Classification: Organic Peroxide; Label: Organic Peroxide (UN2159); Flammable Liquid; Label: Flammable Liquid, Organic Peroxide

CONSENSUS REPORTS: Reported in EPA TSCA Inventory.

SAFETY PROFILE: Moderately toxic by intraperitoneal route. A peroxide. Handle carefully. When heated to decomposition it emits acrid smoke and irritating vapors.

DRQ650   CAS:51-82-1   HR: 3
N,N-DIMETHYLDITHIOCARBAMIC ACID DIMETHYLAMINOMETHYL ESTER
mf: $C_6H_{14}N_2S_2$ mw: 178.34

SYNS: CARBAMIC ACID, DITHIO-, N,N-DIMETHYL-, DIMETHYLAMINOMETHYL ESTER ◇ N,N-DIMETHYL-DITHIOCARBAMINSAEURE-DIMETHYLAMINOMETHYL-ESTER

TOXICITY DATA with REFERENCE
ipr-rat LD50:230 mg/kg   ARZNAD 16,734,66
ipr-mus LD50:410 mg/kg   ARZNAD 16,734,66

CONSENSUS REPORTS: Reported in EPA TSCA Inventory.

SAFETY PROFILE: Poison by intraperitoneal route. When heated to decomposition it emits toxic vapors of NO_x and SO_x.

DSI850   CAS:91-55-4   HR: D
2,3-DIMETHYLINDOLE
mf: $C_{10}H_{11}N$ mw: 145.22

SYNS: 2,3-DIMETHYL-1H-INDOLE ◇ INDOLE, 2,3-DIMETHYL- ◇ 1H-INDOLE, 2,3-DIMETHYL-(9CI)

TOXICITY DATA with REFERENCE
mmo-sat 2500 nmol/plate   TXCYAC 18,219,80
mma-sat 3 µmol/plate   TXCYAC 23,1,82

CONSENSUS REPORTS: Reported in EPA TSCA Inventory.

DSK950   CAS:300-87-8   HR: 2
3,5-DIMETHYLISOXAZOLE
mf: C_5H_7NO mw: 97.13

SYNS: DMI ◊ 3,5-DWUMETYLOIZOKSAZOLU ◊ ISOXAZOLE, 3,5-DIMETHYL- ◊ U 21221

TOXICITY DATA with REFERENCE
ipr-mus LD50:880 mg/kg   DIPHAH 18,19,66

CONSENSUS REPORTS: Reported in EPA TSCA Inventory.

SAFETY PROFILE: Moderately toxic by intraperitoneal route. When heated to decomposition it emits toxic vapors of NO_x.

DTD800   CAS:105-87-3   HR: 1
trans-3,7-DIMETHYL-2,6-OCTADIEN-1-OL ACETATE
mf: $C_{12}H_{20}O_2$ mw: 196.32

PROP: Colorless, sweet, clear liquid; odor of lavender. D: 0.907–0.918 @ 15°, refr index: 1.458–1.464, bp: 128–129° @ 16 mm, flash p: 219°F. Sol in alc, fixed oils, ether; sltly sol in propylene glycol; insol in water and glycerol.

SYNS: ACETIC ACID GERANIOL ESTER ◊ 3,7-DIMETHYL-2-trans-6-OCTADIENYL ACETATE ◊ trans-3,7-DIMETHYL-2,6-OCTADIEN-1-YL ACETATE ◊ trans-2,6-DIMETHYL-2,6-OCTADIEN-8-YL ETHANOATE ◊ FEMA No. 2509 ◊ GERANIOL ACETATE ◊ GERANYL ACETATE (FCC) ◊ NCI-C54728

TOXICITY DATA with REFERENCE
skn-man 16 mg/48H MLD   CTOIDG 94(8),41,79
skn-rbt 100 mg/24H SEV   CTOIDG 94(8),41,79
skn-gpg 100 mg/24H MOD   CTOIDG 94(8),41,79
pic-esc 25 μg/well   MUREAV 260,349,91
mma-mus:lyms 18 mg/L   MUREAV 196,61,88
sce-ham:ovr 70 mg/L   EMMUEG 10(Suppl 10),1,87
orl-rat LD50:6330 mg/kg   FCTXAV 2,327,64

CONSENSUS REPORTS: NTP Carcinogenesis Studies (gavage): No Evidence: mouse,rat NTPTR* NTP-TR-252,87. Reported in EPA TSCA Inventory.

SAFETY PROFILE: Mildly toxic by ingestion. A skin irritant. Mutation data reported. Combustible liquid. When heated to decomposition it emits acrid smoke and irritating fumes. See also ESTERS.

DTF410   CAS:106-22-9   HR: 2
2,6-DIMETHYL-2-OCTEN-8-OL
mf: $C_{10}H_{20}O$ mw: 156.30

SYNS: CEPHROL ◊ CITRONELLOL ◊ 3,7-DIMETHYL-6-OCTEN-1-OL ◊ 6-OCTEN-1-OL, 3,7-DIMETHYL- ◊ RHODINOL ◊ RODINOL

TOXICITY DATA with REFERENCE
skn-man 16 mg/48H MOD   CTOIDG 94(8),41,79
skn-rbt 100 mg/24H SEV   CTOIDG 94(8),41,79
skn-gpg 100 mg/24H SEV   CTOIDG 94(8),41,79
orl-rat LD50:3450 mg/kg   FCTXAV 13,757,75
scu-mus LD50:880 mg/kg   SIZSAR 3,73,52
ims-mus LD50:4 g/kg   JSICAZ 21,342,62
skn-rbt LD50:2650 mg/kg   FCTXAV 13,757,75

CONSENSUS REPORTS: Reported in EPA TSCA Inventory.

SAFETY PROFILE: Moderately toxic by ingestion, subcutaneous, and skin contact routes. A human skin irritant. When heated to decomposition it emits acrid smoke and irritating vapors.

DTF850   CAS:78-66-0   HR: 2
3,6-DIMETHYL-OCTYN-4-DIOL-(3,6)
mf: $C_{10}H_{18}O_2$ mw: 170.28

SYN: 4-OCTYN-3,6-DIOL, 3,6-DIMETHYL-

TOXICITY DATA with REFERENCE
orl-mus LD50:825 mg/kg   ARZNAD 4,477,54

CONSENSUS REPORTS: Reported in EPA TSCA Inventory.

SAFETY PROFILE: Moderately toxic by ingestion. When heated to decomposition it emits acrid smoke and irritating vapors.

DTU850   CAS:67-51-6   HR: 2
3,5-DIMETHYLPYRAZOLE
mf: $C_5H_8N_2$ mw: 96.15

SYNS: DMP ◊ 3,5-DWUMETYLOPIRAZOLU ◊ PYRAZOLE, 3,5-DIMETHYL- ◊ TH 564 ◊ U 6245

TOXICITY DATA with REFERENCE
ipr-mus LD50:570 mg/kg   DIPHAH 18,19,66

CONSENSUS REPORTS: Reported in EPA TSCA Inventory.

SAFETY PROFILE: Moderately toxic by intraperitoneal route. When heated to decomposition it emits toxic vapors of NO_x.

DTY700   CAS:2379-55-7   HR: 3
2,3-DIMETHYLQUINOXALINE
mf: $C_{10}H_{10}N_2$ mw: 158.22

SYN: QUINOXALINE, 2,3-DIMETHYL-

TOXICITY DATA with REFERENCE
ivn-mus LD50:180 mg/kg   CSLNX* NX#00809

CONSENSUS REPORTS: Reported in EPA TSCA Inventory.

SAFETY PROFILE: Poison by intravenous route. When heated to decomposition it emits toxic vapors of NO_x.

DUG450   CAS:631-67-4   *HR: 2*
N,N-DIMETHYLTHIOACETAMIDE
mf: C_4H_9NS mw: 103.20

SYNS: ACETAMIDE, N,N-DIMETHYLTHIO- ◊ N,N-DIMETHYL-ETHANETHIOAMIDE ◊ DIMETHYLTHIOACETAMID ◊ DIMETHYLTHIOACETAMIDE ◊ ETHANETHIOAMIDE, N,N-DIMETHYL-(9CI)

TOXICITY DATA with REFERENCE
ipr-mus LD50:500 mg/kg   AEPPAE 233,376,58

CONSENSUS REPORTS: Reported in EPA TSCA Inventory.

SAFETY PROFILE: Moderately toxic by intraperitoneal route. When heated to decomposition it emits toxic vapors of NO_x and SO_x.

DUG700   CAS:531-53-3   *HR: 3*
DIMETHYLTHIONINE
mf: $C_{14}H_{14}N_3S \cdot Cl$ mw: 291.82

SYNS: 3-AMINO-7-(DIMETHYLAMINO)PHENOTHIAZIN-5-IUM CHLORIDE ◊ AZURE A ◊ C.I. 52005 ◊ asym-DIMETHYL-3,7-DIAMINOPHENAZATHIONIUM CHLORIDE ◊ PHENOTHIAZIN-5-IUM, 3-AMINO-7-(DIMETHYLAMINO)-, CHLORIDE

TOXICITY DATA with REFERENCE
cyt-ham:ovr 20 µmol/L/5H-C   ENMUDM 1,27,79
ivn-rat LD50:37,720 µg/kg   SMBUA9 9,96,51
ivn-mus LD50:59,040 µg/kg   SMBUA9 9,96,51
ivn-rbt LD50:19,340 µg/kg   SMBUA9 9,96,51

CONSENSUS REPORTS: Reported in EPA TSCA Inventory.

SAFETY PROFILE: Poison by intravenous route. Mutation data reported. When heated to decomposition it emits toxic vapors of NO_x, SO_x, and Cl^-.

DUG825   CAS:753-73-1   *HR: 3*
DIMETHYLTIN DICHLORIDE
mf: $C_2H_6Cl_2Sn$ mw: 219.67

SYNS: DICHLORID DIMETHYLCINICITY ◊ DICHLORODIMETHYLSTANNANE ◊ DICHLORODIMETHYLTIN ◊ DIMETHYLDICHLOROSTANNANE ◊ DIMETHYLDICHLOROTIN ◊ STANNANE, DICHLORODIMETHYL- ◊ TIN, DIMETHYL-, DICHLORIDE

TOXICITY DATA with REFERENCE
orl-rat LDLo:160 mg/kg   BJIMAG 15,15,58
ivn-rat LDLo:40 mg/kg   BJIMAG 15,15,58
ivn-mus LD50:56 mg/kg   CSLNX* NX#02187
orl-rbt LDLo:50 mg/kg   SAIGBL 15,3,73

OSHA PEL: TWA 0.1 mg(Sn)/m^3
ACGIH TLV: TWA 0.1 mg(Sn)/m^3; STEL 0.2 mg/m^3 (skin)
NIOSH REL: (organotin compounds): TWA 0.1 mg(Sn)/m^3

CONSENSUS REPORTS: Reported in EPA TSCA Inventory.

SAFETY PROFILE: Poison by ingestion and intravenous routes. When heated to decomposition it emits toxic vapors of Sn and Cl^-.

DUM150   CAS:598-94-7   *HR: 1*
1,1-DIMETHYLUREA
mf: $C_3H_8N_2O$ mw: 88.13

SYN: UREA, 1,1-DIMETHYL-

TOXICITY DATA with REFERENCE
ipr-mus LDLo:6610 mg/kg   JPETAB 54,188,35

CONSENSUS REPORTS: Reported in EPA TSCA Inventory.

SAFETY PROFILE: Mildly toxic by intraperitoneal route. When heated to decomposition it emits toxic vapors of NO_x.

DUP100   CAS:2379-81-9   *HR: 1*
16H-DINAPHTHO(2,3-a:2',3'-i)CARBAZOLE-5,10,15,17-TETRAONE, 6,9-DIBENZAMIDO-
mf: $C_{42}H_{23}N_3O_6$ mw: 665.68

SYNS: AHCOVAT OLIVE ARN ◊ AHCOVAT OLIVE R ◊ AMANTHRENE OLIVE R ◊ ATIC VAT OLIVE R ◊ BENZADONE OLIVE R ◊ CALCOLOID OLIVE R ◊ CALCOLOID OLIVE RC ◊ CALCOLOID OLIVE RL ◊ CALEDONE OLIVE RP ◊ CALEDON OLIVE R ◊ CARBANTHRENE OLIVE R ◊ CERN KYPOVA 27 ◊ C.I. 69005 ◊ CIBANONE OLIVE F2R ◊ CIBANONE OLIVE 2R ◊ C.I. VAT BLACK 27 ◊ FENANTHREN OLIVE R ◊ INDANTHRENE OLIVE R ◊ INDANTHREN OLIVE R ◊ MAYVAT OLIVE AR ◊ MIKETHRENE OLIVE R ◊ NIHONTHRENE OLIVE R ◊ NYANTHRENE OLIVE R ◊ OLIV OSTANTHRENOVY R ◊ OSTANTHREN OLIVE R ◊ PALANTHRENE OLIVE R ◊ PARADONE OLIVE R ◊ PERNITHRENE OLIVE R ◊ PONSOL OLIVE AR ◊ PONSOL OLIVE ARD ◊ ROMANTRENE OLIVE FR ◊ SANDOTHRENE OLIVE N2R ◊ SOLANTHRENE OLIVE R ◊ TINON OLIVE 2R ◊ TYRIAN OLIVE I-R

TOXICITY DATA with REFERENCE
eye-rbt 500 mg/24H MLD   85JCAE -,1324,86

CONSENSUS REPORTS: Reported in EPA TSCA Inventory.

SAFETY PROFILE: An eye irritant. When heated to decomposition it emits toxic vapors of NO_x.

DUR550   CAS:528-45-0   *HR: D*
3,4-DINITROBENZOIC ACID
mf: $C_7H_4N_2O_6$ mw: 212.13

SYN: BENZOIC ACID, 3,4-DINITRO-

TOXICITY DATA with REFERENCE
mmo-sat 500 µg/plate   SAIGBL 29,34,87
mma-sat 1 mg/plate   SAIGBL 29,34,87

CONSENSUS REPORTS: Reported in EPA TSCA Inventory.

SAFETY PROFILE: Mutation data reported. When heated to decomposition it emits toxic vapors of NO_x.

DUR600   CAS:99-34-3   **HR: D**
3,5-DINITROBENZOIC ACID
mf: $C_7H_4N_2O_6$ mw: 212.13

SYNS: BENZOIC ACID, 3,5-DINITRO- ◇ DNBA

TOXICITY DATA with REFERENCE
mmo-sat 100 µg/plate   GDIKAN 29,278,81
mma-sat 100 µg/plate   GDIKAN 29,278,81

CONSENSUS REPORTS: Reported in EPA TSCA Inventory.

SAFETY PROFILE: Mutation data reported. When heated to decomposition it emits toxic vapors of NO_x.

DUR850   CAS:99-33-2   **HR: D**
3,5-DINITROBENZOYL CHLORIDE
mf: $C_7H_3ClN_2O_5$ mw: 230.57

SYNS: BENZOYL CHLORIDE, 3,5-DINITRO- ◇ 3,5-DINITROBENZOIC ACID CHLORIDE

TOXICITY DATA with REFERENCE
mmo-sat 100 µg/plate   SAIGBL 29,34,87
mma-sat 100 µg/plate   SAIGBL 29,34,87

CONSENSUS REPORTS: Reported in EPA TSCA Inventory.

SAFETY PROFILE: Mutation data reported. When heated to decomposition it emits toxic vapors of NO_x and Cl^-.

DUV100   CAS:961-68-2   **HR: 3**
2,4-DINITRODIPHENYLAMINE
mf: $C_{12}H_9N_3O_4$ mw: 259.24

SYN: DIPHENYLAMINE, 2,4-DINITRO-

TOXICITY DATA with REFERENCE
ivn-mus LD50:180 mg/kg   CSLNX* NX#06394

CONSENSUS REPORTS: Reported in EPA TSCA Inventory.

SAFETY PROFILE: Poison by intravenous route. When heated to decomposition it emits toxic vapors of NO_x.

DUV600   CAS:1582-09-8   **HR: 2**
2,6-DINITRO-N,N-DIPROPYL-4-(TRIFLUOROMETHYL)BENZENAMINE
mf: $C_{13}H_{16}F_3N_3O_4$ mw: 335.32

PROP: Technical product contains 84–88 ppm dipropylnitrosoamine NCITR* NCI-CG-TR-34,78.

SYNS: AGREFLAN ◇ AGRIFLAN 24 ◇ CRISALIN ◇ DIGERMIN ◇ 2,6-DINITRO-N,N-DI-N-PROPYL-α,α,α-TRIFLURO-p-TOLUIDINE ◇ 2,6-DINITRO-4-TRIFLUORMETHYL-N,N-DIPROPYLANILIN (GERMAN) ◇ 4-(DI-N-PROPYLAMINO)-3,5-DINITRO-1-TRIFLUOROMETHYLBENZENE ◇ N,N-DI-N-PROPYL-2,6-DINITRO-4-TRIFLUOROMETHYLANILINE ◇ N,N-DIPROPYL-4-TRIFLUOROMETHYL-2,6-DINITROANILINE ◇ ELANCOLAN ◇ L-36352 ◇ LILLY 36,352 ◇ M.T.F. ◇ NCI-C00442 ◇ NITRAN ◇ OLITREF ◇ SUPER-TREFLAN ◇ SU SEGURO CARPIDOR ◇ SYNFLORAN ◇ TREFANOCIDE ◇ TREFICON ◇ TREFLAM ◇ TREFLAN ◇ TREFLANO-CIDE ELANCOLAN ◇ TRI-4 ◇ TRIFLORAN ◇ TRIFLURALIN (USDA) ◇ α,α,α-TRIFLUORO-2,6-DINITRO-N,N-DIPROPYL-p-TOLUIDINE ◇ 4-(TRIFLUOROMETHYL)-2,6-DINITRO-N,N-DIPROPYLANILINE ◇ TRIFLURALIN ◇ TRIFLURALINA 600 ◇ TRIFLURALINE ◇ TRIFUREX ◇ TRIKEPIN ◇ TRIM ◇ TRISTAR

TOXICITY DATA with REFERENCE
mma-sat 1 mg/plate   ENMUDM 8(Suppl 7),1,86
mrc-asn 100 µg/plate   AISSAW 18,123,82
cyt-hmn:lym 2 ppm   PATHAB 73,707,81
sce-hmn:lym 1 mg/L   BSIBAC 60,2149,84
cyt-mus-ipr 200 mg/kg   EESADV 4,263,80
orl-mus TDLo:10 mg/kg (6–15D preg):TER   TJADAB 15,15A,77
ipr-mus TDLo:200 mg/kg (1D male):TER   EESADV 4,263,80
ipr-mus TDLo:200 mg/kg (1D male):TER   EESADV 4,263,80
ipr-mus TDLo:200 mg/kg (1D male):REP   EESADV 4,263,80
orl-mus TDLo:180 g/kg/78W-C:CAR   NCITR* NCI-CG-TR-34,78
ipr-mus TDLo:2600 µg/kg/39D-I:ETA   PATHAB 73,707,81
scu-mus TDLo:2600 µg/kg/39D-I:ETA   PATHAB 73,707,81
orl-mus TD:340 g/kg/78W-C:CAR   NCITR* NCI-CG-TR-34,78
orl-rat LD50:>10 g/kg   PEMNDP 9,851,91
ihl-rat LC50:2800 mg/m^3/1H   NNGADV 16,557,91
skn-rat LD50:>5 g/kg   WRPCA2 9,119,70
orl-mus LD50:3197 mg/kg   NNGADV 16,557,91
ipr-mus LDLo:1500 mg/kg   BECTA6 20,554,78
orl-dog LD50:>2 g/kg   PEMNDP 9,851,91
orl-rbt LD50:>2 g/kg   PEMNDP 9,851,91

CONSENSUS REPORTS: NCI Carcinogenesis Bioassay (feed); Clear Evidence: mouse NCITR* NCI-CG-TR-34,78; No Evidence: rat NCITR* NCI-CG-TR-34,78. EPA Genetic Toxicology Program. Community Right-To-Know List.

SAFETY PROFILE: Moderately toxic by ingestion and intraperitoneal routes. Experimental teratogenic and reproductive effects. Questionable carcinogen with experimental carcinogenic and tumorigenic data. Human mutation data reported. When heated to decomposition it emits very toxic fumes of F^- and NO_x. See also FLUORIDES and DIPROPYLNITROSO AMINE.

DUV700   CAS:69-78-3   **HR: 2**
2, 2'-DINITRO-5,5'-DITHIODIBENZOIC ACID
mf: $C_{14}H_8N_2O_8S_2$ mw: 396.36

SYNS: BENZOIC ACID, 3,3'-DITHIOBIS(6-NITRO- ◇ 2,2'-DINITRO-5,5'-DITHIODIBENZOESAEURE ◇ 3,3'-DITHIOBIS(6-NITROBENZOIC ACID)

TOXICITY DATA with REFERENCE
ipr-mus LD50:2080 mg/kg   ARZNAD 21,284,71

CONSENSUS REPORTS: Reported in EPA TSCA Inventory.

SAFETY PROFILE: Moderately toxic by intraperitoneal

DUW500   CAS:119-15-3   HR: 2
2,4-DINITRO-p-HYDROXYDIPHENYLAMINE
mf: $C_{12}H_9N_3O_5$ mw: 275.24

SYNS: ACETAMINE YELLOW 2R ◇ ACETOQUINONE LIGHT YELLOW 2RZ ◇ AMACEL YELLOW RR ◇ CELLITON FAST YELLOW RR ◇ C.I. 10345 ◇ C.I. DISPERSE YELLOW 1 ◇ CILLA FAST YELLOW RR ◇ C.I. SOLVENT YELLOW 52 ◇ DISPERSE FAST YELLOW 2K ◇ DISPERSE YELLOW R ◇ DISPERSE YELLOW STABLE 2K ◇ DISPERSOL FAST YELLOW A ◇ DISPERSOL PRINTING YELLOW A ◇ DISPERSOL YELLOW B-A ◇ FAST DISPERSE YELLOW 2K ◇ FENACET FAST YELLOW 2R ◇ KAYALON FAST YELLOW RR ◇ MICROSETILE YELLOW 2R ◇ NYLOQUINONE YELLOW 2R ◇ PERLITON YELLOW RR ◇ PERMANENT YELLOW 2K ◇ PHENOL, p-(2,4-DINITROANILINO)- ◇ RELITON YELLOW R ◇ SERISOL FAST YELLOW A ◇ SETACYL YELLOW P-BS ◇ SRA GOLDEN YELLOW VIII ◇ SUPRACET FAST YELLOW 2R ◇ SUPRACET YELLOW RR ◇ SYNTEN YELLOW P 2R

TOXICITY DATA with REFERENCE
mmo-sat 33 µg/plate   EMMUEG 11(Suppl 12),1,88
mma-sat 33 µg/plate   EMMUEG 11(Suppl 12),1,88
ipr-rat LD50:5230 mg/kg   GISAAA 52(11),94,87
orl-mus LD50:6550 mg/kg   GISAAA 52(11),94,87
ipr-uns LD50:2500 mg/kg   GISAAA 40(10),114,75

CONSENSUS REPORTS: Reported in EPA TSCA Inventory.

SAFETY PROFILE: Moderately toxic by intraperitoneal route. Mildly toxic by ingestion. Mutation data reported. When heated to decomposition it emits toxic vapors of NO_x.

DUW503   CAS:2536-18-7   HR: 3
1,3-DINITRO-2-IMIDAZOLIDONE
mf: $C_3H_4N_4O_5$ mw: 176.11

SYNS: 1,3-DINITRO-2-IMIDAZOLIDINONE ◇ 2-IMIDAZOLIDINONE, 1,3-DINITRO-

TOXICITY DATA with REFERENCE
ivn-mus LD50:18 mg/kg   CSLNX* NX#04408

CONSENSUS REPORTS: Reported in EPA TSCA Inventory.

SAFETY PROFILE: Poison by intravenous route. When heated to decomposition it emits toxic vapors of NO_x.

DUX710   CAS:602-38-0   HR: D
1,8-DINITRONAPHTHALENE
mf: $C_{10}H_6N_2O_4$ mw: 218.18

SYN: NAPHTHALENE, 1,8-DINITRO-

TOXICITY DATA with REFERENCE
mmo-sat 100 µg/plate   MUREAV 91,321,81
mma-sat 100 µg/plate   MUREAV 91,321,81

CONSENSUS REPORTS: Reported in EPA TSCA Inventory.

SAFETY PROFILE: Mutation data reported. When heated to decomposition it emits toxic vapors of NO_x.

DVB850   CAS:89-37-2   HR: 3
2,4-DINITROPHENYL-DIMETHYL-DITHIOCARBAMATE
mf: $C_9H_9N_3O_4S_2$ mw: 287.33

SYNS: CARBAMIC ACID, DIMETHYLDITHIO-, 2,4-DINITROPHENYL ESTER ◇ USAF SN-31

TOXICITY DATA with REFERENCE
ipr-mus LD50:200 mg/kg   NTIS** AD277-689

CONSENSUS REPORTS: Reported in EPA TSCA Inventory.

SAFETY PROFILE: Poison by intraperitoneal route. When heated to decomposition it emits toxic vapors of NO_x and SO_x.

DVC300   CAS:1945-92-2   HR: D
N-2,4-DINITROPHENYLETHANOLAMINE
mf: $C_8H_9N_3O_5$ mw: 227.20

SYNS: 2-(2,4-DINITROANILINO)ETHANOL ◇ ETHANOL, 2-(2,4-DINITROANILINO)- ◇ ETHANOL, 2-((2,4-DINITROPHENYL)AMINO)-

TOXICITY DATA with REFERENCE
cyt-rat-ipr 100 mg/kg   BJPCAL 6,357,51

CONSENSUS REPORTS: Reported in EPA TSCA Inventory.

SAFETY PROFILE: Mutation data reported. When heated to decomposition it emits toxic vapors of NO_x.

DVE260   CAS:105-12-4   HR: 3
p-DINITROSOBENZENE
mf: $C_6H_4N_2O_2$ mw: 136.12

SYNS: BENZENE, p-DINITROSO- ◇ BENZENE, 1,4-DINITROSO-(9CI) ◇ 1,4-DINITROSOBENZENE

TOXICITY DATA with REFERENCE
orl-rat LD50:1020 mg/kg   KCRZAE (11),38,85
ihl-rat LCLo:200 mg/m^3/2H   KCRZAE (11),38,85
orl-mus LD50:1230 mg/kg   KCRZAE (11),38,85
ihl-mus LCLo:200 mg/m^3/2H   KCRZAE (11),38,85

CONSENSUS REPORTS: Reported in EPA TSCA Inventory.

SAFETY PROFILE: Poison by inhalation route. Moderately toxic by ingestion. When heated to decomposition it emits toxic vapors of NO_x.

DVG600   CAS:25321-14-6   HR: 3
DINITROTOLUENE
DOT: UN 1600/UN 2038
mf: $C_7H_6N_2O_4$ mw: 182.15

SYNS: DINITROPHENYLMETHANE ◇ ar,ar-DINITROTOLUENE ◇ DINITROTOLUENE, liquid (DOT) ◇ DINITROTOLUENE, molten (DOT) ◇ DINITROTOLUENE, solid (DOT) ◇ METHYLDINITROBENZENE

TOXICITY DATA with REFERENCE
dns-rat-orl 100 mg/kg   CRNGDP 3,241,82
orl-rat TDLo:196 mg/kg (7–20D preg):REP   CIIT** DOCKET #10992/82
orl-rat TDLo:1050 mg/kg (7–20D preg):REP   CIIT** DOCKET #1099/82
orl-rat TDLo:2100 mg/kg (female 7–20D post):REP   FAATDF 5,948,85
orl-rat TDLo:1050 mg/kg (7–20D preg):TER   CIIT** DOCKET #1099/82
orl-rat TDLo:12,775 mg/kg/Y-C:ETA   PAACA3 24,91,83

CONSENSUS REPORTS: Reported in EPA TSCA Inventory. EPA Genetic Toxicology Program.

OSHA PEL: TWA 1.5 mg/m^3 (skin)
ACGIH TLV: TWA 0.15 mg/m^3 (skin); Suspected Human Carcinogen
DFG MAK: Animal Carcinogen, Suspected Human Carcinogen.
NIOSH REL: (Dinitrotoluene): Reduce to lowest level.

DOT Classification: Poison B; Label: Poison

SAFETY PROFILE: Confirmed carcinogen with experimental tumorigenic and teratogenic data. A poison. Experimental reproductive effects. Mutation data reported. Flammable. When heated to decomposition it emits toxic fumes of NO$_x$. See also 2,4-DINITROTOLUENE.

DVO100   CAS:582-52-5   HR: 1
1:2,5:6-DI-O-ISOPROPYLIDENE-α-D-GLUCOFURANOSE
mf: $C_{12}H_{20}O_6$ mw: 260.32

SYNS: 1:2,5:6-DI-O-ISOPROPYLIDEN-α-D-GLUCOFURANOSE ◇ GLUCOFURANOSE, 1:2,5:6-DI-O-ISOPROPYLIDENE-, α-D-

TOXICITY DATA with REFERENCE
orl-mus LD50:4 g/kg   ARZNAD 29,986,79

CONSENSUS REPORTS: Reported in EPA TSCA Inventory.

SAFETY PROFILE: Mildly toxic by ingestion. When heated to decomposition it emits acrid smoke and irritating vapors.

DVW750   CAS:621-09-0   HR: 3
N,N'-DIPHENYLACETAMIDINE
mf: $C_{14}H_{14}N_2$ mw: 210.30

SYN: ACETAMIDINE, N,N'-DIPHENYL-

TOXICITY DATA with REFERENCE
orl-mus LD50:610 mg/kg   THERAP 21,1327,66
scu-mus LD50:250 mg/kg   APFRAD 40,231,82

CONSENSUS REPORTS: Reported in EPA TSCA Inventory.

SAFETY PROFILE: Poison by subcutaneous route. Moderately toxic by ingestion. When heated to decomposition it emits toxic vapors of NO$_x$.

DVY925   CAS:140-22-7   HR: D
1,5-DIPHENYLCARBAZIDE
mf: $C_{13}H_{14}N_4O$ mw: 242.31

SYNS: CARBOHYDRAZIDE, 1,5-DIPHENYL- ◇ CARBONIC DIHYDRAZIDE, 2,2'-DIPHENYL-(9CI) ◇ DIPHENYLCARBAZIDE ◇ N,N'-DIPHENYLCARBAZIDE ◇ sym-DIPHENYLCARBAZIDE ◇ 2,2'-DIPHENYLCARBAZIDE ◇ 1,5-DIPHENYLCARBOHYDRAZIDE ◇ 2,2'-DIPHENYLCARBONIC DIHYDRAZIDE ◇ DPC

TOXICITY DATA with REFERENCE
dnr-esc 200 μg/well   MUREAV 133,161,84

CONSENSUS REPORTS: Reported in EPA TSCA Inventory.

SAFETY PROFILE: Mutation data reported. When heated to decomposition it emits toxic vapors of NO$_x$.

DVY950   CAS:538-62-5   HR: D
1,5-DIPHENYLCARBAZONE
mf: $C_{13}H_{12}N_4O$ mw: 240.29

SYNS: DIAZENECARBOXYLIC ACID, PHENYL-, 2-PHENYLHYDRAZIDE (9CI) ◇ DIPHENYLCARBAZONE ◇ s-DIPHENYLCARBAZONE ◇ FORMIC ACID, (PHENYLAZO)-, 2-PHENYLHYDRAZIDE ◇ PHENYLDIAZENECARBOXYLIC ACID 2-PHENYLHYDRAZIDE

TOXICITY DATA with REFERENCE
dnr-esc 1200 ng/well   MUREAV 133,161,84

CONSENSUS REPORTS: Reported in EPA TSCA Inventory.

SAFETY PROFILE: Mutation data reported. When heated to decomposition it emits toxic vapors of NO$_x$.

DWB400   CAS:150-61-8   HR: 2
N,N'-DIPHENYLETHYLENEDIAMINE
mf: $C_{14}H_{16}N_2$ mw: 212.32

SYNS: N,N'-DIFENYLETHYLENDIAMIN ◇ ETHYLENEDIAMINE, N,N'-DIPHENYL-

TOXICITY DATA with REFERENCE
orl-rat LDLo:500 mg/kg   JPETAB 90,260,47

CONSENSUS REPORTS: Reported in EPA TSCA Inventory.

SAFETY PROFILE: Moderately toxic by ingestion. When heated to decomposition it emits toxic vapors of NO$_x$.

DWX800   CAS:231-36-7   HR: 3
DIQUAT
mf: $C_{12}H_{12}N_2$•2Br   mw: 344.08

PROP: Yellow crystals. Mp: 355°. Sol in water.

SYNS: AQUACIDE ◇ DEIQUAT ◇ DEXTRONE ◇ 9,10-DIHYDRO-8a,10,-DIAZONIAPHENANTHRENE DIBROMIDE ◇ 9,10-DIHYDRO-8a,10 a-DIAZONIAPHENANTHRENE(1,1'-ETHYLENE-2,2'-BIPYRIDYLIUM)DIBROMIDE ◇ 5,6-DIHYDRO-DIPYRIDO(1,2a;2,1c)PYRAZINIUM DIBROMIDE ◇ 6,7-DIHYDROPYRIDO(1,2a;2',1'-C)PYRAZINEDIUM DIBROMIDE ◇ DIQUAT DIBROMIDE ◇ 1,1'-ETHYLENE-2,2'-BIPYRIDYLIUM DIBROMIDE ◇ ETHYLENE DIPYRIDYLIUM DIBROMIDE ◇ 1,1-ETHYLENE 2,2-DIPYRIDYLIUM DIBROMIDE ◇ 1,1'-ETHYLENE-2,2'-DIPYRIDYLIUM DIBROMIDE ◇ FB/2 ◇ FEGLOX ◇ PREEGLONE ◇ REGLON ◇ REGLONE ◇ WEEDTRINE-D

TOXICITY DATA with REFERENCE

skn-rbt 400 mg/kg/20D MLD BJIMAG 27,51,70
eye-rbt 10 mg MLD BJIMAG 27,51,70
mmo-sat 100 nmol/plate TOLED5 3,169,79
dns-hmn:fbr 1 µmol/L MUREAV 42,161,77
ipr-rat TDLo:7 mg/kg (7D preg):TER 26UZAB 6,257,68
ivn-rat TDLo:15 mg/kg (female 17D post):REP TXAPA9 33,450,75
ipr-rat TDLo:7 mg/kg (7D preg):TER 26UZAB 6,257,68
orl-rat LD50:120 mg/kg PRKHDK 1,31,75
skn-rat LD50:433 mg/kg FAATDF 7,299,86
ipr-rat LDLo:500 mg/kg PAREAQ 14,225,62
scu-rat LD50:20 mg/kg PAREAQ 14,225,62
orl-mus LD50:233 mg/kg BJIMAG 27,51,70
orl-dog LDLo:187 mg/kg BJIMAG 27,51,70
orl-rbt LD50:188 mg/kg BJIMAG 27,51,70

CONSENSUS REPORTS: EPA Genetic Toxicology Program.

OSHA PEL: TWA 0.5 mg/m^3
ACGIH TLV: TWA 0.5 mg/m^3; (Proposed: Total Dust: TWA 0.5 mg/m^3; Respirable Dust: 0.1 mg/m^3 (skin))

SAFETY PROFILE: Poison by ingestion, subcutaneous, intravenous, and intraperitoneal routes. Experimental teratogenic and reproductive effects. A skin and eye irritant. Human mutation data reported. When heated to decomposition it emits very toxic fumes of NO$_x$ and Br$^-$. See also PARAQUAT.

DXB450 CAS:133-66-4 **HR: 2**
DISODIUM-4,4'-BIS((4,6-DIANILINO-1,3,5-TRIAZIN-2-YL)AMINO)STILBENE-2,2'-DISULFONATE
mf: C$_{44}$H$_{36}$N$_{12}$O$_6$S$_2$•2Na   mw: 939.02

SYNS: BELOPHOR OD ◇ 4,4'-BIS((4,6-DIANILINO-s-TRIAZIN-2-YL)AMINO)-2,2'-STILBENEDISULFONIC ACID DISODIUM SALT ◇ BLANKOPHOR HZPA ◇ CALCOFLUOR WHITE MR ◇ CELLU-BRITE ◇ C.I. 40621 ◇ C.I. FLUORESCENT BRIGHTENER 9 ◇ COMPOUND 19-28 ◇ OZP 9 ◇ 2,2'-STILBENEDISULFONIC ACID, 4,4'-BIS((4,6-DIANILINO-s-TRIAZIN-2-YL)AMINO)-, DISODIUM SALT

TOXICITY DATA with REFERENCE

orl-rat LD50:>10 g/kg TXAPA9 5,176,63
ipr-rat LD50:1090 mg/kg GISAAA 49(1),85,84
orl-rbt LD50:>10 g/kg TXAPA9 5,176,63
orl-gpg LD50:>7 g/kg TXAPA9 5,176,63

CONSENSUS REPORTS: Reported in EPA TSCA Inventory.

SAFETY PROFILE: Moderately toxic by intraperitoneal route. Slightly toxic by ingestion. When heated to decomposition it emits toxic vapors of NO$_x$ and SO$_x$.

DXG035 CAS:1330-43-4 **HR: 1**
DISODIUM TETRABORATE
mf: B$_4$Na$_2$O$_7$ mw: 201.22

SYNS: ANHYDROUS BORAX ◇ BORATES, TETRA, SODIUM SALT, anhydrous (OSHA) ◇ BORAX GLASS ◇ BORIC ACID, DISODIUM SALT ◇ FR 28 ◇ FUSED BORAX ◇ RASORITE 65 ◇ SODIUM BIBORATE ◇ SODIUM TETRABORATE ◇ SODIUM TETRABORATE (Na$_2$B$_4$O$_7$)

CONSENSUS REPORTS: Reported in EPA TSCA Inventory.

OSHA PEL: TWA 8H TWA 10 mg/m^3

SAFETY PROFILE: A nuisance dust. When heated to decomposition it emits toxic vapors of B.

DXG625 CAS:107-64-2 **HR: 1**
DISTEARYL DIMETHYLAMMONIUM CHLORIDE
mf: C$_{38}$H$_{80}$N•Cl   mw: 586.64

SYNS: ALIQUAT 207 ◇ AMMONIUM, DIMETHYLDIOCTADECYL-, CHLORIDE ◇ AROSURF TA 100 ◇ ARQUAD R 40 ◇ DIMETHYLDIOCTADECYLAMMONIUM CHLORIDE ◇ N,N-DIMETHYL-N-OCTADECYL-1-OCTADECANAMINIUM CHLORIDE ◇ GENAMIN DSAC ◇ KD 83 ◇ 1-OCTADECANAMINIUM, N,N-DIMETHYL-N-OCTADECYL-, CHLORIDE (9CI) ◇ Q-D 86P ◇ QUATERNIUM 5 ◇ TALOFLOC ◇ VARISOFT 100

TOXICITY DATA with REFERENCE

orl-rat LD50:11,300 mg/kg ESKHA5 (101),152,83
scu-rat LD50:11,300 mg/kg SCIEAS 36(1-4),10,89

CONSENSUS REPORTS: Reported in EPA TSCA Inventory.

SAFETY PROFILE: Mildly toxic by ingestion and subcutaneous routes. When heated to decomposition it emits toxic vapors of NO$_x$ and Cl$^-$.

DXI550 CAS:505-29-3 **HR: 2**
p-DITHIANE
mf: C$_4$H$_8$S$_2$ mw: 120.24

SYNS: 1,4-DITHIACYCLOHEXANE ◇ 1,4-DITHIANE

TOXICITY DATA with REFERENCE

orl-rat LD50:2768 mg/kg NTIS** AD-A172-647

CONSENSUS REPORTS: Reported in EPA TSCA Inventory.

SAFETY PROFILE: Moderately toxic by ingestion. When heated to decomposition it emits toxic vapors of SO$_x$.

DXO850   CAS:992-59-6   HR: D
DITOLYLBIS(AZONAPHTHIONIC ACID)
mf: $C_{34}H_{26}N_6O_6S_2 \cdot 2Na$ mw: 724.76

SYNS: AMANIL PURPURINE 4B ◊ ATUL DIRECT RED 4B ◊ AZAMIN 4B ◊ AZOCARD RED 4B ◊ BENCIDAL PURPLE 4B ◊ BENZANIL PURPURINE 4B ◊ BENZOPURPURIN 4B ◊ BENZOPURPURINE 4B ◊ BENZOPURPURINE 4BKX ◊ BENZOPURPURINE 4BX ◊ BRASILAMINA RED 4B ◊ CALCOMINE RED 4BX ◊ CERVEN PRIMA 2 ◊ CHROME LEATHER RED 4B ◊ C.I. 23500 ◊ C.I. DIRECT RED 2 ◊ COTTON RED 4B ◊ DIACOTTON BENZOPURPURINE 4B ◊ DIAMINE PURPURINE 4B ◊ DIAPHTAMINE PURPURINE ◊ DIAZAMINE PURPURINE 4B ◊ DIAZINE RED 4B ◊ DIAZOL PURPURINE 4B ◊ DIPHENYL RED 4B ◊ DIPHENYL RED 4BS ◊ DIRECT PURPURINE 4B ◊ DIRECT PURPURINE M4B ◊ DIRECT RED 2 ◊ DIRECT RED 4A ◊ DIRECT RED 4B ◊ DIRECT RED DCB ◊ ECLIPSE RED ◊ ERIE BENZO 4BP ◊ ERIE RED 4B ◊ FAST SCARLET ◊ HISPAMIN RED 4B ◊ KAYAKU BENZOPURPURINE 4B ◊ MITSUI BENZOPURPURINE 4BX ◊ 1-NAPHTHALENESULFONIC ACID, 3,3'-((3,3'-DIMETHYL(1,1'-BIPHENYL)-4,4'-DIYL)BIS(AZO))BIS(4-AMINO-), DISODIUM SALT ◊ PAPER RED 4B ◊ PHENAMINE PURPURINE 4B ◊ PURPURIN 4B ◊ PURPURINE 4B ◊ TERTRODIRECT RED 4B

TOXICITY DATA with REFERENCE
mma-sat 30 µg/plate   JTEHD6 18,111,86

CONSENSUS REPORTS: Reported in EPA TSCA Inventory.

SAFETY PROFILE: Mutation data reported. When heated to decomposition it emits toxic vapors of NO_x and SO_x.

DXQ740   CAS:1321-74-0   HR: 2
DIVINYLBENZENE
mf: $C_{10}H_{10}$ mw: 130.20

SYNS: BENZENE, DIVINYL- ◊ VINYLSTYRENE

TOXICITY DATA with REFERENCE
orl-rat LDLo:4644 mg/kg   AMIHAB 19,403,59
NIOSH REL: (divinyl benzene) TWA 10 ppm

CONSENSUS REPORTS: Reported in EPA TSCA Inventory.

SAFETY PROFILE: Mildly toxic by ingestion. An eye irritant. Combustible. When heated to decomposition it emits acrid smoke and irritating fumes.

DXS800   CAS:141-63-9   HR: 1
DODECAMETHYLPENTASILOXANE
mf: $C_{12}H_{36}O_4Si_5$ mw: 384.93

SYN: PENTASILOXANE, DODECAMETHYL-

TOXICITY DATA with REFERENCE
orl-gpg LDLo:50 g/kg   JIDHAN 30,332,48

CONSENSUS REPORTS: Reported in EPA TSCA Inventory.

SAFETY PROFILE: Slightly toxic by ingestion. When heated to decomposition it emits acrid smoke and irritating vapors.

DXW050   CAS:2016-56-0   HR: 2
1-DODECYLAMINE ACETATE
mf: $C_{12}H_{27}N \cdot C_2H_4O_2$ mw: 245.46

SYNS: DODECANAMINE ACETATE ◊ DODECYLAMINE, ACETATE

TOXICITY DATA with REFERENCE
orl-mus LD50:1750 mg/kg   CHTPBA 1,11,65

CONSENSUS REPORTS: Reported in EPA TSCA Inventory.

SAFETY PROFILE: Moderately toxic by ingestion. When heated to decomposition it emits toxic vapors of NO_x.

DKW100   CAS:929-73-7   HR: 2
DODECYLAMINE, HYDROCHLORIDE
mf: $C_{12}H_{27}N \cdot ClH$ mw: 221.86

SYNS: DODECANAMINE HYDROCHLORIDE ◊ 1-DODECANAMINE, HYDROCHLORIDE (9CI) ◊ n-DODECYLAMINE HYDROCHLORIDE ◊ DODECYLAMMONIUM CHLORIDE ◊ n-DODECYLAMMONIUM CHLORIDE ◊ LAURYLAMINE HYDROCHLORIDE ◊ LAURYLAMMONIUM HYDROCHLORIDE

TOXICITY DATA with REFERENCE
orl-mus LD50:755 mg/kg   PCJOAU 15,383,81

CONSENSUS REPORTS: Reported in EPA TSCA Inventory.

SAFETY PROFILE: Moderately toxic by ingestion. When heated to decomposition it emits toxic vapors of HCl and Cl^-.

DXY725   CAS:104-74-5   HR: 3
1-DODECYLPYRIDINIUM CHLORIDE
mf: $C_{17}H_{30}N \cdot Cl$ mw: 283.93

SYNS: C 2 ◊ DEHYQUART C ◊ DODECYLPYRIDINIUM CHLORIDE ◊ N-DODECYLPYRIDINIUM CHLORIDE ◊ DPC ◊ ELTREN ◊ LAURYLPYRIDINIUM CHLORIDE ◊ 1-LAURYLPYRIDINIUM CHLORIDE ◊ LPC ◊ QUATERNARIO LPC ◊ PYRIDINIUM, 1-DODECYL-, CHLORIDE

TOXICITY DATA with REFERENCE
unr-mus LD50:119 mg/kg   PHARAT 40,273,85

CONSENSUS REPORTS: Reported in EPA TSCA Inventory.

SAFETY PROFILE: Poison by an unspecified route. When heated to decomposition it emits toxic vapors of NO_x and Cl^-.

DYA810   CAS:1119-94-4   HR: 3
DODECYLTRIMETHYLAMMONIUM BROMIDE
mf: $C_{15}H_{34}N \cdot Br$ mw: 308.41

SYN: AMMONIUM, DODECYLTRIMETHYL-, BROMIDE

TOXICITY DATA with REFERENCE
ivn-rat LD50:6800 µg/kg   APTOA6 47,17,80
ivn-mus LD50:5200 µg/kg   APTOA6 47,17,80

CONSENSUS REPORTS: Reported in EPA TSCA Inventory.

SAFETY PROFILE: Poison by intravenous route. When heated to decomposition it emits toxic vapors of NO_x and Br^-.

DYC900　　　　CAS:544-85-4　　　　**HR: 3**
DOTRIACONTANE
mf: $C_{32}H_{66}$　　mw: 450.98

TOXICITY DATA with REFERENCE
ivn-mus LD50:100 mg/kg　　CSLNX* NX#00741

CONSENSUS REPORTS: Reported in EPA TSCA Inventory.

SAFETY PROFILE: Poison by intravenous route. When heated to decomposition it emits acrid smoke and irritating vapors.

E

EAW100   CAS:90-81-3   **HR: 3**
dl-EPHEDRINE
mf: $C_{10}H_{15}NO$ mw: 165.26

SYNS: BENZENEMETHANOL, α-(1-(METHYLAMINO)ETHYL)-, (R*,S*)-(+ –)-(9CI) ◇ (+ –)-EPHEDRINE ◇ EPHEDRINE, (+ –)- ◇ RACEPHEDRINE

TOXICITY DATA with REFERENCE
ipr-rat LDLo:170 mg/kg   AEPPAE 195,647,40

CONSENSUS REPORTS: Reported in EPA TSCA Inventory.

SAFETY PROFILE: Poison by intraperitoneal route. When heated to decomposition it emits toxic vapors of NO_x.

EAW200   CAS:321-98-2   **HR: 3**
(+)-EPHEDRINE
mf: $C_{10}H_{15}NO$ mw: 165.26

SYNS: BENZENEMETHANOL, α-(1-(METHYLAMINO)ETHYL)-, (S-(R*,S*))-(9CI) ◇ d-EPHEDRINE ◇ EPHEDRINE, (+)- ◇ l-(+)-EPHEDRINE

TOXICITY DATA with REFERENCE
ipr-mus LD50:255 mg/kg   JPMSAE 53,987,64
ivn-mus LD50:105 mg/kg   JPETAB 148,158,65

CONSENSUS REPORTS: Reported in EPA TSCA Inventory.

SAFETY PROFILE: Poison by intravenous and intraperitoneal routes. When heated to decomposition it emits toxic vapors of NO_x.

EBO050   CAS:2386-87-0   **HR: 3**
3,4-EPOXYCYCLOHEXYLMETHYL 3,4-EPOXYCYCLOHEXANE CARBOXYLATE
mf: $C_{14}H_{20}O_4$ mw: 252.34

SYNS: ERL-4221 ◇ 7-OXABICYCLO(4.1.0)HEPTANE-3-CARBOXYLIC ACID, 7-OXABICYCLO(4.1.0)HEPT-3-YLMETHYL ESTER

TOXICITY DATA with REFERENCE
orl-rat LD50:4490 mg/kg   AIHAAP 24,305,63
skn-rbt LD50:20 mg/kg   38MKAJ 2A,2242,81

CONSENSUS REPORTS: Reported in EPA TSCA Inventory.

SAFETY PROFILE: Poison by skin contact. Mildly toxic by ingestion. When heated to decomposition it emits acrid smoke and irritating vapors.

EBO100   CAS:285-67-6   **HR: D**
1,2-EPOXYCYCLOPENTANE
mf: C_5H_8O mw: 84.13

SYNS: CYCLOPENTANE, 1,2-EPOXY- ◇ CYCLOPENTANE OXIDE ◇ CYCLOPENTANE EPOXIDE ◇ CYCLOPENTANEOXIDE ◇ 6-OXABICYCLO(3.1.0)HEXANE

TOXICITY DATA with REFERENCE
mmo-sat 15 μmol/plate   MUREAV 90,67,81
mmo-klp 2 mmol/L   MUREAV 89,269,81
sce-ham:lng 40 mmol/L   MUREAV 249,55,91

CONSENSUS REPORTS: Reported in EPA TSCA Inventory.

SAFETY PROFILE: Mutation data reported. When heated to decomposition it emits acrid smoke and irritating vapors.

ECE525   CAS:1003-14-1   **HR: D**
1,2-EPOXYPENTANE
mf: $C_5H_{10}O$ mw: 86.15

SYNS: PENTANE, 1,2-EPOXY- ◇ PROPYLOXIRANE

TOXICITY DATA with REFERENCE
mmo-klp 1 mmol/L   MUREAV 89,269,81

CONSENSUS REPORTS: Reported in EPA TSCA Inventory.

SAFETY PROFILE: Mutation data reported. When heated to decomposition it emits acrid smoke and irritating vapors.

EEA600   CAS:2224-15-9   **HR: 2**
1,2-ETHANEDIOL DIGLYCIDYL ETHER
mf: $C_8H_{14}O_4$ mw: 174.22

SYNS: 1,2-BIS(GLYCIDYLOXY)ETHANE ◇ DIGLYCIDYLETHYLENE GLYCOL ◇ 1,2-DIGLYCIDYLOXYETHANE ◇ ETHANE, 1,2-BIS(2,3-EPOXYPROPOXY)- ◇ 2,2′-(1,2-ETHANEDIYLBIS(OXYMETHYLENE)) BISOXIRANE ◇ ETHYLENE DIGLYCIDYL ETHER ◇ ETHYLENE GLYCOL DIGLYCIDYL ETHER ◇ ETHYLENGLYKOLDIGLYCIDYLETHER ◇ GLYCOL DIGLYCIDYL ETHER ◇ OXIRANE, 2,2′-(1,2-ETHANEDIYLBIS(OXYMETHYLENE))BIS-(9CI)

TOXICITY DATA with REFERENCE
mmo-sat 300 nmol/plate   MUREAV 231,205,90
oth-esc 1 mmol/L   MUREAV 231,205,90
sce-ham:lng 6250 nmol/L   MUREAV 249,55,91
orl-mus LD50:460 mg/kg   85JCAE -,775,86

CONSENSUS REPORTS: Reported in EPA TSCA Inventory.

SAFETY PROFILE: Moderately toxic by ingestion. Mutation data reported. When heated to decomposition it emits acrid smoke and irritating vapors.

EEB100　　CAS:2001-94-7　　HR: 1
N,N'-1,2-ETHANEDIYLBIS(N-(CARBOXYMETHYL)GLYCINE) DIPOTASSIUM SALT
mf: $C_{10}H_{14}N_2O_8 \cdot 2K$　　mw: 368.46

SYNS: ACETIC ACID, (ETHYLENEDINITRILO)TETRA-, DIPOTASSIUM SALT ◇ DIPOTASSIUM ETHYLENEDIAMINETETRAACETATE ◇ (ETHYLENEDINITRILO)TETRAACETATE DIPOTASSIUM SALT

TOXICITY DATA with REFERENCE
skn-rbt 500 mg MLD　　FCTOD7 20,563,82
eye-rbt 100 mg　　FCTOD7 20,573,82
eye-rbt 100 mg/30S RNS MLD　　FCTOD7 20,573,82

CONSENSUS REPORTS: Reported in EPA TSCA Inventory.

SAFETY PROFILE: A skin and eye irritant. When heated to decomposition it emits toxic vapors of NO_x.

EEC700　　CAS:2002-24-6　　HR: 1
ETHANOLAMINE HYDROCHLORIDE
mf: $C_2H_7NO \cdot ClH$　　mw: 97.56

SYNS: β-AMINOETHANOL HYDROCHLORIDE ◇ 2-AMINOETHANOL HYDROCHLORIDE ◇ COLAMINE HYDROCHLORIDE ◇ ETHANOLAMINE CHLORIDE ◇ ETHANOL, 2-AMINO-, HYDROCHLORIDE ◇ MEA HYDROCHLORIDE ◇ MONOETHANOLAMINE HYDROCHLORIDE

TOXICITY DATA with REFERENCE
scu-mus LD50:4053 mg/kg　　ARZNAD 4,649,54

CONSENSUS REPORTS: Reported in EPA TSCA Inventory.

SAFETY PROFILE: Mildly toxic by subcutaneous route. When heated to decomposition it emits toxic vapors of NO_x, HCl, and Cl^-.

EFU050　　CAS:563-43-9　　HR: 1
ETHYL ALUMINUM DICHLORIDE
mf: $C_2H_5AlCl_2$　　mw: 126.95

SYNS: ALUMINUM, DICHLOROETHYL- ◇ DICHLOROETHYLALUMINUM ◇ DICHLOROMONOETHYLALUMINUM ◇ ETHYLDICHLOROALUMINUM

ACGIH TLV: TWA 2 mg(Al)/m³

CONSENSUS REPORTS: Reported in EPA TSCA Inventory.

SAFETY PROFILE: Mildly toxic by inhalation. When heated to decomposition it emits toxic vapors of Cl^-.

EFU400　　CAS:75-04-7　　HR: 3
ETHYLAMINE
DOT: UN 1036/UN 2270
mf: C_2H_7N　　mw: 45.10

PROP: Colorless gas or liquid, strong ammoniacal odor. Bp: 16.6°, flammable, lel: 4.95%, uel: 20.75%, fp: −80.6°, flash p: −0.4°F, d: 0.662 @ 20°/4°, autoign temp: 725°F, vap d: 1.56. vap press: 400 mm @ 20°. Misc with water, alc, and ether.

SYNS: AETHYLAMINE (GERMAN) ◇ AMINOETHANE ◇ 1-AMINOETHANE ◇ ETHANAMINE ◇ ETILAMINA (ITALIAN) ◇ ETYLOAMINA (POLISH) ◇ MONOETHYLAMINE (DOT) ◇ MONOETHYLAMINE, anhydrous (DOT)

TOXICITY DATA with REFERENCE
skn-rbt 500 mg/24H MLD　　85JCAE -,429,86
eye-rbt 50 ppm/10D-I SEV　　AMIHBC 3,287,51
eye-rbt 250 µg/24H SEV　　85JCAE-,429,86
orl-rat LD50:400 mg/kg　　AMIHBC 10,61,54
ihl-rat LCLo:3000 ppm/4H　　AEHLAU 1,343,60
skn-rbt LD50:390 mg/kg　　AEHLAU 1,343,60
ivn-rbt LDLo:350 mg/kg　　HBAMAK 4,1295,35
ihl-uns LC50:2300 mg/m³　　TPKVAL 14,80,75

CONSENSUS REPORTS: Reported in EPA TSCA Inventory.

OSHA PEL: TWA 10 ppm
ACGIH TLV: TWA 10 ppm (Proposed: TWA 5 ppm; 15 ppm STEL)
DFG MAK: 10 ppm (18 mg/m³)

DOT Classification: Flammable Liquid; Label: Flammable Liquid (UN1036, UN2270); Flammable Gas; Label: Flammable Gas (UN1036); Flammable or Combustible Liquid; Label: Flammable Liquid (UN2270).

SAFETY PROFILE: A poison by ingestion, skin contact, and intravenous routes. Moderately toxic by inhalation. A severe eye irritant. A very dangerous fire hazard when exposed to heat or flame. Moderate explosion hazard when exposed to spark or flame. Keep away from heat and open flame, can react vigorously with oxidizing materials. To fight fire, stop flow of gas, use alcohol foam, dry chemical. Incompatible with cellulose nitrate or oxidizers. When heated to decomposition it emits toxic fumes of NO_x. See also AMINES.

EGF100　　CAS:120-37-6　　HR: D
3-ETHYLAMINO-4-METHYLPHENOL
mf: $C_9H_{13}NO$　　mw: 151.23

SYN: PHENOL, 3-(ETHYLAMINO)-4-METHYL-

TOXICITY DATA with REFERENCE
mma-sat 100 µg/plate　　EMMUEG 11(Suppl 12),1,88

CONSENSUS REPORTS: Reported in EPA TSCA Inventory.

SAFETY PROFILE: Mutation data reported. When heated to decomposition it emits toxic vapors of NO_x.

EGR600   CAS:94-02-0   **HR: 1**
ETHYL BENZOYL ACETATE
mf: $C_{11}H_{12}O_3$ mw: 192.23

SYNS: ACETIC ACID, BENZOYL-, ETHYL ESTER ◊ BENZENEPROPANOIC ACID, β-OXO-, ETHYL ESTER (9CI) ◊ BENZOYLACETIC ACID ETHYL ESTER ◊ ETHYL β-OXOBENZENEPROPANOATE

TOXICITY DATA with REFERENCE
orl-mus LD50:6800 mg/kg   85GMAT -,65,82

CONSENSUS REPORTS: Reported in EPA TSCA Inventory.

SAFETY PROFILE: Mildly toxic by ingestion. When heated to decomposition it emits acrid smoke and irritating vapors.

EGV400   CAS:74-96-4   **HR: 3**
ETHYL BROMIDE
DOT: UN 1891
mf: C_2H_5Br mw: 108.98

PROP: Colorless, volatile liquid. Mp: −119°, bp: 38.4°, lel: 6.7%, uel: 11.3%, flash p: <−4°F, d: 1.451 @ 20°/4°, autoign temp: 952°F, vap press: 400 mm @ 21°, vap d 3.76.

SYNS: BROMOETHANE ◊ BROMURE d'ETHYLE ◊ ETYLU BROMEK (POLISH) ◊ HALON 2001 ◊ MONOBROMOETHANE ◊ NCI-C55481

TOXICITY DATA with REFERENCE
ipr-rat LD50:1750 mg/kg   JPCEAO 320,133,78
ihl-mus LC50:16230 ppm/1H   AMRL** TR-72-62/72
ipr-mus LD50:2850 mg/kg   JPCEAO 320,133,78
orl-rat LD50:1350 mg/kg   85GMAT-,65,82
ihl-rat LDLo:148000 ppm/15M   AMIHBC 6,435,52

CONSENSUS REPORTS: NTP Carcinogenesis Studies (inhalation); Clear Evidence: Mouse NTPTR* NTP-TR-363,89; (Inhalation); Some Evidence: Rat NTPTR* NTP-TR-363,89. EPA Genetic Toxicology Program. Reported in EPA TSCA Inventory.

OSHA PEL: (Transitional: TWA 200 ppm) TWA 200 ppm; STEL 250 ppm
ACGIH TLV: TWA 5 ppm (skin); Suspected Human Carcinogen
DFG MAK: 200 ppm (890 mg/m^3)
DOT Classification: Poison B; Label: Poison

SAFETY PROFILE: A poison. Suspected human carcinogen. Moderately toxic by ingestion and intraperitoneal routes. Mildly toxic by inhalation. An eye and skin irritant. Physiologically, it is an anesthetic and narcotic. Its vapors are markedly irritating to the lungs on inhalation for even short periods. It can produce acute congestion and edema. Liver and kidney damage in humans has been reported. It is much less toxic than methyl bromide, but more toxic than ethyl chloride. It is a preparative hazard. Dangerously flammable by heat, open flame (sparks), oxidizers. Moderately explosive when exposed to flame. Reacts with water or steam to produce toxic and corrosive fumes. Vigorous reaction with oxidizing materials. To fight fire, use CO_2, dry chemical. Readily decomposes when heated to emit toxic fumes of Br$^-$. See also BROMIDES.

EHG050   CAS:120-43-4   **HR: 3**
ETHYLCARBONYL PIPERAZINE
mf: $C_7H_{14}N_2O_2$ mw: 158.23

SYNS: 1-PIPERAZINECARBOXYLIC ACID, ETHYL ESTER ◊ PIPERAZINE ETHYLCARBOXYLATE

TOXICITY DATA with REFERENCE
ipr-mus LD50:138 mg/kg   OYYAA2 33,825,87

CONSENSUS REPORTS: Reported in EPA TSCA Inventory.

SAFETY PROFILE: Poison by intraperitoneal route. When heated to decomposition it emits toxic vapors of NO_x.

EHH100   CAS:638-07-3   **HR: 3**
ETHYL 4-CHLOROACETOACETATE
mf: $C_6H_9ClO_3$ mw: 164.60

SYNS: ACETOACETIC ACID, 4-CHLORO-, ETHYL ESTER ◊ BUTANOIC ACID, 4-CHLORO-3-OXO-, ETHYL ESTER (9CI) ◊ ETHYL γ-CHLOROACETOACETATE ◊ ETHYL 4-CHLORO-3-OXOBUTANOATE

TOXICITY DATA with REFERENCE
ipr-rat LD50:108 mg/kg   OYYAA2 33,695,87
ipr-mus LD50:88 mg/kg   OYYAA2 33,695,87

CONSENSUS REPORTS: Reported in EPA TSCA Inventory.

SAFETY PROFILE: Poison by intraperitoneal route. When heated to decomposition it emits toxic vapors of Cl$^-$.

EHH200   CAS:687-46-7   **HR: 3**
ETHYL 2-CHLOROACRYLATE
mf: $C_5H_7ClO_2$ mw: 134.57

SYNS: ACRYLIC ACID, 2-CHLORO-, ETHYL ESTER ◊ 2-CHLOROACRYLIC ACID ETHYL ESTER ◊ ETHYL α-CHLOROACRYLATE ◊ 2-PROPENOIC ACID, 2-CHLORO-, ETHYL ESTER

TOXICITY DATA with REFERENCE
ivn-mus LD50:180 mg/kg   CSLNX* NX#02176

CONSENSUS REPORTS: Reported in EPA TSCA Inventory.

SAFETY PROFILE: Poison by intravenous route. When heated to decomposition it emits toxic vapors of Cl$^-$.

EHJ600   CAS:92-49-9   **HR: 3**
ETHYL(CHLOROETHYL)ANILINE
mf: $C_{10}H_{14}ClN$ mw: 183.70

SYNS: ANILINE, N-(2-CHLOROETHYL)-N-ETHYL- ◊ BENZENAMINE,

N-(2-CHLOROETHYL)-N-ETHYL-(9CI) ◇ N-(2-CHLOROETHYL)-N-ETHYLANILINE ◇ N-(2-CHLOROETHYL)-N-ETHYLBENZENAMINE ◇ EMERY 5770

TOXICITY DATA with REFERENCE
orl-rat LD50:616 mg/kg   EPASR* 8EHQ-0578-0169S
ipr-rat LD50:200 mg/kg   JMCMAR 8,167,65
ipr-mus LD50:325 mg/kg   JMCMAR 8,167,65
skn-rbt LD50:200 mg/kg   EPASR* 8EHQ-0578-0169S

CONSENSUS REPORTS: Reported in EPA TSCA Inventory.

SAFETY PROFILE: Poison by intraperitoneal and skin contact routes. Moderately toxic by ingestion. When heated to decomposition it emits toxic vapors of NO_x and Cl^-.

EHM100   CAS:97-41-6   **HR: 2**
ETHYL CHRYSANTHEMUMATE
mf: $C_{12}H_{20}O_2$ mw: 196.32

SYNS: CYCLOPROPANECARBOXYLIC ACID, 2,2-DIMETHYL-3-(2-METHYLPROPENYL)-, ETHYL ESTER (8CI) ◇ CYCLOPROPANECARBOXYLIC ACID, 2,2-DIMETHYL-3-(2-METHYL-1-PROPENYL)-, ETHYLESTER ◇ CYCLOPROPANECARBOXYLIC ACID, 2,2-DIMETHYL-3-(2-METHYL-1-PROPENYL)-, ETHYLESTER (9CI) ◇ 2,2-DIMETHYL-3-(2-METHYL-PROPENYL)CYCLOPROPANECARBOXYLIC ACID ETHYL ESTER ◇ ETHYL CHRYSANTHEMATE

TOXICITY DATA with REFERENCE
mrc-smc 1 pph   NTIS** PB85-193761
orl-rat LD50:2600 mg/kg   GISAAA 51(1),16,86   GTPZAB 31(7),53,87
orl-mus LD50:2600 mg/kg   GISAAA 51(1),16,86
skn-mus LD50:>5 g/kg   GISAAA 51(1),16,86
orl-gpg LD50:1900 mg/kg   GISAAA 51(1),16,86

CONSENSUS REPORTS: Reported in EPA TSCA Inventory.

SAFETY PROFILE: Moderately toxic by ingestion. Mildly toxic by skin contact. Mutation data reported. When heated to decomposition it emits acrid smoke and irritating vapors.

EIT100   CAS:882-35-9   **HR: 3**
1,1′-ETHYLENEBIS(PYRIDINIUM)BROMIDE
mf: $C_{12}H_{14}N_2 \cdot 2Br$ mw: 346.10

SYNS: 1,1′-ETHYLENEDIPYRIDINIUM DIBROMIDE ◇ G.L. 102 ◇ P.M. 346 ◇ PYRIDINIUM, 1,1′-ETHYLENEDI-, DIBROMIDE

TOXICITY DATA with REFERENCE
ivn-mus LD50:180 mg/kg   CSLNX* NX#05031

CONSENSUS REPORTS: Reported in EPA TSCA Inventory.

SAFETY PROFILE: Poison by intravenous route. When heated to decomposition it emits toxic vapors of NO_x.

EIV100   CAS:1170-02-1   **HR: 3**
ETHYLENEDIAMINE-DI(o-HYDROXYPHENYL)ACETIC ACID
mf: $C_{18}H_{20}N_2O_6$ mw: 360.40

SYNS: CHEL 138 ◇ EDBPHA ◇ EDDHA ◇ EDHPA ◇ N,N′-ETHYLENEBIS(2-(o-HYDROXYPHENYL)GLYCINE) ◇ ETHYLENEDIAMINE-N,N′-BIS(2-HYDROXYPHENYLACETIC ACID) ◇ ETHYLENEDIAMINE-DI(2-HYDROXYPHENYL)ACETIC ACID ◇ GLYCINE, N,N′-ETHYLENEBIS(2-(o-HYDROXYPHENYL))-

TOXICITY DATA with REFERENCE
ipr-rat LD50:175 mg/kg   JMCMAR 29,1231,86
ivn-rat LD50:47 mg/kg   NTIS** PB82-163692
ipr-mus LD50:350 mg/kg   NTIS** AD691-490
ivn-mus LD50:53 mg/kg   FAATDF 6,292,86

CONSENSUS REPORTS: Reported in EPA TSCA Inventory.

SAFETY PROFILE: Poison by intravenous and intraperitoneal routes. When heated to decomposition it emits toxic vapors of NO_x.

EJC100   CAS:1852-14-8   **HR: 1**
1,1′-ETHYLENEDIUREA
mf: $C_4H_{10}N_4O_2$ mw: 146.18

SYNS: ETHANEDIUREA ◇ N,N″-1,2-ETHANEDIYLBISUREA ◇ 1,1′-ETHYLENEBISUREA ◇ ETHYLENEDIUREA ◇ MONOETHYLENEDIUREA ◇ UREA, N,N″-1,2-ETHANEDIYLBIS-(9CI) ◇ UREA, 1,1′-ETHYLENEDI-

TOXICITY DATA with REFERENCE
ipr-mus LDLo:13140 mg/kg   JPETAB 54,188,35

CONSENSUS REPORTS: Reported in EPA TSCA Inventory.

SAFETY PROFILE: Slightly toxic by intraperitoneal route. When heated to decomposition it emits toxic vapors of NO_x.

EJQ100   CAS:822-38-8   **HR: 2**
ETHYLENE TRITHIOCARBONATE
mf: $C_3H_4S_3$ mw: 136.25

SYNS: CARBONIC ACID, TRITHIO-, CYCLIC ETHYLENE ESTER ◇ CYCLIC ETHYLENE TRITHIOCARBONATE ◇ 1,3-DITHIOLANE-2-THIONE ◇ TRITHIOCARBONIC ACID, CYCLIC ETHYLENE ESTER

TOXICITY DATA with REFERENCE
ipr-mus LD50:500 mg/kg   EJMCA5 17,235,82

CONSENSUS REPORTS: Reported in EPA TSCA Inventory.

SAFETY PROFILE: Moderately toxic by intraperitoneal route. When heated to decomposition it emits toxic vapors of SO_x.

EKK600   CAS:627-45-2   **HR: 2**
N-ETHYLFORMAMIDE
mf: C_3H_7NO mw: 73.11

SYNS: N-AETHYLFORMAMID ◊ ETHYLFORMAMIDE ◊ FORMAMIDE, N-ETHYL- ◊ N-FORMYLETHYLAMINE

TOXICITY DATA with REFERENCE
unr-rat LD50:3000 mg/kg   ARZNAD 18,645,68

CONSENSUS REPORTS: Reported in EPA TSCA Inventory.

SAFETY PROFILE: Moderately toxic by an unspecified route. When heated to decomposition it emits toxic vapors of NO_x.

EKM100   CAS:831-61-8   **HR: 1**
ETHYL GALLATE
mf: $C_9H_{10}O_5$ mw: 198.19

SYNS: BENZOIC ACID, 3,4,5-TRIHYDROXY-, ETHYL ESTER (9CI) ◊ ETHYLESTER KYSELINY GALLOVE ◊ ETHYL 3,4,5-TRIHYDROXY-BENZOATE ◊ GALLIC ACID, ETHYL ESTER ◊ NIPAGALLIN A ◊ NIPA NO. 48 ◊ PHYLLEMBLIN ◊ PROGALLIN A

TOXICITY DATA with REFERENCE
orl-mus LD50:5810 mg/kg   85JCAE -,668,86

CONSENSUS REPORTS: Reported in EPA TSCA Inventory.

SAFETY PROFILE: Mildly toxic by ingestion. When heated to decomposition it emits acrid smoke and irritating vapors.

EKO600   CAS:760-67-8   **HR: 2**
2-ETHYLHEXANOIC ACID CHLORIDE
mf: $C_8H_{15}ClO$ mw: 162.68

SYNS: 2-ETHYLCAPROYL CHLORIDE ◊ 2-ETHYLHEXANOYL CHLORIDE ◊ HEXANOYL CHLORIDE, 2-ETHYL-

TOXICITY DATA with REFERENCE
ihl-rat LC50:1260 mg/m^3 EPASR* 8EHQ-0387-0656
orl-uns LD50:1500 mg/kg   EPASR* 8EHQ-0387-0656
skn-uns LD50:>2 g/kg   EPASR* 8EHQ-0387-0656

CONSENSUS REPORTS: Reported in EPA TSCA Inventory.

SAFETY PROFILE: Moderately toxic by ingestion and inhalation. Slightly toxic by skin contact. When heated to decomposition it emits toxic vapors of Cl^-.

EKU100   CAS:2350-24-5   **HR: 1**
2-ETHYLHEXYL-6-CHLORIDE
mf: $C_8H_{17}Cl$ mw: 148.70

SYN: HEPTANE, 1-CHLORO-5-METHYL-

TOXICITY DATA with REFERENCE
skn-rbt 10 mg/24H open MLD   AMIHBC 4,119,51
eye-rbt 500 mg open   AMIHBC 4,119,51
orl-rat LD50:7340 mg/kg   AMIHBC 4,119,51
ihl-rat LCLo:4000 ppm/4H   JIDHAN 31,343,49

CONSENSUS REPORTS: Reported in EPA TSCA Inventory.

SAFETY PROFILE: Mildly toxic by ingestion and inhalation routes. A skin and eye irritant. When heated to decomposition it emits toxic vapors of Cl^-.

ELC600   CAS:626-86-8   **HR: 1**
ETHYL HYDROGEN ADIPATE
mf: $C_8H_{14}O_4$ mw: 174.22

SYNS: ADIPIC ACID, MONOETHYL ESTER ◊ HEXANOIC ACID, MONOETHYL ESTER (9CI) ◊ MONOETHYL ADIPATE ◊ MONOETHYLADIPIC ACID ESTER ◊ MONOETHYL HEXANEDIOATE

TOXICITY DATA with REFERENCE
orl-uns LD50:4100 mg/kg   GTPZAB 21(10),39,77

CONSENSUS REPORTS: Reported in EPA TSCA Inventory.

SAFETY PROFILE: Mildly toxic by ingestion. When heated to decomposition it emits acrid smoke and irritating vapors.

ELH700   CAS:80-55-7   **HR: 2**
ETHYL 2-HYDROXYISOBUTYRATE
mf: $C_6H_{12}O_3$ mw: 132.18

SYNS: ETHYL α-HYDROXYISOBUTYRATE ◊ ETHYL 2-HYDROXY-2-METHYLPROPANOATE ◊ ETHYL 2-METHYLLACTATE ◊ LACTIC ACID, 2-METHYL-, ETHYL ESTER ◊ 2-METHYLLACTIC ACID ETHYL ESTER ◊ PROPANOIC ACID, 2-HYDROXY-2-METHYL-, ETHYL ESTER (9CI)

TOXICITY DATA with REFERENCE
ims-gpg LDLo:2200 mg/kg   JPETAB 76,189,42

SAFETY PROFILE: Moderately toxic by intramuscular route. When heated to decomposition it emits acrid smoke and irritating vapors.

ELX525   CAS:542-85-8   **HR: D**
ETHYL ISOTHIOCYANATE
mf: C_3H_5NS mw: 87.15

SYNS: ETHANE, ISOTHIOCYANATO-(9CI) ◊ ETHYL MUSTARD OIL ◊ ISOTHIOCYANATOETHANE ◊ ISOTHIOCYANIC ACID, ETHYL ESTER

TOXICITY DATA with REFERENCE
mmo-sat 100 μg/plate   ABCHA6 44,3017,80

CONSENSUS REPORTS: Reported in EPA TSCA Inventory.

SAFETY PROFILE: Mutation data reported. When heated to decomposition it emits toxic vapors of NO_x and SO_x.

ENL900   CAS:2122-70-5   **HR: 2**
ETHYL 1-NAPHTHYLACETATE
mf: $C_{14}H_{14}O_2$ mw: 214.28

SYNS: ETHYL 1-NAPHTHALENEACETATE ◊ 1-NAPHTHALENEACETIC ACID, ETHYL ESTER

TOXICITY DATA with REFERENCE
orl-rat LD50:3580 mg/kg   PESTC* 9,10,80

CONSENSUS REPORTS: Reported in EPA TSCA Inventory.

SAFETY PROFILE: Moderately toxic by ingestion. When heated to decomposition it emits acrid smoke and irritating vapors.

ENN100   CAS:626-35-7   **HR: 2**
ETHYL NITROACETATE
mf: $C_4H_7NO_4$ mw: 133.12

SYN: ACETIC ACID, NITRO-, ETHYL ESTER

TOXICITY DATA with REFERENCE
ipr-rat LD50:2275 mg/kg   VINIT* #6802-83

CONSENSUS REPORTS: Reported in EPA TSCA Inventory.

SAFETY PROFILE: Moderately toxic by intraperitoneal route. When heated to decomposition it emits toxic vapors of NO_x.

ENO100   CAS:838-57-3   **HR: 1**
ETHYL (4-NITROBENZOYL)ACETATE
mf: $C_{11}H_{11}NO_5$ mw: 237.23

SYNS: ACETIC ACID, (p-NITROBENZOYL)-, ETHYL ESTER ◇ BENZENEPROPANOIC ACID, 4-NITRO-β-OXO-, ETHYL ESTER (9CI) ◇ ETHYL (p-NITROBENZOYL)ACETATE ◇ ETHYL 4-NITRO-β-OXOBENZENEPROPANOATE

TOXICITY DATA with REFERENCE
orl-mus LD50:8620 mg/kg   GISAAA 51(1),79,86

CONSENSUS REPORTS: Reported in EPA TSCA Inventory.

SAFETY PROFILE: Mildly toxic by ingestion. When heated to decomposition it emits toxic vapors of NO_x.

EOB300   CAS:609-27-8   **HR: 3**
3-ETHYL-2-PENTANOL
mf: $C_7H_{16}O$ mw: 116.23

SYN: 2-PENTANOL, 3-ETHYL-

TOXICITY DATA with REFERENCE
ivn-mus LD50:180 mg/kg   CSLNX* NX#03027

CONSENSUS REPORTS: Reported in EPA TSCA Inventory.

SAFETY PROFILE: Poison by intravenous route. When heated to decomposition it emits acrid smoke and irritating vapors.

EOE100   CAS:123-07-9   **HR: 3**
4-ETHYLPHENOL
mf: $C_8H_{10}O$ mw: 122.18

SYN: PHENOL, p-ETHYL-

TOXICITY DATA with REFERENCE
ipr-mus LD50:138 mg/kg   JMCMAR 18,868,75

CONSENSUS REPORTS: Reported in EPA TSCA Inventory.

SAFETY PROFILE: Poison by intravenous route. When heated to decomposition it emits acrid smoke and irritating vapors.

EOP600   CAS:993-43-1   **HR: 2**
ETHYL PHOSPHONOTHIOIC DICHLORIDE
DOT: NA 1760
mf: $C_2H_5Cl_2PS$ mw: 163.00

SYNS: DICHLOROETHYLPHOSPHINE SULFIDE ◇ ETHYL PHOSPHONOTHIOIC DICHLORIDE, anhydrous (DOT) ◇ ETHYLPHOSPHONOTHIONIC DICHLORIDE ◇ ETHYL PHOSPHONOTHIOYL DICHLORIDE ◇ ETHYLTHIONOPHOSPHONYL DICHLORIDE ◇ ETHYLTHIOPHOSPHONIC DICHLORIDE ◇ PHOSPHONOTHIOIC DICHLORIDE, ETHYL-

DOT Classification: Corrosive Material; Label: Corrosive

CONSENSUS REPORTS: Reported in EPA TSCA Inventory.

SAFETY PROFILE: A corrosive. When heated to decomposition it emits toxic vapors of SO_x, PO_x, and Cl^-.

EPD600   CAS:606-55-3   **HR: 3**
1-ETHYLQUINALDINIUM IODIDE
mf: $C_{12}H_{14}N•I$ mw: 299.17

SYN: QUINALDINIUM, 1-ETHYL-, IODIDE

TOXICITY DATA with REFERENCE
ivn-mus LD50:18 mg/kg   CSLNX* NX#01580

CONSENSUS REPORTS: Reported in EPA TSCA Inventory.

SAFETY PROFILE: Poison by intravenous route. When heated to decomposition it emits toxic vapors of NO_x and I^-.

EPF700   CAS:111-61-5   **HR: 1**
ETHYL STEARATE
mf: $C_{20}H_{40}O_2$ mw: 312.60

SYNS: ETHYL OCTADECANOATE ◇ ETHYL n-OCTADECANOATE ◇ STEARIC ACID, ETHYL ESTER

TOXICITY DATA with REFERENCE
skn-rbt 500 mg/24H MOD   FCTXAV 17,781,79

CONSENSUS REPORTS: Reported in EPA TSCA Inventory.

SAFETY PROFILE: A skin irritant. When heated to decomposition it emits acrid smoke and irritating vapors.

EQC600 CAS:51-93-4 *HR: 3*
ETHYLTRIMETHYLAMMONIUM IODIDE
mf: $C_5H_{14}N \cdot I$ mw: 215.10

SYNS: AMMONIUM, ETHYLTRIMETHYL-, IODIDE ◇ ETHANAMINIUM, N,N,N-TRIMETHYL-, IODIDE (9CI) ◇ IODURE d'ETHYL-TRIMETHYL-AMMONIUM ◇ N,N,N-TRIMETHYLETHANAMINIUM IODIDE

TOXICITY DATA with REFERENCE
ipr-mus LD50:40 mg/kg   85IXA4 -,675,48

CONSENSUS REPORTS: Reported in EPA TSCA Inventory.

SAFETY PROFILE: Poison by intraperitoneal route. When heated to decomposition it emits toxic vapors of NO_x and I^-.

EQD200 CAS:692-86-4 *HR: 1*
ETHYL 10-UNDECENOATE
mf: $C_{13}H_{24}O_2$ mw: 212.37

SYNS: ETHYL 10-HENDECENOATE ◇ ETHYL UNDECENOATE ◇ ETHYL UNDECYLENATE ◇ 10-UNDECENOIC ACID, ETHYL ESTER

TOXICITY DATA with REFERENCE
skn-rbt 500 mg/24H MLD   FCTOD7 20,687,82
orl-rat LD50:>5 g/kg   FCTOD7 20,687,82
skn-rbt LD50:>5 g/kg   FCTOD7 20,687,82

CONSENSUS REPORTS: Reported in EPA TSCA Inventory.

SAFETY PROFILE: Slightly toxic by ingestion. A skin irritant. When heated to decomposition it emits acrid smoke and irritating vapors.

EQG600 CAS:151-01-9 *HR: D*
ETHYL XANTHATE
mf: $C_3H_6OS_2$ mw: 122.21

SYNS: CARBONIC ACID, DITHIO-, O-ETHYL ESTER ◇ CARBONODITHIOIC ACID, O-ETHYL ESTER ◇ O-ETHYL DITHIOCARBAMATE ◇ O-ETHYL DITHIOCARBONATE ◇ ETHYLXANTHIC ACID ◇ ETHYL XANTHOGENATE ◇ O-ISOBUTYL POTASSIUM XANTHATE ◇ XANTHATE ◇ XANTHOGENIC ACID

TOXICITY DATA with REFERENCE
cyt-rat-ihl 13050 µg/m^3/16W-I   GTPZAB 24(9),33,80

CONSENSUS REPORTS: Reported in EPA TSCA Inventory.

SAFETY PROFILE: Mutation data reported. When heated to decomposition it emits toxic vapors of SO_x.

EQM600 CAS:2028-63-9 *HR: 3*
1-ETHYNYLETHANOL
mf: C_4H_6O mw: 70.10

SYN: 1-BUTYN-3-OL

TOXICITY DATA with REFERENCE
orl-mus LD50:30 mg/kg   ARZNAD 7,85,57

CONSENSUS REPORTS: Reported in EPA TSCA Inventory.

SAFETY PROFILE: Poison by ingestion. When heated to decomposition it emits acrid smoke and irritating vapors.

EQN230 CAS:127-66-2 *HR: 2*
α-ETHYNYL-α-METHYLBENZYL ALCOHOL
mf: $C_{10}H_{10}O$ mw: 146.20

SYNS: 3-BUTYN-2-OL, 2-PHENYL- ◇ 3-PHENYL-BUTIN-1-OL-(3)

TOXICITY DATA with REFERENCE
orl-mus LD50:620 mg/kg   ARZNAD 4,477,54

CONSENSUS REPORTS: Reported in EPA TSCA Inventory.

SAFETY PROFILE: Moderately toxic by ingestion. When heated to decomposition it emits acrid smoke and irritating vapors.

F

FAB000 CAS:88-27-7 HR: 2
F 1 (antioxidant)
mf: $C_{17}H_{29}NO$ mw: 263.47

SYNS: AGIDOL 3 ◇ p-CRESOL, 2,6-DI-tert-BUTYL-α-(DIMETHYLAMINO)- ◇ 2,6-DI-tert-BUTYL-α-(DIMETHYLAMINO)-p-CRESOL ◇ ETHYL 703 ◇ ETHYL ANTIOXIDANT 703 ◇ F 1 ◇ OMI ◇ PHENOL, 4-((DIMETHYL-AMINO)METHYL)-2,6-BIS(1,1-DIMETHYLETHYL)-(9CI)

TOXICITY DATA with REFERENCE
orl-rat LD50:1030 mg/kg   IPSTB3 3,93,76

CONSENSUS REPORTS: Reported in EPA TSCA Inventory.

SAFETY PROFILE: Moderately toxic by ingestion. When heated to decomposition it emits toxic vapors of NO_x.

FAS700 CAS:1185-57-5 HR: 1
FERRIC AMMONIUM CITRATE
mf: $C_6H_8O_7 \bullet xFe \bullet xH_4N$

PROP: A complex salt of undetermined structure. Transparent green scales, granules, powder, or crystals; ammoniacal odor, mild iron-metallic taste. Sol in water; insol in alc. Deliquescent.

SYNS: CITRIC ACID, AMMONIUM IRON(3+) SALT ◇ FAC ◇ FERRIC AMMONIUM CITRATE, GREEN ◇ IRON(III) AMMONIUM CITRATE

ACGIH TLV: TWA 1 mg(Fe)/m^3

CONSENSUS REPORTS: Reported in EPA TSCA Inventory.

SAFETY PROFILE: When heated to decomposition it emits acrid smoke and irritating fumes.

FBB000 CAS:9007-73-2 HR: 2
FERRITIN

PROP: Prepared from rat liver protein by precipitation with a cadmium salt (BECCAN 39,74,61).

TOXICITY DATA with REFERENCE
cyt-ham:ovr 27 mg/L   CNREA8 41,1628,81
scu-rat TDLo:224 mg/kg/15W-I:NEO   BJCAAI 18,667,64

CONSENSUS REPORTS: Reported in EPA TSCA Inventory.

SAFETY PROFILE: Questionable carcinogen with experimental neoplastigenic data. Mutation data reported. When heated to decomposition it emits acrid smoke and irritating fumes.

FBW150 CAS:487-26-3 HR: 3
FLAVANONE
mf: $C_{15}H_{12}O_2$ mw: 224.27

SYNS: 4H-1-BENZOPYRAN-4-ONE, 2,3-DIHYDRO-2-PHENYL-(9CI) ◇ 2,3-DIHYDRO-2-PHENYL-4H-1-BENZOPYRAN-4-ONE ◇ 4-FLAVANONE

TOXICITY DATA with REFERENCE
orl-brd LD50:75 mg/kg   AECTCV 12,355,83

CONSENSUS REPORTS: Reported in EPA TSCA Inventory.

SAFETY PROFILE: Poison by ingestion. When heated to decomposition it emits acrid smoke and irritating vapors.

FBZ100 CAS:483-84-1 HR: D
FLAVIANIC ACID
mf: $C_{10}H_6N_2O_8S$ mw: 314.24

SYNS: 2,4-DINITRONAPHTHOLSULFONIC ACID ◇ D,4-DINITRO-1-NAPHTHOL-7-SULFONIC ACID ◇ DNNS ◇ 8-HYDROXY-5,7-DINITRO-2-NAPHTHALENESULFONIC ACID ◇ 2-NAPHTHALENESULFONIC ACID, 5,7-DINITRO-8-HYDROXY- ◇ 2-NAPHTHALENESULFONIC ACID, 8-HYDROXY-5,7-DINITRO-(8CI)

TOXICITY DATA with REFERENCE
mmo-sat 10 ~2mol/plate   MUREAV 58,11,78

CONSENSUS REPORTS: Reported in EPA TSCA Inventory.

SAFETY PROFILE: Mutation data reported. When heated to decomposition it emits toxic vapors of NO_x and SO_x.

FCA100 CAS:81-11-8 HR: 1
FLAVONIC ACID
mf: $C_{14}H_{14}N_2O_6S_2$ mw: 370.42

SYNS: AMSONIC ACID ◇ BENZENESULFONIC ACID, 2,2'-(1,2-ETHYLENEDIYL)BIS(5-AMINO-(9CI) ◇ DASD ◇ 4,4'-DIAMINO-2,2'-STILBENEDISULFONIC ACID ◇ 2,2'-(1,2-ETHYLENEDIYL)BIS(5-AMINOBENZENESULFONIC ACID) ◇ NCI-C60162 ◇ 2,2'-STILBENE-DISULFONIC ACID, 4,4'-DIAMINO- ◇ TINOPAL BHS

TOXICITY DATA with REFERENCE
orl-gpg LD50:47 g/kg   GISAAA 45(3),73,80

CONSENSUS REPORTS: Reported in EPA TSCA Inventory.

SAFETY PROFILE: Slightly toxic by ingestion. When heated to decomposition it emits toxic vapors of NO_x and SO_x.

FDI100   CAS:86-73-7   *HR: 1*
9H-FLUORENE
mf: $C_{13}H_{10}$ mw: 166.23

SYNS: o-BIPHENYLENEMETHANE ◇ o-BIPHENYLMETHANE ◇ DIPHENYLENEMETHANE ◇ FLUORENE ◇ 2,2′-METHYLENEBIPHENYL

TOXICITY DATA with REFERENCE
mma-mus:lyms 19,500 nmol/L   MUTAEX 3,193,88
otr-mus:mmr 1 µg/L   CNREA8 39,1784,79
dnd-mus:lyms 150 µmol/L   MUREAV 203,155,88
msc-mus:lyms 584 µmol/L   MUTAEX 3,193,88
cyt-ham:lng 25 mg/L   MUREAV 259,103,91
ipr-mus LD50:2 g/kg   RPTOAN 48,143,85
par-mus LD50:>2 g/kg   RPTOAN 52,112,89

CONSENSUS REPORTS: IARC Cancer Review: Group 3 IMEMDT 7,56,87; Animal Inadequate Evidence IMEMDT 32,365,83; Human No Adequate Data IMEMDT 32,365,83

CONSENSUS REPORTS: Reported in EPA TSCA Inventory.

SAFETY PROFILE: Slightly toxic by intraperitoneal and parenteral route. Mutation data reported. When heated to decomposition it emits acrid smoke and irritating vapors.

FGH100   CAS:1194-02-1   *HR: 2*
4-FLUOROBENZONITRILE

SYNS: BENZONITRILE, p-FLUORO- ◇ p-CYANOFLUOROBENZENE ◇ p-FLUOROBENZONITRILE

TOXICITY DATA with REFERENCE
orl-mus LD50:>300 mg/kg   JMCMAR 21,906,78
ipr-mus LD50:1 g/kg   FRPSAX 41,41,86

CONSENSUS REPORTS: On Community Right-To-Know List. Reported in EPA TSCA Inventory.

SAFETY PROFILE: Moderately toxic by ingestion. Mildly toxic by intraperitoneal route. When heated to decomposition it emits toxic vapors of CN^- and F^-.

FJT050   CAS:452-71-1   *HR: D*
4-FLUORO-2-METHYLBENZENAMINE
mf: C_7H_8FN mw: 125.16

SYNS: BENZENAMINE, 4-FLUORO-2-METHYL- ◇ 2-METHYL-4-FLUOROANILINE

TOXICITY DATA with REFERENCE
mma-sat 2 µmol/plate   MUREAV 77,317,80

CONSENSUS REPORTS: Reported in EPA TSCA Inventory.

SAFETY PROFILE: Mutation data reported. When heated to decomposition it emits toxic vapors of NO_x and F^-.

FKK100   CAS:364-76-1   *HR: 2*
4-FLUORO-3-NITROANILINE
mf: $C_6H_5FN_2O_2$ mw: 156.13

SYNS: ANILINE, 4-FLUORO-3-NITRO- ◇ BENZENAMINE, 4-FLUORO-3-NITRO- ◇ 4-F-3NA

TOXICITY DATA with REFERENCE
orl-rat LD50:1100 mg/kg   TXAPA9 72,400,84

CONSENSUS REPORTS: Reported in EPA TSCA Inventory.

SAFETY PROFILE: Moderately toxic by ingestion. When heated to decomposition it emits toxic vapors of NO_x and F^-.

FKT100   CAS:367-12-4   *HR: 2*
2-FLUOROPHENOL
mf: C_6H_5FO mw: 112.11

SYNS: o-FLUOROPHENOL ◇ PHENOL, o-FLUORO-

TOXICITY DATA with REFERENCE
ipr-mus LD50:537 mg/kg   JMCMAR 18,868,75

CONSENSUS REPORTS: Reported in EPA TSCA Inventory.

SAFETY PROFILE: Moderately toxic by intraperitoneal route. When heated to decomposition it emits toxic vapors of F^-.

FLD100   CAS:51-65-0   *HR: D*
4-FLUORO-dl-PHENYLALANINE
mf: $C_9H_{10}FNO_2$ mw: 183.20

SYNS: ALANINE, 3-(p-FLUOROPHENYL)-, dl- ◇ ALNASID ◇ FLUOROPHENYLALANINE ◇ d,l-FLUOROPHENYLALANINE ◇ d,l-p-FLUOROPHENYLALANINE ◇ dl-4-FLUOROPHENYLALANINE ◇ p-FLUOROPHENYLALANINE ◇ FPA ◇ dl-PHENYLALANINE, 4-FLUORO- (9CI)

TOXICITY DATA with REFERENCE
mmo-omi 50 µmol/L   GENRA8 23,47,74
mmo-nsc 1 mg/L   MUREAV 46,345,77
mrc-smc 1200 ppm   HEREAY 59,197,68
sln-smc 1200 ppm   HEREAY 59,197,68
cyt-hmn:lyms 50 µmol/L   HEREAY 88,197,78
cyt-ham:lng 550 µmol/L   HEREAY 95,25,81
dni-rbt:kdy 500 µmol/L   ECREAL 36,92,64

CONSENSUS REPORTS: Reported in EPA TSCA Inventory.

SAFETY PROFILE: Human mutation data reported. When heated to decomposition it emits toxic vapors of NO_x and F^-.

FLY200   CAS:2489-52-3   *HR: 3*
m-FLUOROSULFONYLBENZENESULFONYL CHLORIDE
mf: $C_6H_4ClFO_4S_2$ mw: 258.67

FLZ100 o-FLUOROTOLUENE

SYNS: BENZENE-1,3-DISULFONYL CHLORIDE FLUORIDE ◊ BENZENESULFONYL CHLORIDE, m-(FLUOROSULFONYL)-

TOXICITY DATA with REFERENCE
ivn-mus LD50:56 mg/kg   CSLNX* NX#04699

CONSENSUS REPORTS: Reported in EPA TSCA Inventory.

SAFETY PROFILE: Poison by intravenous route. When heated to decomposition it emits toxic vapors of NO_x, SO_x, F^-, and Cl^-.

FLZ100   CAS:95-52-3   HR: 3
o-FLUOROTOLUENE
mf: C_7H_7F mw: 110.14

SYNS: BENZENE, 1-FLUORO-2-METHYL-(9CI) ◊ 1-FLUORO-2-METHYL-BENZENE ◊ 2-FLUOROTOLUENE ◊ TOLUENE, o-FLUORO-

TOXICITY DATA with REFERENCE
orl-brd LD50:100 mg/kg   AECTCV 12,355,83

CONSENSUS REPORTS: Reported in EPA TSCA Inventory.

SAFETY PROFILE: Poison by ingestion. When heated to decomposition it emits toxic vapors of F^-.

FMV000   CAS:50-00-0   HR: 3
FORMALDEHYDE
DOT: UN 1198/UN 2209
mf: CH_2O mw: 30.03

PROP: Clear, water-white, very sltly acid gas or liquid; pungent odor. Pure formaldehyde is not available commercially because of its tendency to polymerize. It is sold as aqueous solns containing from 37 to 50% formaldehyde by weight and varying amounts of methanol. Some alcoholic solns are used industrially and the physical properties and hazards may be greatly influenced by the solvent. Lel: 7.0%, uel: 73.0%, autoign temp: 806°F, d: 1.0, bp: −3°F, flash p: (37%, methanol-free): 185°F, flash p: (15%, methanol-free): 122°F.

SYNS: ALDEHYDE FORMIQUE (FRENCH) ◊ ALDEIDE FORMICA (ITALIAN) ◊ BFV ◊ FA ◊ FANNOFORM ◊ FORMALDEHYD (CZECH, POLISH) ◊ FORMALDEHYDE, solution (DOT) ◊ FORMALIN ◊ FORMALIN 40 ◊ FORMALIN (DOT) ◊ FORMALINA (ITALIAN) ◊ FORMALINE (GERMAN) ◊ FORMALIN-LOESUNGEN (GERMAN) ◊ FORMALITH ◊ FORMIC ALDEHYDE ◊ FORMOL ◊ FYDE ◊ HOCH ◊ IVALON ◊ KARSAN ◊ LYSOFORM ◊ METHANAL ◊ METHYL ALDEHYDE ◊ METHYLENE GLYCOL ◊ METHYLENE OXIDE ◊ MORBOCID ◊ NCI-C02799 ◊ OPLOSSINGEN (DUTCH) ◊ OXOMETHANE ◊ OXYMETHYLENE ◊ PARAFORM ◊ POLY-OXYMETHYLENE GLYCOLS ◊ RCRA WASTE NUMBER U122 ◊ SUPER-LYSOFORM

TOXICITY DATA with REFERENCE
skn-hmn 150 μg/3D-I MLD   85DKA8 -,127,77
eye-hmn 4 ppm/5M   IAPWAR 4,79,61
eye-hmn 1 ppm/6M nse MLD   AIHAAP 44,463,83
skn-rbt 500 mg/24H SEV   28ZPAK -,40,72
skn-rbt 540 mg open MLD   UCDS** 4/21/67
skn-rbt 50 mg/24H MOD   TXAPA9 21,369,72
eye-rbt 50 μg/24H SEV   28ZPAK -,40,72
eye-rbt 10 mg SEV   TXAPA9 55,501,80
mma-sat 5 μL/plate   BIMADU 6,129,85
dni-esc 5 mmol/L   MUREAV 156,153,85
dnd-hmn:fbr 100 μmol/L   ENMUDM 7,267,85
ihl-rat TCLo:50 μg/m^3/4H (female 1–19D post):REP   TPKVAL 12,78,71
ihl-rat TCLo:12 μg/m^3/24h (female 1–22D post):REP   HYSAAV 33(7-9),112,68
ihl-rat TCLo:12 μg/m^3/24H (female 15D pre):REP   HYSAAV 33(1-3),327,68
ihl-rat TCLo:12 μg/m^3/24H (female 15D pre):REP   HYSAAV 33(1-3),327,68
ihl-rat TCLo:1 mg/m^3/24H (1–22D preg):TER   HYSAAV 34(5),266,69
ipr-mus TDLo:240 mg/kg (female 7–14D post):TER   TJADAB 30(1),34A,84
ipr-mus TDLo:240 mg/kg (female 7–14D post):TER   TJADAB 28,37A,83
itt-rat TDLo:400 mg/kg (male 1D pre):REP   FESTAS 24,884,73
ims-mus TDLo:259 mg/kg (female 11D post):REP   ANREAK 142,479,62
ihl-rat TCLo:35 μg/m^3/8H (male 60D pre):REP   PRKHDK 4,101,79
ipr-rat TDLo:80 mg/kg (male 10D pre):REP   JRBED2 7,42,87
orl-rat TDLo:200 mg/kg (1D male):REP   TJADAB 26(3),14A,82
ipr-rat TDLo:80 mg/kg (male 10D pre):REP   JRBED2 7,42,87
ipr-mus TDLo:240 mg/kg (female 7–14D post):TER   TJADAB 28,37A,83
ipr-mus TDLo:240 mg/kg (female 7–14D post):TER   TJADAB 28,37A,83
ipr-mus TDLo:160 mg/kg (female 7–14D post):TER   TJADAB 30(1),34A,84
ihl-rat TCLo:14300 ppb/6H/2Y-I:CAR   CNREA8 43,4382,83
scu-rat TDLo:1170 mg/kg/65W-I:ETA   GANNA2 45,451,54
ihl-mus TCLo:14300 ppm/6H/2Y-I:ETA   CNREA8 43,4382,83
ihl-rat TC:15 ppm/6H/78W-I:CAR   CNREA8 49,3398,80
scu-rat TD:350 mg/kg/78W-I:ETA   FAONAU 50A,77,72
ihl-rat TC:6 ppm/6H/2Y-I:ETA   EVSRBT 25,353,82
ihl-rat TC:15 ppm/6H/86W-I:CAR   TXAPA9 81,401,85
ihl-rat TC:14 ppm/6H/84W-I:CAR   JJIND8 68,597,82
ihl-rat TC:18,750 μg/m^3/2Y-I:ETA   GISAAA 48(4),60,83
ihl-mus TC:15 ppm/6H/104W-I:ETA   EVSRBT 25,353,82
ihl-rat TC:15 ppm/6H/2Y-I:CAR   CIIT** DOCKET #10992,82
ihl-rat TC:5600 ppb/6H/2Y-I:ETA   CNREA8 43,4382,83
ihl-rat TC:14,300 ppb/6H/2Y-I:ETA   50EXAK -,111,83
orl-wmn LDLo:108 mg/kg   29ZWAE -,328,68
ihl-hmn TCLo:17 mg/m^3/30M:EYE,PUL   JAMAAP 165,1908,57
ihl-man TCLo:300 μg/m^3:NOSE,CNS   GTPZAB 12(7),20,68
unr-man LDLo:477 mg/kg   85DCAI 2,73,70
orl-rat LD50:800 mg/kg   JIHTAB 23,259,41
ihl-rat LC50:590 mg/m^3 GISAAA 41(6),103,76
scu-rat LD50:420 mg/kg   APTOA6 6,299,50

ivn-rat LD50:87 mg/kg   AEPPAE 221,166,54
orl-mus LD50:42 mg/kg   NTIS** AD-A125-539
ihl-mus LC50:400 mg/m^3/2H   85GMAT -,69,82
ipr-mus LDLo:16 mg/kg   TXAPA9 23,288,72
scu-mus LD50:300 mg/kg   APTOA6 6,299,50
scu-dog LDLo:350 mg/kg   IPSTB3 3,93,76
ihl-cat LCLo:400 mg/m^3/2H   85GMAT -,69,82
skn-rbt LD50:270 mg/kg   UCDS** 4/21/67
scu-rbt LDLo:240 mg/kg   JAMAAP 62,984,14
orl-gpg LD50:260 mg/kg   JIHTAB 23,259,41

CONSENSUS REPORTS: NTP Fifth Annual Report on Carcinogens. IARC Cancer Review: Group 2A IMEMDT 7,211,87; Human Inadequate Evidence IMEMDT 29,345,82; Animal Sufficient Evidence IMEMDT 29,345,82. EPA Genetic Toxicology Program. Reported in EPA TSCA Inventory.

OSHA PEL: TWA 0.75 ppm; STEL 2 ppm
ACGIH TLV: TWA 1 ppm; Suspected Human Carcinogen (Proposed: CL 0.3 ppm; Suspected Human Carcinogen)
DFG MAK: 0.5 ppm (0.6 mg/m^3); Suspected Carcinogen.
NIOSH REL: (Formaldehyde) Limit to lowest feasible level.

DOT Classification: Combustible Liquid; Label: None; ORM-A; Label: None; Flammable or Combustible Liquid; Label: Flammable Liquid

SAFETY PROFILE: Confirmed carcinogen with experimental carcinogenic, tumorigenic, and teratogenic data. Human poison by ingestion. Experimental poison by ingestion, skin contact, inhalation, intravenous, intraperitoneal, and subcutaneous routes. Human systemic effects by inhalation: lacrimation, olfactory changes, aggression, and pulmonary changes. Experimental reproductive effects. Human mutation data reported. A human skin and eye irritant. If swallowed it causes violent vomiting and diarrhea that can lead to collapse. Frequent or prolonged exposure can cause hypersensitivity leading to contact dermatitis, possibly of an eczematoid nature. An air concentration of 20 ppm is quickly irritating to eyes. A common air contaminant.

Combustible liquid when exposed to heat or flame; can react vigorously with oxidizers. A moderate explosion hazard when exposed to heat or flame. The gas is a more dangerous fire hazard than the vapor. Should formaldehyde be involved in a fire, irritating gaseous formaldehyde may be evolved. When aqueous formaldehyde solutions are heated above their flash points, a potential for an explosion hazard exists. High formaldehyde concentration or methanol content lowers the flash point. Reacts with NO$_x$ at about 180°; the reaction becomes explosive. Also reacts violently with perchloric acid + aniline; performic acid; nitromethane; magnesium carbonate; H$_2$O$_2$. Moderately dangerous because of irritating vapor that may exist in toxic concentrations locally if storage tank is ruptured. To fight fire, stop flow of gas (for pure form); alcohol foam for 37% methanol-free form. When heated to decomposition it emits acrid smoke and fumes. See also ALDEHYDES.

FMW400   CAS:75-17-2   HR: D
FORMALDOXIME
mf: CH$_3$NO   mw: 45.05

SYNS: FORMALDEHYDE, OXIME ◇ FORMOXIME ◇ METHYLENE-AMINE N-OXIDE

TOXICITY DATA with REFERENCE
mma-esc 10 μmol/plate   MUREAV 164,263,86

CONSENSUS REPORTS: Reported in EPA TSCA Inventory.

SAFETY PROFILE: Mutation data reported. When heated to decomposition it emits toxic vapors of NO$_x$.

FNK010   CAS:119-67-5   HR: 2
2-FORMYLBENZOIC ACID
mf: C$_8$H$_6$O$_3$ mw: 150.14

SYNS: BENZOIC ACID, 2-FORMYL-(9CI) ◇ o-CARBOXYBENZALDEHYDE ◇ 2-CARBOXYBENZALDEHYDE ◇ o-FORMYLBENZOIC ACID ◇ PHTHALALDEHYDIC ACID

TOXICITY DATA with REFERENCE
orl-rat LD50:7500 mg/kg   JANTAJ 27,665,74
scu-rat LD50:2430 mg/kg   JANTAJ 27,665,74
orl-mus LD50:4480 mg/kg   JANTAJ 27,665,74
scu-mus LD50:1860 mg/kg   JANTAJ 27,665,74

CONSENSUS REPORTS: Reported in EPA TSCA Inventory.

SAFETY PROFILE: Moderately toxic by subcutaneous route. Mildly toxic by ingestion. When heated to decomposition it emits acrid smoke and irritating vapors.

FNK150   CAS:564-94-3   HR: 3
2-FORMYL-6,6-DIMETHYLBICYCLO(3.1.1)HEPT-2-ENE
mf: C$_{10}$H$_{14}$O   mw: 150.24

SYNS: BENIHINAL ◇ BICYCLO(3.1.1)HEPT-2-ENE-2-CARBOXALDEHYDE, 6,6-DIMETHYL- ◇ 6,6-DIMETHYLBICYCLO(3.1.1)HEPT-2-ENE-2-CARBOXALDEHYDE ◇ MYRTENAL ◇ 2-NORPINENE-2-CARBOXALDEHYDE, 6,6-DIMETHYL-

TOXICITY DATA with REFERENCE
orl-rat LD50:2300 mg/kg   FCTOD7 26,329,88
ivn-mus LD50:170 mg/kg   FCTOD7 26,329,88
skn-rbt LD50:>5 g/kg   FCTOD7 26,329,88

CONSENSUS REPORTS: Reported in EPA TSCA Inventory.

SAFETY PROFILE: Poison by intravenous route. Moderately toxic by ingestion. Slightly toxic by skin contact. When heated to decomposition it emits acrid smoke and irritating vapors.

FNO100 CAS:487-89-8 HR: 2
3-FORMYLINDOLE
mf: C_9H_7NO mw: 145.17

SYNS: INDOLE-3-ALDEHYDE ◇ INDOLE-3-CARBALDEHYDE ◇ INDOLE-3-CARBOXALDEHYDE ◇ 1H-INDOLE-3-CARBOXALDEHYDE (9CI) ◇ β-INDOLYLALDEHYDE

TOXICITY DATA with REFERENCE
ipr-mus LDLo:600 mg/kg   PCJOAU 6,33,72

CONSENSUS REPORTS: Reported in EPA TSCA Inventory.

SAFETY PROFILE: Moderately toxic by intraperitoneal route. When heated to decomposition it emits toxic vapors of NO_x.

FOE100 CAS:100-83-4 HR: D
m-FORMYLPHENOL
mf: $C_7H_6O_2$ mw: 122.13

SYNS: BENZALDEHYDE, m-HYDROXY- ◇ BENZALDEHYDE, 3-HYDROXY-(9CI) ◇ 3-FORMYLPHENOL ◇ m-HYDROXYBENZALDEHYDE ◇ meta-HYDROXYBENZALDEHYDE ◇ 3-HYDROXYBENZALDEHYDE

TOXICITY DATA with REFERENCE
sce-hmn:lyms 1 mmol/L   MUREAV 206,17,88

CONSENSUS REPORTS: Reported in EPA TSCA Inventory.

SAFETY PROFILE: Human mutation data reported. When heated to decomposition it emits acrid smoke and irritating vapors.

FOO525 CAS:124-73-2 HR: 1
FREON 114B2
mf: $C_2Br_2F_4$ mw: 259.84

SYNS: 1,2-DIBROMOPERFLUOROETHANE ◇ sym-DIBROMOTETRAFLUOROETHANE ◇ 1,2-DIBROMOTETRAFLUOROETHANE ◇ 1,2-DIBROMO-1,1,2,2-TETRAFLUOROETHANE ◇ ETHANE, 1,2-DIBROMOTETRAFLUORO- ◇ ETHANE, 1,2-DIBROMO-1,1,2,2-TETRAFLUORO-(9CI) ◇ F-114B2 ◇ FC 114B2 ◇ FLUOBRENE ◇ HALON 2402 ◇ KHLADON 114B2 ◇ R 114B2

TOXICITY DATA with REFERENCE
ihl-rat LC50:869 g/m^3/2H   GISAAA 55(2),17,90
ihl-mus LC50:300 g/m^3/2H   85JCAE -,136,86

CONSENSUS REPORTS: Reported in EPA TSCA Inventory.

SAFETY PROFILE: Slightly toxic by inhalation. When heated to decomposition it emits toxic vapors of Br^- and F^-.

FOO550 CAS:1717-00-6 HR: 1
FREON 141
mf: $C_2H_3Cl_2F$ mw: 116.95

SYNS: 1,1-DICHLORO-1-FLUOROETHANE ◇ ETHANE, 1,1-DICHLORO-1-FLUORO-

TOXICITY DATA with REFERENCE
ihl-rat LD50:240 g/m^3/2H   85GMAT-,46,82
ihl-mus LC50:151 g/m^3/2H   85JCAE-,134,86

SAFETY PROFILE: Slightly toxic by inhalation. When heated to decomposition it emits toxic vapors of F^- and Cl^-.

FPK050 CAS:539-47-9 HR: 3
2-FURANACRYLIC ACID
mf: $C_7H_6O_3$ mw: 138.13

TOXICITY DATA with REFERENCE
mmo-sat 10 μg/plate   JOPHDQ 1,15,78
ipr-mus LD50:276 mg/kg   YKKZAJ 104,793,84

CONSENSUS REPORTS: Reported in EPA TSCA Inventory.

SAFETY PROFILE: Poison by intravenous route. Mutation data reported. When heated to decomposition it emits acrid smoke and irritating vapors.

FPM100 CAS:623-17-6 HR: D
2-FURANMETHYL ACETATE
mf: $C_7H_8O_3$ mw: 140.15

SYNS: ACETIC ACID FURFURYL ESTER ◇ 2-ACETOXYMETHYLFURAN ◇ 2-FURANMETHANOL, ACETATE (9CI) ◇ FURFURYL ACETATE ◇ FURFURYL ALCOHOL, ACETATE

TOXICITY DATA with REFERENCE
mmo-sat 1 mg/plate   ENMUDM 8(Suppl 7),1,86
mma-sat 2500 μg/plate   ENMUDM 8(Suppl 7),1,86

CONSENSUS REPORTS: Reported in EPA TSCA Inventory.

SAFETY PROFILE: Mutation data reported. When heated to decomposition it emits acrid smoke and irritating vapors.

FPT100 CAS:525-79-1 HR: 2
N-FURFURYLADENINE
mf: $C_{10}H_9N_5O$ mw: 215.24

SYNS: ADENINE, N-FURFURYL- ◇ FAP ◇ N^6-FURFURYLADENINE ◇ 6-FURFURYLADENINE ◇ N^6-(FURFURYLAMINO)PURINE ◇ 6-(FURFURYLAMINO)PURINE ◇ KINETIN ◇ KINETIN (PLANT HORMONE)

TOXICITY DATA with REFERENCE
dns-hmn:leu 1 ~2mol/L   EXPEAM 32,29,76
oth-hmn:leu 1 ~2mol/L   EXPEAM 32,29,76
dni-hmn:leu 100 ~2mol/L   EXPEAM 32,29,76
ipr-mus LD50:450 mg/kg   NYKZAU 61(2),43S,65

CONSENSUS REPORTS: Reported in EPA TSCA Inventory.

SAFETY PROFILE: Moderately toxic by intraperitoneal route. Human mutation data reported. When heated to decomposition it emits toxic vapors of NO_x.

FPX100 CAS:1438-94-4 *HR: 3*
N-FURFURYL PYRROLE
mf: C_9H_9NO mw: 147.19

SYNS: 1-FURFURYLPYRROLE ◇ N-(2-FURFURYL)PYRROLE ◇ PYRROLE, 1-FURFURYL-

TOXICITY DATA with REFERENCE

orl-mus LD50:380 mg/kg   DCTODJ 3,249,80

CONSENSUS REPORTS: Reported in EPA TSCA Inventory.

SAFETY PROFILE: Poison by ingestion. When heated to decomposition it emits toxic vapors of NO_x.

G

GCE600 CAS:83-81-8 **HR: 3**
GEASTIGMOL
mf: $C_{16}H_{24}N_2O_2$ mw: 276.42

SYNS: ANALETIL ◊ 1,2-BENZENEDICARBOXAMIDE, N,N,N',N'-TETRA-ETHYL-(9CI) ◊ BIS-DIETHYLAMID KYSELINY FTALOVE ◊ neo-CARDIAMINE ◊ CARDIOVITAL ◊ CORETONIN ◊ GEASTIMOL ◊ NEO-CARDIAMINE ◊ NEOSPIRAN ◊ PHTHALAMIDE, N,N,N',N'-TETRAETHYL- ◊ PHTHALETHAMIDE ◊ o-PHTHALIC ACID BIS(DIETHYLAMIDE) ◊ o-PHTHALYLBIS(DIETHYLAMIDE) ◊ N,N,N',N'-TETRAETHYL-1,2-BENZENEDICARBOXAMIDE ◊ N,N,N',N'-TETRAETHYLPHTHALAMIDE ◊ UNISPIRAN

TOXICITY DATA with REFERENCE
scu-rbt LD50:30 mg/kg   85JCAE -,344,86

CONSENSUS REPORTS: Reported in EPA TSCA Inventory.

SAFETY PROFILE: Poison by subcutaneous route. When heated to decomposition it emits toxic vapors of NO_x.

GCU100 CAS:141-27-5 **HR: 2**
GERANALDEHYDE
mf: $C_{10}H_{16}O$ mw: 152.26

SYNS: CITRAL α ◊ α-CITRAL ◊ (E)-CITRAL ◊ trans-CITRAL ◊ trans-3,7-DIMETHYL-2,6-OCTADIENAL ◊ GERANIAL ◊ 2,6-OCTADIENAL, 3,7-DIMETHYL-, (E)-

TOXICITY DATA with REFERENCE
orl-rat LD50:500 mg/kg   FRXXBL #2448856

CONSENSUS REPORTS: Reported in EPA TSCA Inventory.

SAFETY PROFILE: Moderately toxic by ingestion. When heated to decomposition it emits acrid smoke and irritating vapors.

GDE810 CAS:106-29-6 **HR: 1**
GERANYL BUTANOATE
mf: $C_{14}H_{24}O_2$ mw: 224.38

SYNS: BUTANOIC ACID, 3,7-DIMETHYL-2,6-OCTADIENYL ESTER, (E)-(9CI) ◊ BUTYRIC ACID, 3,7-DIMETHYL-2,6-OCTADIENYL ESTER, (E)- ◊ trans-3,7-DIMETHYL-2,6-OCTADIEN-1-YL BUTYRATE ◊ GERANIOL BUTYRATE ◊ GERANYL BUTYRATE ◊ GERANYL n-BUTYRATE

TOXICITY DATA with REFERENCE
orl-rat LD50:10,660 mg/kg   FCTXAV 2,327,64
skn-rbt LD50:5 g/kg   FCTXAV 12,889,74

CONSENSUS REPORTS: Reported in EPA TSCA Inventory.

SAFETY PROFILE: Mildly toxic by ingestion and skin contact. When heated to decomposition it emits acrid smoke and irritating vapors.

GDG200 CAS:51-77-4 **HR: 2**
GERANYL FARNESYL ACETATE
mf: $C_{27}H_{44}O_2$ mw: 400.71

SYNS: DA-688 ◊ GEFARNATE ◊ GEFARNIL ◊ GEFARNYL ◊ 4,8,12-TETRADECATRIENOIC ACID, 5,9,13-TRIMETHYL-, 3,7-DIMETHYL-2,6-OCTADIENYLESTER, (E,E,E)- ◊ (E,E,E)-5,9,13-TRIMETHYL-4,8,12-TETRADECATRIENOIC ACID 3,7-DIMETHYL-2,6-OCTADIENYL ESTER

TOXICITY DATA with REFERENCE
orl-rat LD50:>9 g/kg   DRUGAY 6,266,82
ivn-rat LD50:2040 mg/kg   DRUGAY 6,266,82
ims-rat LD50:>13,500 mg/kg   DRUGAY 6,266,82
orl-mus LD50:>8 g/kg   JMCMAR 6,457,63
ipr-mus LD50:>4 g/kg   JMCMAR 6,457,63
ivn-mus LD50:2821 mg/kg   JMCMAR 6,457,63
ims-mus LD50:>13500 mg/kg   DRUGAY 6,266,82

CONSENSUS REPORTS: Reported in EPA TSCA Inventory.

SAFETY PROFILE: Moderately toxic by intravenous route. Mildly toxic by other routes. When heated to decomposition it emits acrid smoke and irritating vapors.

GFM300 CAS:617-65-2 **HR: 1**
dl-GLUTAMIC ACID (9CI)
mf: $C_5H_9NO_4$ mw: 147.15

SYNS: GLUTAMIC ACID, dl- ◊ (+ –)-GLUTAMIC ACID

TOXICITY DATA with REFERENCE
orl-hmn TDLo:71 mg/kg   SCIEAS 163,826,69

CONSENSUS REPORTS: Reported in EPA TSCA Inventory.

SAFETY PROFILE: Human systemic effects by ingestion: headache. When heated to decomposition it emits toxic vapors of NO_x.

GFY200 CAS:56-82-6 **HR: 1**
dl-GLYCERALDEHYDE
mf: $C_3H_6O_3$ mw: 90.09

SYNS: GLYCERALDEHYDE, (+ –)- ◊ dl-GLYCERIC ALDEHYDE

TOXICITY DATA with REFERENCE
mmo-sat 100 μg/plate   ABCHA6 47,2461,83
ipr-rat LD50:2 g/kg   JPPMAB 17,814,65

CONSENSUS REPORTS: Reported in EPA TSCA Inventory.

SAFETY PROFILE: Mildly toxic by intraperitoneal route. Mutation data reported.

GGA915   CAS:544-62-7   HR: 2
GLYCEROL MONOOCTADECYL ETHER
mf: $C_{21}H_{44}O_3$ mw: 344.65

SYNS: BATILOL ◊ BATYL ALCOHOL ◊ MONOOCTADECYL ETHER of GLYCEROL ◊ α-OCTADECYLETHER of GLYCEROL ◊ 1-O-OCTA-DECYLGLYCEROL ◊ 3-(OCTADECYLOXY)-1,2-PROPANEDIOL ◊ 1,2-PROPANEDIOL, 3-(OCTADECYLOXY)-

TOXICITY DATA with REFERENCE
ipr-mus LD50:750 mg/kg   NTIS** AD691-490

CONSENSUS REPORTS: Reported in EPA TSCA Inventory.

SAFETY PROFILE: Moderately toxic by intraperitoneal route. When heated to decomposition it emits acrid smoke and irritating vapors.

GGW600   CAS:930-37-0   HR: D
GLYCIDOL METHYL ETHER
mf: $C_4H_8O_2$ mw: 88.12

SYNS: 1,2-EPOXY-3-METHOXYPROPANE ◊ GLYCIDYL METHYL ETHER ◊ (METHOXYMETHYL)OXIRANE ◊ 3-METHOXYPROPYLENE OXIDE ◊ METHYL GLYCIDYL ETHER ◊ OXIRANE, (METHOXYMETHYL)-(9CI) ◊ PROPANE, 1,2-EPOXY-3-METHOXY-

TOXICITY DATA with REFERENCE
mmo-sat 100 µg/plate   MUREAV 172,105,86
mma-sat 33 µg/plate   MUREAV 172,105,86
oth-esc 3300 µmol/L   MUREAV 231,205,90
mmo-klp 200 µmol/L   MUREAV 89,269,81
sce-ham:lng 1250 µmol/L   MUREAV 249,55,91

CONSENSUS REPORTS: Reported in EPA TSCA Inventory.

SAFETY PROFILE: Mutation data reported. When heated to decomposition it emits acrid smoke and irritating vapors.

GGY100   CAS:2461-15-6   HR: 1
GLYCIDYL 2-ETHYLHEXYL ETHER
mf: $C_{11}H_{22}O_2$ mw: 186.33

SYNS: 2-ETHYLHEXYL GLYCIDYL ETHER ◊ (((2-ETHYL-HEXYL)OXY)METHYL)OXIRANE ◊ OXIRANE, (((2-ETHYL-HEXYL)OXY)METHYL)-(9CI) ◊ PROPANE, 1,2-EPOXY-3-((2-ETHYL-HEXYL)OXY)-

TOXICITY DATA with REFERENCE
mma-sat 100 µg/plate   MUREAV 172,105,86
orl-rat LD50:7800 mg/kg   38MKAJ 2A,2210,81

CONSENSUS REPORTS: Reported in EPA TSCA Inventory.

SAFETY PROFILE: Slightly toxic by ingestion. Mutation data reported. When heated to decomposition it emits acrid smoke and irritating vapors.

GGY150   CAS:2451-62-9   HR: D
GLYCIDYL ISOCYANURATE
mf: $C_{12}H_{15}N_3O_6$ mw: 297.30

SYNS: TGT ◊ s-TRIAZINE-2,4,6(1H,3H,5H)-TRIONE, TRIS(2,3-EPOXYPROPYL)- ◊ s-TRIAZINE-2,4,6(1H,3H,5H)-TRIONE, 1,3,5-TRIS(2,3-EPOXYPROPYL)- ◊ s-TRIAZINE-2,4,6(1H,3H,5H)-TRIONE, 1,3,5-TRIS(OX-IRANYLMETHYL)-(9CI) ◊ TRIGLYCIDYL ISOCYANURATE ◊ N,N′,N″-TRIGLYCIDYL ISOCYANURATE ◊ 1,3,5-TRIGLYCIDYL ISOCYANURATE ◊ 1,3,5-TRIGLYCIDYLISOCYANURIC ACID ◊ TRIS(EPOXYPROPYL)ISOCYANURATE ◊ TRIS(2,3-EPOXYPROPYL)ISOCYANURATE

TOXICITY DATA with REFERENCE
mmo-sat 333 µg/plate   EMMUEG 19(Suppl 21),2,92
mma-sat 2 mg/plate   EMMUEG 19(Suppl 21),2,92
cyt-mus-orl 4 g/kg/5D-C   EPASR* 8EHQ-0491-0490
cyt-ham:lng 1200 µg/L   MUREAV 241,175,90

SAFETY PROFILE: Mutation data reported. When heated to decomposition it emits toxic vapors of NO_x and CN^-.

GGY175   CAS:2186-24-5   HR: D
GLYCIDYL p-TOLYL ETHER
mf: $C_{10}H_{12}O_2$ mw: 164.22

SYNS: p-CRESOL GLYCIDYL ETHER ◊ p-CRESYL GLYCIDYL ETHER ◊ 1,2-EPOXY-3-(p-TOLYLOXY)PROPANE ◊ GLYCIDYL 4-METHYL-PHENYL ETHER ◊ ((4-METHYLPHENOXY)METHYL)OXIRANE ◊ METHYLPHENYL GLYCIDYL ETHER ◊ OXIRANE, ((4-METHYL-PHENOXY)METHYL)-(9CI) ◊ PROPANE, 1,2-EPOXY-3-(p-TOLYLOXY)-

TOXICITY DATA with REFERENCE
mmo-sat 66 nmol/plate   MUREAV 93,297,82
mma-sat 100 µg/plate   MUREAV 172,105,86

SAFETY PROFILE: Mutation data reported. When heated to decomposition it emits acrid smoke and irritating vapors.

GHA050   CAS:107-43-7   HR: 2
GLYCINE BETAINE
mf: $C_5H_{11}NO_2$ mw: 117.17

SYNS: ABROMINE ◊ BETAINE ◊ (CAR-BOXYMETHYL)TRIMETHYLAMMONIUM HYDROXIDE, inner salt ◊ α-EAR-LEINE ◊ GLYCOCOLL BETAINE ◊ GLYCYLBETAINE ◊ GLYKOKOLLBE-TAIN ◊ JORTAINE ◊ LORAMINE AMB 13 ◊ LYCINE ◊ OXYNEURINE ◊ RUBRINE C ◊ TRIMETHYLGLYCINE ◊ TRIMETHYLGLYCOCOLL

TOXICITY DATA with REFERENCE
scu-mus LD50:10,800 mg/kg   ABMGAJ 3,28,59
ivn-mus LD50:830 mg/kg   MPHEAE 16,529,67

CONSENSUS REPORTS: Reported in EPA TSCA Inventory.

SAFETY PROFILE: Moderately toxic by intravenous route. Mildly toxic by subcutaneous route. When heated to decomposition it emits toxic vapors of NO_x.

GHA100 CAS:623-33-6   *HR: 2*
GLYCINE, ETHYL ESTER, HYDROCHLORIDE
mf: $C_4H_9NO_2 \cdot ClH$ mw: 139.60

SYN: USAF DO-10

TOXICITY DATA with REFERENCE
ipr-mus LD50:750 mg/kg   NTIS** AD277-689

CONSENSUS REPORTS: Reported in EPA TSCA Inventory.

SAFETY PROFILE: Moderately toxic by intraperitoneal route. When heated to decomposition it emits toxic vapors of NO_x, HCl, and Cl^-.

GIK100 CAS:1769-41-1   *HR: D*
GLYOXANILIDE OXIME
mf: $C_8H_8N_2O_2$ mw: 164.18

SYNS: ACETAMIDE, 2-(HYDROXYIMINO)-N-PHENYL-(9CI) ◇ GLYOXYLANILIDE, OXIME ◇ GLYOXYLANILIDE, 2-OXIME ◇ 2-(HYDROXYIMINO)-N-PHENYLACETAMIDE ◇ ISONITROSOACETANILIDE ◇ 2-ISONITROSOACETANILIDE ◇ ISONITROSOACETYLANILINE

TOXICITY DATA with REFERENCE
mmo-sat 3333 µg/plate   MUREAV 204,149,88
msc-mus:lyms 613 mg/L   MUREAV 204,149,88

CONSENSUS REPORTS: Reported in EPA TSCA Inventory.

SAFETY PROFILE: Mutation data reported. When heated to decomposition it emits toxic vapors of NO_x.

GJI400 CAS:477-73-6   *HR: 3*
GOSSYPIMINE
mf: $C_{20}H_{19}N_4 \cdot Cl$ mw: 350.88

SYNS: BASIC RED 2 ◇ BRILLIANT SAFRANINE BR ◇ BRILLIANT SAFRANINE G ◇ BRILLIANT SAFRANINE GR ◇ CALCOZINE RED Y ◇ CERVEN ZASADITA 2 ◇ C.I. 50240 ◇ C.I. BASIC RED 2 ◇ 2,8-DIMETHYLPHENOSAFRANINE ◇ HIDACO SAFRANINE ◇ LEATHER RED HT ◇ MITSUI SAFRANINE ◇ NIPPON KAGAKU SAFRANINE GK ◇ NIPPON KAGAKU SAFRANINE T ◇ PHENAZINIUM, 3,7-DIAMINO-2,8-DIMETHYL-5-PHENYL-, CHLORIDE ◇ SAFRANIN ◇ SAFRANINE ◇ SAFRANINE A ◇ SAFRANINE B ◇ SAFRANINE G ◇ SAFRANINE GF ◇ SAFRANINE J ◇ SAFRANINE O ◇ SAFRANINE OK ◇ SAFRANINE SUPERFINE G ◇ SAFRANINE T ◇ SAFRANINE TH ◇ SAFRANINE TN ◇ SAFRANINE Y ◇ SAFRANINE YN ◇ SAFRANINE ZH ◇ SAFRANIN T ◇ TOLUSAFRANINE

TOXICITY DATA with REFERENCE
mmo-sat 16 µg/plate   TRENAF 27,153,76
mma-sat 16 µg/plate   TRENAF 27,153,76
dnr-esc 4 µg/well   ENMUDM 3,429,81
dnr-bcs 2 mg/disc   TRENAF 27,153,76
ivn-rat LD50:28,740 µg/kg   SMBUA9 9,96,51
orl-mus LDLo:1600 mg/kg   JPMSAE 69,327,80
ivn-mus LD50:24,020 µg/kg   SMBUA9 9,96,51
ivn-rbt LD50:26,940 µg/kg   SMBUA9 9,96,51

CONSENSUS REPORTS: Reported in EPA TSCA Inventory.

SAFETY PROFILE: Poison by intravenous route. Moderately toxic by ingestion. Mutation data reported. When heated to decomposition it emits toxic vapors of NO_x and Cl^-.

GKW100 CAS:593-85-1   *HR: 3*
GUANIDINE CARBONATE
mf: $CH_5N_3 \cdot 1/2CH_2O_3$ mw: 493.30

SYNS: AI3-14631 ◇ BISGUANIDINIUM CARBONATE ◇ CARBONIC ACID, compd. with GUANIDINE (1:2) ◇ DIGUANIDINIUM CARBONATE ◇ GUANIDINIUM CARBONATE

TOXICITY DATA with REFERENCE
orl-mus LD50:350 mg/kg   CKFRAY 1,434,52
scu-rbt LDLo:500 mg/kg   HBAMAK 4,1352,35

CONSENSUS REPORTS: Reported in EPA TSCA Inventory.

SAFETY PROFILE: Poison by ingestion. Moderately toxic by subcutaneous route. When heated to decomposition it emits toxic vapors of NO_x.

GLS750 CAS:85-32-5   *HR: 2*
5'-GUANYLIC ACID
mf: $C_{10}H_{14}N_5O_8P$ mw: 363.26

SYNS: GMP ◇ 5'-GMP ◇ GUANIDINE MONOPHOSPHATE ◇ GUANOSINE MONOPHOSPHATE ◇ GUANOSINE 5'-MONOPHOSPHATE ◇ GUANOSINE 5'-MONOPHOSPHORIC ACID ◇ GUANOSINE 5'-PHOSPHATE ◇ GUANYLIC ACID

TOXICITY DATA with REFERENCE
oth-hmn:oth 1 mmol/L   JIDEAE 65,52,75
ipr-mus LDLo:1500 mg/kg   ANYAA9 60,251,54

CONSENSUS REPORTS: Reported in EPA TSCA Inventory.

SAFETY PROFILE: Moderately toxic by intraperitoneal route. Human mutation data reported. When heated to decomposition it emits toxic vapors of NO_x and PO_x.

H

HAO600 CAS:475-25-2 **HR: D**
HEMATEIN
mf: $C_{16}H_{12}O_6$ mw: 300.28

SYNS: BENZ(b)INDENO(1,2-d)PYRAN-9(6H)-ONE,6a,7-DIHYDRO-3,4,6a,10-TETRAHYDROXY- ◇ HAEMATEIN ◇ HEMATINE ◇ 3,4,6a,10-TETRAHYDROXY-6a,7-DIHYDROBENZ(b)INDENO(1,2-d)PYRAN-9(6H)-ONE

TOXICITY DATA with REFERENCE
cyt-ham:ovr 20 ~2mol/L/5H-C ENMUDM 1,27,79

CONSENSUS REPORTS: Reported in EPA TSCA Inventory.

SAFETY PROFILE: Mutation data reported. When heated to decomposition it emits acrid smoke and irritating vapors.

HAQ600 CAS:71-67-0 **HR: 3**
HEPATOSULFALEIN
mf: $C_{20}H_8Br_4O_{10}S_2 \cdot 2Na$ mw: 838.02

SYNS: BROMOSULFALEIN ◇ BROMOSULFOPHTHALEIN ◇ BROMOSULPHALEIN ◇ BROMOSULPHTHALEIN ◇ BROMOTALEINA ◇ BROMSULFALEIN ◇ BROMSULFAN ◇ BROMSULFOPHTHALEIN ◇ BROMSULFTHALEIN ◇ BROMSULPHTHALEIN ◇ BROMSULPHTHALEIN ◇ BROM-TETRAGNOST ◇ BROMTHALEIN ◇ BSF ◇ BSF SIMES ◇ BSP ◇ BSP SODIUM ◇ CBSP ◇ DISODIUM BROMSULFOPHTHALEIN ◇ HEPARTEST ◇ HEPARTESTABROME ◇ PHENOLPHTHALEIN, 4,5,6,7-TETRABROMO-3',3''-DISULFO-, DISODIUM SALT ◇ PHENOLTETRABROMOPHTHALEINSULFONATE ◇ SODIUM BROMOSULFALEIN ◇ SODIUM BROMOSULFOPHTHALEIN ◇ SODIUM BROMSULPHALEIN ◇ SODIUM BROMSULPHTHALEIN ◇ SODIUM PHENOL TETRABROMOPHTHALEIN ◇ SODIUM SULFOBROMOPHTHALEIN ◇ SODIUM SULPHOBROMOPHTHALEIN ◇ SULFOBROMOPHTHALEIN ◇ SULFOBROMOPHTHALEIN SODIUM ◇ SULFOBROMPHTHALEIN ◇ SULPHOBROMOPHTHALEIN ◇ SULPHOBROMOPHTHALEIN SODIUM ◇ TETRABROMOPHENOLSULFOPHTHALEIN ◇ TETRABROMOSULFOPHTHALEIN ◇ TETRABROMSULFTHALEIN

TOXICITY DATA with REFERENCE
ivn-mus LD50:334 mg/kg DRUGAY 6,394,82

CONSENSUS REPORTS: Reported in EPA TSCA Inventory.

SAFETY PROFILE: Poison by intravenous route. When heated to decomposition it emits toxic vapors of SO_x and Br^-.

HAS100 CAS:629-78-7 **HR: 1**
HEPTADECANE
mf: $C_{17}H_{36}$ mw: 240.53

SYN: n-HEPTADECANE

TOXICITY DATA with REFERENCE
ivn-mus LDLo:9821 mg/kg APTOA6 37,56,75

CONSENSUS REPORTS: Reported in EPA TSCA Inventory.

SAFETY PROFILE: Mildly toxic by intravenous route. When heated to decomposition it emits acrid smoke and irritating vapors.

HAW100 CAS:375-01-9 **HR: 2**
2,2,3,3,4,4,4-HEPTAFLUOROBUTANOL
mf: $C_4H_3F_7O$ mw: 200.07

SYNS: 1-BUTANOL, 2,2,3,3,4,4,4-HEPTAFLUORO- ◇ α-α-DIHYDROPERFLUOROBUTANOL ◇ 1,1-DIHYDROPERFLUOROBUTANOL ◇ 1,1-H,H-HEPTAFLUOROBUTANOL

TOXICITY DATA with REFERENCE
orl-rat LD50:3630 mg/kg GTPZAB 13(10),29,69

CONSENSUS REPORTS: Reported in EPA TSCA Inventory.

SAFETY PROFILE: Moderately toxic by ingestion. When heated to decomposition it emits toxic vapors of F^-.

HBL100 CAS:2499-58-3 **HR: 2**
HEPTYL ACRYLATE
mf: $C_{10}H_{18}O_2$ mw: 170.28

SYNS: ACRYLIC ACID, HEPTYL ESTER ◇ ENT 15748 ◇ 2-PROPENOIC ACID, HEPTYL ESTER

TOXICITY DATA with REFERENCE
orl-rat LD50:4500 mg/kg GTPZAB 26(9),52,82
orl-mus LD50:3300 mg/kg GTPZAB 26(9),52,82
ihl-mus LC50:1020 mg/m^3 GTPZAB 20(11),41,76

CONSENSUS REPORTS: Reported in EPA TSCA Inventory.

SAFETY PROFILE: Moderately toxic by ingestion and inhalation routes. When heated to decomposition it emits acrid smoke and irritating vapors.

HBL600 CAS:111-68-2 **HR: 3**
1-HEPTYLAMINE
mf: $C_7H_{17}N$ mw: 115.25

SYNS: 1-AMINOHEPTANE ◇ 1-HEPTANAMINE ◇ HEPTYLAMINE ◇ n-HEPTYLAMINE

TOXICITY DATA with REFERENCE
ipr-rat LDLo:75 mg/kg FATOAO 31,238,68
ipr-mus LD50:100 mg/kg JAPMA8 30,623,41

CONSENSUS REPORTS: Reported in EPA TSCA Inventory.

SAFETY PROFILE: Poison by intraperitoneal route. When heated to decomposition it emits toxic vapors of NO_x.

HBM490   CAS:123-82-0   *HR: 3*
2-HEPTYLAMINE
mf: $C_7H_{17}N$ mw: 115.25

SYNS: dl-2-AMINOHEPTANE ◊ ARMEEN L-7 ◊ HEPTAMINE ◊ 2-HEPTANAMINE ◊ HEPTEDRINE ◊ 1-METHYLHEXYLAMINE ◊ RINEPTIL ◊ TUAMINE ◊ TUAMINOHEPTANE

TOXICITY DATA with REFERENCE
scu-rat LD50:130 mg/kg   JPETAB 85,119,45
ipr-mus LDLo:60 mg/kg   JAPMA8 30,623,41
scu-mus LD50:115 mg/kg   FEPRA7 4,139,45

CONSENSUS REPORTS: Reported in EPA TSCA Inventory.

SAFETY PROFILE: Poison by subcutaneous and intraperitoneal routes. When heated to decomposition it emits toxic fumes of NO_x.

HBN100   CAS:629-04-9   *HR: 2*
n-HEPTYL BROMIDE
mf: $C_7H_{15}Br$ mw: 179.13

SYNS: 1-BROMOHEPTANE ◊ HEPTANE, 1-BROMO- ◊ HEPTYL BROMIDE

TOXICITY DATA with REFERENCE
ipr-mus LD50:2440 mg/kg   GTPZAB 20(12),52,76
ihl-uns LC50:12 g/m³   GTPZAB 18(4),55,74

CONSENSUS REPORTS: Reported in EPA TSCA Inventory.

SAFETY PROFILE: Moderately toxic by intraperitoneal route. Slightly toxic by inhalation. When heated to decomposition it emits toxic vapors of Br^-.

HBN200   CAS:104-67-6   *HR: 1*
γ-HEPTYLBUTYROLACTONE
mf: $C_{11}H_{20}O_2$ mw: 184.31

SYNS: ALDEHYDE C-14 ◊ ALDEHYDE C-14 PEACH ◊ 2(3H)-FURANONE, 5-HEPTYLDIHYDRO- ◊ γ-n-HEPTYLBUTYROLACTONE ◊ 4-HYDROXYUNDECANOIC ACID LACTONE ◊ 4-HYDROXYUNDECANOIC ACID, γ-LACTONE ◊ PEACH ALDEHYDE ◊ PEACH LACTONE ◊ PERSICOL ◊ γ-UNDECALACTONE ◊ UNDECANOIC ACID, 4-HYDROXY-, γ-LACTONE ◊ γ-UNDECANOLACTONE ◊ γ-UNDECANOLIDE ◊ 1,4-UNDECANOLIDE ◊ 4-UNDECANOLIDE ◊ γ-UNDEKALAKTON

TOXICITY DATA with REFERENCE
skn-rbt 100 mg/24H SEV   CTOIDG 94(8),41,79
skn-gpg 100 mg/24H MOD   CTOIDG 94(8),41,79
dnr-bcs 10 mg/disc   OIGZSE 34,267,85
orl-rat LD50:18500 mg/kg   FCTXAV 2,327,64

CONSENSUS REPORTS: Reported in EPA TSCA Inventory.

SAFETY PROFILE: Slightly toxic by ingestion. A severe skin irritant. Mutation data reported. When heated to decomposition it emits acrid smoke and irritating vapors.

HBO650   CAS:1085-12-7   *HR: 2*
HEPTYL 4-HYDROXYBENZOATE
mf: $C_{14}H_{20}O_3$ mw: 236.34

SYNS: BENZOIC ACID, p-HYDROXY-, HEPTYL ESTER ◊ BENZOIC ACID, 4-HYDROXY-, HEPTYL ESTER (9CI) ◊ HEPTYL p-HYDROXYBENZOATE ◊ HEPTYL PARABEN ◊ p-HYDROXYBENZOIC ACID HEPTYL ESTER ◊ NIPAHEPTYL ◊ p-OXYBENZOESAEUREHEPTYLESTER ◊ STAYPRO WS 7

TOXICITY DATA with REFERENCE
scu-mus LD50:4000 mg/kg   AIPTAK 128,135,60

CONSENSUS REPORTS: Reported in EPA TSCA Inventory.

SAFETY PROFILE: Moderately toxic by subcutaneous route. When heated to decomposition it emits acrid smoke and irritating vapors.

HCI000   CAS:67-72-1   *HR: 3*
HEXACHLOROETHANE
DOT: NA 9037
mf: C_2Cl_6 mw: 236.72

PROP: Rhombic, triclinic, or cubic crystals, colorless, camphor-like odor. Mp: 186.6° (subl), d: 2.091, vap press: 1 mm @ 32.7°, bp: 186.8° (triple point). Sol in alc, benzene, chloroform, ether, oils; insol in water.

SYNS: AVLOTANE ◊ CARBON HEXACHLORIDE ◊ DISTOKAL ◊ DISTOPAN ◊ DISTOPIN ◊ EGITOL ◊ ETHANE HEXACHLORIDE ◊ ETHYLENE HEXACHLORIDE ◊ FALKITOL ◊ FASCIOLIN ◊ HEXACHLOR-AETHAN (GERMAN) ◊ 1,1,1,2,2,2-HEXACHLOROETHANE ◊ HEXACHLOROETHYLENE ◊ MOTTENHEXE ◊ NCI-C04604 ◊ PERCHLOROETHANE ◊ PHENOHEP ◊ RCRA WASTE NUMBER U131

TOXICITY DATA with REFERENCE
orl-rat TDLo:5500 mg/kg (6–16D preg):REP   AIHAAP 40,187,79
orl-rat TDLo:5500 mg/kg (6–16D preg):REP   AIHAAP 40,187,79
orl-mus TDLo:230 g/kg/78W-I:CAR   NCITR* NCI-CG-TR-68,78
orl-mus TD:460 g/kg/78W-I:CAR   NCITR* NCI-CG-TR-68,78
orl-rat LD50:4460 mg/kg   AIHAAP 40,187,79
ipr-rat LDLo:2900 mg/kg   AIHAAP 40,187,79
ipr-mus LD50:4500 mg/kg   ARZNAD 11,902,61
ivn-dog LDLo:325 mg/kg   QJPPAL 7,205,34
skn-rbt LD50:32 g/kg   AIHAAP 40,187,79
scu-rbt LDLo:4000 mg/kg   QJPPAL 7,205,34
orl-gpg LD50:4970 mg/kg   AIHAAP 40,187,79

CONSENSUS REPORTS: IARC Cancer Review: Group 3 IMEMDT 7,56,87; Animal Limited Evidence IMEMDT

20,467,79. NCI Carcinogenesis Bioassay (gavage); Clear Evidence: mouse NCITR* NCI-CG-TR-68,78. NCI Carcinogenesis Bioassay (gavage); No Evidence: rat NCITR* NCI-CG-TR-68,78. Community Right-To-Know List. Reported in EPA TSCA Inventory. EPA Genetic Toxicology Program.

OSHA PEL: TWA 1 ppm (skin)
ACGIH TLV: TWA 1 ppm; Suspected Human Carcinogen
DFG MAK: 1 ppm (10 mg/m^3)
NIOSH REL: (Hexachloroethane) Reduce to lowest level.
DOT Classification: ORM-A; Label: None

SAFETY PROFILE: Suspected carcinogen with experimental carcinogenic data. A poison by intravenous route. Moderately toxic by intraperitoneal route. Mildly toxic by ingestion. Experimental reproductive effects. Liver injury has resulted from exposure to this material. An insecticide. Slightly explosive by spontaneous chemical reaction. Dehalogenation of this material by reaction with alkalies, metals, etc., will produce spontaneous explosive chloroacetylenes. When heated to decomposition it emits highly toxic fumes of Cl$^-$ and phosgene. See also CHLORINATED HYDROCARBONS, ALIPHATIC.

HCO600 CAS:544-76-3 **HR: 2**
HEXADECANE
mf: $C_{16}H_{34}$ mw: 226.50

SYNS: CETANE ◇ n-CETANE ◇ n-HEXADECANE

TOXICITY DATA with REFERENCE
skn-man 50 mg/48H SEV   CTOIDG 94(8),41,79
skn-rat 100 mg/24H SEV   CTOIDG 94(8),41,79
skn-rbt 100 mg/24H SEV   CTOIDG 94(8),41,79
skn-pig 50 mg/48H SEV   CTOIDG 94(8),41,79
skn-gpg 100 mg/24H SEV   CTOIDG 94(8),41,79
ivn-mus LDLo:9821 mg/kg   APTOA6 37,56,75

CONSENSUS REPORTS: Reported in EPA TSCA Inventory.

SAFETY PROFILE: Slightly toxic by intravenous route. A severe human skin irritant. When heated to decomposition it emits acrid smoke and irritating vapors.

HCP100 CAS:629-70-9 **HR: 1**
HEXADECYL ACETATE
mf: $C_{18}H_{36}O_2$ mw: 284.54

SYNS: 1-ACETOXYHEXADECANE ◇ CETYL ACETATE ◇ ENT 1025 ◇ 1-HEXADECANOL, ACETATE ◇ PALMITYL ACETATE

TOXICITY DATA with REFERENCE
skn-rbt 500 mg/24H MLD   FCTOD7 21,663,83
orl-rat LD50:>5 g/kg   FCTOD7 21,663,83
skn-rbt LD50:>5 g/kg   FCTOD7 21,663,83

CONSENSUS REPORTS: Reported in EPA TSCA Inventory.

SAFETY PROFILE: Slightly toxic by ingestion and skin contact. A skin irritant. When heated to decomposition it emits acrid smoke and irritating vapors.

HCP525 CAS:112-69-6 **HR: 1**
HEXADECYLDIMETHYLAMINE
mf: $C_{18}H_{39}N$ mw: 269.58

SYNS: ARMEEN DM16D ◇ BAIRDCAT B16 ◇ CETYLDIMETHYLAMINE ◇ DIMETHYLCETYLAMINE ◇ N,N-DIMETHYLCETYLAMINE ◇ N,N-DIMETHYLHEXADECYLAMINE ◇ DIMETHYLPALMITYLAMINE ◇ GENAMIN 16R302D ◇ HEXADECYLAMINE, N,N-DIMETHYL- ◇ PALMITYLDIMETHYLAMINE

TOXICITY DATA with REFERENCE
orl-mus LD50:>3 g/kg   OYYAA2 12,171,76
scu-mus LD50:>3 g/kg   OYYAA2 12,171,76

SAFETY PROFILE: Slightly toxic by ingestion and subcutaneous routes. When heated to decomposition it emits toxic vapors of NO_x.

HCP700 CAS:540-10-3 **HR: 1**
HEXADECYL PALMITATE
mf: $C_{32}H_{64}O_2$ mw: 480.96

SYNS: CETIN ◇ CETYL PALMITATE ◇ HEXADECANOIC ACID, HEXADECYL ESTER ◇ PALMITIC ACID, HEXADECYL ESTER ◇ PALMITYL PALMITATE ◇ STANDAMUL 1616

TOXICITY DATA with REFERENCE
skn-rbt 500 mg/24H MLD   JACTDZ 1(2),13,82

CONSENSUS REPORTS: Reported in EPA TSCA Inventory.

SAFETY PROFILE: A skin irritant. When heated to decomposition it emits acrid smoke and irritating vapors.

HCP800 CAS:140-72-7 **HR: 3**
HEXADECYLPYRIDINE BROMIDE
mf: $C_{21}H_{38}N•Br$ mw: 384.51

SYNS: ACETOQUAT CPB ◇ BROMOCET ◇ CETAPHARM ◇ CETASOL ◇ CETAZOL ◇ CETYLPYRIDINIUM BROMIDE ◇ N-CETYLPYRIDINIUM BROMIDE ◇ 1-CETYLPYRIDINIUM BROMIDE ◇ FIXANOL C ◇ HEXADECYLPYRIDINIUM BROMIDE ◇ N-HEXADECYLPYRIDINIUM BROMIDE ◇ 1-HEXADECYLPYRIDINIUM BROMIDE ◇ MORPAN CBP ◇ NITROGENOL ◇ PYRIDINIUM, 1-HEXADECYL-, BROMIDE ◇ SEPRISAN ◇ STEROGENOL ◇ TsPB

TOXICITY DATA with REFERENCE
dni-hmn:lyms 100 mg/L   BECTA6 28,504,82
ipr-mus LDLo:50 mg/kg   ARZNAD 21,121,71

CONSENSUS REPORTS: Reported in EPA TSCA Inventory.

SAFETY PROFILE: Poison by intraperitoneal route. Human mutation data reported. When heated to decomposition it emits toxic vapors of NO_x and Br$^-$.

HCP900   CAS:1120-01-0   *HR: 3*
HEXADECYL SODIUM SULFATE
mf: $C_{16}H_{33}O_4S$•Na   mw: 344.54

SYNS: AVITEX C ◇ AVITEX SF ◇ CETYL SODIUM SULFATE ◇ CETYL SULFATE SODIUM SALT ◇ CONCO SULFATE C ◇ 1-HEXADECANOL, HYDROGEN SULFATE, SODIUM SALT ◇ NIKKOL S.C.S ◇ SHS ◇ SODIUM CETYL SULFATE ◇ SODIUM HEXADECYL SULFATE ◇ SODIUM MONOHEXADECYL SULFATE ◇ SODIUM PALMITYL SULFATE ◇ TERGITOL ANIONIC 7

TOXICITY DATA with REFERENCE
ipr-mus LD50:356 mg/kg   JAPMA8 42,283,53

CONSENSUS REPORTS: Reported in EPA TSCA Inventory.

SAFETY PROFILE: Poison by intraperitoneal route. When heated to decomposition it emits toxic vapors of SO_x.

HDC450   CAS:382-10-5   *HR: 3*
HEXAFLUOROISOBUTYLENE
mf: $C_4H_2F_6$ mw: 164.06

SYNS: 1,1-BIS(TRIFLUOROMETHYL)ETHENE ◇ 3,3,3,4,4,4-HEXAFLUOROISOBUTYLENE ◇ PROPENE, 3,3,3-TRIFLUORO-2-(TRIFLUOROMETHYL)- ◇ 3,3,3-TRIFLUORO-2-(TRIFLUOROMETHYL) PROPENE

TOXICITY DATA with REFERENCE
ihl-rat LC50:1425 ppm/4H   EPASR* 8EHQ-0683-0476S
ihl-rat TCLo:53 ppm/6H/2W-I   TXAPA9 86,327,86
ihl-rat TCLo:30 ppm/6H/13W-I   TXAPA9 86,327,86

CONSENSUS REPORTS: Reported in EPA TSCA Inventory.

SAFETY PROFILE: Poison by inhalation. When heated to decomposition it emits toxic vapors of F^-.

HDF050   CAS:428-59-1   *HR: D*
HEXAFLUOROPROPENE EPOXIDE
DOT: NA 1956
mf: C_3F_6O mw: 166.03

SYNS: HEXAFLUOROEPOXYPROPANE ◇ HEXAFLUORO-1,2-EPOXYPROPANE ◇ HEXAFLUOROPROPENE OXIDE ◇ HEXAFLUOROPROPYLENE OXIDE (DOT) ◇ OXIRANE, TRIFLUORO(TRIFLUOROMETHYL)- ◇ PERFLUORO(METHYLOXIRANE) ◇ PERFLUOROPROPYLENE OXIDE ◇ PROPANE, 1,2-EPOXY-1,1,2,3,3,3-HEXAFLUORO- ◇ PROPYLENE OXIDE HEXAFLUORIDE ◇ (TRIFLUOROMETHYL)TRIFLUOROOXIRANE ◇ TRIFLUORO(TRIFLUOROMETHYL)OXIRANE

DOT Classification: Nonflammable Gas; Label: Nonflammable Gas

CONSENSUS REPORTS: Reported in EPA TSCA Inventory.

SAFETY PROFILE: Nonflammable gas shipped under pressure. When heated to decomposition it emits toxic vapors of F^-.

HDF300   CAS:77-99-6   *HR: 1*
HEXAGLYCERINE
mf: $C_6H_{14}O_3$ mw: 134.20

SYNS: ETHRIOL ◇ ETHYLTRIMETHYLOLMETHANE ◇ ETRIOL ◇ ETTRIOL ◇ 1,3-PROPANEDIOL, 2-ETHYL-2-(HYDROXYMETHYL)- ◇ TMP ◇ TMP (ALCOHOL) ◇ 1,1,1-TRI(HYDROXYMETHYL)PROPANE ◇ TRIMETHYLOLPROPANE ◇ 1,1,1-TRIMETHYLOLPROPANE ◇ TRIS(HYDROXYMETHYL)PROPANE ◇ 1,1,1-TRIS(HYDROXYMETHYL)PROPANE

TOXICITY DATA with REFERENCE
orl-rat LD50:14,100 mg/kg   HYSAAV 32(5),288,67
orl-mus LD50:13,700 mg/kg   HYSAAV 32(5),288,67

CONSENSUS REPORTS: Reported in EPA TSCA Inventory.

SAFETY PROFILE: Mildly toxic by ingestion. When heated to decomposition it emits acrid smoke and irritating vapors.

HDH100   CAS:98-89-5   *HR: 2*
HEXAHYDROBENZOIC ACID
mf: $C_7H_{12}O_2$ mw: 128.19

SYNS: CARBOXYCYCLOHEXANE ◇ CYCLOHEXANECARBOXYLIC ACID ◇ CYCLOHEXANOIC ACID ◇ CYCLOHEXYLCARBOXYLIC ACID

TOXICITY DATA with REFERENCE
orl-rat LD50:3265 mg/kg   DCTODJ 3,249,80

CONSENSUS REPORTS: Reported in EPA TSCA Inventory.

SAFETY PROFILE: Moderately toxic by ingestion. When heated to decomposition it emits acrid smoke and irritating vapors.

HDH200   CAS:100-49-2   *HR: 3*
HEXAHYDROBENZYL ALCOHOL
mf: $C_7H_{14}O$ mw: 114.21

SYNS: BENZYL ALCOHOL, HEXAHYDRO- ◇ CYCLOHEXANECARBINOL ◇ CYCLOHEXANEMETHANOL ◇ CYCLOHEXYLCARBINOL ◇ CYCLOHEXYLMETHANOL ◇ HYDROXYMETHYLCYCLOHEXANE ◇ METHANOL, CYCLOHEXYL- ◇ USAF DO-49

TOXICITY DATA with REFERENCE
ipr-mus LD50:250 mg/kg   NTIS** AD277-689

CONSENSUS REPORTS: Reported in EPA TSCA Inventory.

SAFETY PROFILE: Poison by intraperitoneal route. When heated to decomposition it emits acrid smoke and irritating vapors.

HDS300   CAS:535-75-1   *HR: 2*
HEXAHYDROPICOLINIC ACID
mf: $C_6H_{11}NO_2$ mw: 129.18

SYNS: ACIDE PIPECOLIQUE ◊ ACIDE PIPERIDINE-CARBOXYLIQUE-2 ◊ DIHYDROBAIKIANE ◊ HOMOPROLINE ◊ PIPECOLATE ◊ PIPECOLIC ACID ◊ PIPECOLINIC ACID ◊ α-PIPECOLINIC ACID ◊ 2-PIPERIDINECARBOXYLIC ACID (9CI) ◊ PIPEROLINIC ACID

TOXICITY DATA with REFERENCE
ivn-mus LD50:2200 mg/kg   THERAP 23,1343,68

CONSENSUS REPORTS: Reported in EPA TSCA Inventory.

SAFETY PROFILE: Moderately toxic by intravenous route. When heated to decomposition it emits toxic vapors of NO_x.

HDW100   CAS:108-74-7   **HR: 3**
HEXAHYDRO-1,3,5-TRIMETHYL-s-TRIAZINE
mf: $C_6H_{15}N_3$ mw: 129.24

SYNS: F 7771 ◊ s-TRIAZINE, HEXAHYDRO-1,3,5-TRIMETHYL- ◊ 1,3,5-TRIAZINE, HEXAHYDRO-1,3,5-TRIMETHYL- ◊ 1,3,5-TRIMETHYLHEXAHYDRO-sym-TRIAZINE ◊ 1,3,5-TRIMETHYLHEXAHYDRO-1,3,5-TRIAZINE

TOXICITY DATA with REFERENCE
ivn-mus LD50:100 mg/kg   CSLNX* NX#03990

CONSENSUS REPORTS: Reported in EPA TSCA Inventory.

SAFETY PROFILE: Poison by intravenous route. When heated to decomposition it emits toxic vapors of NO_x.

HDY600   CAS:695-06-7   **HR: 1**
γ-HEXALACTONE
mf: $C_6H_{10}O_2$ mw: 114.16

SYNS: γ-CAPROLACTONE ◊ 6-CAPROLACTONE ◊ γ-ETHYLBUTYROLACTONE ◊ γ-ETHYL-n-BUTYROLACTONE ◊ 2(3H)-FURANONE, 5-ETHYLDIHYDRO- ◊ γ-HEXANOLACTONE ◊ HEXANOLIDE-1,4 ◊ 4-HYDROXYHEXANOIC ACID LACTONE ◊ TONKALIDE ◊ TOUKALIDE

TOXICITY DATA with REFERENCE
skn-rbt 500 mg/24H MLD   FCTXAV 17,791,79

CONSENSUS REPORTS: Reported in EPA TSCA Inventory.

SAFETY PROFILE: A skin irritant. When heated to decomposition it emits acrid smoke and irritating vapors.

HEF400   CAS:871-67-0   **HR: 3**
HEXAMETHYLENEBIS(DITHIOCARBAMIC ACID) DISODIUM SALT
mf: $C_8H_{14}N_2S_4 \cdot 2Na$ mw: 312.46

SYN: CARBAMIC ACID, HEXAMETHYLENEBIS(DITHIO-, DISODIUM SALT

TOXICITY DATA with REFERENCE
ipr-mus LDLo:300 mg/kg   ARZNAD 21,121,71

CONSENSUS REPORTS: Reported in EPA TSCA Inventory.

SAFETY PROFILE: Poison by intraperitoneal route. When heated to decomposition it emits toxic vapors of NO_x and SO_x.

HEG200   CAS:629-09-4   **HR: D**
HEXAMETHYLENE DIIODIDE
mf: $C_6H_{12}I_2$ mw: 337.98

SYNS: 1,6-DIIODOHEXANE ◊ HEXANE, 1,6-DIIODO-

TOXICITY DATA with REFERENCE
mmo-sat 10 μmol/plate   MUREAV 141,11,84

CONSENSUS REPORTS: Reported in EPA TSCA Inventory.

SAFETY PROFILE: Mutation data reported. When heated to decomposition it emits toxic vapors of I^-.

HEK100   CAS:1608-26-0   **HR: D**
HEXAMETHYLPHOSPHORUS TRIAMIDE
mf: $C_6H_{18}N_3P$ mw: 163.24

SYN: PHOSPHORUS TRIAMIDE, HEXAMETHYL-

TOXICITY DATA with REFERENCE
sln-oin-dmg-orl 1 mmol/L   MUREAV 212,193,89
sln-oin-dmg-par 3 mmol/L   MUREAV 212,193,89
otr-ham:kdy 2500 μg/L   BJCAAI 38,418,78

CONSENSUS REPORTS: Reported in EPA TSCA Inventory.

SAFETY PROFILE: Mutation data reported. When heated to decomposition it emits toxic vapors of NO_x and PO_x.

HEO000   CAS:124-09-4   **HR: 3**
1,6-HEXANEDIAMINE
DOT: UN 1783/UN 2280
mf: $C_6H_{16}N_2$ mw: 116.24

PROP: Colorless leaflets; odor of piperidine. Mp: 39–42°, bp: 205°. Absorbs water and CO_2 from air; very sol in water; sltly sol in alc, benzene.

SYNS: 1,6-DIAMINOHEXANE ◊ 1,6-HEXAMETHYLENEDIAMINE ◊ HEXAMETHYLENE DIAMINE, solid (DOT) ◊ HMDA ◊ NCI-C61405

TOXICITY DATA with REFERENCE
orl-rat TDLo:3 g/kg (female 6–15D post):TER   JJATDK 7,259,87
orl-rat TDLo:3 g/kg (female 6–15D post):TER   JJATDK 7,259,87
ipr-mus TDLo:414 mg/kg (female 10D post):TER   TJADAB 28,237,83
orl-rat TDLo:1840 mg/kg (female 6–15D post):TER   JJATDK 7,259,87
orl-rat LD50:750 mg/kg   TXAPA9 42,417,77
ihl-mus LCLo:750 mg/m^3/10M   NDRC** NDCrc-132,Sept,42
ipr-mus LD50:320 mg/kg   GISAAA 43(11),110,78
scu-mus LD50:1300 mg/kg   85GMAT -,74,82

ivn-mus LD50:180 mg/kg   CSLNX* NX#02313
skn-rbt LD50:1110 mg/kg   TXAPA9 42,417,77

CONSENSUS REPORTS: Reported in EPA TSCA Inventory.

DOT Classification: Corrosive Material; Label: Corrosive (UN1783, UN2280); Corrosive Material; Label: Corrosive, Poison (UN1783)
ACGIH TLV: TWA 0.5 ppm

SAFETY PROFILE: Poison by intravenous and intraperitoneal routes. Moderately toxic by ingestion, inhalation, and skin contact. An experimental teratogen. A corrosive irritant to skin, eyes, and mucous membranes. Combustible when exposed to heat or flame; can react with oxidizing materials. See also AMINES.

HFE200   CAS:2497-21-4   HR: 2
2-HEXEN-4-ONE
mf: $C_6H_{10}O$ mw: 98.16

SYN: 2-HEXENE-4-ONE

TOXICITY DATA with REFERENCE
orl-mus LD50:780 mg/kg   DCTODJ 3,249,80

CONSENSUS REPORTS: Reported in EPA TSCA Inventory.

SAFETY PROFILE: Moderately toxic by ingestion. When heated to decomposition it emits acrid smoke and irritating vapors.

HFS600   CAS:1117-55-1   HR: 1
HEXYL OCTANOATE
mf: $C_{14}H_{28}O_2$ mw: 228.42

SYNS: HEXYL CAPRYLATE ◇ n-HEXYL CAPRYLATE ◇ OCTANOIC ACID, HEXYL ESTER

TOXICITY DATA with REFERENCE
skn-rbt 500 mg/24H MLD   FCTOD7 20,713,82
orl-rat LD50:>5 g/kg   FCTOD7 20,713,82
skn-rbt LD50:>5 g/kg   FCTOD7 20,713,82

CONSENSUS REPORTS: Reported in EPA TSCA Inventory.

SAFETY PROFILE: Mildly toxic by ingestion and skin contact. A skin irritant. When heated to decomposition it emits acrid smoke and irritating vapors.

HGB300   CAS:495-69-2   HR: 2
HIPPURIC ACID
mf: $C_9H_9NO_3$ mw: 179.19

SYNS: ACIDO IPPURICO ◇ BENZAMIDOACETIC ACID ◇ BENZOYLGLYCINE ◇ GLYCINE, N-BENZOYL- ◇ PHENYLCARBONYLAMINOACETIC ACID

TOXICITY DATA with REFERENCE
ipr-rat LD50:1500 mg/kg   BCFAAI 112,53,73

CONSENSUS REPORTS: Reported in EPA TSCA Inventory.

SAFETY PROFILE: Moderately toxic by intraperitoneal route. When heated to decomposition it emits toxic vapors of NO_x.

HGH150   CAS:51-56-9   HR: 3
HOMATROPINE HYDROBROMIDE
mf: $C_{16}H_{21}NO_3 \cdot BrH$ mw: 356.30

SYNS: (+ −)-HOMATROPINE BROMIDE ◇ 1-α-H,5-α-H-TROPAN-3-α-OL, MANDELATE (ester), HYDROBROMIDE

TOXICITY DATA with REFERENCE
ipr-rat LD50:154 mg/kg   DRUGAY 6,789,82
ivn-mus LD50:107 mg/kg   JPETAB 96,1,49

CONSENSUS REPORTS: Reported in EPA TSCA Inventory.

SAFETY PROFILE: Poison by intravenous and intraperitoneal route. When heated to decomposition it emits toxic vapors of NO_x and Br^-.

HGI100   CAS:848-53-3   HR: 3
HOMOCHLOROCYCLIZINE
mf: $C_{19}H_{23}ClN_2$ mw: 314.89

SYNS: CUROSAJIN ◇ 1H-1,4-DIAZEPINE, HEXAHYDRO-1-(p-CHLORO-α-PHENYLBENZYL)-4-METHYL- ◇ HOMOCHLORCYCLIZINE ◇ HOMOCLOMINE ◇ HOMODAMON ◇ HOMORESTAR ◇ LYSILAN ◇ SA 97 ◇ SANKUMIN ◇ WICRON

TOXICITY DATA with REFERENCE
orl-rat LD50:490 mg/kg   YAKUD5 22,375,80
ipr-rat LD50:81 mg/kg   YAKUD5 22,375,80
ivn-rat LD50:36 mg/kg   YAKUD5 22,375,80
orl-mus LD50:343 mg/kg   YAKUD5 22,375,80
ipr-mus LD50:98 mg/kg   YKKZAJ 92,1339,72
scu-mus LD50:164 mg/kg   YAKUD5 22,375,80
ivn-mus LD50:47800 μg/kg   YAKUD5 22,375,80
orl-dog LD50:>200 mg/kg   OYYAA2 1,168,67
ipr-dog LD50:50 mg/kg   OYYAA2 1,168,67
ivn-dog LD50:20 mg/kg   OYYAA2 1,168,67
orl-gpg LD50:370 mg/kg   OYYAA2 1,168,67
ipr-gpg LD50:76 mg/kg   OYYAA2 1,168,67
ivn-gpg LD50:19 mg/kg   OYYAA2 1,168,67

CONSENSUS REPORTS: Reported in EPA TSCA Inventory.

SAFETY PROFILE: Poison by ingestion, intraperitoneal, intravenous, and subcutaneous routes. When heated to decomposition it emits toxic vapors of NO_x and Cl^-.

HGI600   CAS:1516-27-4   ***HR: 3***
HOMONEURINE CHLORIDE
mf: $C_6H_{14}N \cdot Cl$ mw: 135.66

SYNS: ALLYLTRIMETHYLAMMONIUM CHLORIDE ◊ AMMONIUM, AL-LYLTRIMETHYL-, CHLORIDE

TOXICITY DATA with REFERENCE
scu-mus LDLo:180 mg/kg   JPETAB 6,477,14/15

CONSENSUS REPORTS: Reported in EPA TSCA Inventory.

SAFETY PROFILE: Poison by subcutaneous route. When heated to decomposition it emits toxic vapors of NO_x and Cl^-.

HGK600   CAS:93-40-3   ***HR: 3***
HOMOVERATRIC ACID
mf: $C_{10}H_{12}O_4$ mw: 196.22

SYNS: ACETIC ACID, (3,4-DIMETHOXYPHENYL)- ◊ BENZENEACETIC ACID, 3,4-DIMETHOXY-(9CI) ◊ 3,4-DIMETHOXYBENZENEACETIC ACID ◊ 3,4-DIMETHOXYPHENYL ACETIC ACID

TOXICITY DATA with REFERENCE
dni-hmn:lyms 1 mmol/L   BCPCA6 29,1275,80
ivn-mus LD50:180 mg/kg   CSLNX* NX#04366

CONSENSUS REPORTS: Reported in EPA TSCA Inventory.

SAFETY PROFILE: Poison by intravenous route. Human mutation data reported. When heated to decomposition it emits acrid smoke and irritating vapors.

HGR600   CAS:1123-85-9   ***HR: 2***
HYDRATROPIC ALCOHOL
mf: $C_9H_{12}O$ mw: 136.21

SYNS: BENZENEETHANOL, β-METHYL-(9CI) ◊ HYDRATROPYL ALCOHOL ◊ β-METHYLBENZENEETHANOL ◊ β-METHYLPHENETHYL ALCOHOL ◊ α-METHYL PHENYLETHYL ALCOHOL ◊ PHENETHYL ALCOHOL, β-METHYL- ◊ 2-PHENYLPROPAN-1-OL ◊ β-PHENYLPROPYL ALCOHOL ◊ 2-PHENYLPROPYL ALCOHOL

TOXICITY DATA with REFERENCE
orl-rat LD50:2300 mg/kg   FCTXAV 13,547,75

CONSENSUS REPORTS: Reported in EPA TSCA Inventory.

SAFETY PROFILE: Moderately toxic by ingestion. When heated to decomposition it emits acrid smoke and irritating vapors.

HJH100   CAS:1333-39-7   ***HR: 3***
HYDROXYBENZENESULFONIC ACID
mf: $C_6H_6O_4S$ mw: 174.18

SYNS: BENZENESULFONIC ACID, HYDROXY- ◊ PHENOLSULFONIC ACID ◊ SULFOCARBOLIC ACID

TOXICITY DATA with REFERENCE
orl-rat LD50:1900 mg/kg   SIGAAE 36,317,73
ipr-rat LD50:165 mg/kg   SIGAAE 36,317,73
scu-rat LD50:4 g/kg   SIGAAE 36,317,73
orl-mus LD50:1500 mg/kg   SIGAAE 36,317,73
ipr-mus LD50:140 mg/kg   SIGAAE 36,317,73
scu-mus LD50:4200 mg/kg   SIGAAE 36,317,73

CONSENSUS REPORTS: Reported in EPA TSCA Inventory.

SAFETY PROFILE: Poison by subcutaneous and intraperitoneal routes. Moderately toxic by ingestion. When heated to decomposition it emits toxic vapors of SO_x.

HJL100   CAS:936-02-7   ***HR: 3***
2-HYDROXYBENZOIC ACID HYDRAZIDE
mf: $C_7H_8N_2O_2$ mw: 152.17

SYNS: BENZOIC ACID, 2-HYDROXY-, HYDRAZIDE ◊ o-HYDROXYBENZHYDRAZIDE ◊ 2-HYDROXYBENZOHYDRAZIDE ◊ o-HYDROXYBENZOIC ACID HYDRAZIDE ◊ o-HYDROXYBENZOYLHYDRAZIDE ◊ 2-HYDROXYBENZOYLHYDRAZIDE ◊ o-HYDROXYBENZOYLHYDRAZINE ◊ 2-HYDROXYBENZOYLHYDRAZINE ◊ o-HYDROXYLBENZHYDRAZIDE ◊ SALICOYL HYDRAZIDE ◊ SALICYL HYDRAZIDE ◊ SALICYLIC ACID, HYDRAZIDE ◊ SALICYLIC HYDRAZIDE ◊ SALICYCLOHYDRAZINE

TOXICITY DATA with REFERENCE
ipr-mus LD50:100 mg/kg   JMPCAS 4,259,61

CONSENSUS REPORTS: Reported in EPA TSCA Inventory.

SAFETY PROFILE: Poison by intraperitoneal route. When heated to decomposition it emits toxic vapors of NO_x.

HJS900   CAS:1214-47-7   ***HR: 2***
2′-HYDROXYCHALCONE
mf: $C_{15}H_{12}O_2$ mw: 224.27

SYNS: ACRYLOPHENONE, 2′-HYDROXY-3-PHENYL- ◊ 1-(2-HYDROXYPHENYL)-3-PHENYL-2-PROPEN-1-ONE ◊ 2-PROPEN-1-ONE, 1-(2-HYDROXYPHENYL)-3-PHENYL-

TOXICITY DATA with REFERENCE
unr-uns LD50:620 mg/kg   PCJOAU 20,406,86

CONSENSUS REPORTS: Reported in EPA TSCA Inventory.

SAFETY PROFILE: Moderately toxic by an unspecified route. When heated to decomposition it emits acrid smoke and irritating vapors.

HJY100   CAS:93-35-6   ***HR: D***
7-HYDROXYCOUMARIN
mf: $C_9H_6O_3$ mw: 162.15

SYNS: 2H-1-BENZOPYRAN-2-ONE, 7-HYDROXY- ◊ COUMARIN, 7-HYDROXY- ◊ HYDRANGIN ◊ HYDRANGINE ◊ 7-OXYCOUMARIN ◊ SKIMMETIN ◊ SKIMMETINE ◊ UMBELLIFERON ◊ UMBELLIFERONE

TOXICITY DATA with REFERENCE
mmo-klp 800 mg/L   JIMMBG 36,55,80

SAFETY PROFILE: Mutation data reported. When heated to decomposition it emits acrid smoke and irritating vapors.

HKE600   CAS:609-99-4   **HR: 3**
2-HYDROXY-3,5-DINITROBENZOIC ACID
mf: $C_7H_4N_2O_7$ mw: 228.13

SYNS: BENZOIC ACID, 2-HYDROXY-3,5-DINITRO-(9CI) ◇ 3,5-DINITROSALICYLIC ACID ◇ SALICYLIC ACID, 3,5-DINITRO-

TOXICITY DATA with REFERENCE
orl-rat LD50:860 mg/kg   GISAAA 51(1),85,86
orl-mus LD50:270 mg/kg   GISAAA 51(1),85,86

CONSENSUS REPORTS: Reported in EPA TSCA Inventory.

SAFETY PROFILE: Poison by ingestion. When heated to decomposition it emits toxic vapors of NO_x.

HKF300   CAS:91-01-0   **HR: 1**
HYDROXYDIPHENYLMETHANE
mf: $C_{13}H_{12}O$ mw: 184.25

SYNS: BENZHYDROL ◇ BENZHYDRYL ALCOHOL ◇ BENZOHYDROL ◇ DIPHENYL CARBINOL ◇ DIPHENYLMETHANOL ◇ DIPHENYLMETHYL ALCOHOL

TOXICITY DATA with REFERENCE
orl-rat LD50:5000 mg/kg   FCTXAV 17,713,79
scu-gpg LDLo:2 g/kg   JPETAB 24,405,24

CONSENSUS REPORTS: Reported in EPA TSCA Inventory.

SAFETY PROFILE: Mildly toxic by ingestion and subcutaneous routes. When heated to decomposition it emits acrid smoke and irritating vapors.

HKS100   CAS:91-88-3   **HR: 2**
N-HYDROXYETHYL-N-ETHYL-m-TOLUIDINE
mf: $C_{11}H_{17}NO$ mw: 179.29

SYNS: EMERY 5714 ◇ ETHANOL, 2-(N-ETHYL-m-TOLUIDINO)- ◇ N-ETHYL-N-(2-HYDROXYETHYL)-m-TOLUIDINE ◇ 2-(N-ETHYL-m-TOLUIDINO)ETHANOL

TOXICITY DATA with REFERENCE
orl-rat LD50:1370 mg/kg   LONZA# 02JUN80

CONSENSUS REPORTS: Reported in EPA TSCA Inventory.

SAFETY PROFILE: Moderately toxic by ingestion. When heated to decomposition it emits toxic vapors of NO_x.

HLC600   CAS:577-85-5   **HR: 3**
3-HYDROXYFLAVONE
mf: $C_{15}H_{10}O_3$ mw: 238.25

SYN: FLAVONE, 3-HYDROXY-

TOXICITY DATA with REFERENCE
ivn-mus LD50:56 mg/kg   CSLNX* NX#02588

CONSENSUS REPORTS: Reported in EPA TSCA Inventory.

SAFETY PROFILE: Poison by intravenous route. When heated to decomposition it emits acrid smoke and irritating vapors.

HLF600   CAS:613-03-6   **HR: 2**
HYDROXYHYDROQUINONE TRIACETATE
mf: $C_{12}H_{12}O_6$ mw: 252.24

SYNS: 1,2,4-BENZENETRIOL, TRIACETATE ◇ 2-HYDROXYHYDROQUINONE TRIACETATE ◇ TRIACETATE d'HYDROXYHYDROQUINONE ◇ 1,2,4-TRIACETOXYBENZENE

TOXICITY DATA with REFERENCE
ipr-mus LDLo:500 mg/kg   RBPMAZ 22,1,52

CONSENSUS REPORTS: Reported in EPA TSCA Inventory.

SAFETY PROFILE: Moderately toxic by intraperitoneal route. When heated to decomposition it emits acrid smoke and irritating vapors.

HLI100   CAS:2538-85-4   **HR: 3**
3-HYDROXY-4-(2-HYDROXY-1-NAPHTHYLAZO)-1-NAPHTHALENESULFONIC ACID, SODIUM
mf: $C_{20}H_{13}N_2O_5S \cdot Na$ mw: 416.40

SYNS: CALCON ◇ 1-NAPHTHALENESULFONIC ACID, 3-HYDROXY-4-(2-HYDROXY-1-NAPHTHYLAZO)-, SODIUM SALT

TOXICITY DATA with REFERENCE
ivn-mus LD50:180 mg/kg   CSLNX* NX#04757

CONSENSUS REPORTS: Reported in EPA TSCA Inventory.

SAFETY PROFILE: Poison by intravenous route. When heated to decomposition it emits toxic vapors of NO_x and SO_x.

HLR600   CAS:673-22-3   **HR: D**
2-HYDROXY-4-METHOXYBENZALDEHYDE
mf: $C_8H_8O_3$ mw: 152.16

SYNS: p-ANISALDEHYDE, 2-HYDROXY- ◇ BENZALDEHYDE, 2-HYDROXY-4-METHOXY-(9CI) ◇ 4-METHOXYSALICYLALDEHYDE

TOXICITY DATA with REFERENCE
sce-hmn:lyms 500 µmol/L   MUREAV 206,17,88

CONSENSUS REPORTS: Reported in EPA TSCA Inventory.

SAFETY PROFILE: Human mutation data reported. When

heated to decomposition it emits acrid smoke and irritating vapors.

HLT100 CAS:13925-12-7 *HR: 3*
1-HYDROXY-6-METHOXYPHENAZINE 5,10-DIOXIDE
mf: $C_{13}H_{10}N_2O_4$ mw: 258.25

SYNS: 3C ANTIBIOTIC ◇ 6-METHOXY-1-PHENAZINOL 5,10-DIOXIDE ◇ MYXIN ◇ 1-PHENAZINOL, 6-METHOXY-, 5,10-DIOXIDE

TOXICITY DATA with REFERENCE
mma-pro-eug 10 mg/L EXPEAM 27,586,71
ipr-mus LD50:133 mg/kg CMTRAG 12,272,67

SAFETY PROFILE: Poison by intraperitoneal route. Mutation data reported. When heated to decomposition it emits toxic vapors of NO_x.

HLV100 CAS:91-04-3 *HR: 2*
2-HYDROXY-5-METHYL-1,3-BENZENEDIMETHANOL
mf: $C_9H_{12}O_3$ mw: 168.21

SYNS: 1,3-BENZENEDIMETHANOL, 2-HYDROXY-5-METHYL- ◇ α^1,α^3-MESITYLENEDIOL, 2-HYDROXY-(7CI,8CI) ◇ α^1,α^3,2-TRIHYDROXYMESITYLENE

TOXICITY DATA with REFERENCE
par-frg LDLo:1250 mg/kg JPETAB 26,123,26

CONSENSUS REPORTS: Reported in EPA TSCA Inventory.

SAFETY PROFILE: Moderately toxic by parenteral routes. When heated to decomposition it emits acrid smoke and irritating vapors.

HMQ600 CAS:826-81-3 *HR: D*
8-HYDROXY-2-METHYLQUINOLINE
mf: $C_{10}H_9NO$ mw: 159.20

SYNS: HYDROXYQUINALDINE ◇ 8-HYDROXYQUINALDINE ◇ 2-METHYL-8-HYDROXYQUINOLINE ◇ 2-METHYLOXINE ◇ 2-METHYL-8-QUINOLINOL ◇ 8-QUINOLINOL, 2-METHYL-

TOXICITY DATA with REFERENCE
mma-sat 500 nmol/plate MUREAV 42,335,77

CONSENSUS REPORTS: Reported in EPA TSCA Inventory.

SAFETY PROFILE: Mutation data reported. When heated to decomposition it emits toxic vapors of NO_x.

HMR600 CAS:583-91-5 *HR: 2*
2-HYDROXY-4-(METHYLTHIO)BUTANOIC ACID
mf: $C_5H_{10}O_3S$ mw: 150.21

SYNS: ALIMET ◇ BUTANOIC ACID, 2-HYDROXY-4-(METHYLTHIO)- (9CI) ◇ BUTYRIC ACID, 2-HYDROXY-4-(METHYLTHIO)- ◇ METHIONINE HYDROXY ANALOG ◇ MHA ACID ◇ MHA-FA

TOXICITY DATA with REFERENCE
orl-rat LD50:3478 mg/kg TOLED5 31(Suppl),54,86

CONSENSUS REPORTS: Reported in EPA TSCA Inventory.

SAFETY PROFILE: Moderately toxic by ingestion. When heated to decomposition it emits toxic vapors of SO_x.

HMU200 CAS:708-06-5 *HR: 2*
2-HYDROXYNAPHTHALDEHYDE
mf: $C_{11}H_8O_2$ mw: 172.19

SYNS: β-HYDROXYNAPHTHALDEHYDE ◇ 2-HYDROXY-1-NAPHTHALENECARBOXALDEHYDE ◇ 1-NAPHTHALDEHYDE, 2-HYDROXY- ◇ 1-NAPHTHALENECARBOXALDEHYDE, 2-HYDROXY-(9CI)

TOXICITY DATA with REFERENCE
ipr-rat LD50:710 mg/kg GISAAA 51(1),85,86
ipr-mus LD50:1170 mg/kg GISAAA 51(1),85,86

CONSENSUS REPORTS: Reported in EPA TSCA Inventory.

SAFETY PROFILE: Moderately toxic by intraperitoneal route. When heated to decomposition it emits acrid smoke and irritating vapors.

HMX520 CAS:92-70-6 *HR: 3*
3-HYDROXY-2-NAPHTHOIC ACID
mf: $C_{11}H_8O_3$ mw: 188.19

SYNS: BON ◇ BONA ◇ BON ACID ◇ C.I. DEVELOPER 8 ◇ C.I. DEVELOPER 20 (obs.) ◇ DEVELOPER BON ◇ 3-HYDROXY-2-NAPHTHALENECARBOXYLIC ACID ◇ KYSELINA 3-HYDROXY-2-NAFTOOVA ◇ MIKETAZOL DEVELOPER ONS ◇ 2-NAPHTHALENECARBOXYLIC ACID, 3-HYDROXY-(9CI) ◇ 2-NAPHTHOIC ACID, 3-HYDROXY- ◇ NAPHTHOL B.O.N.

TOXICITY DATA with REFERENCE
scu-rat LDLo:376 mg/kg AIPTAK 176,193,68
orl-mus LD50:800 mg/kg 85JCAE -,663,86

CONSENSUS REPORTS: Reported in EPA TSCA Inventory.

SAFETY PROFILE: Poison by subcutaneous route. Moderately toxic by ingestion. When heated to decomposition it emits acrid smoke and irritating vapors.

HMX600 CAS:83-72-7 *HR: 3*
2-HYDROXYNAPHTHOQUINONE
mf: $C_{10}H_6O_3$ mw: 174.16

SYNS: C.I. 75480 ◇ C.I. NATURAL ORANGE 6 ◇ FLOWER of PARADISE ◇ HANA ◇ HENNA ◇ 2-HYDROXY-1,4-NAPHTHALENEDIONE ◇ 2-HYDROXY-1,4-NAPHTHOQUINONE ◇ LAWSONE ◇ 1,4-NAPHTHALENEDIONE, 2-HYDROXY-(9CI) ◇ 1,4-NAPHTHOQUINONE, 2-HYDROXY- ◇ MEHENDI ◇ MENDI

TOXICITY DATA with REFERENCE
mma-sat 33 μg/plate ENMUDM 8(Suppl 7),1,86
ipr-mus LD50:100 mg/kg JMCMAR 26,570,83

CONSENSUS REPORTS: Reported in EPA TSCA Inventory.

SAFETY PROFILE: Poison by intraperitoneal route. Mutation data reported. When heated to decomposition it emits acrid smoke and irritating vapors.

HMY075 CAS:99-42-3 HR: 2
4-HYDROXY-3-NITROBENZOIC ACID METHYL ESTER
mf: $C_8H_7NO_5$ mw: 197.16

SYN: BENZOIC ACID, 4-HYDROXY-3-NITRO-, METHYL ESTER

TOXICITY DATA with REFERENCE
orl-rat LDLo:2500 mg/kg   EPASR* 8EHQ-0287-0657S
skn-rat LD50:>2 g/kg   EPASR* 8EHQ-0287-0657S

CONSENSUS REPORTS: Reported in EPA TSCA Inventory.

SAFETY PROFILE: Moderately toxic by ingestion. Slightly toxic by contact. When heated to decomposition it emits toxic vapors of NO_x.

HND100 CAS:1843-05-6 HR: 1
2-HYDROXY-4-(OCTYLOXY)BENZOPHENONE
mf: $C_{21}H_{26}O_3$ mw: 326.47

SYNS: ADUVEX 248 ◇ ADVASTAB 46 ◇ ANTI-UV P ◇ BENZON 00 ◇ BENZOPHENONE 12 ◇ BENZOPHENONE, 2-HYDROXY-4-(OCTYLOXY)- ◇ BIOSORB 130 ◇ CARSTAB 700 ◇ CHIMASSORB 81 ◇ CYASORB UV 531 ◇ (2-HYDROXY-4-(OCTYLOXY)PHENYL)PHENYLMETHANONE ◇ 2-HYDROXY-4-OKTYLOXYBENZOFENON ◇ METHANONE, (2-HYDROXY-4-(OCTYLOXY)PHENYL)PHENYL-(9CI) ◇ OCTABENZONE ◇ SPECTRA-SORB UV 531 ◇ SUMISORB 130 ◇ UF 4 ◇ UV 1 ◇ UV 531 ◇ UVA 1 ◇ UVINUL 408

TOXICITY DATA with REFERENCE
orl-rat LD50:>10 g/kg   FCTXAV 6,199,68
orl-mus LD50:13 g/kg   85JCAE -,648,86
skn-rbt LD50:>10 g/kg   FCTXAV 6,199,68

CONSENSUS REPORTS: Reported in EPA TSCA Inventory.

SAFETY PROFILE: Slightly toxic by ingestion and skin contact. When heated to decomposition it emits acrid smoke and irritating vapors.

HNG600 CAS:156-38-7 HR: 2
(4-HYDROXYPHENYL)ACETIC ACID
mf: $C_8H_8O_3$ mw: 152.16

SYNS: ACETIC ACID, (p-HYDROXYPHENYL)- ◇ BENZENEACETIC ACID, 4-HYDROXY-(9CI) ◇ 4-HYDROXYBENZENEACETIC ACID ◇ (p-HYDROXYPHENYL)ACETIC ACID

TOXICITY DATA with REFERENCE
ipr-mus LD50:3500 mg/kg   FRPSAX 13,286,58

CONSENSUS REPORTS: Reported in EPA TSCA Inventory.

SAFETY PROFILE: Moderately toxic by intraperitoneal route. When heated to decomposition it emits acrid smoke and irritating vapors.

HNK560 CAS:99-07-0 HR: D
(3-HYDROXYPHENYL)DIMETHYLAMINE
mf: $C_8H_{11}NO$ mw: 137.20

SYNS: m-(DIMETHYLAMINO)PHENOL ◇ 3-(DIMETHYLAMINO)PHENOL ◇ PHENOL, m-(DIMETHYLAMINO)-

TOXICITY DATA with REFERENCE
mma-sat 33 µg/plate   EMMUEG 11(Suppl 12),1,88

CONSENSUS REPORTS: Reported in EPA TSCA Inventory.

SAFETY PROFILE: Mutation data reported. When heated to decomposition it emits toxic vapors of NO_x.

HNL600 CAS:98-67-9 HR: 1
4-HYDROXYPHENYLSULFONIC ACID
mf: $C_6H_6O_4S$ mw: 174.18

SYN: BENZENESULFONIC ACID, p-HYDROXY-

TOXICITY DATA with REFERENCE
orl-mus LD50:6400 mg/kg   JAFCAU 15,845,67

CONSENSUS REPORTS: Reported in EPA TSCA Inventory.

SAFETY PROFILE: Mildly toxic by ingestion. When heated to decomposition it emits toxic vapors of SO_x.

HNS600 CAS:524-38-9 HR: 3
N-HYDROXYPHTHALIMIDE
mf: $C_8H_5NO_3$ mw: 163.14

SYN: PHTHALIMIDE, N-HYDROXY-

TOXICITY DATA with REFERENCE
ivn-mus LD50:178 mg/kg   CSLNX* NX#00747

CONSENSUS REPORTS: Reported in EPA TSCA Inventory.

SAFETY PROFILE: Poison by intravenous route. When heated to decomposition it emits toxic vapors of NO_x.

HNV500 CAS:923-26-2 HR: 1
2-HYDROXYPROPYL METHACRYLATE
mf: $C_7H_{12}O_3$ mw: 144.19

SYNS: β-HYDROXYPROPYL METHACRYLATE ◇ 2-HYDROXYPROPYL 2-METHYL-2-PROPENOATE ◇ METHACRYLIC ACID, 2-HYDROXYPROPYL ESTER ◇ 2-PROPENOIC ACID, 2-METHYL-, 2-HYDROXYPROPYL ESTER (9CI)

TOXICITY DATA with REFERENCE
orl-mus LD50:7964 mg/kg   TOLED5 11,125,82

CONSENSUS REPORTS: Reported in EPA TSCA Inventory.

SAFETY PROFILE: Mildly toxic by ingestion. When heated to decomposition it emits acrid smoke and irritating vapors.

HOB100 CAS:626-64-2 **HR: 2**
4-HYDROXYPYRIDINE
mf: C_5H_5NO mw: 95.11

SYNS: γ-HYDROXYPYRIDINE ◊ 4-PYRIDINOL

TOXICITY DATA with REFERENCE
ipr-mus LD50:923 mg/kg TOXIA6 23,815,85

CONSENSUS REPORTS: Reported in EPA TSCA Inventory.

SAFETY PROFILE: Moderately toxic by intraperitoneal route. When heated to decomposition it emits toxic vapors of NO_x.

HOE100 CAS:86-75-9 **HR: D**
8-HYDROXYQUINOLINE BENZOATE
mf: $C_{16}H_{11}NO_2$ mw: 249.28

SYNS: BENZOIC ACID, 8-QUINOLYL ESTER ◊ BENZOXIQUINE ◊ BENZOXYLINE ◊ BENZOXYQUINE ◊ 8-BENZOYLOXYQUINOLINE ◊ DIOXYLINE ◊ OXYQUINOLINE BENZOATE ◊ 8-QUINOLINOL, BENZOATE (ester)

TOXICITY DATA with REFERENCE
mmo-bcs 10 mmol/L FAVUAI 6,118,74

CONSENSUS REPORTS: Reported in EPA TSCA Inventory.

SAFETY PROFILE: Mutation data reported. When heated to decomposition it emits toxic vapors of NO_x.

HOI200 CAS:453-20-3 **HR: 2**
3-HYDROXYTETRAHYDROFURAN
mf: $C_4H_8O_2$ mw: 88.12

SYNS: 3-FURANOL, TETRAHYDRO- ◊ TETRAHYDRO-3-FURANOL

TOXICITY DATA with REFERENCE
ivn-mus LD50:3850 mg/kg JPPMAB 22,694,70

CONSENSUS REPORTS: Reported in EPA TSCA Inventory.

SAFETY PROFILE: Moderately toxic by intravenous route. When heated to decomposition it emits acrid smoke and irritating vapors.

HOI300 CAS:2226-96-2 **HR: D**
4-HYDROXY-2,2,6,6-TETRAMETHYL-1-PIPERIDINYLOXY
mf: $C_9H_{18}NO_2$ mw: 172.28

SYNS: 4-OXYPIPERIDOL ◊ PIPERIDINOOXY, 4-HYDROXY-2,2,6,6-TETRAMETHYL- ◊ 1-PIPERIDINYLOXY, 4-HYDROXY-2,2,6,6-TETRAMETHYL-(9CI) ◊ TANOL ◊ TEMPOL ◊ TETRAMETHYLPIPERIDINOL N-OXYL ◊ TMPN

TOXICITY DATA with REFERENCE
mmo-sat 24 ~2mol/plate ABBIA4 251,393,86

CONSENSUS REPORTS: Reported in EPA TSCA Inventory.

SAFETY PROFILE: Mutation data reported. When heated to decomposition it emits toxic vapors of NO_x.

HOM300 CAS:791-31-1 **HR: 3**
HYDROXYTRIPHENYLSILANE
mf: $C_{18}H_{16}OSi$ mw: 276.43

SYN: SILANE, HYDROXYTRIPHENYL-

TOXICITY DATA with REFERENCE
ivn-mus LD50:180 mg/kg CSLNX* NX#04018

CONSENSUS REPORTS: Reported in EPA TSCA Inventory.

SAFETY PROFILE: Poison by intravenous route. When heated to decomposition it emits acrid smoke and irritating vapors.

HOR470 CAS:2192-20-3 **HR: 3**
HYDROXYZINE DIHYDROCHLORIDE
mf: $C_{21}H_{27}ClN_2O_2 \cdot 2ClH$ mw: 447.87

SYNS: ALAMON ◊ ATARAX ◊ ATARAX DIHYDROCHLORIDE ◊ ATARAXOID DIHYDROCHLORIDE ◊ ATERAX DIHYDROCHLORIDE ◊ 2-(2-(4-(p-CHLORO-α-PHENYLBENZYL)-1-PIPERAZINYL)ETHOXY)ETHANOLDIHYDROCHLORIDE ◊ DICHLORHYDRATE de1-p.CHLORBENZHYDRYL-4-(2-(2-HYDROXYETHOXY)ETHYL)PIPERAZINE ◊ ETHANOL, 2-(2-(4-(p-CHLORO-α-PHENYLBENZYL)-1-PIPERAZINYL)ETHOXY)-, DIHYDROCHLORIDE ◊ HYDROXYZINE HYDROCHLORIDE ◊ ORGATRAX ◊ TRAN-Q DIHYDROCHLORIDE ◊ TRANQUIZINE DIHYDROCHLORIDE

TOXICITY DATA with REFERENCE
orl-rat LD50:950 mg/kg TXAPA9 18,185,71
ipr-rat LD50:126 mg/kg TXAPA9 18,185,71
ipr-mus LD50:122 mg/kg YKYUA6 24,100,73
ivn-mus LD50:48,900 μg/kg YKYUA6 24,100,73

CONSENSUS REPORTS: Reported in EPA TSCA Inventory.

SAFETY PROFILE: Poison by intraperitoneal and intravenous routes. Moderately toxic by ingestion. When heated to decomposition it emits toxic vapors of NO_x, HCl, and Cl^-.

HOT600 CAS:620-61-1 **HR: 3**
HYOSCYAMINE SULFATE
mf: $C_{34}H_{46}N_2O_6 \cdot H_2O_4S$ mw: 676.90

SYNS: (−)-HYOSCYAMINE SULFATE ◊ 1-α-H,5-α-H-TROPAN-3-α-OL, (−)-TROPATE (ester), SULFATE (2:1)

TOXICITY DATA with REFERENCE
ipr-mus LD50:210 mg/kg   JPMSAE 55,849,66

CONSENSUS REPORTS: Reported in EPA TSCA Inventory.

SAFETY PROFILE: Poison by intraperitoneal route. When heated to decomposition it emits toxic vapors of NO_x and SO_x.

I

IAL100 CAS:72-40-2 **HR: D**
1H-IMIDAZOLE-4-CARBOXAMIDE, 5-AMINO-, MONOHYDROCHLORIDE
mf: $C_4H_6N_4O \cdot ClH$ mw: 162.60

SYNS: 4-AMINO-5-IMIDAZOLECARBOXAMIDE HYDROCHLORIDE ◇ 5-IMIDAZOLECARBOXAMIDE, 4-AMINO-, HYDROCHLORIDE

TOXICITY DATA with REFERENCE
mmo-klp 10 mmol/L/20H   MUREAV 66,207,79

CONSENSUS REPORTS: Reported in EPA TSCA Inventory.

SAFETY PROFILE: Mutation data reported. When heated to decomposition it emits toxic vapors of NO_x, HCl, and Cl^-.

IBB100 CAS:628-87-5 **HR: 3**
1,1'-IMIDODIACETONITRILE
mf: $C_4H_5N_3$ mw: 95.12

SYNS: ACETONITRILE, 2,2'-IMINOBIS-(9CI) ◇ ACETONITRILE, IMINODI- ◇ DI(CIANOMETIL)AMMINA ◇ 2,2'-IMINOBISACETONITRILE ◇ IMINODIACETONITRILE ◇ 2406 I.S.

TOXICITY DATA with REFERENCE
ipr-mus LD50:200 mg/kg   FRPSAX 17,753,62

CONSENSUS REPORTS: Reported in EPA TSCA Inventory.

SAFETY PROFILE: Poison by intraperitoneal route. When heated to decomposition it emits toxic vapors of NO_x.

IBJ100 CAS:1752-24-5 **HR: 3**
4,4'-IMINODIPHENOL
mf: $C_{12}H_{11}NO_2$ mw: 201.24

SYNS: BIS(p-HYDROXYPHENYL)AMINE ◇ BIS(4-HYDROXYPHENYL)AMINE ◇ 4,4'-DIHYDROXYDIPHENYLAMINE ◇ 4,4'-IMINOBISPHENOL ◇ LEUCOINDOPHENOL ◇ PHENOL, 4,4'-IMINOBIS-(9CI) ◇ PHENOL, 4,4'-IMINODI-

TOXICITY DATA with REFERENCE
orl-mus LD50:380 mg/kg   ARZNAD 12,1123,62

CONSENSUS REPORTS: Reported in EPA TSCA Inventory.

SAFETY PROFILE: Poison by ingestion. When heated to decomposition it emits toxic vapors of NO_x.

IBV050 CAS:81-77-6 **HR: 3**
INDANTHRENE BLUE
mf: $C_{28}H_{14}N_2O_4$ mw: 442.44

SYNS: ANTHRAQUINONE BLUE ◇ ANTHRAQUINONE DEEP BLUE ◇ 5,9,14,18-ANTHRAZINETETRONE, 6,15-DIHYDRO- ◇ ATIC VAT BLUE XRN ◇ BENZADONE BLUE RS ◇ BLEU SOLANTHRENE ◇ BLUE ANTHRAQUINONE PIGMENT ◇ BLUE O ◇ CALCOLOID BLUE RS ◇ CALEDON BLUE RN ◇ CALEDON BLUE XRN ◇ CALEDON BRILLIANT BLUE RN ◇ CALEDON PAPER BLUE RN ◇ CALEDON PRINTING BLUE RN ◇ CALEDON PRINTING BLUE XRN ◇ CARBANTHRENE BLUE 2R ◇ CARBANTHRENE BLUE RS ◇ CARBANTHRENE BLUE RSP ◇ CELLITON BLUE RN ◇ C.I. 1106 ◇ C.I. 69800 ◇ CIBANONE BLUE FRS ◇ CIBANONE BLUE FRSN ◇ CIBANONE BLUE RS ◇ CIBANONE BRILLIANT BLUE FR ◇ C.I. PIGMENT BLUE 60 ◇ C.I. VAT BLUE 4 ◇ CROMOPHTAL BLUE A 3R ◇ N,N-DIHYDRO-1,1,1',2'-ANTHRAQUINONE-AZINE ◇ E 130 ◇ FENAN BLUE RSN ◇ FENANTHREN BLUE RS ◇ FOOD BLUE 4 ◇ GRAPHTOL BLUE RL ◇ HELIANTHRENE BLUE RS ◇ HELIOGEN BLUE 6470 ◇ INDANTHREN BLUE ◇ INDANTHREN BLUE GP ◇ INDANTHREN BLUE GPT ◇ INDANTHREN BLUE RPT ◇ INDANTHREN BLUE RS ◇ INDANTHREN BLUE RSN ◇ INDANTHREN BLUE RSP ◇ INDANTHREN BRILLIANT BLUE R ◇ INDANTHRENE ◇ INDANTHRENE BLUE GP ◇ INDANTHRENE BLUE RP ◇ INDANTHRENE BLUE RS ◇ INDANTHRENE BLUE RSA ◇ INDANTHRENE BLUE RSN ◇ INDANTHREN PRINTING BLUE FRS ◇ INDANTHREN PRINTING BLUE KRS ◇ INDANTHRONE ◇ LAKE FAST BLUE BS ◇ LAKE FAST BLUE GGS ◇ LATEXOL FAST BLUE SD ◇ L-BLAU 1 ◇ LIONOGEN BLUE R ◇ LUTETIA FAST BLUE RS ◇ MEDIUM BLUE ◇ MIKETHRENE BLUE RSN ◇ MIKETHRENE BRILLIANT BLUE R ◇ MODR KYPOVA 4 ◇ MODR PIGMENT 60 ◇ MODR POTRAVINARSKA 4 ◇ MONOLITE FAST BLUE 3R ◇ MONOLITE FAST BLUE 3RD ◇ MONOLITE FAST BLUE RV ◇ MONOLITE FAST BLUE SRS ◇ NAVINON BLUE RSN ◇ NAVINON BLUE RSN REDDISH SPECIAL ◇ NIHONTHRENE BLUE RSN ◇ NIHONTHRENE BRILLIANT BLUE RP ◇ NIHONTHRENE BRILLIANT BLUE RS ◇ OSTANTHREN BLUE RSN ◇ OSTANTHREN BLUE RSZ ◇ OSTANTHRENE BLUE RS ◇ PALANTHRENE BLUE GPT ◇ PALANTHRENE BLUE GPZ ◇ PALANTHRENE BLUE RPT ◇ PALANTHRENE BLUE RPZ ◇ PALANTHRENE BLUE RSN ◇ PALANTHRENE BRILLIANT BLUE R ◇ PALANTHRENE PRINTING BLUE KRS ◇ PARADONE BLUE RS ◇ PARADONE BRILLIANT BLUE R ◇ PARADONE PRINTING BLUE FRS ◇ PERNITHRENE BLUE RS ◇ PIGMENT ANTHRAQUINONE DEEP BLUE ◇ PIGMENT BLUE 60 ◇ PIGMENT BLUE ANTHRAQUINONE ◇ PIGMENT BLUE ANTHRAQUINONE V ◇ PIGMENT DEEP BLUE ANTHRAQUINONE ◇ POLYMON BLUE 3R ◇ PONSOL BLUE GZ ◇ PONSOL BLUE RCL ◇ PONSOL BLUE RPC ◇ PONSOL BRILLIANT BLUE R ◇ PONSOL RP ◇ ROMANTHRENE BLUE FRS ◇ ROMANTRENE BLUE FRS ◇ ROMANTRENE BLUE GGSL ◇ ROMANTRENE BLUE RSZ ◇ ROMANTRENE BRILLIANT BLUE FR ◇ ROMANTRENE BRILLIANT BLUE R ◇ SANDOTHRENE BLUE NRSC ◇ SANDOTHRENE BLUE NRSN ◇ SANYO THRENE BLUE IRN ◇ SCHULTZ No. 1228 ◇ SOLANTHRENE BLUE RS ◇ SOLANTHRENE BLUE RSN ◇ SOLANTHRENE R FOR SUGAR ◇ SYMULER FAST BLUE 6011 ◇ TINON BLUE RS ◇ TINON BLUE RSN ◇ TYRIAN BLUE I-RSN ◇ TYRIAN BRILLIANT BLUE I-R ◇ VAT BLUE 4 ◇ VAT BLUE O ◇ VAT BLUE OD ◇ VAT FAST BLUE R ◇ VERSAL BLUE GGSL ◇ VULCAFIX FAST BLUE SD ◇ VULCAFOR FAST BLUE 3R ◇ VULCANOSINE FAST BLUE GG ◇ VULCOL FAST BLUE S ◇ VYNAMON BLUE 3R

TOXICITY DATA with REFERENCE
orl-rat LD50:2 g/kg   FAONAU 38B,60,66
itr-rat LD50:250 mg/kg   85JCAE -,1327,86

CONSENSUS REPORTS: Reported in EPA TSCA Inventory.

SAFETY PROFILE: Poison by intratracheal route. When heated to decomposition it emits toxic vapors of NO_x.

ICS200   CAS:830-96-6   HR: 3
1H-INDOLE-3-PROPIONIC ACID
mf: $C_{11}H_{11}NO_2$ mw: 189.23

SYNS: INDOLEPROPIONIC ACID ◇ β-INDOLEPROPIONIC ACID ◇ 3-(3-INDOLYL)PROPANOIC ACID

TOXICITY DATA with REFERENCE
ipr-mus LDLo:100 mg/kg   PSEBAA 34,138,36

CONSENSUS REPORTS: Reported in EPA TSCA Inventory.

SAFETY PROFILE: Poison by intraperitoneal route. When heated to decomposition it emits toxic vapors of NO_x.

IDA600   CAS:608-08-2   HR: 2
INDOXYLACETATE
mf: $C_{10}H_9NO_2$ mw: 175.20

SYNS: ACETIC ACID, 3-INDOLYL ESTER ◇ 3-ACETOXYINDOLE ◇ INDOLE, 3-ACETATO- ◇ INDOL-3-OL, ACETATE (ester) (8CI) ◇ 1H-INDOL-3-OL, ACETATE (ester) (9CI) ◇ INDOXYL-O-ACETATE

TOXICITY DATA with REFERENCE
ipr-mus LD50:600 mg/kg   NYKZAU 55,1514,59

CONSENSUS REPORTS: Reported in EPA TSCA Inventory.

SAFETY PROFILE: Moderately toxic by intraperitoneal route. When heated to decomposition it emits toxic vapors of NO_x.

IEE025   CAS:619-58-9   HR: 2
p-IODOBENZOIC ACID
mf: $C_7H_5IO_2$ mw: 248.02

SYNS: BENZOIC ACID, p-IODO- ◇ 4-JODBENZOESAEURE

TOXICITY DATA with REFERENCE
ivn-mus LD50:2500 mg/kg   PHARAT 12,415,57

CONSENSUS REPORTS: Reported in EPA TSCA Inventory.

SAFETY PROFILE: Moderately toxic by intravenous route. When heated to decomposition it emits toxic vapors of I^-.

IET500   CAS:628-17-1   HR: 2
1-IODOPENTANE
mf: $C_5H_{11}I$ mw: 198.06

SYNS: AMYL IODIDE ◇ n-AMYL IODIDE ◇ 1-JODPENTAN ◇ PENTANE, 1-IODO- ◇ PENTYL IODIDE ◇ n-PENTYL IODIDE ◇ 1-PENTYL IODIDE

TOXICITY DATA with REFERENCE
ipr-rat LD50:948 mg/kg   85GMAT-,21,82
ipr-mus LD50:489 mg/kg   85GMAT-,21,82

CONSENSUS REPORTS: Reported in EPA TSCA Inventory.

SAFETY PROFILE: Moderately toxic by intraperitoneal route. When heated to decomposition it emits toxic vapors of I^-.

IEU075   CAS:423-62-1   HR: 1
1-IODOPERFLUORODECANE
mf: $C_{10}F_{21}I$ mw: 268.21

SYNS: DECANE, HENEICOSAFLUORO-1-IODO- ◇ DECANE, 1-IODO-1,1,2,2,3,3,4,4,5,5,6,6,7,7,8,8,9,9,10,10,10-HENEICOSAFLUORO- ◇ HENEICOSAFLUORO-1-IODODECANE ◇ PERFLUORODECYL IODIDE ◇ PERFLUORO-1-IODODECANE

TOXICITY DATA with REFERENCE
ivn-mus LD50:11100 mg/kg   MIVRA6 8,320,74

CONSENSUS REPORTS: Reported in EPA TSCA Inventory.

SAFETY PROFILE: Mildly toxic by intravenous route. When heated to decomposition it emits toxic vapors of I^- and F^-.

IEV010   CAS:626-02-8   HR: 2
3-IODOPHENOL
mf: C_6H_5IO mw: 220.01

SYNS: m-HYDROXYIODOBENZENE ◇ m-IODOPHENOL ◇ 3-JODPHENOL ◇ PHENOL, m-IODO- ◇ PHENOL, 3-IODO-

TOXICITY DATA with REFERENCE
orl-mus LD50:2900 mg/kg   PHARAT 18,642,63

CONSENSUS REPORTS: Reported in EPA TSCA Inventory.

SAFETY PROFILE: Moderately toxic by ingestion. When heated to decomposition it emits toxic vapors of I^-.

IGM000   CAS:10102-50-8   HR: 3
IRON(II) ARSENATE (3582)
DOT: UN 1608
mf: $As_2O_8 \bullet 3Fe$ mw: 445.39

SYNS: ARSENATE of IRON, FERROUS ◇ FERROUS ARSENATE (DOT) ◇ FERROUS ARSENATE, solid (DOT) ◇ IRON ARSENATE (DOT)

CONSENSUS REPORTS: Arsenic and its compounds are on the Community Right-To-Know List.

OSHA PEL: Cancer Hazard
ACGIH TLV: TWA 0.2 mg(As)/m^3 (Proposed: 0.01 mg(As)/m^3; Human Carcinogen); TWA 1 mg(Fe)/m^3
NIOSH REL: (Inorganic Arsenic) CL 0.002 mg(As)/m^3/15M

DOT Classification: Poison B; Label: Poison

SAFETY PROFILE: Confirmed human carcinogen. A deadly poison by various routes. A pesticide. When heated to decomposition it emits toxic fumes of As. See also ARSENIC COMPOUNDS and IRON COMPOUNDS.

IGN000 CAS:10102-49-5 HR: 3
IRON(III) ARSENATE (1581)
DOT: UN 1606
mf: $AsO_4 \cdot Fe$ mw: 194.77

SYNS: ARSENATE of IRON, FERRIC ◇ FERRIC ARSENATE, solid (DOT)

CONSENSUS REPORTS: Arsenic and its compounds are on the Community Right-To-Know List.

OSHA PEL: Cancer Hazard
ACGIH TLV: TWA 0.2 mg(As)/m^3 (Proposed: 0.01 mg(As)/m^3; Human Carcinogen); TWA 1 mg(Fe)/m^3
NIOSH REL: (Inorganic Arsenic) CL 0.002 mg(As)/m^3/15M

DOT Classification: Poison B; Label: Poison

SAFETY PROFILE: Confirmed human carcinogen. A deadly poison. A pesticide. When heated to decomposition it emits toxic fumes of As. See also ARSENIC COMPOUNDS and IRON COMPOUNDS.

IGO000 CAS:63989-69-5 HR: 3
IRON(III)-o-ARSENITE PENTAHYDRATE
DOT: UN 1607
mf: $As_2Fe_2O_6 \cdot Fe_2O_3 \cdot 5H_2O$ mw: 607.34

PROP: Brown-yellow powder.

SYNS: FERRIC ARSENITE, BASIC ◇ FERRIC ARSENITE, solid (DOT)

CONSENSUS REPORTS: Arsenic and its compounds are on the Community Right-To-Know List.

OSHA PEL: TWA 0.01 mg(As)/m^3; Cancer Hazard
ACGIH TLV: TWA 0.2 mg(As)/m^3 (Proposed: 0.01 mg(As)/m^3; Human Carcinogen); TWA 1 mg(Fe)/m^3
NIOSH REL: (Inorganic Arsenic) CL 0.002 mg(As)/m^3/15M

DOT Classification: Poison B; Label: Poison

SAFETY PROFILE: Confirmed human carcinogen. A deadly poison. When heated to decomposition it emits toxic fumes of As. See also ARSENIC COMPOUNDS and IRON COMPOUNDS.

IHC550 CAS:1309-38-2 HR: 3
IRON(II,III) OXIDE
mf: Fe_3O_4 mw: 231.54

$FeO \cdot Fe_2O_3$

SYN: 11557 BLACK ◇ BLACK GOLD F 89 ◇ BLACK IRON BM ◇ EPT 500 ◇ H 3S ◇ IRON BLACK ◇ KN 320 ◇ MAGNETIC BLACK ◇ MAGNETIC OXIDE ◇ MAGNETITE ◇ MERAMEC M 25 ◇ RB-BL ◇ TRIIRON TETRAOXIDE

CONSENSUS REPORTS: Reported in EPA TSCA Inventory.

SAFETY PROFILE: Mixtures with aluminum + calcium silicide + sodium nitrate may explode if ignited. Mixtures with aluminum + sulfur react violently if heated. Ignites on contact with hydrogen trisulfide. See also IRON and IRON COMPOUNDS.

IHG100 CAS:1332-37-2 HR: 3
IRON OXIDE, spent
DOT: NA 1383/UN 1376

SYNS: FERROUS FERRITE ◇ IRON MASS, not properly oxidized (NA1383) (DOT) ◇ IRON MASS, spent (UN1376) (DOT) ◇ IRON OXIDE, spent (UN1376) (DOT) ◇ IRON OXIDE RED 130B ◇ IRON SPONGE, not properly oxidized (NA1383) (DOT) ◇ IRON SPONGE, spent (UN1376) (DOT) ◇ MIO 40GN ◇ SIFERRIT ◇ SPENT IRON MASS (UN1376) (DOT) ◇ SPENT IRON SPONGE (UN1376) (DOT)

DOT Classification: Flammable Solid; Label: Flammable Solid

CONSENSUS REPORTS: Reported in EPA TSCA Inventory.

SAFETY PROFILE: Flammable solid. Keep away from sparks and flames.

IHU200 CAS:541-28-6 HR: 2
ISOAMYL IODIDE
mf: $C_5H_{11}I$ mw: 198.06

SYNS: BUTANE, 1-IODO-3-METHYL- ◇ 1-IODO-3-METHYLBUTANE ◇ ISOPENTYL IODIDE ◇ 1-JOD-3-METHYLBUTAN ◇ 3-METHYLBUTYL IODIDE

TOXICITY DATA with REFERENCE
orl-rat LD50:1424 mg/kg   85JCAE -,127,86
ipr-rat LD50:1424 mg/kg   34ZIAG -,756,69
ipr-mus LD50:503 mg/kg   85GMAT -,77,82

CONSENSUS REPORTS: Reported in EPA TSCA Inventory.

SAFETY PROFILE: Moderately toxic by ingestion and intraperitoneal routes. When heated to decomposition it emits toxic vapors of I^-.

IIQ600 CAS:558-30-5 HR: D
ISOBUTYLENEOXIDE
mf: C_4H_8O mw: 72.12

SYN: PROPANE, 1,2-EPOXY-2-METHYL-

TOXICITY DATA with REFERENCE
mmo-klp 5 mmol/L   MUREAV 89,269,81

CONSENSUS REPORTS: Reported in EPA TSCA Inventory.

SAFETY PROFILE: Mutation data reported. When heated to decomposition it emits acrid smoke and irritating vapors.

IJN100 CAS:646-13-9 **HR: 1**
ISOBUTYL STEARATE
mf: $C_{22}H_{44}O_2$ mw: 340.66

SYN: STEARIC ACID, ISOBUTYL ESTER

TOXICITY DATA with REFERENCE
skn-rbt 500 mg MLD   JACTDZ 4(5),107,85

CONSENSUS REPORTS: Reported in EPA TSCA Inventory.

SAFETY PROFILE: A skin irritant. When heated to decomposition it emits acrid smoke and irritating vapors.

IKC100 CAS:470-67-7 **HR: 2**
ISOCINEOLE
mf: $C_{10}H_{18}O$ mw: 154.28

SYNS: 1,4-CINEOL ◇ 1,4-CINEOLE ◇ 1,4-EPOXY-p-MENTHANE ◇ p-MENTHANE, 1,4-EPOXY- ◇ 1-METHYL-4-(1-METHYLETHYL)-7-OXABICYCLO(2.2.1)HEPTANE ◇ 7-OXABICYCLO(2.2.1)HEPTANE, 1-ISOPROPYL-4-METHYL-(6CI) ◇ 7-OXABICYCLO(2.2.1)HEPTANE, 1-METHYL-4-(1-METHYLETHYL)-(9CI)

TOXICITY DATA with REFERENCE
orl-rat LD50:3100 mg/kg   FCTOD7 26,291,88
skn-rbt LD50:>5 g/kg   FCTOD7 26,291,88

CONSENSUS REPORTS: Reported in EPA TSCA Inventory.

SAFETY PROFILE: Moderately toxic by ingestion. Mildly toxic by skin contact. When heated to decomposition it emits acrid smoke and irritating vapors.

IKG800 CAS:2094-99-7 **HR: 3**
1-(1-ISOCYANATO-1-METHYLETHYL)-3-(1-METHYLETHENYL)BENZENE
mf: $C_{13}H_{15}NO$ mw: 201.29

SYNS: BENZENE, 1-(1-ISOCYANATO-1-METHYLETHYL)-3-(1-METHYLETHENYL)- ◇ α-α-DIMETHYL-m-ISOPROPENYL BENZYL ISOCYANATE ◇ ISOCYANIC ACID, m-ISOPROPENYL-α-α-DIMETHYL BENZYL ESTER ◇ m-TMI

TOXICITY DATA with REFERENCE
orl-rat LD50:3410 mg/kg   JACTDZ 1,43,90
ihl-gpg LC50:750 mg/m³/1H   JACTDZ 1,43,90

SAFETY PROFILE: Poison by inhalation. Moderately toxic by ingestion. When heated to decomposition it emits toxic vapors of NO_x.

IKN300 CAS:480-63-7 **HR: 3**
β-ISODURYLIC ACID
mf: $C_{10}H_{12}O_2$ mw: 164.22

SYNS: BENZOIC ACID, 2,4,6-TRIMETHYL- ◇ MESITOIC ACID ◇ 2,4,6-TRIMETHYLBENZOIC ACID

TOXICITY DATA with REFERENCE
ipr-mus LD50:562 mg/kg   JMCMAR 11,1020,68

CONSENSUS REPORTS: Reported in EPA TSCA Inventory.

SAFETY PROFILE: Poison by intraperitoneal route. When heated to decomposition it emits acrid smoke and irritating vapors.

IKO100 CAS:2539-53-9 **HR: 2**
ISOETHYLVANILLIN
mf: $C_9H_{10}O_3$ mw: 166.19

SYNS: BENZALDEHYDE, 4-ETHOXY-3-HYDROXY- ◇ 4-ETHOXY-3-HYDROXYBENZALDEHYDE ◇ ETHYLISOVANILLIN

TOXICITY DATA with REFERENCE
sce-hmn:lyms 500 ~2mol/L   MUREAV 206,17,88
ivn-dog LDLo:1330 mg/kg   APFRAD 14,456,56

CONSENSUS REPORTS: Reported in EPA TSCA Inventory.

SAFETY PROFILE: Moderately toxic by intravenous route. Human mutation data reported. When heated to decomposition it emits acrid smoke and irritating vapors.

IKZ100 CAS:872-85-5 **HR: 2**
ISONICOTINALDEHYDE
mf: C_6H_5NO mw: 107.12

SYNS: 4-FORMYLPYRIDINE ◇ ISONICOTINIC ALDEHYDE ◇ p-PYRIDINEALDEHYDE ◇ 4-PYRIDINEALDEHYDE ◇ PYRIDINE-4-CARBALDEHYDE ◇ 4-PYRIDINECARBOXALDEHYDE (9CI)

TOXICITY DATA with REFERENCE
orl-mus LDLo:1600 mg/kg   AECTCV 14,111,85

CONSENSUS REPORTS: Reported in EPA TSCA Inventory.

SAFETY PROFILE: Moderately toxic by ingestion. When heated to decomposition it emits toxic vapors of NO_x.

ILC100 CAS:1453-82-3 **HR: D**
ISONICOTINIC ACID AMIDE
mf: $C_6H_6N_2O$ mw: 122.14

SYNS: 4-CARBAMOYLPYRIDINE ◇ ISONICOTINAMIDE ◇ γ-PYRIDINECARBOXAMIDE ◇ 4-PYRIDINECARBOXAMIDE (9CI)

TOXICITY DATA with REFERENCE
dni-rat:lvr 20 mmol/L   JJIND8 69,1353,82

CONSENSUS REPORTS: Reported in EPA TSCA Inventory.

SAFETY PROFILE: Mutation data reported. When heated to decomposition it emits toxic vapors of NO_x.

ILG100   CAS:498-94-2   **HR: 2**
ISONIPECOTIC ACID
mf: $C_6H_{11}NO_2$ mw: 129.18

SYNS: ACIDE ISONIPECOTIQUE ◇ ACIDE PIPERIDINE-CARBOXYLIQUE-4 ◇ 4-HEXAHYDROISONICOTINIC ACID ◇ 4-PIPERIDINECARBOXYLIC ACID (9CI)

TOXICITY DATA with REFERENCE
ivn-mus LD50:2100 mg/kg   THERAP 23,1343,68

CONSENSUS REPORTS: Reported in EPA TSCA Inventory.

SAFETY PROFILE: Moderately toxic by intravenous route. When heated to decomposition it emits toxic vapors of NO_x.

ILW100   CAS:543-87-3   **HR: 2**
ISOPENTYL NITRATE
mf: $C_5H_{11}NO_3$ mw: 133.17

SYNS: 1-BUTANOL, 3-METHYL-, NITRATE (9CI) ◇ ISO-AMYL NITRATE ◇ ISOPENTYL ALCOHOL, NITRATE ◇ 3-METHYL-1-BUTANOL NITRATE ◇ NITRITO D'AMILE

TOXICITY DATA with REFERENCE
ipr-mus LD50:480 mg/kg   FRPSAX 11,855,56

CONSENSUS REPORTS: Reported in EPA TSCA Inventory.

SAFETY PROFILE: Moderately toxic by intraperitoneal route. When heated to decomposition it emits toxic vapors of NO_x.

IMK100   CAS:636-53-3   **HR: 2**
ISOPHTHALIC ACID, DIETHYL ESTER
mf: $C_{12}H_{14}O_4$ mw: 222.26

SYN: DIETHYL ISOPHTHALATE

TOXICITY DATA with REFERENCE
ipr-mus LDLo:1111 mg/kg   JPMSAE 56,1446,67

CONSENSUS REPORTS: Reported in EPA TSCA Inventory.

SAFETY PROFILE: Moderately toxic by intraperitoneal route. When heated to decomposition it emits acrid smoke and irritating vapors.

IQN100   CAS:2027-17-0   **HR: 1**
2-ISOPROPYLNAPHTHALENE
mf: $C_{13}H_{14}$ mw: 170.27

SYN: NAPHTHALENE, 2-ISOPROPYL-

TOXICITY DATA with REFERENCE
orl-mus LD50:5300 mg/kg   85JCAE -,50,86

CONSENSUS REPORTS: Reported in EPA TSCA Inventory.

SAFETY PROFILE: Mildly toxic by ingestion. When heated to decomposition it emits acrid smoke and irritating vapors.

IQX090   CAS:618-45-1   **HR: 2**
m-ISOPROPYLPHENOL
mf: $C_9H_{12}O$ mw: 136.21

SYNS: m-CUMENOL ◇ 3-ISOPROPYLPHENOL ◇ PHENOL, m-ISOPROPYL- ◇ PHENOL, 3-(1-METHYLETHYL)-

TOXICITY DATA with REFERENCE
orl-mus LD50:1630 mg/kg   GISAAA 46(1),94,81

CONSENSUS REPORTS: Reported in EPA TSCA Inventory.

SAFETY PROFILE: Moderately toxic by ingestion. When heated to decomposition it emits acrid smoke and irritating vapors.

IQX100   CAS:88-69-7   **HR: 3**
o-ISOPROPYLPHENOL
mf: $C_9H_{12}O$ mw: 136.21

SYNS: 2-ISOPROPYLPHENOL ◇ 2-(1-METHYLETHYL)PHENOL ◇ PHENOL, o-ISOPROPYL- ◇ PHENOL, 2-(1-METHYLETHYL)-(9CI) ◇ PRODOX 131

TOXICITY DATA with REFERENCE
ivn-mus LD50:100 mg/kg   JMCMAR 23,1350,80

CONSENSUS REPORTS: Reported in EPA TSCA Inventory.

SAFETY PROFILE: Poison by intravenous route. When heated to decomposition it emits acrid smoke and irritating vapors.

IRG050   CAS:140-92-1   **HR: 3**
ISOPROPYL POTASSIUM XANTHATE
mf: $C_4H_7OS_2 \cdot K$ mw: 174.33

SYNS: CARBONIC ACID, DITHIO-, O-ISOPROPYL ESTER, POTASSIUM SALT ◇ CARBONODITHIOIC ACID, O-(1-METHYLETHYL) ESTER, POTASSIUM SALT (9CI) ◇ DITHIOCARBONIC ACID O-ISOPROPYL ESTER POTASSIUM SALT ◇ POTASSIUM ISOPROPYL XANTHANATE ◇ POTASSIUM ISOPROPYL XANTHATE ◇ POTASSIUM ISOPROPYL XANTHOGENATE ◇ XANTHIC ACID, ISOPROPYL-, POTASSIUM SALT

TOXICITY DATA with REFERENCE
ivn-mus LD50:207 mg/kg   AIPTAK 135,330,62

CONSENSUS REPORTS: Reported in EPA TSCA Inventory.

SAFETY PROFILE: Poison by intravenous route. When heated to decomposition it emits toxic vapors of SO_x.

IRL100   CAS:112-10-7   ***HR: 1***
ISOPROPYL STEARATE
mf: $C_{21}H_{42}O_2$ mw: 326.63

SYNS: 1-METHYLETHYL OCTADECANOATE ◇ OCTADECANOIC ACID, 1-METHYLETHYL ESTER (9CI) ◇ STEARIC ACID, ISOPROPYL ESTER ◇ WICKENOL 127

TOXICITY DATA with REFERENCE
skn-rbt 500 mg MLD   JACTDZ 4(5),107,85
orl-rat LDLo:8 g/kg   JACTDZ 4(5),107,85

CONSENSUS REPORTS: Reported in EPA TSCA Inventory.

SAFETY PROFILE: Mildly toxic by ingestion. A skin irritant. When heated to decomposition it emits acrid smoke and irritating vapors.

IRN200   CAS:546-68-9   ***HR: 1***
ISOPROPYL TITANATE(IV)
mf: $C_3H_8O \cdot 1/4Ti$ mw: 72.07

SYNS: ISOPROPYL ALCOHOL, TITANIUM(4+) SALT ◇ ISOPROPYL ORTHOTITANATE ◇ TETRAISOPROPOXIDE TITANIUM ◇ TETRAISOPROPOXYTITANIUM ◇ TETRAISOPROPYL ORTHOTITANATE ◇ TETRAISOPROPYL TITANATE ◇ TETRAKIS(ISOPROPOXY)TITANIUM ◇ TITANIUM(4+) ISOPROPOXIDE ◇ TITANIUM ISOPROPYLATE ◇ TITANIUM TETRAISOPROPOXIDE ◇ TITANIUM TETRAISOPROPYLATE ◇ TITANIUM TETRA-n-PROPOXIDE ◇ TYZOR TPt

TOXICITY DATA with REFERENCE
skn-rbt 500 mg/24H MLD   85JCAE -,1263,86
eye-rbt 20 mg/24H MOD   85JCAE -,1263,86
orl-rat LD50:7460 mg/kg   TXAPA9 28,313,74

CONSENSUS REPORTS: Reported in EPA TSCA Inventory.

SAFETY PROFILE: Mildly toxic by ingestion. A skin and eye irritant. When heated to decomposition it emits acrid smoke and irritating vapors.

IRR100   CAS:691-60-1   ***HR: 1***
1-ISOPROPYLUREA
mf: $C_4H_{10}N_2O$ mw: 102.16

SYNS: ISOPROPYLUREA ◇ N-ISOPROPYLUREA ◇ (1-METHYLETHYL)UREA ◇ UREA, ISOPROPYL- ◇ UREA, (1-METHYLETHYL)-(9CI)

TOXICITY DATA with REFERENCE
par-mus LDLo:5107 mg/kg   JPETAB 52,216,34

CONSENSUS REPORTS: Reported in EPA TSCA Inventory.

SAFETY PROFILE: Mildly toxic by parenteral route. When heated to decomposition it emits toxic vapors of NO_x.

ISD100   CAS:592-82-5   ***HR: D***
1-ISOTHIOCYANATOBUTANE
mf: C_5H_9NS mw: 115.21

SYNS: BUTANE, 1-ISOTHIOCYANATO-(9CI) ◇ n-BUTYL ISOTHIOCYANATE ◇ BUTYL MUSTARD OIL ◇ ISOTHIOCYANIC ACID, BUTYL ESTER

TOXICITY DATA with REFERENCE
mmo-sat 100 μg/plate   ABCHA6 44,3017,80

CONSENSUS REPORTS: Reported in EPA TSCA Inventory.

SAFETY PROFILE: Mutation data reported. When heated to decomposition it emits toxic vapors of SO_x.

J

JCA100 CAS:488-10-8 **HR: 1**
cis-JASMONE
mf: $C_{11}H_{16}O$ mw: 164.27

SYNS: 2-CYCLOPENTEN-1-ONE, 3-METHYL-2-(2-PENTENYL)-, (Z)- ◇ JASMONE ◇ (Z)-JASMONE ◇ 3-METHYL-2-(cis-2-PENTEN-1-YL)-2-CYCLOPENTEN-1-ONE

TOXICITY DATA with REFERENCE
skn-rbt 500 mg/24H MOD   FCTXAV 17,845,79
orl-rat LD50:5000 mg/kg   FCTXAV 17,845,79

CONSENSUS REPORTS: Reported in EPA TSCA Inventory.

SAFETY PROFILE: Mildly toxic by ingestion. A skin irritant. When heated to decomposition it emits acrid smoke and irritating vapors.

K

KBB600 CAS:1332-58-7 **HR: 2**
KAOLIN

PROP: Fine white to light-yellow powder; earth taste. Insol in ether, alc, dil acids, and alkali solutions.

SYN: ALTOWHITES ◊ BENTONE ◊ CONTINENTAL ◊ DIXIE ◊ EMATH-LITE ◊ FITROL ◊ FITROL DESICCATE 25 ◊ GLOMAX ◊ HYDRITE ◊ KAOPAOUS ◊ KAOPHILLS-2 ◊ LANGFORD ◊ MCNAMEE ◊ PARCLAY ◊ PEERLESS ◊ SNOW TEX

TOXICITY DATA with REFERENCE
orl-rat TDLo:590 g/kg (female 37D pre):REP JONUAI 107,2020,77

OSHA PEL: (Transitional: TWA Respirable Fraction: 15 mg/m^3; Respirable Fraction: 5 mg/m^3) TWA Total Dust: 10 mg/m^3; Respirable Fraction: 5 mg/m^3
ACGIH TLV: TWA 2 mg/m^3; Respirable Fraction

SAFETY PROFILE: A nuisance dust.

KDA050 CAS:89-63-4 **HR: 3**
KAYAKU FAST RED 3GL BASE
mf: $C_6H_5ClN_2O_2$ mw: 172.58

SYNS: ANILINE, 4-CHLORO-2-NITRO- ◊ AZOENE FAST RED 3GL BASE ◊ AZOIC DIAZO COMPONENT 9 ◊ BENZENAMINE, 4-CHLORO-2-NITRO- (9CI) ◊ p-CHLORO-o-NITROANILINE ◊ 4-CHLORO-2-NITROANILINE ◊ 4-CHLORO-2-NITROBENZENAMINE ◊ C.I. 37040 ◊ C.I. AZOIC DIAZO COMPONENT 9 ◊ DAITO RED BASE 3GL ◊ DEVOL RED F ◊ DIAZO FAST RED 3GL ◊ FAST RED BASE 3GL SPECIAL ◊ FAST RED BASE 3JL ◊ FAST RED 3GL SPECIAL BASE ◊ FAST RED 3GL BASE ◊ FAST RED 2NC BASE ◊ HILTONIL FAST RED 3GL BASE ◊ MITSUI RED 3GL BASE ◊ NAPHTHANIL RED 3G BASE ◊ NAPHTOELAN FAST RED 3GL BASE ◊ NCI-C60355 ◊ 2-NITRO-4-CHLOROANILINE ◊ PCON ◊ PCONA ◊ RED BASE CIBA VI ◊ RED BASE 3 GL ◊ RED BASE IRGA VI ◊ RED 3G BASE ◊ SHINNIPPON FAST RED 3GL BASE

TOXICITY DATA with REFERENCE
mma-sat 33 μg/plate ENMUDM 5(Suppl 1),3,83
cyt-ham:ovr 302 mg/L EMMUEG 10(Suppl 10),1,87
sce-ham:ovr 20 mg/L EMMUEG 10(Suppl 10),1,87
orl-rat LD50:400 mg/kg TSCAT* OTS 206512
ipr-rat LD50:200 mg/kg TSCAT* OTS 206512
orl-mus LD50:800 mg/kg TSCAT* OTS 206512
ipr-mus LD50:200 mg/kg TSCAT* OTS 206512
ivn-mus LD50:63 mg/kg CBCCT* 6,139,54

CONSENSUS REPORTS: Reported in EPA TSCA Inventory.

SAFETY PROFILE: Poison by ingestion, intraperitoneal, and intravenous routes. Mutation data reported. When heated to decomposition it emits toxic vapors of NO_x and Cl^-.

KHU000 CAS:74278-22-1 **HR: 3**
KROMAD

PROP: Contains 5% cadmium sebacate, 5% potassium chromate, 1% malachite green, and 16% thiram (FMCHA2-, D176,80).

TOXICITY DATA with REFERENCE
orl-rat LD50:400 mg/kg FMCHA2 -,D176,80
skn-rbt LD50:1 g/kg FMCHA2 -,D176,80

CONSENSUS REPORTS: Cadmium and its compounds, as well as chromium and its compounds, are on the Community Right-To-Know List.

OSHA PEL: TWA 5 μg(Cd)/m^3
ACGIH TLV: TWA 0.05 mg(Cd)/m^3 (Proposed: TWA 0.01 mg(Cd)/m^3 (dust), Suspected Human Carcinogen; 0.002 mg(Cd)/m^3 (respirable dust), Suspected Human Carcinogen); BEI: 10 μg/g creatinine in urine; 10 μg/L in blood
DFG BAT: Blood 1.5 μg/dL; Urine 15 μg/dL, Suspected Carcinogen
NIOSH REL: (Cadmium) Reduce to lowest feasible level

SAFETY PROFILE: Confirmed human carcinogen. Poison by ingestion. Moderately toxic by skin contact. When heated to decomposition it emits toxic fumes of K_2O, Cd, and Cr. See also CADMIUM COMPOUNDS, POTASSIUM CHROMATE, and BIS(DIMETHYLTHIOCARBAMYL) DISULFIDE (thiram).

L

LAG010   CAS:79-33-4   **HR: 2**
l-(+)-LACTIC ACID
mf: $C_3H_6O_3$ mw: 90.09

SYNS: ESPIRITIN ◊ (S)-2-HYDROXYPROPANOIC ACID ◊ (S)-2-HYDROXYPROPIONIC ACID ◊ (+)-LACTIC ACID ◊ d-LACTIC ACID ◊ (S)-LACTIC ACID ◊ (S)-(+)-LACTIC ACID ◊ PARALACTIC ACID ◊ PROPANOIC ACID, 2-HYDROXY-, (S)-(9CI) ◊ SARCOLACTIC ACID ◊ TISULAC

TOXICITY DATA with REFERENCE
ipr-mus LD50:3194 mg/kg   TXCYAC 62,203,90

CONSENSUS REPORTS: Reported in EPA TSCA Inventory.

SAFETY PROFILE: Moderately toxic by intraperitoneal route. When heated to decomposition it emits acrid smoke and irritating vapors.

LBO050   CAS:141-92-4   **HR: 3**
LAURINE DIMETHYL ACETAL
mf: $C_{12}H_{26}O_3$ mw: 218.38

SYNS: 8,8-DIMETHOXY-2,6-DIMETHYL-2-OCTANOL ◊ HYDROXYCITRONELLAL DIMETHYL ACETAL ◊ HYDROXYCITRONELLAL DMA ◊ OCTANAL, 7-HYDROXY-3,7-DIMETHYL-, DIMETHYL ACETAL ◊ 2-OCTANOL, 8,8-DIMETHOXY-2,6-DIMETHYL-(9CI)

TOXICITY DATA with REFERENCE
skn-rbt 100 mg/24H SEV   CTOIDG 94(8),41,79
skn-gpg 100 mg/24H MLD   CTOIDG 94(8),41,79

CONSENSUS REPORTS: Reported in EPA TSCA Inventory.

SAFETY PROFILE: A severe skin irritant. When heated to decomposition it emits acrid smoke and irritating vapors.

LBX050   CAS:104-74-5   **HR: 3**
LAURYLPYRIDINIUM CHLORIDE
mf: $C_{17}H_{30}N \bullet Cl$ mw: 283.93

SYNS: C 2 ◊ DEHYQUART C ◊ DODECYLPYRIDINIUM CHLORIDE ◊ N-DODECYLPYRIDINIUM CHLORIDE ◊ 1-DODECYLPYRIDINIUM CHLORIDE ◊ DPC ◊ ELTREN ◊ 1-LAURYLPYRIDINIUM CHLORIDE ◊ LPC ◊ PYRIDINIUM, 1-DODECYL-, CHLORIDE ◊ QUATERNARIO LPC

TOXICITY DATA with REFERENCE
unr-mus LD50:119 mg/kg   PHARAT 40,273,85

CONSENSUS REPORTS: Reported in EPA TSCA Inventory.

SAFETY PROFILE: Poison by an unspecified route. When heated to decomposition it emits toxic vapors of NO_x and Cl^-.

LBX100   CAS:90-42-6   **HR: 1**
LAVAMENTHE
mf: $C_{12}H_{20}O$ mw: 180.32

SYNS: (1,1'-BICYCLOHEXYL)-2-ONE ◊ 2-CYCLOHEXYLCYCLOHEXANONE

TOXICITY DATA with REFERENCE
orl-rat LD50:5 g/kg   NPIRI* 1,20,74

CONSENSUS REPORTS: Reported in EPA TSCA Inventory.

SAFETY PROFILE: Mildly toxic by ingestion. When heated to decomposition it emits acrid smoke and irritating vapors.

LEN050   CAS:1940-42-7   **HR: 2**
LEPTOPHOS PHENOL
mf: $C_6H_3BrCl_2O$ mw: 241.90

SYNS: 4-BROMO-2,5-DICHLOROPHENOL ◊ PHENOL, 4-BROMO-2,5-DICHLORO- ◊ PHOSVEL PHENOL

TOXICITY DATA with REFERENCE
orl-rat LD50:1350 mg/kg   NTIS** PB85-143766

CONSENSUS REPORTS: Reported in EPA TSCA Inventory.

SAFETY PROFILE: Moderately toxic by ingestion. When heated to decomposition it emits toxic vapors of Br^- and Cl^-.

LEX200   CAS:476-60-8   **HR: D**
LEUCOQUINIZARIN
mf: $C_{14}H_{10}O_4$ mw: 242.24

SYNS: 1,4,9,10-ANTHRACENETETRAOL ◊ 1,4,9,10-ANTHRACENETETROL (9CI) ◊ 1,4,9,10-TETRAHYDROXYANTHRACENE

TOXICITY DATA with REFERENCE
mmo-sat 100 μg/plate   BCSTB5 5,1489,77
mma-sat 100 μg/plate   MUREAV 40,203,76

CONSENSUS REPORTS: Reported in EPA TSCA Inventory.

SAFETY PROFILE: Mutation data reported. When heated to decomposition it emits acrid smoke and irritating vapors.

LFT800 CAS:474-07-7 HR: 2
LIMAWOOD EXTRACT
mf: $C_{16}H_{14}O_5$ mw: 286.30

SYNS: BENZ(b)INDENO(1,2-d)PYRAN-3,6a,9,10(6H)-TETROL,7,11b-DIHYDRO- ◊ BRASILIN ◊ BRAZILETTO ◊ BRAZILIN ◊ 7,11b-DIHYDROBENZ(b)INDENO(1,2-d)PYRAN-3,6a,9,10(6H)-TETROL ◊ HYPERNIC EXTRACT ◊ PERNAMBUCO EXTRACT ◊ SUPERBRESILINE

TOXICITY DATA with REFERENCE
ipr-mus LD50:1500 mg/kg   85GDA2 8(1),252,82

CONSENSUS REPORTS: Reported in EPA TSCA Inventory.

SAFETY PROFILE: Moderately toxic by intraperitoneal route. When heated to decomposition it emits acrid smoke and irritating vapors.

LFY510 CAS:78-69-3 HR: 1
LINALOOL TETRAHYDRIDE
mf: $C_{10}H_{22}O$ mw: 158.32

SYNS: 3,7-DIMETHYLOCTANOL-3 ◊ 3-OCTANOL, 3,7-DIMETHYL- ◊ TETRAHYDROLINALOOL

TOXICITY DATA with REFERENCE
skn-rbt 500 mg/24H MOD   FCTXAV 17,909,79

CONSENSUS REPORTS: Reported in EPA TSCA Inventory.

SAFETY PROFILE: A skin irritant. When heated to decomposition it emits acrid smoke and irritating vapors.

LGO100 CAS:546-89-4 HR: 2
LITHIUM ACETATE
mf: $C_2H_3O_2 \cdot Li$ mw: 67.00

SYNS: ACETIC ACID, LITHIUM SALT ◊ QUILONE

TOXICITY DATA with REFERENCE
cyt-mus-orl 50 µg/kg   CYTOAN 54,245,89
sce-mus-orl 5 mg/kg   CYTOAN 54,245,89
orl-mus LDLo:1500 mg/kg   PHTXA6 21,419,58
scu-mus LDLo:1500 mg/kg   PHTXA6 21,419,58

CONSENSUS REPORTS: Reported in EPA TSCA Inventory.

SAFETY PROFILE: Moderately toxic by ingestion and subcutaneous routes. Mutation data reported. When heated to decomposition it emits toxic fumes of Li.

LHD150 HR: D
LITHIUM CITRATE HYDRATE
mf: $C_6H_5O_7 \cdot 3Li \cdot xH_2O$ mw: 363.83

SYNS: CITRIC ACID, TRILITHIUM SALT, HYDRATE ◊ 2-HYDROXY-1,2,3-PROPANETRICARBOXYLIC ACID TRILITHIUM SALT HYDRATE ◊ 1,2,3-PROPANETRICARBOXYLIC ACID, 2-HYDROXY-, TRILITHIUM SALT, HYDRATE

TOXICITY DATA with REFERENCE
orl-rat TDLo:554 mg/kg (lactating female 21D postbirth):REP   NURIBL 13,567,76
orl-rat TDLo:554 mg/kg (lactating female 21D postbirth):REP   NURIBL 13,567,76
orl-rat TDLo:529 mg/kg (female 1–22D post):REP   NURIBL 13,567,76
orl-rat TDLo:529 mg/kg (female 1–22D post):REP   NURIBL 13,567,76

SAFETY PROFILE: Experimental reproductive effects. When heated to decomposition it emits toxic fumes of Li.

LHI100 CAS:1310-65-2 HR: 3
LITHIUM HYDROXIDE
mf: HLiO mw: 23.95

SYN: LITHIUM HYDROXIDE (Li(OH)) (9CI)

TOXICITY DATA with REFERENCE
ihl-rat LC50:960 mg/m^3/4H   FAATDF 7,58,86
orl-mus LDLo:200 mg/kg   PHTXA6 21,419,58
scu-mus LDLo:300 mg/kg   PHTXA6 21,419,58

CONSENSUS REPORTS: Reported in EPA TSCA Inventory.

SAFETY PROFILE: Poison by ingestion and subcutaneous routes. Mildly toxic by inhalation. When heated to decomposition it emits toxic fumes of Li.

LIQ500 CAS:106-58-1 HR: 2
LUPETAZINE
mf: $C_6H_{14}N_2$ mw: 114.22

SYNS: N,N'-DIMETHYLPIPERAZINE ◊ 1,4-DIMETHYLPIPERAZINE ◊ PIPERAZINE, 1,4-DIMETHYL-

TOXICITY DATA with REFERENCE
scu-mus LD50:2500 mg/kg   THERAP 9,314,54

CONSENSUS REPORTS: Reported in EPA TSCA Inventory.

SAFETY PROFILE: Moderately toxic by subcutaneous route. When heated to decomposition it emits toxic vapors of NO_x.

LIQ550 CAS:504-03-0 HR: 3
2,6-LUPETIDINE
mf: $C_7H_{15}N$ mw: 113.23

SYNS: 2,6-DIMETHYLPIPERIDINE ◊ LUPETIDINE ◊ NANOFIN ◊ NANOPHYN ◊ PIPERIDINE, 2,6-DIMETHYL-

TOXICITY DATA with REFERENCE
scu-rbt LDLo:400 mg/kg   BDCGAS 34,2408,01

CONSENSUS REPORTS: Reported in EPA TSCA Inventory.

SAFETY PROFILE: Poison by subcutaneous route. When heated to decomposition it emits toxic vapors of NO$_x$.

LIY990   CAS:108-47-4   *HR: 3*
2,4-LUTIDINE
mf: C$_7$H$_9$N   mw: 107.17

SYNS: α-γ-DIMETHYLPYRIDINE ◇ 2,4-DIMETHYLPYRIDINE ◇ PYRIDINE, 2,4-DIMETHYL-(9CI)

TOXICITY DATA with REFERENCE
sln-smc 5000 ppm   MUREAV 163,23,86
orl-rat LD50:200 mg/kg   NTIS** PB85-143766

CONSENSUS REPORTS: Reported in EPA TSCA Inventory.

SAFETY PROFILE: Poison by ingestion. Mutation data reported. When heated to decomposition it emits toxic vapors of NO$_x$.

LJA010   CAS:108-48-5   *HR: 3*
2,6-LUTIDINE
mf: C$_7$H$_9$N   mw: 107.17

SYNS: 2,6-DIMETHYLPYRIDINE ◇ 2,6-DIMETHYPYRIDINE ◇ α-α'-DIMETHYLPYRIDINE ◇ α-α'-LUTIDINE ◇ PYRIDINE, 2,6-DIMETHYL-(9CI)

TOXICITY DATA with REFERENCE
sln-smc 5000 ppm   MUREAV 163,23,86
orl-rat LD50:400 mg/kg   85JCAE -,845,86
ihl-rat LCLo:7500 ppm/1H   85JCAE -,845,86
skn-gpg LD50:2500 mg/kg   85JCAE -,845,86

CONSENSUS REPORTS: Reported in EPA TSCA Inventory.

SAFETY PROFILE: Poison by ingestion. Moderately toxic by skin contact. Mildly toxic by inhalation. Mutation data reported. When heated to decomposition it emits toxic vapors of NO$_x$.

M

MAP750   CAS:7439-96-5   **HR: 3**
MANGANESE
af: Mn   aw: 54.94

PROP: Reddish-grey or silvery, brittle, metallic element. Mp: 1260°, bp: 1900°, d: 7.20, vap press: 1 mm @ 1292°.

SYNS: COLLOIDAL MANGANESE ◇ MANGACAT ◇ MANGAN (POLISH) ◇ MANGAN NITRIDOVANY (CZECH) ◇ TRONAMANG

TOXICITY DATA with REFERENCE
skn-rbt 500 mg/24H MLD   28ZPAK -,21,72
eye-rbt 500 mg/24H MLD   28ZPAK -,21,72
mrc-smc 8 mmol/L/18H   MUREAV 42,343,77
ims-rat TDLo:400 mg/kg/1Y-I:ETA   NCIUS* PH 43-64-886,SEPT,71
ihl-man TCLo:2300 µg/m^3:BRN,CNS   AIHAAP 27,454,66

CONSENSUS REPORTS: Manganese and its compounds are on the Community Right-To-Know List. Reported in EPA TSCA Inventory.

OSHA PEL: Fume: (Transitional: CL 5 mg/m^3) TWA 1 mg/m^3; STEL 3 mg/m^3; Compounds: CL 5 mg/m^3
ACGIH TLV: Fume: 1 mg/m^3; STEL 3 mg/m^3; Dust and Compounds: TWA 5 mg/m^3 (Proposed: TWA 0.2 mg/m^3)
DFG MAK: 5 mg/m^3

SAFETY PROFILE: Human systemic effects by inhalation: degenerative brain changes, change in motor activity, muscle weakness. A skin and eye irritant. Questionable carcinogen with experimental tumorigenic data. Mutation data reported. Flammable and moderately explosive in the form of dust or powder when exposed to flame. The dust may be pyrophoric in air and may explode when heated in carbon dioxide. Mixtures of aluminum dust and manganese dust may explode in air. Mixtures with ammonium nitrate may explode when heated. The powdered metal ignites on contact with fluorine; chlorine + heat; hydrogen peroxide; bromine pentafluoride; sulfur dioxide + heat. Violent reaction with NO$_2$ and oxidants. Incandescent reaction with phosphorus; nitryl fluoride; nitric acid. Will react with water or steam to produce hydrogen; can react with oxidizing materials. To fight fire, use special dry chemical. See also MANGANESE COMPOUNDS.

MCE100   CAS:498-81-7   **HR: 1**
p-MENTHAN-8-OL
mf: C$_{10}$H$_{20}$O   mw: 156.30

SYNS: CYCLOHEXANEMETHANOL, α-α-4-TRIMETHYL-(9CI) ◇ DIHYDRO-α-TERPINEOL ◇ 1-METHYL-4-ISOPROPYLCYCLOHEXANE-8-OL ◇ α-α-4-TRIMETHYLCYCLOHEXANEMETHANOL

TOXICITY DATA with REFERENCE
skn-rbt 500 mg/24h MOD   FCTXAV 12,529,74
orl-rat LD50:>5 g/kg   FCTXAV 12,529,74
skn-rbt LD50:>5 g/kg   FCTXAV 12,529,74

CONSENSUS REPORTS: Reported in EPA TSCA Inventory.

SAFETY PROFILE: Mildly toxic by ingestion and skin contact. A skin irritant. When heated to decomposition it emits acrid smoke and irritating vapors.

MCK900   CAS:2382-96-9   **HR: 3**
2-MERCAPTOBENZOXAZOLE
mf: C$_7$H$_5$NOS   mw: 151.19

SYN: 2-BENZOXAZOLETHIOL

TOXICITY DATA with REFERENCE
ivn-mus LD50:180 mg/kg   CSLNX* NX#04482

CONSENSUS REPORTS: Reported in EPA TSCA Inventory.

SAFETY PROFILE: Poison by intravenous route. When heated to decomposition it emits toxic vapors of SO$_x$ and NO$_x$.

MCQ700   CAS:1121-31-9   **HR: 3**
2-MERCAPTOPYRIDINE MONOXIDE
mf: C$_5$H$_5$NOS   mw: 127.17

SYNS: OMADINE ◇ 2-PYRIDINETHIOL, 1-OXIDE

TOXICITY DATA with REFERENCE
orl-qal LD50:178 mg/kg   JRPFA4 48,371,76

CONSENSUS REPORTS: Reported in EPA TSCA Inventory.

SAFETY PROFILE: Poison by ingestion. When heated to decomposition it emits toxic fumes of NO$_x$.

MDJ740   CAS:527-60-6   **HR: 1**
MESITYL ALCOHOL
mf: C$_9$H$_{12}$O   mw: 136.21

SYNS: 2-HYDROXYMESITYLENE ◇ MESITOL ◇ PHENOL, 2,4,6-TRIMETHYL-(9CI) ◇ 2,4,6-TRIMETHYLPHENOL ◇ 2,4,6-TRIMETYLOFENOL

TOXICITY DATA with REFERENCE
orl-mus LD50:10 g/kg   BCTKAG 14,301,81

CONSENSUS REPORTS: Reported in EPA TSCA Inventory.

SAFETY PROFILE: Slightly toxic by ingestion. When heated to decomposition it emits acrid smoke and irritating vapors.

MDJ745 CAS:487-68-3 **HR: D**
MESITYL ALDEHYDE
mf: $C_{10}H_{12}O$ mw: 148.22

SYNS: BENZALDEHYDE, 2,4,6-TRIMETHYL- ◇ MESITALDEHYDE ◇ MESITYLENECARBOXALDEHYDE ◇ 2-MESITYLENECARBOX-ALDEHYDE ◇ 2,4,6-TRIMETHYLBENZALDEHYDE

TOXICITY DATA with REFERENCE
sce-hmn:lyms 500 µmol/L   MUREAV 206,17,88

CONSENSUS REPORTS: Reported in EPA TSCA Inventory.

SAFETY PROFILE: Human mutation data reported. When heated to decomposition it emits acrid smoke and irritating vapors.

MDJ748 CAS:499-06-9 **HR: 2**
MESITYLENIC ACID
mf: $C_9H_{10}O_2$ mw: 150.19

SYNS: BENZOIC ACID, 3,5-DIMETHYL- ◇ 3,5-DIMETHYLBENZOIC ACID

TOXICITY DATA with REFERENCE
ipr-mus LD50:750 mg/kg   JMCMAR 11,1020,68

CONSENSUS REPORTS: Reported in EPA TSCA Inventory.

SAFETY PROFILE: Moderately toxic by intraperitoneal route. When heated to decomposition it emits acrid smoke and irritating vapors.

MDM760 CAS:98-18-0 **HR: 1**
METANILAMIDE
mf: $C_6H_8N_2O_2S$ mw: 172.22

SYNS: m-AMINOBENZENESULFONAMIDE ◇ 3-AMINOBENZENE-SULFONAMIDE ◇ m-AMINOBENZENESULPHONAMIDE ◇ BENZENESULFONAMIDE, 3-AMINO-

TOXICITY DATA with REFERENCE
orl-mus LDLo:4500 mg/kg   QJPPAL 10,319,37

CONSENSUS REPORTS: Reported in EPA TSCA Inventory.

SAFETY PROFILE: Slightly toxic by ingestion. When heated to decomposition it emits toxic vapors of NO_x and SO_x.

MDN510 CAS:109-16-0 **HR: 1**
METHACRYLIC ACID, DIESTER with TRIETHYLENE GLYCOL
mf: $C_{14}H_{22}O_6$ mw: 286.36

SYNS: NK ESTER 3G ◇ POLYESTER TGM 3 ◇ 2-PROPENOIC ACID, 2-METHYL-, 1,2-ETHANEDIYLBIS(OXY-2,1-ETHANEDIYL) ESTER(9CI) ◇ TEDMA ◇ TGM 3 ◇ TGM 3PC ◇ TGM 3S ◇ TRIETHYLENE GLYCOL DIMETHACRYLATE

TOXICITY DATA with REFERENCE
mma-sat 4 mg/plate   TOLED5 31(Suppl),214,86
orl-rat LD50:10,837 mg/kg   GISAAA 47(4),17,82
orl-mus LD50:10,750 mg/kg   GISAAA 47(4),17,82

CONSENSUS REPORTS: Reported in EPA TSCA Inventory.

SAFETY PROFILE: Slightly toxic by ingestion. Mutation data reported. When heated to decomposition it emits acrid smoke and irritating vapors.

MEA650 CAS:82-39-3 **HR: D**
1-METHOXYANTHRAQUINONE
mf: $C_{15}H_{10}O_3$ mw: 238.25

SYNS: 9,10-ANTHRACENEDIONE, 1-METHOXY-(9CI) ◇ ANTHRAQUINONE, 1-METHOXY- ◇ 1-METHOXY-9,10-ANTHRACENEDIONE

TOXICITY DATA with REFERENCE
mmo-sat 500 µg/plate   MUREAV 40,203,76
mma-sat 500 µg/plate   MUREAV 40,203,76

CONSENSUS REPORTS: Reported in EPA TSCA Inventory.

SAFETY PROFILE: Mutation data reported. When heated to decomposition it emits acrid smoke and irritating vapors.

MEK300 CAS:531-59-9 **HR: 1**
7-METHOXYCOUMARIN
mf: $C_{10}H_8O_3$ mw: 176.18

SYNS: 2H-1-BENZOPYRAN-2-ONE, 7-METHOXY- ◇ AYAPANIN ◇ COUMARIN, 7-METHOXY-(8CI) ◇ HERNIARIN (6CI) ◇ 7-METHOXY-2H-1-BENZOPYRAN-2-ONE ◇ METHYLUMBELLIFERONE

TOXICITY DATA with REFERENCE
orl-rat LD50:4300 mg/kg   FCTOD7 26,375,88
skn-gpg LD50:>5 g/kg   FCTOD7 26,375,88

CONSENSUS REPORTS: Reported in EPA TSCA Inventory.

SAFETY PROFILE: Slightly toxic by ingestion and skin contact. When heated to decomposition it emits acrid smoke and irritating vapors.

MEY800 CAS:2348-82-5 **HR: 3**
2-METHOXY-1,4-NAPHTHALENEDIONE
mf: $C_{11}H_8O_3$ mw: 188.19

SYNS: 2-METHOXYNAPHTHOQUINONE ◇ 2-METHOXY-1,4-

NAPHTHOQUINONE ◇ 1,4-NAPHTHALENEDIONE, 2-METHOXY-(9CI) ◇ 1,4-NAPHTHOQUINONE, 2-METHOXY-

TOXICITY DATA with REFERENCE
ipr-mus LD50:320 mg/kg   JMCMAR 26,570,83
orl-brd LD50:316 mg/kg   AECTCV 12,355,83

CONSENSUS REPORTS: Reported in EPA TSCA Inventory.

SAFETY PROFILE: Poison by ingestion and intraperitoneal route. When heated to decomposition it emits acrid smoke and irritating vapors.

MFL300   CAS:116-11-0   HR: 2
2-METHOXYPROPENE
mf: C_4H_8O mw: 72.12

SYN: 1-PROPENE, 2-METHOXY-

TOXICITY DATA with REFERENCE
orl-rat LD50:1870 mg/kg   AIHAAP 30,470,69
ihl-rat LCLo:64,000 ppm/4H   AIHAAP 30,470,69

CONSENSUS REPORTS: Reported in EPA TSCA Inventory.

SAFETY PROFILE: Moderately toxic by ingestion. Slightly toxic by inhalation.

MGC250   CAS:74-89-5   HR: 3
METHYLAMINE
DOT: UN 1061/UN 1235
mf: CH_5N mw: 31.07

PROP: Colorless gas or liquid; powerful ammonia-like odor. Bp: 6.3°, lel: 4.95%, uel: 20.75%, mp: −93.5°, flash p: 32°F (CC), d: 0.662 @ 20°/4°, autoign temp: 806°F, vap d: 1.07. Fuming liquid when liquefied: d: 0.699 @ −10.8°/4°. Sol in alc; misc with ether.

SYNS: AMINOMETHANE ◇ CARBINAMINE ◇ MERCURIALIN ◇ METHANAMINE (9CI) ◇ METHYLAMINEN (DUTCH) ◇ METILAMINE (ITALIAN) ◇ METYLOAMINA (POLISH) ◇ MONOMETHYLAMINE

TOXICITY DATA with REFERENCE
skn-gpg 100 mg open SEV   CODEDG 6,140,80
dlt-rat-ihl 10 µg/m^3 GISAAA 46(5),7,81
scu-rat LDLo:200 mg/kg   HBAMAK 4,1289,35
ihl-mus LC50:2400 mg/m^3/2H   85GMAT -,80,82
scu-mus LDLo:2500 mg/kg   MEIEDD 11,949,89
scu-gpg LDLo:200 mg/kg   HBAMAK 4,1289,35
scu-frg LDLo:2000 mg/kg   27ZWAY 1,250,23
ihl-mam LC50:2400 mg/m^3 TPKVAL 14,80,75

CONSENSUS REPORTS: Reported in EPA TSCA Inventory.

OSHA PEL: TWA 10 ppm
ACGIH TLV: TWA 5 ppm; STEL 15 ppm
DFG MAK: 10 ppm (12 mg/m^3)

DOT Classification: Flammable Gas; Label: Flammable Gas, Flammable Liquid.

SAFETY PROFILE: Poison by subcutaneous route. Moderately toxic by inhalation. A severe skin irritant. Mutation data reported. A strong base. Flammable gas at ordinary temperature and pressure. Very dangerous fire hazard when exposed to heat, flame, or sparks. Explosive when exposed to heat or flame. To fight fire, stop flow of gas. Forms an explosive mixture with nitromethane. When heated to decomposition it emits toxic fumes of NO_x. See also AMINES.

MGH800   CAS:1668-54-8   HR: 2
2-METHYL-4-AMINO-6-METHOXY-s-TRIAZINE
mf: $C_5H_8N_4O$ mw: 140.17

SYNS: CV 399 ◇ 4-METHOXY-6-METHYL-1,3,5-TRIAZIN-2-AMINE ◇ s-TRIAZINE, 2-AMINO-4-METHOXY-6-METHYL- ◇ 1,3,5-TRIAZINE-2-AMINE, 4-METHOXY-6-METHYL-(9CI)

TOXICITY DATA with REFERENCE
orl-mus LD50:1010 mg/kg   NYKZAU 66,64,70
ipr-mus LD50:880 mg/kg   NYKZAU 66,64,70

CONSENSUS REPORTS: Reported in EPA TSCA Inventory.

SAFETY PROFILE: Moderately toxic by ingestion and intraperitoneal routes. When heated to decomposition it emits toxic vapors of NO_x.

MHB300   CAS:80-18-2   HR: 3
METHYL BENZENESULFONATE
mf: $C_7H_8O_3S$ mw: 172.21

SYNS: BENZENESULFONIC ACID, METHYL ESTER ◇ 20ND3-5

TOXICITY DATA with REFERENCE
orl-rat LD50:740 mg/kg   GISAAA 51(1),82,86
orl-mus LD50:250 mg/kg   GISAAA 51(1),82,86

CONSENSUS REPORTS: Reported in EPA TSCA Inventory.

SAFETY PROFILE: Poison by ingestion. When heated to decomposition it emits toxic vapors of SO_x.

MHJ300   CAS:92-36-4   HR: 2
p-(6-METHYLBENZOTHIAZOL-2-YL)ANILINE
mf: $C_{14}H_{12}N_2S$ mw: 240.34

SYNS: 2-(p-AMINOPHENYL)-6-METHYLBENZOTHIAZOLE ◇ BENZENAMINE, 4-(6-METHYL-2-BENZOTHIAZOLYL)-(9CI) ◇ BENZOTHIAZOLE, 2-(p-AMINOPHENYL)-6-METHYL- ◇ DEHYDRO-p-TOLUIDINE ◇ DHPT ◇ 4-(6-METHYL-2-BENZOTHIAZOLYL)BENZENAMINE

TOXICITY DATA with REFERENCE
ihl-rat LCLo:3 g/m^3/4H   FCTOD7 22,289,84

CONSENSUS REPORTS: Reported in EPA TSCA Inventory.

SAFETY PROFILE: Moderately toxic by inhalation. When heated to decomposition it emits toxic vapors of NO_x and SO_x.

MHM510 CAS:89-93-0 HR: 3
2-METHYLBENZYLAMINE
mf: $C_8H_{11}N$ mw: 121.20

SYN: BENZYLAMINE, o-METHYL-

TOXICITY DATA with REFERENCE
ivn-mus LD50:56 mg/kg   CSLNX* NX#05187

CONSENSUS REPORTS: Reported in EPA TSCA Inventory.

SAFETY PROFILE: Poison by intravenous route. When heated to decomposition it emits toxic vapors of NO_x.

MHN350 CAS:699-10-5 HR: 2
METHYL BENZYL DISULFIDE
mf: $C_8H_{10}S_2$ mw: 170.30

SYN: DISULFIDE, BENZYL METHYL

TOXICITY DATA with REFERENCE
orl-mus LD50:1080 mg/kg   DCTODJ 3,249,80

CONSENSUS REPORTS: Reported in EPA TSCA Inventory.

SAFETY PROFILE: Moderately toxic by ingestion. When heated to decomposition it emits toxic vapors of SO_x.

MHO100 CAS:103-79-7 HR: 2
METHYL BENZYL KETONE
mf: $C_9H_{10}O$ mw: 134.19

SYNS: BENZYL METHYL KETONE ◊ PHENYLACETONE ◊ α-PHENYL-ACETONE ◊ PHENYLMETHYL METHYL KETONE ◊ 1-PHENYL-2-PROPANONE ◊ 2-PROPANONE, 1-PHENYL-

TOXICITY DATA with REFERENCE
ipr-mus LD50:540 mg/kg   JPMSAE 60,799,71

CONSENSUS REPORTS: Reported in EPA TSCA Inventory.

SAFETY PROFILE: Moderately toxic by intraperitoneal route. When heated to decomposition it emits acrid smoke and irritating vapors.

MHR050 CAS:1817-68-1 HR: 1
4-METHYL-2,6-BIS(1-PHENYLETHYL)PHENOL
mf: $C_{23}H_{24}O$ mw: 316.47

SYNS: ALKOFEN MBP ◊ 2,6-BIS(1-PHENYLETHYL)-4-METHYLPHENOL ◊ p-CRESOL, 2,6-BIS(α-METHYLBENZYL)- ◊ IONOL 6 ◊ PHENOL, 4-METHYL-2,6-BIS(1-PHENYLETHYL)-(9CI)

TOXICITY DATA with REFERENCE
orl-mus LD50:4300 mg/kg   85JCAE -,232,86

CONSENSUS REPORTS: Reported in EPA TSCA Inventory.

SAFETY PROFILE: Slightly toxic by ingestion. When heated to decomposition it emits acrid smoke and irritating vapors.

MHR790 CAS:1117-71-1 HR: 3
METHYL 4-BROMOCROTONATE
mf: $C_5H_7BrO_2$ mw: 179.03

SYNS: 2-BUTENOIC ACID, 4-BROMO-, METHYL ESTER (9CI) ◊ CROTONIC ACID, 4-BROMO-, METHYL ESTER ◊ METHYL 4-BROMO-2-BUTENOATE ◊ METHYL BROMOCROTONATE ◊ METHYL γ-BROMOCROTONATE

TOXICITY DATA with REFERENCE
ivn-mus LD50:56 mg/kg   CSLNX* NX#00811

CONSENSUS REPORTS: Reported in EPA TSCA Inventory.

SAFETY PROFILE: Poison by intravenous route. When heated to decomposition it emits toxic vapors of Br^-.

MHY550 CAS:7568-37-8 HR: 3
METHYL CADMIUM AZIDE
mf: CH_3CdN_3 mw: 97.13

CONSENSUS REPORTS: Cadmium and its compounds are on the Community Right-To-Know List.

OSHA PEL: TWA 5 μg(Cd)/m^3
ACGIH TLV: TWA 0.05 mg(Cd)/m^3 (Proposed: TWA 0.01 mg(Cd)/m^3 (dust), Suspected Human Carcinogen; 0.002 mg(Cd)/m^3 (respirable dust), Suspected Human Carcinogen); BEI: 10 μg/g creatinine in urine; 10 μg/L in blood
DFG BAT: Blood 1.5 μg/dL; Urine 15 μg/dL, Suspected Carcinogen
NIOSH REL: (Cadmium) Reduce to lowest feasible level.

SAFETY PROFILE: Confirmed human carcinogen. Hydrolysis reaction in the presence of moisture forms the explosive hydrogen azide gas. When heated to decomposition it emits toxic fumes of Cd and NO_x. See also CADMIUM COMPOUNDS and AZIDES.

MHY650 CAS:110-42-9 HR: 3
METHYL CAPRATE
mf: $C_{11}H_{22}O_2$ mw: 186.33

SYNS: CAPRIC ACID METHYL ESTER ◊ DECANOIC ACID, METHYL ESTER ◊ METHOLENE 2095 ◊ METHYL n-CAPRATE ◊ METHYL CAPRINATE ◊ METHYL DECANOATE ◊ METHYL n-DECANOATE ◊ UNIPHAT A30

TOXICITY DATA with REFERENCE
ivn-mus LDLo:48 mg/kg   RESJAS 3,250,66

SAFETY PROFILE: Poison by intravenous route. When heated to decomposition it emits acrid smoke and irritating vapors.

MHY700   CAS:106-70-7   **HR: 3**
METHYL CAPROATE
mf: $C_7H_{14}O_2$ mw: 130.21

SYNS: HEXANOIC ACID, METHYL ESTER ◊ METHYL CAPRONATE ◊ METHYL HEXANOATE ◊ METHYL n-HEXANOATE ◊ METHYL HEXOATE ◊ METHYL HEXYLATE

TOXICITY DATA with REFERENCE
orl-rat LD50:>5 g/kg   FCTOD7 20,745,82
ihl-mus LC50:14 g/m^3/2H   85GMAT -,82,82
ivn-mus LDLo:48 mg/kg   RESJAS 3,250,66
skn-gpg LD50:>5 g/kg   FCTOD7 20,745,82
ihl-uns LC50:11500 mg/m^3 GISAAA 51(5),61,86

CONSENSUS REPORTS: Reported in EPA TSCA Inventory.

SAFETY PROFILE: Poison by intravenous route. Slightly toxic by ingestion, inhalation, and skin contact routes. When heated to decomposition it emits acrid smoke and irritating vapors.

MHY800   CAS:111-11-5   **HR: 3**
METHYL CAPRYLATE
mf: $C_9H_{18}O_2$ mw: 158.27

SYNS: CAPRYLIC ACID, METHYL ESTER ◊ METHYL OCTANOATE ◊ OCTANOIC ACID, METHYL ESTER ◊ UNIPHAT A20

TOXICITY DATA with REFERENCE
ivn-mus LDLo:48 mg/kg   RESJAS 3,250,66

SAFETY PROFILE: Poison by intravenous route. When heated to decomposition it emits acrid smoke and irritating vapors.

MIM300   CAS:2382-43-6   **HR: 2**
β-METHYLCHOLINE CHLORIDE
mf: $C_6H_{16}NO•Cl$ mw: 153.68

SYNS: AMMONIUM, (2-HYDROXYPROPYL)TRIMETHYL-, CHLORIDE ◊ (2-HYDROXYPROPYL)TRIMETHYLAMMONIUM CHLORIDE ◊ 2-HYDROXY-N,N,N-TRIMETHYL-1-PROPANAMINIUM CHLORIDE ◊ 1-PROPANAMINIUM, 2-HYDROXY-N,N,N-TRIMETHYL-, CHLORIDE (9CI)

TOXICITY DATA with REFERENCE
scu-mus LDLo:630 mg/kg   JPETAB 1,303,09

CONSENSUS REPORTS: Reported in EPA TSCA Inventory.

SAFETY PROFILE: Moderately toxic by subcutaneous route. When heated to decomposition it emits toxic vapors of NO_x and Cl^-.

MJH905   CAS:644-08-6   **HR: 2**
4-METHYLDIPHENYL
mf: $C_{13}H_{12}$ mw: 168.25

SYNS: BIPHENYL, 4-METHYL- ◊ FEMA 3186 ◊ p-METHYLBIPHENYL ◊ 4-METHYL-1,1'-BIPHENYL ◊ p-METHYLDIPHENYL

TOXICITY DATA with REFERENCE
orl-rat LD50:2570 mg/kg   FCTXAV 13,487,75

CONSENSUS REPORTS: Reported in EPA TSCA Inventory.

SAFETY PROFILE: Moderately toxic by ingestion. When heated to decomposition it emits acrid smoke and irritating vapors.

MJH910   CAS:781-35-1   **HR: 2**
METHYL DIPHENYLMETHYL KETONE
mf: $C_{15}H_{14}O$ mw: 210.29

SYNS: BENZHYDRYL METHYL KETONE ◊ 1,1-DIPHENYL ACETONE ◊ 2-PROPANONE, 1,1-DIPHENYL-

TOXICITY DATA with REFERENCE
ipr-mus LD50:500 mg/kg   KHFZAN 21,1326,87

CONSENSUS REPORTS: Reported in EPA TSCA Inventory.

SAFETY PROFILE: Moderately toxic by intraperitoneal route. When heated to decomposition it emits acrid smoke and irritating vapors.

MJM200   CAS:101-14-4   **HR: 3**
4,4'-METHYLENE BIS(2-CHLOROANILINE)
mf: $C_{13}H_{12}Cl_2N_2$ mw: 267.17

PROP: Tan solid.

SYNS: BIS AMINE ◊ CURALIN M ◊ CURENE 442 ◊ CYANASET ◊ DI(-4-AMINO-3-CHLOROPHENYL)METHANE ◊ DI-(4-AMINO-3-CLOROFENIL)METANO (ITALIAN) ◊ 4,4'-DIAMINO-3,3'-DICHLORODIPHENYLMETHANE ◊ 3,3'-DICHLOR-4,4'-DIAMINODIPHENYLMETHAN (GERMAN) ◊ 3,3'-DICHLORO-4,4'-DIAMINODIPHENYLMETHANE ◊ 3,3'-DICLORO-4,4'-DIAMINODIFENILMETANO (ITALIAN) ◊ MBOCA ◊ METHYLENE-4,4'-BIS(o-CHLOROANILINE) ◊ p,p'-METHYLENEBIS(α-CHLOROANILINE) ◊ p,p'-METHYLENEBIS(o-CHLOROANILINE) ◊ 4,4'-METHYLENE(BIS)-CHLOROANILINE ◊ 4,4'-METHYLENEBIS(o-CHLOROANILINE) ◊ 4,4'-METHYLENEBIS-2-CHLOROBENZENAMINE ◊ METHYLENE-BIS-ORTHOCHLOROANILINE ◊ 4,4-METILENE-BIS-o-CLOROANILINA (ITALIAN) ◊ MOCA ◊ RCRA WASTE NUMBER U158

TOXICITY DATA with REFERENCE
otr-mus:fbr 10 μg/L   JJIND8 67,1303,81
sce-ham:ovr 500 μg/L   ENMUDM 7,1,85
orl-rat TDLo:4050 mg/kg/77W-C:CAR   JEPTDQ 2(1),149,78
scu-rat TDLo:25 g/kg/89W-C:CAR   NATWAY 58,578,71
orl-rat TD:27 g/kg/79W-C:ETA   NATWAY 58,578,71
orl-rat TD:27 g/kg/78W-C:CAR   TXAPA9 31,159,75
orl-rat TD:11 g/kg/2Y-C:CAR   JEPTDQ 2,149,78
orl-rat TD:8100 mg/kg/77W-C:CAR   JEPTDQ 2(1),149,78
orl-rat TD:16 g/kg/77W-C:CAR   JEPTDQ 2(1),149,78
orl-rat TD:27 g/kg/65W-C:CAR   JEPTDQ 2(1),149,78
orl-rat TD:34 g/kg/80W-C:CAR   TXAPA9 31,159,75
orl-rat TD:21 g/kg/60W-C:CAR   TXAPA9 31,159,75
orl-rat TD:24 g/kg/57W-C:CAR   TXAPA9 31,159,75
orl-rat LD50:2100 mg/kg   KCRZAE 26(9),28,67

orl-mus LD50:880 mg/kg   KCRZAE 26(9),28,67
ipr-mus LD50:64 mg/kg   PMRSDJ 1,682,81

CONSENSUS REPORTS: NTP Fifth Annual Report on Carcinogens. IARC Cancer Review: Group 2A IMEMDT 7,246,87; Animal Sufficient Evidence IMEMDT 4,65,74. EPA Genetic Toxicology Program. Community Right-To-Know List. Reported in EPA TSCA Inventory.

OSHA PEL: TWA 0.02 ppm (skin)
ACGIH TLV: TWA 0.02 ppm (skin); Suspected Human Carcinogen; (Proposed: 0.01 ppm; Suspected Human Carcinogen)
DFG MAK: Animal Carcinogen, Suspected Human Carcinogen
NIOSH REL: (MOCA) Lowest detectable limit.

SAFETY PROFILE: Confirmed carcinogen with experimental carcinogenic and tumorigenic data. Poison by intraperitoneal route. Moderately toxic by ingestion. Mutation data reported. Flammable liquid. Reactive with active metals such as sodium, potassium, magnesium, or zinc. When heated to decomposition it emits very toxic fumes of Cl⁻ and NO_x. See also AMINES.

MJT100   CAS:1660-94-2   HR: 3
METHYLENEDI(PHOSPHONIC ACID) TETRAETHYL ESTER
mf: $C_9H_{22}O_6P_2$ mw: 288.25

SYNS: METHYLENEBIS(DIETHYLPHOSPHONATE) ◊ PHOSPHONIC ACID, METHYLENEDI-, TETRAETHYL ESTER

TOXICITY DATA with REFERENCE
ivn-mus LD50:180 mg/kg   CSLNX* NX#03139

CONSENSUS REPORTS: Reported in EPA TSCA Inventory.

SAFETY PROFILE: Poison by intravenous route. When heated to decomposition it emits toxic vapors of PO_x.

MJV800   CAS:529-84-0   HR: 2
4-METHYLESCULETIN
mf: $C_{10}H_8O_4$ mw: 192.18

SYNS: 2H-1-BENZOPYRAN-2-ONE,6,7-DIHYDROXY-4-METHYL-(9CI) ◊ COUMARIN, 6,7-DIHYDROXY-4-METHYL- ◊ 6,7-DIHYDROXY-4-METHYL-2H-1-BENZOPYRAN-2-ONE ◊ 6,7-DIHYDROXY-4-METHYL-COUMARIN ◊ 4-METHYLAESCULETIN ◊ METHYLESCULETIN ◊ 4-METHYLESCULETOL

TOXICITY DATA with REFERENCE
ipr-rat LD50:5130 mg/kg   THERAP 20,879,65
orl-mus LD50:3200 mg/kg   MPHEAE 17,497,67

CONSENSUS REPORTS: Reported in EPA TSCA Inventory.

SAFETY PROFILE: Moderately toxic by ingestion. Slightly toxic by intraperitoneal route. When heated to decomposition it emits acrid smoke and irritating vapors.

MJW100   CAS:109-81-9   HR: 3
N-METHYLETHANEDIAMINE
mf: $C_3H_{10}N_2$ mw: 74.15

SYNS: 2-AMINOETHYLMETHYLAMINE ◊ 1,2-ETHANEDIAMINE, N-METHYL-(9CI) ◊ ETHYLENEDIAMINE, N-METHYL- ◊ 2-(METHYLAMINO)ETHYLAMINE ◊ N-METHYLDIAMINOETHANE ◊ N-METHYL-1,2-ETHANEDIAMINE ◊ N-METHYLETHYLENEDIAMINE ◊ N-METHYLETHYLIDENEDIAMINE

TOXICITY DATA with REFERENCE
ipr-mus LD50:50 mg/kg   EJMCA5 17,235,82

CONSENSUS REPORTS: Reported in EPA TSCA Inventory.

SAFETY PROFILE: Poison by intraperitoneal route. When heated to decomposition it emits toxic vapors of NO_x.

MKH550   CAS:620-02-0   HR: 2
5-METHYL FURFURAL
mf: $C_6H_6O_2$ mw: 110.12

SYNS: 2-FORMYL-5-METHYLFURAN ◊ 2-FURALDEHYDE, 5-METHYL- ◊ 5-METHYL-2-FURALDEHYDE ◊ 5-METHYL-2-FURANCARBOXALDEHYDE ◊ 5-METHYL-2-FURFURAL ◊ 5-METHYLFURFURALDEHYDE

TOXICITY DATA with REFERENCE
dnr-bcs 55 µg/disc   DFSCDX 13,353,86
cyt-ham:ovr 37,700 ~2mol/L   CALEDQ 13,89,81
orl-rat LD50:2200 mg/kg   FCTOD7 20,751,82

CONSENSUS REPORTS: Reported in EPA TSCA Inventory.

SAFETY PROFILE: Moderately toxic by ingestion. Mutation data reported. When heated to decomposition it emits acrid smoke and irritating vapors.

MKL350   CAS:597-96-6   HR: 3
3-METHYL-3-HEXANOL
mf: $C_7H_{16}O$ mw: 116.23

SYNS: 3-HEXANOL, 3-METHYL- ◊ 3-METHYL-HEXANOL-(3)

TOXICITY DATA with REFERENCE
orl-mus LD50:785 mg/kg   ARZNAD 4,477,54
scu-mus LD50:680 mg/kg   ARZNAD 5,161,55
ivn-mus LD50:56 mg/kg   CSLNX* NX#04237

CONSENSUS REPORTS: Reported in EPA TSCA Inventory.

SAFETY PROFILE: Poison by intravenous route. Moderately toxic by ingestion and subcutaneous routes. When heated to decomposition it emits acrid smoke and irritating vapors.

MLC600   CAS:547-64-8   *HR: 1*
METHYL LACTATE
mf: $C_4H_8O_3$ mw: 104.12

SYN: LACTIC ACID, METHYL ESTER

TOXICITY DATA with REFERENCE
ipr-rat LDLo:>2 g/kg   JPPMAB 11,150,59

CONSENSUS REPORTS: Reported in EPA TSCA Inventory.

SAFETY PROFILE: Slightly toxic by intraperitoneal route. When heated to decomposition it emits acrid smoke and irritating vapors.

MLC800   CAS:111-82-0   *HR: 3*
METHYL LAURATE
mf: $C_{13}H_{26}O_2$ mw: 214.39

SYNS: LAURIC ACID, METHYL ESTER ◇ METHOLENE 2296 ◇ METHYL DODECANOATE ◇ METHYL DODECYLATE ◇ METHYL LAURINATE ◇ STEPAN C40 ◇ UNIPHAT A40

TOXICITY DATA with REFERENCE
ivn-mus LDLo:48 mg/kg   RESJAS 3,250,66

SAFETY PROFILE: Poison by intravenous route. When heated to decomposition it emits acrid smoke and irritating vapors.

MLH800   CAS:606-45-1   *HR: 2*
METHYL 2-METHOXYBENZOATE
mf: $C_9H_{10}O_3$ mw: 166.19

SYNS: o-ANISIC ACID, METHYL ESTER (7CI,8CI) ◇ BENZOIC ACID, 2-METHOXY-, METHYL ESTER ◇ DIMETHYL SALICYLATE ◇ o-METHOXYBENZOIC ACID METHYL ESTER ◇ o-METHOXY METHYL BENZOATE ◇ METHYL o-ANISATE ◇ METHYL o-METHOXYBENZOATE ◇ METHYLSALICYLATE METHYL ETHER

TOXICITY DATA with REFERENCE
orl-rat LD50:3800 mg/kg   FCTOD7 26,385,88
skn-rbt LD50:>5 g/kg   FCTOD7 26,385,88

CONSENSUS REPORTS: Reported in EPA TSCA Inventory.

SAFETY PROFILE: Moderately toxic by ingestion. Slightly toxic by skin contact. When heated to decomposition it emits acrid smoke and irritating vapors.

MLI800   CAS:1080-12-2   *HR: 2*
METHYL-3-METHOXY-4-HYDROXY STYRYL KETONE
mf: $C_{11}H_{12}O_3$ mw: 192.23

SYNS: 3-BUTEN-2-ONE, 4-(4-HYDROXY-3-METHOXYPHENYL)- ◇ 4-(4-HYDROXY-3-METHOXYPHENYL)-3-BUTEN-2-ONE ◇ MHSK

TOXICITY DATA with REFERENCE
orl-mus LDLo:>2 g/kg   ARZNAD 37,708,87
ipr-mus LD50:610 mg/kg   ARZNAD 37,708,87

CONSENSUS REPORTS: Reported in EPA TSCA Inventory.

SAFETY PROFILE: Moderately toxic by intraperitoneal route. Slightly toxic by ingestion. When heated to decomposition it emits acrid smoke and irritating vapors.

MLJ100   CAS:1210-56-6   *HR: 3*
1-METHYL-6-METHOXY-1,2,3,4-TETRAHYDRO-β-CARBOLINE
mf: $C_{13}H_{16}N_2O$ mw: 216.31

SYNS: 6-METHOXY-1-METHYL-1,2,3,4-TETRAHYDRO-β-CARBOLINE ◇ 9H-PYRIDO(3,4-b)INDOLE, 1,2,3,4-TETRAHYDRO-6-METHOXY-1-METHYL-

TOXICITY DATA with REFERENCE
ipr-mus LD50:350 mg/kg   FATOAO 27,165,64
ivn-mus LD50:56 mg/kg   CSLNX* NX#12476

CONSENSUS REPORTS: Reported in EPA TSCA Inventory.

SAFETY PROFILE: Poison by intravenous and intraperitoneal routes. When heated to decomposition it emits toxic vapors of NO_x.

MLW750   CAS:1646-87-3   *HR: 3*
2-METHYL-2-(METHYLSULFINYL)PROPANAL O-((METHYLAMINO)CARBONYL)OXIME
mf: $C_7H_{14}N_2O_3S$ mw: 206.29

SYNS: ALDICARB SULFOXIDE ◇ 2-METHYL-2-(METHYLSULFINYL)PROPIONALDEHYDE O-(METHYLCARBAMOYL)OXIME ◇ PROPANAL, 2-METHYL-2-(METHYLSULFINYL)-, O-((METHYLAMINO)CARBONYL)OXIME ◇ PROPIONALDEHYDE, 2-METHYL-2-(METHYLSULFINYL)-, O-(METHYLCARBAMOYL)OXIME ◇ TEMIK SULFOXIDE

TOXICITY DATA with REFERENCE
orl-wmn TDLo:54 µg/kg   JAMAAP 256,3218,86
orl-rat LD50:490 µg/kg   TSCAT* FYI-OTS-0885-0443
ipr-rat LD50:470 µg/kg   TSCAT* FYI-OTS-0885-0443
ivn-rat LD50:370 µg/kg   TSCAT* FYI-OTS-0885-0443

SAFETY PROFILE: Poison by ingestion, intraperitoneal, and intravenous routes. Human systemic efects by ingestion: fasciculations, hypermotility or diarrhea, nausea or vomiting. When heated to decomposition it emits toxic vapors of NO_x and SO_x.

MLX800   CAS:1646-75-9   *HR: 3*
2-METHYL-2-(METHYLTHIO)PROPANAL OXIME

SYNS: ALDICARB OXIME ◇ PROPANAL, 2-METHYL-2-(METHYLTHIO)-, OXIME ◇ PROPIONALDEHYDE, 2-METHYL-2-(METHYLTHIO)-, OXIME ◇ TEMIK OXIME

TOXICITY DATA with REFERENCE
msc-mus:lyms 1600 mg/L   MUREAV 204,149,88
orl-rat LD50:742 mg/kg   TSCAT* FYI-OTS-0885-0443
ihl-rat LC50:1230 mg/m³/4H   TSCAT* FYI-OTS-0885-0443

ipr-mus LDLo:100 mg/kg   TSCAT* FYI-OTS-0885-0443
skn-rbt LD50:1900 mg/kg   TSCAT* FYI-OTS-0885-0443

CONSENSUS REPORTS: Reported in EPA TSCA Inventory.

SAFETY PROFILE: Poison by intraperitoneal route. Moderately toxic by ingestion, skin contact, and inhalation routes. Mutation data reported. When heated to decomposition it emits toxic vapors of SO_x.

MMF780   CAS:99-52-5   **HR: D**
2-METHYL-4-NITROANILINE
mf: $C_7H_8N_2O_2$ mw: 152.17

SYNS: ANILINE, 2-METHYL-4-NITRO- ◊ ANSIBASES RED RL ◊ AZOGENE FAST RED GL BASE ◊ AZOGENE FAST RED NRL SALT ◊ AZOGENE FAST RED RL ◊ BENZENAMINE, 2-METHYL-4-NITRO- ◊ C.I. 37100 ◊ DAITO RED BASE RL ◊ DEVOL RED SALT E ◊ DIABASE RED RL ◊ DIAZO FAST RED RL ◊ FAST RED BASE RL ◊ FAST RED 5NT ◊ FAST RED 5NT SALT ◊ FAST RED RL BASE ◊ FAST RED SALT RL ◊ HILTONIL FAST RED RL BASE ◊ HILTOSAL FAST RED RL SALT ◊ KAKO RED RL BASE ◊ KAYAKU RED RL BASE ◊ MEISI FAST RED RL BASE ◊ MITSUI RED RL BASE ◊ NAPHTHOELAN RED RL BASE ◊ RED BASE CIBA X ◊ RED BASE IRGA X ◊ RED BASE NRL ◊ RED RL BASE ◊ SANYO FAST RED RL BASE ◊ SANYO FAST RED SALT RL ◊ SPECTROLENE RED RL ◊ SYMULON RED RL BASE ◊ o-TOLUIDINE, 4-NITRO- ◊ TULABASE FAST RED RL ◊ YAMADA FAST RED RL BASE

TOXICITY DATA with REFERENCE
mmo-sat 2500 µg/plate   SAIGBL 29,34,87
mma-sat 1 mg/plate   54DEAI -,497,84

CONSENSUS REPORTS: Reported in EPA TSCA Inventory.

SAFETY PROFILE: Mutation data reported. When heated to decomposition it emits toxic vapors of NO^x.

MMN800   CAS:881-03-8   **HR: D**
2-METHYL-1-NITRO-NAPHTHALENE
mf: $C_{11}H_9NO_2$ mw: 187.21

TOXICITY DATA with REFERENCE
mma-sat 50 µg/plate   GDIKAN 29,278,81

CONSENSUS REPORTS: Reported in EPA TSCA Inventory.

SAFETY PROFILE: Mutation data reported. When heated to decomposition it emits toxic vapors of NO_x.

MNK100   CAS:625-27-4   **HR: 1**
2-METHYL-2-PENTENE
mf: C_6H_{12} mw: 84.18

SYNS: 2-METHYL-PENTENE-2 ◊ 2-PENTENE, 2-METHYL-

TOXICITY DATA with REFERENCE
ihl-rat LC50:114 g/m^3/4H   RPTOAN 31,162,68
ihl-mus LC50:130 g/m^3/2H   RPTOAN 31,162,68

SAFETY PROFILE: Slightly toxic by inhalation. When heated to decomposition it emits acrid smoke and irritating vapors.

MNM450   CAS:1599-49-1   **HR: 1**
4-METHYL-2-PENTYL-DIOXOLANE
mf: $C_9H_{18}O_2$ mw: 158.27

SYN: 1,3-DIOXOLANE, 4-METHYL-2-PENTYL-

TOXICITY DATA with REFERENCE
orl-mus LD50:6802 mg/kg   DCTODJ 3,249,80

CONSENSUS REPORTS: Reported in EPA TSCA Inventory.

SAFETY PROFILE: Slightly toxic by ingestion. When heated to decomposition it emits acrid smoke and irritating vapors.

MNN520   CAS:2531-84-2   **HR: D**
2-METHYLPHENANTHRENE
mf: $C_{15}H_{12}$ mw: 192.27

SYN: PHENANTHRENE, 2-METHYL-

TOXICITY DATA with REFERENCE
mma-sat 10 µg/plate   MUREAV 156,61,85

SAFETY PROFILE: Mutation data reported. When heated to decomposition it emits acrid smoke and irritating vapors.

MNV800   CAS:707-61-9   **HR: 2**
3-METHYL-1-PHENYL-2-PHOSPHOLENE 1-OXIDE
mf: $C_{11}H_{13}OP$ mw: 192.21

SYN: 2-PHOSPHOLENE, 3-METHYL-1-PHENYL-, 1-OXIDE

TOXICITY DATA with REFERENCE
ipr-mus LD50:840 mg/kg   DANKAS 160,826,65

CONSENSUS REPORTS: Reported in EPA TSCA Inventory.

SAFETY PROFILE: Moderately toxic by intraperitoneal route. When heated to decomposition it emits toxic vapors of PO_x.

MOS900   CAS:1187-58-2   **HR: 2**
N-METHYLPROPIONAMIDE
mf: C_4H_9NO mw: 87.14

SYNS: N-METHYLPROPANAMIDE ◊ N-METHYLPROPIONIC ACID AMIDE ◊ N-METHYLPROPIONSAEUREAMID ◊ PROPANAMIDE, N-METHYL-(9CI) ◊ PROPIONAMIDE, N-METHYL-

TOXICITY DATA with REFERENCE
unr-rat LD50:1700 mg/kg   ARZNAD 18,645,68

CONSENSUS REPORTS: Reported in EPA TSCA Inventory.

SAFETY PROFILE: Moderately toxic by an unspecified

MOX010   CAS:1453-58-3   ***HR: 3***
5-METHYLPYRAZOLE
mf: $C_4H_6N_2$ mw: 82.12

SYNS: 3-METHYLPYRAZOLE ◇ 3(5)-METHYLPYRAZOLE ◇ PYRAZOLE, 3-METHYL- ◇ PYRAZOLE, 5-METHYL-

TOXICITY DATA with REFERENCE
orl-rat LD50:1 g/kg   GISAAA 51(6),29,86
orl-mus LD50:240 mg/kg   GISAAA 51(6),29,86
ipr-mus LD50:900 mg/kg   FATOAO 25,27,62

SAFETY PROFILE: Poison by ingestion. Moderately toxic by intraperitoneal route. When heated to decomposition it emits toxic vapors of NO_x.

MOX100   CAS:108-26-9   ***HR: 2***
3-METHYL-2-PYRAZOLIN-5-ONE
mf: $C_4H_6N_2O$ mw: 98.12

SYNS: 3-METHYL-PYRAZOLON-(5) ◇ 2-PYRAZOLIN-5-ONE, 3-METHYL-

TOXICITY DATA with REFERENCE
unr-rat LDLo:600 mg/kg   BCPCA6 14,1325,65

CONSENSUS REPORTS: Reported in EPA TSCA Inventory.

SAFETY PROFILE: Moderately toxic by unspecified route. When heated to decomposition it emits toxic vapors of NO_x.

MOY550   CAS:1003-73-2   ***HR: 2***
3-METHYLPYRIDINE-1-OXIDE
mf: C_6H_7NO mw: 109.14

SYN: PYRIDINE, 3-METHYL-, 1-OXIDE

TOXICITY DATA with REFERENCE
ipr-mus LD50:1689 mg/kg   TOXIA6 23,815,85

CONSENSUS REPORTS: Reported in EPA TSCA Inventory.

SAFETY PROFILE: Moderately toxic by intraperitoneal route. When heated to decomposition it emits toxic vapors of NO_x.

MOY790   CAS:1003-67-4   ***HR: 2***
4-METHYLPYRIDINE 1-OXIDE
mf: C_6H_7NO mw: 109.14

SYN: PYRIDINE, 4-METHYL-, 1-OXIDE

TOXICITY DATA with REFERENCE
ipr-mus LD50:1242 mg/kg   TOXIA6 23,815,85
orl-qal LD50:750 mg/kg   AECTCV 12,355,83
orl-brd LD50:1 g/kg   AECTCV 12,355,83

CONSENSUS REPORTS: Reported in EPA TSCA Inventory.

SAFETY PROFILE: Moderately toxic by ingestion, intraperitoneal routes. When heated to decomposition it emits toxic vapors of NO_x.

MPA075   CAS:694-85-9   ***HR: 2***
N-METHYLPYRIDONE
mf: C_6H_7NO mw: 109.14

SYNS: 1-METHYL-2(1H)-PYRIDINONE ◇ 1-METHYL-2(1H)-PYRIDONE ◇ 2(1H)-PYRIDINONE, 1-METHYL-(9CI) ◇ 2(1H)-PYRIDONE, 1-METHYL-

TOXICITY DATA with REFERENCE
orl-qal LD50:421 mg/kg   JRPFA4 48,371,76

CONSENSUS REPORTS: Reported in EPA TSCA Inventory.

SAFETY PROFILE: Moderately toxic by ingestion. When heated to decomposition it emits toxic vapors of NO_x.

MPI650   CAS:2288-13-3   ***HR: 2***
1-METHYLSILATRANE
mf: $C_7H_{15}NO_3Si$ mw: 189.32

SYNS: METHYLSILATRAN ◇ METHYLSILATRANE ◇ 1-METHYL-2,8,9-TRIOXO-5-AZA-1-SILABICYCLO(3.3.3)UNDECANE ◇ 2,8,9-TRIOXA-5-AZA-1-SILABICYCLO(3.3.3)UNDECANE, 1-METHYL-

TOXICITY DATA with REFERENCE
orl-mus LD50:1800 mg/kg   PHARAT 26,224,70
ipr-mus LD50:840 mg/kg   PHARAT 26,224,70

CONSENSUS REPORTS: Reported in EPA TSCA Inventory.

SAFETY PROFILE: Moderately toxic by ingestion and intraperitoneal routes. When heated to decomposition it emits toxic vapors of NO_x.

MPV300   CAS:1073-72-9   ***HR: 3***
4-(METHYLTHIO)PHENOL
mf: C_7H_8OS mw: 140.21

SYNS: 4-METHYLMERCAPTOPHENOL ◇ p-(METHYLTHIO)PHENOL ◇ PHENOL, p-(METHYLTHIO)-

TOXICITY DATA with REFERENCE
ice-mus LD50:60 mg/kg   PCBPBS 8,302,78

CONSENSUS REPORTS: Reported in EPA TSCA Inventory.

SAFETY PROFILE: Poison by intracerebral route. When heated to decomposition it emits toxic vapors of SO_x.

MPV790   CAS:2260-00-6   ***HR: 3***
2-METHYL-2-THIOPSEUDOUREA SULFATE
mf: $C_2H_6N_2S \cdot xH_2O_4S$ mw: 776.72

SYNS: CARBAMIMIDOTHIOIC ACID, METHYL ESTER, SULFATE (9CI)

◇ S-METHYLISOTHIURONIUM SULFATE ◇ S-METHYLTHIOPSEUDO-UREA SULFATE ◇ S-METHYLTHIOURONIUM SULFATE ◇ PSEUDOUREA, 2-METHYL-2-THIO-, SULFATE

TOXICITY DATA with REFERENCE
orl-rat LDLo:250 mg/kg   JAFCAU 2,1176,54

CONSENSUS REPORTS: Reported in EPA TSCA Inventory.

SAFETY PROFILE: Poison by ingestion. When heated to decomposition it emits toxic vapors of NO_x and SO_x.

MPW650   CAS:1076-56-8   **HR: 2**
METHYL THYMOL ETHER
mf: $C_{11}H_{16}O$ mw: 164.27

SYNS: ANISOLE, 2-ISOPROPYL-5-METHYL- ◇ O-METHYLTHYMOL ◇ METHYL THYMYL ETHER ◇ THYMOL METHYL ETHER ◇ THYMYL METHYL ETHER

TOXICITY DATA with REFERENCE
ivn-dog LDLo:1850 mg/kg   THERAP 3,109,48

CONSENSUS REPORTS: Reported in EPA TSCA Inventory.

SAFETY PROFILE: Moderately toxic by intravenous route. When heated to decomposition it emits acrid smoke and irritating vapors.

MQF300   CAS:1916-07-0   **HR: 3**
METHYL 3,4,5-TRIMETHOXYBENZOATE
mf: $C_{11}H_{14}O_5$ mw: 226.25

SYN: BENZOIC ACID, 3,4,5-TRIMETHOXY-, METHYL ESTER ◇ 3,4,5-TRIMETHOXYBENZOIC ACID, METHYL ESTER

TOXICITY DATA with REFERENCE
ivn-mus LD50:100 mg/kg   CSLNX* NX#07253

CONSENSUS REPORTS: Reported in EPA TSCA Inventory.

SAFETY PROFILE: Poison by intravenous route. When heated to decomposition it emits acrid smoke and irritating vapors.

MQM800   CAS:1759-28-0   **HR: 3**
4-METHYL-5-VINYL THIAZOLE
mf: C_6H_7NS mw: 125.20

SYN: THIAZOLE, 4-METHYL-5-VINYL-

TOXICITY DATA with REFERENCE
orl-mus LD50:400 mg/kg   DCTODJ 3,249,80

CONSENSUS REPORTS: Reported in EPA TSCA Inventory.

SAFETY PROFILE: Poison by ingestion. When heated to decomposition it emits toxic vapors of NO_x and SO_x.

MQV750   CAS:8012-95-1   **HR: 2**
MINERAL OIL

PROP: Colorless, oily liquid; practically tasteless and odorless. D: 0.83–0.86 (light), 0.875–0.905 (heavy), flash p: 444°F (OC), ULC: 10–20. Insol in water and alc; sol in benzene, chloroform, and ether. A mixture of liquid hydrocarbons from petroleum.

SYNS: ADEPSINE OIL ◇ ALBOLINE ◇ BAYOL F ◇ BLANDLUBE ◇ CRYSTOSOL ◇ DRAKEOL ◇ FONOLINE ◇ GLYMOL ◇ KAYDOL ◇ KONDREMUL ◇ MINERAL OIL, WHITE (FCC) ◇ MOLOL ◇ NEO-CULTOL ◇ NUJOL ◇ OIL MIST, MINERAL (OSHA, ACGIH) ◇ PARAFFIN OIL ◇ PAROL ◇ PAROLEINE ◇ PENETECK ◇ PENRECO ◇ PERFECTA ◇ PETROGALAR ◇ PETROLATUM, liquid ◇ PRIMOL 335 ◇ PROTOPET ◇ SAXOL ◇ TECH PET F ◇ WHITE MINERAL OIL

TOXICITY DATA with REFERENCE
skn-rbt 100 mg/24H MLD   CTOIDG 94(8),41,79
eye-rbt 250 mg/5D MLD   AMIHAB 14,265,56
skn-gpg 100 mg/24H MLD   CTOIDG 94(8),41,79
ihl-man TCLo:5 mg/m^3/5Y-I:CAR,GIT,TER   JOCMA7 23,333,81
skn-mus TDLo:332 g/mg/20W-I:ETA   ANYAA9 132,439,65
ipr-mus TDLo:14 g/kg:ETA   NATUAS 193,1086,62
ipr-mus TD:60 g/kg/17W-I:ETA   CNREA8 38,703,78
ipr-mus TD:60 g/kg/17W-I:ETA   IMMUAM 17,481,69
ipr-mus TD:50 g/kg/9W-I:ETA   IJCNAW 6,422,70
ipr-mus TD:72 g/kg/26W-I:ETA   JOIMA3 92,747,62
orl-mus LD50:22 g/kg   ATXKA8 30,243,73

CONSENSUS REPORTS: Reported in EPA TSCA Inventory.

OSHA PEL: Oil Mist: TWA 5 mg/m^3
ACGIH TLV: TWA 5 mg/m^3; STEL 10 mg/m^3

SAFETY PROFILE: A human teratogen by inhalation that causes testicular tumors in the fetus. Inhalation of vapor or particulates can cause aspiration pneumonia. A skin and eye irritant. Highly purified food grades are of low toxicity. Questionable human carcinogen producing gastrointestinal tumors. Combustible liquid when exposed to heat or flame. To fight fire, use dry chemical, CO_2, foam. When heated to decomposition it emits acrid smoke and fumes.

MQV875   CAS:8042-47-5   **HR: 1**
MINERAL OIL, SLAB OIL

SYNS: SLAB OIL (9CI) ◇ WHITE MINERAL OIL

CONSENSUS REPORTS: IARC Cancer Review: Group 3 IMSULD 7,252,87; Animal Inadequate Evidence IMEMDT 33,87,84. Reported in EPA TSCA Inventory.

SAFETY PROFILE: Highly purified food grades are of low toxicity. Questionable carcinogen. When heated to decomposition it emits acrid smoke and irritating fumes.

MRR090 CAS:729-46-4 HR: 2
4-MORPHOLINETHIOCARBONYL DISULFIDE
mf: $C_{10}H_{16}N_2O_2S_4$ mw: 324.52

SYNS: BIS(MORPHOLINOTHIOCARBONYL)DISULFIDE ◊ DIMORPHOLINETHIURAM DISULFIDE ◊ DISULFIDE, BIS(MORPHOLINOTHIOCARBONYL) ◊ 4,4′-(DITHIO-DICARBONOTHIOYL)BISMORPHOLINE ◊ MORPHOLINE, 4,4′-(DITHIODICARBONOTHIOYL)BIS-(9CI)

TOXICITY DATA with REFERENCE
orl-mus LD50:3250 mg/kg   AIPTAK 112,36,57

CONSENSUS REPORTS: Reported in EPA TSCA Inventory.

SAFETY PROFILE: Moderately toxic by ingestion. When heated to decomposition it emits toxic vapors of NO_x and SO_x.

MRU300 CAS:116-66-5 HR: 1
MOSKENE
mf: $C_{14}H_{18}N_2O_4$ mw: 278.34

SYNS: INDAN, 4,6-DINITRO-1,1,3,3,5-PENTAMETHYL- ◊ 1,1,3,3,5-PENTAMETHYL-4,6-DINITROINDANE

TOXICITY DATA with REFERENCE
skn-rbt 500 mg/24H MOD   FCTXAV 17,885,79

CONSENSUS REPORTS: Reported in EPA TSCA Inventory.

SAFETY PROFILE: A skin irritant. When heated to decomposition it emits toxic vapors of NO_x.

MRW272 CAS:145-39-1 HR: 1
MUSK TIBETENE
mf: $C_{13}H_{18}N_2O_4$ mw: 266.33

SYNS: BENZENE, 1-tert-BUTYL-2,6-DINITRO-3,4,5-TRIMETHYL- ◊ BENZENE, 1-(1,1-DIMETHYLETHYL)-2,6-DINITRO-3,4,5-TRIMETHYL- ◊ 5-tert-BUTYL-1,2,3-TRIMETHYL-4,6-DINITROBENZENE

TOXICITY DATA with REFERENCE
eye-rbt 100 mg/24H MLD   FCTXAV 13,879,75

CONSENSUS REPORTS: Reported in EPA TSCA Inventory.

SAFETY PROFILE: An eye irritant. When heated to decomposition it emits toxic vapors of NO_x.

MSB775 CAS:8016-37-3 HR: 2
MYRRH OIL

PROP: From steam distillation of myrrh gum from *Commiphora* (Fam. Burseraceae). Light-brown to green liquid; characteristic odor. Sol in fixed oils, sltly sol in mineral oil; insol in glycerin, propylene glycol.

SYNS: OIL of HEERABOL-MYRRH

TOXICITY DATA with REFERENCE
orl-rat LD50:1650 mg/kg   FCTXAV 14,621,76

CONSENSUS REPORTS: Reported in EPA TSCA Inventory.

SAFETY PROFILE: Moderately toxic by ingestion. When heated to decomposition it emits acrid smoke and irritating fumes.

MSC050 CAS:1079-01-2 HR: 2
MYRTENYL ACETATE
mf: $C_{12}H_{18}O_2$ mw: 194.30

SYNS: BICYCLO(3.1.1)HEPT-2-ENE-2-METHANOL, 6,6-DIMETHYL-, ACETATE, (1S)- ◊ (1S)-6,6-DIMETHYLBICYCLO(3.1.1)HEPT-2-ENE-2-METHANOL ACETATE ◊ (+)-MYRTENYL ACETATE ◊ 2-PINEN-10-OL, ACETATE (6CI,7CI,8CI)

TOXICITY DATA with REFERENCE
orl-rat LD50:2600 mg/kg   FCTOD7 26,389,88
skn-rbt LDLo:5 g/kg   FCTOD7 26,389,88

CONSENSUS REPORTS: Reported in EPA TSCA Inventory.

SAFETY PROFILE: Moderately toxic by ingestion. Slightly toxic by skin contact. When heated to decomposition it emits acrid smoke and irritating vapors.

N

NAK550 CAS:61-31-4 **HR: 2**
1-NAPHTHALENEACETIC ACID, SODIUM SALT
mf: $C_{12}H_9O_2 \cdot Na$ mw: 208.20

TOXICITY DATA with REFERENCE
orl-mus LD50:670 mg/kg   PEMNDP 9,607,91

CONSENSUS REPORTS: Reported in EPA TSCA Inventory.

SAFETY PROFILE: Moderately toxic by ingestion. When heated to decomposition it emits acrid smoke and irritating vapors.

NAN505 CAS:83-56-7 **HR: D**
1,5-NAPHTHALENEDIOL
mf: $C_{10}H_8O_2$ mw: 160.18

SYNS: C.I. 76625 ◊ 1,5-DIHYDROXYNAPHTHALENE ◊ 1,5-DIHYDROXYNAPTHALENE ◊ DURAFUR DEVELOPER E ◊ OXIDATION BASE

TOXICITY DATA with REFERENCE
dlt-oin-oin-par 10 ppm   EVETBX 2,1029,73
dlt-oin-oin-skn 10 ppm   EVETBX 2,1029,73

CONSENSUS REPORTS: Reported in EPA TSCA Inventory.

SAFETY PROFILE: Mutation data reported. When heated to decomposition it emits acrid smoke and irritating vapors.

NAN510 CAS:92-44-4 **HR: 3**
2,3-NAPHTHALENEDIOL
mf: $C_{10}H_8O_2$ mw: 160.18

TOXICITY DATA with REFERENCE
ivn-mus LD50:56 mg/kg   CSLNX* NX#03884

CONSENSUS REPORTS: Reported in EPA TSCA Inventory.

SAFETY PROFILE: Poison by intravenous route. When heated to decomposition it emits acrid smoke and irritating vapors.

NAP100 CAS:1248-18-6 **HR: D**
1-NAPHTHALENESULFONIC ACID, 2-((2-HYDROXY-1-NAPHTHALENYL)AZO)-, MONOSODIUM SALT
mf: $C_{20}H_{13}N_2O_4S \cdot Na$ mw: 400.40

SYNS: ATUL PIGMENT RED RS SODIUM SALT ◊ BRASILACA RED R ◊ BRITONE RED Y ◊ CARNELIO PALE LITHOL RED ◊ CERTIQUAL LITHOL RED ◊ C.I. 15630 ◊ C.I. PIGMENT RED 49 ◊ C.I. PIGMENT RED 49, MONOSODIUM SALT ◊ D and C RED No. 10 ◊ ELJON LITHOL RED No. 10 ◊ GRAPHIC RED Y ◊ IRGALITE RED RL ◊ KROMON SODIUM LITHOL ◊ LAKE RED R ◊ LAKE RED RL ◊ LIGHT RED RS ◊ LITHOL RED ◊ LITHOL RED 3580 ◊ LITHOL RED 17676 ◊ LITHOL RED B ◊ LITHOL RED 3GS ◊ LITHOL RED LAKE ◊ LITHOL RED R ◊ LITHOL RED RB EXTRA ◊ LITHOL RED RL 151 ◊ LITHOL RED RS ◊ LITHOL RED SODIUM SALT ◊ LITHOL RED TONER ◊ LITHOL TONER ◊ LITHOL TONER EXTRA LIGHT 5000 ◊ LITHOL TONER SODIUM SALT RT 314 ◊ LITHOL TONER YA 8003 ◊ LUTETIA RED R ◊ MONOLITE RED R ◊ NEW YORK RED ◊ OHIO RED ◊ ORALITH RED SR WATER SOLUBLE ◊ PLASTORESIN RED SR ◊ RECOLITE RED LYS ◊ 11935 RED ◊ RED No. 205 ◊ RESAMINE RED RB ◊ RESAMINE RED RC ◊ SEGNALE RED R ◊ SIGNAL RED ◊ SILOTON RED R ◊ SODIUM LITHOL ◊ SODIUM LITHOL RED ◊ SODIUM LITHOL RED 20-4018 ◊ SUNBURST RED ◊ TERTROPIGMENT RED LR ◊ UNDEVELOPED LITHOL TONER ◊ VULCAFIX RED R ◊ ULCOL SCARLET 2G

TOXICITY DATA with REFERENCE
mmo-sat 500 μg/plate   AEMIDF 42,641,81
mma-sat 1 mg/plate   AEMIDF 42,641,81

CONSENSUS REPORTS: Reported in EPA TSCA Inventory.

SAFETY PROFILE: Mutation data reported. When heated to decomposition it emits toxic vapors of NO_x and SO_x.

NAP250 CAS:128-97-2 **HR: 2**
1,4,5,8-NAPHTHALENETETRACARBOXYLIC ACID
mf: $C_{14}H_8O_8$ mw: 304.22

TOXICITY DATA with REFERENCE
orl-rat LD50:7500 mg/kg   85GMAT -,90,82
orl-mus LD50:3800 mg/kg   85GMAT -,90,82

CONSENSUS REPORTS: Reported in EPA TSCA Inventory.

SAFETY PROFILE: Moderately toxic by ingestion. When heated to decomposition it emits acrid smoke and irritating vapors.

NAU600 CAS:238-84-6 **HR: D**
α-NAPHTHOFLUORENE
mf: $C_{17}H_{12}$ mw: 216.29

SYNS: BENZO(a)FLUORENE ◊ 11H-BENZO(a)FLUORENE ◊ 1,2-BENZOFLUORENE ◊ CHRYSOFLUORENE

TOXICITY DATA with REFERENCE
mma-sat 50 μg/plate   MUREAV 174,247,86

CONSENSUS REPORTS: IARC Cancer Review: Group 3 IMEMDT 7,56,87; Animal Inadequate Evidence IMEMDT 32,177,83; Human No Adequate Data IMEMDT 32,177,83

SAFETY PROFILE: Mutation data reported. When heated to decomposition it emits acrid smoke and irritating vapors.

NAV490　　　CAS:86-55-5　　　**HR: 2**
1-NAPHTHOIC ACID
mf: $C_{11}H_8O_2$　　mw: 172.19

SYNS: 1-CARBOXYNAPHTHALENE ◇ NAPHTHALENE-α-CARBOXYLIC ACID ◇ 1-NAPHTHALENECARBOXYLIC ACID (9CI) ◇ α-NAPHTHOIC ACID ◇ α-NAPHTHYLCARBOXYLIC ACID

TOXICITY DATA with REFERENCE
orl-mus LD50:2370 mg/kg　　NTIS** PB85-143766

CONSENSUS REPORTS: Reported in EPA TSCA Inventory.

SAFETY PROFILE: Moderately toxic by ingestion. When heated to decomposition it emits acrid smoke and irritating vapors.

NAX000　　　CAS:135-19-3　　　**HR: 3**
2-NAPHTHOL
mf: $C_{10}H_8O$　　mw: 144.18

PROP: White to yellowish-white crystals; slt phenolic odor. Flash p: 307°F, vap press: 10 mm @ 145.5°, vap d: 4.97 mp: 121–123°, bp: 285–286°, d: 1.22. Darkens with age or exposure to light. Subl when heated; distills in vacuum. Sltly sol in water; more sol in boiling water, glycerol, olive oil, solns of alkali hydroxides. Very sol in alc, ether; sol in chloroform.

SYNS: AZOGEN DEVELOPER A ◇ C.I. 37500 ◇ C.I. AZOIC COUPLING COMPONENT 1 ◇ C.I. DEVELOPER 5 ◇ DEVELOPER A ◇ DEVELOPER AMS ◇ DEVELOPER BN ◇ DEVELOPER SODIUM ◇ β-HYDROXY-NAPHTHALENE ◇ 2-HYDROXYNAPHTHALENE ◇ ISONAPHTHOL ◇ β-MONOXYNAPHTHALENE ◇ β-NAFTOL (DUTCH) ◇ 2-NAFTOL (DUTCH) ◇ β-NAFTOLO (ITALIAN) ◇ 2-NAFTOLO (ITALIAN) ◇ 2-NAPHTHALENOL ◇ NAPHTHOL B ◇ β-NAPHTHOL ◇ β-NAPHTHYL ALCOHOL ◇ β-NAPHTHYL HYDROXIDE ◇ β-NAPHTOL (GERMAN) ◇ 2-NAPHTOL (FRENCH)

TOXICITY DATA with REFERENCE
skn-rbt 500 mg/24H MLD　　BIOFX* 26-4/73
eye-rbt 100 mg MOD　　BIOFX* 26-4/73
dnr-esc 2 mg/disc　　MUREAV 97,1,82
dnr-bcs 5 g/L　　MUREAV 42,19,77
orl-rat LD50:1960 mg/kg　　GTPPAF 8,145,72
ihl-rat LC50:>20 mg/m³　　GISAAA 38(10),15,73
scu-rat LDLo:2940 mg/kg　　AEPPAE 186,195,37
orl-mus LDLo:100 mg/kg　　HBAMAK 4,1289,35
ipr-mus LD50:97500 mg/kg　　EJMCA5 13,381,78
scu-mus LDLo:100 mg/kg　　ZHINAV 64,113,1909
orl-cat LDLo:100 mg/kg　　HBAMAK 4,1289,35
scu-rbt LDLo:3 g/kg　　HBAMAK 4,1289,35
scu-gpg LDLo:2670 mg/kg　　AEPPAE 186,195,37

CONSENSUS REPORTS: Reported in EPA TSCA Inventory. EPA Genetic Toxicology Program.

SAFETY PROFILE: Poison by ingestion, inhalation, and subcutaneous routes. Mutation data reported. A skin and eye irritant. Combustible when exposed to heat or flame. To fight fire, use CO_2 dry chemical. Incompatible with antipyrine; camphor; phenol; ferric salts; menthol; potassium permanganate and other oxidizing materials; urethane. See also α-NAPHTHOL.

NAX100　　　CAS:130-00-7　　　**HR: 2**
1,8-NAPHTHOLACTAM
mf: $C_{11}H_7NO$　　mw: 169.19

SYNS: BENZ(cd)INDOL-2(1H)-ONE ◇ NAPHTHOLACTAM ◇ NAPHTHOSTYRIL ◇ 2(1H)-PERINAPHTHAZOLONE

TOXICITY DATA with REFERENCE
orl-mus LD50:1 g/kg　　CCCCAK 50,1888,85
orl-uns LD50:2700 mg/kg　　GISAAA 51(1),82,86

CONSENSUS REPORTS: Reported in EPA TSCA Inventory.

SAFETY PROFILE: Moderately toxic by an unspecified route. Slightly toxic by ingestion. When heated to decomposition it emits toxic vapors of NO_x.

NAY090　　　CAS:86-53-3　　　**HR: D**
1-NAPHTHONITRILE
mf: $C_{11}H_7N$　　mw: 153.19

SYNS: α-CYANONAPHTHALENE ◇ 1-CYANONAPHTHALENE ◇ 1-NAPHTHALENECARBONITRILE ◇ 1-NAPHTHALENENITRILE ◇ α-NAPHTHONITRILE ◇ α-NAPHTHYLNITRILE ◇ 1-NAPHTHYLNITRILE

TOXICITY DATA with REFERENCE
mmo-sat 10 mg/L　　50NNAZ-,827,85

CONSENSUS REPORTS: Reported in EPA TSCA Inventory.

SAFETY PROFILE: Mutation data reported. When heated to decomposition it emits toxic vapors of NO_x.

NAY100　　　CAS:613-46-7　　　**HR: D**
2-NAPHTHONITRILE
mf: $C_{11}H_7N$　　mw: 153.19

SYNS: β-CYANONAPHTHALENE ◇ 2-CYANONAPHTHALENE ◇ 2-NAPHTHALENECARBONITRILE (9CI) ◇ 2-NAPHTHALENENITRILE ◇ β-NAPHTHONITRILE

TOXICITY DATA with REFERENCE
mmo-sat 10 mg/L　　50NNAZ -,827,85

CONSENSUS REPORTS: Reported in EPA TSCA Inventory.

SAFETY PROFILE: Mutation data reported. When heated to decomposition it emits toxic vapors of NO_x.

NBI100　　　CAS:604-59-1　　　**HR: D**
α-NAPHTHYLFLAVONE
mf: $C_{19}H_{12}O_2$　　mw: 272.31

SYNS: BENZO(h)FLAVONE ◇ 7,8-BENZOFLAVONE (7CI) ◇ α-NAPHTHOFLAVONE ◇ 4H-NAPHTHO(1,2-b)PYRAN-4-ONE, 2-PHENYL- ◇ 2-PHENYL-4H-NAPHTHO(1,2-b)PYRAN-4-ONE

TOXICITY DATA with REFERENCE
cyt-ham:ovr 20 ~2mol/L   CNREA8 47,3662,87
sce-ham:ovr 40 ~2mol/L   CNREA8 47,3662,87

CONSENSUS REPORTS: Reported in EPA TSCA Inventory.

SAFETY PROFILE: Mutation data reported. When heated to decomposition it emits acrid smoke and irritating vapors.

NBJ100   CAS:120-23-0   **HR: 2**
(2-NAPHTHYLOXY)ACETIC ACID

SYNS: ACETIC ACID, (2-NAPHTHYLOXY)- ◇ ACIDE NAPHTHYLOXYACETIQUE ◇ BETAPAL ◇ BETOXON ◇ GERLACH 1396 ◇ 2-NAPHTHALENOXYACETIC ACID ◇ (β-NAPHTHALENYLOXY)ACETIC ACID ◇ β-NAPHTHOXYACETIC ACID ◇ 2-NAPHTHOXYACETIC ACID ◇ O-(2-NAPHTHYL)GLYCOLIC ACID ◇ NOXA ◇ 2-NOXA

TOXICITY DATA with REFERENCE
orl-rat LD50:600 mg/kg   PEMNDP 9,609,91

CONSENSUS REPORTS: Reported in EPA TSCA Inventory.

SAFETY PROFILE: Moderately toxic by ingestion. When heated to decomposition it emits acrid smoke and irritating vapors.

NCR040   CAS:94-44-0   **HR: 3**
NIACIN BENZYL ESTER
mf: $C_{13}H_{11}NO_2$ mw: 213.25

SYNS: BENZYL NICOTINATE ◇ BENZYL PYRIDINE-3-CARBOXYLATE ◇ ESTRU BENZYLOWEGO KWASU NIKOTYNOWEGO ◇ NICOTINIC ACID, BENZYL ESTER ◇ PYCARIL ◇ PYKARYL ◇ 3-PYRIDINECARBOXYLIC ACID, PHENYLMETHYL ESTER ◇ RUBRIMENT

TOXICITY DATA with REFERENCE
orl-rat LD50:2188 mg/kg   BCTKAG 12,79,79
ivn-mus LD50:100 mg/kg   PHARAT 11,242,56

CONSENSUS REPORTS: Reported in EPA TSCA Inventory.

SAFETY PROFILE: Poison by intravenous route. Moderately toxic by ingestion. When heated to decomposition it emits toxic vapors of NO_x.

NDM100   CAS:500-22-1   **HR: 2**
NICOTINALDEHYDE
mf: C_6H_5NO mw: 107.12

SYNS: 3-FORMYLPYRIDINE ◇ NICOTINEALDEHYDE ◇ NICOTINIC ALDEHYDE ◇ 3-PYRIDINALDEHYDE ◇ 3-PYRIDINEALDEHYDE ◇ PYRIDINE-3-CARBALDEHYDE ◇ β-PYRIDINECARBONALDEHYDE ◇ 3-PYRIDINECARBOXALDEHYDE (9CI) ◇ 3-PYRIDYLALDEHYDE ◇ 3-PYRIDYLCARBOXALDEHYDE ◇ ROWALIND

TOXICITY DATA with REFERENCE
ipr-mus LD25:720 mg/kg   JPMSAE 64,528,75

CONSENSUS REPORTS: Reported in EPA TSCA Inventory.

SAFETY PROFILE: Moderately toxic by intraperitoneal route. When heated to decomposition it emits toxic vapors of NO_x.

NDW510   CAS:100-55-0   **HR: 2**
NICOTINIC ALCOHOL
mf: C_6H_7NO mw: 109.14

SYNS: 3-(HYDROXYMETHYL)PYRIDINE ◇ NICOTINYL ALCOHOL ◇ NU-2121 ◇ β-PICOLYL ALCOHOL ◇ PYRIDINE-3-CARBINOL ◇ 3-PYRIDINEMETHANOL ◇ 3-PYRIDYLCARBINOL ◇ 3-PYRIDYLMETHANOL ◇ RO-1-5155 ◇ RONIACOL

TOXICITY DATA with REFERENCE
ivn-mus LD50:1000 mg/kg   PHARAT 11,242,56

CONSENSUS REPORTS: Reported in EPA TSCA Inventory.

SAFETY PROFILE: Moderately toxic by intravenous route. When heated to decomposition it emits toxic vapors of NO_x.

NDW515   CAS:100-54-9   **HR: 2**
NICOTINONITRILE
mf: $C_6H_4N_2$ mw: 104.12

SYNS: 3-AZABENZONITRILE ◇ 3-CYANOPYRIDINE ◇ 3-CYJANOPIRYDYNA ◇ NICOTINIC ACID NITRILE ◇ NITRYL KWASU NIKOTYNOWEGO ◇ 3-PYRIDINECARBONITRILE (9CI) ◇ 3-PYRIDINENITRILE ◇ 3-PYRIDYLCARBONITRILE

TOXICITY DATA with REFERENCE
orl-rat LD50:1185 mg/kg   MEPAAX 30,109,79
ipr-mus LD25:810 mg/kg   JPMSAE 64,528,75
skn-rbt LDLo:4000 mg/kg   MEPAAX 30,109,79

CONSENSUS REPORTS: Reported in EPA TSCA Inventory.

SAFETY PROFILE: Moderately toxic by ingestion, intraperitoneal, and skin contact routes. When heated to decomposition it emits toxic vapors of NO_x.

NEK100   CAS:119-81-3   **HR: 3**
2-NITRO-p-ACETANISIDIDE
mf: $C_9H_{10}N_2O_4$ mw: 210.21

SYNS: 4-ACETAMIDO-3-NITROANISOLE ◇ p-ACETANISIDIDE, 2-NITRO-

TOXICITY DATA with REFERENCE
ivn-mus LD50:180 mg/kg   CSLNX* NX#00426

CONSENSUS REPORTS: Reported in EPA TSCA Inventory.

SAFETY PROFILE: Poison by intravenous route. When heated to decomposition it emits toxic vapors of NO_x.

NET550 CAS:480-68-2 HR: 3
5-NITROBARBITURIC ACID
mf: $C_4H_3N_3O_5$ mw: 173.10

SYNS: BARBITURIC ACID, 5-NITRO- ◊ DILITURIC ACID ◊ 5-NITRO-2,4,6(1H,3H,5H)-PYRIMIDINETRIONE ◊ 2,4,6(1H,3H,5H)-PYRIMIDINETRIONE, 5-NITRO-(9CI)

TOXICITY DATA with REFERENCE
ivn-mus LD50:180 mg/kg   CSLNX* NX#01678

CONSENSUS REPORTS: Reported in EPA TSCA Inventory.

SAFETY PROFILE: Poison by intravenous route. When heated to decomposition it emits toxic vapors of NO_x.

NEV525 CAS:619-80-7 HR: 2
p-NITROBENZAMIDE
mf: $C_7H_6N_2O_3$ mw: 166.15

TOXICITY DATA with REFERENCE
skn-gpg 2500 mg/10D-I MLD   EPASR* 8EHQ-0183-0462
orl-rat LD50:476 mg/kg   EPASR* 8EHQ-0183-0462

CONSENSUS REPORTS: Reported in EPA TSCA Inventory.

SAFETY PROFILE: Moderately toxic by ingestion. A skin irritant. When heated to decomposition it emits toxic vapors of NO_x.

NEZ200 CAS:456-27-9 HR: D
4-NITROBENZENEDIAZONIUM FLUOBORATE
mf: $C_6H_4N_3O_2 \cdot BF_4$ mw: 236.94

SYNS: BENZENEDIAZONIUM, p-NITRO-, TETRAFLUOROBORATE(1-) ◊ BENZENEDIAZONIUM, 4-NITRO-, TETRAFLUOROBORATE(1-) (9CI) ◊ p-NITROBENZENEDIAZONIUM TETRAFLUOROBORATE ◊ 4-NITROBENZENEDIAZONIUM TETRAFLUOROBORATE(1-) ◊ p-NITROPHENYLDIAZONIUM FLUOBORATE

TOXICITY DATA with REFERENCE
mmo-sat 40 µg/plate   FCTOD7 25,669,87

CONSENSUS REPORTS: Reported in EPA TSCA Inventory.

SAFETY PROFILE: Mutation data reported. When heated to decomposition it emits toxic vapors of boron, NO_x, and F^-.

NFI010 CAS:619-72-7 HR: 3
p-NITROBENZONITRILE
mf: $C_7H_4N_2O_2$ mw: 148.13

SYNS: BENZONITRILE, p-NITRO- ◊ BENZONITRILE, 4-NITRO-(9CI) ◊ p-CYANONITROBENZENE ◊ 4-CYANONITROBENZENE ◊ 4-NITROBENZONITRILE

TOXICITY DATA with REFERENCE
dnr-bcs 50 µg/disc   MUREAV 170,11,86
orl-rat LD50:30 mg/kg   ZAARAM 19,225,69
orl-mus LD50:140 mg/kg   APFRAD 41,391,83
ipr-mus LD50:100 mg/kg   ZAARAM 19,225,69

CONSENSUS REPORTS: Reported in EPA TSCA Inventory.

SAFETY PROFILE: Poison by ingestion and intraperitoneal routes. Mutation data reported. When heated to decomposition it emits toxic vapors of NO_x.

NFM100 CAS:619-90-9 HR: D
4-NITROBENZYL ACETATE
mf: $C_9H_9NO_4$ mw: 195.19

SYNS: BENZENEMETHANOL, 4-NITRO-, ACETATE (ester) ◊ BENZYL ALCOHOL, p-NITRO-, ACETATE (ester) ◊ p-NITROBENZYLACETATE

TOXICITY DATA with REFERENCE
mmo-sat 2600 nmol/L   ENMUDM 3,11,81

CONSENSUS REPORTS: Reported in EPA TSCA Inventory.

SAFETY PROFILE: Mutation data reported. When heated to decomposition it emits toxic vapors of NO_x.

NFN010 CAS:619-23-8 HR: D
m-NITROBENZYL CHLORIDE
mf: $C_7H_6ClNO_2$ mw: 171.59

SYNS: BENZENE, 1-(CHLOROMETHYL)-3-NITRO-(9CI) ◊ 1-(CHLOROMETHYL)-3-NITROBENZENE ◊ α-CHLORO-m-NITROTOLUENE ◊ 3-NITROBENZYL CHLORIDE ◊ TOLUENE, α-CHLORO-m-NITRO-

TOXICITY DATA with REFERENCE
mmo-sat 100 µg/plate   MUREAV 170,11,86
mma-sat 200 µg/plate   ENMUDM 9(Suppl 9),1,87
dnr-bcs 500 µg/disc   MUREAV 170,11,86

CONSENSUS REPORTS: Reported in EPA TSCA Inventory.

SAFETY PROFILE: Mutation data reported. When heated to decomposition it emits toxic vapors of NO_x and Cl^-.

NFN600 CAS:610-66-2 HR: D
2-NITROBENZYL NITRILE
mf: $C_8H_6N_2O_2$ mw: 162.16

SYNS: ACETONITRILE, (o-NITROPHENYL)- ◊ BENZENEACETONITRILE, 2-NITRO-(9CI) ◊ o-NITROBENZACETONITRILE ◊ 2-NITROBENZENEACETONITRILE ◊ (o-NITROPHENYL)ACETONITRILE ◊ (2-NITROPHENYL)ACETONITRILE

TOXICITY DATA with REFERENCE
mmo-sat 100 µg/plate   EMMUEG 11(Suppl 12),1,88
mma-sat 333 µg/plate   EMMUEG 11(Suppl 12),1,88

CONSENSUS REPORTS: Reported in EPA TSCA Inventory.

SAFETY PROFILE: Mutation data reported. When heated to decomposition it emits toxic vapors of NO_x.

NFO750 CAS:1083-48-3 **HR: 3**
1-(p-NITROBENZYL)PYRIDINE
mf: $C_{12}H_{10}N_2O_2$ mw: 214.24

SYN: PYRIDINE, 4-(p-NITROBENZYL)-

TOXICITY DATA with REFERENCE
ivn-mus LD50:89 mg/kg   CSLNX* NX#12116

CONSENSUS REPORTS: Reported in EPA TSCA Inventory.

SAFETY PROFILE: Poison by intravenous route. When heated to decomposition it emits toxic vapors of NO_x.

NFQ080 CAS:577-19-5 **HR: D**
2-NITROBROMOBENZENE
mf: $C_6H_4BrNO_2$ mw: 202.02

SYNS: BENZENE, 1-BROMO-2-NITRO- ◇ o-BROMONITROBENZENE ◇ 2-BROMONITROBENZENE ◇ o-NITROBROMOBENZENE

TOXICITY DATA with REFERENCE
mma-sat 250 µg/plate   SAIGBL 29,34,87

CONSENSUS REPORTS: Reported in EPA TSCA Inventory.

SAFETY PROFILE: Mutation data reported. When heated to decomposition it emits toxic vapors of NO_x and Br^-.

NFQ090 CAS:585-79-5 **HR: D**
3-NITROBROMOBENZENE
mf: $C_6H_4BrNO_2$ mw: 202.02

SYNS: BENZENE, 1-BROMO-3-NITRO- ◇ m-BROMONITROBENZENE ◇ 3-BROMONITROBENZENE ◇ m-NITROBROMOBENZENE

TOXICITY DATA with REFERENCE
mma-sat 500 µg/plate   SAIGBL 29,34,87

CONSENSUS REPORTS: Reported in EPA TSCA Inventory.

SAFETY PROFILE: Mutation data reported. When heated to decomposition it emits toxic vapors of NO_x and Br^-.

NFQ100 CAS:586-78-7 **HR: D**
4-NITROBROMOBENZENE

SYNS: BENZENE, 1-BROMO-4-NITRO- ◇ p-BROMONITROBENZENE ◇ 4-BROMONITROBENZENE ◇ p-NITROBROMOBENZENE

TOXICITY DATA with REFERENCE
mma-sat 500 µg/plate   SAIGBL 29,34,87

CONSENSUS REPORTS: Reported in EPA TSCA Inventory.

SAFETY PROFILE: Mutation data reported. When heated to decomposition it emits toxic vapors of NO_x.

NFS505 CAS:1222-98-6 **HR: 1**
4-NITROCHALCONE
mf: $C_{15}H_{11}NO_3$ mw: 253.27

SYNS: CHALCONE, 4-NITRO- ◇ p-NITROBENZYLIDENEACETOPHENONE ◇ 4-NITROCALONE ◇ 3-(4-NITROPHENYL)-1-PHENYL-2-PROPEN-1-ONE ◇ p-NITROSTYRYL PHENYL KETONE ◇ 2-PROPEN-1-ONE, 3-(4-NITROPHENYL)-1-PHENYL-(9CI)

TOXICITY DATA with REFERENCE
mma-sat 10 µg/plate   MUREAV 188,267,87
orl-mus LD50:>3 g/kg   FRPSAX 25,860,70

CONSENSUS REPORTS: Reported in EPA TSCA Inventory.

SAFETY PROFILE: Slightly toxic by ingestion. Mutation data reported. When heated to decomposition it emits toxic vapors of NO_x.

NFT440 CAS:619-89-6 **HR: 3**
4-NITROCINNAMIC ACID
mf: $C_9H_7NO_4$ mw: 193.17

SYNS: CINNAMIC ACID, p-NITRO- ◇ p-NITROCINNAMIC ACID ◇ 3-(4-NITROPHENYL)ACRYLIC ACID ◇ 3-(4-NITROPHENYL)PROPENOIC ACID ◇ 2-PROPENOIC ACID, 3-(4-NITROPHENYL)-

TOXICITY DATA with REFERENCE
mmo-sat 5 µg/plate   MUREAV 170,11,86
mma-sat 10 µg/plate   MUREAV 188,267,87
dnr-bcs 500 µg/disc   MUREAV 170,11,86
ipr-mus LD: >250 mg/kg   IJSIDW 49,77,87

CONSENSUS REPORTS: Reported in EPA TSCA Inventory.

SAFETY PROFILE: Poison by intraperitoneal route. Mutation data reported. When heated to decomposition it emits toxic vapors of NO_x.

NFV010 CAS:99-53-6 **HR: 3**
4-NITRO-o-CRESOL
mf: $C_7H_7NO_3$ mw: 153.15

SYNS: o-CRESOL, 4-NITRO- ◇ 2-METHYL-4-NITROPHENOL

TOXICITY DATA with REFERENCE
orl-mus LD50:390 mg/kg   PHARAT 32,171,77

SAFETY PROFILE: Poison by ingestion. When heated to decomposition it emits toxic vapors of NO_x.

NFW600 CAS:610-17-3 **HR: D**
2-NITRO-N,N-DIMETHYLANILINE
mf: $C_8H_{10}N_2O_2$ mw: 166.20

SYNS: ANILINE, N,N-DIMETHYL-o-NITRO- ◇ BENZENAMINE, N,N-DIMETHYL-2-NITRO-(9CI) ◇ o-(DIMETHYLAMINO)NITROBENZENE ◇ N,N-DIMETHYL-o-NITROANILINE ◇ N,N-DIMETHYL-2-NITROBENZENAMINE

NGB800 5-NITROFURALDEHYDE DIACETATE

TOXICITY DATA with REFERENCE
mma-sat 2500 µg/plate   FCTOD7 23,695,85

CONSENSUS REPORTS: Reported in EPA TSCA Inventory.

SAFETY PROFILE: Mutation data reported. When heated to decomposition it emits toxic vapors of NO_x.

NGB800   CAS:92-55-7   HR: D
5-NITROFURALDEHYDE DIACETATE
mf: $C_9H_9NO_7$ mw: 243.19

SYNS: 2-FURANMETHANEDIOL, 5-NITRO-, DIACETATE ◊ 5-NITRO-2-FURAN METHANEDIOL DIACETATE ◊ 5-NITROFURFURAL DIACETATE ◊ 5-NITROFURFURYLIDENE DIACETATE

TOXICITY DATA with REFERENCE
mmo-sat 5 µg/plate   MUREAV 192,15,87
mma-sat 5 µg/plate   MUREAV 192,15,87
mmo-esc 10 µg/plate   MUREAV 26,3,74

CONSENSUS REPORTS: Reported in EPA TSCA Inventory.

SAFETY PROFILE: Mutation data reported. When heated to decomposition it emits toxic vapors of NO_x.

NHI600   CAS:99-69-4   HR: 3
3,3'-NITROIMINODIPROPIONIC ACID
mf: $C_6H_{10}N_2O_6$ mw: 206.18

SYNS: N-(2-CARBOXYETHYL)-N-NITRO-β-ALANINE ◊ PROPIONIC ACID, 3,3'-NITROIMINODI-

TOXICITY DATA with REFERENCE
ipr-mus LD50:355 mg/kg   PCJOAU 10,1504,76

CONSENSUS REPORTS: Reported in EPA TSCA Inventory.

SAFETY PROFILE: Poison by intraperitoneal route. When heated to decomposition it emits toxic vapors of NO_x.

NHX100   CAS:103-94-6   HR: 2
4'-NITROOXANILIC ACID
mf: $C_8H_6N_2O_5$ mw: 210.16

SYNS: ACETIC ACID, ((4-NITROPHENYL)AMINO)OXO-(9CI) ◊ KYSELINA N-(4-NITROFENYL)OXAMOVA ◊ ((4-NITROPHENYL)AMINO)OXOACETIC ACID ◊ N-(4-NITROPHENYL)OXAMIC ACID

TOXICITY DATA with REFERENCE
orl-mus LDLo:500 mg/kg   85JCAE -,741,86

CONSENSUS REPORTS: Reported in EPA TSCA Inventory.

SAFETY PROFILE: Moderately toxic by ingestion. When heated to decomposition it emits toxic vapors of NO_x.

NIC990   CAS:610-67-3   HR: D
o-NITROPHENETOLE
mf: $C_8H_9NO_3$ mw: 167.18

SYNS: 1-ETHOXY-2-NITROBENZENE ◊ PHENETOLE, p-NITRO-

TOXICITY DATA with REFERENCE
mmo-sat 1 ~2mol/plate   MUREAV 58,11,78
dnr-bcs 500 nL/disc   MUREAV 170,11,86

CONSENSUS REPORTS: Reported in EPA TSCA Inventory.

SAFETY PROFILE: Mutation data reported. When heated to decomposition it emits toxic vapors of NO_x.

NII510   CAS:104-03-0   HR: 2
(p-NITROPHENYL)ACETIC ACID
mf: $C_8H_7NO_4$ mw: 181.16

SYNS: ACETIC ACID, (p-NITROPHENYL)- ◊ BENZENEACETIC ACID, 4-NITRO-(9CI) ◊ 4-NITROBENZENEACETIC ACID ◊ (4-NITROPHENYL)ACETIC ACID ◊ 2-(p-NITROPHENYL)ACETIC ACID

TOXICITY DATA with REFERENCE
dnr-bcs 500 µg/disc   MUREAV 170,11,86
ipr-mus LD50:830 mg/kg   FRPSAX 13,286,58

CONSENSUS REPORTS: Reported in EPA TSCA Inventory.

SAFETY PROFILE: Moderately toxic intraperitoneal route. Mutation data reported. When heated to decomposition it emits toxic vapors of NO_x.

NIM600   CAS:101-63-3   HR: D
p-NITROPHENYL ETHER
mf: $C_{12}H_8N_2O_5$ mw: 260.22

SYNS: BENZENE, 1,1'-OXYBIS(4-NITRO-(9CI) ◊ BIS(p-NITROPHENYL) ETHER ◊ BIS(4-NITROPHENYL)ETHER ◊ p,p'-DINITRODIPHENYL ETHER ◊ 4,4'-DINITRODIPHENYL ETHER ◊ 4,4'-DINITRODIPHENYL OXIDE ◊ DI-4-NITROPHENYL ETHER ◊ ETHER, BIS(p-NITROPHENYL) ◊ OXYBIS(4-NITROBENZENE) ◊ 1,1'-OXYBIS(4-NITROBENZENE)

TOXICITY DATA with REFERENCE
mmo-sat 50 µg/plate   CRNGDP 6,1811,85

CONSENSUS REPORTS: Reported in EPA TSCA Inventory.

SAFETY PROFILE: Mutation data reported. When heated to decomposition it emits toxic vapors of NO_x.

NIN100   CAS:619-91-0   HR: 2
N-(p-NITROPHENYL)GLYCINE
mf: $C_8H_8N_2O_4$ mw: 196.18

SYN: GLYCINE, N-(p-NITROPHENYL)-

TOXICITY DATA with REFERENCE
ipr-mus LD50:632 mg/kg   JAPMA8 48,419,59

CONSENSUS REPORTS: Reported in EPA TSCA Inventory.

SAFETY PROFILE: Moderately toxic by intraperitoneal route. When heated to decomposition it emits toxic vapors of NO_x.

NIQ550   CAS:2719-13-3   **HR: 3**
2-(p-NITROPHENYL)HYDRAZIDEACETIC ACID
mf: $C_8H_9N_3O_3$ mw: 195.20

TOXICITY DATA with REFERENCE
orl-mus LD50:400 mg/kg   PCJOAU 14,162,80

CONSENSUS REPORTS: Reported in EPA TSCA Inventory.

SAFETY PROFILE: Poison by ingestion. When heated to decomposition it emits toxic vapors of NO_x.

NIX100   CAS:89-40-7   **HR: D**
4-NITROPHTHALIMIDE
mf: $C_8H_4N_2O_4$ mw: 192.14

SYNS: 1H-ISOINDOLE-1,3(2H)-DIONE, 5-NITRO- ◊ PHTHALIMIDE, 4-NITRO-

TOXICITY DATA with REFERENCE
mmo-sat 333 µg/plate   EMMUEG 11(Suppl 12),1,88
mma-sat 1666 µg/plate   EMMUEG 11(Suppl 12),1,88

CONSENSUS REPORTS: Reported in EPA TSCA Inventory.

SAFETY PROFILE: Mutation data reported. When heated to decomposition it emits toxic vapors of NO_x.

NJF100   CAS:619-19-2   **HR: 3**
4-NITROSALICYLIC ACID
mf: $C_7H_5NO_5$ mw: 183.13

SYNS: BENZOIC ACID, 2-HYDROXY-4-NITRO- ◊ p-NITROSALICYLIC ACID ◊ 4-NITRO-SALICYLSAEURE ◊ SALICYLIC ACID, 4-NITRO-

TOXICITY DATA with REFERENCE
orl-mus LD50:650 mg/kg   ZNTFA2 6B,183,51
ipr-mus LD50:375 mg/kg   ZNTFA2 6B,183,51

CONSENSUS REPORTS: Reported in EPA TSCA Inventory.

SAFETY PROFILE: Poison by intraperitoneal route. Moderately toxic by ingestion. When heated to decomposition it emits toxic vapors of NO_x.

NJW600   CAS:120-22-9   **HR: D**
p-NITROSODIETHYLANILINE
mf: $C_{10}H_{14}N_2O$ mw: 178.26

SYNS: ANILINE, N,N-DIETHYL-p-NITROSO- ◊ BENZENAMINE, N,N-DIETHYL-4-NITROSO-(9CI) ◊ 4-DIETHYLAMINONITROSOBENZENE ◊ N,N-DIETHYL-4-NITROSOBENZENAMINE

TOXICITY DATA with REFERENCE
mmo-sat 3 µmol/L   XENOBH 16,587,86

CONSENSUS REPORTS: Reported in EPA TSCA Inventory.

SAFETY PROFILE: Mutation data reported. When heated to decomposition it emits toxic vapors of NO_x.

NMC100   CAS:102-96-5   **HR: 2**
β-NITROSTYRENE
mf: $C_8H_7NO_2$ mw: 149.16

SYNS: BNS ◊ NCI-C02211 ◊ γ-NITROSTYRENE ◊ STYRENE, β-NITRO-

TOXICITY DATA with REFERENCE
orl-mus LDLo:710 mg/kg   AECTCV 14,111,85

CONSENSUS REPORTS: Reported in EPA TSCA Inventory.

SAFETY PROFILE: Moderately toxic by ingestion. When heated to decomposition it emits toxic vapors of NO_x.

NMO700   CAS:455-90-3   **HR: D**
4-NITRO-o-TOLUENEDIAZONIUM TETRAFLUOROBORATE
mf: $C_7H_6N_3O_2 \bullet BF_4$ mw: 250.97

SYNS: BENZENEDIAZONIUM, 2-METHYL-4-NITRO-, TETRAFLUOROBORATE(1–) (9CI) ◊ DIAZOLE RED 4S ◊ DIAZOL RED S ◊ o-TOLUENEDIAZONIUM, 4-NITRO-, TETRAFLUOROBORATE(1–)

TOXICITY DATA with REFERENCE
mmo-omi 400 ppm/2H-C   SOGEBZ 4,1466,68

CONSENSUS REPORTS: Reported in EPA TSCA Inventory.

SAFETY PROFILE: Mutation data reported. When heated to decomposition it emits toxic vapors of boron, NO_x, and F^-

NMQ100   CAS:400-99-7   **HR: 2**
2-NITRO-α-α-α-TRIFLUORO-p-CRESOL
mf: $C_7H_4F_3NO_3$ mw: 207.12

SYN: p-CRESOL, 2-NITRO-α-α-α-TRIFLUORO-

TOXICITY DATA with REFERENCE
ipr-mus LD50:800 mg/kg   JPMSAE 57,1763,68

CONSENSUS REPORTS: Reported in EPA TSCA Inventory.

SAFETY PROFILE: Moderately toxic by intraperitoneal route. When heated to decomposition it emits toxic vapors of NO_x and F^-.

NMS520   CAS:89-58-7   **HR: 2**
NITRO-p-XYLENE
mf: $C_8H_9NO_2$ mw: 151.18

SYNS: BENZENE, 1,4-DIMETHYL-2-NITRO- ◇ 1,4-DIMETHYL-2-NITRO-BENZENE ◇ p-XYLENE, 2-NITRO-

TOXICITY DATA with REFERENCE
mma-sat 250 µg/plate   SAIGBL 29,34,87
orl-rat LD50:2440 mg/kg   85JCAE -,420,86

CONSENSUS REPORTS: Reported in EPA TSCA Inventory.

SAFETY PROFILE: Moderately toxic by ingestion. Mutation data reported. When heated to decomposition it emits toxic vapors of NO_x.

NNA100   CAS:762-13-0   HR: 1
NONANOYL PEROXIDE
DOT: UN 2130
mf: $C_{18}H_{34}O_4$ mw: 314.52

SYNS: DIPELARGONYL PEROXIDE ◇ PELARGONOYL PEROXIDE ◇ PELARGONYL PEROXIDE ◇ PELARGONYL PEROXIDE, technically pure (DOT) ◇ PEROXIDE, BIS(1-OXONONYL)

DOT Classification: Organic Peroxide; Label: Organic Peroxide

CONSENSUS REPORTS: Reported in EPA TSCA Inventory.

SAFETY PROFILE: A peroxide. Handle with care. When heated to decomposition it emits acrid smoke and irritating vapors.

NNC510   CAS:104-40-5   HR: 2
4-NONYLPHENOL
mf: $C_{15}H_{24}O$ mw: 220.39

SYNS: PARA-NONYL PHENOL ◇ PHENOL, p-NONYL-

TOXICITY DATA with REFERENCE
orl-rat LD50:1620 mg/kg   NTIS** PB85-143766

CONSENSUS REPORTS: Reported in EPA TSCA Inventory.

SAFETY PROFILE: Moderately toxic by ingestion. When heated to decomposition it emits acrid smoke and irritating vapors.

NNM510   CAS:492-39-7   HR: 3
psi-NOREPHEDRINE
mf: $C_9H_{13}NO$ mw: 151.23

SYNS: threo-2-AMINO-1-HYDROXY-1-PHENYLPROPANE ◇ CATHINE ◇ KATINE ◇ NOR-psi-EPHEDRINE ◇ NORPSEUDOEPHEDRINE, (+)- ◇ d-NOR-psi-EPHEDRINE ◇ d-NORPSEUDOEPHEDRINE ◇ threo-1-PHENYL-1-HYDROXY-2-AMINOPROPANE ◇ PSEUDONOREPHEDRINE

TOXICITY DATA with REFERENCE
scu-mus LD50:275 mg/kg   ARZNAD 5,367,55

CONSENSUS REPORTS: Reported in EPA TSCA Inventory.

SAFETY PROFILE: Poison by subcutaneous route. When heated to decomposition it emits toxic vapors of NO_x.

NNR400   CAS:597-71-7   HR: 2
NORMOSTEROL
mf: $C_{13}H_{20}O_8$ mw: 304.33

SYNS: 2,2-BIS((ACETYLOXY)METHYL)-1,3-PROPANEDIOL DIACETATE ◇ NORMO-LEVEL ◇ PAG ◇ PENTAERYTHRITOL, TETRAACETATE ◇ 1,3-PROPANEDIOL, 2,2-BIS((ACETYLOXY)METHYL)-, DIACETATE (9CI) ◇ T.A.P.E. ◇ TETRAACETIL PENTOETRIOL ◇ TETRAACETYL PENTESTRIOL

TOXICITY DATA with REFERENCE
orl-mus LD50:3500 mg/kg   RMNIBN 86,322,82
ipr-mus LD50:4850 mg/kg   RMNIBN 86,322,82

CONSENSUS REPORTS: Reported in EPA TSCA Inventory.

SAFETY PROFILE: Moderately toxic by ingestion. Slightly toxic by intraperitoneal route. When heated to decomposition it emits acrid smoke and irritating vapors.

NOG500   CAS:8008-45-5   HR: 2
NUTMEG OIL, EAST INDIAN

PROP: Major components are α- and β-pinene, camphene, myristicin, dipentene, and sabanene. Found in fruit of *Myristica fragrans Houttuyn* (Fam. Myristicaceae). Prepared by steam distillation of dried nutmeg (FCTXAV 14,601,76). Colorless to pale-yellow liquid; odor and taste of nutmeg. East Indian: d: 0.880–0.910, refr index: 1.474–1.488; West Indian: d: 0.854–0.880, refr index: 1.469–1.476 @20°. Sol in fixed oils, mineral oil; sltly sol in cold alc; very sol in hot alc, chloroform, ether; insol in glycerin, propylene glycol.

SYNS: MYRISTICA OIL ◇ NUTMEG OIL ◇ OIL of MYRISTICA ◇ OIL of NUTMEG

TOXICITY DATA with REFERENCE
skn-rbt 500 mg/24H MOD   FCTXAV 14,631,76
dnr-bcs 20 mg/disc   TOFOD5 8,91,85
orl-mus TDLo:4 g/kg (40D male):TER   TOLED5 7,239,81
orl-mus TDLo:2400 mg/kg (40D male):REP   TOLED5 7,239,81
orl-rat LD50:2620 mg/kg   FCTXAV 14,631,76
skn-rbt LD50:>10 g/kg   FCTXAV 14,631,76

CONSENSUS REPORTS: Reported in EPA TSCA Inventory.

SAFETY PROFILE: Moderately toxic by ingestion. Low toxicity by skin contact. An experimental teratogen. Experimental reproductive effects. Mutation data reported. A skin irritant. When heated to decomposition it emits acrid smoke and irritating fumes.

O

OAF100   CAS:126-14-7   *HR: 1*
OCTAACETYLSUCROSE
mf: $C_{28}H_{38}O_{19}$ mw: 678.66

SYNS: α-D-GLUCOPYRANOSIDE, 1,3,4,6-TETRA-O-ACETYL-β-D-FRUCTOFURANOSYL-, TETRAACETATE (9CI) ◇ SUCROSE, OCTAACETATE

TOXICITY DATA with REFERENCE
skn-rbt 500 mg/24H MLD   FCTOD7 20,827,82
orl-rat LD50:>5 g/kg   FCTOD7 20,827,82
skn-rbt LD50:>5 g/kg   FCTOD7 20,827,82

CONSENSUS REPORTS: Reported in EPA TSCA Inventory.

SAFETY PROFILE: Slightly toxic by ingestion and skin contact. A skin irritant. When heated to decomposition it emits acrid smoke and irritating vapors.

OAP100   CAS:307-34-6   *HR: 3*
OCTADECAFLUOROOCTANE
mf: C_8F_{18} mw: 438.08

SYNS: OCTANE, OCTADECAFLUORO- ◇ PERFLUOROOCTANE ◇ n-PERFLUOROOCTANE

TOXICITY DATA with REFERENCE
ivn-mus LD50:240 mg/kg   MIVRA6 8,320,74

CONSENSUS REPORTS: Reported in EPA TSCA Inventory.

SAFETY PROFILE: Poison by intravenous route. When heated to decomposition it emits toxic vapors of F^-.

OAP300   CAS:2190-04-7   *HR: 2*
OCTADECANAMINE ACETATE
mf: $C_{18}H_{39}N \cdot C_2H_4O_2$ mw: 329.64

SYNS: ARMAC 18D ◇ ARMAC OD ◇ OCTADECYLAMINE, ACETATE ◇ 1-OCTADEDCYLAMINE ACETATE

TOXICITY DATA with REFERENCE
orl-mus LD50:1000 mg/kg   CHTPBA 1,11,65

CONSENSUS REPORTS: Reported in EPA TSCA Inventory.

SAFETY PROFILE: Moderately toxic by ingestion. When heated to decomposition it emits toxic vapors of NO_x.

OAT000   CAS:2223-93-0   *HR: 3*
OCTADECANOIC ACID, CADMIUM SALT
mf: $C_{36}H_{72}O_4 \cdot Cd$ mw: 681.48

SYNS: CADMIUM STEARATE ◇ KADMIUMSTEARAT (GERMAN) ◇ STEARIC ACID, CADMIUM SALT

TOXICITY DATA with REFERENCE
ihl-wmn TCLo:147 mg/m^3/35M:CNS,GIT   ZHYGAM 17,308,71
ihl-hmn TCLo:1800 μg/m^3/2Y:CVS,CNS,MET   HYSAAV 33,187,68
orl-rat LD50:1125 mg/kg   GISAAA 35(2),98,70
ihl-rat LC50:130 mg/m^3/2H   ZHYGAM 17,308,71
orl-mus LD50:590 mg/kg   HYSAAV 33,187,68

CONSENSUS REPORTS: EPA Extremely Hazardous Substances List. Cadmium and its compounds are on the Community Right-To-Know List. Reported in EPA TSCA Inventory.

OSHA PEL: TWA 5 μg(Cd)/m^3
ACGIH TLV: TWA 0.05 mg(Cd)/m^3 (Proposed: TWA 0.01 mg(Cd)/m^3 (dust), Suspected Human Carcinogen; 0.002 mg(Cd)/m^3 (respirable dust), Suspected Human Carcinogen); BEI: 10 μg/g creatinine in urine; 10 μg/L in blood
DFG BAT: Blood 1.5 μg/dL; Urine 15 μg/dL, Suspected Carcinogen
NIOSH REL: (Cadmium) Reduce to lowest feasible level.

SAFETY PROFILE: Confirmed human carcinogen. Poison by inhalation. Moderately toxic by ingestion. Human systemic effects by inhalation: hallucinations or distorted perceptions; nausea or vomiting, other gastrointestinal effects; weight loss or decreased weight gain; cardiac effects. When heated to decomposition it emits toxic fumes of Cd. See also CADMIUM COMPOUNDS.

OBG100   CAS:1120-04-3   *HR: 3*
OCTADECYL SODIUM SULFATE
mf: $C_{18}H_{37}O_4S \cdot Na$ mw: 372.60

SYNS: SODIUM MONOOCTADECYL SULFATE ◇ SODIUM MONOSTEARYL SULFATE ◇ SODIUM OCTADECYL SULFATE ◇ SODIUM STEARYL SULFATE ◇ SULFURIC ACID, MONOOCTADECYL ESTER, SODIUM SALT

TOXICITY DATA with REFERENCE
ipr-mus LD50:477 mg/kg   JAPMA8 42,283,53

CONSENSUS REPORTS: Reported in EPA TSCA Inventory.

SAFETY PROFILE: Poison by intraperitoneal route. When heated to decomposition it emits toxic vapors of SO_x.

OBM000   CAS: 382-21-8   *HR: 3*
OCTAFLUORO-sec-BUTENE
mf: C_4F_8 mw: 200.04

SYNS: OCTAFLUOROISOBUTENE ◇ OCTAFLUOROISOBUTYLENE ◇ PERFLUOROISOBUTYLENE (ACGIH) ◇ PFIB

TOXICITY DATA with REFERENCE
ihl-rat LC50:500 ppb/6H   34ZIAG -,310,69
ihl-mus LCLo:10 mg/m^3/2H   GTPZAB 5(3),3,61
ihl-rbt LC50:1200 ppb/2H   BIMEA5 57,247,68
ihl-gpg LC50:1050 ppb/2H   BIMEA5 57,247,68
ACGIH TLV: CL 0.01 ppm

SAFETY PROFILE: A deadly poison by inhalation. A skin, eye, and mucous membrane irritant. Human acute exposure causes marked irritation of conjunctivae, throat, and lungs. When heated to decomposition it emits toxic fumes of F$^-$. See also FLUORIDES.

OCE100   CAS: 556-67-2   *HR: 1*
OCTAMETHYLCYCLOTETRASILOXANE
mf: $C_8H_{24}O_4Si_4$ mw: 296.68

SYNS: CYCLOTETRASILOXANE, OCTAMETHYL- ◇ OKTAMETHYLCYKLOTETRASILOXAN

TOXICITY DATA with REFERENCE
skn-rbt 500 mg/24H MLD   85JCAE -,1230,86
eye-rbt 500 mg/24H MLD   85JCAE -,1230,86

CONSENSUS REPORTS: Reported in EPA TSCA Inventory.

SAFETY PROFILE: A skin irritant. When heated to decomposition it emits acrid smoke and irritating vapors.

OCW050   CAS: 629-40-3   *HR: 3*
OCTANEDINITRILE (9CI)
mf: $C_8H_{12}N_2$ mw: 136.22

SYNS: 1,6-DICYANOHEXANE ◇ SUBERONITRILE

TOXICITY DATA with REFERENCE
orl-mus LD50:307 mg/kg   ARTODN 57,88,85

CONSENSUS REPORTS: Reported in EPA TSCA Inventory.

SAFETY PROFILE: Poison by ingestion. When heated to decomposition it emits toxic vapors of NO$_x$ and CN$^-$.

OCW100   CAS: 124-12-9   *HR: 2*
OCTANENITRILE
mf: $C_8H_{15}N$ mw: 125.24

SYNS: ARNEEL 8 ◇ CAPRYLNITRILE ◇ CAPRYLONITRILE ◇ OCTANONITRILE

TOXICITY DATA with REFERENCE
orl-mus LD50:1764 mg/kg   ARTODN 55,47,84

CONSENSUS REPORTS: Reported in EPA TSCA Inventory.

SAFETY PROFILE: Moderately toxic by ingestion. When heated to decomposition it emits toxic vapors of NO$_x$ and CN$^-$.

OCY090   CAS: 123-96-6   *HR: 1*
2-OCTANOL
mf: $C_8H_{18}O$ mw: 130.26

SYN: CAPRYL ALCOHOL

TOXICITY DATA with REFERENCE
unr-uns LD50:6934 mg/kg   GISAAA 51(5),61,86

CONSENSUS REPORTS: Reported in EPA TSCA Inventory.

SAFETY PROFILE: Slightly toxic by an unspecified route. When heated to decomposition it emits acrid smoke and irritating vapors.

ODW028   CAS: 2442-10-6   *HR: 2*
1-OCTEN-3-OL ACETATE
mf: $C_{10}H_{18}O_2$ mw: 170.28

SYNS: AMYL VINYL CARBINOL ACETATE ◇ AMYL VINYL CARBINYL ACETATE ◇ OCTENYL ACETATE ◇ 1-PENTYLALLYL ACETATE

TOXICITY DATA with REFERENCE
orl-rat LD50:850 mg/kg   FCTOD7 20,641,82
skn-rbt LD50:>5 g/kg   FCTOD7 20,641,82

CONSENSUS REPORTS: Reported in EPA TSCA Inventory.

SAFETY PROFILE: Moderately toxic by ingestion. Slightly toxic by skin contact. When heated to decomposition it emits acrid smoke and irritating vapors.

OEK010   CAS: 693-16-3   *HR: 3*
2-OCTYLAMINE
mf: $C_8H_{19}N$ mw: 129.28

SYNS: 2-AMINOOCTANE ◇ CAPRYLAMINE ◇ HEPTYLAMINE, 1-METHYL- ◇ 1-METHYLHEPTYLAMINE ◇ 2-OCTANAMINE

TOXICITY DATA with REFERENCE
ivn-mus LD50:23 mg/kg   BJPCAL 7,42,52

CONSENSUS REPORTS: Reported in EPA TSCA Inventory.

SAFETY PROFILE: Poison by intravenous route. When heated to decomposition it emits toxic vapors of NO$_x$.

OES100   CAS: 2305-05-7   *HR: 1*
γ-n-OCTYL-γ-n-BUTYROLACTONE
mf: $C_{12}H_{22}O_2$ mw: 198.34

SYNS: γ-DODECALACTONE ◇ DODECANOLIDE-1,4 ◇ 2(3H)-FURANONE, DIHYDRO-5-OCTYL-

TOXICITY DATA with REFERENCE
skn-rbt 500 mg/24H MOD   FCTXAV 14,751,76

CONSENSUS REPORTS: Reported in EPA TSCA Inventory.

SAFETY PROFILE: A skin irritant. When heated to decomposition it emits acrid smoke and irritating vapors.

OES300   CAS:2243-27-8   HR: 2
1-OCTYL CYANIDE
mf: $C_9H_{17}N$ mw: 139.27

SYNS: 1-CYANOOCTANE ◊ NONANENITRILE ◊ n-NONANENITRILE ◊ NONANONITRILE ◊ n-NONANONITRILE ◊ OCTYL CYANIDE ◊ n-OCTYL CYANIDE ◊ PELARGONITRILE ◊ PELARGONONITRILE

TOXICITY DATA with REFERENCE
orl-mus LD50:2059 mg/kg   ARTODN 55,47,84

CONSENSUS REPORTS: Reported in EPA TSCA Inventory.

SAFETY PROFILE: Moderately toxic by ingestion. When heated to decomposition it emits toxic vapors of NO_x and CN^-.

OFA100   CAS:629-27-6   HR: 2
1-OCTYL IODIDE
mf: $C_8H_{17}I$ mw: 240.15

SYNS: 1-IODOOCTANE ◊ 1-JODOKTAN ◊ OCTANE, 1-IODO- ◊ OCTYL IODIDE ◊ n-OCTYL IODIDE ◊ 1-n-OCTYL IODIDE

TOXICITY DATA with REFERENCE
ipr-rat LD50:1982 mg/kg   85GMAT -,95,82
ipr-mus LD50:1416 mg/kg   85GMAT -,95,82

CONSENSUS REPORTS: Reported in EPA TSCA Inventory.

SAFETY PROFILE: Moderately toxic by intraperitoneal route. When heated to decomposition it emits toxic vapors of I^-.

OFE050   CAS:693-54-9   HR: 1
OCTYL METHYL KETONE
mf: $C_{10}H_{20}O$ mw: 156.30

SYNS: 2-DECANONE ◊ METHYL OCTYL KETONE ◊ METHYL n-OCTYL KETONE

TOXICITY DATA with REFERENCE
orl-mus LD50:7936 mg/kg   TOLED5 30,13,86

CONSENSUS REPORTS: Reported in EPA TSCA Inventory.

SAFETY PROFILE: Slightly toxic by ingestion. When heated to decomposition it emits acrid smoke and irritating vapors.

OFU200   CAS:142-31-4   HR: 3
OCTYL SODIUM SULFATE
mf: $C_8H_{17}O_4S \cdot Na$ mw: 232.30

SYNS: CYCLORYL OS ◊ DUPONOL 80 ◊ SIPEX OLS ◊ SODIUM CAPRYL SULFATE ◊ SODIUM OCTYL SULFATE ◊ SODIUM OCTYL SULPHATE ◊ SOS ◊ SULFURIC ACID, MONOOCTYL ESTER, SODIUM SALT

TOXICITY DATA with REFERENCE
orl-rat LD50:3200 mg/kg   JAPMA8 42,283,53
ipr-mus LD50:396 mg/kg   JAPMA8 42,283,53

CONSENSUS REPORTS: Reported in EPA TSCA Inventory.

SAFETY PROFILE: Poison by intraperitoneal route. Moderately toxic by ingestion. When heated to decomposition it emits toxic vapors of SO_x.

OGI025   CAS:629-05-0   HR: D
1-OCTYNE
mf: C_8H_{14} mw: 110.22

SYN: HEXYLACETYLENE

TOXICITY DATA with REFERENCE
dns-ham:lvr 100 ~2mol/L   CRNGDP 6,1201,85

CONSENSUS REPORTS: Reported in EPA TSCA Inventory.

SAFETY PROFILE: Mutation data reported. When heated to decomposition it emits acrid smoke and irritating vapors.

OGI050   CAS:818-72-4   HR: 2
1-OCTYN-3-OL
mf: $C_8H_{14}O$ mw: 126.22

TOXICITY DATA with REFERENCE
orl-mus LD50:460 mg/kg   THERAP 11,692,56

CONSENSUS REPORTS: Reported in EPA TSCA Inventory.

SAFETY PROFILE: Moderately toxic by ingestion. When heated to decomposition it emits acrid smoke and irritating vapors.

OJW200   CAS:289-80-5   HR: 2
ORTHODIAZINE
mf: $C_4H_4N_2$ mw: 80.10

SYNS: 1,2-DIAZABENZENE ◊ 1,2-DIAZINE ◊ PYRIDAZINE

TOXICITY DATA with REFERENCE
ipr-mus LD50:2650 mg/kg   PBPHAW 1,542,65

CONSENSUS REPORTS: Reported in EPA TSCA Inventory.

SAFETY PROFILE: Moderately toxic by intraperitoneal route. When heated to decomposition it emits toxic vapors of NO_x.

OKU100 CAS:123-69-3 **HR: 3**
OXACYCLOHEPTADEC-8-EN-2-ONE, (Z)-
mf: $C_{16}H_{28}O_2$ mw: 252.44

SYNS: AMBRETTOLID ◇ AMBRETTOLIDE ◇ MUSK AMBRETTE ◇ MUSK AMBRETTE (NATURAL) ◇ MUSK NATURAL ◇ NATURAL MUSK AMBRETTE

TOXICITY DATA with REFERENCE
orl-rat LD50:339 mg/kg   YAKUD5 22,1513,80
orl-mus LDLo:1600 mg/kg   AECTCV 14,111,85

CONSENSUS REPORTS: Reported in EPA TSCA Inventory.

SAFETY PROFILE: Poison by ingestion. When heated to decomposition it emits acrid smoke and irritating vapors.

OMC300 CAS:96-26-4 **HR: D**
OXANTIN
mf: $C_3H_6O_3$ mw: 90.09

SYNS: DIHYDROXYACETONE ◇ NSC-24343 ◇ OXATONE ◇ CHROMELIN ◇ 1,3-DIHYDROXYACETONE ◇ 1,3-DIHYDROXYPROPANONE ◇ DIHYXAL ◇ OTAN ◇ 2-PROPANONE, 1,3-DIHYDROXY- ◇ SOLEAL ◇ TRIULOSE ◇ VITICOLOR

TOXICITY DATA with REFERENCE
mmo-sat 150 μg/plate   ABCHA6 47,2461,83
oth-esc 3300 μmol/L   MUREAV 203,81,88
dnr-bcs 2 pph   JEPTDQ 3(1-2),227,79

CONSENSUS REPORTS: Reported in EPA TSCA Inventory.

SAFETY PROFILE: Mutation data reported. When heated to decomposition it emits acrid smoke and irritating vapors.

OMY910 CAS:57-71-6 **HR: 3**
2-OXIMINO-3-BUTANONE
mf: $C_4H_7NO_2$ mw: 101.12

SYNS: BIACETYLMONOXIME ◇ 2,3-BUTANEDIONE, MONOOXIME ◇ 2,3-BUTANEDIONE 2-OXIME ◇ DAM ◇ DIACETYLMONOOXIME ◇ DIACETYLMONOXIME

TOXICITY DATA with REFERENCE
ipr-mus LD50:51 mg/kg   JPMSAE 53,1143,64

CONSENSUS REPORTS: Reported in EPA TSCA Inventory.

SAFETY PROFILE: Poison by intraperitoneal route. When heated to decomposition it emits toxic vapors of NO_x.

ONI100 CAS:615-16-7 **HR: 2**
2-OXOBENZIMIDAZOLE
mf: $C_7H_6N_2O$ mw: 134.15

SYN: 2-BENZIMIDAZOLINONE

TOXICITY DATA with REFERENCE
scu-mus LD50:620 mg/kg   ARZNAD 8,42,58

CONSENSUS REPORTS: Reported in EPA TSCA Inventory.

SAFETY PROFILE: Moderately toxic subcutaneous route. When heated to decomposition it emits toxic vapors of NO_x.

ONQ100 CAS:110-99-6 **HR: 2**
OXODIACETIC ACID
mf: $C_4H_6O_5$ mw: 134.10

SYNS: ACETIC ACID, OXYDI- ◇ ACETIC ACID, 2,2'-OXYBIS-(9CI) ◇ BIS(CARBOXYMETHYL)ETHER ◇ DIGLYCOLIC ACID (6CI) ◇ 3-OXAPENTANEDIOIC ACID ◇ OXYBISACETIC ACID ◇ 2,2'-OXYBISACETIC ACID ◇ OXYDIACETIC ACID ◇ 2,2'-OXYDIACETIC ACID ◇ OXYDIETHANOLIC ACID

TOXICITY DATA with REFERENCE
orl-rat LD50:500 mg/kg   ACIEAY 14,94,75

CONSENSUS REPORTS: Reported in EPA TSCA Inventory.

SAFETY PROFILE: Moderately toxic by ingestion. When heated to decomposition it emits acrid smoke and irritating vapors.

OOI100 CAS:97-36-9 **HR: 2**
3-OXO-N-(2,4-METHYLPHENYL)BUTANAMIDE
mf: $C_{12}H_{15}NO_2$ mw: 205.28

SYNS: ACETOACETANILIDE, 2',4'-DIMETHYL- ◇ ACETOACET-2,4-DIMETHYLPHENYL ◇ 2,4-ACETOACETOXYLIDIDE ◇ ACETOACETO-m-XYLIDIDE ◇ 2',4'-ACETOACETOXYLIDIDE ◇ ACETOACET-m-XYLIDIDE ◇ ACETOACETYL-m-XYLIDIDE ◇ BUTANAMIDE, N-(2,4-DIMETHYLPHENYL)-3-OXO-(9CI) ◇ 2',4'-DIMETHYLACETOACETANILIDE ◇ N-(2,4-DIMETHYLPHENYL)-3-OXOBUTANAMIDE

TOXICITY DATA with REFERENCE
orl-rat LD50:3000 mg/kg   LONZA# 13JUL81

CONSENSUS REPORTS: Reported in EPA TSCA Inventory.

SAFETY PROFILE: Moderately toxic by ingestion. When heated to decomposition it emits toxic vapors of NO_x.

OOI200 CAS:734-88-3 **HR: 2**
3-OXO-3H-NAPHTHO(2,1-b)PYRAN-2-CARBOXYLIC ACID ETHYL ESTER
mf: $C_{15}H_{12}O_4$ mw: 256.27

SYNS: BELOPHAR KLA ◇ 5,6-BENZOCOUMARIN-3-CARBONIC ACID ETHYL ETHER ◇ COMPOUND 13-61 ◇ 5,6-BENZOCOUMARIN-3-CARBOXYLIC ACID ETHYL ESTER ◇ OPTICAL BLEACH 13-61 ◇ 3H-NAPHTHO(2,1-b)PYRAN-2-CARBOXYLIC ACID, 3-OXO-, ETHYL ESTER

TOXICITY DATA with REFERENCE
orl-rat LD50:1580 mg/kg   GISAAA 49(1),85,84
orl-mus LD50:590 mg/kg   GISAAA 49(1),85,84

CONSENSUS REPORTS: Reported in EPA TSCA Inventory.

SAFETY PROFILE: Moderately toxic by ingestion. When

heated to decomposition it emits acrid smoke and irritating vapors.

OOO200   CAS:142-08-5   ***HR: 2***
2-OXOPYRIDINE
mf: C_5H_5NO mw: 95.11

SYNS: 2-HYDROXYPYRIDINE ◇ 2-PYRIDINOL ◇ 2-PYRIDINONE ◇ 2(1H)-PYRIDONE ◇ α-PYRIDONE ◇ PYRIDONE-2 ◇ 2-PYRIDONE ◇ 2(1H)-PYRIDINONE (9CI)

TOXICITY DATA with REFERENCE
ipr-mus LD50:410 mg/kg   TOXIA6 23,815,85
ivn-mus LD50:750 mg/kg   AIPTAK 93,143,53

CONSENSUS REPORTS: Reported in EPA TSCA Inventory.

SAFETY PROFILE: Moderately toxic by intraperitoneal and intravenous routes. When heated to decomposition it emits toxic vapors of NO_x.

OPC100   CAS:2469-55-8   ***HR: 3***
OXYBIS((3-AMINOPROPYL)DIMETHYLSILANE)
mf: $C_{10}H_{28}N_2OSi_2$ mw: 248.58

SYNS: 1,3-BIS(3-AMINOPROPYL)-1,1,3,3-TETRAMETHYLDISILOXANE ◇ DISILOXANE, 1,3-BIS(3-AMINOPROPYL)-1,1,3,3-TETRAMETHYL- ◇ SILANE, OXYBIS((3-AMINOPROPYL)DIMETHYL)-

TOXICITY DATA with REFERENCE
ipr-mus LD50:80 mg/kg   RCRVAB 38,975,69

CONSENSUS REPORTS: Reported in EPA TSCA Inventory.

SAFETY PROFILE: Poison by intraperitoneal route. When heated to decomposition it emits toxic vapors of NO_x.

P

PAH260 CAS:985-13-7 **HR: 3**
PAPETHERINE HYDROCHLORIDE
mf: $C_{24}H_{29}NO_4 \cdot ClH$ mw: 432.00

SYNS: BARBONIN HYDROCHLORIDE ◇ 1-(3,4-DIETHOXYBENZYL)-6,7-DIETHOXYISOQUINOLINE HYDROCHLORIDE ◇ ETHAVERINE HYDROCHLORIDE ◇ ISOQUINOLINE, 1-(3,4-DIETHOXYBENZYL)-6,7-DIETHOXY-, HYDROCHLORIDE ◇ ISOQUINOLINE, 1-((3,4-DIETHOXYPHENYL)METHYL)-6,7-DIETHOXY-, HYDROCHLORIDE ◇ PERPARINE HYDROCHLORIDE ◇ PERPARIN HYDROCHLORIDE ◇ PERPERINE HYDROCHLORIDE

TOXICITY DATA with REFERENCE
scu-mus LD50:660 mg/kg   THERAP 5,69,50
ivn-mus LD50:86 mg/kg   APFRAD 9,585,51

CONSENSUS REPORTS: Reported in EPA TSCA Inventory.

SAFETY PROFILE: Poison by intravenous route. Moderately toxic by subcutaneous route. When heated to decomposition it emits toxic vapors of NO_x, HCl, and Cl^-.

PAQ200 CAS:54-35-3 **HR: 2**
PENICILLIN G PROCAINE
mf: $C_{16}H_{18}N_2O_4S \cdot C_{13}H_{20}N_2O_2$ mw: 570.77

SYNS: BENZYLPENICILLIN NOVOCAINE SALT ◇ BENZYLPENICILLIN PROCAINE ◇ DEPOCILLIN ◇ DUPHAPEN ◇ HOSTACILLIN ◇ HYDRACILLIN ◇ JENACILLIN O ◇ MICRO-PEN ◇ NOPCAINE ◇ PENZAL N 300 ◇ PROCAINE BENZYLPENICILLINATE ◇ PROCAINE PENICILLIN G ◇ RETARDILLIN ◇ 4-THIA-1-AZABICYCLO(3.2.0)HEPTANE-2-CARBOXYLIC ACID, 3,3-DIMETHYL-7-OXO-6-(2-PHENYLACETAMIDO)-, compd. with 2-(DIETHYLAMINO)ETHYL p-AMINOBENZOATE (1:1) ◇ VETSPEN ◇ VITABLEND

TOXICITY DATA with REFERENCE
orl-rat LD50:>2 g/kg   ABANAE 3,534,55/56
ipr-rat LD50:420 mg/kg   ABANAE 3,534,55/56
ipr-mus LD50:371 mg/kg   ABANAE 3,534,55/56
scu-mus LDLo:4 g/kg   NYKZAU 55,23,59
ipr-gpg LD50:518 mg/kg   ABANAE 3,534,55/56
ims-gpg LD50:1194 mg/kg   ABANAE 3,534,55/56

CONSENSUS REPORTS: Reported in EPA TSCA Inventory.

SAFETY PROFILE: Moderately toxic by intraperitoneal and intramuscular routes. Slightly toxic by ingestion and subcutaneous routes. When heated to decomposition it emits toxic vapors of NO_x, and SO_x.

PBD275 CAS:602-94-8 **HR: 3**
PENTAFLUOROBENZOIC ACID
mf: $C_7HF_5O_2$ mw: 212.08

SYNS: BENZOIC ACID, PENTAFLUORO- ◇ PERFLUOROBENZOIC ACID

TOXICITY DATA with REFERENCE
ipr-mus LD50:178 mg/kg   JMCMAR 11,1020,68

CONSENSUS REPORTS: Reported in EPA TSCA Inventory.

SAFETY PROFILE: Poison by intraperitoneal route. When heated to decomposition it emits toxic vapors of F^-.

PBE100 CAS:344-07-0 **HR: D**
PENTAFLUOROPHENYL CHLORIDE
mf: C_6ClF_5 mw: 202.51

SYNS: BENZENE, CHLOROPENTAFLUORO- ◇ CHLOROPENTAFLUOROBENZENE ◇ CHLOROPERFLUOROBENZENE ◇ PENTAFLUOROCHLOROBENZENE

TOXICITY DATA with REFERENCE
dns-rat:lvr 10 mg/L   NTIS** AD-A165-849
otr-mus:oth 100 mg/L   NTIS** AD-A187-233
cyt-ham:ovr 30 mg/L   NTIS** AD-A165-849
sce-ham:ovr 62500 µg/L   NTIS** AD-A187-233
orl-rat TDLo:10500 mg/kg (female 6-15D post):REP
   TOXID9 12,105,92

CONSENSUS REPORTS: Reported in EPA TSCA Inventory.

SAFETY PROFILE: Experimental teratogenic effects reported. Mutation data reported. When heated to decomposition it emits toxic vapors of F^- and Cl^-.

PBH125 CAS:628-77-3 **HR: D**
PENTAMETHYLENE DIIODIDE
mf: $C_5H_{10}I_2$ mw: 323.95

SYNS: 1,5-DIIODOPENTANE ◇ PENTANE, 1,5-DIIODO-

TOXICITY DATA with REFERENCE
mmo-sat 10 µmol/plate   MUREAV 141,11,84

CONSENSUS REPORTS: Reported in EPA TSCA Inventory.

SAFETY PROFILE: Mutation data reported. When heated to decomposition it emits toxic vapors of I^-.

PBV505 CAS:110-58-7 **HR: 3**
1-PENTYLAMINE
DOT: UN 1106
mf: C$_5$H$_{13}$N   mw: 87.19

SYNS: 1-AMINOPENTANE ◊ AMYLAMINE ◊ AMYLAMINE (DOT) ◊ n-AMYLAMINE ◊ MONOAMYLAMINE ◊ NORLEUCAMINE ◊ 1-PENTANAMINE ◊ PENTYLAMINE ◊ n-PENTYLAMINE

TOXICITY DATA with REFERENCE
ipr-rat LDLo:37,500 µg/kg   FATOAO 31,238,68

DOT Classification: Flammable Liquid; Label: Flammable Liquid

CONSENSUS REPORTS: Reported in EPA TSCA Inventory.

SAFETY PROFILE: Poison by intraperitoneal route. A flammable liquid. When heated to decomposition it emits toxic vapors of NO$_x$.

PBV800 CAS:2049-96-9 **HR: 1**
n-PENTYL BENZOATE
mf: C$_{12}$H$_{16}$O$_2$ mw: 192.28

SYNS: AMYL BENZOATE ◊ BENZOIC ACID, PENTYL ESTER ◊ PENTYL BENZOATE

TOXICITY DATA with REFERENCE
skn-rbt 500 mg/24H MLD   FCTXAV 11,1079,73

CONSENSUS REPORTS: Reported in EPA TSCA Inventory.

SAFETY PROFILE: A skin irritant. When heated to decomposition it emits acrid smoke and irritating vapors.

PCF275 CAS:127-18-4 **HR: 3**
PERCHLOROETHYLENE
DOT: UN 1897
mf: C$_2$Cl$_4$ mw: 165.82

PROP: Colorless liquid; chloroform-like odor. Mp: −23.35°, bp: 121.20°, d: 1.6311 @ 15°/4°, vap press: 15.8 mm @ 22°, vap d: 5.83.

SYNS: ANKILOSTIN ◊ ANTISOL 1 ◊ CARBON BICHLORIDE ◊ CARBON DICHLORIDE ◊ CZTEROCHLOROETYLEN (POLISH) ◊ DIDAKENE ◊ DOW-PER ◊ ENT 1,860 ◊ ETHYLENE TETRACHLORIDE ◊ FEDAL-UN ◊ NCI-C04580 ◊ NEMA ◊ PERAWIN ◊ PERCHLOORETHYLEEN, PER (DUTCH) ◊ PERCHLOR ◊ PERCHLORAETHYLEN, PER (GERMAN) ◊ PERCHLORETHYLENE ◊ PERCHLORETHYLENE, PER (FRENCH) ◊ PERCLENE ◊ PERCLOROETILENE (ITALIAN) ◊ PERCOSOLVE ◊ PERK ◊ PERKLONE ◊ PERSEC ◊ RCRA WASTE NUMBER U210 ◊ TETLEN ◊ TETRACAP ◊ TETRACHLOORETHEEN (DUTCH) ◊ TETRACHLORAETHEN (GERMAN) ◊ TETRACHLOROETHENE ◊ TETRACHLOROETHYLENE (DOT) ◊ 1,1,2,2-TETRACHLOROETHYLENE ◊ TETRACLOROETENE (ITALIAN) ◊ TETRALENO ◊ TETRALEX ◊ TETRAVEC ◊ TETROGUER ◊ TETROPIL

TOXICITY DATA with REFERENCE
skn-rbt 810 mg/24H SEV   JETOAS 9,171,76
eye-rbt 162 mg MLD   JETOAS 9,171,76
dns-hmn:lng 100 mg/L   NTIS** PB82-185075
otr-rat:emb 97 µmol/L   ITCSAF 14,290,78
ihl-rat TCLo:900 ppm/7H (7–13D preg):REP   TJADAB 19,41A,79
ihl-rat TCLo:900 ppm/7H (7–13D preg):REP   TJADAB 19,41A,79
ihl-rat TCLo:900 ppm/7H (7–13D preg):REP   TJADAB 19,41A,79
ihl-rat TCLo:1000 ppm/24H (14D pre/1–22D preg):TER   APTOD9 19,A21,80
ihl-mus TCLo:300 ppm/7H (6–15D preg):REP   TXAPA9 32,84,75
ihl-mus TCLo:500 ppm/7H (male 5D pre):REP   NTIS** PB82-185075
ihl-mus TCLo:300 ppm/7H (6–15D preg):TER   TXAPA9 32,84,75
ihl-rat TCLo:1000 ppm/24H (14D pre/1–22D preg):TER   APTOD9 19,A21,80
ihl-rat TCLo:200 ppm/6H/2Y-I:CAR   NTPTR* NTP-TR-311,86
orl-mus TDLo:195 g/kg/50W-I:CAR   NCITR* NCI-TR-13,77
ihl-mus TCLo:100 ppm/6H/2Y-I:CAR   NTPTR* NTP-TR-311,86
orl-mus TD:240 g/kg/62W-I:CAR   NCITR* NCI-TR-13,77
ihl-rat TC:200 ppm/6H/2Y-I:NEO   TOLED5 31(Suppl),16,86
ihl-mus TC:100 ppm/6H/2Y-I:NEO   TOLED5 31(Suppl),16,86
ihl-hmn TCLo:96 ppm/7H:PNS,EYE,CNS   NTIS** PB257-185
ihl-man TCLo:280 ppm/2H:EYE,CNS   AMIHBC 5,566,52
ihl-man TCLo:600 ppm/10M:EYE,CNS   AMIHBC 5,566,52
ihl-man LDLo:2857 mg/kg:CNS,PUL   MLDCAS 5,152,72
orl-rat LD50:2629 mg/kg   AIHAAP 20,364,59
ihl-rat LC50:34200 mg/m^3/8H   AIHAAP 20,364,59
orl-mus LD50:8100 mg/kg   NTIS** PB257-185
ihl-mus LC50:5200 ppm/4H   APTOA6 9,303,53
orl-dog LDLo:4000 mg/kg   AJHYA2 9,430,29
ipr-dog LD50:2100 mg/kg   TXAPA9 10,119,67
ivn-dog LDLo:85 mg/kg   QJPPAL 7,205,34
orl-cat LDLo:4000 mg/kg   AJHYA2 9,430,29

CONSENSUS REPORTS: NTP Fifth Annual Report on Carcinogens. IARC Cancer Review: Group 2B IMEMDT 7,355,87; Animal Limited Evidence IMEMDT 20,491,79. NCI Carcinogenesis Bioassay (gavage); Clear Evidence: mouse NCITR* NCI-CG-TR-13,77 (inhalation); Clear Evidence: mouse, rat NTPTR* NTP-TR-311,86 (gavage) Inadequate Studies: rat NCITR* NCI-CG-TR-13,77. Reported in EPA TSCA Inventory. EPA Genetic Toxicology Program. Community Right-To-Know List.

OSHA PEL: (Transitional: TWA 100 ppm; CL 200 ppm; Pk 600 ppm/5M)TWA 25 ppm
ACGIH TLV: TWA 50 ppm; STEL 200 ppm (Proposed: TWA 25 ppm; Animal Carcinogen); BEI: 7 mg/L trichloroacetic acid in urine at end of work week.
DFG MAK: 50 ppm (345 mg/m^3); BAT: blood 100 µg/dl
NIOSH REL: (Tetrachloroethylene) Minimize workplace exposure.

DOT Classification: Poison B; Label: St. Andrews Cross; ORM-A; Label: None.

SAFETY PROFILE: Confirmed carcinogen with experimental carcinogenic, neoplastigenic, and teratogenic data. Experimental poison by intravenous route. Moderately toxic to humans by inhalation, with the following effects: local anesthetic, conjunctiva irritation, general anesthesia, hallucinations, distorted perceptions, coma, and pulmonary changes. Moderately experimentally toxic by ingestion, inhalation, intraperitoneal, and subcutaneous routes. An experimental teratogen. Experimental reproductive effects. Human mutation data reported. An eye and severe skin irritant. The liquid can cause injuries to the eyes; however, with proper precautions it can be handled safely. The symptoms of acute intoxication from this material are the result of its effects upon the nervous system. Can cause dermatitis, particularly after repeated or prolonged contact with the skin. Irritates the gastrointestinal tract upon ingestion. It may be handled in the presence or absence of air, water, and light with any of the common construction materials at temperatures up to 140°. This material is extremely stable and resists hydrolysis. A common air contaminant. Reacts violently under the proper conditions with Ba; Be; Li; N_2O_4; metals; NaOH. When heated to decomposition it emits highly toxic fumes of Cl^-. See also CHLORINATED HYDROCARBONS, ALIPHATIC.

PCG700　　　CAS:306-94-5　　　HR: 3
PERFLUORODECALIN
mf: $C_{10}F_{18}$　　mw: 462.10

SYNS: NAPHTHALENE, DECAHYDROOCTADECAFLUORO- ◇ NAPHTHALENE, OCTADECAFLUORODECAHYDRO- ◇ OCTADECAFLUORODECAHYDRONAPHTHALENE ◇ PERFLUORODECAHYDRONAPHTHALENE ◇ PFD ◇ PP 5

TOXICITY DATA with REFERENCE
ivn-mus LD50:50 mg/kg　　PGPKA8 27(10),8,82

CONSENSUS REPORTS: Reported in EPA TSCA Inventory.

SAFETY PROFILE: Poison by intravenous route. When heated to decomposition it emits toxic vapors of F^-.

PCH050　　　CAS:335-67-1　　　HR: 3
PERFLUOROHEPTANECARBOXYLIC ACID
mf: $C_8HF_{15}O_2$　　mw: 414.09

SYNS: OCTANOIC ACID, PENTADECAFLUORO- ◇ PENTADECAFLUOROOCTANOIC ACID ◇ PENTADECAFLUORO-n-OCTANOIC ACID ◇ PERFLUOROCAPRYLIC ACID ◇ PERFLUOROCTANOIC ACID ◇ PERFLUOROOCTANOIC ACID ◇ PFOA

TOXICITY DATA with REFERENCE
dnd-esc 50 µmol　　MUREAV 89,95,81
dnd-rat-ipr 100 mg/kg　　CALEDQ 57,55,91
dnd-rat-orl 168 mg/kg/2W-C　　CALEDQ 57,55,91
ipr-rat LD50:189 mg/kg　　TXAPA9 70,362,83

CONSENSUS REPORTS: Reported in EPA TSCA Inventory.

SAFETY PROFILE: Poison by intraperitoneal route. Mutation data reported. When heated to decomposition it emits toxic vapors of F^-.

PCH325　　　CAS:507-63-1　　　HR: 1
PERFLUOROOCTYL IODIDE
mf: $C_8F_{17}I$　　mw: 545.98

SYNS: HEPTADECAFLUORO-1-IODOOCTANE ◇ 1-IODOPERFLUOROOCTANE ◇ OCTANE, HEPTADECAFLUORO-1-IODO- ◇ OCTANE, 1,1,1,2,2,3,3,4,4,5,5,6,6,7,7,8,8-HEPTADECAFLUORO-8-IODO-(9CI) ◇ PERFLUORO-1-IODOOCTANE

TOXICITY DATA with REFERENCE
ivn-mus LD50:7500 mg/kg　　MIVRA6 8,320,74

CONSENSUS REPORTS: Reported in EPA TSCA Inventory.

SAFETY PROFILE: Slightly toxic by intravenous route. When heated to decomposition it emits toxic vapors of F^- and I^-.

PCM100　　　CAS:2407-94-5　　　HR: 1
1,1'-PEROXYDICYCLOHEXANOL
DOT: UN 2148
mf: $C_{12}H_{22}O_4$　　mw: 230.34

SYNS: BIS(1-HYDROXYCYCLOHEXYL)PEROXIDE ◇ CYCLOHEXANOL, 1,1'-DIOXYDI- ◇ CYCLOHEXANOL, 1,1'-DIOXYBIS-(9CI) ◇ DIHYDROXYCYCLOHEXYLPEROXIDE ◇ DI-(1-HYDROXYCYCLOHEXYL)PEROXIDE, technically pure (DOT) ◇ 1,1'-DIOXYBISCYCLOHEXANOL

DOT Classification: Organic Peroxide; Label: Organic Peroxide

CONSENSUS REPORTS: Reported in EPA TSCA Inventory.

SAFETY PROFILE: A peroxide. Handle carefully. When heated to decomposition it emits acrid smoke and irritating vapors.

PCR150　　　CAS:1607-17-6　　　HR: 3
PETRIN

TOXICITY DATA with REFERENCE
ipr-mus LD50:200 mg/kg　　85ERAY 3,1636,78

CONSENSUS REPORTS: Reported in EPA TSCA Inventory.

SAFETY PROFILE: Poison by intraperitoneal route. When heated to decomposition it emits acrid smoke and irritating vapors.

PDK800 CAS:621-33-0 **HR: D**
m-PHENETIDINE
mf: $C_8H_{11}NO$ mw: 137.20

SYNS: BENZENAMINE, 3-ETHOXY-(9CI) ◇ m-ETHOXYANILINE ◇ 3-ETHOXYANILINE ◇ 3-ETHOXYBENZENAMINE

TOXICITY DATA with REFERENCE
mma-sat 1 mg/plate   EMMUEG 11(Suppl 12),1,88

CONSENSUS REPORTS: Reported in EPA TSCA Inventory.

SAFETY PROFILE: Mutation data reported. When heated to decomposition it emits toxic vapors of NO_x.

PDK819 CAS:94-70-2 **HR: 2**
o-PHENETIDINE
mf: $C_8H_{11}NO$ mw: 137.20

SYNS: 2-AMINOPHENETOLE ◇ o-ETHOXYANILINE ◇ 2-ETHOXYANILINE ◇ 2-PHENETIDINE

TOXICITY DATA with REFERENCE
orl-mus LDLo:600 mg/kg   NYKZAU 52,215,56
orl-rbt LD50:600 mg/kg   NYKZAU 52,215,56

CONSENSUS REPORTS: Reported in EPA TSCA Inventory.

SAFETY PROFILE: Moderately toxic by ingestion. When heated to decomposition it emits toxic vapors of NO_x.

PDK890 CAS:156-43-4 **HR: 3**
4-PHENETIDINE
DOT: UN 2311
mf: $C_8H_{11}NO$ mw: 137.20

SYNS: AMINOPHENETOLE ◇ p-AMINOPHENETOLE ◇ p-ETHOXYANILINE ◇ 4-ETHOXYANILINE ◇ p-PHENETIDINE (DOT)

TOXICITY DATA with REFERENCE
skn-rbt 500 mg MLD   FCTOD7 20,563,82
eye-rbt 100 mg MOD   FCTOD7 20,563,82
eye-rbt 100 mg/45 rns MLD   FCTOD7 20,563,82
mma-sat 1 µmol/plate   CPBTAL 33,2877,85
orl-rat LD50:580 mg/kg   GISAAA 35(8),28,70
ihl-rat LCLo:250 mg/m^3 GISAAA 35(8),28,70
orl-mus LD50:530 mg/kg   GTPZAB 25(8),50,81
ipr-mus LD50:692 mg/kg   JMCMAR 17,900,74

CONSENSUS REPORTS: Reported in EPA TSCA Inventory.

DOT Classification: Poison B; Label: St. Andrews Cross

SAFETY PROFILE: Poison by inhalation. Moderately toxic by ingestion and intraperitoneal routes. Caution: It can be absorbed through the skin. A skin and eye irritant. Mutation data reported. When heated to decomposition it emits toxic fumes of NO_x.

PDO300 CAS:1464-44-4 **HR: 2**
PHENOL GLUCOSIDE
mf: $C_{12}H_{16}O_6$ mw: 256.28

SYNS: GLUCOPYRANOSIDE, PHENYL-, β-D- ◇ β-D-GLUCOPYRANOSIDE, PHENYL- ◇ PHENYL β-D-GLUCOPYRANOSIDE ◇ PHENYL β-D-GLUCOSIDE

TOXICITY DATA with REFERENCE
ipr-mus LD50:980 mg/kg   NYKZAU 49,84,53

CONSENSUS REPORTS: Reported in EPA TSCA Inventory.

SAFETY PROFILE: Moderately toxic by intraperitoneal route. When heated to decomposition it emits acrid smoke and irritating vapors.

PDO800 CAS:143-74-8 **HR: 2**
PHENOLSULFONPHTHALEIN
mf: $C_{19}H_{14}O_5S$ mw: 354.39

SYNS: FENOLIPUNA ◇ PHENOL, 4,4′-(3H-2,1-BENZOXATHIOL-3-YLIDENE)BIS-, S,S-DIOXIDE (9CI) ◇ PHENOL, 4,4′-(3H-2,1-BENZOXATHIOL-3-YLIDENE)DI-, S,S-DIOXIDE ◇ PHENOL RED ◇ PHENOLSULFONEPHTHALEIN ◇ PHENOLSULPHONPHTHALEIN ◇ PSP ◇ PSP (INDICATOR) ◇ SULFONPHTHAL ◇ SULPHENTAL ◇ SULPHONTHAL

TOXICITY DATA with REFERENCE
mmo-sat 1 mg/plate   AEMIDF 42,641,81
mma-sat 500 µg/plate   AEMIDF 42,641,81
orl-rat LD50:>600 mg/kg   CTOXAO 4(2),185,71
ivn-rat LD50:752 mg/kg   DRUGAY -,930,90
ivn-mus LD50:1368 mg/kg   DRUGAY -,930,90

CONSENSUS REPORTS: Reported in EPA TSCA Inventory.

SAFETY PROFILE: Moderately toxic by ingestion and intravenous routes. Mutation data reported. When heated to decomposition it emits toxic vapors of SO_x.

PDR490 CAS:2688-84-8 **HR: 3**
o-PHENOXYANILINE
mf: $C_{12}H_{11}NO$ mw: 185.24

SYNS: 2-ADE ◇ 2-AMINODIPHENYL ETHER ◇ o-AMINOPHENYL PHENYL ETHER ◇ 2-AMINOPHENYL PHENYL ETHER ◇ ANILINE, 2-PHENOXY- ◇ ANILINE, o-PHENOXY-(8CI) ◇ BENZENAMINE, 2-PHENOXY-(9CI) ◇ 2-PHENOXYANILINE ◇ 2-PHENOXYBENZENAMINE

TOXICITY DATA with REFERENCE
ivn-mus LD50:212 mg/kg   TXCYAC 8,347,77

CONSENSUS REPORTS: Reported in EPA TSCA Inventory.

SAFETY PROFILE: Poison by intravenous route. When heated to decomposition it emits toxic vapors of NO_x.

PDV725 CAS:940-31-8 **HR: 3**
2-PHENOXYPROPIONIC ACID
mf: $C_9H_{10}O_3$ mw: 166.19

SYNS: ACIDE PHENOXY-2-PROPIONIQUE ◇ PROPIONIC ACID, 2-PHENOXY-

TOXICITY DATA with REFERENCE
ipr-rat LD50:391 mg/kg   TSPMA6 34,121,74

CONSENSUS REPORTS: Reported in EPA TSCA Inventory.

SAFETY PROFILE: Poison by intraperitoneal route. When heated to decomposition it emits acrid smoke and irritating vapors.

PEI800   CAS:101-75-7   *HR: 3*
4-(PHENYLAZO)DIPHENYLAMINE
mf: $C_{18}H_{15}N_3$ mw: 273.36

SYNS: AZOBENZENE, 4-ANILINO- ◇ DIPHENYLAMINE, 4-(PHENYLAZO)-

TOXICITY DATA with REFERENCE
ivn-mus LD50:56 mg/kg   CSLNX* NX#03760

CONSENSUS REPORTS: Reported in EPA TSCA Inventory.

SAFETY PROFILE: Poison by intravenous route. When heated to decomposition it emits toxic vapors of NO_x.

PEJ600   CAS:131-22-6   *HR: D*
4-PHENYLAZO-1-NAPHTHYLAMINE
mf: $C_{16}H_{13}N_3$ mw: 247.32

SYNS: C.I. 11350 ◇ C.I. SOLVENT YELLOW 4 ◇ α-NAPHTHYL RED ◇ 1-NAPHTHALENAMINE, 4-(PHENYLAZO)- ◇ NUBIAN YELLOW TB ◇ PHENYLAZO α-NAPHTHYLAMINE

TOXICITY DATA with REFERENCE
mma-sat 500 μg/plate   AEMIDF 42,641,81

CONSENSUS REPORTS: Reported in EPA TSCA Inventory.

SAFETY PROFILE: Mutation data reported. When heated to decomposition it emits toxic vapors of NO_x.

PER700   CAS:525-82-6   *HR: 3*
2-PHENYLCHROMONE
mf: $C_{15}H_{10}O_2$ mw: 222.25

SYNS: ASMACORIL ◇ 4H-1-BENZOPYRAN-4-ONE, 2-PHENYL- ◇ CHROMOCOR ◇ CROMARIL ◇ FLAVONE ◇ 2-PHENYL-4H-1-BENZOPYRAN-4-ONE ◇ PHENYLCHROMONE

TOXICITY DATA with REFERENCE
dni-hmn:fbr 100 mg/L   BCPCA6 33,3823,84
orl-mus LD50:2500 mg/kg   JHTCAO 21,193,84
ipr-mus LD50:380 mg/kg   FRPSAX 13,561,58
scu-mus LD >400 mg/kg   ANTCAO 3,603,53

CONSENSUS REPORTS: Reported in EPA TSCA Inventory.

SAFETY PROFILE: Poison by intraperitoneal route. Moderately toxic by ingestion and subcutaneous routes. Mutation data reported. When heated to decomposition it emits acrid smoke and irritating vapors.

PEW725   CAS:2488-01-9   *HR: 2*
1,4-PHENYLENEBIS(DIMETHYLSILANE)
mf: $C_{10}H_{18}Si_2$ mw: 194.46

SYNS: 1,4-BIS(DIMETHYLSILYL)BENZENE ◇ p-PHENYLENE-BIS(DIMETHYLSILANE) ◇ SILANE, p-PHENYLENEBIS(DIMETHYL)- ◇ SILANE, 1,4-PHENYLENEBIS(DIMETHYL-(9CI))

TOXICITY DATA with REFERENCE
orl-mus LD50:1535 mg/kg   85JCAE -,1235,86
skn-mus LD50:3 g/kg   85JCAE -,1235,86

CONSENSUS REPORTS: Reported in EPA TSCA Inventory.

SAFETY PROFILE: Moderately toxic by ingestion. Slightly toxic by skin contact. When heated to decomposition it emits acrid smoke and irritating vapors.

PEY800   CAS:88-45-9   *HR: D*
p-PHENYLENEDIAMINESULFONIC ACID
mf: $C_6H_8N_2O_3S$ mw: 188.22

SYNS: BENZENESULFONIC ACID, 2,5-DIAMINO- ◇ 2,5-DIAMINOBENZENE SULFONIC ACID

TOXICITY DATA with REFERENCE
dns-rat:lvr 100 μmol/L   MUREAV 206,183,88

CONSENSUS REPORTS: Reported in EPA TSCA Inventory.

SAFETY PROFILE: Mutation data reported. When heated to decomposition it emits toxic vapors of NO_x and SO_x.

PFD400   CAS:1988-89-2   *HR: 3*
4-(1-PHENYLETHYL)PHENOL
mf: $C_{14}H_{14}O$ mw: 198.28

SYNS: 4-(1-FENYLETHYL)FENOL ◇ p-(α-METHYLBENZYL)PHENOL ◇ 4-(α-METHYLBENZYL)PHENOL ◇ PHENOL, p-(α-METHYLBENZYL)- ◇ PHENOL, 4-(1-PHENYLETHYL)- ◇ 4-(PHENYLETHYLIDENE)PHENOL ◇ STYROLFENOL

TOXICITY DATA with REFERENCE
orl-frg LD50:330 mg/kg   GTPZAB 12(12),44,68

CONSENSUS REPORTS: Reported in EPA TSCA Inventory.

SAFETY PROFILE: Poison by ingestion. When heated to decomposition it emits acrid smoke and irritating vapors.

PFJ300   CAS:936-49-2   *HR: 3*
2-PHENYL-2-IMIDAZOLINE
mf: $C_9H_{10}N_2$ mw: 146.21

SYNS: 1H-IMIDAZOLE, 4,5-DIHYDRO-2-PHENYL- ◇ 2-IMIDAZOLINE, 2-PHENYL-

TOXICITY DATA with REFERENCE
ivn-uns LDLo:67 mg/kg   AEPPAE 192,141,39

CONSENSUS REPORTS: Reported in EPA TSCA Inventory.

SAFETY PROFILE: Poison by intravenous route. When heated to decomposition it emits toxic vapors of NO_x.

PFP100   CAS:104-60-9   HR: 3
PHENYLMERCURY OLEATE
mf: $C_{24}H_{38}HgO_2$ mw: 559.21

SYNS: MERCURY, (9-OCTADECENOATO-O)PHENYL-, (Z)-(9CI) ◇ MERCURY, (OLEATO)PHENYL- ◇ PHENYLMERCURIC OLEATE ◇ PMO 10

TOXICITY DATA with REFERENCE
orl-rat LD50:48400 µg/kg   ESKHA5 (101),152,83
NIOSH REL: (Mercury, aryl, and inorganic): CL 0.1 mg/m³ (skin)

CONSENSUS REPORTS: Reported in EPA TSCA Inventory.

SAFETY PROFILE: Poison by ingestion. When heated to decomposition it emits very toxic vapors of Hg.

PFR325   CAS:101-71-3   HR: 2
4-(PHENYLMETHYL)PHENOL CARBAMATE
mf: $C_{14}H_{13}NO_2$ mw: 227.28

SYNS: p-BENZYLPHENYL CARBAMATE ◇ BUTOLAN ◇ BUTOLEN ◇ CARBAURINE ◇ CARPHENOL ◇ p-CRESOL, α-PHENYL-, CARBAMATE ◇ DIPHENAN ◇ DIPHENAN (pharmaceutical) ◇ DIPHENANE ◇ OXYBULAN ◇ OXYLAN ◇ PALAFUGE ◇ PARABENCIL ◇ PARABENCILFENOL ◇ PHENOL, 4-(PHENYLMETHYL)-, CARBAMATE

TOXICITY DATA with REFERENCE
orl-mus LD50:2350 mg/kg   ANTCAO 4,917,54

SAFETY PROFILE: Moderately toxic by ingestion. When heated to decomposition it emits toxic vapors of NO_x.

PGA600   CAS:698-87-3   HR: 2
1-PHENYL-2-PROPANOL
mf: $C_9H_{12}O$ mw: 136.21

SYNS: BENZENEETHANOL, α-METHYL-(9CI) ◇ BENZYL METHYL CARBINOL ◇ α-METHYLBENZENEETHANOL ◇ α-METHYL-PHENETHYL ALCOHOL ◇ PHENETHYL ALCOHOL, α-METHYL-

TOXICITY DATA with REFERENCE
ipr-mus LD50:520 mg/kg   JPMSAE 60,799,71

CONSENSUS REPORTS: Reported in EPA TSCA Inventory.

SAFETY PROFILE: Moderately toxic by intraperitoneal route. When heated to decomposition it emits acrid smoke and irritating vapors.

PGE760   CAS:91-02-1   HR: 2
PHENYL 2-PYRIDYL KETONE
mf: $C_{12}H_9NO$ mw: 183.22

SYNS: 2-BENZOYLPYRIDINE ◇ KETONE, PHENYL 2-PYRIDYL ◇ PYRIDINE, 2-BENZOYL-

TOXICITY DATA with REFERENCE
ipr-mus LD50:475 mg/kg   JMCMAR 14,551,71

CONSENSUS REPORTS: Reported in EPA TSCA Inventory.

SAFETY PROFILE: Moderately toxic by intraperitoneal route. When heated to decomposition it emits toxic vapors of NO_x.

PGL270   CAS:1527-12-4   HR: 2
2-(PHENYLTHIO)BENZOIC ACID
mf: $C_{13}H_{10}O_2S$ mw: 230.29

SYNS: BENZOIC ACID, o-(PHENYLTHIO)- ◇ BENZOIC ACID, 2-(PHENYLTHIO)-(9CI)

TOXICITY DATA with REFERENCE
ipr-mus LDLo:1000 mg/kg   JMCMAR 19,798,76

CONSENSUS REPORTS: Reported in EPA TSCA Inventory.

SAFETY PROFILE: Moderately toxic by intraperitoneal route. When heated to decomposition it emits toxic vapors of SO_x.

PHC800   CAS:813-78-5   HR: 1
PHOSPHORIC ACID, DIMETHYL ESTER
mf: $C_2H_7O_4P$ mw: 126.06

SYNS: DIMETHYL HYDROGEN PHOSPHATE ◇ O,O-DIMETHYL HYDROGEN PHOSPHATE ◇ DIMETHYL PHOSPHATE

TOXICITY DATA with REFERENCE
unr-rat LD50:8714 mg/kg   GISAAA 50(3),57,85

CONSENSUS REPORTS: Reported in EPA TSCA Inventory.

SAFETY PROFILE: Slightly toxic by an unspecified route. When heated to decomposition it emits toxic vapors of PO_x.

PHE300   CAS:838-85-7   HR: D
PHOSPHORIC ACID, DIPHENYL ESTER
mf: $C_{12}H_{11}O_4P$ mw: 250.20

SYNS: DIPHENYL HYDROGEN PHOSPHATE ◇ DIPHENYL PHOSPHATE ◇ PHENYL PHOSPHATE (($PhO)_2$(HO)PO)

TOXICITY DATA with REFERENCE
pic-esc 25 µg/well   MUREAV 260,349,91

SAFETY PROFILE: Mutation data reported. When heated to decomposition it emits toxic vapors of PO_x.

PHW550 CAS:2432-90-8 HR: 2
PHTHALIC ACID, DIDODECYL ESTER
mf: $C_{32}H_{54}O_4$ mw: 502.86

SYNS: 1,2-BENZENEDICARBOXYLIC ACID, DIDODECYL ESTER (9CI) ◇ DIDODECYL PHTHALATE ◇ DI-n-DODECYL PHTHALATE ◇ DILAURYL PHTHALATE

TOXICITY DATA with REFERENCE
orl-rat LDLo:1500 mg/kg   85GMAT -,49,82
orl-mus LDLo:1500 mg/kg   85GMAT -,49,82

CONSENSUS REPORTS: Reported in EPA TSCA Inventory.

SAFETY PROFILE: Moderately toxic by ingestion. When heated to decomposition it emits acrid smoke and irritating vapors.

PHW600 CAS:605-45-8 HR: 2
PHTHALIC ACID, DIISOPROPYL ESTER
mf: $C_{14}H_{18}O_4$ mw: 250.32

SYNS: 1,2-BENZENEDICARBOXYLIC ACID, BIS(1-METHYLETHYL) ESTER ◇ DIISOPROPYL PHTHALATE

TOXICITY DATA with REFERENCE
ipr-mus LDLo:1251 mg/kg   JPMSAE 56,1446,67

CONSENSUS REPORTS: Reported in EPA TSCA Inventory.

SAFETY PROFILE: Moderately toxic by intraperitoneal route. When heated to decomposition it emits acrid smoke and irritating vapors.

PIE530 CAS:88-88-0 HR: D
PICRYL CHLORIDE
mf: $C_6H_2ClN_3O_6$ mw: 247.56

SYNS: BENZENE, 2-CHLORO-1,3,5-TRINITRO- ◇ 2-CHLORO-1,3,5-TRINITROBENZENE ◇ TNCB ◇ 2,4,6-TRINITROCHLOROBENZENE

TOXICITY DATA with REFERENCE
mmo-sat 1600 ng/plate   EMMUEG 11(Suppl 12),1,88
dnd-rat:lvr 5 μmol/L   MUREAV 131,215,84
dnd-mus-ipr 15 mg/kg   MUREAV 116,239,83

CONSENSUS REPORTS: Reported in EPA TSCA Inventory.

SAFETY PROFILE: Mutation data reported. When heated to decomposition it emits toxic vapors of NO_x and Cl^-.

PIH245 CAS:1330-16-1 HR: 3
PINENE (DOT)
DOT: UN 2368
mf: $C_{10}H_{16}$ mw: 136.26

SYN: BICYCLO(3.1.1)HEPTANE, 2,6,6-TRIMETHYL-, DIDEHYDRO deriv.

DOT Classification: Flammable Liquid; Label: Flammable Liquid

CONSENSUS REPORTS: Reported in EPA TSCA Inventory.

SAFETY PROFILE: Flammable liquid. When heated to decomposition it emits acrid smoke and irritating vapors.

PIL525 CAS:2158-03-4 HR: 2
1-PIPERIDINECARBOXAMIDE
mf: $C_6H_{12}N_2O$ mw: 128.20

SYNS: N,N-PENTAMETHYLENEUREA ◇ PIPERIDINE-N-CARBONIC ACID AMIDE

TOXICITY DATA with REFERENCE
ipr-mus LDLo:2050 mg/kg   JPETAB 61,175,37

CONSENSUS REPORTS: Reported in EPA TSCA Inventory.

SAFETY PROFILE: Moderately toxic by intraperitoneal route. When heated to decomposition it emits toxic vapors of NO_x.

PIY500 CAS:98-77-1 HR: 3
PIP-PIP
mf: $C_{11}H_{22}N_2S_2$ mw: 246.47

SYNS: PENTAMETHYLENEDITHIOCARBAMATE ◇ 1-PIPERIDINECARBODITHIOIC ACID, compounded with PIPERIDINE ◇ PIPERIDINIUM ◇ "522" RUBBER ACCELERATOR

TOXICITY DATA with REFERENCE
skn-hmn 500 mg/48H MLD   AMIHBC 5,311,52
ipr-uns LD50:250 mg/kg   AMIHBC 5,311,52

CONSENSUS REPORTS: Reported in EPA TSCA Inventory.

SAFETY PROFILE: Poison by intraperitoneal route. A human skin irritant. An allergen. When heated to decomposition it emits very toxic fumes of NO_x and SO_x. See also CARBAMATES.

PJQ100 CAS:1328-53-6 HR: D
POLYCHLORO COPPER PHTHALOCYANINE

SYNS: ACCOSPERSE CYAN GREEN G ◇ BRILLIANT GREEN PHTHALOCYANINE ◇ CALCOTONE GREEN G ◇ CERES GREEN 3B ◇ CHROMATEX GREEN G ◇ COLANYL GREEN GG ◇ C.I. 74260 ◇ C.I. PIGMENT GREEN 7 ◇ C.I. PIGMENT GREEN 42 ◇ COPPER PHTHALOCYANINE GREEN ◇ CROMOPHTHAL GREEN GF ◇ CYAN GREEN 15-3100 ◇ CYANINE GREEN GP ◇ CYANINE GREEN NB ◇ CYANINE GREEN T ◇ CYANINE GREEN TONER ◇ DAINICHI CYANINE GREEN FG ◇ DAINICHI CYANINE GREEN FGH ◇ DALTOLITE FAST GREEN GN ◇ DURATINT GREEN 1001 ◇ FASTOGEN GREEN 5005 ◇ FASTOGEN GREEN B ◇ FASTOLUX GREEN ◇ FENALAC GREEN G ◇ FENALAC GREEN G DISP ◇ GRANADA GREEN LAKE GL ◇ GRAPHTOL GREEN 2GLS ◇ HELIOGEN GREEN 8680 ◇ HELIOGEN GREEN 8730 ◇ HELIOGEN GREEN A ◇ HELIOGEN GREEN G ◇ HELIOGEN GREEN GA ◇ HELIOGEN GREEN GN ◇ HELIOGEN GREEN GNA ◇ HELIOGEN GREEN GTA ◇ HELIOGEN GREEN GV ◇ HELIOGEN GREEN GWS ◇ HELIOGEN GREEN 8681K ◇ HELIOGEN GREEN 8682T ◇ HOSTAPERM GREEN GG ◇ IRGALITE FAST BRILLIANT GREEN GL ◇ IRGALITE FAST BRILLIANT GREEN 3GL ◇ IRGALITE GREEN GLN ◇ KLON-

DIKE YELLOW X-2261 ◇ LUTETIA FAST EMERALD J ◇ MICROLITH GREEN G-FP ◇ MONARCH GREEN WD ◇ MONASTRAL FAST GREEN BGNA ◇ MONASTRAL FAST GREEN G ◇ MONASTRAL FAST GREEN GD ◇ MONASTRAL FAST GREEN GF ◇ MONASTRAL FAST GREEN GFNP ◇ MONASTRAL FAST GREEN GN ◇ MONASTRAL FAST GREEN GNA ◇ MONASTRAL FAST GREEN GTP ◇ MONASTRAL FAST GREEN GV ◇ MONASTRAL FAST GREEN GWD ◇ MONASTRAL FAST GREEN 2GWD ◇ MONASTRAL FAST GREEN GX ◇ MONASTRAL FAST GREEN GXB ◇ MONASTRAL FAST GREEN GYH ◇ MONASTRAL FAST GREEN LGNA ◇ MONASTRAL GREEN B ◇ MONASTRAL GREEN B PIGMENT ◇ MONASTRAL GREEN G ◇ MONASTRAL GREEN GFN ◇ MONASTRAL GREEN GH ◇ MONASTRAL GREEN GN ◇ MONOLITE FAST GREEN GVSA ◇ NCI-C54637 ◇ NON-FLOCCULATING GREEN G 25 ◇ OPALINE GREEN G 1 ◇ PERMANENT GREEN TONER GT-376 ◇ PHTHALOCYANINE BRILLIANT GREEN ◇ PHTHALOCYANINE GREEN ◇ PHTHALOCYANINE GREEN LX ◇ PHTHALOCYANINE GREEN V ◇ PHTHALOCYANINE GREEN VFT 1080 ◇ PHTHALOCYANINE GREEN WDG 47 ◇ PIGMENT FAST GREEN G ◇ PIGMENT FAST GREEN GN ◇ PIGMENT GREEN 7 ◇ PIGMENT GREEN PHTHALOCYANINE ◇ PIGMENT GREEN PHTHALOCYANINE V ◇ POLYMO GREEN FBH ◇ POLYMO GREEN FGH ◇ POLYMON GREEN G ◇ POLYMON GREEN 6G ◇ POLYMON GREEN GN ◇ PV-FAST GREEN G ◇ RAMAPO ◇ SANYO CYANINE GREEN ◇ SANYO PHTHALOCYANINE GREEN FB PURE ◇ SANYO PHTHALOCYANINE GREEN F6G ◇ SEGNALE LIGHT GREEN G ◇ SHERWOOD GREEN A 4436 ◇ SIEGLE FAST GREEN G ◇ SOLFAST GREEN ◇ SOLFAST GREEN 63102 ◇ SYNTHALINE GREEN ◇ TERMOSOLIDO GREEN FG SUPRA ◇ THALO GREEN No.1 ◇ VERSAL GREEN G ◇ VULCAL FAST GREEN F2G ◇ VULCANOSINE FAST GREEN G ◇ VULCOL FAST GREEN F2G ◇ VYNAMON GREEN BE ◇ VYNAMON GREEN BES ◇ VYNAMON GREEN GNA

TOXICITY DATA with REFERENCE
mma-sat 3333 μg/plate   EMMUEG 11(Suppl 12),1,88

CONSENSUS REPORTS: Reported in EPA TSCA Inventory.

SAFETY PROFILE: Mutation data reported. When heated to decomposition it emits acrid smoke and irritating vapors.

PKV500   CAS:10124-50-2   **HR: 3**
POTASSIUM ARSENITE
DOT: UN 1678
mf: $AsH_3O_3 \cdot xK$ mw: 399.65

PROP: White, hygroscopic powder. Sol in water.

SYNS: ARSENENOUS ACID, POTASSIUM SALT ◇ ARSENITE de POTASSIUM (FRENCH) ◇ ARSONIC ACID, POTASSIUM SALT ◇ KALIUMARSENIT (GERMAN) ◇ NSC 3060 ◇ POTASSIUM METAARSENITE

TOXICITY DATA with REFERENCE
cyt-hmn:leu 1 μmol/L/48H   CNREA8 25,980,65
orl-man TDLo:214 mg/kg/15Y-C:CAR,LIV   GASTAB 68,1582,75
skn-mus TDLo:576 mg/kg/12W-I:ETA   BMJOAE 2,1107,22
orl-man TD:7560 mg/kg/26W-C:CAR,SKN   ANSUA5 99,348,34
orl-man TD:441 mg/kg/3W-C:CAR,SKN   ANSUA5 99,348,34
orl-chd TD:390 mg/kg/3Y-C:CAR,SKN   AIMEAS 61,296,64
orl-hmn TDLo:74 mg/kg:LIV,SKN   LANCAO 1,269,53
orl-rat LD50:14 mg/kg   AFDOAQ 15,122,51
skn-rat LD50:150 mg/kg   PHJOAV 185,361,60
scu-mus LDLo:16 mg/kg   HBAMAK 4,1307,35
orl-dog LDLo:3 mg/kg   HBAMAK 4,1307,35
scu-dog LDLo:700 μg/kg   HBAMAK 4,1307,35
ivn-dog LDLo:2 mg/kg   HBAMAK 4,1307,35
scu-cat LDLo:5 mg/kg   HBAMAK 4,1307,35
scu-rbt LDLo:8 mg/kg   HBAMAK 4,1307,35
ivn-rbt LDLo:6 mg/kg   HBAMAK 4,1307,35
scu-gpg LDLo:9 mg/kg   HBAMAK 4,1307,35
scu-pgn LDLo:12 mg/kg   FDWU** -,-,31

CONSENSUS REPORTS: IARC Cancer Review: Human Sufficient Evidence IMEMDT 23,39,80; Animal Inadequate Evidence IMEMDT 23,39,80, IMEMDT 2,48,73. EPA Extremely Hazardous Substances List. Arsenic and its compounds are on the Community Right-To-Know List.

OSHA PEL: TWA 0.01 mg(As)/m^3; Cancer Hazard
ACGIH TLV: TWA 0.2 mg(As)/m^3 (Proposed: 0.01 mg(As)/m^3; Human Carcinogen)
NIOSH REL: CL (Inorganic Arsenic) 0.002 mg(As)/m^3/15M

DOT Classification: Poison B; Label: Poison

SAFETY PROFILE: Confirmed human carcinogen producing skin and liver tumors. Poison by ingestion, skin contact, subcutaneous, and intravenous routes. Human mutation data reported. Human systemic effects: dermatitis, liver changes. When heated to decomposition it emits toxic fumes of As and K_2O. Used in veterinary medicine and for chronic dermatitis in humans. See also ARSENIC COMPOUNDS.

PLG810   CAS:1319-69-3   **HR: 3**
POTASSIUM GLYCEROPHOSPHATE
mf: $C_3H_7O_6P \cdot 2K$ mw: 248.27

PROP: Pale yellow syrupy liquid. Sol in water.

TOXICITY DATA with REFERENCE
ipr-rat LD50:935 mg/kg   KSRNAM 21,3321,87
ivn-rat LD50:286 mg/kg   KSRNAM 21,3321,87
ipr-mus LD50:1044 mg/kg   KSRNAM 21,3321,87
ivn-mus LD50:407 mg/kg   KSRNAM 21,3321,87

CONSENSUS REPORTS: Reported in EPA TSCA Inventory.

SAFETY PROFILE: Poison by intravenous route. Moderately toxic by intraperitoneal route. When heated to decomposition it emits acrid smoke and irritating fumes.

PLK650   CAS:996-31-6   **HR: 1**
POTASSIUM LACTATE
mf: $C_3H_5O_3K$ mw: 128.17

SYNS: MONOPOTASSIUM 2-HYDROXYPROPANOATE ACID ◇ POTASSIUM α-HYDROXYPROPIONATE ◇ PROPANOIC ACID, 2-HYDROXY-, MONOPOTASSIUM SALT (9CI)

TOXICITY DATA with REFERENCE
eye-rbt 100 mg MLD   FCTOD7 20,573,82

PLL500 POTASSIUM NITRATE

CONSENSUS REPORTS: Reported in EPA TSCA Inventory.

SAFETY PROFILE: An eye irritant. When heated to decomposition it emits acrid smoke and irritating fumes.

PLL500 CAS:7757-79-1 **HR: 3**
POTASSIUM NITRATE
DOT: UN 1486
mf: KNO_3 mw: 140.21

PROP: Transparent, colorless or white crystalline powder or crystals; odorless with a cooling, pungent, salty taste. Mp: 334°, bp: decomp @ 400°, d: 2.109 @ 16°. Sol in glycerin, water; moderately sol in alc.

SYNS: KALIUMNITRAT (GERMAN) ◊ NITER ◊ NITRE ◊ NITRIC ACID, POTASSIUM SALT ◊ SALTPETER ◊ VICKNITE

TOXICITY DATA with REFERENCE
mrc-esc 5 pph   JGMIAN 8,45,53
orl-rat TDLo:598 mg/kg (female 1–22D post):REP
   JANSAG 74,69,89
orl-gpg TDLo:15 g/kg (female 24W pre):REP   TXAPA9 12,179,68
orl-gpg TDLo:1670 g/kg (female 29W pre):REP   TXAPA9 12,179,68
orl-rbt TDLo:6505 mg/kg (female 23–27D post):REP
   SOVEA7 27,246,74
orl-gpg TDLo:1670 g/kg (female 29W pre):REP   TXAPA9 12,179,68
orl-rat TDLo:22 g/kg (female 1–22D post):REP   JANSAG 15,1291,56
orl-rat LD50:3750 mg/kg   NYKZAU 81,469,83
ivn-cat LDLo:100 mg/kg   HBAMAK 4,1289,35
orl-rbt LD50:1901 mg/kg   SOVEA7 27,246,74

CONSENSUS REPORTS: Reported in EPA TSCA Inventory.

DOT Classification: Oxidizer; Label: Oxidizer

SAFETY PROFILE: Poison by intravenous route. Moderately toxic by ingestion. An experimental teratogen. Experimental reproductive effects. Mutation data reported. Ingestion of large quantities may cause gastroenteritis. Chronic exposure can cause anemia, nephritis, and methemoglobinemia. When heated, reaction with calcium hydroxide + polychlorinated phenols forms extremely toxic chlorinated benzodioxins.

A powerful oxidizer. Gunpowder is a mixture of potassium nitrate + sulfur + charcoal. Explosive reaction with aluminum + barium nitrate + potassium perchlorate + water (in storage), boron + laminac + trichloroethylene. Forms explosive mixtures with lactose, powdered metals (e.g., titanium, antimony, germanium), metal sulfides (e.g., antimony trisulfide, barium sulfide, calcium sulfide, germanium monosulfide, titanium disulfide, arsenic disulfide, molybdenum disulfide), nonmetals (e.g., boron, carbon, white phosphorus, arsenic), organic materials, phosphides (e.g., copper(II) phosphide, copper monophosphide), reducing agents (e.g., sodium phosphinate, sodium thiosulfate), sodium acetate. Can react violently under the appropriate conditions with 1,3-bis(trichloromethyl)benzene, boron phosphide, F_2, calcium silicide, charcoal, chromium nitride, Na hypophosphite, (Na_2O_2 + dextrose), red phosphorus, (S + As_2S_3), thorium dicarbide, trichloroethylene, zinc, zirconium. When heated to decomposition it emits very toxic fumes of NO_x and K_2O. See also NITRATES.

PLN300   CAS:583-52-8   **HR: 2**
POTASSIUM OXALATE
mf: $C_2H_2O_4$•2K   mw: 168.24

SYNS: DIPOTASSIUM OXALATE ◊ ETHANEDIOIC ACID, DIPOTASSIUM SALT (9CI) ◊ OXALIC ACID, DIPOTASSIUM SALT ◊ POTASSIUM NEUTRAL OXALATE

TOXICITY DATA with REFERENCE
orl-wmn LDLo:1 g/kg   MLDCAS 4,178,71

CONSENSUS REPORTS: Reported in EPA TSCA Inventory.

SAFETY PROFILE: Human systemic effects by ingestion: cardiac arrhythmias, shock, gastrointestinal changes. When heated to decomposition it emits acrid smoke and irritating vapors.

PLP250   CAS:17014-71-0   **HR: 3**
POTASSIUM PEROXIDE
DOT: UN 1491
mf: K_2O_2 mw: 110.2

PROP: Yellow, amorph mass or white crystals. Mp: 490°.

CONSENSUS REPORTS: Reported in EPA TSCA Inventory.

DOT Classification: Oxidizer; Label: Oxidizer

SAFETY PROFILE: Dangerous fire hazard by spontaneous chemical reaction. It is a very powerful oxidizer. Fires of this material should be handled like sodium peroxide fires. Moderate explosion hazard by spontaneous chemical reaction. Explodes on contact with water. Violent reactions with air; Sb; As; O_2; K. Vigorous reaction on contact with reducing materials. On contact with acid or acid fumes, it can emit toxic fumes. Incompatible with carbon; diselenium dichloride; ethanol; hydrocarbons; metals. When heated to decomposition it emits toxic fumes of K_2O. See also PEROXIDES, INORGANIC.

PLY275   CAS:125-02-0   **HR: 3**
PREDNISOLONE 21-PHOSPHATE DISODIUM
mf: $C_{21}H_{27}O_8P$•2Na   mw: 484.43

SYNS: CODELSOL ◊ DISODIUM PREDNISOLONE 21-PHOSPHATE

◇ HYDELTRASOL ◇ INFLAMASE ◇ OPTIVAL ◇ PHORTISOLONE ◇ PREDNESOL ◇ PREDNISOLONE DISODIUM PHOSPHATE ◇ PREDNISOLONE SODIUM PHOSPHATE ◇ PREDSOL ◇ PREDSOLAN ◇ PREGNA-1,4-DIENE-3,20-DIONE, 11-β,17,21-TRIHYDROXY-, 21-(DIHYDROGEN PHOSPHATE), DISODIUM SALT ◇ PROZORIN ◇ SODIUM PREDNISOLONE PHOSPHATE ◇ SOLUCORT

TOXICITY DATA with REFERENCE
ivn-rbt LD50:360 mg/kg   CPBTAL 22,1439,74

CONSENSUS REPORTS: Reported in EPA TSCA Inventory.

SAFETY PROFILE: Poison by intravenous route. When heated to decomposition it emits toxic vapors of PO_x.

PML300   CAS:814-67-5   ***HR: 3***
1,2-PROPANEDITHIOL
mf: $C_3H_8S_2$ mw: 108.23

SYNS: 1,2-DIMERCAPTOPROPANE ◇ 2,3-DIMERCAPTOPROPANE ◇ 1,2-DITHIOLPROPANE

TOXICITY DATA with REFERENCE
orl-mus LD50:153 mg/kg   DCTODJ 3,249,80

CONSENSUS REPORTS: Reported in EPA TSCA Inventory.

SAFETY PROFILE: Poison by ingestion. When heated to decomposition it emits toxic vapors of SO_x.

PNF100   CAS:105-66-8   ***HR: 1***
PROPYL BUTANOATE
mf: $C_7H_{14}O_2$ mw: 130.21

SYNS: BUTANOIC ACID, PROPYL ESTER (9CI) ◇ BUTYRIC ACID, PROPYL ESTER ◇ PROPYL BUTYRATE ◇ PROPYLESTER KYSELINY MASELNE

TOXICITY DATA with REFERENCE
orl-rat LD50:15 g/kg   FCTXAV 2,327,64
orl-rbt LD50:5729 mg/kg   IMSUAI 41,31,72

CONSENSUS REPORTS: Reported in EPA TSCA Inventory.

SAFETY PROFILE: Slightly toxic by ingestion. When heated to decomposition it emits acrid smoke and irritating vapors.

PNL100   CAS:1331-12-0   ***HR: 1***
PROPYLENE GLYCOL MONOACETATE
mf: $C_5H_{10}O_3$ mw: 118.15

TOXICITY DATA with REFERENCE
orl-rat LD50:18 g/kg   JIDHAN 23,259,41

CONSENSUS REPORTS: Reported in EPA TSCA Inventory.

SAFETY PROFILE: Slightly toxic by ingestion. When heated to decomposition it emits acrid smoke and irritating vapors.

PNL265   CAS:108-65-6   ***HR: 2***
PROPYLENE GLYCOL MONOMETHYL ETHER ACETATE
mf: $C_6H_{12}O_3$ mw: 132.18

SYNS: ACETIC ACID, 2-METHOXY-1-METHYLETHYL ESTER ◇ DOWANOL (R) PMA GLYCOL ETHER ACETATE ◇ 1-METHOXY-2-ACETOXYPROPANE

TOXICITY DATA with REFERENCE
orl-rat LD50:8532 mg/kg   DOWCC* MSD-1582
ipr-mus LD50:750 mg/kg   NTIS** AD691-490
skn-rbt LD50:>5 g/kg   DOWCC* MSD-1582

CONSENSUS REPORTS: Reported in EPA TSCA Inventory.

SAFETY PROFILE: Moderately toxic by intraperitoneal route. Slightly toxic by ingestion and skin contact. When heated to decomposition it emits acrid smoke and irritating vapors.

PNX550   CAS:627-06-5   ***HR: 1***
1-PROPYLUREA
mf: $C_4H_{10}N_2O$ mw: 102.16

SYNS: PROPYLUREA ◇ N-PROPYLUREA ◇ UREA, PROPYL-

TOXICITY DATA with REFERENCE
par-mus LDLo:4086 mg/kg   JPETAB 52,216,34

CONSENSUS REPORTS: Reported in EPA TSCA Inventory.

SAFETY PROFILE: Slightly toxic by parenteral route. When heated to decomposition it emits toxic vapors of NO_x.

POE200   CAS:99-50-3   ***HR: 2***
PROTOCATECHUIC ACID
mf: $C_7H_6O_4$ mw: 154.13

SYNS: BENZOIC ACID, 3,4-DIHYDROXY- ◇ 3,4-DIHYDROXYBENZOIC ACID

TOXICITY DATA with REFERENCE
dni-hmn:lyms 1 mmol/L   BCPCA6 29,1275,80
msc-mus:lyms 300 mg/L   EMMUEG 11,523,88
cyt-ham:ovr 3 g/L   CALEDQ 14,251,81
ipr-mus LD50:>800 mg/kg   JPETAB 196,478,76

SAFETY PROFILE: Moderately toxic by intraperitoneal route. Human mutation data reported. When heated to decomposition it emits acrid smoke and irritating vapors.

POH525   CAS:141-10-6   ***HR: 1***
PSEUDOIONONE
mf: $C_{13}H_{20}O$ mw: 192.33

SYNS: CITRYLIDENEACETONE ◇ 2,6-DIMETHYLUNDECA-2,6,8-TRIENE-10-ONE ◇ 6,10-DIMETHYL-3,5,9-UNDECATRIEN-2-ONE ◇ 3,5,9-UNDECATRIEN-2-ONE, 6,10-DIMETHYL-

POI615   CAS:89-82-7   HR: 3
d-PULEGONE
mf: $C_{10}H_{16}O$ mw: 152.26

SYNS: CYCLOHEXANONE, 5-METHYL-2-(1-METHYLETHYLIDENE)-, (R)- ◊ p-MENTH-4(8)-EN-3-ONE, (R)-(+)- ◊ PULEGON ◊ PULEGONE ◊ (+)-PULEGONE ◊ (+)-(R)-PULEGONE ◊ (R)-(+)-PULEGONE

TOXICITY DATA with REFERENCE
orl-rat LD50:>5 g/kg   FCTOD7 26,311,88
skn-rbt LDLo:5 g/kg   FCTOD7 26,311,88

CONSENSUS REPORTS: Reported in EPA TSCA Inventory.

SAFETY PROFILE: Slightly toxic by ingestion and skin contact routes. When heated to decomposition it emits acrid smoke and irritating vapors.

POI615   CAS:89-82-7   HR: 3
d-PULEGONE
mf: $C_{10}H_{16}O$ mw: 152.26

SYNS: CYCLOHEXANONE, 5-METHYL-2-(1-METHYLETHYLIDENE)-, (R)- ◊ p-MENTH-4(8)-EN-3-ONE, (R)-(+)- ◊ PULEGON ◊ PULEGONE ◊ (+)-PULEGONE ◊ (+)-(R)-PULEGONE ◊ (R)-(+)-PULEGONE

TOXICITY DATA with REFERENCE
ivn-dog LDLo:330 mg/kg   COREAF 236,633,53

CONSENSUS REPORTS: Reported in EPA TSCA Inventory.

SAFETY PROFILE: Poison by intravenous route. When heated to decomposition it emits acrid smoke and irritating vapors.

POL590   CAS:290-37-9   HR: 2
PYRAZINE
mf: $C_4H_4N_2$ mw: 80.10

SYNS: 1,4-DIAZABENZENE ◊ p-DIAZINE ◊ 1,4-DIAZINE ◊ PARADIAZINE ◊ PIAZINE

TOXICITY DATA with REFERENCE
mmo-smc 60 g/L   FCTXAV 18,581,80
cyt-ham:ovr 2500 mg/L   FCTXAV 18,581,80
ipr-mus LD50:2730 mg/kg   PBPHAW 1,542,65

CONSENSUS REPORTS: Reported in EPA TSCA Inventory.

SAFETY PROFILE: Moderately toxic by intraperitoneal route. Mutation data reported. When heated to decomposition it emits toxic vapors of NO_x.

PON340   CAS:1785-51-9   HR: D
1,6-PYRENEDIONE
mf: $C_{16}H_8O_2$ mw: 232.24

SYNS: 3,8-PYRENEDIONE ◊ 1,6-PYRENEQUINONE ◊ 3,8-PYRENEQUINONE

TOXICITY DATA with REFERENCE
mmo-sat 2 µg/plate   MUREAV 156,61,85
mma-sat 2 µg/plate   MUREAV 156,61,85

CONSENSUS REPORTS: Reported in EPA TSCA Inventory.

SAFETY PROFILE: Mutation data reported. When heated to decomposition it emits acrid smoke and irritating vapors.

PON350   CAS:2304-85-0   HR: D
1,8-PYRENEDIONE
mf: $C_{16}H_8O_2$ mw: 232.24

SYNS: 3,10-PYRENEDIONE ◊ 1,8-PYRENEQUINONE ◊ 3,10-PYRENEQUINONE

TOXICITY DATA with REFERENCE
mmo-sat 2 µg/plate   MUREAV 156,61,85
mma-sat 2 µg/plate   MUREAV 156,61,85

CONSENSUS REPORTS: Reported in EPA TSCA Inventory.

SAFETY PROFILE: Mutation data reported. When heated to decomposition it emits acrid smoke and irritating vapors.

POR790   CAS:2044-73-7   HR: 2
2-PYRIDINEMETHANETHIOL
mf: C_6H_7NS mw: 125.20

SYNS: 2-(MERCAPTOMETHYL)PYRIDINE ◊ PYRIDINE-2-THIOCARBINOL ◊ 2-PYRIDYLMETHYL MERCAPTAN

TOXICITY DATA with REFERENCE
orl-mus LD50:600 mg/kg   EJMCA5 18,515,83
ipr-mus LD50:500 mg/kg   EJMCA5 18,515,83

CONSENSUS REPORTS: Reported in EPA TSCA Inventory.

SAFETY PROFILE: Moderately toxic by ingestion and intraperitoneal routes. When heated to decomposition it emits toxic vapors of SO_x and NO_x.

POR810   CAS:586-95-8   HR: 2
4-PYRIDINEMETHANOL
mf: C_6H_7NO mw: 109.14

SYNS: 4-(HYDROXYMETHYL)PYRIDINE ◊ γ-PICOLYL ALCOHOL ◊ 4-PYRIDYLCARBINOL ◊ 4-PYRIDYLMETHANOL

TOXICITY DATA with REFERENCE
ivn-mus LD50:1000 mg/kg   PHARAT 11,242,56
orl-brd LD50:422 mg/kg   AECTCV 12,355,83

CONSENSUS REPORTS: Reported in EPA TSCA Inventory.

SAFETY PROFILE: Moderately toxic by ingestion and intravenous routes. When heated to decomposition it emits toxic vapors of NO_x.

PPH025   CAS:109-00-2   HR: 2
3-PYRIDINOL
mf: C_5H_5NO mw: 95.11

SYNS: β-HYDROXYPYRIDINE ◊ 3-HYDROXYPYRIDINE ◊ 3-PYRIDOL

TOXICITY DATA with REFERENCE
ipr-mus LD50:1822 mg/kg   TOXIA6 23,815,85
orl-brd LD50:750 mg/kg   AECTCV 12,355,83

CONSENSUS REPORTS: Reported in EPA TSCA Inventory.

SAFETY PROFILE: Moderately toxic by ingestion and intraperitoneal routes. When heated to decomposition it emits toxic vapors of NO_x.

PPH300   CAS:109-12-6   *HR: 2*
2-PYRIDIYLAMINE
mf: $C_4H_5N_3$ mw: 95.12

SYNS: 2-AMINOPYRIMIDINE ◇ 2-PYRIMIDINAMINE (9CI) ◇ PYRIMIDINE, 2-AMINO-

TOXICITY DATA with REFERENCE
mmo-omi 50 mg/L   SOGEBZ 6,1623,70
ivn-rat LD50:790 mg/kg   AEPPAE 211,367,50

CONSENSUS REPORTS: Reported in EPA TSCA Inventory.

SAFETY PROFILE: Moderately toxic by intraperitoneal route. Mutation data reported. When heated to decomposition it emits toxic vapors of NO_x.

PPN100   CAS:603-50-9   *HR: 3*
4,4′-(2-PYRIDYLMETHYLENE)DIPHENOL DIACETATE
mf: $C_{22}H_{19}NO_4$ mw: 361.42

SYNS: BICOL ◇ BIS(p-ACETOXYPHENYL)-2-PYRIDYLMETHANE ◇ BISACODYL ◇ BROXALAX ◇ CONTALAX ◇ DEFICOL ◇ 2-(4,4′-DIACETOXYDIPHENYLMETHYL)PYRIDINE ◇ 4,4′-DIACETOXYDIPHENYL-PYRID-2-YLMETHANE ◇ (4,4′-DIACETOXYDIPHENYL)(2-PYRIDYL)METHANE ◇ DI-(p-ACETOXYPHENYL)-2-PYRIDYLMETHANE ◇ DI-(4-ACETOXYPHENYL)-2-PYRIDYLMETHANE ◇ DULCOLAN ◇ DULCOLAX ◇ DUROLAX ◇ ENDOKOLAT ◇ EULAXAN ◇ GODALAX ◇ LA96A ◇ LACO ◇ LAXADIN ◇ LAXAGETTEN ◇ LAXANIN N ◇ LAXANS ◇ LAXOREX ◇ NIGALAX ◇ PERILAX ◇ PHENOL, 4,4′-(2-PYRIDINYLMETHYLENE)BIS-, DIACETATE (ESTER) ◇ PHENOL, 4,4′-(2-PYRIDYLMETHYLENE)DI-, DIACETATE (ester) ◇ PYRILAX ◇ SK-BISACODYL ◇ STADALAX ◇ TELEMIN ◇ THERALAX ◇ ULCOLAX

TOXICITY DATA with REFERENCE
orl-rat LD50:4320 mg/kg   TXAPA9 2,243,60
orl-mus LD50:17,500 mg/kg   TXAPA9 2,243,60
ivn-mus LDLo:80 mg/kg   TXAPA9 2,243,60
orl-dog LD50:>15 g/kg   DRUGAY -,854,90

CONSENSUS REPORTS: Reported in EPA TSCA Inventory.

SAFETY PROFILE: Poison by intravenous route. Slightly toxic by ingestion. When heated to decomposition it emits toxic vapors of NO_x.

PPP100   CAS:1193-24-4   *HR: D*
4,6-PYRIMIDINEDIOL
mf: $C_4H_4N_2O_2$ mw: 112.10

SYNS: 4,6-DIHYDROXYPYRIMIDIN ◇ 6-HYDROXY-4(1H)-PYRIMIDINONE ◇ 4(1H)-PYRIMIDINONE, 6-HYDROXY-

TOXICITY DATA with REFERENCE
pic-esc 1 g/L   ZAPOAK 12,583,72

CONSENSUS REPORTS: Reported in EPA TSCA Inventory.

SAFETY PROFILE: Mutation data reported. When heated to decomposition it emits toxic vapors of NO_x.

PPQ630   CAS:89-05-4   *HR: 3*
PYROMELLITIC ACID
mf: $C_{10}H_6O_8$ mw: 254.16

SYNS: 1,2,4,5-BENZENETETRACARBOXYLIC ACID ◇ 1,2,4,5-TETRACARBOXYBENZENE ◇ USAF XR-20

TOXICITY DATA with REFERENCE
ipr-mus LD50:300 mg/kg   NTIS** AD277-689

CONSENSUS REPORTS: Reported in EPA TSCA Inventory.

SAFETY PROFILE: Poison by intraperitoneal route. When heated to decomposition it emits acrid smoke and irritating vapors.

PPQ635   CAS:89-32-7   *HR: 3*
PYROMELLITIC ACID ANHYDRIDE
mf: $C_{10}H_2O_6$ mw: 218.12

SYNS: 1,2,4,5 BENZENETETRACARBOXYLIC 1,2:4,5 DIANHYDRIDE ◇ 1H,3H-BENZO(1,2-c:4,5-c′)DIFURAN-1,3,5,7-TETRONE ◇ PYROMELLITIC ACID DIANHYDRIDE ◇ PYROMELLITIC ANHYDRIDE ◇ PYROMELLITIC DIANHYDRIDE

TOXICITY DATA with REFERENCE
orl-rat LD50:2250 mg/kg   GISAAA 51(11),79,86
ihl-rat LCLo:150 mg/m^3/4H   85GMAT -,105,82
orl-mus LD50:2400 mg/kg   TPKVAL 14,125,75
orl-gpg LD50:1595 mg/kg   GISAAA 51(11),79,86

CONSENSUS REPORTS: Reported in EPA TSCA Inventory.

SAFETY PROFILE: Poison by inhalation. Moderately toxic by ingestion. When heated to decomposition it emits acrid smoke and irritating vapors.

Q

QFS100   CAS:747-45-5   *HR: 1*
QUINIDINE SULFATE (1:1) (salt)
mf: $C_{20}H_{24}N_2O_2 \cdot H_2O_4S$ mw: 422.54

SYNS: BIQUIN DURULES ◇ CHINIDIN DURULES ◇ CHINIDIN VUFB ◇ CINCHONAN-9-OL, 6′-METHOXY-, (9S)-, SULFATE (1:1) (SALT) (9CI) ◇ (9S)-6′-METHOXYCINCHONAN-9-OL SULFATE (1:1) (SALT) ◇ KINIDIN DURETTER ◇ KINIDIN DURULES ◇ KINILENTIN ◇ OPTOCHINIDIN ◇ QUINIDINE BISULFATE

TOXICITY DATA with REFERENCE
orl-man TDLo:225 mg/kg/3W-I   AIMDAP 145,2051,85
orl-wmn TDLo:210 mg/kg/3W-I   AIMEAS 108,369,88

CONSENSUS REPORTS: Reported in EPA TSCA Inventory.

SAFETY PROFILE: Human systemic effects by ingestion: vascular changes, joint changes. When heated to decomposition it emits toxic vapors of NO_x and SO_x.

QOJ100   CAS:2637-37-8   *HR: 2*
2-QUINOLINETHIOL
mf: C_9H_7NS mw: 161.23

SYNS: CARBOSTYRIL, THIO- ◇ 2-MERCAPTOQUINOLINE ◇ 2(1H)-QUINOLINETHIONE ◇ THIOCARBOSTYRIL

TOXICITY DATA with REFERENCE
ipr-mus LD50:450 mg/kg   RAREAE 7,13,57

CONSENSUS REPORTS: Reported in EPA TSCA Inventory.

SAFETY PROFILE: Moderately toxic by intraperitoneal route. When heated to decomposition it emits toxic vapors of NO_x and SO_x.

QSJ100   CAS:1196-57-2   *HR: 3*
2-QUINOXALINOL
mf: $C_8H_6N_2O$ mw: 146.16

SYNS: 2(1H)-QUINOXALINONE

TOXICITY DATA with REFERENCE
ivn-mus LD50:178 mg/kg   CSLNX* NX#00285

CONSENSUS REPORTS: Reported in EPA TSCA Inventory.

SAFETY PROFILE: Poison by intravenous route. When heated to decomposition it emits toxic vapors of NO_x.

QUJ990   CAS:1619-34-7   *HR: D*
3-QUINUCLIDINOL
mf: $C_7H_{13}NO$ mw: 127.21

SYNS: 1-AZABICYCLO(2.2.2)OCTAN-3-OL (9CI) ◇ 3-HYDROXYQUINUCLIDINE

TOXICITY DATA with REFERENCE
cyt-ham-ipr 10 mg/kg   ACNSAX 17,253,75

CONSENSUS REPORTS: Reported in EPA TSCA Inventory.

SAFETY PROFILE: Mutation data reported. When heated to decomposition it emits toxic vapors of NO_x.

R

REF025   CAS:151-10-0   HR: 2
RESORCINOL DIMETHYL ETHER
mf: $C_8H_{10}O_2$ mw: 138.18

SYNS: BENZENE, m-DIMETHOXY- ◇ BENZENE, 1,3-DIMETHOXY- ◇ m-DIMETHOXYBENZENE ◇ 1,3-DIMETHOXYBENZENE ◇ DIMETHYL-ETHER RESORCINOL ◇ 3-METHOXYANISOLE

TOXICITY DATA with REFERENCE
ipr-mus LD50:900 mg/kg   JAPMA8 46,185,57

CONSENSUS REPORTS: Reported in EPA TSCA Inventory.

SAFETY PROFILE: Moderately toxic by intraperitoneal route. When heated to decomposition it emits acrid smoke and irritating vapors.

REF070   CAS:2479-46-1   HR: 2
RESORCINOL OXYDIANILINE
mf: $C_{18}H_{16}N_2O_2$ mw: 292.36

SYNS: ANILINE, p,p'-(m-PHENYLENEDIOXY)DI- ◇ BENZENAMINE, 4,4'-(1,3-PHENYLENEBIS(OXY))BIS- ◇ 1,3-BIS(4-AMINOPHENOXY)BENZENE ◇ 1,3-PHENYLENE-DI-4-AMINOPHENYL ETHER ◇ 4,4'-(m-PHENYLENEDIOXY)DIANILINE

TOXICITY DATA with REFERENCE
mma-sat 100 μg/plate   SAIGBL 24,498,82
dns-rat:lvr 10 μmol/L   MUREAV 204,683,88
orl-rat LDLo:690 mg/kg   EPASR* 8EHQ-0790-0898S

SAFETY PROFILE: Moderately toxic by ingestion. Mutation data reported. When heated to decomposition it emits toxic vapors of NO_x.

REF100   CAS:95-01-2   HR: 3
β-RESORCYLALDEHYDE
mf: $C_7H_6O_3$ mw: 138.13

SYNS: BENZALDEHYDE, 2,4-DIHYDROXY- ◇ 2,4-DIHYDROXYBENZALDEHYDE ◇ 2,4-DIHYDROXYBENZENECARBONAL ◇ 4-FORMYLRESORCINOL ◇ 4-HYDROXYSALICYLALDEHYDE ◇ β-RESORCINALDEHYDE ◇ β-RESORCYLIC ALDEHYDE ◇ β-RESORCALDEHYDE

TOXICITY DATA with REFERENCE
orl-mus LD50:1380 mg/kg   IJANDP 10,741,87
ipr-mus LD50:200 mg/kg   IJOCAP 14,449,76

CONSENSUS REPORTS: Reported in EPA TSCA Inventory.

SAFETY PROFILE: Poison by intraperitoneal route. Moderately toxic by ingestion. When heated to decomposition it emits acrid smoke and irritating vapors.

REF200   CAS:99-10-5   HR: 2
α-RESORCYLIC ACID
mf: $C_7H_6O_4$ mw: 154.13

SYNS: BENZOIC ACID, 3,5-DIHYDROXY-(9CI) ◇ 5-CARBOXYRESORCINOL ◇ 3,5-DIHYDROXYBENZOIC ACID

TOXICITY DATA with REFERENCE
ivn-mus LD50:2000 mg/kg   PMDCAY 5,59,67

CONSENSUS REPORTS: Reported in EPA TSCA Inventory.

SAFETY PROFILE: Moderately toxic by intravenous. When heated to decomposition it emits acrid smoke and irritating vapors.

RIF100   CAS:146-14-5   HR: 2
RIBOFLAVIN-ADENINE DINUCLEOTIDE
mf: $C_{27}H_{33}N_9O_{15}P_2$ mw: 785.63

SYNS: ADENINE-FLAVIN DINUCLEOTIDE ◇ ADENINE-FLAVINE DINUCLEOTIDE ◇ ADENINE-RIBOFLAVIN DINUCLEOTIDE ◇ ADENINE-RIBOFLAVINE DINUCLEOTIDE ◇ ADENOSINE 5'-(TRIHYDROGEN PYROPHOSPHATE), 5'-5'-ESTER with RIBOFLAVINE ◇ FAD ◇ FLAVIN ADENIN DINUCLEOTIDE ◇ FLAVIN ADENINE DINUCLEOTIDE ◇ FLAVINAT ◇ FLAVINE-ADENINE DINUCLEOTIDE ◇ FLAVINE ADENOSINE DIPHOSPHATE ◇ FLAVITAN ◇ ISOALLOXAZINE-ADENINE DINUCLEOTIDE ◇ RIBOFLAVINE-ADENINE DINUCLEOTIDE ◇ RIBOFLAVIN 5'-(TRIHYDROGEN DIPHOSPHATE), 5'-5'-ESTER with ADENOSINE (9CI)

TOXICITY DATA with REFERENCE
ivn-mus LD50:700 mg/kg   DRUGAY 6,689,82

CONSENSUS REPORTS: Reported in EPA TSCA Inventory.

SAFETY PROFILE: Moderately toxic by intravenous route. When heated to decomposition it emits toxic vapors of NO_x and PO_x.

RPB200   CAS:584-09-8   HR: 2
RUBIDIUM CARBONATE
mf: $CO_3 \cdot 2Rb$ mw: 230.95

SYNS: DIRUBIDIUM CARBONATE ◇ CARBONIC ACID, DIRUBIDIUM SALT ◇ DIRUBIDIUM MONOCARBONATE

RPB200 RUBIDIUM CARBONATE

TOXICITY DATA with REFERENCE
orl-rat LD50:2625 mg/kg   GTPZAB 31(9),55,87
ipr-rat LD50:450 mg/kg   GTPZAB 31(9),55,87

CONSENSUS REPORTS: Reported in EPA TSCA Inventory.

SAFETY PROFILE: Moderately toxic by ingestion and intraperitoneal routes. When heated to decomposition it emits toxic fumes of Rb.

S

SAA025   CAS:89-65-6   *HR: D*
SACCHAROSONIC ACID
mf: $C_6H_8O_6$ mw: 176.14

SYNS: ARABOASCORBIC ACID ◇ d-ARABOASCORBIC ACID ◇ ERYCORBIN ◇ ERYTHORBIC ACID ◇ d-ERYTHORBIC ACID ◇ GLUCOSACCHARONIC ACID ◇ d-erythro-HEX-2-ENONIC ACID, γ-LACTONE ◇ ISOASCORBIC ACID ◇ d-ISOASCORBIC ACID ◇ ISOVITAMIN C ◇ MERCATE 5 ◇ NEO-CEBICURE

TOXICITY DATA with REFERENCE
mma-sat 50 mg/plate   FCTOD7 22,623,84
dnd-uns:lyms 5 mmol/L   JNSVA5 21,237,75

CONSENSUS REPORTS: Reported in EPA TSCA Inventory.

SAFETY PROFILE: Mutation data reported. When heated to decomposition it emits acrid smoke and irritating vapors.

SAU480   CAS:89-28-1   *HR: 1*
SANTOFLEX DD
mf: $C_{24}H_{39}N$ mw: 341.64

SYNS: QUINOLINE, 6-DODECYL-1,2-DIHYDRO-2,2,4-TRIMETHYL- ◇ 6-DODECYL-2,2,4-TRIMETHYL-1,2-DIHYDROQUINOLINE

TOXICITY DATA with REFERENCE
orl-rat LDLo:40 g/kg   IPSTB3 3,93,76
orl-rbt LDLo:10 g/kg   IPSTB3 3,93,76
skn-rbt LDLo:>10 g/kg   RCTEA4 45,627,72

CONSENSUS REPORTS: Reported in EPA TSCA Inventory.

SAFETY PROFILE: Slightly toxic by ingestion and skin contact. When heated to decomposition it emits toxic vapors of NO_x.

SCF550   CAS:287-29-6   *HR: 2*
SILACYCLOBUTANE
mf: C_3H_8Si mw: 72.20

SYN: SILICETANE

TOXICITY DATA with REFERENCE
orl-rat LD50:1539 mg/kg   EPASR* 8EHQ-0990-1084

SAFETY PROFILE: Moderately toxic by ingestion. When heated to decomposition it emits acrid smoke and irritating vapors.

SCH000   CAS:69012-64-2   *HR: 2*
SILICA, AMORPHOUS FUMED
mf: O_2Si mw: 60.09

PROP: A finely powdered microcellular silica foam with minimum SiO_2 content of 89.5%. Insol in water; sol in hydrofluoric acid.

SYNS: ACTICEL ◇ AEROSIL ◇ AMORPHOUS SILICA DUST ◇ AQUAFIL ◇ CAB-O-GRIP II ◇ CAB-O-SIL ◇ CAB-O-SPERSE ◇ CATALOID ◇ COLLOIDAL SILICA ◇ COLLOIDAL SILICON DIOXIDE ◇ DAVISON SG-67 ◇ DICALITE ◇ DRI-DIE INSECTICIDE 67 ◇ ENT 25,550 ◇ FLO-GARD ◇ FOSSIL FLOUR ◇ FUMED SILICA ◇ FUMED SILICON DIOXIDE ◇ HI-SEL ◇ LO-VEL ◇ LUDOX ◇ NALCOAG ◇ NYACOL ◇ NYACOL 830 ◇ NYACOL 1430 ◇ SANTOCEL ◇ SG-67 ◇ SILICA AEROGEL ◇ SILICA, AMORPHOUS ◇ SILICIC ANHYDRIDE ◇ SILICON DIOXIDE (FCC) ◇ SILIKILL ◇ SYNTHETIC AMORPHOUS SILICA ◇ VULKASIL

TOXICITY DATA with REFERENCE
dns-rat-itr 120 mg/kg   ENVRAL 41,61,86
bfa-rat:lng 120 mg/kg   ENVRAL 41,61,86
ihl-rat TCLo:50 mg/m^3/6H/2Y-I:CAR   CNREA8 2,255,86
orl-rat LD50:3160 mg/kg   ARSIM* 20,9,66
ipr-rat LDLo:50 mg/kg   AHBAAM 136,1,52
ivn-rat LD50:15 mg/kg   BSIBAC 44,1685,68
itr-rat LDLo:10 mg/kg   AHBAAM 136,1,52
ipr-gpg LDLo:120 mg/kg   BJEPA5 3,75,22

CONSENSUS REPORTS: IARC Cancer Review: Group 3 IMEMDT 7,341,87; Animal Inadequate Evidence IMEMDT 42,209,88; Human Inadequate Evidence IMEMDT 42,209,88. Reported in EPA TSCA Inventory.

OSHA PEL: (Transitional: TWA 80 mg/m^3/%SiO_2) TWA 6 mg/m^3
ACGIH TLV: TWA 2 mg/m^3 (Respirable Dust)

SAFETY PROFILE: Poison by intraperitoneal, intravenous, and intratracheal routes. Moderately toxic by ingestion. An inhalation hazard. Much less toxic than crystalline forms. Questionable carcinogen with experimental carcinogenic data. Mutation data reported. Does not cause silicosis. See also other silica entries.

SCK600   CAS:60676-86-0   *HR: 3*
SILICA, FUSED
mf: O_2Si mw: 60.09

PROP: Made up of spherical submicroscopic particles under 0.1 micron in size (AMIHBC 9,389,54).

SYNS: AMORPHOUS FUSED SILICA ◇ FUSED QUARTZ ◇ FUSED SILICA (ACGIH) ◇ QUARTZ GLASS ◇ SILICA, AMORPHOUS FUSED ◇ SILICA, VITREOUS ◇ SILICON DIOXIDE ◇ VITREOUS QUARTZ

TOXICITY DATA with REFERENCE
imp-rat TDLo:400 mg/kg:ETA   NATWAY 41,534,54
ipr-rat LDLo:400 mg/kg   AMIHBC 9,389,54
itr-rat LDLo:120 mg/kg   AMIHBC 9,389,54
ipr-mus LDLo:40 mg/kg   BJEPA5 3,75,22
ivn-cat LDLo:15 mg/kg   JLCMAK 26,774,41
ivn-rbt LDLo:35 mg/kg   BJEPA5 3,75,22

CONSENSUS REPORTS: IARC Cancer Review: Group 3 IMEMDT 7,341,87; Animal Inadequate Evidence IMEMDT 42,39,87; Human Inadequate Evidence IMEMDT 42,39,87. Reported in EPA TSCA Inventory.

OSHA PEL: (Transitional: TWA Respirable: 10 mg/m^3/ 2(%SiO$_2$+2); Total Dust: TWA 30 mg/m^3/2(%SiO$_2$+2)) TWA 0.1 mg/m^3
ACGIH TLV: TWA 0.1 mg/m^3 (Respirable Fraction)
NIOSH REL: (Silica, Crystalline): TWA 0.05 mg/m^3

SAFETY PROFILE: An inhalation hazard. Questionable carcinogen with experimental tumorigenic data. Poison by intraperitoneal, intravenous, and intratracheal routes. See also other silica entries.

SDI800   CAS:563-63-3   *HR: 3*
SILVER(I) ACETATE
mf: C$_2$H$_3$O$_2$•Ag   mw: 166.92

SYNS: ACETIC ACID, SILVER(1+) SALT ◊ SILVER ACETATE ◊ SILVER(1+) ACETATE ◊ SILVER MONOACETATE

TOXICITY DATA with REFERENCE
ipr-mus LD50:34 mg/kg   DCTODJ 6,267,83
ACGIH TLV: TWA 0.01 mg(Ag)/m^3

CONSENSUS REPORTS: Reported in EPA TSCA Inventory.

SAFETY PROFILE: Poison by intraperitoneal route. When heated to decomposition it emits acrid smoke and irritating vapors.

SDX250   CAS:694-53-1   *HR: 2*
SILYLBENZENE
mf: C$_6$H$_8$Si   mw: 108.23

SYNS: FENYLSILAN ◊ PHENYLSILANE ◊ SILANE, PHENYL-

TOXICITY DATA with REFERENCE
orl-mus LD50:700 mg/kg   85JCAE -,1227,86
ihl-mus LC50:14,500 mg/m^3/2H   85JCAE -,1227,86

CONSENSUS REPORTS: Reported in EPA TSCA Inventory.

SAFETY PROFILE: Moderately toxic by ingestion. Mildly toxic by inhalation. When heated to decomposition it emits acrid smoke and irritating vapors.

SEH475   CAS:130-22-3   *HR: 3*
SODIUM ALIZARIN-3-SULFONATE
mf: C$_{14}$H$_7$O$_7$S•Na   mw: 342.26

SYNS: ACID RED ALIZARINE ◊ AHCOQUINONE RED S ◊ ALIZARIN CARMINE (BIOLOGICAL STAIN) ◊ ALIZARINE CARMINE INDICATOR ◊ ALIZARINE RED A ◊ ALIZARINE RED AS ◊ ALIZARINE RED INDICATOR ◊ ALIZARINE RED S (BIOLOGICAL STAIN) ◊ ALIZARINE RED S SODIUM SALT ◊ ALIZARINE RED SW ◊ ALIZARINE RED SZ ◊ ALIZARINE RED W ◊ ALIZARINE RED WA ◊ ALIZARINE RED for WOOL ◊ ALIZARINE RED WS ◊ ALIZARINE S ◊ ALIZARINE S EXTRA CONC. A EXPORT ◊ ALIZARINE S EXTRA PURE A ◊ ALIZARIN RED S ◊ ALIZARINROT-S ◊ ALIZARIN S ◊ ALIZARINSULFONATE ◊ 2-ANTHRACENESULFONIC ACID, 9,10-DIHYDRO-3,4-DIHYDROXY-9,10-DIOXO-, MONOSODIUM SALT ◊ CALCOCHROME ALIZARINE RED SC ◊ CARNELIO RUBINE LAKE ◊ CHROME RED ALIZARINE ◊ C.I. 58005 ◊ C.I. MORDANT RED 3 ◊ DIAMOND RED W ◊ 9,10-DIHYDRO-3,4-DIHYDROXY-9,10-DIOXO-2-ANTHRACENESULFONIC ACID MONOSODIUM SALT ◊ EXT. D and C RED NO. 7 ◊ FENAKROM RED W ◊ MITSUI ALIZARINE RED S ◊ OXANAL FAST RED SW ◊ SODIUM ALIZARINESULFONATE

TOXICITY DATA with REFERENCE
ivn-mus LD50:70 mg/kg   EXPEAM 28,180,72

CONSENSUS REPORTS: Reported in EPA TSCA Inventory.

SAFETY PROFILE: Poison by intravenous route. When heated to decomposition it emits toxic vapors of SO$_x$.

SEY500   CAS:7784-46-5   *HR: 3*
SODIUM ARSENITE
mf: AsO$_2$•Na   mw: 129.91
DOT: UN 1686/UN 2027

PROP: White or grayish-white powder. Commercially: 95–98% pure. Very sol in water; sltly sol in alc.

SYNS: ARSENIOUS ACID, SODIUM SALT ◊ ARSENITE de SODIUM (FRENCH) ◊ ARSENOUS ACID, SODIUM SALT (9CI) ◊ ATLAS "A" ◊ CHEM PELS C ◊ CHEM-SEN 56 ◊ KILL-ALL ◊ PENITE ◊ PRODALUMNOL ◊ PRODALUMNOL DOUBLE ◊ SODANIT ◊ SODIUM ARSENITE, liquid (solution) (DOT) ◊ SODIUM ARSENITE, solid (DOT) ◊ SODIUM METAARSENITE

TOXICITY DATA with REFERENCE
dnr-esc 63 µg/well   ENMUDM 3,429,81
mmo-smc 100 mmol/L   MUREAV 117,149,83
oms-hmn:lym 700 nmol/L   SWEHDO 7,277,81
cyt-hmn:lym 1 mg/L   ENMUDM 3,597,81
orl-ham TDLo:25 mg/kg (female 12D post):TER   BECTA6 29,671,82
ipr-rat TDLo:11 mg/kg (7D preg):TER   TJADAB 23,66A,81
orl-ham TDLo:5 mg/kg (female 9D post):REP   TJADAB 23,40A,81
ipr-rat TDLo:11 mg/kg (7D preg):TER   TJADAB 23,66A,81
ipr-mus TDLo:10 mg/kg (female 9D post):TER   BECTA6 7,216,72
ipr-rat TDLo:11 mg/kg (7D preg):TER   TJADAB 23,66A,81
ipr-rat TDLo:11 mg/kg (7D preg):TER   TJADAB 23,66A,81
ipr-mus TDLo:10 mg/kg (female 9D post):TER   EVHPAZ 19,219,77

orl-chd TDLo:1 mg/kg   CTOXAO 10,477,77
orl-chd LDLo:2 mg/kg   CTOXAO 10,477,77
orl-rat LD50:41 mg/kg   AIHAAP 30,470,69
skn-rat LD50:150 mg/kg   PHJOAV 185,361,60
ivn-rat LDLo:6 mg/kg   JPETAB 33,270,28
ipr-mus LD50:1170 µg/kg   COREAF 257,791,63
ims-mus LD50:14 mg/kg   EXMDA4 (440),312,78
orl-rbt LDLo:12 mg/kg   JPETAB 33,270,28
ivn-rbt LD50:7600 µg/kg   TXCYAC 51,213,88

CONSENSUS REPORTS: NTP Fifth Annual Report on Carcinogens. IARC Cancer Review: Group 1 IMEMDT 7,100,87; Animal Inadequate Evidence IMEMDT 23,39,80; Human Sufficient Evidence IMEMDT 23,39,80; Animal No Evidence IMEMDT 2,48,73. Arsenic and its compounds are on the Community Right-To-Know List. Reported in EPA TSCA Inventory. EPA Genetic Toxicology Program. EPA Extremely Hazardous Substances List.

OSHA PEL: TWA 0.01 mg(As)/m^3; Cancer Hazard
ACGIH TLV: TWA 0.2 mg(As)/m^3 (Proposed: 0.01 mg(As)/m^3; Human Carcinogen)
NIOSH REL: CL (Inorganic Arsenic) 0.002 mg(As)/m^3/15M

DOT Classification: Poison B; Label: Poison

SAFETY PROFILE: Confirmed human carcinogen. Human poison by ingestion. Experimental poison by ingestion, skin contact, intravenous, intramuscular, and intraperitoneal routes. An experimental teratogen. Experimental reproductive effects. Human mutation data reported. Used as a herbicide and pesticide. When heated to decomposition it emits toxic fumes of As and Na$_2$O. See also ARSENIC COMPOUNDS.

SEZ000　　　CAS:7784-46-5　　　**HR: 3**
SODIUM ARSENITE (liquid)

CONSENSUS REPORTS: Arsenic and its compounds are on the Community Right-To-Know List. Reported in EPA TSCA Inventory.

OSHA PEL: TWA 0.01 mg(As)/m^3; Cancer Hazard
ACGIH TLV: TWA 0.2 mg(As)/m^3 (Proposed: 0.01 mg(As)/m^3; Human Carcinogen)
NIOSH REL: CL (Inorganic Arsenic) 0.002 mg(As)/m^3/15M

DOT Classification: Poison B; Label: Poison

SAFETY PROFILE: Confirmed human carcinogen. A deadly poison. When heated to decomposition it emits toxic fumes of As and Na$_2$O. See also SODIUM ARSENITE.

SFA200　　　CAS:114-70-5　　　**HR: 2**
SODIUM BENZENEACETATE
mf: C$_8$H$_7$O$_2$•Na　　mw: 158.14

SYNS: ACETIC ACID, PHENYL-, SODIUM SALT ◇ BENZENEACETIC ACID, SODIUM SALT (9CI) ◇ PHENYLACETATE SODIUM SALT ◇ PHENYLACETIC ACID SODIUM SALT ◇ PHENYLESSIGSAEURE NATRIUM-SALZ ◇ SODIUM PHENYLACETATE

TOXICITY DATA with REFERENCE
ipr-mus LD50:2710 mg/kg   DECRDP 12(Suppl 1),11,86
ivn-mus LD50:1500 mg/kg   ARZNAD 12,751,62

CONSENSUS REPORTS: Reported in EPA TSCA Inventory.

SAFETY PROFILE: Moderately toxic by intraperitoneal and intravenous routes. When heated to decomposition it emits acrid smoke and irritating vapors.

SGB550　　　CAS:2624-17-1　　　**HR: 2**
SODIUM CYANURATE
mf: C$_3$H$_3$N$_3$O$_3$•Na　　mw: 152.08

SYNS: ACOVENOSIDE B ◇ MONOSODIUM CYANURATE ◇ NCI-C56542 ◇ SODIUM ISOCYANURATE ◇ s-TRIAZINE-2,4,6(1H,3H,5H)-TRIONE, MONOSODIUM SALT ◇ 1,3,5-TRIAZINE-2,4,6(1H,3H,5H)-TRIONE, MONO-SODIUM SALT (9CI)

TOXICITY DATA with REFERENCE
orl-rat LDLo:>7500 mg/kg   GISAAA 27(12),13,62
ivn-cat LD50:2144 mg/kg   JPETAB 103,420,51
orl-rat TDLo:560 g/kg/20W-C   TXAPA9 7,667,65
orl-dog TDLo:2912 g/kg/2Y-C   TXAPA9 7,667,65

CONSENSUS REPORTS: Reported in EPA TSCA Inventory.

SAFETY PROFILE: Moderately toxic by intravenous route. Slightly toxic by ingestion. When heated to decomposition it emits toxic vapors of NO$_x$.

SGB600　　　CAS:1002-62-6　　　**HR: D**
SODIUM DECANOATE
mf: C$_{10}$H$_{19}$O$_2$•Na　　mw: 194.28

SYNS: CAPRINIC ACID, SODIUM SALT ◇ DECANOIC ACID, SODIUM SALT ◇ SODIUM CAPRINATE ◇ SODIUM CAPRATE ◇ SODIUM-n-DECANOATE ◇ SODIUM DECANOIC ACID

TOXICITY DATA with REFERENCE
dni-gpg:kdy 400 µmol/L   FCTXAV 14,431,76

CONSENSUS REPORTS: Reported in EPA TSCA Inventory.

SAFETY PROFILE: Mutation data reported. When heated to decomposition it emits acrid smoke and irritating vapors.

SGG650　　　CAS:620-45-1　　　**HR: 3**
SODIUM 2,6-DICHLOROINDOPHENOL
mf: C$_{12}$H$_6$Cl$_2$NO$_2$•Na　　mw: 290.08

SYNS: 2,5-CYCLOHEXADIEN-1-ONE,2,6-DICHLORO-4-((p-HYDROXYPHENYL)IMINO)-, SODIUM SALT ◇ 2,6-DICHLORO-4-((p-HYDROXYPHENYL)IMINO)-2,5-CYCLOHEXADIEN-1-ONE SODIUM SALT ◇ 2,6-DICHLOROINDOPHENOL, SODIUM SALT ◇ INDOPHENOL,

2,6-DICHLORO-, SODIUM SALT ◇ SODIUM 2,6-DICHLOROINDOPHENOLATE

TOXICITY DATA with REFERENCE
ipr-mus LD50:75 mg/kg   EJMCA5 17,349,82
ivn-mus LD50:180 mg/kg   CSLNX* NX#04765

CONSENSUS REPORTS: Reported in EPA TSCA Inventory.

SAFETY PROFILE: Poison by intravenous and intraperitoneal routes. When heated to decomposition it emits toxic vapors of NO_x and Cl^-.

SHG500   CAS:62-74-8   **HR: 3**
SODIUM FLUOROACETATE
DOT: UN 2629
mf: $C_2H_2FO_2 \cdot Na$ mw: 100.03

PROP: Fine, white powder. Sol in water.

SYNS: 1080 ◇ COMPOUND No. 1080 ◇ FLUORACETATO di SODIO (ITALIAN) ◇ FLUOROACETIC ACID, SODIUM SALT ◇ FLUORESSIGSAEURE (GERMAN) ◇ FRATOL ◇ FURATOL ◇ MONOFLUORESSIGSAEURE, NATRIUM (GERMAN) ◇ NATRIUMFLUORACETAAT (DUTCH) ◇ NATRIUMFLUORACETAT (GERMAN) ◇ RATBANE 1080 ◇ RCRA WASTE NUMBER P058 ◇ SODIO, FLUORACETATO di (ITALIAN) ◇ SODIUM FLUOACETATE ◇ SODIUM FLUOACETIC ACID ◇ SODIUM FLUORACETATE de (FRENCH) ◇ SODIUM MONOFLUOROACETATE ◇ TL 869 ◇ YASOKNOCK

TOXICITY DATA with REFERENCE
orl-rat TDLo:210 µg/kg (3D male):REP   JRPFA4 56,201,79
orl-hmn LDLo:714 µg/kg   34ZIAG -,542,69
unr-man LDLo:5 mg/kg   AJPEAG 36,1427,46
orl-rat LD50:100 µg/kg   AJPEAG 36,1427,46
ipr-rat LDLo:3 mg/kg   JAPMA8 36,59,47
orl-mus LD50:500 µg/kg   JAPMA8 37,307,48
ipr-mus LD50:7 mg/kg   NEZAAQ 34,193,79
scu-mus LDLo:7200 µg/kg   NDRC** 30101,2,45
orl-dog LD50:66 µg/kg   JPETAB 101,82,50
ivn-dog LD50:60 µg/kg   YKYUA6 31,1385,80
ivn-mky LD50:5 mg/kg   AJPEAG 36,1427,46

CONSENSUS REPORTS: Reported in EPA TSCA Inventory. EPA Extremely Hazardous Substances List.

OSHA PEL: (Transitional: TWA 0.05 mg/m^3 (skin)) TWA 0.05 mg/m^3 (skin); STEL 0.15 mg/m^3 (skin)
ACGIH TLV: TWA 0.05 mg/m^3 (skin); STEL 0.15 mg/m^3 (skin) (Proposed: TWA 0.05 mg/m^3 (skin))
DFG MAK: 0.05 mg/m^3

DOT Classification: Poison B: Label: Poison

SAFETY PROFILE: A deadly human poison by ingestion. Experimental poison by ingestion, intraperitoneal, subcutaneous, and intravenous routes. A very highly toxic water-soluble salt used mainly as an immediate-action rodenticide. It is absorbed rapidly by the gastrointestinal tract but slowly by the skin unless the skin is abraded or cut. It operates by blocking the Krebs cycle by formation of fluorocitric acid, which inhibits aconitase. It has an effect on either the cardiovascular or nervous system, or both, in all species and, in some species, the skeletal muscles. Humans have mixed responses, with the cardiac feature predominating. By a direct action on the heart, contractile power is lost, which leads to declining blood pressure. Ventricular premature contractions and arrhythmias are seen in all species, including humans. The central nervous system is directly attacked by sodium fluoroacetate. In humans, the action on the central nervous system produces epileptiform convulsive seizures followed by severe depression. The dangerous dose for humans is 0.5–2 mg/kg. Other species vary considerably in their response to this material, with primates and birds being the most resistant and carnivora and rodents being the most susceptible. Most domestic animals show a susceptibility falling between the two extremes indicated above. Experimental reproductive effects. When heated to decomposition it emits highly toxic fumes of Na_2O and F^-.

SHM000   CAS:12005-86-6   **HR: 3**
SODIUM HEXAFLUOROARSENATE
mf: $AsF_6 \cdot Na$ mw: 211.91

TOXICITY DATA with REFERENCE
ivn-mus LD50:56 mg/kg   CSLNX* NX#00127

CONSENSUS REPORTS: Arsenic and its compounds are on the Community Right-To-Know List.

OSHA PEL: Cancer Hazard
ACGIH TLV: TWA 0.2 mg(As)/m^3 (Proposed: 0.01 mg(As)/m^3; Human Carcinogen)
NIOSH REL: (Arsenic, Inorganic) CL 2 µg/m^3/15M

SAFETY PROFILE: Confirmed human carcinogen. Poison by intravenous route. When heated to decomposition it emits very toxic fumes of As, F^-, and Na_2O. See also FLUORIDES and ARSENIC COMPOUNDS.

SIB700   CAS:1847-58-1   **HR: 2**
SODIUM LAURYL SULFOACETATE
mf: $C_{14}H_{27}O_5S \cdot Na$ mw: 330.46

SYNS: ACETIC ACID, SULFO-, DODECYL ESTER, S-SODIUM SALT (7CI) ◇ ACETIC ACID, SULFO-, 1-DODECYL ESTER, SODIUM SALT ◇ DODECYL SODIUM SULFOACETATE ◇ LATHANOL ◇ LATHANOL LAL ◇ LATHANOL-LAL 70 ◇ NACCONOL LAL ◇ SULFOACETIC ACID DODECYL ESTER S-SODIUM SALT ◇ SULFOACETIC ACID 1-DODECYL ESTER, SODIUM SALT

TOXICITY DATA with REFERENCE
skn-rbt 500 mg/24H MOD   JACTDZ 6(3),261,87
eye-rbt 35 mg MLD   JACTDZ 6(3),261,87
orl-rat LD50:700 mg/kg   JACTDZ 6(3),261,87
ipr-rat LD50:980 mg/kg   JADSAY 26,1461,39

CONSENSUS REPORTS: Reported in EPA TSCA Inventory.

SAFETY PROFILE: Moderately toxic by ingestion and intraperitoneal routes. A skin and eye irritant. When heated to decomposition it emits toxic vapors of SO_x.

SIM100   CAS:1121-70-6   **HR: 3**
SODIUM-4-METHYLPHENOXIDE
mf: C_7H_7O•Na   mw: 130.13

$NaOC_6H_4CH_3$

SYNS: 4-METHYLPHENOL, SODIUM SALT ◊ PHENOL, 4-METHYL-, SODIUM SALT (9CI) ◊ SODIUM p-CRESOLATE ◊ SODIUM p-CRESOXIDE ◊ SODIUM p-METHYLPHENOLATE ◊ SODIUM 4-METHYLPHENOLATE ◊ SODIUM p-METHYLPHENOXIDE

TOXICITY DATA with REFERENCE
scu-mus LDLo:150 mg/kg   AEXPBL 52,220,05
scu-frg LDLo:150 mg/kg   AEXPBL 52,220,05

CONSENSUS REPORTS: Reported in EPA TSCA Inventory.

SAFETY PROFILE: Poison by subcutaneous route. Ignites when heated in air even when moist. Oxidizes vigorously when heated in air. When heated to decomposition it emits toxic fumes of Na_2O. See also SODIUM COMPOUNDS.

SIN900   CAS:822-12-8   **HR: 3**
SODIUM MYRISTATE
mf: $C_{14}H_{28}O_2$•Na   mw: 251.41

SYNS: MYRISTIC ACID, SODIUM SALT ◊ SODIUM TETRADECANOATE ◊ TETRADECANOIC ACID, SODIUM SALT (9CI)

TOXICITY DATA with REFERENCE
dni-gpg:kdy 200 µmol/L   FCTXAV 14,431,76
ivn-dog LDLo:10 mg/kg   JCINAO 42,860,63

CONSENSUS REPORTS: Reported in EPA TSCA Inventory.

SAFETY PROFILE: Poison by intravenous route. Mutation data reported. When heated to decomposition it emits acrid smoke and irritating vapors.

SIT850   CAS:584-42-9   **HR: D**
SODIUM m-NITROBENZENEAZOSALICYLATE
mf: $C_{13}H_8N_3O_5$•Na   mw: 309.23

SYNS: ACID CHROME YELLOW GG ◊ ACID CHROME YELLOW 2GW ◊ ALIZARINE GR ◊ ALIZARINE YELLOW AGP ◊ ALIZARINE YELLOW G ◊ ALIZARINE YELLOW 2G ◊ ALIZARINE YELLOW GG ◊ ALIZARINE YELLOW GGW ◊ ALIZARINE YELLOW GM ◊ ALIZARIN YELLOW ◊ ALIZARIN YELLOW G ◊ ALIZARIN YELLOW GG ◊ ALIZARIN YELLOW G SODIUM SALT ◊ ALIZAROL YELLOW GW ◊ ANTHRANOL CHROME YELLOW 2G ◊ ANTHRANOL CHROME YELLOW 5GS ◊ ATLANTICHROME YELLOW 2G ◊ AZOCHROMOL YELLOW 5G ◊ BENZOIC ACID, 2-HYDROXY-5-((3-NITROPHENYL)AZO)-, MONOSODIUM SALT ◊ CALCOCHROME YELLOW 2G ◊ CHROMAL YELLOW M ◊ CHROME YELLOW 2G ◊ CHROME YELLOW 2GR ◊ CHROMOL YELLOW G ◊ CHROMOL YELLOW N ◊ C.I. 14025 ◊ C.I. MORDANT YELLOW 1 ◊ C.I. MORDANT YELLOW 1, MONOSODIUM SALT ◊ ENIACROMO YELLOW G ◊ ERIOCHROMAL YELLOW G ◊ ERIOCHROME YELLOW 2G ◊ ERIOCHROME YELLOW GS ◊ FENAKROM YELLOW R ◊ HIDACHROME YELLOW 2G ◊ HISPACROM YELLOW 2G ◊ HISPACROM YELLOW 2GR ◊ JAVA CHROME YELLOW GT ◊ JAVA UNICHROME YELLOW GT ◊ KAYAKU MORDANT YELLOW GG ◊ MAGRACROM YELLOW GG ◊ METACHROME YELLOW ◊ METACHROME YELLOW RA ◊ METOMEGA CHROME YELLOW GM ◊ MITSUI CHROME YELLOW GG ◊ MONOCHROME YELLOW MG ◊ ORTHOCHROME YELLOW GGW ◊ PONTACHROME YELLOW GS ◊ RV 1 ◊ SALICYL YELLOW ◊ SHOWA CHROME YELLOW GG ◊ SOLOCHROME YELLOW WN ◊ SUNCHROMINE YELLOW GG ◊ YUDOCHROME YELLOW GGN

TOXICITY DATA with REFERENCE
mmo-sat 500 µg/plate   MUREAV 56,249,78
mma-sat 500 µg/plate   MUREAV 56,249,78

CONSENSUS REPORTS: Reported in EPA TSCA Inventory.

SAFETY PROFILE: Mutation data reported. When heated to decomposition it emits toxic vapors of NO_x.

SIX600   CAS:1984-06-1   **HR: D**
SODIUM n-OCTANOATE
mf: $C_8H_{16}O_2$•Na   mw: 167.23

SYNS: CAPRYLIC ACID SODIUM SALT ◊ OCTANOIC ACID, SODIUM SALT ◊ SODIUM CAPRYLATE ◊ SODIUM OCTANOATE

TOXICITY DATA with REFERENCE
dni-gpg:kdy 600 µmol/L   FCTXAV 14,431,76

CONSENSUS REPORTS: Reported in EPA TSCA Inventory.

SAFETY PROFILE: Mutation data reported. When heated to decomposition it emits acrid smoke and irritating vapors.

SIY600   CAS:2013-26-5   **HR: 2**
SODIUM 2-OXOBUTYRATE
mf: $C_4H_5O_3$•Na   mw: 124.08

SYNS: BUTANOIC ACID, 2-OXO-, SODIUM SALT (9CI) ◊ BUTYRIC ACID, 2-OXO-, SODIUM SALT ◊ SODIUM-α-KETOBUTYRATE ◊ SODIUM 2-OXOBUTANOATE

TOXICITY DATA with REFERENCE
scu-mus LD50:2960 mg/kg   FATOAO 42,221,79

CONSENSUS REPORTS: Reported in EPA TSCA Inventory.

SAFETY PROFILE: Moderately toxic by subcutaneous route. When heated to decomposition it emits acrid smoke and irritating vapors.

SJB400   CAS:3313-92-6   **HR: 2**
SODIUM PERCARBONATE
mf: $C_2H_2O_6$•2Na   mw: 168.02

SYNS: DISODIUM PEROXYDICARBONATE ◊ PEROXYDICARBONIC ACID, DISODIUM SALT ◊ SODIUM CARBONATE ◊ SODIUM PEROXYDICARBONATE

TOXICITY DATA with REFERENCE
orl-mus LD50:2050 mg/kg   SKEZAP 27,553,86

CONSENSUS REPORTS: Reported in EPA TSCA Inventory.

SAFETY PROFILE: Moderately toxic by ingestion. When heated to decomposition it emits acrid smoke and irritating vapors.

SJH050   CAS:515-42-4   HR: 1
SODIUM PHENYLSULFONATE
mf: $C_6H_5O_3S \cdot Na$ mw: 180.16

SYNS: BENZENESULFONIC ACID, SODIUM SALT ◇ SODIUM BENZENESULFONATE ◇ SODIUM BENZOSULFONATE

TOXICITY DATA with REFERENCE
orl-mus LD50:9378 mg/kg   SKEZAP 4,15,63

CONSENSUS REPORTS: Reported in EPA TSCA Inventory.

SAFETY PROFILE: Slightly toxic by ingestion. When heated to decomposition it emits toxic vapors of SO_x.

SJW100   CAS:150-90-3   HR: 1
SODIUM SUCCINATE
mf: $C_4H_4O_4 \cdot 2Na$ mw: 162.06

SYNS: BUTANEDIOIC ACID, DISODIUM SALT (9CI) ◇ DISODIUM SUCCINATE ◇ JANTARAN SODNY ◇ SODUXIN ◇ SUCCINIC ACID, DISODIUM SALT

TOXICITY DATA with REFERENCE
ivn-mus LD50:4500 mg/kg   85ESA3 11,1368,89

CONSENSUS REPORTS: Reported in EPA TSCA Inventory.

SAFETY PROFILE: Slightly toxic by intravenous route. When heated to decomposition it emits acrid smoke and irritating vapors.

SLG600   CAS:334-50-9   HR: 2
SPRIDINE TRIHYDROCHLORIDE
mf: $C_7H_{19}N_3 \cdot 3ClH$ mw: 254.67

SYNS: N-(3-AMINOPROPYL)-1,4-BUTANEDIAMINETRIHYDROCHLORIDE ◇ 1,4-BUTANEDIAMINE, N-(3-AMINOPROPYL)-, TRIHYDROCHLORIDE ◇ SPRIDINE HYDROCHLORIDE

TOXICITY DATA with REFERENCE
ipr-mus LD50:870 mg/kg   FEPRA7 41,1575,82

CONSENSUS REPORTS: Reported in EPA TSCA Inventory.

SAFETY PROFILE: Moderately toxic by intraperitoneal route. When heated to decomposition it emits toxic vapors of NO_x, HCl, and Cl^-.

SLG700   CAS:111-01-3   HR: 1
SQUALANE
mf: $C_{30}H_{62}$ mw: 422.92

SYNS: COSBIOL ◇ 2,6,10,15,19,23-HEXAMETHYLTETRACOSANE ◇ TETRACOSANE, 2,6,10,15,19,23-HEXAMETHYL-

TOXICITY DATA with REFERENCE
skn-rbt 100 mg/24H MLD   CTOIDG 94(8),41,79
skn-gpg 100 mg/24H MOD   CTOIDG 94(8),41,79

CONSENSUS REPORTS: Reported in EPA TSCA Inventory.

SAFETY PROFILE: A skin irritant. When heated to decomposition it emits acrid smoke and irritating vapors.

SLG800   CAS:111-02-4   HR: 2
trans-SQUALENE

SYNS: SPINACEN ◇ SPINACENE ◇ SQUALEN ◇ SQUALENE ◇ (E,E,E,E)-SQUALENE ◇ 2,6,10,14,18,22-TETRACOSAHEXENE, 2,6,10,15,19,23-HEXAMETHYL-, (all-E)-

TOXICITY DATA with REFERENCE
orl-mus LD50:5 g/kg   JJCREP 76,1021,85
ivn-mus LD50:1800 mg/kg   JJCREP 76,1021,85

CONSENSUS REPORTS: Reported in EPA TSCA Inventory.

SAFETY PROFILE: Moderately toxic by intravenous route. Slightly toxic by ingestion. When heated to decomposition it emits acrid smoke and irritating vapors.

SLJ500   CAS:9005-25-8   HR: 2
STARCH DUST

SYNS: AMAIZO W 13 ◇ AMYLOMAIZE VII ◇ AMYLUM ◇ AQUAPEL (POLYSACCHARIDE) ◇ ARGO BRAND CORN STARCH ◇ ARROWROOT STARCH ◇ CLARO 5591 ◇ CLEARJEL ◇ CLEAREL ◇ CORN PRODUCTS ◇ CPC 3005 ◇ CPC 6448 ◇ FARINEX 100 ◇ GALACTASOL A ◇ GENVIS ◇ HRW 13 ◇ KEESTAR ◇ MAIZENA ◇ MARANTA ◇ MELOGEL ◇ MELUNA ◇ OK PRE-GEL ◇ PENFORD GUM 380 ◇ REMYLINE Ac ◇ RICE STARCH ◇ SORGHUM GUM ◇ STARAMIC 747 ◇ STARCH ◇ α-STARCH ◇ STARCH, CORN ◇ STARCH (OSHA) ◇ STA-RX 1500 ◇ TAPIOCA STARCH ◇ TAPON ◇ TROGUM ◇ W-GUM ◇ W-13 STABILIZER

TOXICITY DATA with REFERENCE
skn-hmn 300 μg/3D-I MLD   85DKA8 -,127,77
ipr-mus LD50:6600 mg/kg   PCJOAU 15,139,81

CONSENSUS REPORTS: Reported in EPA TSCA Inventory.

OSHA PEL: Total Dust: 15 mg/m^3; Respirable Fraction: 5 mg/m^3
ACGIH TLV: TWA (nuisance particulate) 10 mg/m^3 of total dust (when toxic impurities are not present, e.g., quartz <1%).
NIOSH REL: (Starch, respirable fraction) 5 mg/m^3; (total dust) 10 mg/m^3

SAFETY PROFILE: A nuisance dust. Mildly toxic by intraperitoneal route. A skin irritant. An allergen. Flammable when exposed to flame, can react with oxidizing materials. Moderately explosive when exposed to flame.

SLR100 CAS:103-30-0 **HR: 2**
trans-STILBENE
mf: $C_{14}H_{12}$ mw: 180.26

SYNS: BENZENE, 1,1′-(1,2-ETHENEDIYL)BIS-, (E)-(9CI) ◊ trans-DIPHENYLETHENE ◊ trans-1,2-DIPHENYLETHENE ◊ trans-α-β-DIPHENYLETHYLENE ◊ (E)-1,2-DIPHENYLETHYLENE ◊ trans-1,2-DIPHENYLETHYLENE ◊ (E)-1,1′-(1,2-ETHENEDIYL)BISBENZENE ◊ STILBENE, (E)- ◊ (E)-STILBENE

TOXICITY DATA with REFERENCE
orl-mus LD50:920 mg/kg   YHHPAL 14,227,79
ipr-mus LD50:6500 mg/kg   YHHPAL 14,227,79

CONSENSUS REPORTS: Reported in EPA TSCA Inventory.

SAFETY PROFILE: Moderately toxic by ingestion. Slightly toxic by intraperitoneal route. When heated to decomposition it emits acrid smoke and irritating vapors.

SMA000 CAS:3930-19-6 **HR: 3**
STREPTONIGRAN
mf: $C_{25}H_{22}N_4O_8$ mw: 506.51

SYNS: BRUNEOMYCIN ◊ NIGRIN ◊ NSC 45383 ◊ 5278 R.P. ◊ RUFOCHROMOMYCIN ◊ RUFOCROMOMYCINE ◊ SN ◊ STREPTONIGRIN ◊ VALACIDIN

TOXICITY DATA with REFERENCE
mmo-sat 100 ng/plate   PNASA6 79,7445,82
mmo-bcs 600 µg/L   CMMUAO 9,165,84
par-rat TDLo:250 µg/kg (female 11D post):TER   JEEMAF 18,215,67
ipr-rat TDLo:100 µg/kg (10D preg):TER   CCROBU 53,23,69
unr-rat TDLo:100 µg/kg (female 11D post):TER   85DJA5 -,95,71
ivn-rbt TDLo:90 µg/kg (female 1D pre):REP   MUREAV 127,73,84
ivn-rbt TDLo:90 µg/kg (female 1D pre):REP   MUREAV 127,73,84
ipr-rat TDLo:200 µg/kg (female 9D post):TER   CCROBU 53,23,69
ipr-rat TDLo:100 µg/kg (10D preg):TER   CCROBU 53,23,69
ipr-rat TDLo:100 µg/kg (10D preg):TER   CCROBU 53,23,69
ipr-rat TDLo:200 µg/kg (female 9D post):TER   CCROBU 53,23,69
ipr-rat TDLo:100 µg/kg (10D preg):TER   CCROBU 53,23,69
unr-rat TDLo:100 µg/kg (female 2D post):TER   85DJA5 -,95,71
ipr-rat TDLo:200 µg/kg (female 9D post):TER   CCROBU 53,23,69
ipr-rat LD50:150 µg/kg   FATOAO 37,197,74
orl-mus LD50:2330 µg/kg   ANTBAL 12,132,67
ipr-mus LD50:520 µg/kg   FATOAO 37,197,74
scu-mus LD50:1000 µg/kg   COREAF 261,4911,65
ivn-mus LD50:500 µg/kg   85ERAY 2,1359,78
orl-dog LD50:500 µg/kg   ANTBAL 11,1090,66
ivn-rat LD50:600 µg/kg   ANTBAL 12,132,67

CONSENSUS REPORTS: Reported in EPA TSCA Inventory.

SAFETY PROFILE: A deadly poison by ingestion, intraperitoneal, subcutaneous, and intravenous routes. An experimental teratogen. Experimental reproductive effects. Human mutation data reported. When heated to decomposition it emits toxic fumes of NO_x.

SMH000 CAS:7789-06-2 **HR: 3**
STRONTIUM CHROMATE (1:1)
mf: $CrO_4 \cdot Sr$ mw: 203.62

PROP: Monoclinic, yellow crystals. D: 3.895 @ 15°.

SYNS: CHROMIC ACID, STRONTIUM SALT (1:1) ◊ C.I. PIGMENT YELLOW 32 ◊ DEEP LEMON YELLOW ◊ STRONTIUM CHROMATE (VI) ◊ STRONTIUM CHROMATE 12170 ◊ STRONTIUM YELLOW

TOXICITY DATA with REFERENCE
mmo-sat 800 ng/plate   MUREAV 156,219,85
sce-ham:ovr 100 µg/L   MUREAV 156,219,85
itr-rat TDLo:40 mg/kg/34W-I:ETA   AEHLAU 5,445,62
imp-rat TDLo:125 mg/kg:ETA   AIHAAP 20,274,59
orl-rat LD50:3118 mg/kg   GISAAA 45(10),76,80
itr-rat LD50:16,600 mg/kg   GISAAA 45(10),76,80

CONSENSUS REPORTS: NTP Fifth Annual Report on Carcinogens. IARC Cancer Review: Group 1 IMEMDT 7,165,87; Animal Sufficient Evidence IMEMDT 2,100,73; IMEMDT 23,205,80; Human Sufficient Evidence IMEMDT 23,205,80. Chromium and its compounds are on the Community Right-To-Know List. Reported in EPA TSCA Inventory.

OSHA PEL: CL 0.1 mg(CrO_3)/m^3
ACGIH TLV: TWA 0.0005 ppm; Suspected Human Carcinogen
DFG TRK: 0.1 mg/m^3; Animal Carcinogen, Suspected Human Carcinogen
NIOSH REL: TWA 0.0001 mg(Cr(VI))/m^3

SAFETY PROFILE: Confirmed human carcinogen with experimental carcinogenic and tumorigenic data. Moderately toxic by ingestion. Mutation data reported. See also CHROMIUM COMPOUNDS and STRONTIUM COMPOUNDS.

SMQ100 CAS:93-56-1 **HR: 1**
STYRENE GLYCOL
mf: $C_8H_{10}O_2$ mw: 138.18

SYNS: α-β-DIHYDROXYETHYLBENZENE ◊ 1,2-DIHYDROXYETHYLBENZENE ◊ 1,2-ETHANEDIOL, PHENYL- ◊ 1,2-ETHANEDIOL, 1-PHENYL- ◊ 1-FENYL-1,2-ETHANDIOL ◊ FENYLGLYCOL ◊ PHENYLETHANEDIOL

◇ PHENYLETHYLENE GLYCOL ◇ 1-PHENYLETHYLENE GLYCOL ◇ PHENYL GLYCOL ◇ STYROLYL ALCOHOL

TOXICITY DATA with REFERENCE
orl-gpg LD50:2 g/kg   85JCAE -,215,86

CONSENSUS REPORTS: Reported in EPA TSCA Inventory.

SAFETY PROFILE: Slightly toxic by ingestion. When heated to decomposition it emits acrid smoke and irritating vapors.

SNB100   CAS:106-65-0   HR: 1
SUCCINIC ACID, DIMETHYL ESTER
mf: $C_6H_{10}O_4$ mw: 146.16

SYNS: BUTANEDIOIC ACID, DIMETHYL ESTER ◇ DIMETHYL BUTANEDIOATE ◇ DIMETHYL SUCCINATE

TOXICITY DATA with REFERENCE
orl-rat LD50:>5 g/kg   FCTXAV 18,677,80
skn-rbt LD50:>5 g/kg   FCTXAV 18,677,80

SAFETY PROFILE: Slightly toxic by ingestion and skin contact. When heated to decomposition it emits acrid smoke and irritating vapors.

SNH050   CAS:126-13-6   HR: 1
SUCROSE ACETATE ISOBUTYRATE
mf: $C_{40}H_{62}O_{19}$ mw: 847.02

SYNS: SACCHAROSE ACETATE ISOBUTYRATE ◇ SAIB ◇ SUCROSE ACETOISOBUTYRATE ◇ SUCROSE, DIACETATE HEXAISOBUTYRATE

TOXICITY DATA with REFERENCE
orl-rat LDLo:25,600 mg/kg   JAFCAU 21,473,73
ipr-rat LDLo:25,600 mg/kg   JAFCAU 21,473,73
orl-mus LDLo:>25,600 mg/kg   JAFCAU 21,473,73
ipr-mus LDLo:25,600 mg/kg   JAFCAU 21,473,73

CONSENSUS REPORTS: Reported in EPA TSCA Inventory.

SAFETY PROFILE: Slightly toxic by ingestion and intraperitoneal routes. When heated to decomposition it emits acrid smoke and irritating vapors.

SNL840   CAS:138-41-0   HR: 3
p-SULFAMOYLBENZOIC ACID
mf: $C_7H_7NO_4S$ mw: 201.21

SYNS: 4-(AMINOSULFONYL)BENZOIC ACID ◇ BENZOIC ACID, 4-(AMINOSULFONYL)-(9CI) ◇ BENZOIC ACID p-SULFAMIDE ◇ BENZOIC ACID, p-SULFAMOYL- ◇ p-CARBOXYBENZENESULFONAMIDE ◇ 4-CARBOXYBENZENESULFONAMIDE ◇ CARZENID ◇ CARZENIDE ◇ DIRNATE ◇ 4-SULFAMOYLBENZOIC ACID ◇ p-SULFAMYLBENZOIC ACID ◇ p-SULFONAMIDOBENZOIC ACID

TOXICITY DATA with REFERENCE
ipr-rat LD50:350 mg/kg   PHARAT 38,102,83

CONSENSUS REPORTS: Reported in EPA TSCA Inventory.

SAFETY PROFILE: Poison by intraperitoneal route. When heated to decomposition it emits toxic vapors of NO_x and SO_x.

SNN600   CAS:121-57-3   HR: 1
SULFANILIC ACID
mf: $C_6H_7NO_3S$ mw: 173.20

SYNS: p-AMINOBENZENESULFONIC ACID ◇ 4-AMINOBENZENESULFONIC ACID ◇ p-AMINOPHENYLSULFONIC ACID ◇ ANILINE-p-SULFONIC ACID ◇ ANILINE-4-SULFONIC ACID ◇ ANILINE-p-SULPHONIC ACID ◇ BENZENESULFONIC ACID, 4-AMINO-(9CI) ◇ KYSELINA SULFANILOVA ◇ SULFANILSAEURE ◇ SULPHANILIC ACID

TOXICITY DATA with REFERENCE
skn-rbt 500 mg/24H MLD   28ZPAK -,180,72
eye-rbt 100 mg/24H MOD   28ZPAK -,180,72
orl-rat LD50:12,300 mg/kg   28ZPAK -,180,72
ivn-rat LD50:6 g/kg   AEPPAE 211,367,50

CONSENSUS REPORTS: Reported in EPA TSCA Inventory.

SAFETY PROFILE: Slightly toxic by ingestion and intravenous routes. A skin irritant. When heated to decomposition it emits toxic vapors of NO_x and SO_x.

SNO100   CAS:88-21-1   HR: D
o-SULFANILIC ACID
mf: $C_6H_7NO_3S$ mw: 173.20

SYNS: o-AMINOBENZENESULFONIC ACID ◇ 2-AMINOBENZENESULFONIC ACID ◇ o-AMINOPHENYLSULFONIC ACID ◇ ANILINO-o-SULFONIC ACID ◇ ANILINO-2-SULFONIC ACID ◇ ANILINO-o-SULPHONIC ACID ◇ BENZENESULFONIC ACID, o-AMINO- ◇ ORTHANILIC ACID

TOXICITY DATA with REFERENCE
mma-sat 6667 μg/plate   EMMUEG 11(Suppl 12),1,88

CONSENSUS REPORTS: Reported in EPA TSCA Inventory.

SAFETY PROFILE: Mutation data reported.

SOB600   CAS:81-69-6   HR: 3
SULFOPARABLUE
mf: $C_{20}H_{15}N_3O_8S_2$ mw: 489.50

SYNS: 1-AMINO-2-SULFO-4-(4'-AMINO-3'-SULFOANILINO)ANTHRAQUINONE ◇ 2-ANTHRACENESULFONIC ACID, 1-AMINO-4-(4-AMINO-3-SULFOANILINO)-9,10-DIHYDRO-9,10-DIOXO- ◇ 2-ANTHRACENESULFONIC ACID,1-AMINO-4-((4-AMINO-3-SULFOPHENYL)AMINO)-9,10-DIHYDRO-9,10-DIOXO- ◇ USAF ND-60

TOXICITY DATA with REFERENCE
ipr-mus LD50:300 mg/kg   NTIS** AD277-689

CONSENSUS REPORTS: Reported in EPA TSCA Inventory.

SAFETY PROFILE: Poison by intraperitoneal route. When heated to decomposition it emits toxic vapors of NO_x and SO_x.

SOD300   CAS:1639-66-3   **HR: 2**
SULFOSUCCINIC ACID 1,4-DIOCTYL ESTER SODIUM SALT
mf: $C_{20}H_{37}O_7S \cdot Na$ mw: 444.62

SYNS: BUTANEDIOIC ACID, SULFO-, 1,4-DIOCTYL ESTER, SODIUM SALT (9CI) ◇ DI-n-OCTYL SODIUM SULFOSUCCINATE ◇ DIOKTYLESTER SULFOJANTARANU SODNEHO ◇ SUCCINIC ACID, SULFO-, 1,4-DIOCTYL ESTER, SODIUM SALT

TOXICITY DATA with REFERENCE
orl-rat LD50:1900 mg/kg   85JCAE -,1065,86
orl-mus LD50:4800 mg/kg   85JCAE -,1065,86

CONSENSUS REPORTS: Reported in EPA TSCA Inventory.

SAFETY PROFILE: Moderately toxic by ingestion. When heated to decomposition it emits toxic vapors of SO_x.

SPE700   CAS:530-57-4   **HR: D**
SYRINGIC ACID
mf: $C_9H_{10}O_5$ mw: 198.19

SYNS: BENZOIC ACID, 4-HYDROXY-3,5-DIMETHOXY- ◇ 4-HYDROXY-3,5-DIMETHOXYBENZOIC ACID

TOXICITY DATA with REFERENCE
cyt-ham:ovr 3 g/L   CALEDQ 14,251,81

CONSENSUS REPORTS: Reported in EPA TSCA Inventory.

SAFETY PROFILE: Mutation data reported. When heated to decomposition it emits acrid smoke and irritating vapors.

T

TAN750   CAS:100-21-0   *HR: 2*
TEREPHTHALIC ACID
mf: $C_8H_6O_4$ mw: 166.14

PROP: White crystals or powder. D: 1.51, sublimes @ 300°. Insol in water, chloroform, ether, acetic acid; sltly sol in alc; sol in alkalies.

SYNS: ACIDE TEREPHTALIQUE (FRENCH) ◊ p-BENZENEDICARBOXYLIC ACID ◊ 1,4-BENZENEDICARBOXYLIC ACID ◊ KYSELINA TERFTALOVA (CZECH) ◊ TA 12 ◊ TA-33MP

TOXICITY DATA with REFERENCE
eye-rbt 500 mg/24H MOD   28ZPAK -,52,72
orl-rat LD50:18,800 mg/kg   28ZPAK -,52,72
orl-mus LDLo:10 g/kg   85GMAT -,107,82
ipr-mus LD50:1430 mg/kg   CPBTAL 16,1655,68
ivn-dog LDLo:767 mg/kg   TXAPA9 18,469,71

CONSENSUS REPORTS: Community Right-To-Know List. Reported in EPA TSCA Inventory. EPA Genetic Toxicology Program.
ACGIH TLV: (Proposed: TWA 10 mg/m^3)

SAFETY PROFILE: Moderately toxic by intravenous and intraperitoneal routes. Mildly toxic by ingestion. An eye irritant. Can explode during preparation. When heated to decomposition it emits acrid smoke and irritating fumes.

TBD775   CAS:138-87-4   *HR: D*
β-TERPINEOL
mf: $C_{10}H_{18}O$ mw: 154.28

SYNS: CYCLOHEXANOL, 1-METHYL-4-(1-METHYLETHENYL)-(9CI) ◊ p-MENTH-8-EN-1-OL ◊ t-MENTH-1-EN-8-OL ◊ 1-METHYL-4-(1-METHYLETHENYL)CYCLOHEXANOL

TOXICITY DATA with REFERENCE
bfa-rat:sat 2500 mg/kg   NUCADQ 1,10,79

CONSENSUS REPORTS: Reported in EPA TSCA Inventory.

SAFETY PROFILE: Mutation data reported. When heated to decomposition it emits acrid smoke and irritating vapors.

TBQ290   CAS:2539-17-5   *HR: 2*
TETRACHLOROGUAIACOL
mf: $C_7H_4Cl_4O_2$ mw: 261.91

SYNS: 2-METHOXYTETRACHLOROPHENOL ◊ 2-METHOXY-3,4,5,6-TETRACHLOROPHENOL ◊ PHENOL, 2-METHOXY-3,4,5,6-TETRACHLORO-

TOXICITY DATA with REFERENCE
orl-rat LD50:1690 mg/kg   TXAPA9 45,295,78

SAFETY PROFILE: Moderately toxic by ingestion. When heated to decomposition it emits toxic vapors of Cl$^-$.

TBT100   CAS:632-58-6   *HR: 1*
TETRACHLOROPHTHALIC ACID
mf: $C_8H_2Cl_4O_4$ mw: 303.90

SYNS: 1,2-BENZENEDICARBOXYLIC ACID, 3,4,5,6-TETRACHLORO-(9CI) ◊ PHTHALIC ACID, TETRACHLORO- ◊ 3,4,5,6-TETRACHLORO-1,2-BENZENEDICARBOXYLIC ACID

TOXICITY DATA with REFERENCE
orl-mus LD50:4640 mg/kg   NNGADV 1,283,76

CONSENSUS REPORTS: Reported in EPA TSCA Inventory.

SAFETY PROFILE: Slightly toxic by ingestion. When heated to decomposition it emits toxic vapors of Cl$^-$.

TCB200   CAS:1119-97-7   *HR: 3*
TETRADECYLTRIMETHYLAMMONIUM BROMIDE
mf: $C_{17}H_{38}N \bullet Br$ mw: 336.47

SYNS: AMMONIUM, TETRADECYLTRIMETHYL-, BROMIDE ◊ AMMONIUM, TRIMETHYLTETRADECYL-, BROMIDE ◊ MORPAN T ◊ MYRISTYLTRIMETHYLAMMONIUM BROMIDE ◊ MYTAB ◊ QUATERNIUM 13 ◊ 1-TETRADECANAMINIUM, N,N,N-TRIMETHYL-, BROMIDE (9CI) ◊ TETRADONIUM BROMIDE ◊ TRIMETHYLMYRISTYLAMMONIUM BROMIDE ◊ N,N,N-TRIMETHYL-1-TETRADECANAMINIUM BROMIDE ◊ TRIMETHYLTETRADECYLAMMONIUM BROMIDE

TOXICITY DATA with REFERENCE
ivn-rat LD50:15 mg/kg   APTOA6 47,17,80
ivn-mus LD50:12 mg/kg   APTOA6 47,17,80

CONSENSUS REPORTS: Reported in EPA TSCA Inventory.

SAFETY PROFILE: Poison by intravenous route. When heated to decomposition it emits toxic vapors of NO$_x$ and Br$^-$.

TCD002   CAS:2567-83-1   *HR: 2*
TETRAETHYLAMMONIUM PERCHLORATE (dry) (DOT)
DOT: NA 1325
mf: $C_8H_{20}N \bullet ClO_4$ mw: 229.74

SYNS: AMMONIUM, TETRAETHYL-, PERCHLORATE ◊ ETHANAMINIUM, N,N,N-TRIETHYL-, PERCHLORATE

CONSENSUS REPORTS: Reported in EPA TSCA Inventory.

DOT Classification: Flammable Solid; Label: Flammable Solid

SAFETY PROFILE: Flammable solid. When heated to decomposition it emits very toxic fumes of NO_x, NH_3, and Cl^-. See also PERCHLORATES.

TCD250 CAS:78-13-7 *HR: 1*
TETRA(2-ETHYLBUTYL) ORTHOSILICATE
mf: $C_{24}H_{52}O_4Si$ mw: 432.85

PROP: D: 0.8920–0.9018 @ 20/20°, mp: <− 100°, bp: 238° @ 50 mm, flash p: 335°F (OC). Insol in water; sltly sol in methanol; misc with most organic solvents.

SYNS: 2-ETHYL-1-BUTANOL, SILICATE ◇ TETRA(2-ETHYLBUTOXY) SILANE

TOXICITY DATA with REFERENCE
skn-rbt 500 mg open MLD   UCDS** 2/15/66
orl-rat LD50:20 g/kg   UCDS** 2/15/66

CONSENSUS REPORTS: Reported in EPA TSCA Inventory.

SAFETY PROFILE: Mildly toxic by ingestion. A skin irritant. Combustible when exposed to heat or flame. To fight fire, use mist, spray, dry chemical. When heated to decomposition it emits acrid smoke and irritating fumes. See also SILICATES and SILANE.

TCE400 CAS:109-17-1 *HR: D*
TETRAETHYLENE GLYCOL DIMETHACRYLATE
mf: $C_{16}H_{26}O_7$ mw: 330.42

SYNS: METHACRYLIC ACID, DIESTER with TETRAETHYLENE GLYCOL ◇ 2-PROPENOIC ACID, 2-METHYL-, OXYBIS(2,1-ETHANEDIYLOXY-2,1-ETHANEDIYL)ESTER ◇ SR 209 ◇ TGM 4

TOXICITY DATA with REFERENCE
skn-rbt 500 mg MLD   JTEHD6 19,149,86
mnt-mus:lyms 475 mg/L   MUTAEX 4,381,89
cyt-mus:lyms 350 mg/L   MUTAEX 4,381,89

CONSENSUS REPORTS: Reported in EPA TSCA Inventory.

SAFETY PROFILE: A skin irritant. Mutation data reported. When heated to decomposition it emits acrid smoke and irritating vapors.

TCS550 CAS:1192-30-9 *HR: 3*
TETRAHYDROFURFURYL BROMIDE
mf: C_5H_9BrO mw: 165.05

SYNS: 2-(BROMOMETHYL)-TETRAHYDROFURAN ◇ FURAN, 2-(BROMOMETHYL)TETRAHYDRO- ◇ FURAN, TETRAHYDRO-2-(BROMOMETHYL)-

TOXICITY DATA with REFERENCE
ivn-mus LD50:180 mg/kg   CSLNX* NX#03416

CONSENSUS REPORTS: Reported in EPA TSCA Inventory.

SAFETY PROFILE: Poison by intravenous route. When heated to decomposition it emits toxic vapors of Br^-.

TCU650 CAS:1200-89-1 *HR: 3*
1,4,4a,8a-TETRAHYDRO-1,4-METHANO-NAPHTHALENE-5,8-DIONE
mf: $C_{11}H_{10}O_2$ mw: 174.21

SYN: 1,4-METHANONAPHTHALENE-5,8-DIONE,1,4,4a,8a-TETRAHYDRO-

TOXICITY DATA with REFERENCE
ipr-mus LDLo:250 mg/kg   CBCCT* 6,223,54

CONSENSUS REPORTS: Reported in EPA TSCA Inventory.

SAFETY PROFILE: Poison by intraperitoneal route. When heated to decomposition it emits acrid smoke and irritating vapors.

TDB100 CAS:85-40-5 *HR: D*
TETRAHYDROPHTHALIC ACID IMIDE
mf: $C_8H_9NO_2$ mw: 151.18

SYNS: 4-CYCLOHEXENE-1,2-DICARBOXIMIDE ◇ 1H-ISOINDOLE-1,3(2H)-DIONE, 3-α-4,7,7-α-TETRAHYDRO-(9CI) ◇ TETRAHYDROPHTHALIMIDE ◇ Δ^4-TETRAHYDROPHTHALIMIDE ◇ 1,2,3,6-TETRAHYDROPHTHALIMIDE

TOXICITY DATA with REFERENCE
ipr-mus TDLo:100 mg/kg (female 9D post):REP   MOPMA3 13,133,77
ipr-mus TDLo:25 mg/kg (female 9D post):REP   MOPMA3 13,133,77
ipr-mus TDLo:25 mg/kg (female 9D post):REP   MOPMA3 13,133,77

CONSENSUS REPORTS: Reported in EPA TSCA Inventory.

SAFETY PROFILE: Experimental reproductive effects. When heated to decomposition it emits toxic vapors of NO_x.

TDC800 CAS:1003-10-7 *HR: 3*
TETRAHYDRO-2-THIOPHENONE
mf: C_4H_6OS mw: 102.16

SYNS: 4-BUTYROTHIOLACTONE ◇ DIHYDRO-2(3H)-THIOPHENONE ◇ 4,5-DIHYDRO-3(2H)-THIOPHENONE ◇ THIACYCLOPENTANONE-2 ◇ THIACYCLOPENTAN-2-ONE ◇ γ-THIOBUTYROLACTONE ◇ 4-THIOBUTYROLACTONE ◇ THIOLAN-2-ONE ◇ 2(3H)-THIOPHENONE, DIHYDRO-(8CI,9CI) ◇ 2-THIOPHENONE, TETRAHYDRO-

TOXICITY DATA with REFERENCE
orl-mus LD50:1860 mg/kg   DCTODJ 3,249,80
ipr-mus LD50:199 mg/kg   JMCMAR 14,846,71

CONSENSUS REPORTS: Reported in EPA TSCA Inventory.

SAFETY PROFILE: Poison by intraperitoneal route. Moderately toxic by ingestion. When heated to decomposition it emits toxic vapors of SO_x.

TDQ400   CAS:628-21-7   **HR: D**
TETRAMETHYLENE DIIODIDE
mf: $C_4H_8I_2$ mw: 309.92

SYNS: BUTANE, 1,4-DIIODO- ◊ 1,4-DIIODOBUTANE ◊ TETRAMETHYLENE IODIDE

TOXICITY DATA with REFERENCE
mmo-sat 10 μmol/plate   MUREAV 141,11,84

CONSENSUS REPORTS: Reported in EPA TSCA Inventory.

SAFETY PROFILE: Mutation data reported. When heated to decomposition it emits toxic vapors of I^-.

TDR100   CAS:111-18-2   **HR: 3**
N,N,N′,N′-TETRAMETHYL-1,6-HEXANEDIAMINE
mf: $C_{10}H_{24}N_2$ mw: 172.36

SYNS: 1,6-HEXANEDIAMINE, N,N,N′,N′-TETRAMETHYL- ◊ N,N,N′,N′-TETRAMETHYLHEXAMETHYLENE DIAMINE

TOXICITY DATA with REFERENCE
ivn-mus LD50:141 mg/kg   IJNEAQ 8,131,69

CONSENSUS REPORTS: Reported in EPA TSCA Inventory.

SAFETY PROFILE: Poison by intravenous route. When heated to decomposition it emits toxic vapors of NO_x.

TDT770   CAS:826-36-8   **HR: 3**
2,2,6,6-TETRAMETHYL-4-PIPERIDINONE
mf: $C_9H_{17}NO$ mw: 155.27

SYNS: 4-OXO-2,2,6,6-TETRAMETHYLPIPERIDINE ◊ 4-PIPERIDONE, 2,2,6,6-TETRAMETHYL- ◊ 4-PIPERIDINONE, 2,2,6,6-TETRAMETHYL-(9CI) ◊ 2,2,6,6-TETRAMETHYLPIPERIDINONE ◊ 2,2,6,6-TETRAMETHYL-PIPERIDONE ◊ 2,2,6,6-TETRAMETHYL-4-PIPERIDONE ◊ TRIACETONAMIN ◊ TRIACETONAMINE ◊ TRIACETONE AMINE ◊ TROJACETONOAMINY ◊ VINCUBINA ◊ VINCUBINE

TOXICITY DATA with REFERENCE
orl-rat LD50:1539 mg/kg   GTPZAB 28(5),53,84
ipr-rat LD50:385 mg/kg   APPHAX 24,652,67

CONSENSUS REPORTS: Reported in EPA TSCA Inventory.

SAFETY PROFILE: Poison by intraperitoneal route. Moderately toxic by ingestion. When heated to decomposition it emits toxic vapors of NO_x.

TDT800   CAS:2564-83-2   **HR: 2**
2,2,6,6-TETRAMETHYLPIPERIDINOOXY
mf: $C_9H_{18}NO$ mw: 156.28

SYNS: PIPERIDINOOXY, 2,2,6,6-TETRAMETHYL- ◊ 1-PIPERIDINYLOXY, 2,2,6,6-TETRAMETHYL-(9CI) ◊ TANAN ◊ TANANE ◊ TEMPO ◊ TETRAMETHYLPIPERIDINE NITROXIDE ◊ 2,2,6,6-TETRAMETHYLPIPERIDINE-1-OXYL ◊ TETRAMETHYLPIPERIDINOOXY ◊ 2,2,6,6-TETRAMETHYLPIPERIDINOOXYL ◊ 2,2,6,6-TETRAMETHYLPIPERIDINYLOXY ◊ 2,2,6,6-TETRAMETHYL-1-PIPERIDINYLOXY ◊ 2,2,6,6-TETRAMETHYLPIPERIDOXYL

TOXICITY DATA with REFERENCE
skn-rbt 500 mg/4H SEV   EPASR* 8EHQ-0786-0607S
eye-rbt 100 mg/24H SEV   EPASR* 8EHQ-0786-0607S
ihl-rat LCLo:4500 mg/m^3/2H   EPASR* 8EHQ-0786-0607S
skn-rbt LDLo:2 g/kg   EPASR* 8EHQ-0786-0607S

SAFETY PROFILE: Slightly toxic by inhalation and skin contact. A severe skin and eye irritant. When heated to decomposition it emits toxic vapors of NO_x.

TDV300   CAS:2154-68-9   **HR: 2**
2,2,5,5-TETRAMETHYL-1-PYRROLIDINYLOXY-3-CARBOXYLIC ACID
mf: $C_9H_{16}NO_3$ mw: 186.26

SYNS: PCA ◊ 1-PYRROLIDINYLOXY, 3-CARBOXY-2,2,5,5-TETRAMETHYL-

TOXICITY DATA with REFERENCE
ivn-rat LD50:2812 mg/kg   INVRAV 19,549,84

CONSENSUS REPORTS: Reported in EPA TSCA Inventory.

SAFETY PROFILE: Moderately toxic by intravenous route. When heated to decomposition it emits toxic vapors of NO_x.

TDY250   CAS:509-14-8   **HR: 3**
TETRANITROMETHANE
DOT: UN 1510
mf: CN_4O_8 mw: 196.05

PROP: Colorless or yellow liquid. Mp: 13°, bp: 125.7°, d: 1.650 @ 13°, vap press: 10 mm @ 22.7°. Insol in water; very sol in alc, ether.

SYNS: NCI-C55947 ◊ RCRA WASTE NUMBER P112 ◊ TNM

TOXICITY DATA with REFERENCE
orl-rat LD50:130 mg/kg   NTIS** AD-A051-334
ihl-rat LC50:18 ppm/4H   AMRL** TR-77-25,77
ivn-rat LD50:12,600 μg/kg   AMRL** TR-77-25,77
orl-mus LD50:375 mg/kg   NTIS** AD-A051-334
ihl-mus LC50:54 ppm/4H   AMRL** TR-77-25,77
ipr-mus LD50:53 mg/kg   NTIS** AD-A051-334
ivn-mus LD50:63,100 μg/kg   AMRL** TR-77-25,77

CONSENSUS REPORTS: EPA Extremely Hazardous Substances List. Reported in EPA TSCA Inventory.

OSHA PEL: TWA 1 ppm
ACGIH TLV: TWA 1 ppm (Proposed: TWA 0.005 ppm; Suspected Human Carcinogen)
DFG MAK: 1 ppm (8 mg/m^3)

DOT Classification: Oxidizer; Label: Oxidizer

SAFETY PROFILE: Poison by ingestion, inhalation, intravenous, and intraperitoneal routes. Irritating to the skin, eyes, mucous membranes, and respiratory passages, and does serious damage to the liver. It occurs as an impurity in crude TNT, and is thought to be mainly responsible for the irritating properties of that material. It can cause pulmonary edema, mild methemoglobinemia, and fatty degeneration of the liver and kidneys.

A powerful oxidizer. A very dangerous fire hazard. A severe explosion hazard when shocked or exposed to heat. May explode during distillation. Potentially explosive reaction with ferrocene, pyridine, sodium ethoxide. Mixtures with amines (e.g., aniline) ignite spontaneously and may explode. Mixtures with cotton or toluene may explode when ignited. Forms sensitive and powerful explosive mixtures with nitrobenzene; 1-nitrotoluene; 4-nitrotoluene; 1,3-dinitrobenzene; 1-nitronaphthalene; other oxygen-deficient explosives; hydrocarbons. Can react vigorously with oxidizing materials. Incompatible with aluminum. When heated to decomposition it emits highly toxic fumes of NO$_x$. Used as an oxidizer in rocket propellants and as an explosive. See also NITRATES, and EXPLOSIVES, HIGH.

TED650   CAS:994-65-0   HR: 1
TETRAPROPYLGERMANE
mf: C$_{12}$H$_{28}$Ge   mw: 244.99

SYNS: GERMANE, TETRAPROPYL- ◊ TETRAPROPYLGERMANIUM

TOXICITY DATA with REFERENCE
orl-rat LD50:>12 g/kg   CHDDAT 262,1302,66
ipr-rat LD50:12 g/kg   85JCAE -,1243,86
orl-mus LD50:>20 g/kg   CHDDAT 262,1302,66
ipr-mus LD50:5640 mg/kg   85JCAE -,1243,86

CONSENSUS REPORTS: Reported in EPA TSCA Inventory.

SAFETY PROFILE: Slightly toxic by ingestion and intraperitoneal routes. When heated to decomposition it emits toxic vapors of Ge$^-$.

TEL040   CAS:1314-12-1   HR: 3
THALLIUM OXIDE
mf: OTl$_2$ mw: 424.74

SYNS: DITHALLIUM OXIDE ◊ THALLOUS OXIDE

TOXICITY DATA with REFERENCE
orl-rat LD50:40,600 µg/kg   85ECAN 2,101,78
ipr-rat LD50:75 mg/kg   85ECAN 2,101,78
orl-dog LDLo:31,200 µg/kg   85ECAN 2,101,78
ipr-dog LDLo:31,200 µg/kg   85ECAN 2,101,78
orl-rbt LDLo:31,200 µg/kg   85ECAN 2,101,78
ipr-rbt LDLo:62,400 µg/kg   85ECAN 2,101,78
ivn-rbt LDLo:40,600 µg/kg   85ECAN 2,101,78

CONSENSUS REPORTS: Reported in EPA TSCA Inventory.

SAFETY PROFILE: Poison by ingestion, intraperitoneal, and intravenous route. When heated to decomposition it emits toxic vapors of Tl.

TES300   CAS:132-65-0   HR: 2
9-THIAFLUORENE
mf: C$_{12}$H$_8$S   mw: 184.26

SYNS: 2,2'-BIPHENYLYLENE SULFIDE ◊ DIBENZOTHIOPHENE ◊ DIBENZO(b,d)THIOPHENE ◊ DIPHENYLENE SULFIDE

TOXICITY DATA with REFERENCE
orl-mus LD50:470 mg/kg   FAATDF 12,787,89

SAFETY PROFILE: Moderately toxic by ingestion. When heated to decomposition it emits toxic vapors of SO$_x$.

TES800   CAS:67-16-3   HR: 2
THIAMINE DISULFIDE
mf: C$_{24}$H$_{34}$N$_8$O$_4$S$_2$ mw: 562.78

SYNS: ALGONEURINA ◊ ALITIA S ◊ ANEURINE DISULFIDE ◊ APREN S ◊ DAIOMIN ◊ DAISAZIN ◊ FEIDMIN 5 ◊ FORMAMIDE, N,N'-(DITHIOBIS(2-(2-HYDROXYETHYL)-1-METHYLVINYLENE))BIS(N-((4-AMINO-2-METHYL-5-PYRIMIDINYL)METHYL)- ◊ NEOLAMIN ◊ SSB$_1$ ◊ TDS ◊ TDS (NEUROTROPE) ◊ THIAMIDIN F ◊ THIAMIN DISULFIDE ◊ VITAMIN B$_1$ DISULFIDE

TOXICITY DATA with REFERENCE
ivn-rat LD50:1250 mg/kg   AEPPAE 236,149,59

CONSENSUS REPORTS: Reported in EPA TSCA Inventory.

SAFETY PROFILE: Moderately toxic by intravenous route. When heated to decomposition it emits toxic vapors of NO$_x$ and SO$_x$.

TEU300   CAS:288-47-1   HR: 2
THIAZOLE
mf: C$_3$H$_3$NS   mw: 85.13

TOXICITY DATA with REFERENCE
orl-mus LD50:983 mg/kg   DCTODJ 3,249,80

CONSENSUS REPORTS: Reported in EPA TSCA Inventory.

SAFETY PROFILE: Moderately toxic by ingestion. When heated to decomposition it emits toxic vapors of NO$_x$ and SO$_x$.

TEV600   CAS:1779-81-3   HR: 3
2-THIAZOLIDINIMINE
mf: $C_3H_6N_2S$ mw: 102.17

SYNS: 2-AMINO-2-THIAZOLINE ◇ 4,5-DIHYDRO-2-THIAZOLAMINE ◇ 2-IMINOTHIAZOLIDINE ◇ 2-THIAZOLAMINE, 4,5-DIHYDRO-(9CI) ◇ 2-THIAZOLINE, 2-AMINO- ◇ USAF PD-57

TOXICITY DATA with REFERENCE
ipr-mus LD50:150 mg/kg   NTIS** AD691-490

CONSENSUS REPORTS: Reported in EPA TSCA Inventory.

SAFETY PROFILE: Poison by intraperitoneal route. When heated to decomposition it emits toxic vapors of NO_x and SO_x.

TFC550   CAS:98-91-9   HR: 3
THIOBENZOIC ACID
mf: C_7H_6OS mw: 138.19

SYNS: ACIDO MERCAPTOBENZOICO ◇ BENZENECARBOTHIOIC ACID ◇ BENZOIC ACID, THIO- ◇ BENZOYL THIOL ◇ MONOTHIOBENZOIC ACID

TOXICITY DATA with REFERENCE
ipr-rat LD50:350 mg/kg   BCFAAI 112,53,73

CONSENSUS REPORTS: Reported in EPA TSCA Inventory.

SAFETY PROFILE: Poison by intraperitoneal route. When heated to decomposition it emits toxic vapors of SO_x.

TFH800   CAS:333-49-3   HR: 3
2-THIOCYTOSINE
mf: $C_4H_5N_3S$ mw: 127.18

SYNS: 2(1H)-PYRIMIDINETHIONE, 4-AMINO- ◇ THIOCYTOSINE

TOXICITY DATA with REFERENCE
dni-hmn:lyms 100 mmol/L   CYTBAI 5,97,72
oth-hmn:lyms 5 mmol/L   CYTBAI 5,97,72
ipr-mus LD50:255 mg/kg   ARZNAD 31,1713,81

CONSENSUS REPORTS: Reported in EPA TSCA Inventory.

SAFETY PROFILE: Poison by intraperitoneal route. Mutation data reported. When heated to decomposition it emits toxic vapors of NO_x and SO_x.

TFM600   CAS:527-72-0   HR: 2
2-THIOPHENECARBOXYLIC ACID
mf: $C_5H_4O_2S$ mw: 128.15

SYNS: 2-CARBOXYTHIOPHENE ◇ 2-THENOIC ACID ◇ α-THIOPHENECARBOXYLIC ACID ◇ 2-THIOPHENIC ACID

TOXICITY DATA with REFERENCE
ivn-mus LD50:1670 mg/kg   THERAP 26,831,71

CONSENSUS REPORTS: Reported in EPA TSCA Inventory.

SAFETY PROFILE: Moderately toxic by intravenous route. When heated to decomposition it emits toxic vapors of SO_x.

TFP300   CAS:2637-34-5   HR: 3
2-THIOPYRIDINE
mf: C_5H_5NS mw: 111.17

SYNS: 2-MERCAPTOPYRIDINE ◇ 2-PYRIDINETHIOL ◇ 2(1H)-PYRIDINETHIONE ◇ THIOPYRIDONE-2

TOXICITY DATA with REFERENCE
ipr-mus LD50:250 mg/kg   NTIS** AD691-490
ivn-mus LD50:250 mg/kg   AIPTAK 93,143,53

CONSENSUS REPORTS: Reported in EPA TSCA Inventory.

SAFETY PROFILE: Poison by intravenous and intraperitoneal routes. When heated to decomposition it emits toxic vapors of NO_x and SO_x.

TFQ650   CAS:636-26-0   HR: 3
2-THIOTHYMINE
mf: $C_5H_6N_2OS$ mw: 142.19

SYNS: 2,3-DIHYDRO-5-METHYL-2-THIOXO-4(1H)-PYRIMIDINONE ◇ 4(1H)-PYRIMIDINONE, 2,3-DIHYDRO-5-METHYL-2-THIOXO-(9CI) ◇ THIOTHYMINE ◇ THYMINE, 2-THIO-

TOXICITY DATA with REFERENCE
ipr-rat LDLo:1 g/kg   JPETAB 97,478,49
unr-rbt LDLo:100 mg/kg   JPETAB 97,478,49

CONSENSUS REPORTS: Reported in EPA TSCA Inventory.

SAFETY PROFILE: Poison by unspecified route. Slightly toxic by intraperitoneal route. When heated to decomposition it emits toxic vapors of NO_x and SO_x.

TFX850   CAS:76-61-9   HR: D
THYMOL, 6,6'-(3H-2,1-BENZOXATHIOL-3-YLIDENE)DI-, S,S-DIOXIDE
mf: $C_{27}H_{30}O_5S$ mw: 466.63

SYNS: PHENOL,4,4'-(3H-2,1-BENZOXATHIOL-3-YLIDENE)BIS(5-METHYL-2-(1-METHYLETHYL)-, S,S-DIOXIDE ◇ THYMOL BLUE ◇ THYMOLSULFONEPHALEIN ◇ THYMOLSULFONEPHTHALEIN ◇ THYMOLSULFOPHTHALEIN ◇ THYMOLSULPHONPHTHALEIN ◇ THYMOSULFONPHTHALEIN

TOXICITY DATA with REFERENCE
dnd-esc 50 μmol/L   MUREAV 89,95,81

CONSENSUS REPORTS: Reported in EPA TSCA Inventory.

SAFETY PROFILE: Mutation data reported. When heated to decomposition it emits toxic vapors of SO_x.

TGB250   CAS:7440-31-5   *HR: 3*
TIN
af: Sn   aw: 118.71

PROP: Cubic, gray, crystalline metallic element. Mp: 231.9°, stabilizes <18°, d: 7.31, vap press: 1 mm @ 1492°, bp: 2507°.

SYNS: SILVER MATT POWDER ◊ TIN (α) ◊ TIN FLAKE ◊ TIN POWDER ◊ ZINN (GERMAN)

TOXICITY DATA with REFERENCE
imp-rat TDLo:395 g/kg:ETA   RCOCB8 18,201,77
imp-mus TDLo:840 g/kg:ETA   RCOCB8 18,201,77

CONSENSUS REPORTS: Reported in EPA TSCA Inventory.

OSHA PEL: Organic Compounds: TWA 0.1 mg(Sn)/m^3 (skin); Inorganic Compounds (except oxides): TWA 2 mg/m^3
ACGIH TLV: TWA metal, oxide and inorganic compounds (except SnH_4) as Sn 2 mg/m^3; organic compounds TWA 0.1 mg(Sn)/m^3; STEL 0.2 mg(Sn)/m^3 (skin)
DFG MAK: Inorganic 2 mg/m^3, organic 0.1 mg/m^3
NIOSH REL: (Organotin Compounds) TWA 0.1 mg(Sn)/m^3

SAFETY PROFILE: An inhalation hazard. Questionable carcinogen with experimental tumorigenic data by implant route. Combustible in the form of dust when exposed to heat or by spontaneous chemical reaction with Br_2; BrF_3; Cl_2; ClF_3; $Cu(NO_3)$; K_2O_2; S. See also POWDERED METALS and TIN COMPOUNDS.

TGK750   CAS:108-88-3   *HR: 3*
TOLUENE
DOT: UN 1294
mf: C_7H_8 mw: 92.15

PROP: Colorless liquid; benzol-like odor. Mp: −95 to −94.5°, bp: 110.4°, flash p: 40°F (CC), ULC: 75–80, lel: 1.27%, uel: 7%, d: 0.866 @ 20°/4°, autoign temp: 996°F, vap press: 36.7 mm @ 30°, vap d: 3.14. Insol in water; sol in acetone; misc in abs alc, ether, chloroform.

SYNS: ANTISAL 1a ◊ BENZENE, METHYL- ◊ METHACIDE ◊ METHANE, PHENYL- ◊ METHYLBENZENE ◊ METHYLBENZOL ◊ NCI-C07272 ◊ PHENYLMETHANE ◊ RCRA WASTE NUMBER U220 ◊ TOLUEEN (DUTCH) ◊ TOLUEN (CZECH) ◊ TOLUOL (DOT) ◊ TOLUOLO (ITALIAN) ◊ TOLU-SOL

TOXICITY DATA with REFERENCE
eye-hmn 300 ppm   JIHTAB 25,282,43
skn-rbt 435 mg MLD   UCDS** 7/23/70
skn-rbt 500 MOD   FCTOD7 20,563,82
eye-rbt 870 μg MLD   UCDS** 7/23/70
eye-rbt 2 mg/24H SEV   28ZPAK -,23,72
eye-rbt 100 mg/30S rns MLD   FCTOD7 20,573,82
oms-grh-ihl 562 mg/L   MUREAV 113,467,83
cyt-rat-scu 12 g/kg/12D-I   GTPZAB 17(3),24,73
ihl-mus TCLo:400 ppm/7H (female 7–16D post):REP   FAATDF 6,145,86
orl-mus TDLo:9 g/kg (female 6–15D post):TER   TJADAB 19,41A,79
ihl-rat TCLo:1500 mg/m^3/24H (1–8D preg):TER   TXCYAC 11,55,78
ihl-rbt TCLo:1 g/m^3/24H (female 7–20D post):REP   ARTODN 8,425,85
ihl-rat TCLo:100 ppm (male 51W pre):REP   SAIGBL 13,501,71
orl-mus TDLo:30 g/kg (female 6–15D post):TER   TJADAB 19,41A,79
ihl-rat TCLo:1500 mg/m^3/24H (1–8D preg):TER   TXCYAC 11,55,78
ihl-mus TCLo:200 ppm/7H (female 7–16D post):TER   FAATDF 6,145,86
orl-hmn LDLo:50 mg/kg   YAKUD5 22,883,80
ihl-hmn TCLo:200 ppm:BRN,CNS,BLD   JAMAAP 123,1106,43
ihl-man TCLo:100 ppm:CNS   WEHRBJ 9,131,72
orl-rat LD50:5000 mg/kg   AMIHAB 19,403,59
ihl-rat LCLo:4000 ppm/4H   AIHAAP 30,470,69
ipr-rat LDLo:800 mg/kg   TXAPA9 1,156,59
ivn-rat LD50:1960 mg/kg   MELAAD 54,486,63
ihl-mus LC50:5320 ppm/8H   JIHTAB 25,366,43
ipr-mus LD50:640 mg/kg   ANYAA9 243,104,75
ihl-rbt LCLo:55,000 ppm/40M   JIDHAN 26,69,44
skn-rbt LD50:12,124 mg/kg   AIHAAP 30,470,69

CONSENSUS REPORTS: Community Right-To-Know List. Reported in EPA TSCA Inventory. EPA Genetic Toxicology Program.

OSHA PEL: (Transitional: TWA 200 ppm; CL 300 ppm; Pk 500 ppm/10M/8H) TWA 100 ppm; STEL 150 ppm
ACGIH TLV: TWA 50 ppm (skin); BEI: 1 mg(toluene)/L in venous blood at end of shift; 20 ppm toluene in end-exhaled air during shift
DFG MAK: 100 ppm (380 mg/m^3); BAT: 340 μg/dL in blood at end of shift.
NIOSH REL: (Toluene) TWA 100 ppm; CL 200 ppm/10M

DOT Classification: Flammable Liquid; Label: Flammable Liquid

SAFETY PROFILE: Poison by intraperitoneal route. Moderately toxic by intravenous and subcutaneous routes. Mildly toxic by inhalation. An experimental teratogen. Human systemic effects by inhalation: CNS recording changes, hallucinations or distorted perceptions, motor activity changes, antipsychotic, psychophysiological test changes, and bone marrow changes. Experimental reproductive effects. Mutation data reported. A human eye irritant. An experimental skin and severe eye irritant.

    Toluene is derived from coal tar, and commercial grades usually contain small amounts of benzene as an impurity. Inhalation of 200 ppm of toluene for 8 hours may cause impairment of coordination and reaction time; with higher concentrations (up to 800 ppm) these effects are increased and are observed in a shorter time. In the few cases of acute toluene poisoning reported, the effect has been that of a narcotic, the workman passing through a stage of intoxication into one of coma. Recovery following removal from exposure has been the rule. An occasional report of chronic poisoning describes an anemia and leukopenia, with biopsy showing a bone marrow hypoplasia. These effects, however, are less common in people working with toluene, and they are not as severe. At 200–500 ppm, headache, nausea, eye irritation, loss of appetite, a bad taste, lassitude, impairment of coordination and reaction time are reported, but are not usually accompanied by any laboratory or physical findings of significance. With higher concentrations, the above complaints are increased and, in addition, anemia, leukopenia, and enlarged liver may be found in rare cases. A common air contaminant, emitted from modern building materials (CENEAR 69,22,91). Used in production of drugs of abuse.

Flammable liquid. A very dangerous fire hazard when exposed to heat, flame, or oxidizers. Explosive in the form of vapor when exposed to heat or flame. Explosive reaction with 1,3-dichloro-5,5-dimethyl-2,4-imidazolididione; dinitrogen tetraoxide; concentrated nitric acid; H_2SO_4 + HNO_3; N_2O_4; $AgClO_4$; BrF_3; UF_6, sulfur dichloride. Forms an explosive mixture with tetranitromethane. Can react vigorously with oxidizing materials. To fight fire, use foam, CO_2, dry chemical. When heated to decomposition it emits acrid smoke and irritating fumes.

TGM450   CAS:459-44-9   HR: D
p-TOLUENEDIAZONIUM FLUOROBORATE
mf: $C_7H_7N_2 \cdot BF_4$ mw: 205.97

SYNS: BENZENEDIAZONIUM, 4-METHYL-, TETRAFLUOROBORATE (1−) (9CI) ◊ 4-METHYLBENZENEDIAZONIUM TETRAFLUOROBORATE (1−) ◊ p-METHYLPHENYLDIAZONIUM FLUOROBORATE ◊ p-TOLUENEDIAZONIUM, TETRAFLUOROBORATE (1−) ◊ p-TOLUENEDIAZONIUM TETRAFLUOROBORATE (7CI)

TOXICITY DATA with REFERENCE
mmo-sat 20 µg/plate   FCTOD7 25,669,87

CONSENSUS REPORTS: Reported in EPA TSCA Inventory.

SAFETY PROFILE: Mutation data reported. When heated to decomposition it emits toxic vapors of NO_x, boron, and Fl⁻.

TGN600   CAS:68-34-8   HR: 3
p-TOLUENESULFONANILIDE
mf: $C_{13}H_{13}NO_2S$ mw: 247.33

SYNS: BENZENESULFONAMIDE, 4-METHYL-N-PHENYL-(9CI) ◊ 4-METHYL-N-PHENYLBENZENESULFONAMIDE ◊ N-PHENYL-p-TOLUENESULFONAMIDE ◊ p-TOLUENESULFANILIDE ◊ p-TOLUENESULFONYLANILIDE ◊ N-TOSYLANILINE

TOXICITY DATA with REFERENCE
ivn-mus LD50:44 mg/kg   CPBTAL 12,1451,64

CONSENSUS REPORTS: Reported in EPA TSCA Inventory.

SAFETY PROFILE: Poison by intravenous route. When heated to decomposition it emits toxic vapors of NO_x and SO_x.

TGO300   CAS:329-98-6   HR: 3
α-TOLUENESULFONYL FLUORIDE
mf: $C_7H_7FO_2S$ mw: 174.20

SYNS: BENZENEMETHANESULFONYL FLUORIDE ◊ BENZYLSULFONYL FLUORIDE ◊ BENZYLSULPHONYL FLUORIDE ◊ PHENYLMETHANESULFONYL FLUORIDE ◊ PHENYLMETHYLSULFONYL FLUORIDE ◊ PMSF

TOXICITY DATA with REFERENCE
ipr-rat LD50:150 mg/kg   NATUAS 173,33,54
ipr-mus LD50:215 mg/kg   LIFSAK 31,1193,82

CONSENSUS REPORTS: Reported in EPA TSCA Inventory.

SAFETY PROFILE: Poison by intraperitoneal route. When heated to decomposition it emits toxic vapors of SO_x and F⁻.

TGO800   CAS:108-40-7   HR: 3
m-TOLUENETHIOL
mf: C_7H_8S mw: 124.21

SYNS: BENZENETHIOL, 3-METHYL-(9CI) ◊ m-MERCAPTOTOLUENE ◊ m-METHYLBENZENETHIOL ◊ 3-METHYLBENZENETHIOL ◊ m-METHYLTHIOPHENOL ◊ 3-METHYLTHIOPHENOL ◊ m-THIOCRESOL ◊ 3-THIOCRESOL ◊ m-TOLYLMERCAPTAN ◊ USAF EK-2680

TOXICITY DATA with REFERENCE
ipr-mus LD50:50 mg/kg   NTIS** AD277-689

CONSENSUS REPORTS: Reported in EPA TSCA Inventory.

SAFETY PROFILE: Poison by intraperitoneal route. When heated to decomposition it emits toxic vapors of SO_x.

TGX550   CAS:93-69-6   HR: 2
o-TOLYLBIGUANIDE
mf: $C_9H_{13}N_5$ mw: 191.27

SYNS: ALIANT ◊ BIGUANIDE, 1-o-TOLYL- ◊ EPONOC B ◊ IMIDODICARBONIMIDIC DIAMIDE, N-(2-METHYLPHENYL)-(9CI) ◊ N-(2-METHYLPHENYL)IMIDODICARBONIMIDIC DIAMIDE ◊ NOCCELER BG ◊ SOPANOX ◊ 1-o-TOLYLBIGUANIDE ◊ o-TOLYLDIGUANIDE ◊ VULKACIT 1000

TOXICITY DATA with REFERENCE
orl-rat LD50:800 mg/kg   IPSTB3 3,93,76

CONSENSUS REPORTS: Reported in EPA TSCA Inventory.

SAFETY PROFILE: Moderately toxic by ingestion. When heated to decomposition it emits toxic vapors of NO_x.

TGY800　　CAS:2687-25-4　　HR: 3
2,3-TOLYLENEDIAMINE
mf: $C_7H_{10}N_2$　　mw: 122.19

SYNS: 1,2-BENZENEDIAMINE, 3-METHYL-(9CI) ◇ 2,3-DIAMINO-TOLUENE ◇ 3-METHYL-1,2-BENZENEDIAMINE ◇ TOLUENE-2,3-DIAMINE ◇ 2,3-TOLUYLENEDIAMINE

TOXICITY DATA with REFERENCE
ipr-mus LD50:286 mg/kg　　GENEA3 26,109,85

CONSENSUS REPORTS: Reported in EPA TSCA Inventory.

SAFETY PROFILE: Poison by intraperitoneal route. When heated to decomposition it emits toxic vapors of NO_x.

THA300　　CAS:86-92-0　　HR: 1
1-(p-TOLYL)-3-METHYLPYRAZOLONE-5
mf: $C_{11}H_{12}N_2O$　　mw: 188.25

SYNS: 3-METHYL-1-p-TOLYL-PYRAZOLIN-5-ONE ◇ 2-PYRAZOLIN-5-ONE, 3-METHYL-1-p-TOLYL-

TOXICITY DATA with REFERENCE
orl-rat LD50:7450 mg/kg　　LONZA# 26APR82

CONSENSUS REPORTS: Reported in EPA TSCA Inventory.

SAFETY PROFILE: Slightly toxic by ingestion. When heated to decomposition it emits toxic vapors of NO_x.

THL800　　CAS:675-10-5　　HR: 2
TRIACETIC ACID LACTONE
mf: $C_6H_6O_3$　　mw: 126.12

SYNS: 4-HYDROXY-6-METHYL-2H-PYRAN-2-ONE ◇ 6-METHYL-4-HYDROXYPYRON-(2) ◇ 2H-PYRAN-2-ONE, 4-HYDROXY-6-METHYL-

TOXICITY DATA with REFERENCE
scu-mus LD50:3200 mg/kg　　AIPTAK 128,126,60

CONSENSUS REPORTS: Reported in EPA TSCA Inventory.

SAFETY PROFILE: Moderately toxic by subcutaneous route.

THM900　　CAS:959-52-4　　HR: 3
1,3,5-TRIACRYLOYLHEXAHYDROTRIAZINE
mf: $C_{12}H_{15}N_3O_3$　　mw: 249.30

SYNS: FIXIERER P ◇ HEXAHYDRO-1,3,5-TRIS(1-OXO-2-PROPENYL)-1,3,5-TRIAZINE ◇ TRIACRYLFORMAL ◇ TRIACRYLOYLHEXAHYDROTRIAZINE ◇ TRI(N-ACRYLOYL)HEXAHYDROTRIAZINE ◇ TRIACRYLOYLHEXAHYDRO-s-TRIAZINE ◇ TRIACRYLOYLPERHYDROTRIAZINE ◇ s-TRIAZINE, HEXAHYDRO-1,3,5-TRIACRYLOYL- ◇ 1,3,5-TRIAZINE, HEXAHYDRO-1,3,5-TRIS(1-OXO-2-PROPENYL)-(9CI) ◇ TRIS(N-ACRYLOYL)HEXAHYDROTRIAZINE ◇ TRIS(ACRYLOYL)HEXAHYDRO-s-TRIAZINE

TOXICITY DATA with REFERENCE
eye-rbt 500 μg SEV　　TXCYAC 40,145,86
mnt-mus-ipr 9300 μg/kg　　TXCYAC 40,145,86
cyt-ham:ovr 700 μg/L　　TXCYAC 40,145,86
orl-rat LD50:350 mg/kg　　TXCYAC 40,145,86
skn-rbt LD50:70 mg/kg　　TXCYAC 40,145,86

CONSENSUS REPORTS: Reported in EPA TSCA Inventory.

SAFETY PROFILE: Poison by ingestion and skin contact routes. Mutation data reported. A severe eye irritant. When heated to decomposition it emits toxic vapors of NO_x.

THN775　　CAS:108-72-5　　HR: D
1,3,5-TRIAMINOBENZENE
mf: $C_6H_9N_3$　　mw: 123.18

SYNS: 1,3,5-BENZENETRIAMINE ◇ s-TRIAMINOBENZENE ◇ sym-TRIAMINOBENZENE

TOXICITY DATA with REFERENCE
mma-sat 5 umol/plate　　JTSCDR 4,317,79

CONSENSUS REPORTS: Reported in EPA TSCA Inventory.

SAFETY PROFILE: Mutation data reported. When heated to decomposition it emits toxic vapors of NO_x.

THS825　　CAS:288-88-0　　HR: 2
s-TRIAZOLE
mf: $C_2H_3N_3$　　mw: 69.08

SYNS: TA ◇ 1H-1,2,4-TRIAZOLE (9CI)

TOXICITY DATA with REFERENCE
skn-rbt 25%　　GISAAA 51(11),65,86
eye-rbt 12%　　GISAAA 51(11),65,86
orl-rat LD50:1750 mg/kg　　GISAAA 51(11),65,86
orl-mus LD50:1350 mg/kg　　GISAAA 51(11),65,86

CONSENSUS REPORTS: Reported in EPA TSCA Inventory.

SAFETY PROFILE: Moderately toxic by ingestion. A skin and eye irritant. When heated to decomposition it emits toxic vapors of NO_x.

THV100　　CAS:598-16-3　　HR: 2
1,1,2-TRIBROMOETHYLENE
mf: C_2HBr_3　　mw: 264.76

SYNS: ETHENE, TRIBROMO- ◇ ETHYLENE, TRIBROMO- ◇ TRIBROMOETHENE ◇ TRIBROMOETHYLENE

TOXICITY DATA with REFERENCE
orl-mus LD50:1100 mg/kg　　85JCAE -,123,86
ihl-mus LC50:3900 mg/m^3/2H　　85JCAE -,123,86

CONSENSUS REPORTS: Reported in EPA TSCA Inventory.

SAFETY PROFILE: Moderately toxic by ingestion and inhalation. When heated to decomposition it emits toxic vapors of Br⁻.

THV500   CAS:73941-35-2   *HR: 3*
2,4,5-TRIBROMOIMIDAZOLE CADMIUM SALT (2:1)
mf: $C_6Br_6N_4 \cdot Cd$ mw: 719.96

SYN: CADMIUM salt of 2,4-5-TRIBROMOIMIDAZOLE

TOXICITY DATA with REFERENCE
ivn-mus LD50:56 mg/kg   CSLNX* NX#06532

CONSENSUS REPORTS: Cadmium and its compounds are on the Community Right-To-Know List.

OSHA PEL: TWA 5 μg(Cd)/m³
ACGIH TLV: TWA 0.05 mg(Cd)/m³ (Proposed: TWA 0.01 mg(Cd)/m³ (dust), Suspected Human Carcinogen; 0.002 mg(Cd)/m³ (respirable dust), Suspected Human Carcinogen); BEI: 10 μg/g creatinine in urine; 10 μg/L in blood
NIOSH REL: (Cadmium) Reduce to lowest feasible level.

SAFETY PROFILE: Confirmed human carcinogen. Poison by intravenous route. When heated to decomposition it emits very toxic fumes of Br⁻, Cd, and NO_x. See also BROMIDES and CADMIUM COMPOUNDS.

TIA300   CAS:998-40-3   *HR: 2*
TRIBUTYLPHOSPHINE
mf: $C_{12}H_{27}P$ mw: 202.36

SYNS: PHOSPHINE, TRIBUTYL- ◇ TRIBUTYLFOSFIN ◇ TRI-n-BUTYLPHOSPHINE

TOXICITY DATA with REFERENCE
orl-rat LD50:750 mg/kg   34ZIAG -,600,69
skn-rbt LD50:2 g/kg   34ZIAG -,600,69

CONSENSUS REPORTS: Reported in EPA TSCA Inventory.

SAFETY PROFILE: Moderately toxic by ingestion. Slightly toxic by skin contact. When heated to decomposition it emits toxic vapors of PO_x.

TIH000   CAS:12380-95-9   *HR: 3*
TRICADMIUM DINITRIDE
mf: Cd_3N_2 mw: 365.21

SYN: CADMIUM NITRIDE

CONSENSUS REPORTS: Cadmium compounds are on the Community Right-To-Know List.

OSHA PEL: TWA 5 μg(Cd)/m³
ACGIH TLV: TWA 0.05 mg(Cd)/m³ (Proposed: TWA 0.01 mg(Cd)/m³ (dust), Suspected Human Carcinogen; 0.002 mg(Cd)/m³ (respirable dust), Suspected Human Carcinogen); BEI: 10 μg/g creatinine in urine; 10 μg/L in blood
NIOSH REL: (Cadium) Reduce to lowest feasible level.

SAFETY PROFILE: Confirmed human carcinogen. Many cadmium compounds are poisons. Explodes violently on shock or heating. Explodes on contact with water, acids, or bases. When heated to decomposition it emits very toxic fumes of NO_x and Cd. See also NITRIDES and CADMIUM COMPOUNDS.

TIK100   CAS:87-61-6   *HR: 2*
1,2,3-TRICHLOROBENZENE
mf: $C_6H_3Cl_3$ mw: 181.44

SYNS: BENZENE, 1,2,3-TRICHLORO- ◇ vic-TRICHLOROBENZENE ◇ 1,2,6-TRICHLOROBENZENE

TOXICITY DATA with REFERENCE
mnt-mus-ipr 250 mg/kg/24H   MUTAEX 2,111,87
ipr-mus LD50:1390 mg/kg   MUTAEX 2,111,87

CONSENSUS REPORTS: Reported in EPA TSCA Inventory.

SAFETY PROFILE: Moderately toxic by intraperitoneal route. Mutation data reported. When heated to decomposition it emits toxic vapors of Cl⁻.

TIK300   CAS:108-70-3   *HR: 2*
1,3,5-TRICHLOROBENZENE
mf: $C_6H_3Cl_3$ mw: 181.44

SYNS: BENZENE, 1,3,5-TRICHLORO- ◇ s-TRICHLOROBENZENE ◇ sym-TRICHLOROBENZENE

TOXICITY DATA with REFERENCE
dlt-oin-oin-par 10 ppm   EVETBX 2,1029,73
dlt-oin-oin-skn 10 ppm   EVETBX 2,1029,73
mnt-mus-ipr 425 mg/kg/24H   MUTAEX 2,111,87
orl-rat LD50:800 mg/kg   48RKAL -,389,81
orl-mus LD50:3350 mg/kg   NTIS** UCRL-13701
ipr-mus LD50:2260 mg/kg   MUTAEX 2,111,87

CONSENSUS REPORTS: Reported in EPA TSCA Inventory.

SAFETY PROFILE: Moderately toxic by ingestion and intraperitoneal routes. Mutation data reported. When heated to decomposition it emits toxic vapors of Cl⁻.

TIO750   CAS:79-01-6   *HR: 3*
TRICHLOROETHYLENE
DOT: UN 1710
mf: C_2HCl_3 mw: 131.38

PROP: Clear, colorless, mobile liquid; characteristic sweet odor of chloroform. D: 1.4649 @ 20°/4°, bp: 86.7°, mp: −73°, fp: −86.8°, autoign temp: 788°F, vap press: 100 mm @ 32°, vap d: 4.53, refr index: 1.477 @ 20°. Immisc with water; misc with alc, ether, acetone, carbon tetrachloride.

SYNS: ACETYLENE TRICHLORIDE ◇ ALGYLEN ◇ ANAMENTH ◇ BENZINOL ◇ BLACOSOLV ◇ CECOLENE ◇ 1-CHLORO-2,2-DICHLOROETHYLENE ◇ CHLORYLEA ◇ CHORYLEN ◇ CIRCOSOLV ◇ CRAWHASPOL ◇ DENSINFLUAT ◇ 1,1-DICHLORO-2-CHLOROETHY-

LENE ◇ DOW-TRI ◇ DUKERON ◇ ETHINYL TRICHLORIDE ◇ ETHYLENE TRICHLORIDE ◇ FLECK-FLIP ◇ FLUATE ◇ GERMALGENE ◇ LANADIN ◇ LETHURIN ◇ NARCOGEN ◇ NARKOSOID ◇ NCI-C04546 ◇ NIALK ◇ PERM-A-CHLOR ◇ PETZINOL ◇ RCRA WASTE NUMBER U228 ◇ THRETHYLENE ◇ TRIAD ◇ TRIASOL ◇ TRICHLOORETHEEN (DUTCH) ◇ TRICHLOORETHYLEEN, (DUTCH) ◇ TRICHLORAETHEN (GERMAN) ◇ TRICHLORAETHYLEN, (GERMAN) ◇ TRICHLORAN ◇ TRICHLORETHENE (FRENCH) ◇ TRICHLORETHYLENE, (FRENCH) ◇ TRICHLOROETHENE ◇ 1,2,2-TRICHLOROETHYLENE ◇ TRI-CLENE ◇ TRICLORETENE (ITALIAN) ◇ TRICLOROETILENE (ITALIAN) ◇ TRIELINA (ITALIAN) ◇ TRILENE ◇ TRIMAR ◇ TRI-PLUS ◇ VESTROL ◇ VITRAN ◇ WESTROSOL

TOXICITY DATA with REFERENCE
skn-rbt 500 mg/24H SEV   28ZPAK -,28,72
eye-rbt 20 mg/24H MOD   28ZPAK -,28,72
mmo-asn 2500 ppm   MUREAV 155,105,85
otr-ham:emb 5 mg/L   CRNGDP 4,291,83
orl-rat TDLo:2688 mg/kg (1–22D preg/21D post):REP   TOXID9 4,179,84
orl-rat TDLo:36 g/kg (female 15D pre):REP   TXCYAC 32,229,84
ihl-rat TCLo:100 ppm/4H (female 6–22D post):TER   JPHYA7 276,24P,78
ihl-rat TCLo:100 ppm/4H (female 6–22D post):REP   JPHYA7 276,24P,78
ihl-mus TCLo:100 ppm/7H (male 5D pre):REP   NTIS** PB82-185075
ihl-mus TCLo:150 ppm/24H (male 4W pre):TER   TOLED5 23,223,84
ihl-rat TCLo:1800 ppm/24H (female 1–20D post):TER   APTOD9 19,A22,80
ihl-rat TCLo:1800 ppm/24H (female 1–20D post):TER   APTOD9 19,A22,80
ihl-rat TCLo:1800 ppm/6H (female 1–20D post):TER   TXCYAC 14,153,79
ihl-rat TCLo:150 ppm/7H/2Y-I:CAR   INHEAO 21,243,83
orl-mus TDLo:455 g/kg/78W-I:CAR   NCITR* NCI-CG-TR-2,76
ihl-mus TCLo:150 ppm/7H/2Y-I:CAR   INHEAO 21,243,83
ihl-ham TCLo:100 ppm/6H/77W-I:ETA   ARTODN 43,237,80
orl-mus TD:912 g/kg/78W-I:CAR   NCITR* NCI-CG-TR-2,76
ihl-mus TC:500 ppm/6H/77W-I:ETA   ARTODN 43,237,80
ihl-mus TC:150 ppm/7H/2Y-I:CAR   INHEAO 21,243,83
orl-man TDLo:2143 mg/kg:GIT   34ZIAG -,602,69
ihl-hmn TCLo:6900 mg/m^3/10M:CNS   AHBAAM 116,131,36
ihl-hmn TCLo:160 ppm/83M:CNS   AIHAAP 23,167,62
ihl-hmn TDLo:812 mg/kg:CNS,GIT,LIV   BMJOAE 2,689,45
ihl-man TCLo:110 ppm/8H:EYE,CNS   BJIMAG 28,293,71
ihl-man LCLo:2900 ppm   NZMJAX 50,119,51
orl-hmn LDLo:7 g/kg   ARTODN 35,295,76
ihl-rat LC50:25,700 ppm/1H   TXAPA9 42,417,77
orl-mus LD50:2402 mg/kg   NTIS** AD-A080-636
ihl-mus LC50:8450 ppm/4H   APTOA6 9,303,53
ivn-mus LD50:33,900 μg/kg   CBCCT* 6,141,54
ipr-dog LD50:1900 mg/kg   TXAPA9 10,119,67
scu-dog LDLo:150 mg/kg   HBTXAC 5,76,59
ivn-dog LDLo:150 mg/kg   QJPPAL 7,205,34
orl-cat LDLo:5864 mg/kg   HBTXAC 5,76,59
orl-rbt LDLo:7330 mg/kg   HBTXAC 5,76,59
scu-rbt LDLo:1800 mg/kg   QJPPAL 7,205,34
ihl-gpg LCLo:37,200 ppm/40M   HBTXAC 5,76,59

CONSENSUS REPORTS: IARC Cancer Review: Group 3 IMEMDT 7,364,87; Animal Limited Evidence IMEMDT 20,545,79; Human Inadequate Evidence IMEMDT 20,545,79; Animal Sufficient Evidence IMEMDT 11,263,76. NCI Carcinogenesis Bioassay (gavage); No Evidence: rat NCITR* NCI-CG-TR-2,76; (gavage); Clear Evidence: mouse NCITR* NCI-CG-TR-2,76. Community Right-To-Know List. Reported in EPA TSCA Inventory. EPA Genetic Toxicology Program.

OSHA PEL: (Transitional: TWA 100 ppm; CL 200 ppm; Pk 300 ppm/5M/2H)TWA 50 ppm; STEL 200 ppm
ACGIH TLV: TWA 50 ppm; STEL 200 ppm (Proposed: TWA 50 ppm; 100 STEL; NOT Suspected as a Human Carcinogen); BEI: 320 mg(trichloroethanol)/g creatinine in urine at end of shift; 0.5 ppm trichloroethylene in end-exhaled air prior to shift and end of work week.
DFG MAK: Suspected Carcinogen; 50 ppm (270 mg/m^3); BAT: 500 μg/dL in blood at end of shift or work week
NIOSH REL: (Trichloroethylene) TWA 250 ppm; (Waste Anesthetic Gases) CL 2 ppm/1H

DOT Classification: ORM-A; Label: None; Poison B; Label: St. Andrews Cross

SAFETY PROFILE: Suspected carcinogen with experimental carcinogenic, tumorigenic, and teratogenic data. Experimental poison by intravenous and subcutaneous routes. Moderately toxic experimentally by ingestion and intraperitoneal routes. Mildly toxic to humans by ingestion and inhalation. Mildly toxic experimentally by inhalation. Human systemic effects by ingestion and inhalation: eye effects, somnolence, hallucinations or distorted perceptions, gastrointestinal changes, and jaundice. Experimental reproductive effects. Human mutation data reported. An eye and severe skin irritant. Inhalation of high concentrations causes narcosis and anesthesia. A form of addiction has been observed in exposed workers. Prolonged inhalation of moderate concentrations causes headache and drowsiness. Fatalities following severe, acute exposure have been attributed to ventricular fibrillation resulting in cardiac failure. There is damage to liver and other organs from chronic exposure. A common air contaminant.

Nonflammable, but high concentrations of trichloroethylene vapor in high-temperature air can be made to burn mildly if plied with a strong flame. Though such a condition is difficult to produce, flames or arcs should not be used in closed equipment that contains any solvent residue or vapor. Reacts with alkali, epoxides [e.g., 1-chloro-2,3-epoxypropane, 1,4-butanediol mono-2,3-epoxypropylether, 1,4-butanediol di-2,3-epoxypropylether, 2,2-bis[(4(2′,3′-epoxypropoxy)phenyl)propane] to form the spontaneously flammable gas dichloroacetylene. Can react violently with Al, Ba, N_2O_4, Li, Mg, liquid O_2, O_3, KOH, KNO_3, Na, NaOH, Ti. Reacts with water under heat and pressure to form HCl gas. When heated to decomposition it emits toxic fumes of Cl^-. See also CHLORINATED HYDROCARBONS, ALIPHATIC.

TIO800   CAS:306-52-5   HR: 2
TRICHLOROETHYL PHOSPHATE
mf: $C_2H_4Cl_3O_4P$ mw: 229.38

SYNS: ETHANOL, 2,2,2-TRICHLORO-, DIHYDROGEN PHOSPHATE ◇ PHOSPHORIC ACID, 2,2,2-TRICHLOROETHYL ESTER ◇ SCH 10159 ◇ 2,2,2-TRICHLOROETHYL PHOSPHATE ◇ TRICLOFOS ◇ TRICLOS

TOXICITY DATA with REFERENCE
orl-rat LD50:1408 mg/kg   TXAPA9 18,185,71
orl-mus LD50:742 mg/kg   GISAAA 54(9),77,89
orl-gpg LD50:1013 mg/kg   GISAAA 54(9),77,89

CONSENSUS REPORTS: Reported in EPA TSCA Inventory.

SAFETY PROFILE: Moderately toxic by ingestion. When heated to decomposition it emits toxic vapors of PO_x and Cl^-.

TIR800   CAS:2000-43-3   HR: 2
α-(TRICHLOROMETHYL)BENZENEMETHANOL
mf: $C_8H_7Cl_3O$ mw: 225.50

SYNS: BENZENEMETHANOL, α-(TRICHLOROMETHYL)-(9CI) ◇ BENZYL ALCOHOL, α-(TRICHLOROMETHYL)- ◇ EFIRAN 99 ◇ PHENYL(TRICHLOROMETHYL)CARBINOL ◇ α-(TRICHLOROMETHYL)BENZYL ALCOHOL ◇ TRICHLOROMETHYLPHENYL CARBINOL

TOXICITY DATA with REFERENCE
scu-rat LDLo:790 mg/kg   JPETAB 24,405,24
scu-rbt LDLo:1670 mg/kg   JPETAB 24,405,24
scu-frg LDLo:200 mg/kg   JPETAB 24,405,24

CONSENSUS REPORTS: Reported in EPA TSCA Inventory.

SAFETY PROFILE: Moderately toxic by subcutaneous route. When heated to decomposition it emits toxic vapors of Cl^-.

TJD600   CAS:2077-46-5   HR: 2
2,3,6-TRICHLOROTOLUENE
mf: $C_7H_5Cl_3$ mw: 195.47

SYNS: BENZENE, 1,2,4-TRICHLORO-3-METHYL- ◇ TOLUENE, 2,3,6-TRICHLORO-

TOXICITY DATA with REFERENCE
orl-mus LD50:2000 mg/kg   GISAAA 45(12),64,80

CONSENSUS REPORTS: Reported in EPA TSCA Inventory.

SAFETY PROFILE: Moderately toxic by ingestion. When heated to decomposition it emits toxic vapors of Cl^-.

TJE100   CAS:354-58-5   HR: 3
1,1,1-TRICHLORO-2,2,2-TRIFLUOROETHANE
mf: $C_2Cl_3F_3$ mw: 187.37

SYNS: FC 113 ◇ FC133a ◇ FREON FT ◇ PRECISION CLEANING AGENT ◇ TF ◇ T-WD602 ◇ TRICHLOROTRIFLUOROETHANE

TOXICITY DATA with REFERENCE
ihl-rat LC50:13 pph/15M   HUTODJ 1,239,82
ipr-mus LD50:8600 mg/kg   EJTXAZ 7,247,74

CONSENSUS REPORTS: Reported in EPA TSCA Inventory.

SAFETY PROFILE: Poison by inhalation. When heated to decomposition it emits toxic vapors of F^- and Cl^-.

TJE200   CAS:2631-68-7   HR: 3
1,3,5-TRICHLORO-2,4,6-TRINITROBENZENE
mf: $C_6Cl_3N_3O_6$ mw: 316.44

SYNS: BENZENE, 1,3,5-TRICHLORO-2,4,6-TRINITRO- ◇ BULBOSAN ◇ TCTNB ◇ sym-TRICHLOROTRINITROBENZENE ◇ TRICHLORO-1,3,5-TRINITROBENZENE

TOXICITY DATA with REFERENCE
ihl-rat LCLo:684 mg/m^3/1H   NTIS** UCRL-13701

CONSENSUS REPORTS: Reported in EPA TSCA Inventory.

SAFETY PROFILE: Poison by inhalation. When heated to decomposition it emits toxic vapors of NO_x and Cl^-.

TJL600   CAS:368-39-8   HR: D
TRIETHOXONIUM FLUOROBORATE
mf: $C_6H_{15}O•BF_4$ mw: 190.02

SYNS: OXONIUM, TRIETHYL-, TETRAFLUOROBORATE(1−) ◇ TEOF ◇ TRIETHYLOXONIUM BOROFLUORIDE ◇ TRIETHYLOXONIUM FLUOBORATE ◇ TRIETHYLOXONIUM FLUOROBORATE ◇ TRIETHYLOXONIUM TETRAFLUOROBORATE

TOXICITY DATA with REFERENCE
dnd-smc 500 mmol/L   CPBTAL 23,2485,75
dnd-ofs-sal:oth 500 mmol/L   CPBTAL 23,2485,75

CONSENSUS REPORTS: Reported in EPA TSCA Inventory.

SAFETY PROFILE: Mutation data reported. When heated to decomposition it emits toxic vapors of boron and F^-.

TJO050   CAS:554-68-7   HR: 2
TRIETHYLAMINE, HYDROCHLORIDE
mf: $C_6H_{15}N•ClH$ mw: 137.68

TOXICITY DATA with REFERENCE
scu-mus LDLo:600 mg/kg   JPETAB 37,309,29

CONSENSUS REPORTS: Reported in EPA TSCA Inventory.

SAFETY PROFILE: Moderately toxic by subcutaneous

route. When heated to decomposition it emits toxic vapors of NO$_x$, HCl, and Cl$^-$.

TJP550   CAS:597-49-9   **HR: 2**
TRIETHYLCARBINOL
mf: C$_7$H$_{16}$O   mw: 116.23

SYNS: 3-AETHYL-PENTANOL-(3) ◇ 3-ETHYL-3-PENTANOL ◇ 3-PENTANOL, 3-ETHYL- ◇ TRIETHYLMETHANOL

TOXICITY DATA with REFERENCE
scu-mus LD50:700 mg/kg   ARZNAD 5,161,55

CONSENSUS REPORTS: Reported in EPA TSCA Inventory.

SAFETY PROFILE: Moderately toxic by subcutaneous route. When heated to decomposition it emits acrid smoke and irritating vapors.

TJY900   CAS:420-46-2   **HR: D**
1,1,1-TRIFLUOROETHANE
mf: C$_2$H$_3$F$_3$ mw: 84.05

SYNS: ETHANE, 1,1,1-TRIFLUORO- ◇ FC143a ◇ FLUOROCARBON FC143a ◇ METHYLFLUOROFORM ◇ R 143a ◇ 1,1,1-TRIFLUOROFORM

TOXICITY DATA with REFERENCE
mma-sat 50 pph/48H   TXAPA9 72,15,84

SAFETY PROFILE: Mutation data reported. When heated to decomposition it emits toxic vapors of F$^-$.

TKB775   CAS:368-53-6   **HR: 2**
5-(TRIFLUOROMETHYL)-1,3-BENZENEDIAMINE
mf: C$_7$H$_7$F$_3$N$_2$ mw: 176.16

SYNS: 1,3-BENZENEDIAMINE, 5-(TRIFLUOROMETHYL)- ◇ 3,5-BENZOTRIFLUORODIAMINE ◇ 3,5-DIAMINOBENZOTRIFLUORIDE ◇ TOLUENE-3,5-DIAMINE, α-α-α-TRIFLUORO-(8CI) ◇ α-α-α-TRIFLUOROTOLUENE-3,5-DIAMINE

TOXICITY DATA with REFERENCE
mmo-sat 500 μg/plate   EPASR* 8EHQ-1190-1113
mma-sat 1 mg/plate   EPASR* 8EHQ-1190-1113
msc-mus:lyms 20 mg/L   EMMUEG 19(Suppl 20),4,92
orl-rat LDLo:500 mg/kg   EPASR* 8EHQ-1190-1113

SAFETY PROFILE: Moderately toxic by ingestion. Mutation data reported. When heated to decomposition it emits toxic vapors of NO$_x$ and F$^-$.

TKE775   CAS:92-30-8   **HR: 3**
2-(TRIFLUOROMETHYL)PHENOTHIAZINE
mf: C$_{13}$H$_8$F$_3$NS   mw: 267.28

SYN: PHENOTHIAZINE, 2-(TRIFLUOROMETHYL)-

TOXICITY DATA with REFERENCE
orl-mus LD50:520 mg/kg   TPKVAL 12,110,71
ivn-mus LD50:32 mg/kg   CSLNX* NX#07426

CONSENSUS REPORTS: Reported in EPA TSCA Inventory.

SAFETY PROFILE: Poison by intravenous route. Moderately toxic by ingestion. When heated to decomposition it emits toxic vapors of NO$_x$, and F$^-$.

TKG800   CAS:1187-93-5   **HR: 3**
TRIFLUOROMETHYL TRIFLUOROVINYL ETHER
mf: C$_3$F$_6$O   mw: 166.03

SYNS: ETHENE, TRIFLUORO(TRIFLUOROMETHOXY)- ◇ ETHER, TRIFLUOROMETHYL TRIFLUOROVINYL

TOXICITY DATA with REFERENCE
ihl-mus LC50:46 μg/m^3/2H   85JCAE -,540,86

CONSENSUS REPORTS: Reported in EPA TSCA Inventory.

SAFETY PROFILE: Poison by inhalation. When heated to decomposition it emits toxic vapors of F$^-$.

TKH015   CAS:677-21-4   **HR: 1**
1,1,1-TRIFLUOROPROPENE
mf: C$_3$H$_3$F$_3$ mw: 96.06

SYNS: PROPENE, 3,3,3-TRIFLUORO- ◇ 1-PROPENE, 3,3,3-TRIFLUORO-(9CI) ◇ TRIFLUOROMETHYLETHYLENE ◇ 3,3,3-TRIFLUOROPROPENE ◇ 3,3,3-TRIFLUORO-1-PROPENE ◇ 3,3,3-TRIFLUOROPROPYLENE

TOXICITY DATA with REFERENCE
ihl-mus LC50:1691 g/m^3/2H   85GMAT -,116,82

CONSENSUS REPORTS: Reported in EPA TSCA Inventory.

SAFETY PROFILE: Slightly toxic by inhalation. When heated to decomposition it emits toxic vapors of F$^-$.

TKH020   CAS:460-40-2   **HR: 3**
3,3,3-TRIFLUOROPROPIONALDEHYDE
mf: C$_3$H$_3$F$_3$O   mw: 112.06

SYNS: PROPIONALDEHYDE, 3,3,3-TRIFLUORO- ◇ TRIFLUOROPROPIONALDEHYDE

TOXICITY DATA with REFERENCE
orl-rat LDLo:25 mg/kg   EPASR* 8EHQ-1186-0644

SAFETY PROFILE: Poison by ingestion. When heated to decomposition it emits toxic vapors of F$^-$.

TKH250   CAS:59544-89-7   **HR: 3**
TRIFLUORO SELENIUM HEXAFLUORO ARSENATE
mf: AsF$_9$Se   mw: 324.9

CONSENSUS REPORTS: Selenium and its compounds, as well as arsenic and its compounds, are on the Community Right-To-Know List.

TKH310 CAS:447-14-3 HR: 2
α-β,β-TRIFLUOROSTYRENE
mf: $C_8H_5F_3$ mw: 158.13

SYNS: BENZENE, (TRIFLUOROETHENYL)-(9CI) ◇ STYRENE, α-β,β-TRIFLUORO- ◇ (TRIFLUOROETHENYL)BENZENE

TOXICITY DATA with REFERENCE
orl-rat LD50:2500 mg/kg 85GMAT -,117,82
ihl-rat LC50:8 g/m³/4H 85GMAT -,117,82

CONSENSUS REPORTS: Reported in EPA TSCA Inventory.

SAFETY PROFILE: Moderately toxic by ingestion. Slightly toxic by inhalation. When heated to decomposition it emits toxic vapors of F^-.

TKN800 CAS:2144-08-3 HR: 2
2,3,4-TRIHYDROXYBENZALDEHYDE
mf: $C_7H_6O_4$ mw: 154.13

SYN: BENZALDEHYDE, 2,3,4-TRIHYDROXY-

TOXICITY DATA with REFERENCE
orl-mus LD50:2195 mg/kg IJANDP 10,741,87

CONSENSUS REPORTS: Reported in EPA TSCA Inventory.

SAFETY PROFILE: Moderately toxic by ingestion. When heated to decomposition it emits acrid smoke and irritating vapors.

TKP100 CAS:2295-58-1 HR: 3
2,4,6-TRIHYDROXYPROPIOPHENONE
mf: $C_9H_{10}O_4$ mw: 182.19

SYNS: ARGOBYL ◇ COSPANON ◇ FLOPROPION ◇ FLOPROPIONE ◇ LABRODA ◇ LABRODAX ◇ LABRODAX SUPANATE ◇ PHLOROPROPIONONE ◇ PHLOROPROPIOPHENONE ◇ 1-PROPANONE, 1-(2,4,6-TRIHYDROXYPHENYL)- ◇ PROPIONYLPHLOROGLUCINOL ◇ PROPIOPHENONE, 2′,4′,6′-TRIHYDROXY- ◇ PROPIOPHLOROGLUCINE ◇ 13907 R.P. ◇ RP 13907 ◇ 2′,4′,6′-TRIHYDROXYPROPIOPHENONE

TOXICITY DATA with REFERENCE
orl-rat LD50:2380 mg/kg DRUGAY 6,734,82
ipr-rat LD50:412 mg/kg DRUGAY 6,734,82
scu-rat LD50:640 mg/kg DRUGAY 6,734,82
ivn-rat LD50:246 mg/kg DRUGAY 6,734,82
orl-mus LD50:2780 mg/kg DRUGAY 6,734,82
ipr-mus LD50:578 mg/kg DRUGAY 6,734,82
scu-mus LD50:804 mg/kg DRUGAY 6,734,82
ivn-mus LD50:300 mg/kg DRUGAY 6,734,82

CONSENSUS REPORTS: Reported in EPA TSCA Inventory.

SAFETY PROFILE: Poison by intravenous route. Moderately toxic by ingestion and subcutaneous routes.

TKP500 CAS:102-71-6 HR: 3
TRIHYDROXYTRIETHYLAMINE
mf: $C_6H_{15}NO_3$ mw: 149.22

PROP: Pale-yellow viscous liquid. Mp: 21.2°, bp: 360°, flash p: 355°F (CC), d: 1.1258 @ 20/20°, vap press: 10 mm @ 205°, vap d: 5.14.

SYNS: DALTOGEN ◇ NITRILO-2,2′,2″-TRIETHANOL ◇ 2,2′,2″-NITRILO-TRIETHANOL ◇ STEROLAMIDE ◇ THIOFACO T-35 ◇ TRIAETHANOL-AMIN-NG ◇ TRIETHANOLAMIN ◇ TRIETHANOLAMINE (ACGIH) ◇ TRIETHYLOLAMINE ◇ TRI(HYDROXYETHYL)AMINE ◇ 2,2′,2″-TRIHYDROXYTRIETHYLAMINE ◇ TRIS(2-HYDROXYETHYL)AMINE ◇ TROLAMINE

TOXICITY DATA with REFERENCE
skn-hmn 15 mg/3D-I MLD 85DKA8 -,127,77
skn-rbt 560 mg/24H MLD TXAPA9 19,276,71
eye-rbt 10 mg MLD TXAPA9 55,501,80
orl-mus TDLo:16 g/kg/64W-C:CAR CNREA8 38,3918,78
orl-mus TD:154 g/kg/61W-C:CAR CNREA8 38,3918,78
orl-rat LD50:8 g/kg NTIS** PB158-507
orl-mus LD50:7400 mg/kg GTPZAB 26(8),53,82
ipr-mus LD50:1450 mg/kg RCRVAB 38,975,69
orl-gpg LD50:5300 mg/kg GISAAA 29(11),25,64

CONSENSUS REPORTS: Reported in EPA TSCA Inventory. EPA Genetic Toxicology Program.
ACGIH TLV: (Proposed: TWA 0.5 mg/m³)

SAFETY PROFILE: Moderately toxic by intraperitoneal route. Mildly toxic by ingestion. Liver and kidney damage has been demonstrated in animals from chronic exposure. A human and experimental skin irritant. An eye irritant. Questionable carcinogen with experimental carcinogenic data. Combustible liquid when exposed to heat or flame; can react vigorously with oxidizing materials. To fight fire, use alcohol foam, CO_2, dry chemical. When heated to decomposition it emits toxic fumes of NO_x and CN^-.

TKU700 CAS:528-44-9 HR: 2
TRIMELLITIC ACID
mf: $C_9H_6O_6$ mw: 210.15

SYNS: 1,2,4-BENZENETRICARBOXYLIC ACID ◇ TMA ◇ 1,2,4-TRICARBOXYBENZENE

[preceding entry continued:]

OSHA PEL: TWA 0.01 mg(As)/m³; Cancer Hazard; TWA 0.2 mg(Se)/m³
ACGIH TLV: TWA 0.2 mg(As)/m³ (Proposed: 0.01 mg(As)/m³; Human Carcinogen); TWA 0.2 mg(Se)/m³
DFG TRK: 0.2 mg/m³ calculated as arsenic in that portion of dust that can possibly be inhaled.
NIOSH REL: CL 2 μg(As)/m³

SAFETY PROFILE: Arsenic compounds are poisons. Violent reaction with water. When heated to decomposition it emits very toxic fumes of As, F^-, and Se. See also FLUORIDES; ARSENIC COMPOUNDS; and SELENIUM COMPOUNDS.

TOXICITY DATA with REFERENCE
orl-mus LD50:2500 mg/kg   GTPZAB 18(7),57,74

CONSENSUS REPORTS: Reported in EPA TSCA Inventory.

SAFETY PROFILE: Moderately toxic by ingestion. When heated to decomposition it emits acrid smoke and irritating vapors.

TKV000   CAS:552-30-7   HR: 3
TRIMELLITIC ANHYDRIDE
mf: $C_9H_4O_5$ mw: 192.13

PROP: Crystals. Mp: 162°, bp: 240–245° @ 14 mm. Sol in acetone, ethyl acetate, dimethylformamide.

SYNS: ANHYDROTRIMELLIC ACID ◇ 1,2,4-BENZENETRICARBOXYLIC ACID ANHYDRIDE ◇ 1,2,4-BENZENETRICARBOXYLIC ACID, CYCLIC 1,2-ANHYDRIDE ◇ 1,2,4-BENZENETRICARBOXYLIC ANHYDRIDE ◇ 4-CARBOXYPHTHALIC ANHYDRIDE ◇ 1,3-DIHYDRO-1,3-DIOXO-5-ISOBENZOFURANCARBOXYLIC ACID ◇ 1,3-DIOXO-5-PHTHALANCARBOXYLIC ACID ◇ DIPHENYLMETHANE-4,4'-DIISOCYANATE-TRIMELLIC ANHYDRIDE-ETHOMID HT POLYMER ◇ NCI-C56693 ◇ TMA ◇ TMAN ◇ TRIMELLIC ACID ANHYDRIDE ◇ TRIMELLIC ACID-1,2-ANHYDRIDE ◇ TRIMELLITIC ACID CYCLIC-1,2-ANHYDRIDE

TOXICITY DATA with REFERENCE
orl-rat LD50:5600 mg/kg   DTLVS* 4,415,80

CONSENSUS REPORTS: Reported in EPA TSCA Inventory.

OSHA PEL: TWA 0.005 ppm
ACGIH TLV: TWA 0.005 ppm; (Proposed: CL 0.04 mg/m^3)
DFG MAK: 0.005 ppm (0.04 mg/m^3)
NIOSH REL: (Trimellitic Anhydride): handle as extremely toxic.

SAFETY PROFILE: Mildly toxic by ingestion. Has caused pulmonary edema from inhalation. Irritant to lungs and air passages. May be a powerful allergen. Typical attack consists of breathlessness, wheezing, cough, running nose, immunological sensitization, and asthma symptoms. When heated to decomposition it emits acrid smoke and irritating fumes. See also ANHYDRIDES.

TKZ100   CAS:102-24-9   HR: 1
TRIMETHOXYBOROXINE
mf: $C_3H_9B_3O_6$ mw: 173.55

SYNS: BOROXIN, TRIMETHOXY- ◇ TRIMETHOXYBOROXIN

TOXICITY DATA with REFERENCE
orl-rat LD50:5160 mg/kg   14KTAK -,693,64

CONSENSUS REPORTS: Reported in EPA TSCA Inventory.

SAFETY PROFILE: Slightly toxic by ingestion. When heated to decomposition it emits toxic fumes of boron.

TLD500   CAS:75-50-3   HR: 3
TRIMETHYLAMINE
DOT: UN 1083/UN 1297
mf: C_3H_9N mw: 59.13

PROP: Colorless gas. Pungent, fishy, ammonia-like odor; saline taste. Bp: 2.87°, lel: 2%, uel: 11.6%, fp: −117.1°, d: 0.662 @ −5°, autoign temp: 374°F, vap d: 2.0, flash p: 20°F (CC). Misc with alc; sol in ether, benzene, toluene, xylene, chloroform.

SYNS: N,N-DIMETHYLMETHANAMINE ◇ TMA ◇ TRIMETHYLAMINE, anhydrous (DOT) ◇ TRIMETHYLAMINE, aqueous solution (DOT) ◇ TRIMETHYLAMINE, aqueous solutions containing not more than 30% of trimethylamine (DOT)

TOXICITY DATA with REFERENCE
ihl-rat LCLo:3500 ppm/4H   TOXID9 4,68,84
ivn-mus LD50:90 mg/kg   MPHEAE 16,529,67
scu-mus LDLo:1000 mg/kg   BBMS** -,-,48
scu-rbt LDLo:800 mg/kg   CRSBAW 83,481,20
ivn-rbt LDLo:400 mg/kg   BBMS** -,-,48
rec-rbt LDLo:800 mg/kg   CRSBAW 83,481,20
scu-frg LDLo:2000 mg/kg   SAPHAO 10,201,1900
ihl-mam LC50:19 g/m^3 TPKVAL 14,80,75

CONSENSUS REPORTS: Reported in EPA TSCA Inventory.

OSHA PEL: TWA 10 ppm; STEL 15 ppm
ACGIH TLV: TWA 5 ppm; STEL 15 ppm

DOT Classification: Flammable Gas; Label: Flammable Gas, Anhydrous; Flammable Liquid; Label: Flammable Liquid

SAFETY PROFILE: Poison by intravenous route. Moderately toxic by subcutaneous and rectal routes. Mildly toxic by inhalation. A very dangerous fire hazard when exposed to heat or flame. Self-reactive. Moderately explosive in the form of vapor when exposed to heat or flame. Can react with oxidizing materials. To fight fire, stop flow of gas. Potentially explosive reaction with bromine + heat; ethylene oxide; triethynylaluminum. When heated to decomposition it emits toxic fumes of NO$_x$. See also AMINES.

TLG100   CAS:121-72-2   HR: 3
N,N,3-TRIMETHYLANILINE
mf: $C_9H_{13}N$ mw: 135.23

SYNS: BENZENAMINE, N,N,3-TRIMETHYL-(9CI) ◇ DIMETHYL-m-TOLUIDINE ◇ DIMETIL-m-TOLUIDINA ◇ m,N,N-TRIMETHYLANILINE ◇ N,N,3-TRIMETHYLBENZENAMINE ◇ m-TOLUIDINE, N,N-DIMETHYL-

TOXICITY DATA with REFERENCE
ipr-mus LD50:300 mg/kg   AISFAR 1,284,51

CONSENSUS REPORTS: Reported in EPA TSCA Inventory.

TLG150   CAS:99-97-8   HR: 3
N,N,4-TRIMETHYLANILINE
mf: $C_9H_{13}N$ mw: 135.23

SYNS: BENZENAMINE, N,N,4-TRIMETHYL- ◊ DIMETHYL-p-TOLUIDINE ◊ DIMETIL-p-TOLUIDINA ◊ p-TOLUIDINE, N,N-DIMETHYL- ◊ p,N,N-TRIMETHYLANILINE

TOXICITY DATA with REFERENCE
ipr-mus LD50:212 mg/kg   AISFAR 1,284,51

CONSENSUS REPORTS: Reported in EPA TSCA Inventory.

SAFETY PROFILE: Poison by intraperitoneal route. When heated to decomposition it emits toxic vapors of NO_x.

TLM100   CAS:122-08-7   HR: 3
N,N,N-TRIMETHYLBENZENEMETHANAMINIUM METHOXIDE
mf: $C_{10}H_{16}N \cdot CH_3O$ mw: 181.31

SYNS: AMMONIUM, BENZYLTRIMETHYL-, METHOXIDE ◊ BENZENEMETHANAMINIUM, N,N,N-TRIMETHYL-, METHOXIDE (9CI) ◊ BENZYLTRIMETHYLAMMONIUM METHOXIDE

TOXICITY DATA with REFERENCE
ivn-mus LD50:5600 µg/kg   CSLNX* NX#03547

CONSENSUS REPORTS: Reported in EPA TSCA Inventory.

SAFETY PROFILE: Poison by intravenous route. When heated to decomposition it emits toxic vapors of NO_x.

TLO510   CAS:933-48-2   HR: 2
cis-3,3,5-TRIMETHYLCYCLOHEXANOL
mf: $C_9H_{18}O$ mw: 142.27

SYN: CYCLOHEXANOL, 3,3,5-TRIMETHYL-, cis-

TOXICITY DATA with REFERENCE
orl-rat LD50:2400 mg/kg   GWXXAW #2757641

CONSENSUS REPORTS: Reported in EPA TSCA Inventory.

SAFETY PROFILE: Moderately toxic by ingestion. When heated to decomposition it emits acrid smoke and irritating vapors.

TLR050   CAS:627-31-6   HR: D
TRIMETHYLENE DIIODIDE
mf: $C_3H_6I_2$ mw: 295.89

SYNS: 1,3-DIIODOPROPANE ◊ PROPANE, 1,3-DIIODO-

TOXICITY DATA with REFERENCE
mmo-sat 10 µmol/plate   MUREAV 141,11,84

CONSENSUS REPORTS: Reported in EPA TSCA Inventory.

SAFETY PROFILE: Mutation data reported. When heated to decomposition it emits toxic vapors of I^-.

TLU200   CAS:118-12-7   HR: 3
1,3,3-TRIMETHYL-2-METHYLENEINDOLINE
mf: $C_{12}H_{15}N$ mw: 173.28

SYNS: INDOLINE, 2-METHYLENE-1,3,3-TRIMETHYL- ◊ 2-METHYLENE-1,3,3-TRIMETHYLINDOLINE

TOXICITY DATA with REFERENCE
ivn-mus LD50:56 mg/kg   CSLNX* NX#07782

CONSENSUS REPORTS: Reported in EPA TSCA Inventory.

SAFETY PROFILE: Poison by intravenous route. When heated to decomposition it emits toxic vapors of NO_x.

TME272   CAS:108-75-8   HR: 3
2,4,6-TRIMETHYLPYRIDINE
mf: $C_8H_{11}N$ mw: 121.20

SYNS: α-γ,α'-COLLIDINE ◊ γ-COLLIDINE ◊ s-COLLIDINE ◊ sym-COLLIDINE ◊ 2,4,6-COLLIDINE ◊ 2,4,6-KOLLIDIN ◊ PYRIDINE, 2,4,6-TRIMETHYL-

TOXICITY DATA with REFERENCE
orl-rat LD50:400 mg/kg   85JCAE -,847,86
ihl-rat LCLo:2500 ppm/2H   85JCAE -,847,86
skn-gpg LD50:1 g/kg   85JCAE -,847,86

CONSENSUS REPORTS: Reported in EPA TSCA Inventory.

SAFETY PROFILE: Poison by ingestion. Moderately toxic by inhalation. When heated to decomposition it emits toxic vapors of NO_x.

TMG800   CAS:1774-47-6   HR: 3
TRIMETHYL SULPHOXONIUM IODIDE
mf: $C_3H_9OS \cdot I$ mw: 220.08

SYNS: SULFONIUM, TRIMETHYL-, IODIDE, OXIDE ◊ SULFOXONIUM, TRIMETHYL-, IODIDE

TOXICITY DATA with REFERENCE
ipr-mus LD50:900 mg/kg   IJRBA3 3,41,61
ivn-mus LD50:180 mg/kg   CSLNX* NX#04108

CONSENSUS REPORTS: Reported in EPA TSCA Inventory.

SAFETY PROFILE: Poison by intravenous route. Moderately toxic by intraperitoneal route. When heated to decomposition it emits toxic vapors of SO_x and I^-.

TMJ100   CAS:1117-41-5   *HR: 1*
3,6,10-TRIMETHYL-3,5,9-UNDECATRIEN-2-ONE
mf: $C_{14}H_{22}O$ mw: 206.36

SYNS: METHYLISOPSEUDOIONONE ◊ PSEUDO-α-ISOMETHYL IONONE ◊ 2,6,9-TRIMETHYLUNDECA-2,6,8-TRIEN-10-ONE ◊ 3,5,9-UNDECATRIEN-2-ONE, 3,6,10-TRIMETHYL-

TOXICITY DATA with REFERENCE
orl-rat LDLo:5 g/kg   FCTOD7 26,413,88
skn-rbt LD50:>2500 mg/kg   FCTOD7 26,413,88

CONSENSUS REPORTS: Reported in EPA TSCA Inventory.

SAFETY PROFILE: Slightly toxic by ingestion and skin contact. When heated to decomposition it emits acrid smoke and irritating vapors.

TMK300   CAS:606-35-9   *HR: D*
2,4,6-TRINITROANISOLE
mf: $C_7H_5N_3O_7$ mw: 243.15

SYNS: ANISOLE, 2,4,6-TRINITRO- ◊ BENZENE, 2-METHOXY-1,3,5-TRINITRO-(9CI) ◊ 2-METHOXY-1,3,5-TRINITROBENZENE ◊ METHYL PICRATE

TOXICITY DATA with REFERENCE
mmo-sat 102 nmol/plate   MUREAV 136,209,84
mma-sat 408 nmol/plate   MUREAV 136,209,84

CONSENSUS REPORTS: Reported in EPA TSCA Inventory.

SAFETY PROFILE: Mutation data reported. When heated to decomposition it emits toxic vapors of NO_x.

TMU800   CAS:76-86-8   *HR: 3*
TRIPHENYLSILYL CHLORIDE
mf: $C_{18}H_{15}ClSi$ mw: 294.87

SYNS: CHLOROTRIPHENYLSILANE ◊ SILANE, CHLOROTRIPHENYL-

TOXICITY DATA with REFERENCE
ivn-mus LD50:56 mg/kg   CSLNX* NX#04717

CONSENSUS REPORTS: Reported in EPA TSCA Inventory.

SAFETY PROFILE: Poison by intravenous route. When heated to decomposition it emits toxic vapors of Cl^-.

TMV850   CAS:379-52-2   *HR: 2*
TRIPHENYLTIN FLUORIDE
mf: $C_{18}H_{15}FSn$ mw: 369.02

SYNS: BIOMET 204 ◊ FLUOROTRIPHENYLSTANNANE ◊ STANNANE, FLUOROTRIPHENYL- ◊ TIN, FLUOROTRIPHENYL-

TOXICITY DATA with REFERENCE
orl-mus LDLo:710 mg/kg   AECTCV 14,111,85

SAFETY PROFILE: Moderately toxic by ingestion. When heated to decomposition it emits toxic vapors of Cl^- and fumes of tin.

TNC100   CAS:2396-43-2   *HR: 1*
2,4,6-TRIPROPYL-s-TRIOXANE
mf: $C_{12}H_{24}O_3$ mw: 216.36

SYNS: PARABUTYRALDEHYDE ◊ 1,3,5-TRIOXANE, 2,4,6-TRIPROPYL- ◊ s-TRIOXANE, 2,4,6-TRIPROPYL-

TOXICITY DATA with REFERENCE
orl-rbt LDLo:5408 mg/kg   JPETAB 48,488,33

CONSENSUS REPORTS: Reported in EPA TSCA Inventory.

SAFETY PROFILE: Slightly toxic by ingestion. When heated to decomposition it emits acrid smoke and irritating vapors.

TNI300   CAS:301-13-3   *HR: 1*
TRIS(2-ETHYLHEXYL)PHOSPHITE
mf: $C_{24}H_{51}O_3P$ mw: 418.72

SYN: PHOSPHOROUS ACID, TRIS(2-ETHYLHEXYL) ESTER

TOXICITY DATA with REFERENCE
orl-rat LD50:10700 mg/kg   ALBRW* #OPB-3,84

CONSENSUS REPORTS: Reported in EPA TSCA Inventory.

SAFETY PROFILE: Slightly toxic by ingestion. When heated to decomposition it emits toxic vapors of PO_x.

TNP600   CAS:596-43-0   *HR: 3*
TRITYL BROMIDE
mf: $C_{19}H_{15}Br$ mw: 323.25

SYNS: BROMOTRIPHENYLMETHANE ◊ METHANE, BROMOTRIPHENYL-

TOXICITY DATA with REFERENCE
ivn-mus LD50:180 mg/kg   CSLNX* NX#04015

CONSENSUS REPORTS: Reported in EPA TSCA Inventory.

SAFETY PROFILE: Poison by intravenous route. When heated to decomposition it emits toxic vapors of Br^-.

U

UVJ440   CAS:104-98-3   ***HR: 3***
UROCANIC ACID
mf: $C_6H_6N_2O_2$ mw: 138.14

SYNS: IMIDAZOLEACRYLIC ACID ◇ IMIDAZOLE-4-ACRYLIC ACID ◇ 5-IMIDAZOLEACRYLIC ACID ◇ 3-(1H-IMIDAZOL-4-YL)-2-PROPENOIC ACID ◇ 2-PROPENOIC ACID, 3-(1H-IMIDAZOL-4-YL)-(9CI) ◇ UROCANINIC ACID

TOXICITY DATA with REFERENCE
dni-hmn:oth 1 mmol/L   JIDEAE 65,400,75
ivn-mus LD50:100 mg/kg   85GDA2 5,81,81

CONSENSUS REPORTS: Reported in EPA TSCA Inventory.

SAFETY PROFILE: Poison by intravenous route. Mutation data reported. When heated to decomposition it emits toxic vapors of NO_x.

V

VAQ100 CAS:624-24-8 **HR: 1**
VALERIC ACID, METHYL ESTER
mf: $C_6H_{12}O_2$ mw: 116.18

SYNS: METHYL PENTANOATE ◇ METHYL VALERATE ◇ METHYL n-VALERATE ◇ METHYL VALERIANATE ◇ PENTANOIC ACID, METHYL ESTER (9CI)

TOXICITY DATA with REFERENCE
ihl-mus LC50:6600 mg/m^3/2H 85GMAT -,88,82
ihl-uns LC50:6600 mg/m^3 GISAAA 51(5),61,86

CONSENSUS REPORTS: Reported in EPA TSCA Inventory.

SAFETY PROFILE: Slightly toxic by inhalation. When heated to decomposition it emits acrid smoke and irritating vapors.

VAV300 CAS:110-59-8 **HR: 3**
VALERONITRILE
mf: C_5H_9N mw: 83.15

SYNS: 1-CYANOBUTANE ◇ PENTANENITRILE (9CI) ◇ n-VALERONITRILE

TOXICITY DATA with REFERENCE
orl-mus LD50:191 mg/kg ARTODN 55,47,84

CONSENSUS REPORTS: Reported in EPA TSCA Inventory.

SAFETY PROFILE: Poison by ingestion. When heated to decomposition it emits toxic vapors of NO_x.

VHP600 CAS:93-07-2 **HR: 2**
VERATRIC ACID
mf: $C_9H_{10}O_4$ mw: 182.19

SYNS: BENZOIC ACID, 3,4-DIMETHOXY- ◇ 3,4-DIMETHOXYBENZOIC ACID ◇ 3,4-DIMETHYLPROTOCATECHUIC ACID

TOXICITY DATA with REFERENCE
dni-hmn:lyms 1 mmol/L BCPCA6 29,1275,80
ipr-mus LD50:>800 mg/kg JPETAB 196,478,76

CONSENSUS REPORTS: Reported in EPA TSCA Inventory.

SAFETY PROFILE: Moderately toxic by intraperitoneal route. Human mutation data reported. When heated to decomposition it emits acrid smoke and irritating vapors.

VLU250 CAS:108-05-4 **HR: 3**
VINYL ACETATE
DOT: UN 1301
mf: $C_4H_6O_2$ mw: 86.10

$H_2C=CHOCO \cdot CH_3$

PROP: Colorless, mobile liquid; polymerizes to solid on exposure to light. Mp:−92.8°, bp: 73°, flash p: 18°F, d: 0.9335 @ 20°, autoign temp: 800°F, vap press: 100 mm @ 21.5°, lel: 2.6%, uel: 13.4%, vap d: 3.0. Misc in alc, ether. Somewhat sol in water.

SYNS: ACETIC ACID ETHENYL ESTER ◇ ACETIC ACID VINYL ESTER ◇ 1-ACETOXYETHYLENE ◇ ETHENYL ACETATE ◇ OCTAN WINYLU (POLISH) ◇ VAC ◇ VINILE (ACETATO di) (ITALIAN) ◇ VINYLACETAAT (DUTCH) ◇ VINYLACETAT (GERMAN) ◇ VINYL A MONOMER ◇ VINYLE (ACETATE de) (FRENCH) ◇ VYAC ◇ ZESET T

TOXICITY DATA with REFERENCE
eye-hmn 22 ppm AIHAAP 30,449,69
skn-rbt 10 mg/24H open JIHTAB 30,63,48
eye-rbt 500 mg open JIHTAB 30,63,48
eye-rbt 500 mg/24H MLD 85JCAE -,354,86
cyt-hmn:lym 250 μmol/L MUREAV 159,109,86
sce-ham:ovr 125 μmol/L CNREA8 45,4816,85
orl-rat TDLo:500 mg/kg/D (multi) :REP EPASR* 8EHQ-0185-0543
orl-rat TDLo:100 g/kg/2Y-C:CAR TXAPA9 68,43,83
ihl-rat TCLo:600 ppm/6H/5D/2Y-I:ETA EPASR* 8EHQ-0187-0650
ihl-mus TCLo:600 ppm/6H/5D/2Y-I:ETA EPASR* 8EHQ-0187-0650
orl-rat LD50:2920 mg/kg UCDS** 4/25/58
ihl-rat LC50:4000 ppm/2H DUPON* ES-3574,75
orl-mus LD50:1613 mg/kg GISAAA 31(8),19,66
ihl-mus LC50:1550 ppm/4H DUPON* ES-3574,75
ihl-rbt LC50:2500 ppm/4H 85INA8 5,621,86
skn-rbt LD50:2335 mg/kg DUPON* ES-3574,75

CONSENSUS REPORTS: IARC Cancer Review: Group 3 IMEMDT 7,56,87; Animal Inadequate Evidence IMEMDT 19,341,79; IMEMDT 39,113,86; Human Inadequate Evidence IMEMDT 39,113,86. Reported in EPA TSCA Inventory. Community Right-To-Know List. EPA Extremely Hazardous Substances List.

OSHA PEL: TWA 10 ppm; STEL 20 ppm
ACGIH TLV: TWA 10 ppm; STEL 20 ppm; (Proposed: 10 ppm; Animal Carcinogen)
DFG MAK: 10 ppm (35 mg/m^3)

NIOSH REL: (Vinyl Acetate) CL 15 mg/m^3/15M

DOT Classification: Label: Flammable Liquid.

SAFETY PROFILE: Moderately toxic by ingestion, inhalation, and intraperitoneal routes. A skin and eye irritant. Questionable carcinogen with experimental carcinogenic and tumorigenic data. Experimental reproductive effects. Human mutation data reported. Highly dangerous fire hazard when exposed to heat, flame, or oxidizers. A storage hazard, it may undergo spontaneous exothermic polymerization. Reaction with air or water to form peroxides that catalyze an exothermic polymerization reaction has caused several large industrial explosions. Reaction with hydrogen peroxide forms the explosive peracetic acid. Reacts with oxygen above 50°C to form an unstable explosive peroxide. Reacts with ozone to form the explosive vinyl acetate ozonide. Solution polymerization of the acetate dissolved in toluene has resulted in large industrial explosions. Polymerization reaction with dibenzoyl peroxide + ethyl acetate may release ignitable and explosive vapors. The vapor may react vigorously with desiccants (e.g., silica gel or alumina). Incompatible (explosive) with 2-amino ethanol; chlorosulfonic acid; ethylenediamine; ethyleneimine; HCl; HF; HNO$_3$; oleum; peroxides; H$_2$SO$_4$. See also ESTERS.

VNK100   CAS:1484-13-5   ***HR: 3***
N-VINYLCARBAZOLE
mf: C$_{14}$H$_{11}$N   mw: 193.26

SYNS: 9H-CARBAZOLE, 9-ETHENYL-(9CI) ◇ CARBAZOLE, 9-VINYL- ◇ N-ETHENYLCARBAZOLE ◇ 9-ETHENYL-9H-CARBAZOLE ◇ VINYLCARBAZOLE ◇ 9-VINYLCARBAZOLE ◇ N-VINYLKARBAZOL

TOXICITY DATA with REFERENCE
orl-mus LD50:50 mg/kg   85JCAE -,824,86
orl-gpg LD50:100 mg/kg   85JCAE -,824,86

CONSENSUS REPORTS: Reported in EPA TSCA Inventory.

SAFETY PROFILE: Poison by ingestion. When heated to decomposition it emits toxic vapors of NO$_x$.

VNZ990   CAS:2622-21-1   ***HR: 1***
1-VINYLCYCLOHEXENE
mf: C$_8$H$_{12}$ mw: 108.20

SYNS: CYCLOHEXENE, 1-VINYL- ◇ 1-ETHENYLCYCLOHEXENE

TOXICITY DATA with REFERENCE
ihl-rat LC50:10,500 mg/m^3/4H   85JCAE -,24,86

CONSENSUS REPORTS: Reported in EPA TSCA Inventory.

SAFETY PROFILE: Slightly toxic by inhalation. When heated to decomposition it emits acrid smoke and irritating vapors.

VQA150   CAS:1464-69-3   ***HR: 2***
2-(VINYLOXY)ETHYL METHACRYLATE
mf: C$_8$H$_{12}$O$_3$ mw: 156.20

SYNS: 2-(ETHENYLOXY)ETHYL 2-METHYL-2-PROPENOATE ◇ METHACRYLIC ACID, 2-(VINYLOXY)ETHYL ESTER ◇ 2-PROPENOIC ACID, 2-METHYL-, 2-(ETHENYLOXY)ETHYL ESTER (9CI)

TOXICITY DATA with REFERENCE
orl-rat LD50:5940 mg/kg   GTPZAB 31(2),45,87
orl-mus LD50:3830 mg/kg   GTPZAB 31(2),45,87
ihl-mus LC50:2115 mg/m^3 GTPZAB 31(2),45,87

CONSENSUS REPORTS: Reported in EPA TSCA Inventory.

SAFETY PROFILE: Moderately toxic by ingestion and inhalation. When heated to decomposition it emits acrid smoke and irritating vapors.

VQK590   CAS:100-43-6   ***HR: 3***
4-VINYLPYRIDINE
mf: C$_7$H$_7$N   mw: 105.15

SYNS: 4-ETHENYLPYRIDINE ◇ PYRIDINE, 4-VINYL-

TOXICITY DATA with REFERENCE
orl-rat LD50:100 mg/kg   85JCAE -,845,86
ihl-rat LCLo:2000 ppm/2H   85JCAE -,845,86
skn-gpg LDLo:500 mg/kg   85JCAE -,845,86
orl-brd LD50:100 mg/kg   AECTCV 12,355,83

CONSENSUS REPORTS: Reported in EPA TSCA Inventory.

SAFETY PROFILE: Poison by ingestion. Slightly toxic by inhalation. When heated to decomposition it emits toxic vapors of NO$_x$.

VQK600   CAS:2097-18-9   ***HR: 2***
1-VINYLSILATRANE
mf: C$_8$H$_{15}$NO$_3$Si   mw: 201.33

SYNS: 1-ETHENYLSILATRANE ◇ 2,8,9-TRIOXA-5-AZA-1-SILABICYCLO(3.3.3)UNDECANE, 1-ETHENYL- ◇ VINYLSILATRAN ◇ VINYLSILATRANE ◇ 1-VINYL-2,8,9-TRIOXA-5-AZA-1-SILABICYCLO(3.3.3)UNDECANE ◇ 2,8,9-TRIOXA-5-AZA-1-SILABICYCLO(3.3.3)UNDECANE, 1-VINYL-

TOXICITY DATA with REFERENCE
orl-mus LD50:1750 mg/kg   PHARAT 26,224,70
ipr-mus LD50:1100 mg/kg   PHARAT 26,224,70

CONSENSUS REPORTS: Reported in EPA TSCA Inventory.

SAFETY PROFILE: Moderately toxic by ingestion and intraperitoneal route. When heated to decomposition it emits toxic vapors of NO$_x$.

VQK700   CAS:622-97-9   **HR: 3**
4-VINYLTOLUENE
mf: C_9H_{10} mw: 118.19

SYNS: BENZENE, 1-ETHENYL-4-METHYL-(9CI) ◊ 1-ETHENYL-4-METHYLBENZENE ◊ p-METHYLSTYRENE ◊ STYRENE, p-METHYL- ◊ 1-p-TOLYLETHENE ◊ p-VINYLTOLUENE

TOXICITY DATA with REFERENCE

cyt-hmn:lyms 330 µmol/L   CRNGDP 2,237,81
sce-hmn:lyms 330 µmol/L   CRNGDP 2,237,81
orl-rat LD50:2255 mg/kg   JACTDZ 1,77,90
ipr-rat LD50:2324 mg/kg   JACTDZ 1,77,90
orl-mus LD50:1072 mg/kg   JACTDZ 1,76,90
ipr-mus LD50:581 mg/kg   JACTDZ 1,76,90
ivn-mus LD50:280 mg/kg   JACTDZ 1,76,90
orl-dog LD50:>5 g/kg   JACTDZ 1,77,90
skn-rbt LD50:>5 g/kg   JACTDZ 1,77,90

CONSENSUS REPORTS: Reported in EPA TSCA Inventory.

SAFETY PROFILE: Poison by intravenous route. Moderately toxic by ingestion. Slightly toxic by skin contact. Human mutation data reported.

VSK985   CAS:116-31-4   **HR: D**
trans-VITAMIN A ALDEHYDE
mf: $C_{20}H_{28}O$ mw: 284.48

SYNS: AXEROPHTHAL ◊ RETINAL (9CI) ◊ RETINAL, all-trans- ◊ all-E-RETINAL ◊ all-trans-RETINAL ◊ E-RETINAL ◊ trans-RETINAL ◊ RETINALDEHYDE ◊ RETINENE ◊ α-RETINENE ◊ RETINENE 1 ◊ VITAMIN A ALDEHYDE ◊ VITAMIN A1 ALDEHYDE

TOXICITY DATA with REFERENCE

dni-rat:mmr 3 µmol/L   JJIND8 70,949,83
dni-mus:lyms 100 µmol/L   ONCOBS 44,356,87

CONSENSUS REPORTS: Reported in EPA TSCA Inventory.

SAFETY PROFILE: Mutation data reported. When heated to decomposition it emits acrid smoke and irritating vapors.

VSZ095   CAS:1406-16-2   **HR: 1**
VITAMIN D

TOXICITY DATA with REFERENCE

orl-wmn TDLo:875 µg/kg/6W-I   AJMEAZ 82,224,87

CONSENSUS REPORTS: Reported in EPA TSCA Inventory.

SAFETY PROFILE: Human systemic effects by ingestion: changes in tubules (including acute renal failure, acute tubular necrosis), depressed renal function tests, calcium level changes. When heated to decomposition it emits acrid smoke and irritating vapors.

X

XQJ300 CAS:58-86-6
(D)-XYLOSE
mf: $C_5H_{10}O_5$ mw: 150.15

SYNS: XYLOSE, D-

TOXICITY DATA with REFERENCE
orl-mus LD50:23 g/kg   YKYUA6 32,1367,81
ivn-mus LD50:11300 mg/kg   YKYUA6 32,1367,81

HR: 1

CONSENSUS REPORTS: Reported in EPA TSCA Inventory.

SAFETY PROFILE: Slightly toxic by ingestion and intravenous routes. When heated to decomposition it emits acrid smoke and irritating vapors.

Z

ZDS000   CAS:10326-24-6   **HR: 3**
ZINC-m-ARSENITE
DOT: UN 1712
mf: $AsHO_2 \cdot 1/2Zn$ mw: 140.61

PROP: A white powder.

SYNS: ARSENIOUS ACID, ZINC SALT ◇ ZINC ARSENITE, solid (DOT) ◇ ZINC METAARSENITE ◇ ZINC METHARSENITE ◇ ZMA

CONSENSUS REPORTS: Arsenic and its compounds, as well as zinc and its compounds, are on the Community Right-To-Know List.

OSHA PEL: TWA 0.01 mg(As)/m^3: Cancer Hazard
ACGIH TLV: TWA 0.2 mg(As)/m^3 (Proposed: 0.01 mg(As)/m^3; Human Carcinogen)
NIOSH REL: (Inorganic Arsenic) CL 0.002 mg(As)/m^3/15M

DOT Classification: Poison B; Label: Poison

SAFETY PROFILE: Confirmed human carcinogen. A poison. When heated to decomposition it emits toxic fumes of As and ZnO. See also ARSENIC COMPOUNDS and ZINC COMPOUNDS.

ZIJ300   CAS:127-82-2   **HR: 3**
ZINC p-HYDROXYBENZENESULFONATE
mf: $C_{12}H_{12}O_8S_2 \cdot Zn$ mw: 413.73

SYNS: BENZENESULFONIC ACID, p-HYDROXY-, ZINC SALT (2:1) ◇ BENZENESULFONIC ACID, 4-HYDROXY-, ZINC SALT (2:1) ◇ p-HYDROXYBENZENESULFONIC ACID ZINC SALT ◇ 1-PHENOL-4-SULFONIC ACID ZINC SALT ◇ PHENOZIN ◇ ZINC PHENOLSULFONATE ◇ ZINC p-PHENOL SULFONATE ◇ ZINC SULFOCARBOLATE ◇ ZINC SULFOPHENATE

TOXICITY DATA with REFERENCE
eye-rbt 3 mg MOD   JACTDZ 5(5),373,86
orl-rat LD50:1800 mg/kg   JACTDZ 5(5),373,86
orl-mus LD50:3 g/kg   JACTDZ 5(5),373,86
ipr-mus LD50:225 mg/kg   JACTDZ 5(5),373,86

CONSENSUS REPORTS: Reported in EPA TSCA Inventory.

SAFETY PROFILE: Poison by intraperitoneal route. Moderately toxic by ingestion. An eye irritant. When heated to decomposition it emits toxic vapors of SO_x and fumes of zinc.

CAS Registry Number Cross-Index

50-00-0 see FMV000	85-40-5 see TDB100	95-73-8 see DGM700	106-22-9 see DTF410	121-72-2 see TLG100
50-84-0 see DER100	85-52-9 see BDL850	95-77-2 see DFY425	106-29-6 see GDE810	122-08-7 see TLM100
51-56-9 see HGH150	86-53-3 see NAY090	95-84-1 see AKY950	106-31-0 see BSW550	122-18-9 see BEL900
51-65-0 see FLD100	86-55-5 see NAV490	96-26-4 see OMC300	106-37-6 see BNV775	122-65-6 see DDF700
51-77-4 see GDG200	86-73-7 see FDI100	97-36-9 see OOI100	106-46-7 see DEP800	122-69-0 see CMQ850
51-82-1 see DRQ650	86-75-9 see HOE100	97-41-6 see EHM100	106-58-1 see LIQ500	122-75-8 see DDF800
51-93-4 see EQC600	86-92-0 see THA300	98-18-0 see MDM760	106-65-0 see SNB100	123-07-9 see EOE100
53-59-8 see CNF400	86-97-5 see ALJ500	98-27-1 see BRU800	106-70-7 see MHY700	123-41-1 see CMF800
53-84-9 see CNF390	86-98-6 see DGJ250	98-50-0 see ARA250	107-01-7 see BOW255	123-69-3 see OKU100
54-35-3 see PAQ200	87-12-7 see BOD600	98-51-1 see BSP500	107-09-5 see BNI650	123-82-0 see HBM490
56-23-5 see CBY000	87-61-6 see TIK100	98-67-9 see HNL600	107-43-7 see GHA050	123-86-4 see BPU750
56-37-1 see BFL300	87-63-8 see CLK227	98-77-1 see PIY500	107-58-4 see BPW050	123-96-6 see OCY090
56-55-3 see BBC250	88-18-6 see BSE460	98-86-2 see ABH000	107-64-2 see DXG625	124-04-9 see AEN250
56-82-6 see GFY200	88-21-1 see SNO100	98-89-5 see HDH100	108-05-4 see VLU250	124-09-4 see HEO000
57-71-6 see OMY910	88-27-7 see FAB000	98-91-9 see TFC550	108-24-7 see AAX500	124-12-9 see OCW100
58-64-0 see AEK100	88-45-9 see PEY800	99-07-0 see HNK560	108-26-9 see MOX100	124-40-3 see DOQ800
58-86-6 see XQJ300	88-58-4 see DEC800	99-10-5 see REF200	108-36-1 see DDK050	124-73-2 see FOO525
61-31-4 see NAK550	88-60-8 see BQV600	99-28-5 see DDQ500	108-40-7 see TGO800	125-02-0 see PLY275
62-33-9 see CAR780	88-64-2 see AHQ300	99-33-2 see DUR850	108-47-4 see LIY990	126-13-6 see SNH050
62-74-8 see SHG500	88-68-6 see AID620	99-34-3 see DUR600	108-48-5 see LAJ010	126-14-7 see OAF100
63-37-6 see CQL300	88-69-7 see IQX100	99-42-3 see HMY075	108-65-6 see PNL265	127-18-4 see PCF275
65-47-4 see CQL400	88-88-0 see PIE530	99-50-3 see POE200	108-70-3 see TIK300	127-66-2 see EQN230
66-23-9 see CMF260	88-96-0 see BBO500	99-52-5 see MMF780	108-72-5 see THN775	127-82-2 see ZIJ300
67-16-3 see TES800	89-05-4 see PPQ630	99-53-6 see NFV010	108-74-7 see HDW100	128-39-2 see DEG100
67-51-6 see DTU850	89-28-1 see SAU480	99-69-4 see NIM600	108-75-8 see TME272	128-94-9 see DBQ250
67-72-1 see HCI000	89-32-7 see PPQ635	99-97-8 see TLG150	108-88-3 see TGK750	128-97-2 see NAP250
68-34-8 see TGN600	89-37-2 see DVB850	100-03-8 see CEJ600	109-00-2 see PPH025	130-00-7 see NAX100
69-78-3 see DUV700	89-40-7 see NIX100	100-09-4 see AOU600	109-04-6 see BOB600	130-16-5 see CLD600
71-30-7 see CQM600	89-52-1 see AAJ150	100-18-5 see DNN830	109-12-6 see PPH300	130-20-1 see DFN425
71-67-0 see HAQ600	89-58-7 see NMS520	100-21-0 see TAN750	109-13-7 see BSC600	130-22-3 see SEH475
72-40-2 see IAL100	89-63-4 see KDA050	100-40-3 see CPD750	109-16-0 see MDN510	131-09-9 see CEI100
74-11-3 see CEL300	89-65-6 see SAA025	100-43-6 see VQK590	109-17-1 see TCE400	131-22-6 see PEJ600
74-89-5 see MGC250	89-68-9 see CLJ800	100-49-2 see HDH200	109-81-9 see MJW100	131-53-3 see DMW250
74-96-4 see EGV400	89-82-7 see POI615	100-54-9 see NDW515	109-89-7 see DHJ200	131-54-4 see BKH200
75-04-7 see EFU400	89-93-0 see MHM510	100-55-0 see NDW510	110-42-9 see MHY650	132-22-9 see CLX300
75-07-0 see AAG250	90-11-9 see BNS200	100-83-4 see FOE100	110-53-2 see AOF800	132-64-9 see DDB500
75-17-2 see FMW400	90-42-6 see LBX100	101-14-4 see MJM200	110-58-7 see PBV505	132-65-0 see TES300
75-34-3 see DFF809	90-44-8 see APM500	101-63-3 see NIM600	110-59-8 see VAV300	133-18-6 see APJ500
75-50-3 see TLD500	90-50-6 see CMQ100	101-70-2 see BKO600	110-64-5 see BOX300	133-66-4 see DXB450
75-65-0 see BPX000	90-81-3 see EAW100	101-71-3 see PFR325	110-99-6 see ONQ100	135-19-3 see NAX000
75-87-6 see CDN550	91-01-0 see HKF300	101-75-7 see PEI800	111-01-3 see SLG700	135-57-9 see BDK800
76-61-9 see TFX850	91-02-1 see PGE760	101-76-8 see BIM800	111-02-4 see SLG800	136-81-2 see AOM325
76-86-8 see TMU800	91-04-3 see HLV100	101-82-6 see BFG600	111-11-5 see MHY800	138-41-0 see SNL840
76-93-7 see BBY990	91-55-4 see DSI850	102-24-9 see TKZ100	111-18-2 see TDR100	138-87-4 see TBD775
77-48-5 see DDI900	91-68-9 see DIO400	102-56-7 see AKD925	111-61-5 see EPF700	140-10-3 see CMP980
77-99-6 see HDF300	91-88-3 see HKS100	102-71-6 see TKP500	111-68-2 see HBL600	140-11-4 see BDX000
78-04-6 see DEJ100	92-30-8 see TKE775	102-96-5 see NMC100	111-69-3 see AER250	140-22-7 see DVY925
78-13-7 see TCD250	92-36-4 see MHJ300	103-00-4 see CPG700	111-82-0 see MLC800	140-28-3 see BHB300
78-18-2 see CPC300	92-39-7 see CJL100	103-14-0 see BDY750	112-10-7 see IRL100	140-53-4 see CEP300
78-66-0 see DTF850	92-44-4 see NAN510	103-30-0 see SLR100	112-29-8 see BNB800	140-72-7 see HCP800
78-69-3 see LFY510	92-49-9 see EHJ600	103-79-7 see MHO100	112-69-6 see HCP525	140-80-7 see DJQ850
79-01-6 see TIO750	92-55-7 see NGB800	103-88-8 see BMR100	114-70-5 see SFA200	140-92-1 see IRG050
79-33-4 see LAG010	92-66-0 see BMW300	103-94-6 see NHX100	115-27-5 see CDS050	141-00-4 see CAI750
80-07-1 see BBS300	92-70-6 see HMX250	104-03-0 see NII150	116-11-0 see MFL300	141-10-6 see POH525
80-18-2 see MHB300	92-81-9 see DKQ675	104-13-2 see AJA550	116-31-4 see VSK985	141-17-3 see DDT500
80-55-7 see ELH700	93-07-2 see VHP600	104-23-4 see AIB350	116-66-5 see MRU300	141-27-5 see GCU100
81-11-8 see FCA100	93-35-6 see HJY100	104-40-5 see NNC510	118-12-7 see TLU200	141-63-9 see DXS800
81-26-5 see DQR350	93-40-3 see HGK600	104-60-9 see HYP100	119-15-3 see DUW500	141-92-4 see LBO050
81-63-0 see DBR450	93-52-7 see DDN900	104-67-6 see HBN200	119-21-1 see CGC100	142-08-5 see OOO200
81-69-6 see SOB600	93-56-1 see SMQ100	104-74-5 see LBX050	119-67-5 see FNK010	142-31-4 see OFU200
81-77-6 see IBV050	93-69-6 see TGX550	104-83-6 see CFB600	119-80-2 see BHM300	142-77-8 see BSB000
82-39-3 see MEA650	93-70-9 see AAY500	104-88-1 see CEI600	119-81-3 see NEK100	143-74-8 see PDO800
82-86-0 see AAF300	93-91-4 see BDJ800	104-92-7 see AOY450	120-22-9 see NJW600	145-39-1 see MRW272
83-07-8 see AIB300	94-01-9 see BHB100	104-96-1 see AMS675	120-23-0 see NBJ100	146-14-5 see RIF100
83-32-9 see AAF275	94-02-0 see EGR600	104-98-3 see UVJ440	120-35-4 see AKM500	150-25-4 see DMT500
83-42-1 see CJG825	94-44-0 see NCR040	105-12-4 see DVE260	120-37-6 see EGF100	150-61-8 see DWB400
83-56-7 see NAN505	94-70-2 see PDK819	105-16-8 see DIB300	120-43-4 see EHG050	150-90-3 see SJW100
83-72-7 see HMX600	95-01-2 see REF100	105-18-0 see BJA250	120-80-9 see CCP850	151-01-9 see EQG600
83-81-8 see GCE600	95-29-4 see DNN900	105-60-2 see CBF700	120-86-5 see BKD800	151-10-0 see REF025
84-61-7 see DGV700	95-46-5 see BOG260	105-66-8 see PNF100	121-05-1 see DNP700	156-38-7 see HNG600
84-89-9 see ALI240	95-50-1 see DEP600	105-77-1 see BSS550	121-50-6 see CEG800	203-12-3 see BCJ200
85-32-5 see GLS750	95-52-3 see FLZ100	105-87-3 see DTD800	121-57-3 see SNN600	205-99-2 see BAW250

231-36-7 see DWX800	505-29-3 see DXI550	588-07-8 see CDZ050	628-20-6 see CEU300	838-85-7 see PHE300
238-84-6 see NAU600	506-82-1 see DQW800	590-46-5 see CCH850	628-21-7 see TDQ400	848-53-3 see HGI100
243-17-4 see BCJ800	507-63-1 see PCH325	591-17-3 see BOG300	628-77-3 see PBH125	860-39-9 see BLA800
275-51-4 see ASP600	509-14-8 see TDY250	591-31-1 see AOT300	628-87-5 see IBB100	871-67-0 see HEF400
283-24-9 see ARX300	513-49-5 see BPY100	591-35-5 see DFY450	629-04-9 see HBN100	872-85-5 see IKZ100
285-67-6 see EBO100	513-78-0 see CAD800	592-82-5 see ISD100	629-05-0 see OGI025	880-52-4 see AEE100
287-29-6 see SCF550	513-79-1 see CNB475	593-85-1 see GKW100	629-09-4 see HEG200	881-03-8 see MMN800
288-47-1 see TEU300	515-42-4 see SJH050	594-34-3 see DDQ150	629-27-6 see OFA100	882-35-9 see EIT100
288-88-0 see THS825	515-69-5 see BGO775	596-27-0 see CNX400	629-40-3 see OCW050	923-26-2 see HNV500
289-80-5 see OJW200	522-75-8 see DNT300	596-43-0 see TNP600	629-70-9 see HCP100	926-57-8 see DEU650
290-37-9 see POL590	524-38-9 see HNS600	597-49-9 see TJP550	629-78-7 see HAS100	927-68-4 see BNI600
300-87-8 see DSK950	525-79-1 see FPT100	597-71-7 see NNR400	630-08-0 see CDW750	929-73-7 see DKW100
300-92-5 see AHA275	525-82-6 see PER700	597-96-6 see MKL350	631-67-4 see DUG450	930-37-0 see GGW600
301-13-3 see TNI300	527-60-6 see MDJ740	598-16-3 see THV100	632-58-6 see TBT100	932-52-5 see AMW100
305-15-7 see DGC850	527-72-0 see TFM600	598-74-3 see AOE200	633-65-8 see BFN600	933-48-2 see TLO510
306-52-5 see TIO800	528-44-9 see TKU700	598-94-7 see DUM150	634-60-6 see AML600	936-02-7 see HJL100
306-94-5 see PCG700	528-45-0 see DUR550	599-64-4 see COF400	634-95-7 see DKD650	936-49-2 see PFJ300
307-34-6 see OAP100	528-94-9 see ANT600	599-71-3 see BDW100	635-93-8 see CLD800	936-52-7 see CPY800
312-30-1 see DKH250	529-84-0 see MJV800	600-05-5 see DDS500	636-09-9 see DKB160	940-31-8 see PDV725
321-14-2 see CLD825	530-57-4 see SPE700	602-09-5 see BGC100	636-26-0 see TFQ650	947-04-6 see COW930
321-30-2 see AEH500	531-53-3 see DUG700	602-38-0 see DUX710	636-53-3 see IMK100	947-42-2 see DMN450
321-98-2 see EAW200	531-59-9 see MEK300	602-94-8 see PBD275	637-12-7 see AHH825	951-78-0 see DAZ050
329-98-6 see TGO300	535-75-1 see HDS300	603-50-9 see PPN100	638-07-3 see EHH100	959-52-4 see THM900
333-49-3 see TFH800	535-80-8 see CEL300	604-59-1 see NBI100	644-08-6 see MJH905	961-68-2 see DUV100
334-50-9 see SLG600	535-87-5 see DBQ190	605-45-8 see PHW600	645-15-8 see BLA600	985-13-7 see PAH260
335-67-1 see PCH050	536-90-3 see AVO890	606-35-9 see TMK300	646-13-9 see IJN100	987-65-5 see AEM100
344-07-0 see PBE100	538-41-0 see ASK925	606-45-1 see MLH800	646-25-3 see DAG650	992-59-6 see DXO850
344-72-9 see AMU550	538-62-5 see DVY950	606-55-3 see EPD600	661-95-0 see DDO450	993-43-1 see EOP600
350-30-1 see CHI950	539-32-2 see BSJ550	608-08-2 see IDA600	673-22-3 see HLR600	994-65-0 see TED650
354-58-5 see TJE100	539-47-9 see FPK050	609-27-8 see EOB300	675-10-5 see THL800	995-33-5 see BHL200
356-18-3 see DFM025	540-10-3 see HCP700	609-86-9 see API800	675-62-7 see DFS700	996-31-6 see PLK650
364-76-1 see FKK100	540-49-8 see DDN950	609-99-4 see HKE600	677-21-4 see TKH015	998-40-3 see TIA300
366-29-0 see BJF600	541-02-6 see DAF350	610-17-3 see NFW500	683-72-7 see DEM300	1002-62-6 see SGB600
367-12-4 see FKT100	541-28-6 see IHU200	610-66-2 see NFN600	686-31-7 see AOM200	1002-89-7 see ANU200
367-25-9 see DKF700	541-33-3 see BRQ050	610-67-3 see NIC990	687-46-7 see EHH200	1003-10-7 see TDC800
368-39-8 see TJL600	542-18-7 see CPI400	613-03-6 see HLF600	689-11-2 see BSS300	1003-14-1 see ECE525
368-53-6 see TKB775	542-58-5 see CGO600	613-91-2 see ABH150	691-60-1 see IRR100	1003-67-4 see MOY790
375-01-9 see HAW100	542-85-8 see ELX525	614-60-8 see CNU850	692-86-4 see EQD200	1003-73-2 see MOY550
379-52-2 see TMV850	543-87-3 see ILW100	614-94-8 see DBO100	693-16-3 see OEK010	1004-40-6 see AMS750
382-10-5 see HDC450	543-90-8 see CAD250	615-16-7 see ONI100	693-54-9 see OFE050	1006-59-3 see DJU700
382-21-8 see OBM000	544-47-8 see CEQ800	615-18-9 see CEM850	694-53-1 see SDX250	1006-99-1 see CIH100
393-75-9 see CGM225	544-62-7 see GGA915	615-48-5 see CHU100	694-85-9 see MPA075	1068-27-5 see DRJ825
400-99-7 see NMQ100	544-76-3 see HCO600	615-58-7 see DDR150	695-06-7 see HDY600	1071-27-8 see COR550
420-46-2 see TJY900	544-85-4 see DYC900	615-74-7 see CFE500	698-87-3 see PGA600	1073-72-9 see MPV300
423-62-1 see IEU075	546-68-9 see IRN200	617-65-2 see GFM300	699-10-5 see MHN350	1075-76-9 see AOT100
428-59-1 see HDF050	546-89-4 see LGO100	618-45-1 see IQX090	706-79-6 see DFM050	1076-56-8 see MPW650
446-35-5 see DKH900	547-64-8 see MLC600	618-56-4 see DBQ200	707-61-9 see MNV800	1079-01-2 see MSC050
447-14-3 see TKH310	552-30-7 see TKV000	618-76-8 see DNF300	708-06-5 see HMU200	1080-12-2 see MLI800
447-31-4 see CFJ100	553-26-4 see BGO600	619-04-5 see DQM850	719-22-2 see DDV500	1083-48-3 see NFO750
452-71-1 see FJT050	554-00-7 see DEO290	619-19-2 see NJF100	719-59-5 see AJE400	1085-12-7 see HBO650
453-20-3 see HOI200	554-68-7 see TJO050	619-23-8 see NFN010	722-27-0 see ALW100	1113-38-8 see ANO750
455-90-3 see NMO700	555-31-7 see AHC600	619-45-4 see AIN150	723-62-6 see APG550	1113-59-3 see BOD550
456-27-9 see NEZ200	555-60-2 see CKA550	619-58-9 see IEE025	728-84-7 see BNH100	1114-51-8 see DJX250
459-44-9 see TGM450	556-18-3 see AIC825	619-72-7 see NFI010	729-46-4 see MRR090	1115-47-5 see ACQ270
460-40-2 see TKH020	556-67-2 see OCE100	619-80-7 see NEV525	734-88-3 see OOI200	1117-41-5 see TMJ100
461-05-2 see CCK655	558-30-5 see IIQ600	619-84-1 see DOU650	747-45-5 see QFS100	1117-55-1 see HFS600
464-07-3 see BRU300	563-43-9 see EUF050	619-89-6 see NFT440	753-73-1 see DUG825	1117-71-1 see MHR790
470-67-7 see IKC100	563-63-3 see SDI800	619-90-9 see NFM100	757-44-8 see DKD500	1118-12-3 see BSS310
471-34-1 see CAT775	564-94-3 see FNK150	619-91-0 see NIN100	760-67-8 see EKO600	1119-94-4 see DYA810
474-07-7 see LFT800	565-74-2 see BNM100	620-02-0 see MKH550	760-79-2 see DQV300	1119-97-7 see TCB200
475-25-2 see HAO600	574-66-3 see BCS400	620-45-1 see SGG650	762-12-9 see DAH475	1120-01-0 see HCP900
476-60-8 see LEX200	576-24-9 see DFX500	620-61-1 see HOT600	762-13-0 see NNA100	1120-04-3 see OBG100
477-73-6 see GJI400	577-19-5 see NFQ080	621-09-0 see DVW750	762-16-3 see CBF705	1121-31-9 see MCQ700
480-63-7 see IKN300	577-85-5 see HLC600	621-33-0 see PDK800	764-41-0 see DEV000	1121-70-6 see SIM100
480-68-2 see NET550	578-57-4 see BMT400	621-79-4 see CMP970	765-09-3 see BOI500	1122-54-9 see ADA365
482-54-2 see CPB120	578-95-0 see ADI825	622-88-8 see BNW550	765-87-7 see CPB200	1122-91-4 see BMT700
482-89-3 see BGB275	579-66-8 see DIS650	622-97-9 see VQK700	766-66-5 see CFF600	1123-54-2 see AIB340
483-84-1 see FBZ100	580-17-6 see AMK725	623-17-6 see FPM100	773-76-2 see DEL300	1123-85-9 see HGR600
485-71-2 see CMP910	580-22-3 see AMK700	623-33-6 see BSG300	780-11-0 see BSG300	1134-35-6 see DQS100
487-26-3 see FBW150	581-08-8 see ABG350	624-24-8 see VAQ100	781-35-1 see MJH910	1170-02-1 see EIV100
487-68-3 see MDJ745	582-52-5 see DVO100	624-72-6 see DKH300	787-84-8 see DDE250	1185-57-5 see FAS700
487-89-8 see FNO100	582-60-5 see DQM100	625-27-4 see MNK100	791-31-1 see HOM300	1187-33-3 see DEH300
488-10-8 see JCA100	583-52-8 see PLN300	626-02-8 see IEV010	813-78-5 see PHC800	1187-58-2 see MOS900
492-39-7 see NNM510	583-68-6 see BOG500	626-35-7 see ENN100	814-67-5 see PML300	1187-93-5 see TKG800
495-69-2 see HGB300	583-78-8 see DFX850	626-64-2 see HOB100	818-72-4 see OGI050	1191-79-3 see BAI800
496-41-3 see CNU875	583-91-5 see HMR600	626-86-8 see ELC600	822-12-8 see SIN900	1192-30-9 see TCS550
498-67-9 see DFL100	584-09-8 see RPB200	627-06-5 see PNX550	822-38-8 see EJQ100	1192-62-7 see ACM200
498-81-7 see MCE100	584-42-9 see SIT850	627-30-5 see CKP725	826-36-8 see TDT770	1193-24-4 see PPP100
498-94-2 see ILG100	585-79-5 see NFQ090	627-31-6 see TLR050	826-81-3 see HMQ600	1194-02-1 see FGH100
499-06-9 see MDJ748	586-76-5 see BMU100	627-45-2 see EKK600	830-96-6 see ICS200	1196-57-2 see QSJ100
500-22-1 see NDM100	586-78-7 see NFQ100	627-75-8 see AQV500	831-61-8 see EKM100	1197-55-3 see AID700
504-03-0 see LIQ550	586-95-8 see POR810	628-17-1 see IET500	838-57-3 see ENO100	1200-89-1 see TCU650

1210-56-6 see MLJ100	1668-54-8 see MGH800	2179-57-9 see AGF300	2539-53-9 see IKO100	8028-73-7 see ARE500
1214-39-7 see BDX090	1670-14-0 see BBM750	2186-24-5 see GGY175	2564-83-2 see TDT800	8042-47-5 see MQV875
1214-47-7 see HJS900	1688-71-7 see AJC750	2190-04-7 see OAP300	2567-83-1 see TCD002	9005-25-8 see SLJ500
1222-98-6 see NFS505	1717-00-6 see FOO550	2191-10-8 see CAD750	2605-44-9 see CAG775	9007-73-2 see FBB000
1248-18-6 see NAP100	1752-24-5 see IBJ100	2192-20-3 see HOR470	2618-77-1 see DRJ100	10022-68-1 see CAH250
1303-18-0 see ARJ750	1759-28-0 see MQM800	2213-63-0 see DGJ950	2621-62-7 see DGB480	10048-95-0 see ARC250
1303-28-2 see ARH500	1769-41-1 see GIK100	2219-82-1 see BRU790	2622-21-1 see VNZ990	10102-49-5 see IGN000
1303-33-9 see ARI000	1774-47-6 see TMG800	2223-93-0 see OAT000	2624-17-1 see SGB550	10102-50-8 see IGM000
1304-76-3 see BKW600	1779-51-7 see BSR900	2224-15-9 see EEA600	2628-16-2 see ABW550	10102-53-1 see ARB000
1304-85-4 see BKW100	1779-81-3 see TEV600	2226-96-2 see HOI300	2631-68-7 see TJE200	10103-50-1 see ARD000
1306-19-0 see CAH500	1783-81-9 see ALX100	2243-27-8 see OES300	2634-33-5 see BCE475	10103-60-3 see ARD600
1306-23-6 see CAJ750	1785-51-9 see PON340	2260-00-6 see MPV790	2637-34-5 see TFP300	10108-64-2 see CAE250
1309-38-2 see IHC550	1817-68-1 see MHR050	2284-30-2 see BFI400	2637-37-8 see QOJ100	10124-36-4 see CAJ000
1310-65-2 see LHI100	1833-31-4 see BEE800	2288-13-3 see MPI650	2638-94-0 see ASL500	10124-50-2 see PKV500
1314-12-1 see TEL040	1843-05-6 see HND100	2295-58-1 see TKP100	2687-25-4 see TGY800	10290-12-7 see CNN500
1317-25-5 see AHA135	1847-58-1 see SIB700	2304-85-0 see PON350	2688-84-8 see PDR490	10325-94-7 see CAH000
1319-69-3 see PLG810	1852-14-8 see EJC100	2305-05-7 see OES100	2719-13-3 see NIQ550	10326-24-6 see ZDS000
1321-74-0 see DXQ740	1852-16-0 see BPM660	2346-00-1 see DLV900	2921-88-2 see CMA100	12005-86-6 see SHM000
1324-55-6 see DFN450	1863-63-4 see ANB100	2348-82-5 see MEY800	3313-92-6 see SJB400	12014-28-7 see CAI125
1328-53-6 see PJQ100	1866-15-5 see ADC300	2350-24-5 see EKU100	3930-19-6 see SMA000	12380-95-9 see TIH000
1330-16-1 see PIH245	1888-91-1 see ACF100	2372-21-6 see BSD150	5743-04-4 see CAD275	13477-17-3 see CAI000
1330-43-4 see DXG035	1916-07-0 see MQF300	2374-05-2 see BOL303	5902-95-4 see CAM000	13477-21-9 see CAJ500
1331-12-0 see PNL100	1928-38-7 see DAA825	2379-55-7 see DTY700	7439-96-5 see MAP750	13925-12-7 see HLT100
1332-37-2 see IHG100	1937-19-5 see AKC800	2379-81-9 see DUP100	7440-31-5 see TGB250	14060-38-9 see ARJ500
1332-58-7 see KBB600	1940-42-7 see LEN050	2379-90-0 see AKO350	7440-38-2 see ARA750	14215-29-3 see CAD350
1333-39-7 see HJH100	1945-92-2 see DVC300	2382-43-6 see MIM300	7440-43-9 see CAD000	14239-68-0 see BJB500
1406-16-2 see VSZ095	1962-75-0 see DEH700	2382-96-9 see MCK900	7440-48-4 see CNA250	14486-19-2 see CAG000
1421-49-4 see DEC900	1984-06-1 see SIX600	2386-87-0 see EBO050	7495-93-4 see CAD500	15120-17-9 see ARD500
1421-89-2 see DOZ100	1988-89-2 see PFD400	2396-43-2 see TNC100	7568-37-8 see MHY550	15954-91-3 see CAF750
1438-94-4 see FPX100	2000-43-3 see TIR800	2406-25-9 see DEF090	7631-89-2 see ARD750	16039-55-7 see CAG750
1453-58-3 see MOX010	2001-94-7 see EEB100	2407-94-5 see PCM100	7757-79-1 see PLL500	17010-21-8 see CAG500
1453-82-3 see ILC100	2002-24-6 see EEC700	2418-14-6 see DNV610	7774-41-6 see ARC500	17014-71-0 see PLP250
1464-44-4 see PDO300	2013-26-5 see SIY600	2429-74-5 see CMO500	7778-39-4 see ARB250	18897-36-4 see CAI350
1464-69-3 see VQA150	2016-56-0 see DXW050	2432-90-8 see PHW550	7778-43-0 see ARC000	19010-79-8 see CAI400
1484-13-5 see VNK100	2027-17-0 see IQN100	2439-35-2 see DPB300	7778-44-1 see ARB750	22298-29-9 see BFV760
1516-27-4 see HGI600	2028-63-9 see EQM600	2442-10-6 see ODW028	7784-33-0 see ARF250	22750-53-4 see CAD325
1518-15-6 see BIY600	2044-73-7 see POR790	2451-62-9 see GGY150	7784-34-1 see ARF500	25321-14-6 see DVG600
1527-12-4 see PGL270	2049-96-9 see PBV800	2460-49-3 see DFL720	7784-35-2 see ARI250	27152-57-4 see CAM500
1561-49-5 see DGV650	2052-15-5 see BRR700	2461-15-6 see GGY100	7784-41-0 see ARD250	35658-65-2 see CAE500
1563-01-5 see DIJ300	2077-46-5 see TJD600	2469-55-8 see OPC100	7784-44-3 see DCG800	59544-89-7 see TKH250
1582-09-8 see DUV600	2082-84-0 see DAJ500	2475-44-7 see BKP500	7784-45-4 see ARG750	60676-86-0 see SCK600
1603-41-4 see AMA010	2094-99-7 see IKG800	2479-46-1 see REF070	7784-46-5 see SEY500	62865-26-3 see CMS205
1607-17-6 see PCR150	2097-18-9 see VQK600	2488-01-9 see PEW725	7784-46-5 see SEZ000	63951-03-1 see ARJ900
1608-26-0 see HEK100	2113-57-7 see BMW290	2489-52-3 see FLY200	7789-06-2 see SMH000	63989-69-5 see IGO000
1619-34-7 see QUJ990	2114-00-3 see BOB550	2494-56-6 see BSY300	7789-42-6 see CAD600	69012-64-2 see SCH000
1639-66-3 see SOD300	2122-70-5 see ENL900	2497-21-4 see HFE200	7790-78-5 see CAE425	72589-96-9 see CAE375
1646-26-0 see ACC100	2144-08-3 see TKN800	2499-58-3 see HBL100	7790-79-6 see CAG250	73419-42-8 see CAK250
1646-75-9 see MLX800	2150-60-9 see APG700	2531-84-2 see MNN520	7790-84-3 see CAJ250	73941-35-2 see THV500
1646-87-3 see MLW750	2154-68-9 see TDV300	2536-18-7 see DUW503	8008-45-5 see NOG500	74278-22-1 see KHU000
1660-94-2 see MJT100	2158-03-4 see PIL525	2538-85-4 see HLI100	8012-95-1 see MQV750	
1665-59-4 see DJV300	2158-76-1 see BKB100	2539-17-5 see TBQ290	8016-37-3 see MSB775	

Synonym Cross-Index

AAoC see AAY600
AAP see AIB300
ABG 3034 see BDX090
ABROMINE see GHA050
ACCOSPERSE CYAN GREEN G see PJQ100
ACENAPHTHENE see AAF275
ACENAPHTHENEDIONE see AAF300
ACENAPHTHYLENE, 1,2-DIHYDRO- see AAF275
1,2-ACENAPHTHYLENEDIONE see AAF300
ACETALDEHYD (GERMAN) see AAG250
ACETALDEHYDE see AAG250
ACETALDEHYDE, TRICHLORO-(9CI) see CDN550
ACETAMIDE, N-(1-ADAMANTYL)- see AEE100
ACETAMIDE, N-(4-BROMOPHENYL)- see BMR100
ACETAMIDE, DICHLORO- see DEM300
ACETAMIDE, 2,2-DICHLORO-(8CI,9CI) see DEM300
ACETAMIDE, N-(2,5-DICHLOROPHENYL)-(9CI) see DGB480
ACETAMIDE, N,N-DIMETHYLTHIO- see DUG450
ACETAMIDE, N-(2-ETHOXYPHENYL)-(9CI) see ABG350
ACETAMIDE, 2-(HYDROXYIMINO)-N-PHENYL-(9CI) see GIK100
ACETAMIDINE, N,N'-DIPHENYL- see DVW750
2-ACETAMIDOBENZOIC ACID see AAJ150
o-ACETAMIDOBENZOIC ACID see AAJ150
4-ACETAMIDO-3-NITROANISOLE see NEK100
ACETAMINE YELLOW 2R see DUW500
ACETANILIDE, p-BROMO- see BMR100
ACETANILIDE, 4'-BROMO- see BMR100
ACETANILIDE, 2',5'-DICHLORO- see DGB480
ACETANILIDE, 2'-ETHOXY- see ABG350
ACETANILIDE-4-SULFONIC ACID, 3-AMINO- see AHQ300
p-ACETANISIDIDE, 2-NITRO- see NEK100
ACETATE de BUTYLE (FRENCH) see BPU750
ACETIC ACID, (p-AMINOPHENYL)- see AID700
ACETIC ACID, ANHYDRIDE see AAX500
ACETIC ACID, BENZOYL-, ETHYL ESTER see EGR600
ACETIC ACID BENZYL ESTER see BDX000
ACETIC ACID n-BUTYL ESTER see BPU750
ACETIC ACID, CADMIUM SALT see CAD250
ACETIC ACID, CADMIUM SALT, DIHYDRATE see CAD275
ACETIC ACID, (1,2-CYCLOHEXYLENEDINITRILO)TETRA- see CPB120
ACETIC ACID, (2,4-DICHLOROPHENOXY)-, METHYL ESTER see DAA825
ACETIC ACID, (3,4-DIMETHOXYPHENYL)- see HGK600
ACETIC ACID, 2-(DIMETHYLAMINO)ETHYL ESTER see DOZ100
ACETIC ACID ETHENYL ESTER see VLU250
ACETIC ACID, (ETHYLENEDINITRILO)TETRA-, CALCIUM DISODIUM SALT see CAR780
ACETIC ACID, (ETHYLENEDINITRILO)TETRA-, DIPOTASSIUM SALT see EEB100
ACETIC ACID FURFURYL ESTER see FPM100
ACETIC ACID GERANIOL ESTER see DTD800
ACETIC ACID, (p-HYDROXYPHENYL)- see HNG600
ACETIC ACID, 3-INDOLYL ESTER see IDA600
ACETIC ACID, LITHIUM SALT see LGO100
ACETIC ACID, 2-METHOXY-1-METHYLETHYL ESTER see PNL265
ACETIC ACID, (2-NAPHTHYLOXY)- see NBJ100
ACETIC ACID, (p-NITROBENZOYL)-, ETHYL ESTER see ENO100
ACETIC ACID, NITRO-, ETHYL ESTER see ENN100
ACETIC ACID, (p-NITROPHENYL)- see NII510
ACETIC ACID, ((4-NITROPHENYL)AMINO)OXO-(9CI) see NHX100
ACETIC ACID, 2,2'-OXYBIS-(9CI) see ONQ100
ACETIC ACID, OXYDI- see ONQ100
ACETIC ACID PHENYLMETHYL ESTER see BDX000
ACETIC ACID, PHENYL-, SODIUM SALT see SFA200
ACETIC ACID, SILVER(1+) SALT see SDI800
ACETIC ACID, SULFO-, 1-DODECYL ESTER, SODIUM SALT see SIB700
ACETIC ACID, SULFO-, DODECYL ESTER, S-SODIUM SALT (7CI) see SIB700
ACETIC ACID VINYL ESTER see VLU250
ACETIC ALDEHYDE see AAG250
ACETIC ANHYDRIDE see AAX500
ACETIC OXIDE see AAX500
ACETOACETANILIDE, o-CHLORO- see AAY600
ACETOACETANILIDE, 2'-CHLORO- see AAY600
ACETOACETANILIDE, 2',4'-DIMETHYL- see OOI100
ACETOACET-o-CHLORANILIDE see AAY600
ACETOACET-o-CHLOROANILIDE see AAY600
ACETOACET-2,4-DIMETHYLPHENYL see OOI100
ACETOACETIC ACID, 4-CHLORO-, ETHYL ESTER see EHH100
o-ACETOACETOCHLORANILIDE see AAY600
ACETOACETOPHENONE see BDJ800
ACETOACETO-m-XYLIDIDE see OOI100
2,4-ACETOACETOXYLIDIDE see OOI100
2',4'-ACETOACETOXYLIDIDE see OOI100
ACETOACET-m-XYLIDIDE see OOI100
ACETOACETYL-2-CHLOROANILIDE see AAY600
ACETOACETYL-m-XYLIDIDE see OOI100
o-ACETOAMINOBENZOIC ACID see AAJ150
ACETONITRILE, (p-CHLOROPHENYL)- see CEP300
ACETONITRILE, 2,2'-IMINOBIS-(9CI) see IBB100
ACETONITRILE, IMINODI- see IBB100
ACETONITRILE, (o-NITROPHENYL)- see NFN600
o-ACETOPHENETIDIDE see ABG350
ACETOPHENONE see ABH000
ACETOPHENONE, α-CHLORO-α-PHENYL- see CFJ100
ACETOPHENONE, 2-CHLORO-2-PHENYL-(8CI) see CFJ100
ACETOPHENONE, OXIME see ABH150
ACETOQUAT CPB see HCP800
ACETOQUINONE LIGHT PINK RLZ see AKO350
ACETOQUINONE LIGHT YELLOW 2RZ see DUW500
ACETOXYETHYL CHLORIDE see CGO600
1-ACETOXYETHYLENE see VLU250
ACETOXYETHYL-TRIMETHYLAMMONIUM BROMIDE see CMF260
1-ACETOXYHEXADECANE see HCP100
3-ACETOXYINDOLE see IDA600
2-ACETOXYMETHYLFURAN see FPM100
p-ACETOXYSTYRENE see ABW550
4-ACETOXYSTYRENE see ABW550
α-ACETOXYTOLUENE see BDX000
α-ACETYLACETOPHENONE see BDJ800
2-ACETYLACETOPHENONE see BDJ800
N-ACETYLAMINOBENZOIC ACID see AAJ150
2-(ACETYLAMINO)BENZOIC ACID see AAJ150
ACETYL ANHYDRIDE see AAX500
ACETYLANTHRANILIC ACID see AAJ150
N-ACETYLANTHRANILIC ACID see AAJ150
ACETYLBENZENE see ABH000
2-ACETYLBENZOFURAN see ACC100
ACETYLBENZOYLMETHANE see BDJ800
ACETYLCAPROLACTAM see ACF100
N-ACETYLCAPROLACTAM see ACF100
ACETYLCHOLINE BROMHYDRATE see CMF260
ACETYLCHOLINE BROMIDE see CMF260
ACETYLCHOLINE HYDROBROMIDE see CMF260
2-ACETYLCOUMARONE see ACC100
ACETYLENE TRICHLORIDE see TIO750
ACETYL ETHER see AAX500
ACETYLFURAN see ACM200
2-ACETYLFURAN see ACM200
ACETYLKAPROLAKTAM see ACF100
ACETYL-dl-METHIONINE see ACQ270
dl-N-ACETYLMETHIONINE see ACQ270
N-ACETYL-dl-METHIONINE see ACQ270
ACETYL OXIDE see AAX500
2-(ACETYLOXY)-N,N,N-TRIMETHYLETHANAMINIUM BROMIDE see CMF260
4-ACETYLPYRIDINE see ADA365
ACETYLTHIOCHOLINE DIIODIDE see ADC300
ACETYLTHIOCHOLINE IODIDE see ADC300
S-ACETYLTHIOCHOLINE IODIDE see ADC300
2-(ACETYLTHIO)-N,N,N-TRIMETHYLETHANAMINIUM IODIDE see ADC300
ACID ALIZARINE SAPPHIRE SE see APG700
ACID ANTHRAQUINONE BLUE see APG700
ACID BLUE 43 see APG700
ACID CHROME YELLOW GG see SIT850

ACID CHROME YELLOW 2GW

ACID CHROME YELLOW 2GW see SIT850
ACID COPPER ARSENITE see CNN500
ACIDE ARSENIQUE LIQUIDE (FRENCH) see ARB250
ACIDE DIPHENYLHYDROXYACETIQUE see BBY990
ACIDE ISONIPECOTIQUE see ILG100
ACIDE NAPHTHYLOXYACETIQUE see NBJ100
ACIDE PHENOXY-2-PROPIONIQUE see PDV725
ACIDE PIPECOLIQUE see HDS300
ACIDE PIPERIDINE-CARBOXYLIQUE-2 see HDS300
ACIDE PIPERIDINE-CARBOXYLIQUE-4 see ILG100
ACIDE TEREPHTHALIQUE (FRENCH) see TAN750
ACID LEATHER BLUE G see APG700
ACIDO m-CLOROBENZOICO see CEL290
ACIDO p-CLOROBENZOICO see CEL300
ACIDO IPPURICO see HGB300
ACIDO MERCAPTOBENZOICO see TFC550
ACID RED ALIZARINE see SEH475
ACIFLOCTIN see AEN250
ACILAN SAPPHIROL SE see APG700
ACINETTEN see AEN250
ACOVENOSIDE B see SGB550
ACRIDAN see ADI775
ACRIDANE see ADI775
ACRIDANONE see ADI825
9-ACRIDANONE see ADI825
ACRIDINE, 9,10-DIHYDRO-(9CI) see ADI775
9(10H)-ACRIDINONE (9CI) see ADI825
ACRIDONE see ADI825
9-ACRIDONE see ADI825
9(10H)-ACRIDONE see ADI825
ACRYLAMIDE, N-BUTOXYMETHYL- see BPM660
ACRYLAMIDE, N-tert-BUTYL- see BPW050
ACRYLIC ACID, 2-CHLORO-, ETHYL ESTER see EHH200
ACRYLIC ACID, 2-(DIMETHYLAMINO)ETHYL ESTER see DPB300
ACRYLIC ACID, HEPTYL ESTER see HBL100
ACRYLOPHENONE, 2′-HYDROXY-3-PHENYL- see HJS900
ACTICEL see SCH000
N-(1-ADAMANTYL)ACETAMIDE see AEE100
ADAME see DPB300
2-ADE see PDR490
ADENINE, N-BENZYL- see BDX090
ADENINE-FLAVIN DINUCLEOTIDE see RIF100
ADENINE-FLAVINE DINUCLEOTIDE see RIF100
ADENINE, N-FURFURYL- see FPT100
ADENINE-NICOTINAMIDE DINUCLEOTIDE see CNF390
ADENINE-RIBOFLAVIN DINUCLEOTIDE see RIF100
ADENINE SULFATE see AEH500
ADENINSULFAT see AEH500
ADENOSINE DIPHOSPHATE see AEK100
ADENOSINE 5′-DIPHOSPHATE see AEK100
ADENOSINE DIPHOSPHORIC ACID see AEK100
ADENOSINE 5′-DIPHOSPHORIC ACID see AEK100
ADENOSINE PYROPHOSPHATE see AEK100
ADENOSINE 5′-PYROPHOSPHATE see AEK100
ADENOSINE 5′-PYROPHOSPHORIC ACID see AEK100
ADENOSINE 5′-(TETRAHYDROGENTRIPHOSPHATE), DISODIUM SALT see AEM100
ADENOSINE, 5′-(TRIHYDROGEN DIPHOSPHATE) (9CI) see AEK100
ADENOSINE, 5′-(TRIHYDROGEN PYROPHOSPHATE) see AEK100
ADENOSINE 5′-(TRIHYDROGEN PYROPHOSPHATE), 5′-5′-ESTER with RIBOFLAVINE see RIF100
ADENOSINE TRIPHOSPHATE DISODIUM see AEM100
5′-ADENYLPHOSPHORIC ACID see AEK100
ADEPSINE OIL see MQV750
ADETPHOS see AEM100
ADILACTETTEN see AEN250
ADIPIC ACID see AEN250
ADIPIC ACID, BIS(2-(2-BUTOXYETHOXY)ETHYL) ESTER see DDT500
ADIPIC ACID DINITRILE see AER250
ADIPIC ACID, MONOETHYL ESTER see ELC600
ADIPIC ACID NITRILE see AER250
ADIPINIC ACID see AEN250
ADIPODINITRILE see AER250
ADIPONITRILE see AER250
ADP see AEK100
5′-ADP see AEK100
ADP (NUCLEOTIDE) see AEK100
ADSORBONAC see CAR780
ADUVEX 248 see HND100
ADVASTAB 46 see HND100
ADVASTAB 47 see DMW250
ADVASTAB DBTM see DEJ100
ADVASTAB T290 see DEJ100
ADVASTAB T340 see DEJ100
AEROMATT see CAT775
AEROSIL see SCH000
AEROSOL of THERMOVACUUM CADMIUM see CAK000
AETHYLAMINE (GERMAN) see EFU400
N-AETHYLFORMAMID see EKK600
AETHYLIDENCHLORID (GERMAN) see DFF809
3-AETHYL-PENTANOL-(3) see TJP550
AGIDOL 3 see FAB000
AGREFLAN see DUV600
AGRIFLAN 24 see DUV600
AGRISYNTH B2D see BOX300
AHCOQUINONE RED S see SEH475
AHCOVAT BLUE BCF see DFN425
AHCOVAT BRILLIANT VIOLET 2R see DFN450
AHCOVAT BRILLIANT VIOLET 4R see DFN450
AHCOVAT OLIVE ARN see DUP100
AHCOVAT OLIVE R see DUP100
AI3-14631 see GKW100
AIREDALE BLUE D see CMO500
AIR-FLO GREEN see CNN500
AIZEN DIRECT SKY BLUE 5BH see CMO500
AKADAMA see CAT775
ALAMON see HOR470
ALANINE, 3-(p-FLUOROPHENYL)-, dl- see FLD100
ALBACAR see CAT775
ALBACAR 5970 see CAT775
ALBAFIL see CAT775
ALBAGLOS see CAT775
ALBAGLOS SF see CAT775
ALBOLINE see MQV750
ALCA see AHA135
ALCLOXA see AHA135
ALCOOL BUTYLIQUE TERTIAIRE (FRENCH) see BPX000
ALDEHYDE ACETIQUE (FRENCH) see AAG250
ALDEHYDE C-14 see HBN200
ALDEHYDE C-14 PEACH see HBN200
ALDEHYDE FORMIQUE (FRENCH) see FMV000
ALDEIDE ACETICA (ITALIAN) see AAG250
ALDEIDE FORMICA (ITALIAN) see FMV000
ALDICARB OXIME see MLX800
ALDICARB SULFOXIDE see MLW750
ALGONEURINA see TES800
ALGYLEN see TIO750
ALIANT see TGX550
ALIMET see HMR600
ALIQUAT 207 see DXG625
ALITIA S see TES800
ALIZANTHRENE BLUE RC see DFN425
ALIZARIN BRILLANT BLUE BS see APG700
ALIZARIN CARMINE (BIOLOGICAL STAIN) see SEH475
ALIZARINE AZUROL SE see APG700
ALIZARINE BLUE SE see APG700
ALIZARINE BRILLANT BLUE BS see APG700
ALIZARINE CARMINE INDICATOR see SEH475
ALIZARINE GR see SIT850
ALIZARINE LIGHT BLUE SE see APG700
ALIZARINE MSE see APG700
ALIZARINE RED A see SEH475
ALIZARINE RED AS see SEH475
ALIZARINE RED INDICATOR see SEH475
ALIZARINE RED S (BIOLOGICAL STAIN) see SEH475
ALIZARINE RED S SODIUM SALT see SEH475
ALIZARINE RED SW see SEH475
ALIZARINE RED SZ see SEH475
ALIZARINE RED W see SEH475
ALIZARINE RED WA see SEH475
ALIZARINE RED for WOOL see SEH475
ALIZARINE RED WS see SEH475
ALIZARINE S see SEH475
ALIZARINE SAPPHIRE SE see APG700
ALIZARINE SAPPHIROL SE see APG700
ALIZARINE S EXTRA CONC. A EXPORT see SEH475
ALIZARINE S EXTRA PURE A see SEH475
ALIZARINE YELLOW 2G see SIT850
ALIZARINE YELLOW AGP see SIT850
ALIZARINE YELLOW G see SIT850
ALIZARINE YELLOW GG see SIT850
ALIZARINE YELLOW GGW see SIT850
ALIZARINE YELLOW GM see SIT850
ALIZARIN LIGHT BLUE SE see APG700

ALIZARIN RED S see SEH475
ALIZARINROT-S see SEH475
ALIZARIN S see SEH475
ALIZARINSULFONATE see SEH475
ALIZARIN YELLOW see SIT850
ALIZARIN YELLOW G see SIT850
ALIZARIN YELLOW GG see SIT850
ALIZARIN YELLOW G SODIUM SALT see SIT850
ALIZAROL YELLOW GW see SIT850
ALKOFEN MBP see MHR050
ALLERGICAN see CLX300
ALLERGISAN see CLX300
ALLIED WHITING see CAT775
ALLYL DISULFIDE see AGF300
ALLYL DISULPHIDE see AGF300
ALLYLTRIMETHYLAMMONIUM CHLORIDE see HGI600
ALNASID see FLD100
ALTOWHITES see KBB600
ALUGEL 34TN see AHH825
ALUMINIUM CHLOROHYDROXYALLANTOINATE see AHA135
ALUMINIUM STEARATE see AHH825
ALUMINUM, CHLORO((2,5-DIOXO-4-IMIDAZOLIDINYL)UREATO)TETRAHYDROXYDI- see AHA135
ALUMINUM CHLOROHYDROXYALLANTOINATE see AHA135
ALUMINUM, CHLOROTETRAHYDROXY((2-HYDROXY-5-OXO-2-IMIDAZOLIN-4-YL)UREATO)DI- see AHA135
ALUMINUM, DICHLOROETHYL- see EFU050
ALUMINUM DISTEARATE see AHA275
ALUMINUM DISTEARATE (ACGIH) see AHA275
ALUMINUM HYDROXIDE DISTEARATE see AHA275
ALUMINUM, HYDROXYBIS(OCTADECANOATO-O)-(9CI) see AHA275
ALUMINUM, HYDROXYBIS(STEARATO)- see AHA275
ALUMINUM HYDROXYDISTEARATE see AHA275
ALUMINUM ISOPROPOXIDE see AHC600
ALUMINUM(II) ISOPROPYLATE see AHC600
ALUMINUM STEARATE see AHH825
ALUMINUM TRISTEARATE see AHH825
AMACEL YELLOW RR see DUW500
AMAIZO W 13 see SLJ500
AMANIL PURPURINE 4B see DXO850
AMANIL SKY BLUE see CMO500
AMANTHRENE BLUE BCL see DFN425
AMANTHRENE BRILLIANT VIOLET RR see DFN450
AMANTHRENE OLIVE R see DUP100
AMBRETTOLID see OKU100
AMBRETTOLIDE see OKU100
3-AMINOACETANILIDE-4-SULFONIC ACID see AHQ300
m-AMINOANISOLE see AVO890
3-AMINOANISOLE see AVO890
4-AMINOANTIPYRENE see AIB300
AMINOANTIPYRIN see AIB300
AMINOANTIPYRINE see AIB300
4-AMINOANTIPYRINE see AIB300
6-AMINO-8-AZAPURINE see AIB340
4′-AMINOAZOBENZENE-4-SULFONIC ACID see AIB350
4-AMINOAZOBENZENE-4′-SULPHONIC ACID see AIB350
AMINOAZOPHENAZONE see AIB300
p-AMINOBENZALDEHYDE see AIC825
4-AMINOBENZALDEHYDE see AIC825
o-AMINOBENZAMIDE see AID620
2-AMINOBENZAMIDE see AID620
4-AMINOBENZENEACETIC ACID see AID700
p-AMINOBENZENEARSONIC ACID see ARA250
4-AMINOBENZENEARSONIC ACID see ARA250
m-AMINOBENZENESULFONAMIDE see MDM760
3-AMINOBENZENESULFONAMIDE see MDM760
2-AMINOBENZENESULFONIC ACID see SNO100
o-AMINOBENZENESULFONIC ACID see SNO100
p-AMINOBENZENESULFONIC ACID see SNN600
4-AMINOBENZENESULFONIC ACID see SNN600
m-AMINOBENZENESULPHONAMIDE see MDM760
p-AMINOBENZOIC ACID METHYL ESTER see AIN150
p-AMINOBUTYLBENZENE see AJA550
1-AMINO-4-BUTYLBENZENE see AJA550
p-AMINOBUTYROPHENONE see AJC750
4′-AMINOBUTYROPHENONE see AJC750
AMINOCAPROIC LACTAM see CBF700
2-AMINO-5-CHLOROBENZOPHENONE see AJE400
3-AMINO-4-CHLOROBENZOTRIFLUORIDE see CEG800
(2-AMINO-5-CHLOROPHENYL)PHENYLMETHANONE see AJE400
2-AMINO-3-CHLOROTOLUENE see CLK227
3-AMINO-4-CHLORO-α-α-α-TRIFLUOROTOLUENE see CEG800

2-AMINO-3,5-DIIODOBENZOIC ACID see API800
3-AMINO-7-(DIMETHYLAMINO)PHENOTHIAZIN-5-IUM CHLORIDE see DUG700
2-AMINODIPHENYL ETHER see PDR490
AMINOETHANE see EFU400
1-AMINOETHANE see EFU400
β-AMINOETHANOL HYDROCHLORIDE see EEC700
2-AMINOETHANOL HYDROCHLORIDE see EEC700
2-AMINOETHYL BROMIDE see BNI650
2-AMINOETHYLMETHYLAMINE see MJW100
AMINOGUANIDINE HYDROCHLORIDE see AKC800
1-AMINOHEPTANE see HBL600
dl-2-AMINOHEPTANE see HBM490
6-AMINOHEXANOIC ACID CYCLIC LACTAM see CBF700
AMINOHYDROQUINONE DIMETHYL ETHER see AKD925
1-AMINO-4-HYDROXY-2-METHOXY-9,10-ANTHRACENEDIONE see AKO350
1-AMINO-4-HYDROXY-2-METHOXYANTHRAQUINONE see AKO350
threo-2-AMINO-1-HYDROXY-1-PHENYLPROPANE see NNM510
4-AMINO-2-HYDROXYPYRIMIDINE see CQM600
4-AMINO-5-IMIDAZOLECARBOXAMIDE HYDROCHLORIDE see IAL100
AMINOMETHANE see MGC250
3-AMINO-4-METHOXY BENZANILIDE see AKM500
1-AMINO-2-METHOXY-4-OXYANTHRAQUINONE see AKO350
2-AMINO-4-METHYLPHENOL see AKY950
2-AMINO-5-METHYLPYRIDINE see AMA010
5-AMINO-1-NAPHTHALENESULFONIC ACID see ALI240
5-AMINO-2-NAPHTHOL see ALJ500
2-AMINOOCTANE see OEK010
1-AMINOPENTANE see PBV505
2-AMINOPENTANE see DHJ200
AMINOPHENAZONE see AIB300
4-AMINOPHENAZONE see AIB300
2-AMINOPHENETOLE see PDK819
(p-AMINOPHENYL)ACETIC ACID see AID700
4-AMINOPHENYLACETIC ACID see AID700
AMINOPHENYLARSINE ACID see ARA250
p-AMINOPHENYLARSINE ACID see ARA250
p-AMINOPHENYLARSINIC ACID see ARA250
4-AMINOPHENYLARSONIC ACID see ARA250
p-((p-AMINOPHENYL)AZO)BENZENESULFONIC ACID see AIB350
1-(4-AMINOPHENYL)-1-BUTANONE see AJC750
4-AMINOPHENYL DISULFIDE see ALW100
2-(p-AMINOPHENYL)-6-METHYLBENZOTHIAZOLE see MHJ300
m-AMINOPHENYL METHYL SULFIDE see ALX100
p-AMINOPHENYL METHYL SULFIDE see AMS675
o-AMINOPHENYL PHENYL ETHER see PDR490
2-AMINOPHENYL PHENYL ETHER see PDR490
o-AMINOPHENYLSULFONIC ACID see SNO100
p-AMINOPHENYLSULFONIC ACID see SNN600
6-AMINO-3-PICOLINE see AMA010
N-(3-AMINOPROPYL)-1,4-BUTANEDIAMINE TRIHYDROCHLORIDE see SLG600
2-AMINOPYRIMIDINE see PPH300
4-AMINO-2(1H)-PYRIMIDINONE see CQM600
2-AMINOQUINOLINE see AMK700
3-AMINOQUINOLINE see AMK725
2-AMINORESORCINOL HYDROCHLORIDE see AML600
1-AMINO-2-SULFO-4-(4′-AMINO-3′-SULFOANILINO)ANTHRAQUINONE see SOB600
4-(AMINOSULFONYL)BENZOIC ACID see SNL840
2-AMINO-2-THIAZOLINE see TEV600
m-AMINOTHIOANISOLE see ALX100
p-AMINOTHIOANISOLE see AMS675
3-AMINOTHIOANISOLE see ALX100
4-AMINOTHIOANISOLE see AMS675
6-AMINO-2-THIOURACIL see AMS750
p-AMINO-α-TOLUIC ACID see AID700
7-AMINO-1H-v-TRIAZOLO(4,5-d)PYRIMIDINE see AIB340
2-AMINO-4-(TRIFLUOROMETHYL)-5-THIAZOLECARBOXYLIC ACID ETHYL ESTER see AMU550
5-AMINOURACIL see AMW100
4-AMMINOANTIPIRINA see AIB300
AMMONIUM ACID ARSENATE see DCG800
AMMONIUM, ALLYLTRIMETHYL-, CHLORIDE see HGI600
AMMONIUM ARSENATE, solid (DOT) see DCG800
AMMONIUM BENZOATE see ANB100
AMMONIUM, BENZYLDIMETHYLHEXADECYL-, CHLORIDE see BEL900
AMMONIUM, BENZYLTRIETHYL-, CHLORIDE see BFL300
AMMONIUM, BENZYLTRIMETHYL-, METHOXIDE see TLM100
AMMONIUM, (2-BUTYRYLOXYETHYL)TRIMETHYL-, IODIDE see BSY300
AMMONIUM CADMIUM CHLORIDE see AND250
AMMONIUM, (3-CARBOXY-2-HYDROXYPROPYL)TRIMETHYL-, CHLORIDE, (+ −)- see CCK655
AMMONIUM, (CARBOXYMETHYL)TRIMETHYL-, CHLORIDE see CCH850
AMMONIUM, DECYLTRIMETHYL-, BROMIDE see DAJ500

AMMONIUM, DIMETHYLDIOCTADECYL-, CHLORIDE

AMMONIUM, DIMETHYLDIOCTADECYL-, CHLORIDE see DXG625
AMMONIUM, DODECYLTRIMETHYL-, BROMIDE see DYA810
AMMONIUM, ETHYLTRIMETHYL-, IODIDE see EQC600
AMMONIUM, (2-HYDROXYPROPYL)TRIMETHYL-, CHLORIDE see MIM300
AMMONIUM, (2-MERCAPTOETHYL)TRIMETHYL-, IODIDE ACETATE see ADC300
AMMONIUM OXALATE see ANO750
AMMONIUM SALICYLATE see ANT600
AMMONIUM STEARATE see ANU200
AMMONIUM, TETRADECYLTRIMETHYL-, BROMIDE see TCB200
AMMONIUM, TETRAETHYL-, PERCHLORATE see TCD002
AMMONIUM, TRIMETHYLTETRADECYL-, BROMIDE see TCB200
AMMONIUM VANADO-ARSENATE see ANY750
1A-2MO-4OA see AKO350
AMORPHOUS FUSED SILICA see SCK600
AMORPHOUS SILICA DUST see SCH000
AMPYRONE see AIB300
AMSONIC ACID see FCA100
AMYLAMINE see PBV505
n-AMYLAMINE see PBV505
AMYLAMINE (DOT) see PBV505
iso-AMYLAMINE see AOE200
AMYL BENZOATE see PBV800
AMYL BROMIDE see AOF800
n-AMYL BROMIDE see AOF800
AMYL IODIDE see IET500
n-AMYL IODIDE see IET500
AMYLOMAIZE VII see SLJ500
tert-AMYL PEROXY-2-ETHYLHEXANOATE, technically pure (DOT) see AOM200
o-AMYLPHENOL see AOM325
AMYLUM see SLJ500
AMYL VINYL CARBINOL ACETATE see ODW028
AMYL VINYL CARBINYL ACETATE see ODW028
ANALETIL see GCE600
ANAMENTH see TIO750
ANEURINE DISULFIDE see TES800
ANHYDRIDE ACETIQUE (FRENCH) see AAX500
ANHYDRIDE ARSENIQUE (FRENCH) see ARH500
ANHYDRID KYSELINY MASELNE see BSW550
ANHYDROTRIMELLIC ACID see TKV000
ANHYDROUS BORAX see DXG035
ANHYDROUS CHLORAL see CDN550
ANIDRIDE ACETICA (ITALIAN) see AAX500
p-ANILINEARSONIC ACID see ARA250
ANILINE, 4,4′-AZODI- see ASK925
ANILINE, 4-BUTYL- see AJA550
ANILINE, N-(2-CHLOROETHYL-)-N-ETHYL- see EHJ600
ANILINE, 5-CHLORO-2-MERCAPTO-, HYDROCHLORIDE see CHU100
ANILINE, 4-CHLORO-2-NITRO- see KDA050
ANILINE, 2,4-DICHLORO- see DEO290
ANILINE, 2,6-DIETHYL- see DIS650
ANILINE, N,N-DIETHYL-p-NITROSO- see NJW600
ANILINE, 2,4-DIFLUORO- see DKF700
ANILINE, 2,5-DIMETHOXY- see AKD925
ANILINE, N,N-DIMETHYL-o-NITRO- see NFW600
ANILINE, p,p′-DITHIODI- see ALW100
ANILINE, 4-FLUORO-3-NITRO- see FKK100
ANILINE, 2-METHYL-4-NITRO- see MMF780
ANILINE, m-(METHYLTHIO)- see ALX100
ANILINE, p-(METHYLTHIO)- see AMS675
ANILINE, 2-PHENOXY- see PDR490
ANILINE, o-PHENOXY-(8CI) see PDR490
ANILINE, p,p′-(m-PHENYLENEDIOXY)DI- see REF070
ANILINE-p-SULFONIC ACID see SNN600
ANILINE-4-SULFONIC ACID see SNN600
ANILINE-p-SULPHONIC ACID see SNN600
β-ANILINOPROPIONITRILE see AOT100
3-ANILINOPROPIONITRILE see AOT100
ANILINO-o-SULFONIC ACID see SNO100
ANILINO-2-SULFONIC ACID see SNO100
ANILINO-o-SULPHONIC ACID see SNO100
m-ANISALDEHYDE see AOT300
p-ANISALDEHYDE, 2-HYDROXY- see HLR600
p-ANISIC ACID see AOU600
4-ANISIC ACID see AOU600
o-ANISIC ACID, METHYL ESTER (7CI,8CI) see MLH800
m-ANISIDINE see AVO890
ANISOLE, o-BROMO- see BMT400
ANISOLE, p-BROMO- see AOY450
ANISOLE, 2-ISOPROPYL-5-METHYL- see MPW650
ANISOLE, 2,4,6-TRINITRO- see TMK300
m-ANISYLAMINE see AVO890
ANISYL BROMIDE see AOY450, BMT400

ANKILOSTIN see PCF275
ANSIBASES RED RL see MMF780
ANTALLIN see CAR780
ANTHRACENE-9-CARBOXYLIC ACID see APG550
9,10-ANTHRACENEDIONE, 1-AMINO-4-HYDROXY-2-METHOXY-(9CI) see AKO350
9,10-ANTHRACENEDIONE, 2-CHLORO- see CEI100
9,10-ANTHRACENEDIONE, 1,4-DIAMINO-2,3-DIHYDRO- see DBR450
9,10-ANTHRACENEDIONE, 1,8-DIAMINO-4,5-DIHYDROXY- see DBQ250
9,10-ANTHRACENEDIONE, 1-METHOXY-(9CI) see MEA650
2-ANTHRACENESULFONIC ACID, 1-AMINO-4-(4-AMINO-3-SULFOANILINO)-9,10-DIHYDRO-9,10-DIOXO- see SOB600
2-ANTHRACENESULFONIC ACID, 1-AMINO-4-((4-AMINO-3-SULFOPHENYL)AMINO)-9,10-DIHYDRO-9,10-DIOXO- see SOB600
2-ANTHRACENESULFONIC ACID, 4,8-DIAMINO-9,10-DIHYDRO-1,5-DIHYDROXY-9,10-DIOXO-, MONOSODIUM SALT see APG700
2-ANTHRACENESULFONIC ACID, 9,10-DIHYDRO-3,4-DIHYDROXY-9,10-DIOXO-, MONOSODIUM SALT see SEH475
1,4,9,10-ANTHRACENETETRAOL see LEX200
1,4,9,10-ANTHRACENETETROL (9CI) see LEX200
9(10H)-ANTHRACENONE see APM500
ANTHRANILAMIDE see AID620
ANTHRANILIC ACID, N-ACETYL- see AAJ150
ANTHRANILIC ACID, 3,5-DIIODO- see API800
ANTHRANILIC ACID, PHENETHYL ESTER see APJ500
ANTHRANILIMIDIC ACID see AID620
ANTHRANOL CHROME YELLOW 2G see SIT850
ANTHRANOL CHROME YELLOW 5GS see SIT850
ANTHRAQUINONE ACID BLUE see APG700
ANTHRAQUINONE, 1-AMINO-4-HYDROXY-2-METHOXY- see AKO350
ANTHRAQUINONE, 1,4-BIS(METHYLAMINO)- see BKP500
ANTHRAQUINONE BLUE see IBV050
ANTHRAQUINONE, 2-CHLORO- see CEI100
ANTHRAQUINONE DEEP BLUE see IBV050
ANTHRAQUINONE, 1,8-DIAMINO-4,5-DIHYDROXY- see DBQ250
ANTHRAQUINONE, 2,3-DIHYDRO-1,4-DIAMINO- see DBR450
ANTHRAQUINONE, 1-METHOXY- see MEA650
5,9,14,18-ANTHRAZINETETRONE, 7,16-DICHLORO-6,15-DIHYDRO- see DFN425
5,9,14,18-ANTHRAZINETETRONE, 6,15-DIHYDRO- see IBV050
9-ANTHROIC ACID see APG550
ANTHRONE see APM500
3C ANTIBIOTIC see HLT100
ANTINOLO RED B see DNT300
ANTISAL 1a see TGK750
ANTISEPSIN see BMR100
ANTISOL 1 see PCF275
ANTI-UV P see HND100
ANTOXYLIC ACID see ARA250
APREN S see TES800
AQUACAT see CNA250
AQUACIDE see DWX800
AQUAFIL see SCH000
AQUAPEL (POLYSACCHARIDE) see SLJ500
ARABOASCORBIC ACID see SAA025
d-ARABOASCORBIC ACID see SAA025
d-ARGININE HYDROCHLORIDE see AQV500
ARGININE, MONOHYDROCHLORIDE, d- see AQV500
d-ARGININE, MONOHYDROCHLORIDE (9CI) see AQV500
ARGO BRAND CORN STARCH see SLJ500
ARGOBYL see TKP100
ARLANTHRENE VIOLET 4R see DFN450
ARMAC 18D see OAP300
ARMAC OD see OAP300
ARMEEN L-7 see HBM490
ARMEEN DM16D see HCP525
ARNEEL 8 see OCW100
AROMANTRENE OLIVE FR see DUP100
AROSURF TA 100 see DXG625
ARQUAD R 40 see DXG625
ARROWROOT STARCH see SLJ500
ARSANILIC ACID see ARA250
4-ARSANILIC ACID see ARA250
p-ARSANILIC ACID see ARA250
ARSENATE see ARB250
ARSENATE of IRON, FERRIC see IGN000
ARSENATE of IRON, FERROUS see IGM000
ARSENENOUS ACID, POTASSIUM SALT see PKV500
ARSENIATE de CALCIUM (FRENCH) see ARB750
ARSENIATE de MAGNESIUM (FRENCH) see ARD000
ARSENIC see ARA750
ARSENIC-75 see ARA750
ARSENIC, metallic (DOT) see ARA750
ARSENIC ACID see ARH500

m-ARSENIC ACID see ARB000
o-ARSENIC ACID see ARB250
ARSENIC ACID, liquid (DOT) see ARB250
ARSENIC ACID, solid (DOT) see ARB250, ARC500
ARSENIC ACID ANHYDRIDE see ARH500
ARSENIC ACID, CALCIUM SALT (2583) see ARB750
ARSENIC ACID, DISODIUM SALT see ARC000
ARSENIC ACID, DISODIUM SALT, HEPTAHYDRATE see ARC250
ARSENIC ACID, HEMIHYDRATE see ARC500
ARSENIC ACID, MAGNESIUM SALT see ARD000
ARSENIC ACID, MONOPOTASSIUM SALT see ARD250
ARSENIC ACID, MONOSODIUM SALT see ARD500, ARD600
ARSENIC ACID, SODIUM SALT see ARD750
ARSENIC ACID, SODIUM SALT (9CI) see ARD500
ARSENICAL DUST see ARE500
ARSENICAL FLUE DUST see ARE500
ARSENICALS see ARA750, ARF750
ARSENIC ANHYDRIDE see ARH500
ARSENIC BLACK see ARA750
ARSENIC(III) BROMIDE see ARF250
ARSENIC BUTTER see ARF500
ARSENIC CHLORIDE see ARF500
ARSENIC(III) CHLORIDE see ARF500
ARSENIC COMPOUNDS see ARF750
ARSENIC FLUORIDE see ARI250
ARSENIC IODIDE see ARG750
ARSENIC OXIDE see ARH500
ARSENIC(V) OXIDE see ARH500
ARSENIC PENTASULFIDE see ARH250
ARSENIC PENTOXIDE see ARH500
ARSENIC SESQUISULFIDE see ARI000
ARSENIC SULFIDE see ARI000
ARSENIC SULFIDE YELLOW see ARI000
ARSENIC SULPHIDE see ARI000
ARSENIC TRIBROMIDE see ARF250
ARSENIC TRIFLUORIDE see ARI250
ARSENIC TRIIODIDE see ARG750
ARSENIC TRISULFIDE see ARI000
ARSENIC YELLOW see ARI000
ARSENIOUS ACID, CALCIUM SALT see CAM500
ARSENIOUS ACID, SODIUM SALT see ARJ500, SEY500
ARSENIOUS ACID, SODIUM SALT POLYMERS see ARJ500
ARSENIOUS ACID, ZINC SALT see ZDS000
ARSENIOUS CHLORIDE see ARF500
ARSENIOUS SULPHIDE see ARI000
ARSENITE de POTASSIUM (FRENCH) see PKV500
ARSENITE de SODIUM (FRENCH) see SEY500
ARSENOMARCASITE see ARJ750
ARSENOPYRITE see ARJ750
ARSENOUS ACID, SODIUM SALT (9CI) see SEY500
ARSENOUS BROMIDE see ARF250
ARSENOUS CHLORIDE see ARF500
ARSENOUS FLUORIDE see ARI250
ARSENOUS IODIDE see ARG750
ARSENOUS SULFIDE see ARI000
ARSENOUS TRIBROMIDE see ARF250
ARSENOUS TRICHLORIDE (9CI) see ARF500
ARSENOUS TRIIODIDE (9CI) see ARG750
ARSENOXIDE SODIUM see ARJ900
ARSONIC ACID, COPPER(2+) SALT (1:1) (9CI) see CNN500
ARSONIC ACID, POTASSIUM SALT see PKV500
ARSONIC ACID, SODIUM SALT (9CI) see ARJ500
ARTISIL BRILLIANT PINK RFS see AKO350
ASEPSIN see BMR100
ASMACORIL see PER700
ATARAX see HOR470
ATARAX DIHYDROCHLORIDE see HOR470
ATARAXOID DIHYDROCHLORIDE see HOR470
ATERAX DIHYDROCHLORIDE see HOR470
ATIC VAT BLUE BC see DFN425
ATIC VAT BLUE XRN see IBV050
ATIC VAT BRILLIANT PURPLE 4R see DFN450
ATIC VAT OLIVE R see DUP100
ATLANTICHROME YELLOW 2G see SIT850
ATLANTIC SKY BLUE A see CMO500
ATLAS "A" see SEY500
ATOMIT see CAT775
ATOMITE see CAT775
ATOXYLIC ACID see ARA250
ATP DISODIUM see AEM100
ATP DISODIUM SALT see AEM100
ATUL DIRECT RED 4B see DXO850

ATUL DIRECT SKY BLUE see CMO500
ATUL PIGMENT RED RS SODIUM SALT see NAP100
AURORA YELLOW see CAJ750
AVITEX C see HCP900
AVITEX SF see HCP900
AVLOTANE see HCI000
AX 363 see CAT775
AXEROPHTHAL see VSK985
AYAPANIN see MEK300
8-AZAADENINE see AIB340
3-AZABENZONITRILE see NDW515
3-AZABICYCLO(3.2.2)NONANE see ARX300
1-AZABICYCLO(2.2.2)OCTAN-3-OL (9CI) see QUJ990
2-AZACYCLOHEPTANONE see CBF700
AZACYCLOTRIDECAN-2-ONE see COW930
2-AZACYCLOTRIDECANONE see COW930
AZAMIN 4B see DXO850
8-AZAPURINE, 6-AMINO- see AIB340
2H-AZEPIN-2-ONE, 1-ACETYLHEXAHYDRO- see ACF100
AZIJNZUURANHYDRIDE (DUTCH) see AAX500
AZINE SKY BLUE 5B see CMO500
p-AZOANILINE see ASK925
AZOBENZENE, 4-ANILINO- see PEI800
4,4′-AZOBISBENZENAMINE see ASK925
4,4′-AZOBIS(4-CYANOPENTANOIC ACID) see ASL500
AZOBIS(CYANOVALERIC ACID) see ASL500
4,4′-AZOBIS(4-CYANOVALERIC ACID) see ASL500
AZOCARD RED 4B see DXO850
AZOCHROMOL YELLOW 5G see SIT850
4,4′-AZODIANILINE see ASK925
AZOENE FAST ORANGE RD SALT see CEG800
AZOENE FAST RED GL BASE see MMF780
AZOENE FAST RED 3GL BASE see KDA050
AZOGEN DEVELOPER A see NAX000
AZOGENE FAST RED NRL SALT see MMF780
AZOGENE FAST RED RL see MMF780
AZOIC DIAZO COMPONENT 9 see KDA050
AZULENE see ASP600
AZURE A see DUG700

BA see BBC250, BDX090
6-BA see BDX090
BA (GROWTH STIMULANT) see BDX090
BAIRDCAT B16 see HCP525
BAP see BDX090
6-BAP see BDX090
BAP (GROWTH STIMULANT) see BDX090
BARBITURIC ACID, 5-NITRO- see NET550
BARBONIN HYDROCHLORIDE see PAH260
BARIUM CADMIUM STEARATE see BAI800
BASIC BISMUTH NITRATE see BKW100
BASIC RED 2 see GJI400
BATILOL see GGA915
BATYL ALCOHOL see GGA915
BAYOL F see MQV750
BELAMINE SKY BLUE A see CMO500
BELOPHAR KLA see OOI200
BELOPHOR OD see DXB450
BENCIDAL PURPLE 4B see DXO850
BENIHINAL see FNK150
BENTONE see KBB600
BENZ(e)ACEPHENANTHRYLENE see BAW250
3,4-BENZ(e)ACEPHENANTHRYLENE see BAW250
BENZADONE BLUE RC see DFN425
BENZADONE BLUE RS see IBV050
BENZADONE BRILLIANT PURPLE 2R see DFN450
BENZADONE BRILLIANT PURPLE 4R see DFN450
BENZADONE OLIVE R see DUP100
BENZALDEHYDE, 4-AMINO- see AIC825
BENZALDEHYDE, p-BROMO- see BMT700
BENZALDEHYDE, 4-BROMO-(9CI) see BMT700
BENZALDEHYDE, p-CHLORO- see CEI600
BENZALDEHYDE, 2,4-DIHYDROXY- see REF100
BENZALDEHYDE, 4-ETHOXY-3-HYDROXY- see IKO100
BENZALDEHYDE, m-HYDROXY- see FOE100
BENZALDEHYDE, 3-HYDROXY-(9CI) see FOE100
BENZALDEHYDE, 2-HYDROXY-4-METHOXY-(9CI) see HLR600
BENZALDEHYDE, 3-METHOXY-(9CI) see AOT300
BENZALDEHYDE, 2,3,4-TRIHYDROXY- see TKN800
BENZALDEHYDE, 2,4,6-TRIMETHYL- see MDJ745
BENZAMIDE, o-AMINO- see AID620
BENZAMIDE, 2-AMINO-(9CI) see AID620

BENZAMIDE, 5-BROMO-N-(4-BROMOPHENYL)-2-HYDROXY-

BENZAMIDE, 5-BROMO-N-(4-BROMOPHENYL)-2-HYDROXY- see BOD600
BENZAMIDE, N,N'-(DITHIODI-2,1-PHENYLENE)BIS- see BDK800
BENZAMIDINE, HYDROCHLORIDE see BBM750
BENZAMIDOACETIC ACID see HGB300
BENZANILIDE, 3-AMINO-4-METHOXY- see AKM500
BENZANILIDE, 2',2'''-DITHIOBIS- see BDK800
BENZANIL PURPURINE 4B see DXO850
BENZANIL SKY BLUE see CMO500
BENZANTHRACENE see BBC250
BENZ(a)ANTHRACENE see BBC250
1,2-BENZANTHRACENE see BBC250
1,2-BENZ(a)ANTHRACENE see BBC250
1,2-BENZANTHRAZEN (GERMAN) see BBC250
BENZANTHRENE see BBC250
1,2-BENZANTHRENE see BBC250
BENZATHINE see BHB300
BENZATIN see BHB300
BENZENAMINE, 4,4'-AZOBIS-(9CI) see ASK925
BENZENAMINE, 4-BUTYL-(9CI) see AJA550
BENZENAMINE, N-(2-CHLOROETHYL)-N-ETHYL-(9CI) see EHJ600
BENZENAMINE, 4-CHLORO-2-NITRO-(9CI) see KDA050
BENZENAMINE, 2,4-DICHLORO-(9CI) see DEO290
BENZENAMINE, 2,6-DIETHYL-(9CI) see DIS650
BENZENAMINE, N,N-DIETHYL-4-NITROSO-(9CI) see NJW600
BENZENAMINE, 2,4-DIFLUORO-(9CI) see DKF700
BENZENAMINE, 2,5-DIMETHOXY-(9CI) see AKD925
BENZENAMINE, N,N-DIMETHYL-2-NITRO-(9CI) see NFW600
BENZENAMINE, 4,4'-DITHIOBIS-(9CI) see ALW100
BENZENAMINE, 3-ETHOXY-(9CI) see PDK800
BENZENAMINE, 4-FLUORO-2-METHYL- see FJT050
BENZENAMINE, 4-FLUORO-3-NITRO- see FKK100
BENZENAMINE, 3-METHOXY-(9CI) see AVO890
BENZENAMINE, 4-METHOXY-N-(4-METHOXYPHENYL)- see BKO600
BENZENAMINE, 4-(6-METHYL-2-BENZOTHIAZOLYL)-(9CI) see MHJ300
BENZENAMINE, 2-METHYL-4-NITRO- see MMF780
BENZENAMINE, 3-(METHYLTHIO)-(9CI) see ALX100
BENZENAMINE, 4-(METHYLTHIO)-(9CI) see AMS675
BENZENAMINE, N,N,4-TRIMETHYL- see TLG150
BENZENAMINE, N,N,3-TRIMETHYL-(9CI) see TLG100
BENZENAMINE, 2-PHENOXY-(9CI) see PDR490
BENZENAMINE, 4,4'-(1,3-PHENYLENEBIS(OXY))BIS- see REF070
BENZENEACETIC ACID, 4-AMINO-(9CI) see AID700
BENZENEACETIC ACID, 3,4-DIMETHOXY-(9CI) see HGK600
BENZENEACETIC ACID, 4-HYDROXY-(9CI) see HNG600
BENZENEACETIC ACID, α-HYDROXY-α-PHENYL-(9CI) see BBY990
BENZENEACETIC ACID, 4-NITRO-(9CI) see NII510
BENZENEACETIC ACID, SODIUM SALT (9CI) see SFA200
BENZENEACETONITRILE, 4-CHLORO- see CEP300
BENZENEACETONITRILE, 2-NITRO-(9CI) see NFN600
BENZENE, 1,4-BIS(1-METHYLETHYL)-(9CI) see DNN830
BENZENE, 1-BROMO-2-METHOXY-(9CI) see BMT400
BENZENE, 1-BROMO-4-METHOXY-(9CI) see AOY450
BENZENE, 1-BROMO-3-METHYL- see BOG300
BENZENE, 1-BROMO-2-METHYL-(9CI) see BOG260
BENZENE, 1-BROMO-2-NITRO- see NFQ080
BENZENE, 1-BROMO-3-NITRO- see NFQ090
BENZENE, 1-BROMO-4-NITRO- see NFQ100
BENZENE, 1-tert-BUTYL-2,6-DINITRO-3,4,5-TRIMETHYL- see MRW272
BENZENECARBOPEROXOIC ACID, 1,1,4,4-TETRAMETHYL-1,4-BUTANEDIYL ESTER see DRJ100
BENZENECARBOTHIOIC ACID see TFC550
BENZENECARBOXIMIDAMIDE HYDROCHLORIDE see BBM750
BENZENE, 2-CHLORO-1,3-DINITRO-5-(TRIFLUOROMETHYL)- see CGM225
BENZENE, 2-CHLORO-1-FLUORO-4-NITRO- see CHI950
BENZENE, 1-CHLORO-2-METHYL-3-NITRO- see CJG825
BENZENE, 1-(CHLOROMETHYL)-3-NITRO-(9CI) see NFN010
BENZENE, 5-CHLORO-1-NITRO-2,4-DIMETHOXY- see CGC100
BENZENE, CHLOROPENTAFLUORO- see PBE100
BENZENE, 2-CHLORO-1,3,5-TRINITRO- see PIE530
1,3-BENZENEDIAMINE, 4-METHOXY-, DIHYDROCHLORIDE see DBO100
1,2-BENZENEDIAMINE, 3-METHYL-(9CI) see TGY800
1,3-BENZENEDIAMINE, 5-(TRIFLUOROMETHYL) see TKB775
BENZENEDIAZONIUM, 2-METHYL-4-NITRO-, TETRAFLUOROBORATE(1⁻) (9CI) see NMO700
BENZENEDIAZONIUM, 4-METHYL-, TETRAFLUOROBORATE (1⁻) (9CI) see TGM450
BENZENEDIAZONIUM, p-NITRO-, TETRAFLUOROBORATE(1⁻) (9CI) see NEZ200
BENZENEDIAZONIUM, 4-NITRO-, TETRAFLUOROBORATE(1⁻) (9CI) see NEZ200
BENZENE, m-DIBROMO- see DDK050
BENZENE, p-DIBROMO- see BNV775
BENZENE, 1,3-DIBROMO-(9CI) see DDK050
BENZENE, 1,4-DIBROMO-(9CI) see BNV775
BENZENE, (1,2-DIBROMOETHYL)- see DDN900
1,2-BENZENEDICARBOXAMIDE see BBO500
1,2-BENZENEDICARBOXAMIDE, N,N,N',N'-TETRAETHYL-(9CI) see GCE600
p-BENZENEDICARBOXYLIC ACID see TAN750
1,4-BENZENEDICARBOXYLIC ACID see TAN750
1,2-BENZENEDICARBOXYLIC ACID, BIS(1-METHYLETHYL) ESTER see PHW600
1,4-BENZENEDICARBOXYLIC ACID, DIBUTYL ESTER see DEH700
1,2-BENZENEDICARBOXYLIC ACID, DICYCLOHEXYL ESTER see DGV700
1,2-BENZENEDICARBOXYLIC ACID, DIDODECYL ESTER (9CI) see PHW550
1,4-BENZENEDICARBOXYLIC ACID, DIETHYL ESTER see DKB160
1,2-BENZENEDICARBOXYLIC ACID, 3,4,5,6-TETRACHLORO- see TBT100
BENZENE, (DICHLOROFLUOROMETHYL)- see DFL100
BENZENE, 2,4-DICHLORO-1-METHYL-(9CI) see DGM700
BENZENE, 2,4-DIFLUORO-1-NITRO- see DKH900
BENZENE, p-DIISOPROPYL- see DNN830
1,3-BENZENEDIMETHANOL, 2-HYDROXY-5-METHYL- see HLV100
BENZENE, m-DIMETHOXY- see REF025
BENZENE, 1,3-DIMETHOXY- see REF025
BENZENE, 1-(1,1-DIMETHYLETHYL)-2,6-DINITRO-3,4,5-TRIMETHYL- MRW272
BENZENE, 1,4-DIMETHYL-2-NITRO- see NMS520
BENZENE, p-DINITROSO- see DVE260
BENZENE, 1,4-DINITROSO-(9CI) see DVE260
o-BENZENEDIOL see CCP850
1,2-BENZENEDIOL see CCP850
1,3-BENZENEDIOL, DIBENZOATE see BHB100
BENZENE-1,3-DISULFONYL CHLORIDE FLUORIDE see FLY200
BENZENE, DIVINYL- see DXQ740
BENZENEETHANOL, α-METHYL-(9CI) see PGA600
BENZENEETHANOL, β-METHYL-(9CI) see HGR600
BENZENE, 1,1'-(1,2-ETHENEDIYL)BIS-, (E)-(9CI) see SLR100
BENZENE, 1-ETHENYL-4-METHYL-(9CI) see VQK700
BENZENE, 1-FLUORO-2-METHYL-(9CI) see FLZ100
BENZENE, 1-(1-ISOCYANATO-1-METHYLETHYL)-3-(1-METHYLETHENYL)- see IKG800
BENZENEMETHANAMINIUM, N-HEXADECYL-N,N-DIMETHYL-, CHLORIDE see BEL900
BENZENEMETHANAMINIUM, N,N,N-TRIETHYL-, CHLORIDE (9CI) see BFL300
BENZENEMETHANAMINIUM, N,N,N-TRIMETHYL-, METHOXIDE (9CI) see TLM100
BENZENEMETHANESULFONYL FLUORIDE see TGO300
BENZENEMETHANOL, α-(1-(METHYLAMINO)ETHYL)-, (R*,S*)-(+ −)-(9CI) see EAW100
BENZENEMETHANOL, α-(1-(METHYLAMINO)ETHYL)-, (S-(R*,S*))-(9CI) see EAW200
BENZENEMETHANOL, 4-NITRO-, ACETATE (ester) see NFM100
BENZENEMETHANOL, α-(TRICHLOROMETHYL)-(9CI) see TIR800
BENZENE, 2-METHOXY-1,3,5-TRINITRO-(9CI) see TMK300
BENZENE, METHYL- see TGK750
BENZENE, 1,1'-OXYBIS(4-NITRO-(9CI) see NIM600
BENZENEPROPANOIC ACID, 4-NITRO-β-OXO-, ETHYL ESTER (9CI) see ENO100
BENZENEPROPANOIC ACID, β-OXO-, ETHYL ESTER (9CI) see EGR600
BENZENESULFINIC ACID, p-CHLORO- see CEJ600
BENZENESULFOHYDRAZIDE see BBS300
BENZENESULFONAMIDE, 3-AMINO- see MDM760
BENZENESULFONAMIDE, 4-METHYL-N-PHENYL-(9CI) see TGN600
BENZENESULFONIC ACID, 4-ACETAMIDO-2-AMINO- see AHQ300
BENZENESULFONIC ACID, o-AMINO- see SNO100
BENZENESULFONIC ACID, 4-AMINO-(9CI) see SNN600
BENZENESULFONIC ACID, p-((p-AMINOPHENYL)AZO)- see AIB350
BENZENESULFONIC ACID, 2,5-DIAMINO- see PEY800
BENZENESULFONIC ACID, 3-(4-DIETHYLAMINO-2-HYDROXYPHENYLAZO)-4-HYDROXY- see DIJ300
BENZENESULFONIC ACID, 2,2'-(1,2-ETHYLENEDIYL)BIS(5-AMINO-(9CI) see FCA100
BENZENESULFONIC ACID, HYDRAZIDE see BBS300
BENZENESULFONIC ACID, HYDROXY- see HJH100
BENZENESULFONIC ACID, p-HYDROXY- see HNL600
BENZENESULFONIC ACID, p-HYDROXY-, ZINC SALT (2:1) see ZIJ300
BENZENESULFONIC ACID, p-HYDROXY-, ZINC SALT (2:1) see ZIJ300
BENZENESULFONIC ACID, METHYL ESTER see MHB300
BENZENESULFONIC ACID, SODIUM SALT see SJH050
BENZENESULFONIC HYDRAZIDE see BBS300
BENZENESULFONOHYDRAZIDE see BBS300
BENZENE, 1,1'-SULFONYLBIS(4-FLUORO-3-NITRO- see DKH250
BENZENESULFONYL CHLORIDE, m-(FLUOROSULFONYL)- see FLY200
BENZENESULFONYL HYDRAZIDE see BBS300
BENZENESULFONYL HYDRAZINE see BBS300
BENZENE SULPHONOHYDRAZIDE see BBS300
1,2,4,5-BENZENETETRACARBOXYLIC ACID see PPQ630
1,2,4,5 BENZENETETRACARBOXYLIC 1,2:4,5 DIANHYDRIDE see PPQ635
BENZENETHIOL, 3-METHYL-(9CI) see TGO800
1,3,5-BENZENETRIAMINE see THN775

1,2,4-BENZENETRICARBOXYLIC ACID see TKU700
1,2,4-BENZENETRICARBOXYLIC ACID ANHYDRIDE see TKV000
1,2,4-BENZENETRICARBOXYLIC ACID, CYCLIC 1,2-ANHYDRIDE see TKV000
1,2,4-BENZENETRICARBOXYLIC ANHYDRIDE see TKV000
BENZENE, 1,2,3-TRICHLORO- see TIK100
BENZENE, 1,3,5-TRICHLORO- see TIK300
BENZENE, 1,2,4-TRICHLORO-3-METHYL- see TJD600
BENZENE, 1,3,5-TRICHLORO-2,4,6-TRINITRO- see TJE200
BENZENE, (TRIFLUOROETHENYL)-(9CI) see TKH310
1,2,4-BENZENETRIOL, TRIACETATE see HLF600
2,3-BENZFLUORANTHENE see BAW250
3,4-BENZFLUORANTHENE see BAW250
BENZHYDROL see HKF300
BENZHYDRYL ALCOHOL see HKF300
BENZHYDRYL METHYL KETONE see MJH910
BENZIDINE, N,N,N′,N′-TETRAMETHYL- see BJF600
BENZILIC ACID see BBY990
BENZIMIDAZOLE, 5,6-DIMETHYL- see DQM100
2-BENZIMIDAZOLINONE see ONI100
BENZ(b)INDENO(1,2-d)PYRAN-9(6H)-ONE, 6a,7-DIHYDRO-3,4,6a,10-TETRAHYDROXY- see HAO600
BENZ(b)INDENO(1,2-d)PYRAN-3,6a,9,10(6H)-TETROL, 7,11b-DIHYDRO- see LFT800
BENZ(cd)INDOL-2(1H)-ONE see NAX100
BENZINOFORM see CBY000
BENZINOL see TIO750
1,2-BENZISOTHIAZOLIN-3-ONE see BCE475
1,2-BENZISOTHIAZOL-3(2H)-ONE see BCE475
BENZOANTHRACENE see BBC250
BENZO(a)ANTHRACENE see BBC250
1,2-BENZOANTHRACENE see BBC250
BENZO(g)(1,3)BENZODIOXOLO(5,6-a)QUINOLIZINIUM,5,6-DIHYDRO-9,10-DIMETHOXY-, CHLORIDE (9CI) see BFN600
5,6-BENZOCOUMARIN-3-CARBOXYLIC ACID ETHYL ESTER see OOI200
5,6-BENZOCUMARIN-3-CARBONIC ACID ETHYL ETHER see OOI200
1H,3H-BENZO(1,2-c:4,5-c′)DIFURAN-1,3,5,7-TETRONE see PPQ635
BENZO(h)FLAVONE see NBI100
7,8-BENZOFLAVONE (7CI) see NBI100
BENZO(b)FLUORANTHENE see BAW250
BENZO(e)FLUORANTHENE see BAW250
2,3-BENZOFLUORANTHENE see BAW250
3,4-BENZOFLUORANTHENE see BAW250
2,13-BENZOFLUORANTHENE see BCJ200
7,10-BENZOFLUORANTHENE see BCJ200
BENZO(ghi)FLUORANTHENE see BCJ200
BENZO(mno)FLUORANTHENE see BCJ200
2,3-BENZOFLUORANTHRENE see BAW250
BENZO(a)FLUORENE see NAU600
1,2-BENZOFLUORENE see NAU600
2,3-BENZOFLUORENE see BCJ800
11H-BENZO(a)FLUORENE see NAU600
11H-BENZO(b)FLUORENE see BCJ800
2-BENZOFURANCARBOXYLIC ACID see CNU875
1-(2-BENZOFURANYL)ETHANONE see ACC100
2-BENZOFURANYL METHYL KETONE see ACC100
BENZO(b)FURAN-2-YL METHYL KETONE see ACC100
BENZOHYDROL see HKF300
BENZOIC ACID, 2-(ACETYLAMINO)-(9CI) see AAJ150
BENZOIC ACID, 2-AMINO-3,5-DIIODO-(9CI) see API800
BENZOIC ACID, p-AMINO-, METHYL ESTER see AIN150
BENZOIC ACID, 2-AMINO-, 2-PHENYLETHYL ESTER see APJ500
BENZOIC ACID, 4-(AMINOSULFONYL)-(9CI) see SNL840
BENZOIC ACID, AMMONIUM SALT see ANB100
BENZOIC ACID, 2-BENZOYL- see BDL850
BENZOIC ACID, o-BENZOYL- see BDL850
BENZOIC ACID, p-BROMO- see BMU100
BENZOIC ACID, 4-BROMO-(9CI) see BMU100
BENZOIC ACID, 3-CHLORO- see CEL290
BENZOIC ACID, m-CHLORO- see CEL290
BENZOIC ACID, p-CHLORO- see CEL300
BENZOIC ACID, 4-CHLORO-(9CI) see CEL300
BENZOIC ACID, 5-CHLORO-2-HYDROXY- see CLD825
BENZOIC ACID, 3,5-DIAMINO- see DBQ190
BENZOIC ACID, 3,5-DIAMINO-, DIHYDROCHLORIDE see DBQ200
BENZOIC ACID, 3,5-DI-tert-BUTYL-4-HYDROXY- see DEC900
BENZOIC ACID, 2,4-DICHLORO- see DER100
BENZOIC ACID, 3,4-DIHYDROXY- see POE200
BENZOIC ACID, 3,5-DIHYDROXY-(9CI) see REF200
BENZOIC ACID, 3,5-DIIODO-4-HYDROXY- see DNF300
BENZOIC ACID, 3,4-DIMETHOXY- see VHP600
BENZOIC ACID, 3,4-DIMETHYL- see DQM850
BENZOIC ACID, 3,5-DIMETHYL- see MDJ748
BENZOIC ACID, p-(DIMETHYLAMINO)- see DOU650
BENZOIC ACID, 3,4-DINITRO- see DUR550
BENZOIC ACID, 3,5-DINITRO- see DUR600
BENZOIC ACID, 2,2′-DITHIOBIS-(9CI) see BHM300
BENZOIC ACID, 3,3′-DITHIOBIS(6-NITRO- see DUV700
BENZOIC ACID, 2,2′-DITHIODI- see BHM300
BENZOIC ACID, 2-FORMYL-(9CI) see FNK010
BENZOIC ACID, 4-HYDROXY-3,5-DIMETHOXY- see SPE700
BENZOIC ACID, 2-HYDROXY-3,5-DINITRO-(9CI) see HKE600
BENZOIC ACID, p-HYDROXY-, HEPTYL ESTER see HBO650
BENZOIC ACID, 4-HYDROXY-, HEPTYL ESTER (9CI) see HBO650
BENZOIC ACID, 2-HYDROXY-, HYDRAZIDE see HJL100
BENZOIC ACID, 2-HYDROXY-4-NITRO- see NJF100
BENZOIC ACID, 4-HYDROXY-3-NITRO-, METHYL ESTER see HMY075
BENZOIC ACID, 2-HYDROXY-5-((3-NITROPHENYL)AZO)-, MONOSODIUM SALT see SIT850
BENZOIC ACID, p-IODO- see IEE025
BENZOIC ACID, 2-METHOXY-, METHYL ESTER see MLH800
BENZOIC ACID, PENTAFLUORO- see PBD275
BENZOIC ACID, PENTYL ESTER see PBV800
BENZOIC ACID, o-(PHENYLTHIO)- see PGL270
BENZOIC ACID, 2-(PHENYLTHIO)-(9CI) see PGL270
BENZOIC ACID, 8-QUINOLYL ESTER see HOE100
BENZOIC ACID p-SULFAMIDE see SNL840
BENZOIC ACID, p-SULFAMOYL- see SNL840
BENZOIC ACID, THIO- see TFC550
BENZOIC ACID, 3,4,5-TRIHYDROXY-, ETHYL ESTER (9CI) see EKM100
BENZOIC ACID, 3,4,5-TRIMETHOXY-, METHYL ESTER see MQF300
BENZOIC ACID, 2,4,6-TRIMETHYL- see IKN300
BENZON 00 see HND100
BENZONITRILE, p-FLUORO- see FGH100
BENZONITRILE, p-NITRO- see NFI010
BENZONITRILE, 4-NITRO-(9CI) see NFI010
BENZO(a)PHENANTHRENE see BBC250
BENZO(b)PHENANTHRENE see BBC250
2,3-BENZOPHENANTHRENE see BBC250
BENZO(rst)PHENANTHRO(10,1,2-cde)PENTAPHENE-9,18-DIONE, DICHLORO- see DFN450
BENZOPHENONE-6 see BKH200
BENZOPHENONE-8 see DMW250
BENZOPHENONE 12 see HND100
BENZOPHENONE-2-CARBOXYLIC ACID see BDL850
BENZOPHENONE, 2,2′-DIHYDROXY-4,4′-DIMETHOXY- see BKH200
BENZOPHENONE, 2,2′-DIHYDROXY-4-METHOXY- see DMW250
BENZOPHENONE, 2-HYDROXY-4-(OCTYLOXY)- see HND100
BENZOPHENONE, OXIME see BCS400
BENZOPHENOXIME see BCS400
BENZOPURPURIN 4B see DXO850
BENZOPURPURINE 4B see DXO850
BENZOPURPURINE 4BKX see DXO850
BENZOPURPURINE 4BX see DXO850
4H-1-BENZOPYRAN-4-ONE, 2,3-DIHYDRO-2-PHENYL-(9CI) see FBW150
2H-1-BENZOPYRAN-2-ONE, 6,7-DIHYDROXY-4-METHYL-(9CI) see MJV800
2H-1-BENZOPYRAN-2-ONE, 7-HYDROXY- see HJY100
2H-1-BENZOPYRAN-2-ONE, 7-METHOXY- see MEK300
4H-1-BENZOPYRAN-4-ONE, 2-PHENYL- see PER700
1,4-BENZOQUINONE, 2-(N,N-BIS(2-HYDROXYETHYL)AMINO)- see BKB100
p-BENZOQUINONE, 2,6-DI-tert-BUTYL- see DDV500
BENZO SKY BLUE A-CF see CMO500
BENZO SKY BLUE S see CMO500
BENZOTHIAZOLE, 2-(p-AMINOPHENYL)-6-METHYL- see MHJ300
BENZOTHIAZOLE, 5-CHLORO-2-METHYL- see CIH100
2-BENZOTHIAZOLESULFENAMIDE, N,N-DIISOPROPYL- see DNN900
BENZOTRIFLUORIDE, 4-CHLORO-3,5-DINITRO- see CGM225
3,5-BENZOTRIFLUORODIAMINE see TKB775
BENZOXAZOLE, 2-CHLORO- see CEM850
2-BENZOXAZOLETHIOL see MCK900
BENZOXIQUINE see HOE100
BENZOXYLINE see HOE100
BENZOXYQUINE see HOE100
BENZOYLACETIC ACID ETHYL ESTER see EGR600
BENZOYL-ACETON see BDJ800
BENZOYLACETONE see BDJ800
o-(BENZOYLAMINO)PHENYL DISULFIDE see BDK800
2-BENZOYLBENZOIC ACID see BDL850
BENZOYL CHLORIDE, 3,5-DINITRO- see DUR850
BENZOYLGLYCINE see HGB300
BENZOYL METHIDE see ABH000
8-BENZOYLOXYQUINOLINE see HOE100
2-BENZOYLPYRIDINE see PGE760
BENZOYL THIOL see TFC550
2,3-BENZPHENANTHRENE see BBC250
BENZSULFOHYDROXAMIC ACID see BDW100

BENZYL ACETATE

BENZYL ACETATE see BDX000
BENZYLADENINE see BDX090
6-BENZYLADENINE see BDX090
N-BENZYLADENINE see BDX090
N⁶-BENZYLADENINE see BDX090
BENZYL ALCOHOL, HEXAHYDRO- see HDH200
BENZYL ALCOHOL, p-NITRO-, ACETATE (ester) see NFM100
BENZYL ALCOHOL, α-(TRICHLOROMETHYL)- see TIR800
BENZYLAMINE, o-METHYL- see MHM510
4-(BENZYLAMINO)PHENOL see BDY750
BENZYLAMINOPURINE see BDX090
6-(BENZYLAMINO)PURINE see BDX090
6-(N-BENZYLAMINO)PURINE see BDX090
N⁶-(BENZYLAMINO)PURINE see BDX090
BENZYLCARBINYL ANTHRANILATE see APJ500
BENZYLCHLORODIMETHYLSILANE see BEE800
BENZYLDIMETHYLCETYLAMMONIUM CHLORIDE see BEL900
BENZYLDIMETHYLHEXADECYLAMMONIUM CHLORIDE see BEL900
BENZYL ETHANOATE see BDX000
2-BENZYLIDENEACETAMIDE see CMP970
BENZYL METHYL CARBINOL see PGA600
BENZYL METHYL KETONE see MHO100
BENZYL NICOTINATE see NCR040
BENZYLPENICILLIN NOVOCAINE SALT see PAQ200
BENZYLPENICILLIN PROCAINE see PAQ200
p-BENZYLPHENYL CARBAMATE see PFR325
2-BENZYLPYRIDINE see BFG600
BENZYL PYRIDINE-3-CARBOXYLATE see NCR040
4-BENZYL RESORCINOL see BFI400
BENZYLSULFONYL FLUORIDE see TGO300
BENZYLSULPHONYL FLUORIDE see TGO300
BENZYLTRIETHYLAMMONIUM CHLORIDE see BFL300
BENZYLTRIMETHYLAMMONIUM METHOXIDE see TLM100
BERBERINE CHLORIDE see BFN600
BERBERINE HYDROCHLORIDE see BFN600
BERBERINIUM CHLORIDE see BFN600
BERBINIUM, 7,8,13,13a-TETRADEHYDRO-9,10-DIMETHOXY-2,3-(METHYLENEDIOXY)-, CHLORIDE see BFN600
BETAINE see GHA050
BETAMETHASONE BENZOATE see BFV760
BETAMETHASONE 17-BENZOATE see BFV760
BETAPAL see NBJ100
BETHAMETHASONE 17-BENZOATE see BFV760
BETOXON see NBJ100
B(b)F see BAW250
BF 200 see CAT775
BFV see FMV000
BIACETYLMONOXIME see OMY910
1,1′-BIANTHRACENE-9,9′,10,10′-TETRAONE, 2,2′-DIMETHYL- see DQR350
(Δ²,²′(3H,3′H)-BIBENZO(b)THIOPHENE)-3,3′-DIONE see DNT300
BICARNESINE see CCK655
BICOL see PPN100
BICYCLO(5.3.0)DECAPENTAENE see ASP600
BICYCLO(0.3.5)DECA-1,3,5,7,9-PENTAENE see ASP600
BICYCLO(5.3.0)-DECA-2,4,6,8,10-PENTAENE see ASP600
BICYCLO(3.1.1)HEPTANE, 2,6,6-TRIMETHYL-, DIDEHYDRO deriv. see PIH245
BICYCLO(3.1.1)HEPT-2-ENE-2-CARBOXALDEHYDE, 6,6-DIMETHYL- see FNK150
BICYCLO(3.1.1)HEPT-2-ENE-2-METHANOL, 6,6-DIMETHYL-, ACETATE, (1S)- see MSC050
(1,1′-BICYCLOHEXYL)-2-ONE see LBX100
BIGUANIDE, 1-o-TOLYL- see TGX550
(2,2′-BIINDOLINE)-3,3′-DIONE see BGB275
(Δ²,²′)-BIINDOLINE)-3,3′-DIONE see BGB275
(1,1′-BINAPHTHALENE)-2,2′-DIOL see BGC100
β-BINAPHTHOL see BGC100
1,1′-BI-2-NAPHTHOL see BGC100
BIOMET 204 see TMV850
BIOSORB 130 see HND100
4-BIPHENYLAMINE, 4,4′-DIMETHOXY- see BKO600
BIPHENYL, 3-BROMO- see BMW290
(1,1′-BIPHENYL)-4,4′-DIAMINE, N,N,N′,N′-TETRAMETHYL- see BJF600
o-BIPHENYLENEMETHANE see FDI100
2,2′-BIPHENYLENE OXIDE see DDB500
o-BIPHENYLMETHANE see FDI100
BIPHENYL, 4-METHYL- see MJH905
2,2′-BIPHENYLYLENE OXIDE see DDB500
2,2′-BIPHENYLYLENE SULFIDE see TES300
(Δ²,²′)-BIPSEUDOINDOXYL see BGB275
4,4′-BIPYRIDINE see BGO600
2,2′-BIPYRIDINE, 4,4′-DIMETHYL- see DQS100
4,4′-BIPYRIDYL see BGO600
4,4′-BIPYRIDYL see BGO600

γ-BIPYRIDYL see BGO600
BIQUIN DURULES see QFS100
BISABOLOL see BGO775
(−)-α-BISABOLOL see BGO775
BIS(ACETOXY)CADMIUM see CAD250
BIS(p-ACETOXYPHENYL)-2-PYRIDYLMETHANE see PPN100
2,2-BIS((ACETYLOXY)METHYL)-1,3-PROPANEDIOL DIACETATE see NNR400
BISACODYL see PPN100
BIS AMINE see MJM200
1,3-BIS(4-AMINOPHENOXY)BENZENE see REF070
1,3-BIS(3-AMINOPROPYL)-1,1,3,3-TETRAMETHYLDISILOXANE see OPC100
BIS(p-ANISYLAMINE) see BKO600
BIS(o-BENZAMIDOPHENYL) DISULFIDE see BDK800
BIS-o-BENZOYLAMINOFENYL-DISULFID see BDK800
BIS(o-BENZAMIDOPHENYL) DISULFIDE see BDK800
1,3-BIS(BENZOYLOXY)BENZENE see BHB100
1,2-BIS(BENZYLAMINO)ETHANE see BHB300
4,4-BIS(tert-BUTYLPEROXY)VALERIC ACID BUTYL ESTER see BHL200
2,6-BIS(tert-BUTYL)PHENOL see DEG100
BIS-BUTYLXANTHOGEN see BSS550
BIS(CARBOXYMETHYL)ETHER see ONQ100
BIS(o-CARBOXYPHENYL) DISULFIDE see BHM300
BIS(2-CARBOXYPHENYL) DISULFIDE see BHM300
BIS(p-CHLOROPHENYL)METHANE see BIM800
4,4′-BIS((4,6-DIANILINO-s-TRIAZIN-2-YL)AMINO)-2,2′-STILBENEDISULFONICACID DISODIUM SALT see DXB450
1,4-BIS(DICYANOMETHYLENE)CYCLOHEXANE see BIY600
BIS-DIETHYLAMID KYSELINY FTALOVE see GCE600
1,4-BIS(DIETHYLAMINO)-2-BUTYNE see BJA250
BIS(DIETHYLDITHIOCARBAMATO)CADMIUM see BJB500
4,4′-BIS(N,N-DIMETHYLAMINO)BIPHENYL see BJF600
1,4-BIS(DIMETHYLSILYL)BENZENE see PEW725
BIS(2-ETHYLHEXYL) ESTER PHOSPHOROUS ACID CADMIUM SALT see CAD500
BIS(4-FLUORO-3-NITROPHENYL)SULFONE see DKH250
1,2-BIS(GLYCIDYLOXY)ETHANE see EEA600
BISGUANIDINIUM CARBONATE see GKW100
BIS(2-HYDROXYBENZOATO-O¹,O²-), (T-4)-CADMIUM (9CI) see CAI400
BIS(1-HYDROXYCYCLOHEXYL)PEROXIDE see PCM100
2-(N,N-BIS(2-HYDROXYETHYL)AMINO)-1,4-BENZOQUINONE see BKB100
N,N′-BIS(2-HYDROXYETHYL)-DITHIOOXAMIDE see BKD800
BIS(2-HYDROXY-4-METHOXYPHENYL)METHANONE see BKH200
BIS(4-HYDROXYPHENYL)AMINE see IBJ100
BIS(p-HYDROXYPHENYL)AMINE see IBJ100
BIS(4-METHOXYPHENYL)AMINE see BKO600
BIS(p-METHOXYPHENYL)AMINE see BKO600
1,4-BIS(METHYLAMINO)-9,10-ANTHRACENEDIONE see BKP500
1,4-BIS(1-METHYLETHYL)BENZENE see DNN830
BIS(MORPHOLINOTHIOCARBONYL)DISULFIDE see MRR090
BISMUTH HYDROXIDE NITRATE OXIDE see BKW100
BISMUTH MAGISTERY see BKW100
BISMUTHOUS OXIDE see BKW600
BISMUTH OXIDE see BKW600
BISMUTH(3+) OXIDE see BKW600
BISMUTH SESQUIOXIDE see BKW600
BISMUTH SUBNITRATE see BKW100
BISMUTH SUBNITRICUM see BKW100
BISMUTH TRIOXIDE see BKW600
BISMUTH WHITE see BKW100
BISMUTH YELLOW see BKW600
BISMUTHYL NITRATE see BKW100
BIS-β-NAPHTHOL see BGC100
BIS(p-NITROPHENYL)ETHER see NIM600
BIS(4-NITROPHENYL)ETHER see NIM600
BIS(p-NITROPHENYL) PHOSPHATE see BLA600
BIS(4-NITROPHENYL) PHOSPHATE see BLA600
BIS(2-NITRO-4-TRIFLUOROMETHYLPHENYL) DISULFIDE see BLA800
2,6-BIS(1-PHENYLETHYL)-4-METHYLPHENOL see MHR050
1,1-BIS(TRIFLUOROMETHYL)ETHENE see HDC450
11557 BLACK see IHC550
BLACK GOLD F 89 see IHC550
BLACK IRON BM see IHC550
BLACOSOLV see TIO750
BLANC DE FARD see BKW100
BLANDLUBE see MQV750
BLANKOPHOR HZPA see DXB450
L-BLAU 1 see IBV050
BLEU SOLANTHRENE see IBV050
11669 BLUE see BGB275
BLUE ANTHRAQUINONE PIGMENT see IBV050
BLUE K see DFN425
BLUE NO. 201 see BGB275
BLUE O see IBV050

BNPP see BLA600
BNS see NMC100
BON see HMX520
BONA see HMX520
BON ACID see HMX520
BORATES, TETRA, SODIUM SALT, anhydrous (OSHA) see DXG035
BORAX GLASS see DXG035
BORIC ACID, DISODIUM SALT see DXG035
BOROXIN, TRIMETHOXY- see TKZ100
BRASILACA RED R see NAP100
BRASILAMINA RED 4B see DXO850
BRASILIN see LFT800
BRAZILETTO see LFT800
BRAZILIN see LFT800
BRILLIANT 15 see CAT775
BRILLIANT GREEN PHTHALOCYANINE see PJQ100
BRILLIANT SAFRANINE BR see GJI400
BRILLIANT SAFRANINE G see GJI400
BRILLIANT SAFRANINE GR see GJI400
BRILLIANT VIOLET K see DFN450
BRITOMYA M see CAT775
BRITONE RED Y see NAP100
BRODAN see CMA100
p-BROMANILID KYSELINY 5-BROMSALICYLOVE see BOD600
p-BROMANISOLE see AOY450
p-BROMOACETANILIDE see BMR100
4-BROMOACETANILIDE see BMR100
4′-BROMOACETANILIDE see BMR100
p-BROMO-N-ACETANILIDE see BMR100
BROMOANILIDE see BMR100
o-BROMOANISOLE see BMT400
p-BROMOANISOLE see AOY450
2-BROMOANISOLE see BMT400
4-BROMOANISOLE see AOY450
BROMOANTIFEBRIN see BMR100
p-BROMOBENZALDEHYDE see BMT700
4-BROMOBENZALDEHYDE see BMT700
p-BROMOBENZOIC ACID see BMU100
4-BROMOBENZOIC ACID see BMU100
3-BROMOBIPHENYL see BMW290
4-BROMOBIPHENYL see BMW300
BROMOCET see HCP800
1-BROMODECANE see BNB800
4-BROMO-2,5-DICHLOROPHENOL see LEN050
4-BROMO-2,6-DIMETHYLPHENOL see BOL303
2-BROMO-1,3-DIPHENYL-1,3-PROPANEDIONE see BNH100
2-BROMOETHANAMINE see BNI650
BROMOETHANE see EGV400
BROMOETHYL ACETATE see BNI600
2-BROMOETHYL ACETATE see BNI600
β-BROMOETHYLAMINE see BNI650
2-BROMOETHYLAMINE see BNI650
1-BROMOHEPTANE see HBN100
3-BROMO-6-HYDROXYBENZ-p-BROMANILIDE see BOD600
α-BROMOISOVALERIC ACID see BNM100
2-BROMOISOVALERIC ACID see BNM100
1-BROMO-2-METHOXYBENZENE see BMT400
1-BROMO-4-METHOXYBENZENE see AOY450
1-BROMO-2-METHYLBENZENE see BOG260
2-BROMO-3-METHYLBUTANOIC ACID see BNM100
2-BROMO-3-METHYLBUTYRIC ACID see BNM100
2-(BROMOMETHYL)-TETRAHYDROFURAN see TCS550
α-BROMONAPHTHALENE see BNS200
1-BROMONAPHTHALENE see BNS200
m-BROMONITROBENZENE see NFQ090
o-BROMONITROBENZENE see NFQ080
2-BROMONITROBENZENE see NFQ080
3-BROMONITROBENZENE see NFQ090
4-BROMONITROBENZENE see NFQ100
p-BROMONITROBENZENE see NFQ100
3-BROMO-2-OXOPROPANOIC ACID see BOD550
1-BROMOPENTANE see AOF800
p-BROMOPHENYL BROMIDE see BNV775
4-BROMOPHENYL HYDRAZINE HYDROCHLORIDE see BNW550
o-BROMOPHENYL METHYL ETHER see BMT400
p-BROMOPHENYL METHYL ETHER see AOY450
α-BROMOPROPIOPHENONE see BOB550
2-BROMOPROPIOPHENONE see BOB550
2-BROMOPYRIDINE see BOB600
3-BROMOPYRUVATE see BOD550
BROMOPYRUVIC ACID see BOD550
β-BROMOPYRUVIC ACID see BOD550

3-BROMOPYRUVIC ACID see BOD550
5-BROMOSALICYL-4-BROMOANILIDE see BOD600
BROMOSULFALEIN see HAQ600
BROMOSULFOPHTHALEIN see HAQ600
BROMOSULPHALEIN see HAQ600
BROMOSULPHTHALEIN see HAQ600
BROMOTALEINA see HAQ600
m-BROMOTOLUENE see BOG300
o-BROMOTOLUENE see BOG260
2-BROMOTOLUENE see BOG260
3-BROMOTOLUENE see BOG300
5-BROMOTOLUENE see BOG300
2-BROMOTOLUIDINE see BOG500
1-BROMOTRIDECANE see BOI500
BROMOTRIPHENYLMETHANE see TNP600
4-BROMO-2,6-XYLENOL see BOL303
BROMSULFALEIN see HAQ600
BROMSULFAN see HAQ600
BROMSULFOPHTHALEIN see HAQ600
BROMSULFTHALEIN see HAQ600
BROMSULPHALEIN see HAQ600
BROMSULPHTHALEIN see HAQ600
BROM-TETRAGNOST see HAQ600
BROMTHALEIN see HAQ600
BROMURE d'ETHYLE see EGV400
BROXALAX see PPN100
BRUNEOMYCIN see SMA000
BSF see HAQ600
BSF SIMES see HAQ600
BSP see HAQ600
BSP SODIUM see HAQ600
BT 31 see DEJ100
BUCACID ALIZARINE LIGHT BLUE SE see APG700
BULBOSAN see TJE200
BURSINE see CMF800
BUTADIENE DIMER see CPD750
BUTANAMIDE, N-(2,4-DIMETHYLPHENYL)-3-OXO-(9CI) see OOI100
2-BUTANAMINE, 3-METHYL-(9CI) see AOE200
BUTANEAMIDE, N-(2-CHLOROPHENYL)-3-OXO- see AAY600
1,4-BUTANEDIAMINE, N-(3-AMINOPROPYL)-, TRIHYDROCHLORIDE see SLG600
1,4-BUTANEDICARBOXYLIC ACID see AEN250
BUTANE, 1,1-DICHLORO- see BRQ050
BUTANE, 1,4-DIIODO- see TDQ400
BUTANEDIOIC ACID, 2,3-DIMERCAPTO-(9CI) see DNV610
BUTANEDIOIC ACID, DIMETHYL ESTER see SNB100
BUTANEDIOIC ACID, DISODIUM SALT (9CI) see SJW100
BUTANEDIOIC ACID, SULFO-, 1,4-DIOCTYL ESTER, SODIUM SALT (9CI) see SOD300
2,3-BUTANEDIONE, MONOOXIME see OMY910
2,3-BUTANEDIONE 2-OXIME see OMY910
1,3-BUTANEDIONE, 1-PHENYL- see BDJ800
BUTANE, 1-IODO-3-METHYL- see IHU200
BUTANE, 1-ISOTHIOCYANATO-(9CI) see ISD100
BUTANENITRILE, 4-CHLORO-(9CI) see CEU300
BUTANOIC ACID, ANHYDRIDE (9CI) see BSW550
BUTANOIC ACID, 2-BROMO-3-METHYL-(9CI) see BNM100
BUTANOIC ACID, 4-CHLORO-3-OXO-, ETHYL ESTER (9CI) see EHH100
BUTANOIC ACID, 3,7-DIMETHYL-2,6-OCTADIENYL ESTER, (E)-(9CI) see GDE810
BUTANOIC ACID, 2-HYDROXY-4-(METHYLTHIO)-(9CI) see HMR600
BUTANOIC ACID, 2-OXO-, SODIUM SALT (9CI) see SIY600
BUTANOIC ACID, PROPYL ESTER (9CI) see PNF100
BUTANOIC ANHYDRIDE see BSW550
tert-BUTANOL see BPX000
2-BUTANOL, 3,3-DIMETHYL- see BRU300
1-BUTANOL, 2,2,3,3,4,4,4-HEPTAFLUORO- see HAW100
1-BUTANOL, 3-METHYL-, NITRATE (9CI) see ILW100
BUTANOL TERTIAIRE (FRENCH) see BPX000
1-BUTANONE, 1-(4-AMINOPHENYL)-(9CI) see AJC750
2-BUTENE see BOW255
2-BUTENE, 1,3-DICHLORO- see DEU650
2-BUTENE, 1,4-DIHYDROXY- see BOX300
2-BUTENE-1,4-DIOL see BOX300
2-BUTENOIC ACID, 4-BROMO-, METHYL ESTER (9CI) see MHR790
3-BUTEN-2-ONE, 4-(4-HYDROXY-3-METHOXYPHENYL)- see MLI800
BUTILE (ACETATI di) (ITALIAN) see BPU750
BUTOLAN see PFR325
BUTOLEN see PFR325
N-(BUTOXYMETHYL)ACRYLAMIDE see BPM660
N-BUTOXYMETHYLAKRYLAMID see BPM660
N-(BUTOXYMETHYL)-2-PROPENAMIDE see BPM660
BUTYLACETAT (GERMAN) see BPU750
BUTYL ACETATE see BPU750

n-BUTYL ACETATE see BPU750
1-BUTYL ACETATE see BPU750
BUTYLACETATEN (DUTCH) see BPU750
N-tert-BUTYLACRYLAMIDE see BPW050
tert-BUTYL ALCOHOL see BPX000
(+)-2-BUTYLAMINE see BPY100
S-2-BUTYLAMINE see BPY100
sec-BUTYLAMINE, (S)- see BPY100
p-n-BUTYLANILINE see AJA550
4-BUTYLBENZENAMINE see AJA550
6-tert-BUTYL-m-CRESOL see BQV600
n-BUTYL-4,4-DI-(tert-BUTYLPEROXY)VALERATE, technically pure (UN2140)(DOT) see BHL200
n-BUTYL-4,4-DI-(tert-BUTYLPEROXY)VALERATE, not >52% with inert solid(UN2141) (DOT) see BHL200
BUTYLE (ACETATE de) (FRENCH) see BPU750
β-BUTYLENE see BOW255
BUTYL ETHANOATE see BPU750
tert-BUTYL HYDROXIDE see BPX000
BUTYLIDENE CHLORIDE see BRQ050
n-BUTYL ISOTHIOCYANATE see ISD100
BUTYL LAEVULINATE see BRR700
n-BUTYL LAEVULINATE see BRR700
BUTYL LEVULINATE see BRR700
n-BUTYL LEVULINATE see BRR700
tert-BUTYL METHYL CARBINOL see BRU300
2-tert-BUTYL-6-METHYLPHENOL see BRU790
4-tert-BUTYL-2-METHYLPHENOL see BRU800
BUTYL MUSTARD OIL see ISD100
BUTYL OLEATE see BSB000
BUTYL 4-OXOPENTANOATE see BRR700
tert-BUTYL PERISOBUTYRATE see BSC600
tert-BUTYL PEROXYISOBUTYRATE see BSC600
tert-BUTYL PEROXYISOBUTYRATE, >77% in solution (DOT) see BSC600
tert-BUTYL PEROXYISOBUTYRATE, not >52% in solution (UN2562) (DOT) see BSC600
tert-BUTYL PEROXYISOBUTYRATE, >52% but not >77% in solution (UN2142)(DOT) see BSC600
tert-BUTYL PEROXYISOPROPYL CARBONATE, technically pure (DOT) see BSD150
2-t-BUTYLPHENOL see BSE460
3-tert-BUTYLPHENYL N-METHYLCARBAMATE see BSG300
3-BUTYLPYRIDINE see BSJ550
3-n-BUTYLPYRIDINE see BSJ550
p-tert-BUTYLTOLUENE see BSP500
5-tert-BUTYL-1,2,3-TRIMETHYL-4,6-DINITROBENZENE see MRW272
n-BUTYLTRIPHENYLPHOSPHONIUM BROMIDE see BSR900
sec-BUTYLUREA see BSS300
tert-BUTYLUREA see BSS310
BUTYLXANTHIC DISULFIDE see BSS550
1-BUTYN-3-OL see EQM600
3-BUTYN-2-OL, 2-PHENYL- see EQN230
2-BUTYNYLENEDIAMINE, N,N,N′N′-TETRAETHYL- see BJA250
BUTYRAMIDE, N,N-DIMETHYL- see DQV300
BUTYRANHYDRID see BSW550
BUTYRIC ACID ANHYDRIDE see BSW550
n-BUTYRIC ACID ANHYDRIDE see BSW550
BUTYRIC ACID, 2-BROMO-3-METHYL- see BNM100
BUTYRIC ACID, 3,7-DIMETHYL-2,6-OCTADIENYL ESTER, (E)- see GDE810
BUTYRIC ACID, 2-HYDROXY-4-(METHYLTHIO)- see HMR600
BUTYRIC ACID, 2-OXO-, SODIUM SALT see SIY600
BUTYRIC ACID, PROPYL ESTER see PNF100
BUTYRIC ANHYDRIDE see BSW550
n-BUTYRIC ANHYDRIDE see BSW550
BUTYRONITRILE, 4-CHLORO- see CEU300
BUTYRONITRILE, 4-(TRICHLOROSILYL)- see COR550
BUTYROPHENONE, 4′-AMINO- see AJC750
4-BUTYROTHIOLACTONE see TDC800
BUTYRYLCHOLINE IODIDE see BSY300
BUTYRYL OXIDE see BSW550
B-3-Zh see CMS205

C 2 see LBX050
C-908 see ABW550
CAB-O-GRIP II see SCH000
CAB-O-SIL see SCH000
CAB-O-SPERSE see SCH000
CADDY see CAE250
CADMINATE see CAI750
CADMIUM see CAD000
CADMIUM(II) ACETATE see CAD250
CADMIUM ACETATE (DOT) see CAD250
CADMIUM ACETATE DIHYDRATE see CAD275

CADMIUM AMIDE see CAD325
CADMIUM AZIDE see CAD350
CADMIUM BARIUM STEARATE see BAI800
CADMIUM BIS(2-ETHYLHEXYL) PHOSPHITE see CAD500
CADMIUM, BIS(1-HYDROXY-2(1H)-PYRIDINETHIONATO)- see CAI350
CADMIUM, BIS(SALICYLATO)- see CAI400
CADMIUM BROMIDE see CAD600
CADMIUM CAPRYLATE see CAD750
CADMIUM CARBONATE see CAD800
CADMIUM CHLORATE see CAE000
CADMIUM CHLORIDE see CAE250
CADMIUM CHLORIDE, DIHYDRATE see CAE375
CADMIUM CHLORIDE, HYDRATE (2:5) see CAE425
CADMIUM CHLORIDE, MONOHYDRATE see CAE500
CADMIUM COMPOUNDS see CAE750
CADMIUM DIACETATE see CAD250
CADMIUM DIACETATE DIHYDRATE see CAD275
CADMIUM DIAMIDE see CAD325
CADMIUM DIAZIDE see CAD350
CADMIUM DIBROMIDE see CAD600
CADMIUM DICHLORIDE see CAE250
CADMIUM DICYANIDE see CAF500
CADMIUM DIETHYL DITHIOCARBAMATE see BJB500
CADMIUM DILAURATE see CAG775
CADMIUM DINITRATE see CAH000
CADMIUM DODECANOATE see CAG775
CADMIUM(II) EDTA COMPLEX see CAF750
CADMIUM FLUOBORATE see CAG000
CADMIUM FLUORIDE see CAG250
CADMIUM FLUOROBORATE see CAG000
CADMIUM FLUORURE (FRENCH) see CAG250
CADMIUM FLUOSILICATE see CAG500
CADMIUM FUME see CAH750
CADMIUM GOLDEN see CMS205
CADMIUM GOLDEN 366 see CAJ750
CADMIUM LACTATE see CAG750
CADMIUM LAURATE see CAG775
CADMIUM LEMON see CMS205
CADMIUM LEMON YELLOW 527 see CAJ750
CADMIUM MONOCARBONATE see CAD800
CADMIUM NITRATE see CAH000
CADMIUM(II) NITRATE TETRAHYDRATE (1:2:4) see CAH250
CADMIUM NITRIDE see TIH000
CADMIUM ORANGE see CAJ750
CADMIUM OXIDE see CAH500
CADMIUM OXIDE FUME see CAH750
CADMIUM PHOSPHATE see CAI000
CADMIUM PHOSPHIDE see CAI125
CADMIUM PRIMROSE see CMS205
CADMIUM PRIMROSE 819 see CAJ750
CADMIUM PROPIONATE see CAI250
CADMIUM PT see CAI350
CADMIUM 2-PYRIDINETHIONE see CAI350
CADMIUM SALICYLATE see CAI400
CADMIUM SELENIDE see CAI500
CADMIUM STEARATE see OAT000
CADMIUM SUCCINATE see CAI750
CADMIUM SULFATE see CAJ000
CADMIUM SULFATE (1:1) see CAJ000
CADMIUM SULFATE (1:1) HYDRATE (3588) see CAJ250
CADMIUM SULFATE OCTAHYDRATE see CAJ250
CADMIUM SULFATE TETRAHYDRATE see CAJ500
CADMIUM SULFIDE see CAJ750
CADMIUM SULFIDE mixed with ZINC SULFIDE (1:1) see CMS205
CADMIUM SULPHATE see CAJ000
CADMIUM SULPHIDE see CAJ750
CADMIUM THERMOVACUUM AEROSOL see CAK000
CADMIUM-THIONEINE see CAK250
CADMIUM salt of 2,4-5-TRIBROMOIMIDAZOLE see THV500
CADMIUM YELLOW see CAJ750
CADMOPUR YELLOW see CAJ750
CALAR see CAM000
CALCENE CO see CAT775
CALCIATE(2⁻), ((ETHYLENEDINITRILO)TETRAACETATO)-, DISODIUM see CAR780
CALCICOLL see CAT775
CALCIDAR 40 see CAT775
CALCILIT 8 see CAT775
CALCITETRACEMATE DISODIUM see CAR780
CALCIUM ACID METHANEARSONATE see CAM000
CALCIUM ACID METHYL ARSONATE see CAM000
CALCIUMARSENAT see ARB750

CALCIUM ARSENATE (MAK) see ARB750
CALCIUM ARSENITE see CAM500
CALCIUM ARSENITE, solid (DOT) see CAM500
CALCIUM CARBONATE (1:1) see CAT775
CALCIUM DISODIUM EDATHAMIL see CAR780
CALCIUM DISODIUM EDETATE see CAR780
CALCIUM DISODIUM EDTA see CAR780
CALCIUM DISODIUM ETHYLENEDIAMINETETRAACETATE see CAR780
CALCIUM DISODIUM (ETHYLENEDINITRILO)TETRAACETATE see CAR780
CALCIUM DISODIUM VERSENATE see CAR780
CALCIUM EDTA see CAR780
CALCIUM HYDROGEN METHANEARSONATE see CAM000
CALCIUM METHANEARSONATE see CAM000
CALCIUM MONOCARBONATE see CAT775
CALCIUM ORTHOARSENATE see ARB750
CALCIUM TITRIPLEX see CAR780
CALCOCHROME ALIZARINE RED SC see SEH475
CALCOCHROME YELLOW 2G see SIT850
CALCOCID ALIZARINE BLUE SE see APG700
CALCOFLUOR WHITE MR see DXB450
CALCOLOID BLUE BLC see DFN425
CALCOLOID BLUE BLD see DFN425
CALCOLOID BLUE BLFD see DFN425
CALCOLOID BLUE BLR see DFN425
CALCOLOID BLUE RS see IBV050
CALCOLOID OLIVE R see DUP100
CALCOLOID OLIVE RC see DUP100
CALCOLOID OLIVE RL see DUP100
CALCOLOID VIOLET 4RD see DFN450
CALCOLOID VIOLET 4RP see DFN450
CALCOMINE RED 4BX see DXO850
CALCON see HLI100
CALCOTONE GREEN G see PJQ100
CALCOZINE RED Y see GJI400
CALEDON BLUE RN see IBV050
CALEDON BLUE XRC see DFN425
CALEDON BLUE XRN see IBV050
CALEDON BRILLIANT BLUE RN see IBV050
CALEDON BRILLIANT PURPLE 4R see DFN450
CALEDON BRILLIANT PURPLE 4RP see DFN450
CALEDONE OLIVE RP see DUP100
CALEDON OLIVE R see DUP100
CALEDON PAPER BLUE RN see IBV050
CALEDON PRINTING BLUE RN see IBV050
CALEDON PRINTING BLUE XRN see IBV050
CALEDON PRINTING PURPLE 4R see DFN450
CALIBRITE see CAT775
CAL-LIGHT SA see CAT775
CALMOS see CAT775
CALMOTE see CAT775
CALOFIL A 4 see CAT775
CALOFORT S see CAT775
CALOFORT U see CAT775
CALOFOR U 50 see CAT775
CALOPAKE F see CAT775
CALOPAKE HIGH OPACITY see CAT775
CALSEEDS see CAT775
CALTEC see CAT775
CAMA see CAM000
CAMEL-CARB see CAT775
CAMEL-TEX see CAT775
CAMEL-WITE see CAT775
CAPRIC ACID METHYL ESTER see MHY650
CAPRINIC ACID SODIUM SALT see SGB600
CAPROLACTAM see CBF700
6-CAPROLACTAM see CBF700
ω-CAPROLACTAM (MAK) see CBF700
γ-CAPROLACTONE see HDY600
6-CAPROLACTONE see HDY600
CAPROLATTAME (FRENCH) see CBF700
CAPROLYL PEROXIDE see CBF705
CAPRYL ALCOHOL see OCY090
CAPRYLAMINE see OEK010
CAPRYLIC ACID METHYL ESTER see MHY800
CAPRYLIC ACID SODIUM SALT see SIX600
CAPRYLNITRILE see OCW100
CAPRYLONITRILE see OCW100
CAPRYL PEROXIDE see CBF705
CAPRYLYL PEROXIDE see CBF705
CAPRYLYL PEROXIDE (DOT) see CBF705
CAPRYLYL PEROXIDE SOLUTION (DOT) see CBF705
CAPSEBON see CAJ750

CARBAMIC ACID, DIMETHYLDITHIO-, 2,4-DINITROPHENYL ESTER see DVB850
CARBAMIC ACID, DITHIO-, N,N-DIMETHYL-, DIMETHYLAMINOMETHYL ESTER see DRQ650
CARBAMIC ACID, HEXAMETHYLENEBIS(DITHIO-, DISODIUM SALT see HEF400
CARBAMIC ACID, METHYL-, 3-tert-BUTYLPHENYL ESTER see BSG300
CARBAMIMIDOTHIOIC ACID, METHYL ESTER, SULFATE (9CI) see MPV790
2-CARBAMOYLANILINE see AID620
4-CARBAMOYLPYRIDINE see ILC100
CARBANTHRENE BLUE 2R see IBV050
CARBANTHRENE BLUE BCF see DFN425
CARBANTHRENE BLUE BCS see DFN425
CARBANTHRENE BLUE RBCF see DFN425
CARBANTHRENE BLUE RCS see DFN425
CARBANTHRENE BLUE RS see IBV050
CARBANTHRENE BLUE RSP see IBV050
CARBANTHRENE BRILLIANT VIOLET 4R see DFN450
CARBANTHRENE OLIVE R see DUP100
CARBANTHRENE VIOLET 2R see DFN450
CARBANTHRENE VIOLET 2RP see DFN450
CARBAURINE see PFR325
CARBAZINE see ADI775
9H-CARBAZOLE, 9-ETHENYL-(9CI) see VNK100
CARBAZOLE, 9-VINYL- see VNK100
CARBINAMINE see MGC250
CARBITAL 90 see CAT775
CARBIUM see CAT775
CARBIUM MM see CAT775
CARBOHYDRAZIDE, 1,5-DIPHENYL- see DVY925
CARBONA see CBY000
CARBON BICHLORIDE see PCF275
CARBON CHLORIDE see CBY000
CARBON DICHLORIDE see PCF275
CARBONE (OXYDE de) (FRENCH) see CBW750
CARBON HEXACHLORIDE see HCI000
CARBONIC ACID, CADMIUM SALT see CAD800
CARBONIC ACID, CALCIUM SALT (1:1) see CAT775
CARBONIC ACID, COBALT(2+) SALT (1:1) see CNB475
CARBONIC ACID, DIRUBIDIUM SALT see RPB200
CARBONIC ACID, DITHIO-, O-ETHYL ESTER see EQG600
CARBONIC ACID, DITHIO-, O-ISOPROPYL ESTER, POTASSIUM SALT see IRG050
CARBONIC ACID, compd. with GUANIDINE (1:2) see GKW100
CARBONIC ACID, TRITHIO-, CYCLIC ETHYLENE ESTER see EJQ100
CARBONIC DIHYDRAZIDE, 2,2′-DIPHENYL-(9CI) see DVY925
CARBONIC OXIDE see CBW750
CARBONIO (OSSIDO di) (ITALIAN) see CBW750
CARBON MONOXIDE see CBW750
CARBON MONOXIDE, CRYOGENIC liquid (DOT) see CBW750
CARBONODITHIOIC ACID, O-ETHYL ESTER see EQG600
CARBONODITHIOIC ACID, O-(1-METHYLETHYL) ESTER, POTASSIUM SALT (9CI) see IRG050
CARBON OXIDE (CO) see CBW750
CARBON TET see CBY000
CARBON TETRACHLORIDE see CBY000
CARBONYL CYANIDE, 3-CHLOROPHENYLHYDRAZONE see CKA550
CARBOREX 2 see CAT775
CARBOSTYRIL, THIO- see QOJ100
CARBOTHRONE see APM500
2-CARBOXYACETANILIDE see AAJ150
9-CARBOXYANTHRACENE see APG550
o-CARBOXYBENZALDEHYDE see FNK010
2-CARBOXYBENZALDEHYDE see FNK010
p-CARBOXYBENZENESULFONAMIDE see SNL840
4-CARBOXYBENZENESULFONAMIDE see SNL840
p-CARBOXYBROMOBENZENE see BMU100
p-CARBOXYCHLOROBENZENE see CEL300
CARBOXYCYCLOHEXANE see HDH100
N-(2-CARBOXYETHYL)-N-NITRO-β-ALANINE see NHI600
(+ −)-(3-CARBOXY-2-HYDROXYPROPYL)TRIMETHYLAMMONIUM CHLORIDE see CCK655
4-CARBOXYMETHYLANILINE see AID700
(CARBOXYMETHYL)TRIMETHYLAMMONIUM CHLORIDE see CCH850
(CARBOXYMETHYL)TRIMETHYLAMMONIUM HYDROXIDE, inner salt see GHA050
1-CARBOXYNAPHTHALENE see NAV490
4-CARBOXYPHTHALIC ANHYDRIDE see TKV000
5-CARBOXYRESORCINOL see REF200
trans-β-CARBOXYSTYRENE see CMP980
2-CARBOXYTHIOPHENE see TFM600
neo-CARDIAMINE see GCE600
CARDIOVITAL see GCE600
CARNELIO PALE LITHOL RED see NAP100
CARNELIO RUBINE LAKE see SEH475
(+ −)-CARNITINE CHLORIDE see CCK655

dl-CARNITINE CHLORIDE see CCK655
(+ −)-CARNITINE HYDROCHLORIDE see CCK655
d,l-CARNITINE HYDROCHLORIDE see CCK655
dl-CARNITINE HYDROCHLORIDE see CCK655
CARPHENOL see PFR325
CARSTAB 700 see HND100
CARUSIS P see CAT775
CARZENID see SNL840
CARZENIDE see SNL840
CATALOID see SCH000
CATECHIN see CCP850
CATECHOL see CCP850
CATHINE see NNM510
CBSP see HAQ600
CCC G-WHITE see CAT775
CCC No.AA OOLITIC see CAT775
CCCP see CKA550
CCP see CKA550
CCR see CAT775
CCW see CAT775
CdPT see CAI350
CDTA see CPB120
CECOLENE see TIO750
CELLITON BLUE RN see IBV050
CELLITON FAST PINK RF see AKO350
CELLITON FAST PINK RFA-CF see AKO350
CELLITON FAST YELLOW RR see DUW500
CELLU-BRITE see DXB450
CELOGEN BSH see BBS300
CEPHROL see DTF410
CERES GREEN 3B see PJQ100
CERN KYPOVA 27 see DUP100
CERTIQUAL LITHOL RED see NAP100
CERVEN DISPERZNI 4 see AKO350
CERVEN PRIMA 2 see DXO850
CERVEN ZASADITA 2 see GJI400
CETANE see HCO600
n-CETANE see HCO600
CETAPHARM see HCP800
CETASOL see HCP800
CETAZOL see HCP800
CETIN see HCP700
CETYL ACETATE see HCP100
CETYLDIMETHYLAMINE see HCP525
CETYL PALMITATE see HCP700
CETYLPYRIDINIUM BROMIDE see HCP800
1-CETYLPYRIDINIUM BROMIDE see HCP800
N-CETYLPYRIDINIUM BROMIDE see HCP800
CETYL SODIUM SULFATE see HCP900
CETYL SULFATE SODIUM SALT see HCP900
CGTA see CPB120
CHALCONE, 4-NITRO- see NFS505
CHEL 138 see EIV100
CHEL 600 see CPB120
CHEMCARB see CAD800, CAT775
CHEM PELS C see SEY500
CHEM-SEN 56 see SEY500
CHIMASSORB 81 see HND100
CHINIDIN DURILES see QFS100
CHINIDIN VUFB see QFS100
ChKhZ 9 see BBS300
CHLOFUCID see DEL300
CHLORAL see CDN550
CHLORAMINE SKY BLUE 4B see CMO500
CHLORAMINE SKY BLUE A see CMO500
p-CHLORBENZOIC ACID see CEL300
4-CHLOR-BENZYL-CYANID see CEP300
CHLORENDIC ANHYDRIDE see CDS050
2-CHLORETHYLACETAT see CGO600
CHLORHYDROXYALUMINUM ALLANTOINATE see AHA135
CHLORINATED HYDROCHLORIC ETHER see DFF809
m-CHLOROACETANILIDE see CDZ050
3′-CHLOROACETANILIDE see CDZ050
o-CHLOROACETOACETANILIDE see AAY600
2′-CHLOROACETOACETANILIDE see AAY600
2-CHLOROACRYLIC ACID ETHYL ESTER see EHH200
4-CHLORO-3-AMINOBENZOTRIFLUORIDE see CEG800
3-CHLORO-2-AMINOTOLUENE see CLK227
2-CHLORO-9,10-ANTHRACENEDIONE see CEI100
2-CHLOROANTHRAQUINONE see CEI100
CHLOROBEN see DEP600
p-CHLOROBENZALDEHYDE see CEI600
4-CHLOROBENZALDEHYDE see CEI600
4-CHLOROBENZENEACETONITRILE see CEP300
p-CHLOROBENZENECARBOXALDEHYDE see CEI600
p-CHLOROBENZENESULFINIC ACID see CEJ600
m-CHLOROBENZOIC ACID see CEL290
p-CHLOROBENZOIC ACID see CEL300
3-CHLOROBENZOIC ACID see CEL290
4-CHLOROBENZOIC ACID see CEL300
2-CHLOROBENZOXAZOLE see CEM850
p-CHLOROBENZYL CYANIDE see CEP300
4-CHLOROBENZYL CYANIDE see CEP300
α-CHLOROBENZYL PHENYL KETONE see CFJ100
p-CHLOROBENZYLPSEUDOTHIURONIUM CHLORIDE see CEQ800
2-(4-CHLOROBENZYL)-2-THIOPSEUDOUREA HYDROCHLORIDE see CEQ800
4-CHLOROBUTANENITRILE see CEU300
γ-CHLOROBUTYRONITRILE see CEU300
4-CHLOROBUTYRONITRILE see CEU300
1-CHLORO-4-CHLOROMETHYLBENZENE see CFB600
6-CHLORO-m-CRESOL see CFE500
α-CHLOROCYCLOHEPTANONE see CFF600
2-CHLOROCYCLOHEPTANONE see CFF600
CHLOROCYCLOHEXANE see CPI400
CHLORODEN see DEP600
α-CHLORODEOXYBENZOIN see CFJ100
1-CHLORO-2,2-DICHLOROETHYLENE see TIO750
1-CHLORO-2,4-DIMETHOXY-5-NITROBENZENE see CGC100
2-(p-CHLORO-α-(2-(DIMETHYLAMINO)ETHYL)BENZYL)PYRIDINE see CLX300
4-CHLORO-3,5-DINITROBENZOTRIFLUORIDE see CGM225
4-CHLORO-3,5-DINITRO-α,α,α-TRIFLUOROTOLUENE see CGM225
2-CHLORO-1,2-DIPHENYLETHANONE see CFJ100
CHLORODRACYLIC ACID see CEL300
2-CHLOROETHANOL ACETATE see CGO600
β-CHLOROETHYL ACETATE see CGO600
2-CHLOROETHYL ACETATE see CGO600
N-(2-CHLOROETHYL)-N-ETHYLANILINE see EHJ600
N-(2-CHLOROETHYL)-N-ETHYLBENZENAMINE see EHJ600
3-CHLORO-4-FLUORONITROBENZENE see CHI950
2-CHLORO-1-FLUORO-4-NITROBENZENE see CHI950
5-CHLORO-8-HYDROXYQUINOLINE see CLD600
5-CHLORO-2-MERCAPTOANILINE HYDROCHLORIDE see CHU100
6-CHLORO-2-METHYLANILINE see CLK227
5-CHLORO-2-METHYLBENZOTHIAZOLE see CIH100
4-CHLORO-5-METHYL-2-(1-METHYLETHYL)PHENOL see CLJ800
1-(CHLOROMETHYL)-3-NITROBENZENE see NFN010
2-CHLORO-5-METHYLPHENOL see CFE500
p-CHLORO-o-NITROANILINE see KDA050
4-CHLORO-2-NITROANILINE see KDA050
4-CHLORO-2-NITROBENZENAMINE see KDA050
α-CHLORO-m-NITROTOLUENE see NFN010
2-CHLORO-6-NITROTOLUENE see CJG825
6-CHLORO-2-NITROTOLUENE see CJG825
CHLOROPENTAFLUOROBENZENE see PBE100
CHLOROPERFLUOROBENZENE see PBE100
4-CHLOROPHENIRAMINE see CLX300
2-CHLOROPHENOTHIAZINE see CJL100
N-(2-CHLOROPHENYL)ACETOACETAMIDE see AAY600
p-CHLOROPHENYLACETONITRILE see CEP300
(4-CHLOROPHENYL)ACETONITRILE see CEP300
2-(4-CHLOROPHENYL)ACETONITRILE see CEP300
2-CHLORO-2-PHENYLACETOPHENONE see CFJ100
2-(2-(4-(p-CHLORO-α-PHENYLBENZYL)-1-PIPERAZINYL)ETHOXY)ETHANOLDIHYDROCH LORIDE see HOR470
p-CHLOROPHENYL CHLORIDE see DEP800
((3-CHLOROPHENYL)HYDRAZONO)PROPANEDINITRILE see CKA550
CHLOROPHENYLPYRIDAMINE see CLX300
1-(p-CHLOROPHENYL)-1-(2-PYRIDYL)-3-DIMETHYLAMINOPROPANE see CLX300
1-(p-CHLOROPHENYL)-1-(2-PYRIDYL)-3-N,N-DIMETHYLPROPYLAMINE see CLX300
CHLOROPIRIL see CLX300
3-CHLOROPROPANOL see CKP725
CHLOROPROPHENPYRIDAMINE see CLX300
5-CHLORO-8-QUINOLINOL see CLD600
5-CHLOROSALICYLALDEHYDE see CLD800
5-CHLOROSALICYLIC ACID see CLD825
CHLOROTETRAHYDROXY((2-HYDROXY-5-OXO-2-IMIDAZOLIN-4-YL)UREATO)DIALUMINUM see AHA135
CHLOROTHYMOL see CLJ800
6-CHLOROTHYMOL see CLJ800
6-CHLORO-o-TOLUIDINE see CLK227
6-CHLORO-2-TOLUIDINE see CLK227
2-CHLORO-5-(TRIFLUOROMETHYL)ANILINE see CEG800
6-CHLORO-α,α,α-TRIFLUORO-m-TOLUIDINE see CEG800

2-CHLORO-1,3,5-TRINITROBENZENE see PIE530
CHLOROTRIPHENYLSILANE see TMU800
CHLOROXINE see DEL300
CHLOROXYQUINOLINE see CLD600, DEL300
CHLORPHENAMINE see CLX300
CHLORPHENIRAMINE see CLX300
3-CHLORPROPAN-1-OL see CKP725
CHLORPROPHENPYRIDAMINE see CLX300
CHLORPYRIFOS see CMA100
CHLORQUINOL see DEL300
CHLORTHYMOL see CLJ800
CHLOR-TRIMETON see CLX300
CHLOR-TRIPOLON see CLX300
CHLORURE d'ARSENIC (FRENCH) see ARF500
CHLORURE ARSENIEUX (FRENCH) see ARF500
CHLORURE d'ETHYLIDENE (FRENCH) see DFF809
CHLORYLEA see TIO750
CHOLINE ACETATE (ESTER), BROMIDE see CMF260
CHOLINE, ACETYL-, BROMIDE see CMF260
CHOLINE, S-ACETYLTHIO-, IODIDE see ADC300
CHOLINE HYDROXIDE see CMF800
CHORYLEN see TIO750
CHQ see DEL300
CHROMAL YELLOW M see SIT850
CHROMATEX GREEN G see PJQ100
CHROME LEATHER PURE BLUE see CMO500
CHROME LEATHER RED 4B see DXO850
CHROMELIN see OMC300
CHROME RED ALIZARINE see SEH475
CHROME YELLOW 2G see SIT850
CHROME YELLOW 2GR see SIT850
CHROMIC ACID, STRONTIUM SALT (1:1) see SMH000
CHROMOCOR see PER700
CHROMOL YELLOW G see SIT850
CHROMOL YELLOW N see SIT850
CHRYSOFLUORENE see NAU600
C.I. 1106 see IBV050
C.I. 10345 see DUW500
C.I. 11350 see PEJ600
C.I. 14025 see SIT850
C.I. 15630 see NAP100
C.I. 23500 see DXO850
C.I. 24400 see CMO500
C.I. 35811 see AKD925
C.I. 37040 see KDA050
C.I. 37050 see CEG800
C.I. 37100 see MMF780
C.I. 37500 see NAX000
C.I. 40621 see DXB450
C.I. 50240 see GJI400
C.I. 52005 see DUG700
C.I. 58005 see SEH475
C.I. 60010 see DFN450
C.I. 60755 see AKO350
C.I. 63000 see APG700
C.I. 69005 see DUP100
C.I. 69800 see IBV050
C.I. 69825 see DFN425
C.I. 73000 see BGB275
C.I. 73300 see DNT300
C.I. 74260 see PJQ100
C.I. 75480 see HMX600
C.I. 76500 see CCP850
C.I. 76625 see NAN505
C.I. 77086 see ARI000
C.I. 77160 see BKW600
C.I. 77169 see BKW100
C.I. 77180 see CAD000
C.I. 77185 see CAD250
C.I. 77199 see CAJ750
C.I. 77205 see CMS205
C.I. 77320 see CNA250
C.I. 77353 see CNB475
C.I. ACID BLUE 43 see APG700
C.I. AZOIC COUPLING COMPONENT 1 see NAX000
C.I. AZOIC DIAZO COMPONENT 9 see KDA050
CIBANONE BLUE FG see DFN425
CIBANONE BLUE FGF see DFN425
CIBANONE BLUE FGL see DFN425
CIBANONE BLUE FRS see IBV050
CIBANONE BLUE FRSN see IBV050
CIBANONE BLUE GF see DFN425

CIBANONE BLUE RS see IBV050
CIBANONE BRILLIANT BLUE FR see IBV050
CIBANONE OLIVE 2R see DUP100
CIBANONE OLIVE F2R see DUP100
CIBANONE VIOLET F 4R see DFN450
CIBANONE VIOLET F 2RB see DFN450
CIBANONE VIOLET 2R see DFN450
CIBANONE VIOLET 4R see DFN450
CIBA PINK B see DNT300
C.I. BASIC RED 2 see GJI400
C.I. DEVELOPER 5 see NAX000
C.I. DEVELOPER 8 see HMX520
C.I. DEVELOPER 20 (obs.) see HMX520
C.I. DIRECT BLUE 15 see CMO500
C.I. DIRECT BLUE 15, TETRASODIUM SALT see CMO500
C.I. DIRECT RED 2 see DXO850
C.I. DISPERSE BLUE 78 see BKP500
C.I. DISPERSE RED 4 see AKO350
C.I. DISPERSE YELLOW 1 see DUW500
C.I. FLUORESCENT BRIGHTENER 9 see DXB450
C.I. FOOD YELLOW 6 see AIB350
CILLA FAST PINK RF see AKO350
CILLA FAST YELLOW RR see DUW500
C.I. MORDANT RED 3 see SEH475
C.I. MORDANT YELLOW 1 see SIT850
C.I. MORDANT YELLOW 1, MONOSODIUM SALT see SIT850
C.I. NATURAL ORANGE 6 see HMX600
CINCHONAN-9-OL, (8-α-9R)-(9CI) see CMP910
CINCHONAN-9-OL, 6′-METHOXY-, (9S)-, SULFATE (1:1) (SALT) (9CI) see QFS100
CINCHONIDINE see CMP910
(−)-CINCHONIDINE see CMP910
(8S,9R)-CINCHONIDINE see CMP910
CINCHOVATINE see CMP910
1,4-CINEOL see IKC100
1,4-CINEOLE see IKC100
CINNAMAMIDE see CMP970
CINNAMIC ACID, (E)- see CMP980
(E)-CINNAMIC ACID see CMP980
trans-CINNAMIC ACID see CMP980
CINNAMIC ACID, CINNAMYL ESTER see CMQ850
CINNAMIC ACID, o-HYDROXY-, (E)- see CNU850
CINNAMIC ACID, p-NITRO- see NFT440
CINNAMIC ACID, 3,4,5-TRIMETHOXY- see CMQ100
CINNAMYL ALCOHOL, CINNAMATE see CMQ850
CINNAMYL CINNAMATE see CMQ850
CINNAMYLESTER KYSELINY SKORICOVE see CMQ850
C.I. OXIDATION BASE 26 see CCP850
C.I. PIGMENT BLUE 60 see IBV050
C.I. PIGMENT GREEN 7 see PJQ100
C.I. PIGMENT GREEN 42 see PJQ100
C.I. PIGMENT ORANGE 20 see CAJ750
C.I. PIGMENT RED 49 see NAP100
C.I. PIGMENT RED 49, MONOSODIUM SALT see NAP100
C.I. PIGMENT VIOLET 31 see DFN450
C.I. PIGMENT WHITE 17 see BKW100
C.I. PIGMENT WHITE 18 see CAT775
C.I. PIGMENT YELLOW 32 see SMH000
C.I. PIGMENT YELLOW 35 see CMS205
C.I. PIGMENT YELLOW 37 see CAJ750
CIRCOSOLV see TIO750
C.I. SOLVENT BLUE 78 see BKP500
C.I. SOLVENT BLUE 93 see BKP500
C.I. SOLVENT YELLOW 4 see PEJ600
C.I. SOLVENT YELLOW 52 see DUW500
(E)-CITRAL see GCU100
CITRAL α see GCU100
α-CITRAL see GCU100
trans-CITRAL see GCU100
CITRIC ACID, AMMONIUM IRON(3+) SALT see FAS700
CITRIC ACID, TRILITHIUM SALT, HYDRATE see LHD150
CITRONELLOL see DTF410
CITRYLIDENEACETONE see POH525
C.I. VAT BLACK 27 see DUP100
C.I. VAT BLUE 1 see BGB275
C.I. VAT BLUE 4 see IBV050
C.I. VAT BLUE 6 see DFN425
C.I. VAT RED 41 see DNT300
C.I. VAT VIOLET 1 (8CI) see DFN450
CLARO 5591 see SLJ500
CLEARJEL see SLJ500
CLEARJREL see SLJ500
CLEFNON see CAT775

CLOFUZID see DEL300
CLORALIO see CDN550
CLORFENIRAMINA see CLX300
CLOROBEN see DEP600
CLOROPIRIL see CLX300
CLORURO di ETILIDENE (ITALIAN) see DFF809
CMP see CQL300
5′-CMP see CQL300
CMP (nucleotide) see CQL300
COBALT see CNA250
COBALT-59 see CNA250
COBALT CARBONATE see CNB475
COBALT CARBONATE (1:1) see CNB475
COBALT(2+) CARBONATE see CNB475
COBALT MONOCARBONATE see CNB475
COBALTOUS CARBONATE see CNB475
CODEHYDRASE I see CNF390
CODEHYDRASE II see CNF400
CODEHYDROGENASE I see CNF390
CODEHYDROGENASE II see CNF400
CODELSOL see PLY275
COENZYME I see CNF390
COENZYME II see CNF400
COLAMINE HYDROCHLORIDE see EEC700
COLANYL GREEN GG see PJQ100
s-COLLIDINE see TME272
γ-COLLIDINE see TME272
sym-COLLIDINE see TME272
α-γ,α′-COLLIDINE see TME272
2,4,6-COLLIDINE see TME272
COLLOIDAL ARSENIC see ARA750
COLLOIDAL CADMIUM see CAD000
COLLOIDAL MANGANESE see MAP750
COLLOIDAL SILICA see SCH000
COLLOIDAL SILICON DIOXIDE see SCH000
COMPLEXON IV see CPB120
COMPOUND 13-61 see OOI200
COMPOUND 19-28 see DXB450
COMPOUND No. 1080 see SHG500
CONCO SULFATE C see HCP900
CONTALAX see PPN100
CONTINENTAL see KBB600
COPPER ARSENITE, solid (DOT) see CNN500
COPPER ORTHOARSENITE see CNN500
COPPER PHTHALOCYANINE GREEN see PJQ100
CORETONIN see GCE600
CORN PRODUCTS see SLJ500
COSBIOL see SLG700
COSMETIC WHITE see BKW100
COSPANON see TKP100
COTTON RED 4B see DXO850
trans-o-COUMARIC ACID see CNU850
COUMARILIC ACID see CNU875
COUMARIN, 6,7-DIHYDROXY-4-METHYL- see MJV800
COUMARIN, 7-HYDROXY- see HJY100
COUMARIN, 7-METHOXY-(8CI) see MEK300
COZYMASE see CNF390
COZYMASE I see CNF390
COZYMASE II see CNF400
CPB see BSS550
CPC 3005 see SLJ500
CPC 6448 see SLJ500
CRAWHASPOL see TIO750
p-CRESOL, 2,6-BIS(α-METHYLBENZYL)- see MHR050
p-CRESOL, 2,6-DI-tert-BUTYL-α-(DIMETHYLAMINO)- see FAB000
p-CRESOL GLYCIDYL ETHER see GGY175
o-CRESOL, 4-NITRO- see NFV010
p-CRESOL, 2-NITRO-α,α,α-TRIFLUORO- see NMQ100
p-CRESOL, α-PHENYL-, CARBAMATE see PFR325
o-CRESOLPHTHALEIN see CNX400
CRESOTINE PURE BLUE see CMO500
p-CRESYL GLYCIDYL ETHER see GGY175
CRISALIN see DUV600
CROMARIL see PER700
CROMOPHTAL BLUE A 3R see IBV050
CROMOPHTAL GREEN GF see PJQ100
CROTONIC ACID, 4-BROMO-, METHYL ESTER see MHR790
CRYSTIC PREFIL S see CAT775
CRYSTOSOL see MQV750
CTP see CQL400
5′-CTP see CQL400
m-CUMENOL see IQX090

p-CUMYLPHENOL see COF400
p-(α-CUMYL)PHENOL see COF400
CUPRIC ARSENITE see CNN500
CUPRIC GREEN see CNN500
CURALIN M see MJM200
CURENE 442 see MJM200
CUROSAJIN see HGI100
CV 399 see MGH800
CYANASET see MJM200
CYAN GREEN 15-3100 see PJQ100
CYANINE GREEN GP see PJQ100
CYANINE GREEN NB see PJQ100
CYANINE GREEN T see PJQ100
CYANINE GREEN TONER see PJQ100
1-CYANOBUTANE see VAV300
N-(CYANOETHYL)ANILINE see AOT100
N-(β-CYANOETHYL)ANILINE see AOT100
N-(2-CYANOETHYL)ANILINE see AOT100
p-CYANOFLUOROBENZENE see FGH100
α-CYANONAPHTHALENE see NAY090
1-CYANONAPHTHALENE see NAY090
4-CYANONITROBENZENE see NFI010
p-CYANONITROBENZENE see NFI010
1-CYANOOCTANE see OES300
(3-CYANOPROPYL) TRICHLOROSILANE see COR550
3-CYANOPYRIDINE see NDW515
CYASORB UV 12 see BKH200
CYASORB UV 24 see DMW250
CYASORB UV 531 see HND100
CYASORB UV 24 LIGHT ABSORBER see DMW250
CYCLIC ETHYLENE TRITHIOCARBONATE see EJQ100
CYCLOBUTANE, 1,2-DICHLOROHEXAFLUORO- see DFM025
CYCLOBUTANE, 1,2-DICHLORO-1,2,3,3,4,4-HEXAFLUORO-(9CI) see DFM025
CYCLODODECALACTAM see COW930
CYCLOHEPTANONE, 2-CHLORO- see CFF600
2,5-CYCLOHEXADIENE-1,4-DIONE, 2-(BIS(2-HYDROXYETHYL)AMINO)- see BKB100
2,5-CYCLOHEXADIEN-1-ONE, 2,6-DICHLORO-4-((p-HYDROXYPHENYL)IMINO)-, SODIUM SALT see SGG650
CYCLOHEXANECARBINOL see HDH200
CYCLOHEXANECARBOXYLIC ACID see HDH100
CYCLOHEXANE, CHLORO- see CPI400
1,2-CYCLOHEXANEDIAMINETETRAACETIC ACID see CPB120
1,2-CYCLOHEXANEDIAMINE-N,N,N′,N′-TETRAACETIC ACID see CPB120
Δ$^{1-α,4-α'}$-CYCLOHEXANEDIMALONONITRILE see BIY600
1,2-CYCLOHEXANEDIONE see CPB200
CYCLOHEXANEMETHANOL see HDH200
CYCLOHEXANEMETHANOL, α,α-4-TRIMETHYL-(9CI) see MCE100
CYCLOHEXANOIC ACID see HDH100
CYCLOHEXANOL, 1,1′-DIOXYBIS-(9CI) see PCM100
CYCLOHEXANOL, 1,1′-DIOXYDI- see PCM100
CYCLOHEXANOL, 1-((1-HYDROPEROXYCYCLOHEXYL)DIOXY)- see CPC300
CYCLOHEXANOL, 1-METHYL-4-(1-METHYLETHENYL)-(9CI) see TBD775
CYCLOHEXANOL, 3,3,5-TRIMETHYL-, cis- see TLO510
CYCLOHEXANONE ISO-OXIME see CBF700
CYCLOHEXANONE, 5-METHYL-2-(1-METHYLETHYLIDENE)-, (R)- see POI615
CYCLOHEXANONE PEROXIDE see CPC300
4-CYCLOHEXENE-1,2-DICARBOXIMIDE see TDB100
CYCLOHEXENE, 1-VINYL- see VNZ990
CYCLOHEXENYLETHYLENE see CPD750
1-CYCLOHEXYLAMINO-2-PROPANOL see CPG700
CYCLOHEXYLCARBINOL see HDH200
CYCLOHEXYLCARBOXYLIC ACID see HDH100
CYCLOHEXYL CHLORIDE see CPI400
2-CYCLOHEXYLCYCLOHEXANONE see LBX100
1,2-CYCLOHEXYLENEDIAMINETETRAACETIC ACID see CPB120
(1,2-CYCLOHEXYLENEDINITRILO)TETRAACETIC ACID see CPB120
CYCLOHEXYLMETHANOL see HDH200
CYCLOPENTACYCLOHEPTENE see ASP600
CYCLOPENTANE, 1,2-EPOXY- see EBO100
CYCLOPENTANE OXIDE see EBO100
CYCLOPENTASILOXANE, DECAMETHYL- see DAF350
CYCLOPENTENE, 1,2-DICHLOROHEXAFLUORO- see DFM050
CYCLOPENTENE, 1,2-DICHLORO-3,3,4,4,5,5-HEXAFLUORO-(9CI) see DFM050
CYCLOPENTENE EPOXIDE see EBO100
CYCLOPENTENEOXIDE see EBO100
2-CYCLOPENTEN-1-ONE, 3-METHYL-2-(2-PENTENYL)-, (Z)- see JCA100
N-(1-CYCLOPENTEN-1-YL)-MORPHOLINE see CPY800
CYCLOPROPANECARBOXYLIC ACID, 2,2-DIMETHYL-3-(2-METHYL-1-PROPENYL)-, ETHYLESTER see EHM100
CYCLOPROPANECARBOXYLIC ACID, 2,2-DIMETHYL-3-(METHYLPROPENYL)-, ETHYL ESTER (8CI) see EHM100

CYCLOPROPANECARBOXYLIC ACID, 2,2-DIMETHYL-3-(2-METHYL-1-PROPENYL)-ETHYLESTER (9CI) see EHM100
CYCLORYL OS see OFU200
CYCLOTETRASILOXANE, OCTAMETHYL- see OCE100
CYDTA see CPB120
3-CYJANOPIRYDYNA see NDW515
CYSTOCEVA see BGB275
CYTIDINE 5'-(DIHYDROGENPHOSPHATE) see CQL300
CYTIDINE MONOPHOSPHATE see CQL300
CYTIDINE 5'-MONOPHOSPHATE see CQL300
CYTIDINE 5'-MONOPHOSPHORIC ACID see CQL300
CYTIDINE 5'-PHOSPHATE see CQL300
CYTIDINE 5'-PHOSPHORIC ACID see CQL300
CYTIDINE, 5'-(TETRAHYDROGEN TRIPHOSPHATE) see CQL400
CYTIDINE-5'-TRIPHOSPHATE see CQL400
CYTIDINE 5'-TRIPHOSPHORIC ACID see CQL400
CYTIDYLIC ACID see CQL300
5'-CYTIDYLIC ACID see CQL300
CYTOSINE (8CI) see CQM600
CYTOSINIMINE see CQM600
CZTEROCHLOREK WEGLA (POLISH) see CBY000
CZTEROCHLOROETYLEN (POLISH) see PCF275

DA-688 see GDG200
DACOTE see CAT775
DAINICHI CYANINE GREEN FG see PJQ100
DAINICHI CYANINE GREEN FGH see PJQ100
DAIOMIN see TES800
DAISAZIN see TES800
DAITO ORANGE SALT RD see CEG800
DAITO RED BASE 3GL see KDA050
DAITO RED BASE RL see MMF780
DAKTOSE B see DIB300
DALTOGEN see TKP500
DALTOLITE FAST GREEN GN see PJQ100
DAM see OMY910
DASD see FCA100
DAVISON SG-67 see SCH000
DBED see BHB300
DBED DIACETATE see DDF800
DBQ see DDV500
DCB see DEP600, DEV000
1,4-DCB see DEV000
D and C BLUE NO. 6 see BGB275
D and C BLUE No. 9 see DFN425
D&C BLUE NO. 6 see BGB275
D and C RED No. 10 see NAP100
DCTA see CPB120
D,4-DINITRO-1-NAPHTHOL-7-SULFONIC ACID see FBZ100
DECAMETHYLCYCLOPENTASILOXANE see DAF350
DECANE, 1-BROMO- see BNB800
1,10-DECANEDIAMINE see DAG650
DECANE, HENEICOSAFLUORO-1-IODO- see IEU075
DECANE, 1-IODO-1,1,2,2,3,3,4,4,5,5,6,6,7,7,8,8,9,9,10,10,10-HENEICOSAFLUORO- see IEU075
DECANOIC ACID, METHYL ESTER see MHY650
DECANOIC ACID, SODIUM SALT see SGB600
2-DECANONE see OFE050
DECANOX see DAH475
DECANOYL PEROXIDE see DAH475
DECANOYL PEROXIDE, technically pure (DOT) see DAH475
DECYL BROMIDE see BNB800
n-DECYL BROMIDE see BNB800
1-DECYL BROMIDE see BNB800
DECYLTRIMETHYLAMMONIUM BROMIDE see DAJ500
DEEP LEMON YELLOW see SMH000
DEFICOL see PPN100
DEHYDRO-p-TOLUIDINE see MHJ300
DEHYQUART C see LBX050
DEIQUAT see DWX800
DEKAMETHYLCYKLOPENTASILOXAN see DAF350
DENSINFLUAT see TIO750
DEOXYRIBOSE URACIL see DAZ050
DEOXYURIDINE see DAZ050
2'-DEOXYURIDINE see DAZ050
DEPOCILLIN see PAQ200
DESICCANT L-10 see ARB250
2'-DESOXYURIDINE see DAZ050
DESYL CHLORIDE see CFJ100
DEVELOPER A see NAX000
DEVELOPER AMS see NAX000
DEVELOPER BN see NAX000
DEVELOPER BON see HMX520
DEVELOPER SODIUM see NAX000
DEVOL RED F see KDA050
DEVOL RED SALT E see MMF780
DEXTRONE see DWX800
DHPT see MHJ300
DIABASE RED RL see MMF780
2-(4,4'-DIACETOXYPHENYLMETHYL)PYRIDINE see PPN100
(4,4'-DIACETOXYDIPHENYL)(2-PYRIDYL)METHANE see PPN100
4,4'-DIACETOXYDIPHENYLPYRID-2-YLMETHANE see PPN100
DI-(p-ACETOXYPHENYL)-2-PYRIDYLMETHANE see PPN100
DI-(4-ACETOXYPHENYL)-2-PYRIDYLMETHANE see PPN100
DIACETYLMONOOXIME see OMY910
DIACETYLMONOXIME see OMY910
DIACOTTON BENZOPURPURINE 4B see DXO850
DIACOTTON SKY BLUE 5B see CMO500
DIAETHYLAMIN (GERMAN) see DHJ200
O,O-DIAETHYL-O-3,5,6-TRICHLOR-2-PYRIDYLMONOTHIOPHOSPHAT (GERMAN) see CMA100
DIALLYL DISULFIDE see AGF300
DIALLYL DISULPHIDE see AGF300
DIAMINE PURPURINE 4B see DXO850
DIAMINE SKY BLUE CI see CMO500
2,4-DIAMINOANISOLE DIHYDROCHLORIDE see DBO100
p-DIAMINOAZOBENZENE see ASK925
4,4'-DIAMINOAZOBENZENE see ASK925
2,5-DIAMINOBENZENE SULFONIC ACID see PEY800
3,5-DIAMINOBENZOIC ACID see DBQ190
3,5-DIAMINOBENZOIC ACID DIHYDROCHLORIDE see DBQ200
3,5-DIAMINOBENZOTRIFLUORIDE see TKB775
DI(4-AMINO-3-CHLOROPHENYL)METHANE see MJM200
4,5-DIAMINOCHRYSAZIN see DBQ250
1,8-DIAMINOCHRYSAZINE see DBQ250
DI-(4-AMINO-3-CLOROFENIL)METANO (ITALIAN) see MJM200
1,2-DIAMINOCYCLOHEXANETETRAACETIC ACID see CPB120
1,2-DIAMINOCYCLOHEXANE-N,N'-TETRAACETIC ACID see CPB120
4,4'-DIAMINO-3,3'-DICHLORODIPHENYLMETHANE see MJM200
1,4-DIAMINO-2,3-DIHYDROANTHRAQUINONE see DBR450
1,8-DIAMINO-4,5-DIHYDROXYANTHRACHINON see DBQ250
1,8-DIAMINO-4,5-DIHYDROXYANTHRAQUINONE see DBQ250
4,5-DIAMINO-1,8-DIHYDROXYANTHRAQUINONE see DBQ250
1,8-DIAMINO-4,5-DIHYDROXY-9,10-ANTHRAQUINONE see DBQ250
1,6-DIAMINOHEXANE see HEO000
4,4'-DIAMINO-2,2'-STILBENEDISULFONIC ACID see FCA100
2,3-DIAMINOTOLUENE see TGY800
DIAMMONIUM ARSENATE see DCG800
DIAMMONIUM HYDROGEN ARSENATE see DCG800
DIAMMONIUM MONOHYDROGEN ARSENATE see DCG800
DIAMOND RED W see SEH475
DI-p-ANISYLAMINE see BKO600
DIANIX FAST PINK R see AKO350
DIAPHTAMINE PURE BLUE see CMO500
DIAPHTAMINE PURPURINE see DXO850
DIARESIN BLUE K see BKP500
DIARSENIC PENTOXIDE see ARH500
DIARSENIC TRISULFIDE see ARI000
1,2-DIAZABENZENE see OJW200
1,4-DIAZABENZENE see POL590
DIAZAMINE PURPURINE 4B see DXO850
DIAZENECARBOXYLIC ACID, PHENYL-, 2-PHENYLHYDRAZIDE (9CI) see DVY950
1H-1,4-DIAZEPINE, HEXAHYDRO-1-(p-CHLORO-α-PHENYLBENZYL)-4-METHYL- see HGI100
p-DIAZINE see POL590
1,2-DIAZINE see OJW200
1,4-DIAZINE see POL590
DIAZINE RED 4B see DXO850
DIAZO FAST ORANGE RD see CEG800
DIAZO FAST RED 3GL see KDA050
DIAZO FAST RED RL see MMF780
DIAZOLE RED 4S see NMO700
DIAZOL PURE BLUE 4B see CMO500
DIAZOL PURPURINE 4B see DXO850
DIAZOL RED S see NMO700
DIBASIC AMMONIUM ARSENATE see DCG800
o,o'-DIBENZAMIDODIPHENYL DISULFIDE see BDK800
DI-o-BENZAMIDOPHENYL DISULPHIDE see BDK800
DIBENZOFURAN see DDB500
DIBENZO(b,d)FURAN see DDB500
DIBENZOTHIOPHENE see TES300
DIBENZO(b,d)THIOPHENE see TES300
2,2'-DIBENZOYLAMINODIPHENYL DISULFIDE see BDK800

1,2-DIBENZOYLHYDRAZINE see DDE250
N,N′-DIBENZYLDITHIOOXAMIDE see DDF700
N,N′-DIBENZYLETHYLENEDIAMINE see BHB300
N,N′-DIBENZYLETHYLENEDIAMINE DIACETATE see DDF800
DIBISMUTH TRIOXIDE see BKW600
DIBROMANTIN see DDI900
DIBROMANTINE see DDI900
m-DIBROMOBENZENE see DDK050
p-DIBROMOBENZENE see BNV775
1,3-DIBROMOBENZENE see DDK050
1,4-DIBROMOBENZENE see BNV775
N,N′-DIBROMODIMETHYLHYDANTOIN see DDI900
1,3-DIBROMO-5,5-DIMETHYL-2,4-IMIDAZOLIDINEDIONE see DDI900
α-β-DIBROMOETHYLBENZENE see DDN900
(1,2-DIBROMOETHYL)BENZENE see DDN900
1,2-DIBROMOETHYLENE see DDN950
1,2-DIBROMOHEXAFLUOROPROPANE see DDO450
1,2-DIBROMO-1,1,2,3,3,3-HEXAFLUOROPROPANE see DDO450
1,2-DIBROMO-2-METHYLPROPANE see DDQ150
2,6-DIBROMO-4-NITROPHENOL see DDQ500
1,2-DIBROMOPERFLUOROETHANE see FOO525
2,4-DIBROMOPHENOL see DDR150
1,2-DIBROMO-1-PHENYLETHANE see DDN900
2,3-DIBROMOPROPANOIC ACID see DDS500
α-β-DIBROMOPROPIONIC ACID see DDS500
2,3-DIBROMOPROPIONIC ACID see DDS500
4′,5-DIBROMOSALICYLANILIDE see BOD600
1,2-DIBROMOTETRAFLUOROETHANE see FOO525
sym-DIBROMOTETRAFLUOROETHANE see FOO525
1,2-DIBROMO-1,1,2,2-TETRAFLUOROETHANE see FOO525
DIBROMSALAN see BOD600
DIBUTOXYETHOXYETHYL ADIPATE see DDT500
DI(BUTOXYTHIOCARBONYL) DISULFIDE see BSS550
2,5-DI-tert-BUTYLBENZENE-1,4-DIOL see DEC800
2,6-DI-tert-BUTYL-p-BENZOQUINONE see DDV500
2,6-DI-tert-BUTYL-α-(DIMETHYLAMINO)-p-CRESOL see FAB000
2,2-DIBUTYL-1,3,2-DIOXASTANNEPIN-4,7-DIONE see DEJ100
DIBUTYL DIXANTHOGEN see BSS550
DIBUTYLDIXANTOGENATE see BSS550
2,5-DI-t-BUTYLHYDROQUINONE see DEC800
3,5-DI-tert-BUTYL-4-HYDROXYBENZOIC ACID see DEC900
DIBUTYL(MALEOYLDIOXY)TIN see DEJ100
DI-tert-BUTYL NITROXIDE see DEF090
2,6-DI-tert-BUTYLPHENOL see DEG100
N,N-DIBUTYLPROPIONAMIDE see DEH300
DIBUTYLSTANNYLENE MALEATE see DEJ100
DIBUTYL TEREPHTHALATE see DEH700
DIBUTYLTIN MALEATE see DEJ100
DIBUTYL XANTHOGEN DISULFIDE see BSS550
DICALITE see SCH000
DICAPRYLYL PEROXIDE see CBF705
p-DICHLOORBENZEEN (DUTCH) see DEP800
1,4-DICHLOORBENZEEN (DUTCH) see DEP800
1,1-DICHLOORETHAAN (DUTCH) see DFF809
2,5-DICHLORACETANILID see DGB480
1,1-DICHLORAETHAN (GERMAN) see DFF809
2,4-DICHLORANILIN see DEO290
o-DICHLORBENZENE see DEP600
o-DICHLOR BENZOL see DEP600
p-DICHLORBENZOL (GERMAN) see DEP800
1,4-DICHLOR-BENZOL (GERMAN) see DEP800
3,3′-DICHLOR-4,4′-DIAMINODIPHENYLMETHAN (GERMAN) see MJM200
DICHLORHYDRATE de l-p.CHLORBENZHYDRYL-4-(2-(2-HYDROXYETHOXY)ETHYL)PIPERAZINE see HOR470
5,7-DICHLOR-8-HYDROXYCHINOLIN see DEL300
DI-CHLORICIDE see DEP800
DICHLORID DIMETHYLCINICITY see DUG825
DICHLOROACETAMIDE see DEM300
2,2-DICHLOROACETAMIDE see DEM300
2′,5′-DICHLOROACETANILIDE see DGB480
2,4-DICHLOROANILINE see DEO290
2,4-DICHLOROBENZENAMINE see DEO290
o-DICHLOROBENZENE see DEP600
p-DICHLOROBENZENE see DEP800
1,2-DICHLOROBENZENE (MAK) see DEP600
1,4-DICHLOROBENZENE (MAK) see DEP800
DICHLOROBENZENE, ORTHO, liquid (DOT) see DEP600
DICHLOROBENZENE, PARA, solid (DOT) see DEP800
2,4-DICHLOROBENZOIC ACID see DER100
p-DICHLOROBENZOL see DEP800
1,1-DICHLOROBUTANE see BRQ050
1,3-DICHLORO-2-BUTENE see DEU650

1,4-DICHLORO-2-BUTENE see DEV000
1,4-DICHLOROBUTENE-2 (MAK) see DEV000
1,1-DICHLORO-2-CHLOROETHYLENE see TIO750
3,3′-DICHLORO-4,4′-DIAMINODIPHENYLMETHANE see MJM200
DICHLORODIMETHYLSTANNANE see DUG825
DICHLORODIMETHYLTIN see DUG825
1,1-DICHLOROETHANE see DFF809
DICHLOROETHYLALUMINUM see EFU050
DICHLOROETHYLPHOSPHINE SULFIDE see EOP600
1,1-DICHLORO-1-FLUOROETHANE see FOO550
(DICHLOROFLUOROMETHYL)BENZENE see DFL100
α-α-DICHLORO-α-FLUOROTOLUENE see DFL100
4,5-DICHLOROGUAIACOL see DFL720
1,2-DICHLOROHEXAFLUOROCYCLOBUTANE see DFM025
1,2-DICHLORO-1,2,3,3,4,4-HEXAFLUOROCYCLOBUTANE see DFM025
1,2-DICHLOROHEXAFLUOROCYCLOPENTENE see DFM050
1,2-DICHLORO-3,3,4,4,5,5-HEXAFLUOROCYCLOPENTENE see DFM050
2,6-DICHLORO-4-((p-HYDROXYPHENYL)IMINO)-2,5-CYCLOHEXADIEN-1-ONE SODIUM SALT see SGG650
DICHLOROHYDROXYQUINOLINE see DEL300
5,7-DICHLORO-8-HYDROXYQUINOLINE see DEL300
DICHLOROINDANTHRONE see DFN425
3,3′-DICHLOROINDANTHRONE see DFN425
7,16-DICHLOROINDANTHRONE see DFN425
2,6-DICHLOROINDOPHENOL, SODIUM SALT see SGG650
DICHLOROISOVIOLANTHRONE see DFN450
4,5-DICHLORO-2-METHOXYPHENOL see DFL720
2,4-DICHLORO-1-METHYLBENZENE see DGM700
DICHLOROMETHYL-3,3,3-TRIFLUOROPROPYLSILANE see DFS700
DICHLOROMONOETHYLALUMINUM see EFU050
5,7-DICHLOROOXINE see DEL300
1,2-DICHLOROPERFLUOROCYCLOBUTANE see DFM025
1,2-DICHLOROPERFLUOROCYCLOPENTENE see DFM050
2,3-DICHLOROPHENOL see DFX500
2,5-DICHLOROPHENOL see DFX850
3,4-DICHLOROPHENOL see DFY425
3,5-DICHLOROPHENOL see DFY450
N-(2,5-DICHLOROPHENYL)ACETAMIDE see DGB480
(2,5-DICHLOROPHENYL)HYDRAZINE see DGC850
DI-(4-CHLOROPHENYL)METHANE see BIM800
DI-(p-CHLOROPHENYL)METHANE see BIM800
4,7-DICHLOROQUINOLINE see DGJ250
DICHLOROQUINOLINOL see DEL300
2,3-DICHLOROQUINOXALINE see DGJ950
2,4-DICHLOROTOLUENE see DGM700
DICHLOROXIN see DEL300
5,7-DICHLOROXINE see DEL300
DI(CIANOMETIL)AMMINA see IBB100
p-DICLOROBENZENE (ITALIAN) see DEP800
1,4-DICLOROBENZENE (ITALIAN) see DEP800
3,3′-DICLORO-4,4′-DIAMINODIFENILMETANO (ITALIAN) see MJM200
1,1-DICLOROETANO (ITALIAN) see DFF809
1,4-DICYANOBUTANE see AER250
1,6-DICYANOHEXANE see OCW050
DICYCLOHEXYL PEROXIDE CARBONATE see DGV650
DICYCLOHEXYL PEROXYDICARBONATE, technically pure (UN2152) (DOT) see DGV650
DICYCLOHEXYL PEROXYDICARBONATE, not >91% with water (UN2153) (DOT) see DGV650
DICYCLOHEXYL PHTHALATE see DGV700
DIDAKENE see PCF275
DIDECANOYL PEROXIDE see DAH475
DIDODECYL PHTHALATE see PHW550
DI-n-DODECYL PHTHALATE see PHW550
1-(3,4-DIETHOXYBENZYL)-6,7-DIETHOXYISOQUINOLINE HYDROCHLORIDE see PAH260
2-(DIETHOXYPHOSPHINYL)ETHYLTRIETHOXYSILANE see DKD500
DIETHYLAMIDE of PROPIONIC ACID see DJX250
DIETHYLAMINE see DHJ200
N,N-DIETHYLAMINE see DHJ200
2-DIETHYLAMINOETHYLESTER KYSELINY METHAKRYLOVE see DIB300
DIETHYLAMINOETHYL METHACRYLATE see DIB300
β-(DIETHYLAMINO)ETHYL METHACRYLATE see DIB300
2-(DIETHYLAMINO)ETHYL METHACRYLATE see DIB300
2-(N,N-DIETHYLAMINO)ETHYL METHACRYLATE see DIB300
3-(4-DIETHYLAMINO-2-HYDROXYPHENYLAZO)-4-HYDROXYBENZENESULFONIC ACID see DIJ300
4-DIETHYLAMINONITROSOBENZENE see NJW600
m-(DIETHYLAMINO)PHENOL see DIO400
3-(DIETHYLAMINO)PHENOL see DIO400
2,6-DIETHYLANILINE see DIS650
2,6-DIETHYLBENZENAMINE see DIS650

DIETHYLCADMIUM see DIV800
DIETHYL ISOPHTHALATE see IMK100
N,N-DIETHYL-4-METHYLTETRAMETHYLENEDIAMINE see DJQ850
N,N-DIETHYL-4-NITROSOBENZENAMINE see NJW600
2,6-DIETHYLPHENOL see DJU700
N,N-DIETHYL-N'-PHENYLETHYLENEDIAMINE see DJV300
N,N-DIETHYLPROPANAMIDE see DJX250
N,N-DIETHYLPROPIONAMIDE see DJX250
DIETHYL TEREPHTHALATE see DKB160
O,O-DIETHYL-O-3,5,6-TRICHLORO-2-PYRIDYL PHOSPHOROTHIOATE see CMA100
DIETHYL (2-(TRIETHOXYSILYL)ETHYL)PHOSPHONIC ACID see DKD500
1,1-DIETHYLUREA see DKD650
N,N-DIETHYLUREA see DKD650
asym-DIETHYLUREA see DKD650
DIETILAMINA (ITALIAN) see DHJ200
DIFENYL-DIHYDROXYSILAN see DMN450
N,N'-DIFENYLETHYLENDIAMIN see DWB400
2,4-DIFLUOROANILINE see DKF700
2,4-DIFLUOROBENZENAMINE see DKF700
p,p'-DIFLUORO-m,m'-DINITRODIPHENYL SULFONE see DKH250
4,4'-DIFLUORO-3,3-DINITRODIPHENYL SULFONE see DKH250
1,2-DIFLUOROETHANE see DKH300
2,4-DIFLUORONITROBENZENE see DKH900
DIGERMIN see DUV600
DIGLYCIDYLETHYLENE GLYCOL see EEA600
1,2-DIGLYCIDYLOXYETHANE see EEA600
DIGLYCOLIC ACID (6CI) see ONQ100
DIGUANIDINIUM CARBONATE see GKW100
9,10-DIHYDROACRIDINE see ADI775
N,N-DIHYDRO-1,1',1',2'-ANTHRAQUINONE-AZINE see IBV050
DIHYDROBAIKIANE see HDS300
7,11b-DIHYDROBENZ(b)INDENO(1,2-d)PYRAN-3,6a,9,10(6H)-TETROL see LFT800
9,10-DIHYDRO-8a,10-DIAZONIAPHENANTHRENE DIBROMIDE see DWX800
9,10-DIHYDRO-8a,10a-DIAZONIAPHENANTHRENE(1,1'-ETHYLENE-2,2'-BIPYRIDYL-IUM)DIBROMIDE see DWX800
9,10-DIHYDRO-3,4-DIHYDROXY-9,10-DIOXO-2-ANTHRACENESULFONIC ACID MONOSODIUM SALT see SEH475
1,3-DIHYDRO-1,3-DIOXO-5-ISOBENZOFURANCARBOXYLIC ACID see TKV000
5,6-DIHYDRO-DIPYRIDO(1,2a;2,1c)PYRAZINIUM DIBROMIDE see DWX800
4,5-DIHYDRO-2-METHYLTHIAZOLE see DLV900
2,3-DIHYDRO-5-METHYL-2-THIOXO-4(1H)-PYRIMIDINONE see TFQ650
9,10-DIHYDRO-9-OXOANTHRACENE see APM500
α-α-DIHYDROPERFLUOROBUTANOL see HAW100
1,1-DIHYDROPERFLUOROBUTANOL see HAW100
2,3-DIHYDRO-2-PHENYL-4H-1-BENZOPYRAN-4-ONE see FBW150
6,7-DIHYDROPYRIDO(1,2a;2',1'-C)PYRAZINEDIUM DIBROMIDE see DWX800
DIHYDRO-α-TERPINEOL see MCE100
4,5-DIHYDRO-2-THIAZOLAMINE see TEV600
DIHYDRO-2(3H)-THIOPHENONE see TDC800
4,5-DIHYDRO-3(2H)-THIOPHENONE see TDC800
DIHYDROXYACETONE see OMC300
1,3-DIHYDROXYACETONE see OMC300
2,4-DIHYDROXYBENZALDEHYDE see REF100
o-DIHYDROXYBENZENE see CCP850
1,2-DIHYDROXYBENZENE see CCP850
2,4-DIHYDROXYBENZENECARBONAL see REF100
3,4-DIHYDROXYBENZOIC ACID see POE200
3,5-DIHYDROXYBENZOIC ACID see REF200
2,2'-DIHYDROXYBINAPHTHALENE see BGC100
1,4-DIHYDROXY-2-BUTENE see BOX300
DIHYDROXYCYCLOHEXYLPEROXIDE see PCM100
DI-(1-HYDROXYCYCLOHEXYL)PEROXIDE, technically pure (DOT) see PCM100
1,8-DIHYDROXY-4,5-DIAMINOANTHRACHINON see DBQ250
2,2'-DIHYDROXYDINAPHTHYL see BGC100
4,4'-DIHYDROXYDIPHENYLAMINE see IBJ100
DIHYDROXYDIPHENYLSILANE see DMN450
DI(2'-HYDROXYETHYL)AMINO-1,4-BENZOQUINONE see BKB100
α-β-DIHYDROXYETHYLBENZENE see SMQ100
1,2-DIHYDROXYETHYLBENZENE see SMQ100
N,N-DIHYDROXYETHYL GLYCINE see DMT500
2,2'-DIHYDROXY-4-METHOXYBENZOPHENONE see DMW250
6,7-DIHYDROXY-4-METHYL-2H-1-BENZOPYRAN-2-ONE see MJV800
6,7-DIHYDROXY-4-METHYLCOUMARIN see MJV800
1,5-DIHYDROXYNAPHTHALENE see NAN505
1,5-DIHYDROXYNAPTHALENE see NAN505
1,3-DIHYDROXYPROPANONE see OMC300
4,6-DIHYDROXYPYRIMIDIN see PPP100
DIHYXAL see OMC300
DIINDOGEN see BGB275
3,5-DIIODOANTHRANILIC ACID see API800
1,4-DIIODOBUTANE see TDQ400
1,6-DIIODOHEXANE see HEG200

3,5-DIIODO-4-HYDROXYBENZOIC ACID see DNF300
1,5-DIIODOPENTANE see PBH125
1,3-DIIODOPROPANE see TLR050
p-DIISOPROPYLBENZENE see DNN830
1,4-DIISOPROPYLBENZENE see DNN830
p-DIISOPROPYLBENZOL see DNN830
N,N-DIISOPROPYL-2-BENZOTHIAZOLESULFENAMIDE see DNN900
N,N-DIISOPROPYL ETHYLENEDIAMINE see DNP700
1:2,5:6-DI-O-ISOPROPYLIDEN-α-D-GLUCOFURANOSE see DVO100
1:2,5:6-DI-O-ISOPROPYLIDENE-α-D-GLUCOFURANOSE see DVO100
DIISOPROPYL PHTHALATE see PHW600
4,4'-DIISOTHIOINDIGO see DNT300
DILANTIN DB see DEP600
DILATIN DB see DEP600
DILAURYL PHTHALATE see PHW550
DILITURIC ACID see NET550
2,3-DIMERCAPTOBUTANEDIOIC ACID see DNV610
1,2-DIMERCAPTOPROPANE see PML300
2,3-DIMERCAPTOPROPANE see PML300
DIMERCAPTOSUCCINIC ACID see DNV610
α-β-DIMERCAPTOSUCCINIC ACID see DNV610
2,3-DIMERCAPTOSUCCINIC ACID see DNV610
2,5-DIMETHOXYANILINE see AKD925
2,5-DIMETHOXYBENZENAMINE see AKD925
m-DIMETHOXYBENZENE see REF025
1,3-DIMETHOXYBENZENE see REF025
3,4-DIMETHOXYBENZENEACETIC ACID see HGK600
3,4-DIMETHOXYBENZOIC ACID see VHP600
8,8-DIMETHOXY-2,6-DIMETHYL-2-OCTANOL see LBO050
p,p'-DIMETHOXYDIPHENYLAMINE see BKO600
4,4'-DIMETHOXYDIPHENYLAMINE see BKO600
3,4-DIMETHOXYPHENYL ACETIC ACID see HGK600
DI-p-METHOXYPHENYLAMINE see BKO600
2',4'-DIMETHYLACETOACETANILIDE see OOI100
DIMETHYLAMINE see DOQ800
DIMETHYLAMINE, anhydrous (DOT) see DOQ800
DIMETHYLAMINE, aqueous solution (DOT) see DOQ800
DIMETHYLAMINE, solution (DOT) see DOQ800
p-DIMETHYLAMINO BENZOIC ACID see DOU650
DIMETHYLAMINOETHANOL ACETATE see DOZ100
2-DIMETHYLAMINOETHANOL ACETATE see DOZ100
DIMETHYLAMINOETHYL ACETATE see DOZ100
2-(DIMETHYLAMINO)ETHYL ACETATE see DOZ100
DIMETHYLAMINOETHYL ACRYLATE see DPB300
o-DIMETHYLAMINONITROBENZENE see NFW600
m-(DIMETHYLAMINO)PHENOL see HNK560
3-(DIMETHYLAMINO)PHENOL see HNK560
5,6-DIMETHYLBENZIMIDAZOLE see DQM100
3,4-DIMETHYLBENZOIC ACID see DQM850
3,5-DIMETHYLBENZOIC ACID see MDJ748
p-(α-α-DIMETHYLBENZYL)PHENOL see COF400
2,2'-DIMETHYL-1,1'-BIANTHRACENE-9,9',10,10'-TETRONE see DQR350
2,2'-DIMETHYL-1,1'-BIANTHRAQUINONE see DQR350
6,6-DIMETHYLBICYCLO(3.1.1)HEPT-2-ENE-2-CARBOXALDEHYDE see FNK150
(1S)-6,6-DIMETHYLBICYCLO(3.1.1)HEPT-2-ENE-2-METHANOL ACETATE see MSC050
4,4'-DIMETHYL-2,2'-BIPYRIDINE see DQS100
DIMETHYL BUTANEDIOATE see SNB100
3,3-DIMETHYL-2-BUTANOL see BRU300
N,N-DIMETHYLBUTYRAMIDE see DQV300
DIMETHYLCADMIUM see DQW800
DIMETHYLCETYLAMINE see HCP525
N,N-DIMETHYLCETYLAMINE see HCP525
asym-DIMETHYL-3,7-DIAMINOPHENAZATHIONIUM CHLORIDE see DUG700
2,5-DIMETHYL-2,5-DI-(BENZOYLPEROXY)HEXANE see DRJ100
2,5-DIMETHYL-2,5-DI-(BENZOYLPEROXY)HEXANE, technically pure (UN2172) (DOT) see DRJ100
2,5-DIMETHYL-2,5-DI-(BENZOYLPEROXY)HEXANE, not >82% with inert solid (UN2173) (DOT) see DRJ100
2,5-DIMETHYL-2,5-DI(tert-BUTYLPEROXY)HEXYNE-3 see DRJ825
2,5-DIMETHYL-2,5-DI-(tert-BUTYLPEROXY)HEXYNE-3, technically pure (UN2158)(DOT) see DRJ825
2,5-DIMETHYL-2,5-DI-(tert-BUTYLPEROXY)HEXYNE-3, not >52% with inert solid(UN2159) (DOT) see DRJ825
DIMETHYLDICHLOROSTANNANE see DUG825
DIMETHYLDICHLOROTIN see DUG825
DIMETHYLDIOCTADECYLAMMONIUM CHLORIDE see DXG625
N,N-DIMETHYLDITHIOCARBAMIC ACID DIMETHYLAMINOMETHYL ESTER see DRQ650
N,N-DIMETHYL-DITHIOCARBAMINSAEURE-DIMETHYLAMINOMETHYL-ESTER see DRQ650
N,N-DIMETHYLETHANETHIOAMIDE see DUG450

1,1-DIMETHYLETHANOL

1,1-DIMETHYLETHANOL see BPX000
DIMETHYLETHER RESORCINOLU see REF025
2-(1,1-DIMETHYLETHYL)-5-METHYLPHENOL see BQV600
N-(1,1-DIMETHYLETHYL)-2-PROPENAMIDE see BPW050
(1,1-DIMETHYLETHYL)UREA see BSS310
N,N-DIMETHYLHEXADECYLAMINE see HCP525
DIMETHYL HYDROGEN PHOSPHATE see PHC800
O,O-DIMETHYL HYDROGEN PHOSPHATE see PHC800
2,3-DIMETHYLINDOLE see DSI850
2,3-DIMETHYL-1H-INDOLE see DSI850
α-α-DIMETHYL-m-ISOPROPENYL BENZYL ISOCYANATE see IKG800
3,5-DIMETHYLISOXAZOLE see DSK950
N,N-DIMETHYLMETHANAMINE see TLD500
α-4-DIMETHYL-α-(4-METHYL-3-PENTENYL)-3-CYCLOHEXENE-1-METHANOL see BGO775
2,2-DIMETHYL-3-(2-METHYLPROPENYL)CYCLOPROPANECARBOXYLIC ACID ETHYL ESTER see EHM100
N,N-DIMETHYL-o-NITROANILINE see NFW600
N,N-DIMETHYL-2-NITROBENZENAMINE see NFW600
1,4-DIMETHYL-2-NITROBENZENE see NMS520
N,N-DIMETHYL-N-OCTADECYL-1-OCTADECANAMINIUM CHLORIDE see DXG625
trans-3,7-DIMETHYL-2,6-OCTADIENAL see GCU100
trans-3,7-DIMETHYL-2,6-OCTADIEN-1-OL ACETATE see DTD800
3,7-DIMETHYL-2-trans-6-OCTADIENYL ACETATE see DTD800
trans-3,7-DIMETHYL-2,6-OCTADIEN-1-YL ACETATE see DTD800
trans-3,7-DIMETHYL-2,6-OCTADIEN-1-YL BUTYRATE see GDE810
trans-2,6-DIMETHYL-2,6-OCTADIEN-8-YL ETHANOATE see DTD800
3,7-DIMETHYLOCTANOL-3 see LFY510
2,6-DIMETHYL-2-OCTEN-8-OL see DTF410
3,7-DIMETHYL-6-OCTEN-1-OL see DTF410
3,6-DIMETHYL-OCTIN-4-DIOL-(3,6) see DTF850
DIMETHYLPALMITYLAMINE see HCP525
3′,3″-DIMETHYLPHENOLPHTHALEIN see CNX400
2,8-DIMETHYLPHENOSAFRANINE see GJI400
1,5-DIMETHYL-2-PHENYL-4-AMINOPYRAZOLINE see AIB300
4-(DIMETHYLPHENYLMETHYL)PHENOL see COF400
N-(2,4-DIMETHYLPHENYL)-3-OXOBUTANAMIDE see OOI100
DIMETHYL PHOSPHATE see PHC800
N,N′-DIMETHYLPIPERAZINE see LIQ500
1,4-DIMETHYLPIPERAZINE see LIQ500
2,6-DIMETHYLPIPERIDINE see LIQ550
1,2-DIMETHYLPROPANAMINE see AOE200
1,2-DIMETHYLPROPYLAMINE see AOE200
3,4-DIMETHYLPROTOCATECHUIC ACID see VHP600
3,5-DIMETHYLPYRAZOLE see DTU850
α-γ-DIMETHYLPYRIDINE see LIY990
2,4-DIMETHYLPYRIDINE see LIY990
2,3-DIMETHYLQUINOXALINE see DTY700
DIMETHYL SALICYLATE see MLH800
DIMETHYL SUCCINATE see SNB100
DIMETHYLTHIOACETAMID see DUG450
DIMETHYLTHIOACETAMIDE see DUG450
N,N-DIMETHYLTHIOACETAMIDE see DUG450
DIMETHYLTHIONINE see DUG700
DIMETHYLTIN DICHLORIDE see DUG825
DIMETHYL-m-TOLUIDINE see TLG100
DIMETHYL-p-TOLUIDINE see TLG150
2,6-DIMETHYLUNDECA-2,6,8-TRIENE-10-ONE see POH525
6,10-DIMETHYL-3,5,9-UNDECATRIEN-2-ONE see POH525
1,1-DIMETHYLUREA see DUM150
DIMETIL-m-TOLUIDINA see TLG100
DIMETIL-p-TOLUIDINA see TLG150
DIMORPHOLINETHIURAM DISULFIDE see MRR090
16H-DINAPHTHO(2,3-a:2′,3′-i)CARBAZOLE-5,10,15,17-TETRAONE, 6,9-DIBENZAMIDO- see DUP100
2,2′-DINAPHTHOL see BGC100
2-(2,4-DINITROANILINO)ETHANOL see DVC300
3,4-DINITROBENZOIC ACID see DUR550
3,5-DINITROBENZOIC ACID see DUR600
3,5-DINITROBENZOIC ACID CHLORIDE see DUR850
3,5-DINITROBENZOYL CHLORIDE see DUR850
3,5-DINITRO-4-CHLORO-α-α-α-TRIFLUOROTOLUENE see CGM225
3,3′-DINITRO-4,4′-DIFLUORODIPHENYL SULFONE see DKH250
2,4-DINITRODIPHENYLAMINE see DUV100
p,p′-DINITRODIPHENYL ETHER see NIM600
4,4′-DINITRODIPHENYL ETHER see NIM600
4,4′-DINITRODIPHENYL OXIDE see NIM600
2,6-DINITRO-N,N-DIPROPYL-4-(TRIFLUOROMETHYL)BENZENAMINE see DUV600
2,6-DINITRO-N,N-DI-N-PROPYL-α,α,α-TRIFLURO-p-TOLUIDINE see DUV600
2,2′-DINITRO-5,5′-DITHIODIBENZOESAEURE see DUV700
2,2′-DINITRO-5,5′-DITHIODIBENZOIC ACID see DUV700
2,4-DINITRO-p-HYDROXYDIPHENYLAMINE see DUW500
1,3-DINITRO-2-IMIDAZOLIDINONE see DUW503
1,3-DINITRO-2-IMIDAZOLIDONE see DUW503
1,8-DINITRONAPHTHALENE see DUX710
2,4-DINITRONAPHTHOLSULFONIC ACID see FBZ100
2,4-DINITROPHENYL-DIMETHYL-DITHIOCARBAMATE see DVB850
N-2,4-DINITROPHENYLETHANOLAMINE see DVC300
DI-4-NITROPHENYL ETHER see NIM600
DINITROPHENYLMETHANE see DVG600
DI-p-NITROPHENYL PHOSPHATE see BLA600
3,5-DINITROSALICYLIC ACID see HKE600
p-DINITROSOBENZENE see DVE260
1,4-DINITROSOBENZENE see DVE260
DINITROTOLUENE see DVG600
ar,ar-DINITROTOLUENE see DVG600
DINITROTOLUENE, liquid (DOT) see DVG600
DINITROTOLUENE, molten (DOT) see DVG600
DINITROTOLUENE, solid (DOT) see DVG600
2,6-DINITRO-4-TRIFLUORMETHYL-N,N-DIPROPYLANILIN (GERMAN) see DUV600
DIOCTANOYL PEROXIDE see CBF705
DI-n-OCTYL SODIUM SULFOSUCCINATE see SOD300
DIOKTYLESTER SULFOJANTARANU SODNEHO see SOD300
1,3,2-DIOXASTANNEPIN-4,7-DIONE, 2,2-DIBUTYL- see DEJ100
1,2-DIOXOCYCLOHEXANE see CPB200
1,3-DIOXO-5-PHTHALANCARBOXYLIC ACID see TKV000
o-DIOXYBENZENE see CCP850
DIOXYBENZON see DMW250
DIOXYBENZONE see DMW250
1,1′-DIOXYBISCYCLOHEXANOL see PCM100
DIOXYLINE see HOE100
DIPAC see DNN900
DIPELARGONYL PEROXIDE see NNA100
DIPHENAN see PFR325
DIPHENAN (pharmaceutical) see PFR325
DIPHENANE see PFR325
o-DIPHENOL see CCP850
N,N-DIPHENYLACETAMIDE see DVW750
1,1-DIPHENYL ACETONE see MJH910
DIPHENYLAMINE, 2,4-DINITRO- see DUV100
DIPHENYLAMINE, 4-(PHENYLAZO)- see PEI800
DIPHENYL BRILLIANT BLUE see CMO500
DIPHENYLCARBAZIDE see DVY925
N,N′-DIPHENYLCARBAZIDE see DVY925
sym-DIPHENYLCARBAZIDE see DVY925
1,5-DIPHENYLCARBAZIDE see DVY925
2,2′-DIPHENYLCARBAZIDE see DVY925
DIPHENYLCARBAZONE see DVY950
s-DIPHENYLCARBAZONE see DVY950
1,5-DIPHENYLCARBAZONE see DVY950
DIPHENYL CARBINOL see HKF300
1,5-DIPHENYLCARBOHYDRAZIDE see DVY925
2,2′-DIPHENYLCARBONIC DIHYDRAZIDE see DVY925
DIPHENYLENEMETHANE see FDI100
DIPHENYLENE OXIDE see DDB500
DIPHENYLENE SULFIDE see TES300
trans-DIPHENYLETHENE see SLR100
trans-1,2-DIPHENYLETHENE see SLR100
(E)-1,2-DIPHENYLETHYLENE see SLR100
trans-1,2-DIPHENYLETHYLENE see SLR100
trans-α-β-DIPHENYLETHYLENE see SLR100
N,N′-DIPHENYLETHYLENEDIAMINE see DWB400
DIPHENYLGLYCOLIC ACID see BBY990
α-α-DIPHENYLGLYCOLIC ACID see BBY990
DIPHENYL HYDROGEN PHOSPHATE see PHE300
DIPHENYLHYDROXYACETIC ACID see BBY990
DIPHENYL KETOXIME see BCS400
DIPHENYLMETHANE-4,4′-DIISOCYANATE-TRIMELLIC ANHYDRIDE-ETHOMID HT POLYMER see TKV000
DIPHENYLMETHANOL see HKF300
DIPHENYLMETHANONE OXIME see BCS400
DIPHENYLMETHYL ALCOHOL see HKF300
(DIPHENYLMETHYLENE)HYDROXYLAMINE see BCS400
DIPHENYL PHOSPHATE see PHE300
DIPHENYL RED 4B see DXO850
DIPHENYL RED 4BS see DXO850
DIPHENYL SKY BLUE 6B see CMO500
DIPHOSPHOPYRIDINE NUCLEOTIDE see CNF390
DIPOTASSIUM ETHYLENEDIAMINETETRAACETATE see EEB100
DIPOTASSIUM OXALATE see PLN300
4-(DI-N-PROPYLAMINO)-3,5-DINITRO-1-TRIFLUOROMETHYLBENZENE see DUV600
N,N-DI-N-PROPYL-2,6-DINITRO-4-TRIFLUOROMETHYLANILINE see DUV600
N,N-DIPROPYL-4-TRIFLUOROMETHYL-2,6-DINITROANILINE see DUV600

4,4′-DIPYRIDINE see BGO600
γ,γ′-DIPYRIDYL see BGO600
4,4-DIPYRIDYL see BGO600
4,4′-DIPYRIDYL see BGO600
DIQUAT see DWX800
DIQUAT DIBROMIDE see DWX800
DIRECT BLUE 15 see CMO500
DIRECT BLUE 10G see CMO500
DIRECT BLUE HH see CMO500
DIRECT PURE BLUE see CMO500
DIRECT PURE BLUE M see CMO500
DIRECT PURPURINE 4B see DXO850
DIRECT PURPURINE M4B see DXO850
DIRECT RED 2 see DXO850
DIRECT RED 4A see DXO850
DIRECT RED 4B see DXO850
DIRECT RED DCB see DXO850
DIRECT SKY BLUE A see CMO500
DIRNATE see SNL840
DIRUBIDIUM CARBONATE see RPB200
DIRUBIDIUM MONOCARBONATE see RPB200
DISILOXANE, 1,3-BIS(3-AMINOPROPYL)-1,1,3,3-TETRAMETHYL- see OPC100
DISODIUM ADENOSINE TRIPHOSPHATE see AEM100
DISODIUM ADENOSINE 5′-TRIPHOSPHATE see AEM100
DISODIUM ARSENATE see ARC000
DISODIUM ARSENATE, HEPTAHYDRATE see ARC250
DISODIUM ARSENIC ACID see ARC000
DISODIUM ATP see AEM100
DISODIUM-4,4′-BIS((4,6-DIANILINO-1,3,5-TRIAZIN-2-YL)AMINO)STILBENE-2,2′-DISULFONATE see DXB450
DISODIUM BROMOSULFOPHTHALEIN see HAQ600
DISODIUM CALCIUM EDTA see CAR780
DISODIUM CALCIUM ETHYLENEDIAMINETETRAACETATE see CAR780
DISODIUM DIHYDROGEN ATP see AEM100
DISODIUM HYDROGEN ARSENATE see ARC000
DISODIUM HYDROGEN ORTHOARSENATE see ARC000
DISODIUM MONOHYDROGEN ARSENATE see ARC000
DISODIUM PEROXYDICARBONATE see SJB400
DISODIUM PREDNISOLONE 21-PHOSPHATE see PLY275
DISODIUM SUCCINATE see SJW100
DISODIUM TETRABORATE see DXG035
DISPERSE BLUE 78 see BKP500
DISPERSE BLUE 110 see BKP500
DISPERSE FAST YELLOW 2K see DUW500
DISPERSE PINK Zh see AKO350
DISPERSE RED-4 see AKO350
DISPERSE ROSE Zh see AKO350
DISPERSE YELLOW R see DUW500
DISPERSE YELLOW STABLE 2K see DUW500
DISPERSOL FAST YELLOW A see DUW500
DISPERSOL PRINTING YELLOW A see DUW500
DISPERSOL YELLOW B-A see DUW500
DISTEARYL DIMETHYLAMMONIUM CHLORIDE see DXG625
DISTOKAL see HCI000
DISTOPAN see HCI000
DISTOPIN see HCI000
DISULFIDE, BENZYL METHYL see MHN350
DISULFIDE, BIS(MORPHOLINOTHIOCARBONYL) see MRR090
DISULFIDE, BIS(2-NITRO-α-α-α-TRIFLUORO-p-TOLYL) see BLA800
DISULFIDE, DI-2-PROPENYL (9CI) see AGF300
DITHALLIUM OXIDE see TEL040
1,4-DITHIACYCLOHEXANE see DXI550
p-DITHIANE see DXI550
1,4-DITHIANE see DXI550
4,5-DITHIA-1,7-OCTADIENE see AGF300
2′,2‴-DITHIOBISBENZANILIDE see BDK800
4,4′-DITHIOBISBENZENAMINE see ALW100
2,2′-DITHIOBIS(BENZOIC ACID) see BHM300
3,3′-DITHIOBIS(6-NITROBENZOIC ACID) see DUV700
DITHIOBIS(THIOFORMIC ACID) O,O-DIBUTYL ESTER see BSS550
DITHIOCARBONIC ACID O-ISOPROPYL ESTER POTASSIUM SALT see IRG050
p,p′-DITHIODIANILINE see ALW100
4,4′-DITHIODIANILINE see ALW100
2′,2‴-DITHIODIBENZANILIDE see BDK800
2,2′-DITHIODIBENZOESAEURE see BHM300
2,2′-DITHIODIBENZOIC ACID see BHM300
4,4′-(DITHIODICARBONOTHIOYL)BISMORPHOLINE see MRR090
N,N′-(DITHIODI-2,1-PHENYLENE)BISBENZAMIDE see BDK800
1,3-DITHIOLANE-2-THIONE see EJQ100
1,2-DITHIOLPROPANE see PML300
DITOLYLBIS(AZONAPHTHIONIC ACID) see DXO850
DIVINYLBENZENE see DXQ740

DIXIE see KBB600
DIZENE see DEP600
DMA see DOQ800
2,4-D METHYL ESTER see DAA825
DMI see DSK950
DMP see DTU850
DNBA see DUR600
DNNS see FBZ100
γ-DODECALACTONE see OES100
DODECAMETHYLPENTASILOXANE see DXS800
DODECANAMINE ACETATE see DXW050
DODECANAMINE HYDROCHLORIDE see DKW100
1-DODECANAMINE, HYDROCHLORIDE (9CI) see DKW100
DODECANOIC ACID, CADMIUM SALT (9CI) see CAG775
DODECANOLIDE-1,4 see OES100
DODECYLAMINE, ACETATE see DXW050
1-DODECYLAMINE ACETATE see DXW050
DODECYLAMINE, HYDROCHLORIDE see DKW100
n-DODECYLAMINE HYDROCHLORIDE see DKW100
DODECYLAMMONIUM CHLORIDE see DKW100
n-DODECYLAMMONIUM CHLORIDE see DKW100
DODECYLPYRIDINIUM CHLORIDE see LBX050
N-DODECYLPYRIDINIUM CHLORIDE see LBX050
1-DODECYLPYRIDINIUM CHLORIDE see LBX050
DODECYL SODIUM SULFOACETATE see SIB700
DODECYLTRIMETHYLAMMONIUM BROMIDE see DYA810
6-DODECYL-2,2,4-TRIMETHYL-1,2-DIHYDROQUINOLINE see SAU480
DOMAR see CAT775
DOTRIACONTANE see DYC900
DOWANOL (R) PMA GLYCOL ETHER ACETATE see PNL265
DOWCO 179 see CMA100
DOW-PER see PCF275
DOWSPRAY 9 see DDN900
DOWTHERM E see DEP600
DOW-TRI see TIO750
DPC see DVY925, LBX050
DPN see CNF390
DRACONIC ACID see AOU600
DRAKEOL see MQV750
DRI-DIE INSECTICIDE 67 see SCH000
DTBN see DEF090
dU see DAZ050
DUKERON see TIO750
DULCOLAN see PPN100
DULCOLAX see PPN100
DUPHAPEN see PAQ200
DUPONOL 80 see OFU200
DURAFUR DEVELOPER C see CCP850
DURAFUR DEVELOPER E see NAN505
DURAMITE see CAT775
DURATINT GREEN 1001 see PJQ100
DURCAL 10 see CAT775
dURD see DAZ050
DURINDONE PRINTING RED B see DNT300
DURINDONE RED B see DNT300
DURINDONE RED BP see DNT300
DUROLAX see PPN100
DURSBAN see CMA100
DURSBAN F see CMA100
DUSICNAN KADEMNATY (CZECH) see CAH250
DWUETYLOAMINA (POLISH) see DHJ200
3,5-DWUMETYLOIZOKSAZOLU see DSK950
3,5-DWUMETYLOPIRAZOLU see DTU850
DXG see BSS550
DYMEX see ABH000

E 130 see IBV050
α-EARLEINE see GHA050
ECLIPSE RED see DXO850
EDATHAMIL CALCIUM DISODIUM see CAR780
EDBPHA see EIV100
EDDHA see EIV100
EDETAMIN see CAR780
EDETAMINE see CAR780
EDETATE CALCIUM see CAR780
EDETIC ACID CALCIUM DISODIUM SALT see CAR780
EDHPA see EIV100
EDTACAL see CAR780
EDTA CALCIUM DISODIUM SALT see CAR780
EFIRAN 99 see TIR800
EGITOL see HCI000
EGRI M 5 see CAT775

ELANCOLAN see DUV600
ELJON LITHOL RED No. 10 see NAP100
ELTREN see LBX050
EMATHLITE see KBB600
EMERY 5714 see HKS100
EMERY 5770 see EHJ600
ENDIARON see DEL300
ENDOKOLAT see PPN100
ENIACROMO YELLOW G see SIT850
ENIANIL PURE BLUE AN see CMO500
ENT 1025 see HCP100
ENT 1,860 see PCF275
ENT 4,705 see CBY000
ENT 15748 see HBL100
ENT 25,550 see SCH000
ENT 27,311 see CMA100
ENZOPRIDE see CNF390
(+−)-EPHEDRINE see EAW100
EPHEDRINE, (+−)- see EAW100
(+)-EPHEDRINE see EAW200
EPHEDRINE, (+)- see EAW200
d-EPHEDRINE see EAW200
dl-EPHEDRINE see EAW100
l-(+)-EPHEDRINE see EAW200
EPONOC B see TGX550
3,4-EPOXYCYCLOHEXYLMETHYL 3,4-EPOXYCYCLOHEXANE CARBOXYLATE see EBO050
1,2-EPOXYCYCLOPENTANE see EBO100
1,4-EPOXY-p-MENTHANE see IKC100
1,2-EPOXY-3-METHOXYPROPANE see GGW600
1,2-EPOXYPENTANE see ECE525
1,2-EPOXY-3-(p-TOLYLOXY)PROPANE see GGY175
EPSYLON KAPROLAKTAM (POLISH) see CBF700
EPT 500 see IHC550
ERADEX see CMA100
E-RETINAL see VSK985
ERGOPLAST.FDC see DGV700
ERIE BENZO 4BP see DXO850
ERIE RED 4B see DXO850
ERIOCHROMAL YELLOW G see SIT850
ERIOCHROME YELLOW 2G see SIT850
ERIOCHROME YELLOW GS see SIT850
ERIO FAST CYANINE SE see APG700
ERL-4221 see EBO050
ERYCORBIN see SAA025
ERYTHORBIC ACID see SAA025
d-ERYTHORBIC ACID see SAA025
ESKALON 100 see CAT775
ESPEROX 24M see BSC600
ESPIRITIN see LAG010
ESSIGSAEUREANHYDRID (GERMAN) see AAX500
ESTEROQUINONE LIGHT PINK RLL see AKO350
ESTRU BENZYLOWEGO KWASU NIKOTYNOWEGO see NCR040
ETHANAL see AAG250
ETHANAMINE see EFU400
ETHANAMINE, 2-BROMO-(9CI) see BNI650
ETHANAMINIUM, 2-(ACETYLOXY)-N,N,N-TRIMETHYL-, BROMIDE (9CI) see CMF260
ETHANAMINIUM, 2-(ACETYLTHIO)-N,N,N-TRIMETHYL-, IODIDE (9CI) see ADC300
ETHANAMINIUM, N,N,N-TRIETHYL-, PERCHLORATE see TCD002
ETHANAMINIUM, N,N,N-TRIMETHYL-, IODIDE (9CI) see EQC600
ETHANE, 1,2-BIS(2,3-EPOXYPROPOXY)- see EEA600
1,2-ETHANEDIAMINE, N-METHYL- see MJW100
ETHANE, 1,2-DIBROMOTETRAFLUORO- see FOO525
ETHANE, 1,2-DIBROMO-1,1,2,2-TETRAFLUORO-(9CI) see FOO525
ETHANE, 1,1-DICHLORO-1-FLUORO- see FOO550
ETHANE, 1,2-DIFLUORO- see DKH300
ETHANEDIOIC ACID DIAMMONIUM SALT see ANO750
ETHANEDIOIC ACID, DIPOTASSIUM SALT (9CI) see PLN300
1,2-ETHANEDIOL DIGLYCIDYL ETHER see EEA600
1,2-ETHANEDIOL, PHENYL- see SMQ100
1,2-ETHANEDIOL, 1-PHENYL- see SMQ100
ETHANEDIUREA see EJC100
N,N′-1,2-ETHANEDIYLBIS(N-(CARBOXYMETHYL)GLYCINE, DIPOTASSIUM SALT see EEB100
2,2′-(1,2-ETHANEDIYLBIS(OXYMETHYLENE))BISOXIRANE see EEA600
N,N″-1,2-ETHANEDIYLBISUREA see EJC100
ETHANE HEXACHLORIDE see HCI000
ETHANE, ISOTHIOCYANATO-(9CI) see ELX525
ETHANETHIOAMIDE, N,N-DIMETHYL-(9CI) see DUG450
ETHANE, 1,1,1-TRIFLUORO- see TJY900

ETHANOIC ANHYDRATE see AAX500
ETHANOLAMINE CHLORIDE see EEC700
ETHANOLAMINE HYDROCHLORIDE see EEC700
ETHANOL, 2-AMINO-, HYDROCHLORIDE see EEC700
ETHANOL, 2-BROMO-, ACETATE see BNI600
ETHANOL, 2-CHLORO-, ACETATE see CGO600
ETHANOL, 2-(2-(4-(p-CHLORO-α-PHENYLBENZYL)-1-PIPERAZINYL)ETHOXY)-, DIHYDROCHLORIDE see HOR470
ETHANOL, 2-(2,4-DINITROANILINO)- DVC300
ETHANOL, 2-((2,4-DINITROPHENYL)AMINO)- see DVC300
ETHANOL, 2-(N-ETHYL-m-TOLUIDINO)- see HKS100
ETHANOL, 2,2,2-TRICHLORO-, DIHYDROGEN PHOSPHATE see TIO800
ETHANONE, 1-(2-BENZOFURANYL)-(9CI) see ACC100
ETHANONE, 1-(2-FURANYL)-(9CI) see ACM200
ETHANONE, 1-PHENYL-, OXIME see ABH150
ETHANOX 701 see DEG100
ETHAVERINE HYDROCHLORIDE see PAH260
(E)-1,1′-(1,2-ETHENEDIYL)BISBENZENE see SLR100
ETHENE, TRIBROMO- see THV100
ETHENE, TRIFLUORO(TRIFLUOROMETHOXY)- see TKG800
ETHENYL ACETATE see VLU250
N-ETHENYLCARBAZOLE see VNK100
9-ETHENYL-9H-CARBAZOLE see VNK100
1-ETHENYLCYCLOHEXENE see VNZ990
4-ETHENYL-1-CYCLOHEXENE see CPD750
1-ETHENYL-4-METHYLBENZENE see VQK700
2-(ETHENYLOXY)ETHYL 2-METHYL-2-PROPENOATE see VQA150
4-ETHENYLPHENOL ACETATE see ABW550
4-ETHENYLPYRIDINE see VQK590
1-ETHENYLSILATRANE see VQK600
ETHER, BIS(p-NITROPHENYL) see NIM600
ETHER, TRIFLUOROMETHYL TRIFLUOROVINYL see TKG800
ETHINYL TRICHLORIDE see TIO750
ETHONE, 2-CHLORO-1,2-DIPHENYL-(9CI) see CFJ100
2-ETHOXYACETANILIDE see ABG350
2′-ETHOXYACETANILIDE see ABG350
m-ETHOXYANILINE see PDK800
o-ETHOXYANILINE see PDK819
2-ETHOXYANILINE see PDK819
3-ETHOXYANILINE see PDK800
3-ETHOXYBENZENAMINE see PDK800
4-ETHOXY-3-HYDROXYBENZALDEHYDE see IKO100
1-ETHOXY-2-NITROBENZENE see NIC990
N-(2-ETHOXYPHENYL)ACETAMIDE see ABG350
ETHRIOL see HDF300
ETHYL 703 see FAB000
ETHYL ALDEHYDE see AAG250
ETHYL ALUMINUM DICHLORIDE see EFU050
ETHYLAMINE see EFU400
ETHYLAMINE, 2-BROMO- see BNI650
3-ETHYLAMINO-4-METHYLPHENOL see EGF100
ETHYL ANTIOXIDANT 703 see FAB000
ETHYL BENZOYL ACETATE see EGR600
ETHYL BROMIDE see EGV400
2-ETHYL-1-BUTANOL, SILICATE see TCD250
γ-ETHYLBUTYROLACTONE see HDY600
γ-ETHYL-n-BUTYROLACTONE see HDY600
ETHYL CADMATE see BJB500
2-ETHYLCAPROYL CHLORIDE see EKO600
ETHYLCARBONYL PIPERAZINE see EHG050
ETHYL γ-CHLOROACETOACETATE see EHH100
ETHYL 4-CHLOROACETOACETATE see EHH100
ETHYL 2-CHLOROACRYLATE see EHH200
ETHYL α-CHLOROACRYLATE see EHH200
ETHYL(CHLOROETHYL)ANILINE see EHJ600
ETHYL 4-CHLORO-3-OXOBUTANOATE see EHH100
ETHYL CHRYSANTHEMATE see EHM100
ETHYL CHRYSANTHEMUMATE see EHM100
ETHYLDICHLOROALUMINUM see EFU050
O-ETHYL DITHIOCARBAMATE see EQG600
O-ETHYL DITHIOCARBONATE see EQG600
1,1′-ETHYLENE-2,2′-BIPYRIDYLIUM DIBROMIDE see DWX800
N,N′-ETHYLENEBIS(2-(o-HYDROXYPHENYL)GLYCINE) see EIV100
1,1′-ETHYLENEBIS(PYRIDINIUM)BROMIDE see EIT100
1,1′-ETHYLENEBISUREA see EJC100
ETHYLENEDIAMINE-N,N′-BIS(2-HYDROXYPHENYLACETIC ACID) see EIV100
ETHYLENEDIAMINE, N,N′-DIBENZYL- see BHB300
ETHYLENEDIAMINE, N,N′-DIBENZYL-, DIACETATE see DDF800
ETHYLENEDIAMINE, N,N-DIETHYL-N′-PHENYL- see DJV300
ETHYLENEDIAMINE-DI(o-HYDROXYPHENYL)ACETIC ACID see EIV100
ETHYLENEDIAMINE-DI(2-HYDROXYPHENYL)ACETIC ACID see EIV100
ETHYLENEDIAMINE, N,N-DIISOPROPYL- see DNP700

ETHYLENEDIAMINE, N,N′-DIPHENYL- see DWB400
ETHYLENEDIAMINE, N-METHYL- see MJW100
ETHYLENEDIAMINETETRAACETIC ACID, CALCIUM DISODIUM CHELATE see CAR780
ETHYLENE, 1,2-DIBROMO- see DDN950
ETHYLENE DIGLYCIDYL ETHER see EEA600
(ETHYLENEDINITRILO)TETRAACETATE DIPOTASSIUM SALT see EEB100
(ETHYLENEDINITRILO)TETRAACETIC ACID CADMIUM(II) COMPLEX see CAF750
1,1′-ETHYLENEDIPYRIDINIUM DIBROMIDE see EIT100
ETHYLENE DIPYRIDYLIUM DIBROMIDE see DWX800
1,1-ETHYLENE 2,2-DIPYRIDYLIUM DIBROMIDE see DWX800
1,1′-ETHYLENE-2,2′-DIPYRIDYLIUM DIBROMIDE see DWX800
ETHYLENEDIUREA see EJC100
1,1′-ETHYLENEDIUREA see EJC100
2,2′-(1,2-ETHYLENEDIYL)BIS(5-AMINOBENZENESULFONIC ACID) see FCA100
ETHYLENE GLYCOL DIGLYCIDYL ETHER see EEA600
ETHYLENE HEXACHLORIDE see HCI000
1,8-ETHYLENENAPHTHALENE see AAF275
ETHYLENE TETRACHLORIDE see PCF275
ETHYLENE, TRIBROMO- see THV100
ETHYLENE TRICHLORIDE see TIO750
ETHYLENE TRITHIOCARBONATE see EJQ100
ETHYLENGLYKOLDIGLYCIDYLETHER see EEA600
ETHYLESTER KYSELINY GALLOVE see EKM100
N-ETHYL-ETHANAMINE see DHJ200
ETHYLFORMAMIDE see EKK600
N-ETHYLFORMAMIDE see EKK600
ETHYL GALLATE see EKM100
ETHYL 10-HENDECENOATE see EQD200
2-ETHYLHEXANOIC ACID CHLORIDE see EKO600
2-ETHYLHEXANOYL CHLORIDE see EKO600
2-ETHYLHEXYL-6-CHLORIDE see EKU100
2-ETHYLHEXYL GLYCIDYL ETHER see GGY100
(((2-ETHYLHEXYL)OXY)METHYL)OXIRANE see GGY100
ETHYL HYDROGEN ADIPATE see ELC600
N-ETHYL-N-(2-HYDROXYETHYL)-m-TOLUIDINE see HKS100
ETHYL α-HYDROXYISOBUTYRATE see ELH700
ETHYL 2-HYDROXYISOBUTYRATE see ELH700
ETHYL 2-HYDROXY-2-METHYLPROPANOATE see ELH700
ETHYLIDENE CHLORIDE see DFF809
ETHYLIDENE DICHLORIDE see DFF809
ETHYL ISOTHIOCYANATE see ELX525
ETHYLISOVANILLINE see IKO100
ETHYL 2-METHYLLACTATE see ELH700
ETHYL MUSTARD OIL see ELX525
ETHYL 1-NAPHTHALENEACETATE see ENL900
ETHYL 1-NAPHTHYLACETATE see ENL900
ETHYL NITROACETATE see ENN100
ETHYL (p-NITROBENZOYL)ACETATE see ENO100
ETHYL (4-NITROBENZOYL)ACETATE see ENO100
ETHYL 4-NITRO-β-OXOBENZENEPROPANOATE see ENO100
ETHYL OCTADECANOATE see EPF700
ETHYL n-OCTADECANOATE see EPF700
ETHYL β-OXOBENZENEPROPANOATE see EGR600
3-ETHYL-2-PENTANOL see EOB300
3-ETHYL-3-PENTANOL see TJP550
2-ETHYLPEROXYHEXANOIC ACID tert-PENTYL ESTER see AOM200
4-ETHYLPHENOL see EOE100
ETHYL PHOSPHONOTHIOIC DICHLORIDE see EOP600
ETHYL PHOSPHONOTHIOIC DICHLORIDE, anhydrous (DOT) see EOP600
ETHYLPHOSPHONOTHIONIC DICHLORIDE see EOP600
ETHYL PHOSPHONOTHIOYL DICHLORIDE see EOP600
1-ETHYLQUINALDINIUM IODIDE see EPD600
ETHYL STEARATE see EPF700
ETHYLTHIONOPHOSPHONYL DICHLORIDE see EOP600
ETHYLTHIOPHOSPHONIC DICHLORIDE see EOP600
2-(N-ETHYL-m-TOLUIDINO)ETHANOL see HKS100
ETHYL 3,4,5-TRIHYDROXYBENZOATE see EKM100
ETHYLTRIMETHYLAMMONIUM IODIDE see EQC600
ETHYLTRIMETHYLOLMETHANE see HDF300
ETHYL TUADS see BJB500
ETHYL UNDECENOATE see EQD200
ETHYL 10-UNDECENOATE see EQD200
ETHYL UNDECYLENATE see EQD200
ETHYL XANTHATE see EQG600
ETHYLXANTHIC ACID see EQG600
ETHYL XANTHOGENATE see EQG600
1-ETHYNYLETHANOL see EQM600
α-ETHYNYL-α-METHYLBENZYL ALCOHOL see EQN230
ETILAMINA (ITALIAN) see EFU400
ETRIOL see HDF300
ETTRIOL see HDF300

ETYLOAMINA (POLISH) see EFU400
ETYLU BROMEK (POLISH) see EGV400
EULAXAN see PPN100
EVOLA see DEP800
EXHAUST GAS see CBW750
EXT. D and C RED NO. 7 see SEH475

F 1 see FAB000
1167 F see DJV300
F 7771 see HDW100
F 1 (antioxidant) see FAB000
FA see FMV000
FAC see FAS700
FAD see RIF100
FAGINE see CMF800
FALKITOL see HCI000
FANNOFORM see FMV000
FAP see FPT100
FARINEX 100 see SLJ500
FASCIOLIN see CBY000, HCI000
FAST DISPERSE YELLOW 2K see DUW500
FASTOGEN GREEN 5005 see PJQ100
FASTOGEN GREEN B see PJQ100
FASTOLUX GREEN see PJQ100
FAST ORANGE RD OIL see CEG800
FAST ORANGE RD SALT see CEG800
FAST ORANGE SALT RD see CEG800
FAST ORANGE SALT RDA see CEG800
FAST ORANGE SALT RDN see CEG800
FAST RED BASE 3JL see KDA050
FAST RED BASE RL see MMF780
FAST RED 2NC BASE see KDA050
FAST RED 3GL BASE see KDA050
FAST RED 3GL SPECIAL BASE see KDA050
FAST RED 5NT see MMF780
FAST RED 5NT SALT see MMF780
FAST RED RL BASE see MMF780
FAST RED SALT RL see MMF780
FAST SCARLET see DXO850
FATSCO ANT POISON see ARD750
FB/2 see DWX800
F-114B2 see FOO525
FC 113 see TJE100
FC143 see DKH300
FC133a see TJE100
FC143a see TJY900
FC 114B2 see FOO525
FEDAL-UN see PCF275
FEGLOX see DWX800
FEIDMIN 5 see TES800
FEMA 3186 see MJH905
FEMA No. 2003 see AAG250
FEMA No. 2009 see ABH000
FEMA No. 2011 see AEN250
FEMA No. 2135 see BDX000
FEMA No. 2174 see BPU750
FEMA No. 2509 see DTD800
FENACET FAST PINK RF see AKO350
FENACET FAST YELLOW 2R see DUW500
FENAKROM RED W see SEH475
FENAKROM YELLOW R see SIT850
FENALAC GREEN G see PJQ100
FENALAC GREEN G DISP see PJQ100
FENAMIN SKY BLUE see CMO500
FENAN BLUE BCS see DFN425
FENAN BLUE RSN see IBV050
FENANTHREN BLUE BC see DFN425
FENANTHREN BLUE BD see DFN425
FENANTHREN BLUE RS see IBV050
FENANTHREN BRILLIANT VIOLET 2R see DFN450
FENANTHREN BRILLIANT VIOLET 4R see DFN450
FENANTHREN OLIVE R see DUP100
FENAZO LIGHT BLUE AE see APG700
FENOLIPUNA see PDO800
1-FENYL-1,2-ETHANDIOL see SMQ100
4-(1-FENYLETHYL)FENOL see PFD400
FENYLGLYCOL see SMQ100
FENYLSILAN see SDX250
FERRIC AMMONIUM CITRATE see FAS700
FERRIC AMMONIUM CITRATE, GREEN see FAS700
FERRIC ARSENATE, solid (DOT) see IGN000
FERRIC ARSENITE, solid (DOT) see IGO000

FERRIC ARSENITE, BASIC

FERRIC ARSENITE, BASIC see IGO000
FERRITIN see FBB000
FERROUS ARSENATE (DOT) see IGM000
FERROUS ARSENATE, solid (DOT) see IGM000
FERROUS FERRITE see IHG100
FERRO YELLOW see CAJ750
FILTEX WHITE BASE see CAT775
FINNCARB 6002 see CAT775
FITROL see KBB600
FITROL DESICCITE 25 see KBB600
FIXANOL C see HCP800
FIXIERER P see THM900
FLAKE WHITE see BKW100
FLAVANONE see FBW150
4-FLAVANONE see FBW150
FLAVIANIC ACID see FBZ100
FLAVIN ADENIN DINUCLEOTIDE see RIF100
FLAVIN ADENINE DINUCLEOTIDE see RIF100
FLAVINAT see RIF100
FLAVINE-ADENINE DINUCLEOTIDE see RIF100
FLAVINE ADENOSINE DIPHOSPHATE see RIF100
FLAVITAN see RIF100
FLAVONE see PER700
FLAVONE, 3-HYDROXY- see HLC600
FLAVONIC ACID see FCA100
FLECK-FLIP see TIO750
FLO-GARD see SCH000
FLOPROPION see TKP100
FLOPROPIONE see TKP100
FLOWER of PARADISE see HMX600
FLUATE see TIO750
FLUE DUST, ARSENIC CONTAINING see ARE500
FLUE GAS see CBW750
FLUKOIDS see CBY000
FLUOBRENE see FOO525
FLUORACETATO di (ITALIAN) see SHG500
FLUORENE see FDI100
9H-FLUORENE see FDI100
FLUORESSIGSAEURE (GERMAN) see SHG500
FLUOROACETIC ACID, SODIUM SALT see SHG500
p-FLUOROBENZONITRILE see FGH100
4-FLUOROBENZONITRILE see FGH100
FLUOROCARBON FC143 see DKH300
FLUOROCARBON FC143a see TJY900
4-FLUORO-2-METHYLBENZENAMINE see FJT050
1-FLUORO-2-METHYLBENZENE see FLZ100
4-FLUORO-3-NITROANILINE see FKK100
o-FLUOROPHENOL see FKT100
2-FLUOROPHENOL see FKT100
FLUOROPHENYLALANINE see FLD100
p-FLUOROPHENYLALANINE see FLD100
d,l-FLUOROPHENYLALANINE see FLD100
4-FLUORO-dl-PHENYLALANINE see FLD100
dl-4-FLUOROPHENYLALANINE see FLD100
d,l-p-FLUOROPHENYLALANINE see FLD100
m-FLUOROSULFONYLBENZENESULFONYL CHLORIDE see FLY200
o-FLUOROTOLUENE see FLZ100
2-FLUOROTOLUENE see FLZ100
FLUOROTRIPHENYLSTANNANE see TMV850
FONOLINE see MQV750
FOOD BLUE 4 see IBV050
FORMALDEHYD (CZECH, POLISH) see FMV000
FORMALDEHYDE see FMV000
FORMALDEHYDE, solution (DOT) see FMV000
FORMALDEHYDE, OXIME see FMW400
FORMALDOXIME see FMW400
FORMALIN see FMV000
FORMALIN 40 see FMV000
FORMALIN (DOT) see FMV000
FORMALINA (ITALIAN) see FMV000
FORMALINE (GERMAN) see FMV000
FORMALIN-LOESUNGEN (GERMAN) see FMV000
FORMALITH see FMV000
FORMAMIDE, N,N'-(DITHIOBIS(2-(2-HYDROXYETHYL)-1-METHYL-VINYLENE))BIS(N-((4-AMINO-2-METHYL-5-PYRIMIDINYL)METHYL)- see TES800
FORMAMIDE, N-ETHYL- see EKK600
FORMIC ACID, DITHIOBIS(THIO-, O,O-DIBUTYL ESTER see BSS550
FORMIC ACID, (PHENYLAZO)-, 2-PHENYLHYDRAZIDE see DVY950
FORMIC ALDEHYDE see FMV000
FORMOL see FMV000
FORMOXIME see FMW400

o-FORMYLBENZOIC ACID see FNK010
2-FORMYLBENZOIC ACID see FNK010
2-FORMYL-6,6-DIMETHYLBICYCLO(3.1.1)HEPT-2-ENE see FNK150
N-FORMYLETHYLAMINE see EKK600
3-FORMYLINDOLE see FNO100
2-FORMYL-5-METHYLFURAN see MKH550
m-FORMYLPHENOL see FOE100
3-FORMYLPHENOL see FOE100
3-FORMYLPYRIDINE see NDM100
4-FORMYLPYRIDINE see IKZ100
4-FORMYLRESORCINOL see REF100
FOSSIL FLOUR see SCH000
FOURAMINE PCH see CCP850
FOURRINE 68 see CCP850
FPA see FLD100
FR 28 see DXG035
FRATOL see SHG500
FREON 141 see FOO550
FREON 152 see DKH300
FREON 114B2 see FOO525
FREON FT see TJE100
FUMED SILICA see SCH000
FUMED SILICON DIOXIDE see SCH000
FUMING LIQUID ARSENIC see ARF500
2-FURALDEHYDE, 5-METHYL- see MKH550
FURAN, 2-ACETYL- see ACM200
2-FURANACRYLIC ACID see FPK050
FURAN, 2-(BROMOMETHYL)TETRAHYDRO- see TCS550
2-FURANMETHANEDIOL, 5-NITRO-, DIACETATE see NGB800
2-FURANMETHANOL, ACETATE (9CI) see FPM100
2-FURANMETHYL ACETATE see FPM100
3-FURANOL, TETRAHYDRO- see HOI200
2(3H)-FURANONE, DIHYDRO-5-OCTYL- see OES100
2(3H)-FURANONE, 5-ETHYLDIHYDRO- see HDY600
2(3H)-FURANONE, 5-HEPTYLDIHYDRO- see HBN200
FURAN, TETRAHYDRO-2-(BROMOMETHYL)- see TCS550
1-(2-FURANYL)ETHANONE see ACM200
FURATOL see SHG500
FURFURYL ACETATE see FPM100
6-FURFURYLADENINE see FPT100
N-FURFURYLADENINE see FPT100
N⁶-FURFURYLADENINE see FPT100
FURFURYL ALCOHOL, ACETATE see FPM100
6-(FURFURYLAMINO)PURINE see FPT100
N⁶-(FURFURYLAMINO)PURINE see FPT100
1-FURFURYLPYRROLE see FPX100
N-FURFURYL PYRROLE see FPX100
N-(2-FURFURYL)PYRROLE see FPX100
2-FURYL METHYL KETONE see ACM200
FUSED BORAX see DXG035
FUSED QUARTZ see SCK600
FUSED SILICA (ACGIH) see SCK600
FYDE see FMV000

GALACTASOL A see SLJ500
GALLIC ACID, ETHYL ESTER see EKM100
GAROLITE SA see CAT775
GEASTIGMOL see GCE600
GEASTIMOL see GCE600
GEFARNATE see GDG200
GEFARNIL see GDG200
GEFARNYL see GDG200
GENAMIN 16R302D see HCP525
GENAMIN DSAC see DXG625
GENITRON BSH see BBS300
GENVIS see SLJ500
GERANALDEHYDE see GCU100
GERANIAL see GCU100
GERANIOL ACETATE see DTD800
GERANIOL BUTYRATE see GDE810
GERANYL ACETATE (FCC) see DTD800
GERANYL BUTANOATE see GDE810
GERANYL BUTYRATE see GDE810
GERANYL n-BUTYRATE see GDE810
GERANYL FARNESYLACETATE see GDG200
GERLACH 1396 see NBJ100
GERMALGENE see TIO750
GERMANE, TETRAPROPYL- see TED650
GILDER'S WHITING see CAT775
G.L. 102 see EIT100
GLOMAX see KBB600
GLUCOFURANOSE, 1,2,5,6-DI-O-ISOPROPYLIDENE-, α-D- see DVO100

GLUCOPYRANOSIDE, PHENYL-, β-D- see PDO300
β-D-GLUCOPYRANOSIDE, PHENYL- see PDO300
α-D-GLUCOPYRANOSIDE, 1,3,4,6-TETRA-O-ACETYL-β-D-FRUCTOFURANOSYL-, TETRAACETATE (9CI) see OAF100
GLUCOSACCHARONIC ACID see SAA025
(+ −)-GLUTAMIC ACID see GFM300
GLUTAMIC ACID, dl- see GFM300
dl-GLUTAMIC ACID (9CI) see GFM300
GLYCERALDEHYDE, (+ −)- see GFY200
dl-GLYCERALDEHYDE see GFY200
dl-GLYCERIC ALDEHYDE see GFY200
GLYCEROL MONOOCTADECYL ETHER see GGA915
GLYCIDOL METHYL ETHER see GGW600
GLYCIDYL 2-ETHYLHEXYL ETHER see GGY100
GLYCIDYL ISOCYANURATE see GGY150
GLYCIDYL METHYL ETHER see GGW600
GLYCIDYL 4-METHYLPHENYL ETHER see GGY175
GLYCIDYL p-TOLYL ETHER see GGY175
GLYCINE, N-BENZOYL- see HGB300
GLYCINE BETAINE see GHA050
GLYCINE, N,N-DIHYDROXYETHYL- see DMT500
GLYCINE, N,N'-ETHYLENEBIS(2-(o-HYDROXYPHENYL)- see EIV100
GLYCINE, ETHYL ESTER, HYDROCHLORIDE see GHA100
GLYCINE, N,N^1-1,2-CYCLOHEXANEDIYLBIS(N-(CARBOXYMETHYL)-(9CI) see CPB120
GLYCINE, N-(p-NITROPHENYL)- see NIN100
GLYCOCOLL BETAINE see GHA050
GLYCOL DIGLYCIDYL ETHER see EEA600
GLYCYLBETAINE see GHA050
GLYKOKOLLBETAIN see GHA050
GLYKOKOLLBETAIN-CHLORID see CCH850
GLYMOL see MQV750
GLYOXANILIDE OXIME see GIK100
GLYOXYLANILIDE, OXIME see GIK100
GLYOXYLANILIDE, 2-OXIME see GIK100
GMP see GLS750
5'-GMP see GLS750
GODALAX see PPN100
GOSSYPIMINE see GJI400
GOSSYPINE see CMF800
GRANADA GREEN LAKE GL see PJQ100
GRAPHIC RED Y see NAP100
GRAPHTOL BLUE RL see IBV050
GRAPHTOL GREEN 2GLS see PJQ100
GRASEX see CDN550
GREENOCKITE see CAJ750
GREY ARSENIC see ARA750
GUANIDINE, AMINO-, HYDROCHLORIDE see AKC800
GUANIDINE CARBONATE see GKW100
GUANIDINE MONOPHOSPHATE see GLS750
GUANIDINIUM CARBONATE see GKW100
GUANOSINE MONOPHOSPHATE see GLS750
GUANOSINE 5'-MONOPHOSPHATE see GLS750
GUANOSINE 5'-MONOPHOSPHORIC ACID see GLS750
GUANOSINE 5'-PHOSPHATE see GLS750
GUANYLHYDRAZINE HYDROCHLORIDE see AKC800
GUANYLIC ACID see GLS750
5'-GUANYLIC ACID see GLS750

H-22 see BSG300
HAEMATEIN see HAO600
HAKUENKA CC see CAT775
HAKUENKA R 06 see CAT775
HALON 2001 see EGV400
HALON 2402 see FOO525
HANA see HMX600
HARMONE B 79 see DFN425
HAYNON see CLX300
HELANTHRENE BLUE BC see DFN425
HELIANE RED 5B see DNT300
HELIANTHRENE BLUE RS see IBV050
HELINDON RED BB see DNT300
HELIOGEN BLUE 6470 see IBV050
HELIOGEN GREEN 8680 see PJQ100
HELIOGEN GREEN 8730 see PJQ100
HELIOGEN GREEN 8681K see PJQ100
HELIOGEN GREEN 8682T see PJQ100
HELIOGEN GREEN A see PJQ100
HELIOGEN GREEN G see PJQ100
HELIOGEN GREEN GA see PJQ100
HELIOGEN GREEN GN see PJQ100
HELIOGEN GREEN GNA see PJQ100
HELIOGEN GREEN GTA see PJQ100
HELIOGEN GREEN GV see PJQ100
HELIOGEN GREEN GWS see PJQ100
HEMATEIN see HAO600
HEMATINE see HAO600
HENEICOSAFLUORO-1-IODODECANE see IEU075
HENNA see HMX600
HEPARTEST see HAQ600
HEPARTESTABROME see HAQ600
HEPATOSULFALEIN see HAQ600
HEPTADECAFLUORO-1-IODOOCTANE see PCH325
HEPTADECANE see HAS100
n-HEPTADECANE see HAS100
2,2,3,3,4,4,4-HEPTAFLUOROBUTANOL see HAW100
HEPTAMINE see HBM490
1-HEPTANAMINE see HBL600
2-HEPTANAMINE see HBM490
HEPTANE, 1-BROMO- see HBN100
HEPTANE, 1-CHLORO-5-METHYL- see EKU100
HEPTEDRINE see HBM490
5-HEPTEN-2-OL, 6-METHYL-2-(4-METHYL-3-CYCLOHEXEN-1-YL)- see BGO775
HEPTYL ACRYLATE see HBL100
HEPTYLAMINE see HBL600
n-HEPTYLAMINE see HBL600
1-HEPTYLAMINE see HBL600
2-HEPTYLAMINE see HBM490
HEPTYLAMINE, 1-METHYL- see OEK010
HEPTYL BROMIDE see HBN100
n-HEPTYL BROMIDE see HBN100
γ-HEPTYLBUTYROLACTONE see HBN200
γ-n-HEPTYLBUTYROLACTONE see HBN200
HEPTYL p-HYDROXYBENZOATE see HBO650
HEPTYL 4-HYDROXYBENZOATE see HBO650
HEPTYL PARABEN see HBO650
HERNIARIN (6CI) see MEK300
HEXACHLOR-AETHAN (GERMAN) see HCI000
HEXACHLOROETHANE see HCI000
1,1,1,2,2,2-HEXACHLOROETHANE see HCI000
HEXACHLOROETHYLENE see HCI000
HEXADECANE see HCO600
n-HEXADECANE see HCO600
HEXADECANOIC ACID, HEXADECYL ESTER see HCP700
1-HEXADECANOL, ACETATE see HCP100
1-HEXADECANOL, HYDROGEN SULFATE, SODIUM SALT see HCP900
HEXADECYL ACETATE see HCP100
HEXADECYLAMINE, N,N-DIMETHYL- see HCP525
HEXADECYLDIMETHYLAMINE see HCP525
HEXADECYL PALMITATE see HCP700
HEXADECYLPYRIDINE BROMIDE see HCP800
HEXADECYLPYRIDINIUM BROMIDE see HCP800
1-HEXADECYLPYRIDINIUM BROMIDE see HCP800
N-HEXADECYLPYRIDINIUM BROMIDE see HCP800
HEXADECYL SODIUM SULFATE see HCP900
HEXAFLUOROEPOXYPROPANE see HDF050
HEXAFLUORO-1,2-EPOXYPROPANE see HDF050
HEXAFLUOROISOBUTYLENE see HDC450
3,3,3,4,4,4-HEXAFLUOROISOBUTYLENE see HDC450
HEXAFLUOROPROPENE EPOXIDE see HDF050
HEXAFLUOROPROPENE OXIDE see HDF050
HEXAFLUOROPROPYLENE OXIDE (DOT) see HDF050
HEXAGLYCERINE see HDF300
HEXAHYDRO-2-AZEPINONE see CBF700
HEXAHYDRO-2H-AZEPIN-2-ONE see CBF700
HEXAHYDROBENZOIC ACID see HDH100
HEXAHYDROBENZYL ALCOHOL see HDH200
4-HEXAHYDROISONICOTINIC ACID see ILG100
HEXAHYDROPICOLINIC ACID see HDS300
HEXAHYDRO-1,3,5-TRIMETHYL-s-TRIAZINE see HDW100
HEXAHYDRO-1,3,5-TRIS(1-OXO-2-PROPENYL)-1,3,5-TRIAZINE see THM900
γ-HEXALACTONE see HDY600
HEXAMETHYLENEBIS(DITHIOCARBAMIC ACID) DISODIUM SALT see HEF400
1,6-HEXAMETHYLENEDIAMINE see HEO000
HEXAMETHYLENE DIAMINE, solid (DOT) see HEO000
HEXAMETHYLENE DIIODIDE see HEG200
HEXAMETHYLPHOSPHOROUS TRIAMIDE see HEK100
2,6,10,15,19,23-HEXAMETHYLTETRACOSANE see SLG700
1,6-HEXANEDIAMINE see HEO000
1,6-HEXANEDIAMINE, N,N,N',N'-TETRAMETHYL- see TDR100
HEXANE, 1,6-DIIODO- see HEG200
HEXANEDINITRILE see AER250
1,6-HEXANEDIOIC ACID see AEN250

HEXANEDIOIC ACID, BIS(2-(2-BUTOXYETHOXY)ETHYL) ESTER (9CI)

HEXANEDIOIC ACID, BIS(2-(2-BUTOXYETHOXY)ETHYL) ESTER (9CI) see DDT500
HEXANEDIOIC ACID DINITRILE see AER250
6-HEXANELACTAM see CBF700
HEXANOIC ACID, METHYL ESTER see MHY700
HEXANOIC ACID, MONOETHYL ESTER (9CI) see ELC600
γ-HEXANOLACTONE see HDY600
HEXANOLIDE-1,4 see HDY600
3-HEXANOL, 3-METHYL- see MKL350
HEXANONE ISOXIME see CBF700
HEXANONISOXIM (GERMAN) see CBF700
HEXANOYL CHLORIDE, 2-ETHYL- see EKO600
2-HEXENE-4-ONE see HFE200
2-HEXEN-4-ONE see HFE200
d-erythro-HEX-2-ENONIC ACID, γ-LACTONE see SAA025
1,6-HEXOLACTAM see CBF700
HEXYLACETYLENE see OGI025
HEXYL CAPRYLATE see HFS600
n-HEXYL CAPRYLATE see HFS600
HEXYL OCTANOATE see HFS600
3-HEXYNE, 2,5-DIMETHYL-2,5-DI(t-BUTYLPEROXY)- see DRJ825
HF 191 see DGV700
HIDACHROME YELLOW 2G see SIT850
HIDACO SAFRANINE see GJI400
HILTONIL FAST RED 3GL BASE see KDA050
HILTONIL FAST RED RL BASE see MMF780
HILTOSAL FAST ORANGE RD SALT see CEG800
HILTOSAL FAST RED RL SALT see MMF780
HIPPURIC ACID see HGB300
HI-SEL see SCH000
HISPACROM YELLOW 2G see SIT850
HISPACROM YELLOW 2GR see SIT850
HISPAMIN RED 4B see DXO850
HISPAMIN SKY BLUE 3B see CMO500
HISTADUR see CLX300
HI-YIELD DESSICANT H-10 see ARB250
HMDA see HEO000
HOCH see FMV000
(+ −)-HOMATROPINE BROMIDE see HGH150
HOMATROPINE HYDROBROMIDE see HGH150
HOMOCAL D see CAT775
HOMOCHLORCYCLIZINE see HGI100
HOMOCHLOROCYCLIZINE see HGI100
HOMOCLOMINE see HGI100
HOMODAMON see HGI100
HOMONEURINE CHLORIDE see HGI600
HOMOPROLINE see HDS300
HOMORESTAR see HGI100
HOMOVERATRIC ACID see HGK600
HOSTACILLIN see PAQ200
HOSTAPERM GREEN GG see PJQ100
HRW 13 see SLJ500
1,1-H,H-HEPTAFLUOROBUTANOL see HAW100
HYDANTOIN, 1,3-DIBROMO-5,5-DIMETHYL- see DDI900
HYDELTRASOL see PLY275
HYDRACILLIN see PAQ200
HYDRANGIN see HJY100
HYDRANGINE see HJY100
HYDRATROPIC ALCOHOL see HGR600
HYDRATROPYL ALCOHOL see HGR600
HYDRAZIDE BSG see BBS300
HYDRAZINE, 1-(p-BROMOPHENYL)-, HYDROCHLORIDE see BNW550
HYDRAZINECARBOXIMIDAMIDE HYDROCHLORIDE see AKC800
HYDRAZINE, 1,2-DIBENZOYL- see DDE250
HYDRAZINE, (2,5-DICHLOROPHENYL)- see DGC850
HYDRITE see KBB600
HYDROCARB 60 see CAT775
1-HYDROPEROXYCYCLOHEXYL-1-HYDROXYCYCLOHEXYL PEROXIDE see CPC300
o-HYDROQUINONE see CCP850
HYDROQUINONE, 2,5-DI-tert-BUTYL- see DEC800
HYDROXAMIC ACID, BENZSULFO- see BDW100
m-HYDROXYBENZALDEHYDE see FOE100
3-HYDROXYBENZALDEHYDE see FOE100
meta-HYDROXYBENZALDEHYDE see FOE100
4-HYDROXYBENZENEACETIC ACID see HNG600
HYDROXYBENZENESULFONIC ACID see HJH100
p-HYDROXYBENZENESULFONIC ACID ZINC SALT see ZIJ300
o-HYDROXYBENZHYDRAZIDE see HJL100
2-HYDROXYBENZOHYDRAZIDE see HJL100
p-HYDROXYBENZOIC ACID HEPTYL ESTER see HBO650
o-HYDROXYBENZOIC ACID HYDRAZIDE see HJL100
2-HYDROXYBENZOIC ACID HYDRAZIDE see HJL100
2-HYDROXYBENZOIC ACID MONOAMMONIUM SALT see ANT600
o-HYDROXYBENZOYLHYDRAZIDE see HJL100
2-HYDROXYBENZOYLHYDRAZIDE see HJL100
o-HYDROXYBENZOYLHYDRAZINE see HJL100
2-HYDROXYBENZOYLHYDRAZINE see HJL100
2′-HYDROXYCHALCONE see HJS900
trans-o-HYDROXYCINNAMIC ACID see CNU850
trans-2-HYDROXYCINNAMIC ACID see CNU850
HYDROXYCITRONELLAL DIMETHYL ACETAL see LBO050
HYDROXYCITRONELLAL DMA see LBO050
7-HYDROXYCOUMARIN see HJY100
4-HYDROXY-3,5-DIMETHOXYBENZOIC ACID see SPE700
2-HYDROXY-3,5-DINITROBENZOIC ACID see HKE600
8-HYDROXY-5,7-DINITRO-2-NAPHTHALENESULFONIC ACID see FBZ100
HYDROXYDIPHENYLACETIC ACID see BBY990
4-HYDROXYDIPHENYLDIMETHYLMETHANE see COF400
HYDROXYDIPHENYLMETHANE see HKF300
N-HYDROXYETHYL-N-ETHYL-m-TOLUIDINE see HKS100
3-HYDROXYFLAVONE see HLC600
4-HYDROXYHEXANOIC ACID LACTONE see HDY600
1-HYDROXY-1-HYDROPEROXYDICYCLOHEXYL PEROXIDE see CPC300
1-HYDROXY-1′-HYDROPEROXYDICYCLOHEXYL PEROXIDE see CPC300
HYDROXYHYDROQUINONE TRIACETATE see HLF600
2-HYDROXYHYDROQUINONE TRIACETATE see HLF600
3-HYDROXY-4-(2-HYDROXY-1-NAPHTHYLAZO)-1-NAPHTHALENESULFONIC ACID, SODIUM see HLI100
2-(HYDROXYIMINO)-N-PHENYLACETAMIDE see GIK100
m-HYDROXYIODOBENZENE see IEV010
o-HYDROXYLBENZHYDRAZIDE see HJL100
2-HYDROXYMESITYLENE see MDJ740
2-HYDROXY-4-METHOXYBENZALDEHYDE see HLR600
1-HYDROXY-6-METHOXYPHENAZINE 5,10-DIOXIDE see HLT100
4-(4-HYDROXY-3-METHOXYPHENYL)-3-BUTEN-2-ONE see MLI800
2-HYDROXY-5-METHYL-1,3-BENZENEDIMETHANOL see HLV100
HYDROXYMETHYLCYCLOHEXANE see HDH200
4-HYDROXY-6-METHYL-2H-PYRAN-2-ONE see THL800
3-(HYDROXYMETHYL)PYRIDINE see NDW510
4-(HYDROXYMETHYL)PYRIDINE see POR810
8-HYDROXY-2-METHYLQUINOLINE see HMQ600
2-HYDROXY-4-(METHYLTHIO)BUTANOIC ACID see HMR600
β-HYDROXYNAPHTHALDEHYDE see HMU200
2-HYDROXYNAPHTHALDEHYDE see HMU200
2-HYDROXYNAPHTHALENE see NAX000
β-HYDROXYNAPHTHALENE see NAX000
2-HYDROXY-1-NAPHTHALENECARBOXALDEHYDE see HMU200
3-HYDROXY-2-NAPHTHALENECARBOXYLIC ACID see HMX520
2-HYDROXY-1,4-NAPHTHALENEDIONE see HMX600
3-HYDROXY-2-NAPHTHOIC ACID see HMX520
2-HYDROXYNAPHTHOQUINONE see HMX600
2-HYDROXY-1,4-NAPHTHOQUINONE see HMX600
4-HYDROXY-3-NITROBENZOIC ACID METHYL ESTER see HMY075
2-HYDROXY-4-(OCTYLOXY)BENZOPHENONE see HND100
(2-HYDROXY-4-(OCTYLOXY)PHENYL)PHENYLMETHANONE see HND100
2-HYDROXY-4-OKTYLOXYBENZOFENON see HND100
o-HYDROXYPHENOL see CCP850
2-HYDROXYPHENOL see CCP850
(p-HYDROXYPHENYL)ACETIC ACID see HNG600
(4-HYDROXYPHENYL)ACETIC ACID see HNG600
α-HYDROXY-α-PHENYLBENZENEACETIC ACID see BBY990
(3-HYDROXYPHENYL)DIMETHYLAMINE see HNK560
1-(2-HYDROXYPHENYL)-3-PHENYL-2-PROPEN-1-ONE see HJS900
(E)-3-(2-HYDROXYPHENYL)-2-PROPENOIC ACID see CNU850
4-HYDROXYPHENYLSULFONIC ACID see HNL600
N-HYDROXYPHTHALIMIDE see HNS600
2-HYDROXY-1,2,3-PROPANETRICARBOXYLIC ACID TRILITHIUM SALT HYDRATE see LHD150
(S)-2-HYDROXYPROPANOIC ACID see LAG010
(S)-2-HYDROXYPROPIONIC ACID see LAG010
β-HYDROXYPROPYL METHACRYLATE see HNV500
2-HYDROXYPROPYL METHACRYLATE see HNV500
2-HYDROXYPROPYL 2-METHYL-2-PROPENOATE see HNV500
(2-HYDROXYPROPYL)TRIMETHYLAMMONIUM CHLORIDE see MIM300
2-HYDROXYPYRIDINE see OOO200
3-HYDROXYPYRIDINE see PPH025
4-HYDROXYPYRIDINE see HOB100
β-HYDROXYPYRIDINE see PPH025
γ-HYDROXYPYRIDINE see HOB100
6-HYDROXY-4(1H)-PYRIMIDINONE see PPP100
8-HYDROXYQUINALDINE see HMQ600
HYDROXYQUINALDINE see HMQ600
8-HYDROXYQUINOLINE BENZOATE see HOE100

3-HYDROXYQUINUCLIDINE see QUJ990
4-HYDROXYSALICYLALDEHYDE see REF100
3-HYDROXYTETRAHYDROFURAN see HOI200
4-HYDROXY-2,2,6,6-TETRAMETHYL-1-PIPERIDINYLOXY see HOI300
2-HYDROXY-N,N,N-TRIMETHYL-1-PROPANAMINIUM CHLORIDE see MIM300
HYDROXYTRIPHENYLSILANE see HOM300
4-HYDROXYUNDECANOIC ACID LACTONE see HBN200
4-HYDROXYUNDECANOIC ACID, γ-LACTONE see HBN200
HYDROXYZINE DIHYDROCHLORIDE see HOR470
HYDROXYZINE HYDROCHLORIDE see HOR470
HYOSCYAMINE SULFATE see HOT600
(-)-HYOSCYAMINE SULFATE see HOT600
HYPERNIC EXTRACT see LFT800
HYPNONE see ABH000

IMIDAZOLEACRYLIC ACID see UVJ440
IMIDAZOLE-4-ACRYLIC ACID see UVJ440
5-IMIDAZOLEACRYLIC ACID see UVJ440
5-IMIDAZOLECARBOXAMIDE, 4-AMINO-, HYDROCHLORIDE see IAL100
1H-IMIDAZOLE-4-CARBOXAMIDE, 5-AMINO-, MONOHYDROCHLORIDE see IAL100
1H-IMIDAZOLE, 4,5-DIHYDRO-2-PHENYL- see PFJ300
2,4-IMIDAZOLIDINEDIONE, 1,3-DIBROMO-5,5-DIMETHYL-(9CI) see DDI900
2-IMIDAZOLIDINONE, 1,3-DINITRO- see DUW503
2-IMIDAZOLINE, 2-PHENYL- see PFJ300
3-(1H-IMIDAZOL-4-YL)-2-PROPENOIC ACID see UVJ440
1,1′-IMIDODIACETONITRILE see IBB100
IMIDODICARBONIMIDIC DIAMIDE, N-(2-METHYLPHENYL)-(9CI) see TGX550
2,2′-IMINOBISACETONITRILE see IBB100
4,4′-IMINOBISPHENOL see IBJ100
IMINODIACETONITRILE see IBB100
4,4′-IMINODIPHENOL see IBJ100
2-IMINOTHIAZOLIDINE see TEV600
INDAN, 4,6-DINITRO-1,1,3,3,5-PENTAMETHYL- see MRU300
INDANTHREN BLUE see IBV050
INDANTHREN BLUE BC see DFN425
INDANTHREN BLUE BCA see DFN425
INDANTHREN BLUE BCS see DFN425
INDANTHREN BLUE GP see IBV050
INDANTHREN BLUE GPT see IBV050
INDANTHREN BLUE RPT see IBV050
INDANTHREN BLUE RS see IBV050
INDANTHREN BLUE RSN see IBV050
INDANTHREN BLUE RSP see IBV050
INDANTHREN BRILLIANT BLUE R see IBV050
INDANTHREN BRILLIANT VIOLET 4R see DFN450
INDANTHREN BRILLIANT VIOLET RR see DFN450
INDANTHRENE see IBV050
INDANTHRENE BLUE see IBV050
INDANTHRENE BLUE BC see DFN425
INDANTHRENE BLUE BCF see DFN425
INDANTHRENE BLUE GP see IBV050
INDANTHRENE BLUE RP see IBV050
INDANTHRENE BLUE RS see IBV050
INDANTHRENE BLUE RSA see IBV050
INDANTHRENE BLUE RSN see IBV050
INDANTHRENE BRILLIANT VIOLET 4R see DFN450
INDANTHRENE BRILLIANT VIOLET RR see DFN450
INDANTHRENE OLIVE R see DUP100
INDANTHREN OLIVE R see DUP100
INDANTHREN PRINTING BLUE FRS see IBV050
INDANTHREN PRINTING BLUE KRS see IBV050
INDANTHREN PRINTING VIOLET F 4R see DFN450
INDANTHRONE see IBV050
INDIGO see BGB275
INDIGO BLUE see BGB275
INDIGO CIBA see BGB275
INDIGO CIBA SL see BGB275
INDIGO J see BGB275
INDIGO N see BGB275
INDIGO NAC see BGB275
INDIGO NACCO see BGB275
INDIGO P see BGB275
INDIGO PLN see BGB275
INDIGO POWDER W see BGB275
INDIGO PURE BASF see BGB275
INDIGO PURE BASF POWDER K see BGB275
INDIGO SYNTHETIC see BGB275
INDIGOTIN see BGB275
INDIGO VS see BGB275
INDO BLUE B-I see DFN425
INDO BLUE WD 279 see DFN425

INDOFAST VIOLET LAKE see DFN450
INDOLE, 3-ACETATO- see IDA600
INDOLE-3-ALDEHYDE see FNO100
INDOLE-3-CARBALDEHYDE see FNO100
INDOLE-3-CARBOXALDEHYDE see FNO100
1H-INDOLE-3-CARBOXALDEHYDE (9CI) see FNO100
INDOLE, 2,3-DIMETHYL- see DSI850
1H-INDOLE, 2,3-DIMETHYL-(9CI) see DSI850
INDOLEPROPIONIC ACID see ICS200
β-INDOLEPROPIONIC ACID see ICS200
1H-INDOLE-3-PROPIONIC ACID see ICS200
INDOLINE, 2-METHYLENE-1,3,3-TRIMETHYL- see TLU200
INDOL-3-OL, ACETATE (ester) (8CI) see IDA600
1H-INDOL-3-OL, ACETATE (ester) (9CI) see IDA600
3H-INDOL-3-ONE, 2(1,3-DIHYDRO-3-OXO-2H-INDOL-2-YLIDENE)-1,2-DIHYDRO-(9CI) see BGB275
β-INDOLYLALDEHYDE see FNO100
3-(3-INDOLYL)PROPANOIC ACID see ICS200
INDOPHENOL, 2,6-DICHLORO-, SODIUM SALT see SGG650
INDOTONER BLUE B 79 see DFN425
INDOXYLACETATE see IDA600
INDOXYL-O-ACETATE see IDA600
INFLAMASE see PLY275
INTERCHEM ACETATE FAST PINK DNA see AKO350
INTRAVAT BLUE GF see DFN425
p-IODOBENZOIC ACID see IEE025
1-IODO-3-METHYLBUTANE see IHU200
1-IODOOCTANE see OFA100
1-IODOPENTANE see IET500
1-IODOPERFLUORODECANE see IEU075
1-IODOPERFLUOROOCTANE see PCH325
m-IODOPHENOL see IEV010
3-IODOPHENOL see IEV010
IODURE d'ETHYL-TRIMETHYL-AMMONIUM see EQC600
IONOL 6 see MHR050
IRGALITE FAST BRILLIANT GREEN GL see PJQ100
IRGALITE FAST BRILLIANT GREEN 3GL see PJQ100
IRGALITE GREEN GLN see PJQ100
IRGALITE RED RL see NAP100
IRGasTAB T 4 see DEJ100
IRGasTAB T 150 see DEJ100
IRGasTAB T 290 see DEJ100
IRON(III) AMMONIUM CITRATE see FAS700
IRON ARSENATE (DOT) see IGM000
IRON(II) ARSENATE (3582) see IGM000
IRON(III) ARSENATE (1581) see IGN000
IRON(III)-o-ARSENITE PENTAHYDRATE see IGO000
IRON BLACK see IHC550
IRON(II,III) OXIDE see IHC550
IRON MASS, spent (UN1376) (DOT) see IHG100
IRON MASS, not properly oxidized (NA1383) (DOT) see IHG100
IRON OXIDE, spent see IHG100
IRON OXIDE, spent (UN1376) (DOT) see IHG100
IRON OXIDE RED 130B see IHG100
IRON SPONGE, spent (UN1376) (DOT) see IHG100
IRON SPONGE, not properly oxidized (NA1383) (DOT) see IHG100
2406 I.S. see IBB100
ISOALLOXAZINE-ADENINE DINUCLEOTIDE see RIF100
ISOAMYL IODIDE see IHU200
ISO-AMYL NITRATE see ILW100
ISOASCORBIC ACID see SAA025
d-ISOASCORBIC ACID see SAA025
ISOBUTYLENEOXIDE see IIQ600
O-ISOBUTYL POTASSIUM XANTHATE see EQG600
ISOBUTYL STEARATE see IJN100
ISOCINEOLE see IKC100
1-(1-ISOCYANATO-1-METHYLETHYL)-3-(1-METHYLETHENYL)BENZENE see IKG800
ISOCYANIC ACID, m-ISOPROPENYL-α-α-DIMETHYL BENZYL ESTER see IKG800
β-ISODURYLIC ACID see IKN300
ISOETHYLVANILLIN see IKO100
1H-ISOINDOLE-1,3(2H)-DIONE, 5-NITRO- see NIX100
1H-ISOINDOLE-1,3(2H)-DIONE, 3-α-4,7,7-α-TETRAHYDRO-(9CI) see TDB100
ISONAPHTHOL see NAX000
ISONICOTINALDEHYDE see IKZ100
ISONICOTINAMIDE see ILC100
ISONICOTINIC ACID AMIDE see ILC100
ISONICOTINIC ALDEHYDE see IKZ100
ISONIPECOTIC ACID see ILG100
ISONITROSOACETANILIDE see GIK100
2-ISONITROSOACETANILIDE see GIK100
ISONITROSOACETYLANILINE see GIK100

ISOPENTYL ALCOHOL, NITRATE see ILW100
ISOPENTYL IODIDE see IHU200
ISOPENTYL NITRATE see ILW100
ISOPHTHALIC ACID, DIETHYL ESTER see IMK100
ISOPROPYL ALCOHOL, TITANIUM(4+) SALT see IRN200
2-ISOPROPYLNAPHTHALENE see IQN100
ISOPROPYL ORTHOTITANATE see IRN200
m-ISOPROPYLPHENOL see IQX090
o-ISOPROPYLPHENOL see IQX100
2-ISOPROPYLPHENOL see IQX100
3-ISOPROPYLPHENOL see IQX090
ISOPROPYL POTASSIUM XANTHATE see IRG050
ISOPROPYL STEARATE see IRL100
ISOPROPYL TITANATE(IV) see IRN200
ISOPROPYLUREA see IRR100
1-ISOPROPYLUREA see IRR100
N-ISOPROPYLUREA see IRR100
ISOQUINOLINE, 1-(3,4-DIETHOXYBENZYL)-6,7-DIETHOXY-, HYDROCHLORIDE see PAH260
ISOQUINOLINE, 1-((3,4-DIETHOXYPHENYL)METHYL)-6,7-DIETHOXY-, HYDROCHLORIDE see PAH260
1-ISOTHIOCYANATOBUTANE see ISD100
ISOTHIOCYANATOETHANE see ELX525
ISOTHIOCYANIC ACID, BUTYL ESTER see ISD100
ISOTHIOCYANIC ACID, ETHYL ESTER see ELX525
ISOTHIOINDIGO see DNT300
ISOVITAMIN C see SAA025
ISOXAZOLE, 3,5-DIMETHYL- see DSK950
IVALON see FMV000

JANTARAN SODNY see SJW100
JASMONE see JCA100
(Z)-JASMONE see JCA100
cis-JASMONE see JCA100
JAVA CHROME YELLOW GT see SIT850
JAVA UNICHROME YELLOW GT see SIT850
JENACILLIN O see PAQ200
4-JODBENZOESAEURE see IEE025
1-JOD-3-METHYLBUTAN see IHU200
1-JODOKTAN see OFA100
1-JODPENTAN see IET500
3-JODPHENOL see IEV010
JORTAINE see GHA050

K 250 see CAT775
KADMIUM (GERMAN) see CAD000
KADMIUMCHLORID (GERMAN) see CAE250
KADMIUMSTEARAT (GERMAN) see OAT000
KADMU TLENEK (POLISH) see CAH500
KAKO RED RL BASE see MMF780
KALIUMARSENIT (GERMAN) see PKV500
KALIUMNITRAT (GERMAN) see PLL500
KALZIUMARSENIAT (GERMAN) see ARB750
KAOLIN see KBB600
KAOPAOUS see KBB600
KAOPHILLS-2 see KBB600
e-KAPROLAKTAM (CZECH) see CBF700
KARSAN see FMV000
KATINE see NNM510
KAYACARBON BIC see BSD150
KAYAKU BENZOPURPURINE 4B see DXO850
KAYAKU DIRECT SKY BLUE 5B see CMO500
KAYAKU FAST RED 3GL BASE see KDA050
KAYAKU MORDANT YELLOW GG see SIT850
KAYAKU RED RL BASE see MMF780
KAYALON FAST YELLOW RR see DUW500
KAYDOL see MQV750
KD 83 see DXG625
KEESTAR see SLJ500
2-KETOHEXAMETHYLENIMINE see CBF700
KETONE, 2-BENZOFURANYL METHYL see ACC100
KETONE, 2-FURYL METHYL see ACM200
KETONE METHYL PHENYL see ABH000
KETONE, METHYL 4-PYRIDYL see ADA365
KETONE, PHENYL 2-PYRIDYL see PGE760
4-KETOPENTANOIC ACID BUTYL ESTER see BRR700
KHLADON 114B2 see FOO525
KILL-ALL see SEY500
KINETIN see FPT100
KINETIN (PLANT HORMONE) see FPT100
KING'S YELLOW see ARI000
KINIDIN DURETTER see QFS100

KINIDIN DURULES see QFS100
KINILENTIN see QFS100
KITON FAST BLUE G see APG700
KLONDIKE YELLOW X-2261 see PJQ100
KN 320 see IHC550
KNOCKBAL see BSG300
KOBALT (GERMAN, POLISH) see CNA250
KOHLENMONOXID (GERMAN) see CBW750
KOHLENOXYD (GERMAN) see CBW750
2,4,6-KOLLIDIN see TME272
KOMPLEXON IV see CPB120
KONDREMUL see MQV750
KOOLMONOXYDE (DUTCH) see CBW750
KOTAMITE see CAT775
KP 201 see DGV700
KREDAFIL 150 EXTRA see CAT775
KREDAFIL RM 5 see CAT775
KROMAD see KHU000
KROMON SODIUM LITHOL see NAP100
KS 4B see DEJ100
KS 1300 see CAT775
KULU 40 see CAT775
KYSELINA ADIPOVA (CZECH) see AEN250
KYSELINA 4,4'-AZO-BIS-(4-KYANVALEROVA) see ASL500
KYSELINA 1,2-CYKLOHEXYLENDIAMINTETRAOCTOVA see CPB120
KYSELINA 3-HYDROXY-2-NAFTOOVA see HMX520
KYSELINA 4-METHOXYBENZOOVA see AOU600
KYSELINA N-(4-NITROFENYL)OXAMOVA see NHX100
KYSELINA SULFANILOVA see SNN600
KYSELINA TERFTALOVA (CZECH) see TAN750

L-36352 see DUV600
LA96A see PPN100
LABRODA see TKP100
LABRODAX see TKP100
LABRODAX SUPANATE see TKP100
LACO see PPN100
(+)-LACTIC ACID see LAG010
d-LACTIC ACID see LAG010
(S)-LACTIC ACID see LAG010
l-(+)-LACTIC ACID see LAG010
(S)-(+)-LACTIC ACID see LAG010
LACTIC ACID, CADMIUM SALT see CAG750
LACTIC ACID, METHYL ESTER see MLC600
LACTIC ACID, 2-METHYL-, ETHYL ESTER see ELH700
LAKE FAST BLUE BS see IBV050
LAKE FAST BLUE GGS see IBV050
LAKE RED R see NAP100
LAKE RED RL see NAP100
LANADIN see TIO750
LANGFORD see KBB600
LATEXOL FAST BLUE SD see IBV050
LATHANOL see SIB700
LATHANOL LAL see SIB700
LATHANOL-LAL 70 see SIB700
LAURIC ACID, CADMIUM SALT (2:1) see CAG775
LAURIC ACID, METHYL ESTER see MLC800
LAURINE DIMETHYL ACETAL see LBO050
LAURYLAMINE HYDROCHLORIDE see DKW100
LAURYLAMMONIUM HYDROCHLORIDE see DKW100
LAURYLPYRIDINIUM CHLORIDE see LBX050
1-LAURYLPYRIDINIUM CHLORIDE see LBX050
LAVAMENTHE see LBX100
LAWSONE see HMX600
LAXADIN see PPN100
LAXAGETTEN see PPN100
LAXANIN N see PPN100
LAXANS see PPN100
LAXOREX see PPN100
LEATHER RED HT see GJI400
LEDCLAIR see CAR780
LEPTOPHOS PHENOL see LEN050
LETHURIN see TIO750
LEUCOINDOPHENOL see IBJ100
LEUCOQUINIZARIN see LEX200
LEVIGATED CHALK see CAT775
LEVULINIC ACID, BUTYL ESTER see BRR700
LIGHT RED RS see NAP100
LILLY 36,352 see DUV600
LIMAWOOD EXTRACT see LFT800
LINALOOL TETRAHYDRIDE see LFY510
LIONOGEN BLUE R see IBV050

LITHIUM ACETATE see LGO100
LITHIUM CITRATE HYDRATE see LHD150
LITHIUM HYDROXIDE see LHI100
LITHIUM HYDROXIDE (Li(OH)) (9CI) see LHI100
LITHOL RED see NAP100
LITHOL RED 3580 see NAP100
LITHOL RED 17676 see NAP100
LITHOL RED B see NAP100
LITHOL RED LAKE see NAP100
LITHOL RED R see NAP100
LITHOL RED RB EXTRA see NAP100
LITHOL RED RL 151 see NAP100
LITHOL RED RS see NAP100
LITHOL RED 3GS see NAP100
LITHOL RED SODIUM SALT see NAP100
LITHOL RED TONER see NAP100
LITHOL TONER see NAP100
LITHOL TONER EXTRA LIGHT 5000 see NAP100
LITHOL TONER SODIUM SALT RT 314 see NAP100
LITHOL TONER YA 8003 see NAP100
LITHOSOL DEEP BLUE V see BGB275
LORAMINE AMB 13 see GHA050
LORSBAN see CMA100
LO-VEL see SCH000
LPC see LBX050
LUDOX see SCH000
LUPERSOL 8 see BSC600
LUPERSOL TBIC see BSD150
LUPERSOL TBIC-M75 see BSD150
LUPETAZINE see LIQ500
LUPETIDINE see LIQ550
2,6-LUPETIDINE see LIQ550
LURIDINE see CMF800
LUTETIA FAST BLUE RS see IBV050
LUTETIA FAST EMERALD J see PJQ100
LUTETIA RED R see NAP100
2,4-LUTIDINE see LIY990
LYCINE see GHA050
LYSILAN see HGI100
LYSOFORM see FMV000

MA300A see DEJ100
MACQUER'S SALT see ARD250
MACROLEX BLUE FR see BKP500
MAGISTERY OF BISMUTH see BKW100
MAGNACAT see MAP750
MAGNESIUM ARSENATE see ARD000
MAGNESIUM ARSENATE PHOSPHOR see ARD000
MAGNETIC BLACK see IHC550
MAGNETIC OXIDE see IHC550
MAGNETITE see IHC550
MAGRACROM YELLOW GG see SIT850
MAIZENA see SLJ500
MANGAN (POLISH) see MAP750
MANGANESE see MAP750
MANGAN NITRIDOVANY (CZECH) see MAP750
MARANTA see SLJ500
MARBLEWHITE 325 see CAT775
MARFIL see CAT775
MARKURE UL2 see DEJ100
MAYVAT OLIVE AR see DUP100
MBOCA see MJM200
MCNAMEE see KBB600
MC-T see CAT775
MEA HYDROCHLORIDE see EEC700
MEDIUM BLUE see IBV050
MEHENDI see HMX600
MEISI FAST RED RL BASE see MMF780
MELOGEL see SLJ500
MELUNA see SLJ500
MENDI see HMX600
p-MENTHANE, 1,4-EPOXY- see IKC100
p-MENTHAN-8-OL see MCE100
p-MENTH-8-EN-1-OL see TBD775
t-MENTH-1-EN-8-OL see TBD775
p-MENTH-4(8)-EN-3-ONE, (R)-(+)- see POI615
MERAMEC M 25 see IHC550
2-MERCAPTOBENZOXAZOLE see MCK900
((2-MERCAPTOETHYL)TRIMETHYLAMMONIUM IODIDE ACETATE) see ADC300
2-(MERCAPTOMETHYL)PYRIDINE see POR790
2-MERCAPTOPYRIDINE see TFP300
2-MERCAPTOPYRIDINE MONOXIDE see MCQ700

2-MERCAPTOQUINOLINE see QOJ100
m-MERCAPTOTOLUENE see TGO800
MERCATE 5 see SAA025
MERCURIALIN see MGC250
MERCURY, (9-OCTADECENOATO-O)PHENYL-, (Z)-(9CI) see PFP100
MERCURY, (OLEATO)PHENYL- see PFP100
MESITALDEHYDE see MDJ745
MESITOIC ACID see IKN300
MESITOL see MDJ740
MESITYL ALCOHOL see MDJ740
MESITYL ALDEHYDE see MDJ745
MESITYLENECARBOXALDEHYDE see MDJ745
2-MESITYLENECARBOXALDEHYDE see MDJ745
α^1,α^3-MESITYLENEDIOL, 2-HYDROXY-(7CI,8CI) see HLV100
MESITYLENIC ACID see MDJ748
MESOXALONITRILE, (m-CHLOROPHENYL)HYDRAZONE (8CI) see CKA550
METAARSENIC ACID see ARB000
METACHROME YELLOW see SIT850
METACHROME YELLOW RA see SIT850
METALLIC ARSENIC see ARA750
METANILAMIDE see MDM760
METAPIRAZONE see AIB300
METASAP XX see AHH825
METHACIDE see TGK750
METHACRYLIC ACID, DIESTER with TETRAETHYLENE GLYCOL see TCE400
METHACRYLIC ACID, DIESTER with TRIETHYLENE GLYCOL see MDN510
METHACRYLIC ACID, 2-(DIETHYLAMINO)ETHYL ESTER see DIB300
METHACRYLIC ACID, 2-HYDROXYPROPYL ESTER see HNV500
METHACRYLIC ACID, 2-(VINYLOXY)ETHYL ESTER see VQA150
METHANAL see FMV000
METHANAMINE (9CI) see MGC250
METHANAMINIUM, 1-CARBOXY-N,N,N-TRIMETHYL-, CHLORIDE see CCH850
METHANE, BIS(4-CHLOROPHENYL)- see BIM800
METHANE, BROMOTRIPHENYL- see TNP600
METHANE, PHENYL- see TGK750
METHANE TETRACHLORIDE see CBY000
METHANOL, CYCLOHEXYL- see HDH200
1,4-METHANONAPHTHALENE-5,8-DIONE, 1,4,4a,8a-TETRAHYDRO- see TCU650
METHANONE, (2-AMINO-5-CHLOROPHENYL)PHENYL- see AJE400
METHANONE, BIS(2-HYDROXY-4-METHOXYPHENYL)-(9CI) see BKH200
METHANONE, DIPHENYL-, OXIME (9CI) see BCS400
METHANONE, (2-HYDROXY-4-METHOXYPHENYL)(2-HYDROXYPHENYL)-(9CI) see DMW250
METHANONE, (2-HYDROXY-4-(OCTYLOXY)PHENYL)PHENYL-(9CI) see HND100
METHIONINE, N-ACETYL-, dl- see ACQ270
dl-METHIONINE, N-ACETYL-(9CI) see ACQ270
METHIONINE HYDROXY ANALOG see HMR600
METHOLENE 2095 see MHY650
METHOLENE 2296 see MLC800
1-METHOXY-2-ACETOXYPROPANE see PNL265
3-METHOXYANILINE see AVO890
3-METHOXYANISOLE see REF025
1-METHOXY-9,10-ANTHRACENEDIONE see MEA650
1-METHOXYANTHRAQUINONE see MEA650
m-METHOXYBENZALDEHYDE see AOT300
3-METHOXYBENZALDEHYDE see AOT300
3-METHOXYBENZENAMINE see AVO890
p-METHOXYBENZOIC ACID see AOU600
4-METHOXYBENZOIC ACID see AOU600
o-METHOXYBENZOIC ACID METHYL ESTER see MLH800
7-METHOXY-2H-1-BENZOPYRAN-2-ONE see MEK300
o-METHOXYBROMOBENZENE see BMT400
p-METHOXYBROMOBENZENE see AOY450
2-METHOXYBROMOBENZENE see BMT400
4-METHOXYBROMOBENZENE see AOY450
(9S)-6'-METHOXYCINCHONAN-9-OL SULFATE (1:1) (SALT) see QFS100
7-METHOXYCOUMARIN see MEK300
o-METHOXY METHYL BENZOATE see MLH800
(METHOXYMETHYL)OXIRANE see GGW600
6-METHOXY-1-METHYL-1,2,3,4-TETRAHYDRO-β-CARBOLINE see MLJ100
4-METHOXY-6-METHYL-1,3,5-TRIAZIN-2-AMINE see MGH800
2-METHOXY-1,4-NAPHTHALENEDIONE see MEY800
2-METHOXYNAPHTHOQUINONE see MEY800
2-METHOXY-1,4-NAPHTHOQUINONE see MEY800
6-METHOXY-1-PHENAZINOL 5,10-DIOXIDE see HLT100
o-METHOXYPHENYL BROMIDE see BMT400
p-METHOXYPHENYL BROMIDE see AOY450
2-METHOXYPHENYL BROMIDE see BMT400
4-METHOXYPHENYL BROMIDE see AOY450
2-METHOXYPROPENE see MFL300
3-METHOXYPROPYLENE OXIDE see GGW600

4-METHOXYSALICYLALDEHYDE

4-METHOXYSALICYLALDEHYDE see HLR600
2-METHOXYTETRACHLOROPHENOL see TBQ290
2-METHOXY-3,4,5,6-TETRACHLOROPHENOL see TBQ290
2-METHOXY-1,3,5-TRINITROBENZENE see TMK300
4-METHYLAESCULETIN see MJV800
METHYL ALDEHYDE see FMV000
METHYLAMINE see MGC250
METHYLAMINEN (DUTCH) see MGC250
METHYL p-AMINOBENZOATE see AIN150
2-(METHYLAMINO)ETHYLAMINE see MJW100
2-METHYL-4-AMINO-6-METHOXY-s-TRIAZINE see MGH800
METHYL o-ANISATE see MLH800
METHYLBENZENE see TGK750
3-METHYL-1,2-BENZENEDIAMINE see TGY800
4-METHYLBENZENEDIAZONIUM TETRAFLUOROBORATE (1-) see TGM450
α-METHYLBENZENEETHANOL see PGA600
β-METHYLBENZENEETHANOL see HGR600
METHYL BENZENESULFONATE see MHB300
m-METHYLBENZENETHIOL see TGO800
3-METHYLBENZENETHIOL see TGO800
METHYLBENZOL see TGK750
p-(6-METHYLBENZOTHIAZOL-2-YL)ANILINE see MHJ300
4-(6-METHYL-2-BENZOTHIAZOLYL)BENZENAMINE see MHJ300
2-METHYLBENZYLAMINE see MHM510
METHYL BENZYL DISULFIDE see MHN350
METHYL BENZYL KETONE see MHO100
p-(α-METHYLBENZYL)PHENOL see PFD400
4-(α-METHYLBENZYL)PHENOL see PFD400
p-METHYLBIPHENYL see MJH905
4-METHYL-1,1′-BIPHENYL see MJH905
4-METHYL-2,6-BIS(1-PHENYLETHYL)PHENOL see MHR050
4-METHYL-2-BROMOANILINE see BOG500
m-METHYLBROMOBENZENE see BOG300
2-METHYLBROMOBENZENE see BOG260
3-METHYLBROMOBENZENE see BOG300
METHYL 4-BROMO-2-BUTENOATE see MHR790
METHYL BROMOCROTONATE see MHR790
METHYL γ-BROMOCROTONATE see MHR790
METHYL 4-BROMOCROTONATE see MHR790
3-METHYL-2-BUTANAMINE see AOE200
3-METHYL-1-BUTANOL NITRATE see ILW100
1-METHYL-4-tert-BUTYLBENZENE see BSP500
p-METHYL-tert-BUTYLBENZENE see BSP500
3-METHYLBUTYLIODIDE see IHU200
METHYL CADMIUM AZIDE see MHY550
METHYL CAPRATE see MHY650
METHYL n-CAPRATE see MHY650
METHYL CAPRINATE see MHY650
METHYL CAPROATE see MHY700
METHYL CAPRONATE see MHY700
METHYL CAPRYLATE see MHY800
β-METHYLCHOLINE CHLORIDE see MIM300
METHYL DECANOATE see MHY650
METHYL n-DECANOATE see MHY650
METHYL 2,4-D ESTER see DAA825
N-METHYLDIAMINOETHANE see MJW100
METHYLDINITROBENZENE see DVG600
p-METHYLDIPHENYL see MJH905
4-METHYLDIPHENYL see MJH905
METHYL DIPHENYLMETHYL KETONE see MJH910
METHYL DODECANOATE see MLC800
METHYL DODECYLATE see MLC800
METHYLENEAMINE N-OXIDE see FMW400
2,2′-METHYLENEBIPHENYL see FDI100
4,4′-METHYLENE(BIS)-CHLOROANILINE see MJM200
p,p′-METHYLENEBIS(o-CHLOROANILINE) see MJM200
p,p′-METHYLENEBIS(α-CHLOROANILINE) see MJM200
4,4′-METHYLENE BIS(2-CHLOROANILINE) see MJM200
METHYLENE-4,4′-BIS(o-CHLOROANILINE) see MJM200
4,4′-METHYLENEBIS(o-CHLOROANILINE) see MJM200
4,4′-METHYLENEBIS-2-CHLOROBENZENAMINE see MJM200
METHYLENEBIS(DIETHYLPHOSPHONATE) see MJT100
METHYLENE-BIS-ORTHOCHLOROANILINE see MJM200
METHYLENEDI(PHOSPHONIC ACID) TETRAETHYL ESTER see MJT100
METHYLENE GLYCOL see FMV000
METHYLENE OXIDE see FMV000
2-METHYLENE-1,3,3-TRIMETHYLINDOLINE see TLU200
METHYLESCULETIN see MJV800
4-METHYLESCULETIN see MJV800
4-METHYLESCULETOL see MJV800
N-METHYLETHANEDIAMINE see MJW100
N-METHYL-1,2-ETHANEDIAMINE see MJW100
N-METHYLETHYLENEDIAMINE see MJW100
N-METHYLETHYLIDENEDIAMINE see MJW100
1-METHYLETHYL OCTADECANOATE see IRL100
2-(1-METHYLETHYL)PHENOL see IQX100
(1-METHYLETHYL)UREA see IRR100
2-METHYL-4-FLUOROANILINE see FJT050
METHYLFLUOROFORM see TJY900
5-METHYL-2-FURALDEHYDE see MKH550
5-METHYL-2-FURANCARBOXALDEHYDE see MKH550
5-METHYL FURFURAL see MKH550
5-METHYL-2-FURFURAL see MKH550
5-METHYLFURFURALDEHYDE see MKH550
METHYL 2-FURYL KETONE see ACM200
METHYL GLYCIDYL ETHER see GGW600
1-METHYLHEPTYLAMINE see OEK010
METHYL HEXANOATE see MHY700
METHYL n-HEXANOATE see MHY700
3-METHYL-HEXANOL-(3) see MKL350
3-METHYL-3-HEXANOL see MKL350
METHYL HEXOATE see MHY700
1-METHYLHEXYLAMINE see HBM490
METHYL HEXYLATE see MHY700
6-METHYL-4-HYDROXYPYRON-(2) see THL800
2-METHYL-8-HYDROXYQUINOLINE see HMQ600
1-METHYL-4-ISOPROPYLCYCLOHEXANE-8-OL see MCE100
METHYLISOPSEUDOIONONE see TMJ100
S-METHYLISOTHIURONIUM SULFATE see MPV790
METHYL LACTATE see MLC600
2-METHYLLACTIC ACID ETHYL ESTER see ELH700
METHYL LAURATE see MLC800
METHYL LAURINATE see MLC800
3-METHYLMERCAPTOANILINE see ALX100
4-METHYLMERCAPTOPHENOL see MPV300
N-METHYLMETHANAMINE see DOQ800
METHYL o-METHOXYBENZOATE see MLH800
METHYL 2-METHOXYBENZOATE see MLH800
METHYL-3-METHOXY-4-HYDROXY STYRYL KETONE see MLI800
1-METHYL-6-METHOXY-1,2,3,4-TETRAHYDRO-β-CARBOLINE see MLJ100
6-METHYL-2-(4-METHYL-3-CYCLOHEXEN-1-YL)-5-HEPTEN-2-OL see BGO775
1-METHYL-4-(1-METHYLETHENYL)CYCLOHEXANOL see TBD775
1-METHYL-4-(1-METHYLETHYL)-7-OXABICYCLO(2.2.1)HEPTANE see IKC100
2-METHYL-2-(METHYLSULFINYL)PROPANAL O-((METHYLAMINO)CARBONYL)OXIME see MLW750
2-METHYL-2-(METHYLSULFINYL)PROPIONALDEHYDE O-(METHYLCARBAMOYL)OXIME see MLW750
2-METHYL-2-(METHYLTHIO)PROPANAL OXIME see MLX800
2-METHYL-4-NITROANILINE see MMF780
2-METHYL-1-NITRO-NAPHTHALENE see MMN800
2-METHYL-4-NITROPHENOL see NFV010
METHYL OCTANOATE see MHY800
METHYL OCTYL KETONE see OFE050
METHYL n-OCTYL KETONE see OFE050
2-METHYLOXINE see HMQ600
METHYL PENTANOATE see VAQ100
2-METHYL-PENTENE-2 see MNK100
2-METHYL-2-PENTENE see MNK100
3-METHYL-2-(cis-2-PENTEN-1-YL)-2-CYCLOPENTEN-1-ONE see JCA100
2-METHYLPHENANTHRENE see MNN520
α-METHYL-PHENETHYL ALCOHOL see PGA600
β-METHYLPHENETHYL ALCOHOL see HGR600
4-(1-METHYL-1-PHENETHYL)PHENOL see COF400
4-METHYLPHENOL SODIUM SALT see SIM100
((4-METHYLPHENOXY)METHYL)OXIRANE see GGY175
4-METHYL-N-PHENYLBENZENESULFONAMIDE see TGN600
o-METHYLPHENYL BROMIDE see BOG260
p-METHYLPHENYLDIAZONIUM FLUOROBORATE see TGM450
α-METHYL PHENYLETHYL ALCOHOL see HGR600
METHYLPHENYL GLYCIDYL ETHER see GGY175
N-(2-METHYLPHENYL)IMIDODICARBONIMIDIC DIAMIDE see TGX550
METHYL PHENYL KETONE see ABH000
3-METHYL-1-PHENYL-2-PHOSPHOLENE 1-OXIDE see MNV800
METHYL PICRATE see TMK300
N-METHYLPROPANAMIDE see MOS900
2-METHYL-2-PROPANOL see BPX000
N-METHYLPROPIONAMIDE see MOS900
N-METHYLPROPIONIC ACID AMIDE see MOS900
N-METHYLPROPIONSAEUREAMID see MOS900
3-METHYLPYRAZOLE see MOX010
5-METHYLPYRAZOLE see MOX010
3(5)-METHYLPYRAZOLE see MOX010
3-METHYL-2-PYRAZOLIN-5-ONE see MOX100
3-METHYL-PYRAZOLON-(5) see MOX100

3-METHYLPYRIDINE-1-OXIDE see MOY550
4-METHYLPYRIDINE 1-OXIDE see MOY790
1-METHYL-2(1H)-PYRIDINONE see MPA075
N-METHYLPYRIDONE see MPA075
1-METHYL-2(1H)-PYRIDONE see MPA075
METHYL 4-PYRIDYL KETONE see ADA365
2-METHYL-8-QUINOLINOL see HMQ600
METHYLSALICYLATE METHYL ETHER see MLH800
METHYLSILATRAN see MPI650
METHYLSILATRANE see MPI650
1-METHYLSILATRANE see MPI650
O-METHYLSINAPIC ACID see CMQ100
p-METHYLSTYRENE see VQK700
METHYL-2Δ-2 THIAZOLINE see DLV900
m-(METHYLTHIO)ANILINE see ALX100
4-(METHYLTHIO)ANILINE see AMS675
3-(METHYLTHIO)BENZENAMINE see ALX100
4-(METHYLTHIO)BENZENAMINE see AMS675
m-METHYLTHIOPHENOL see TGO800
p-(METHYLTHIO)PHENOL see MPV300
3-METHYLTHIOPHENOL see TGO800
4-(METHYLTHIO)PHENOL see MPV300
S-METHYLTHIOPSEUDOUREA SULFATE see MPV790
2-METHYL-2-THIOPSEUDOUREA SULFATE see MPV790
S-METHYLTHIOURONIUM SULFATE see MPV790
O-METHYLTHYMOL see MPW650
METHYL THYMOL ETHER see MPW650
METHYL THYMYL ETHER see MPW650
3-METHYL-1-p-TOLYL-PYRAZOLIN-5-ONE see THA300
METHYL 3,4,5-TRIMETHOXYBENZOATE see MQF300
1-METHYL-2,8,9-TRIOXO-5-AZA-1-SILABICYCLO(3.3.3)UNDECANE see MPI650
METHYLUMBELLIFERONE see MEK300
METHYL VALERATE see VAQ100
METHYL n-VALERATE see VAQ100
METHYL VALERIANATE see VAQ100
4-METHYL-5-VINYL THIAZOLE see MQM800
METILAMINE (ITALIAN) see MGC250
4,4-METILENE-BIS-o-CLOROANILINA (ITALIAN) see MJM200
METOMEGA CHROME YELLOW GM see SIT850
METYLOAMINA (POLISH) see MGC250
MHA ACID see HMR600
MHA-FA see HMR600
MHSK see MLI800
MICROCARB see CAT775
MICROLITH GREEN G-FP see PJQ100
MICROMIC CR 16 see CAT775
MICROMYA see CAT775
MICRO-PEN see PAQ200
MICROSETILE YELLOW 2R see DUW500
MICROWHITE 25 see CAT775
MIKETAZOL DEVELOPER ONS see HMX520
MIKETHRENE BLUE BC see DFN425
MIKETHRENE BLUE BCS see DFN425
MIKETHRENE BLUE RSN see IBV050
MIKETHRENE BRILLIANT BLUE R see IBV050
MIKETHRENE OLIVE R see DUP100
MIKETON FAST PINK RL see AKO350
MIKETON POLYESTER PINK RL see AKO350
MINERAL OIL see MQV750
MINERAL OIL, SLAB OIL see MQV875
MINERAL OIL, WHITE (FCC) see MQV750
MIO 40GN see IHG100
MISPICKEL see ARJ750
MITSUI ALIZARINE RED S see SEH475
MITSUI ALIZARINE SAPHIROL SE see APG700
MITSUI BENZOPURPURINE 4BX see DXO850
MITSUI CHROME YELLOW GG see SIT850
MITSUI DIRECT SKY BLUE 5B see CMO500
MITSUI INDIGO PASTE see BGB275
MITSUI INDIGO PURE see BGB275
MITSUI RED 3GL BASE see KDA050
MITSUI RED RL BASE see MMF780
MITSUI SAFRANINE see GJI400
MOCA see MJM200
MODR KYPOVA 1 see BGB275
MODR KYPOVA 4 see IBV050
MODR PIGMENT 60 see IBV050
MODR POTRAVINARSKA 4 see IBV050
MODR PRIMA 15 see CMO500
MOLOL see MQV750
MOLTEN ADIPIC ACID see AEN250
MONARCH GREEN WD see PJQ100
MONASTRAL FAST GREEN BGNA see PJQ100
MONASTRAL FAST GREEN G see PJQ100
MONASTRAL FAST GREEN GD see PJQ100
MONASTRAL FAST GREEN GF see PJQ100
MONASTRAL FAST GREEN GFNP see PJQ100
MONASTRAL FAST GREEN GN see PJQ100
MONASTRAL FAST GREEN GNA see PJQ100
MONASTRAL FAST GREEN GTP see PJQ100
MONASTRAL FAST GREEN GV see PJQ100
MONASTRAL FAST GREEN GWD see PJQ100
MONASTRAL FAST GREEN GX see PJQ100
MONASTRAL FAST GREEN GXB see PJQ100
MONASTRAL FAST GREEN GYH see PJQ100
MONASTRAL FAST GREEN LGNA see PJQ100
MONASTRAL FAST GREEN 2GWD see PJQ100
MONASTRAL GREEN B see PJQ100
MONASTRAL GREEN B PIGMENT see PJQ100
MONASTRAL GREEN G see PJQ100
MONASTRAL GREEN GFN see PJQ100
MONASTRAL GREEN GH see PJQ100
MONASTRAL GREEN GN see PJQ100
MONOAMYLAMINE see PBV505
MONOBROMOETHANE see EGV400
MONOCALCIUM ARSENITE see CAM500
MONOCALCIUM CARBONATE see CAT775
MONOCALCIUM DISODIUM EDTA see CAR780
MONOCHLOROCYCLOHEXANE see CPI400
MONOCHROME YELLOW MG see SIT850
MONOETHANOLAMINE HYDROCHLORIDE see EEC700
MONOETHYL ADIPATE see ELC600
MONOETHYLADIPIC ACID ESTER see ELC600
MONOETHYLAMINE (DOT) see EFU400
MONOETHYLAMINE, anhydrous (DOT) see EFU400
MONOETHYLENEDIUREA see EJC100
MONOETHYL HEXANEDIOATE see ELC600
MONOFLUORESSIGSAEURE, NATRIUM (GERMAN) see SHG500
MONOLITE FAST BLUE 3R see IBV050
MONOLITE FAST BLUE 3RD see IBV050
MONOLITE FAST BLUE RV see IBV050
MONOLITE FAST BLUE SRS see IBV050
MONOLITE FAST BLUE 2RV see DFN425
MONOLITE FAST BLUE 2RVSA see DFN425
MONOLITE FAST GREEN GVSA see PJQ100
MONOLITE FAST NAVY BLUE BV see BGB275
MONOLITE RED R see NAP100
MONOMETHYLAMINE see MGC250
MONOOCTADECYL ETHER of GLYCEROL see GGA915
MONOPOTASSIUM ARSENATE see ARD250
MONOPOTASSIUM DIHYDROGEN ARSENATE see ARD250
MONOPOTASSIUM 2-HYDROXYPROPANOATE ACID see PLK650
MONOSODIUM ARSENATE see ARD600
MONOSODIUM CYANURATE see SGB550
MONOTHIOBENZOIC ACID see TFC550
β-MONOXYNAPHTHALENE see NAX000
MORBOCID see FMV000
MORPAN CBP see HCP800
MORPAN T FAST see TCB200
MORPHOLINE, 4-(1-CYCLOPENTEN-1-YL)- see CPY800
MORPHOLINE, 4,4'-(DITHIODICARBONOTHIOYL)BIS-(9CI) see MRR090
4-MORPHOLINETHIOCARBONYL DISULFIDE see MRR090
(1-MORPHOLINOCYCLOPENTENE) see CPY800
MOSATIL see CAR780
MOSKENE see MRU300
MOTTENHEXE see HCI000
MS-1112 see BFV760
MSK-C see CAT775
M.T.F. see DUV600
MULTIFLEX MM see CAT775
MUSK AMBRETTE see OKU100
MUSK AMBRETTE (NATURAL) see OKU100
MUSK NATURAL see OKU100
MUSK TIBETENE see MRW272
MYRISTIC ACID, SODIUM SALT see SIN900
MYRISTICA OIL see NOG500
MYRISTYLTRIMETHYLAMMONIUM BROMIDE see TCB200
MYRRH OIL see MSB775
MYRTENAL see FNK150
MYRTENYL ACETATE see MSC050
(+)-MYRTENYL ACETATE see MSC050
MYTAB see TCB200
MYXIN see HLT100

N 34 see CAT775
NACCONOL LAL see SIB700
NAD see CNF390
NAD+ see CNF390
β-NAD see CNF390
NADIDE see CNF390
NADP see CNF400
β-NADP see CNF400
NAD PHOSPHATE see CNF400
β-NAFTOL (DUTCH) see NAX000
2-NAFTOL (DUTCH) see NAX000
β-NAFTOLO (ITALIAN) see NAX000
2-NAFTOLO (ITALIAN) see NAX000
NALCOAG see SCH000
NANOFIN see LIQ550
NANOPHYN see LIQ550
NAPHTAMINE BLUE 10G see CMO500
1-NAPHTHALDEHYDE, 2-HYDROXY- see HMU200
1-NAPHTHALENAMINE, 4-(PHENYLAZO)- see PEJ600
1-NAPHTHALENEACETIC ACID, ETHYL ESTER see ENL900
1-NAPHTHALENEACETIC ACID, SODIUM SALT see NAK550
NAPHTHALENE, 1-BROMO- see BNS200
1-NAPHTHALENECARBONITRILE see NAY090
1-NAPHTHALENECARBOXALDEHYDE, 2-HYDROXY-(9CI) see HMU200
NAPHTHALENE-α-CARBOXYLIC ACID see NAV490
1-NAPHTHALENECARBOXYLIC ACID (9CI) see NAV490
2-NAPHTHALENECARBOXYLIC ACID, 3-HYDROXY-(9CI) see HMX520
NAPHTHALENE, DECAHYDROOCTADECAFLUORO- see PCG700
NAPHTHALENE, 1,8-DINITRO- see DUX710
1,5-NAPHTHALENEDIOL see NAN505
2,3-NAPHTHALENEDIOL see NAN510
1,4-NAPHTHALENEDIONE, 2-HYDROXY-(9CI) see HMX600
1,4-NAPHTHALENEDIONE, 2-METHOXY-(9CI) see MEY800
NAPHTHALENE, 2-ISOPROPYL- see IQN100
1-NAPHTHALENENITRILE see NAY090
NAPHTHALENE, OCTADECAFLUORODECAHYDRO- see PCG700
1-NAPHTHALENESULFONIC ACID, 5-AMINO- see ALI240
1-NAPHTHALENESULFONIC ACID, 3,3′-((3,3′-DIMETHYL(1,1′-BIPHENYL)-4,4′-DIYL)BIS(AZO))BIS (4-AMINO-, DISODIUM SALT see DXO850
2-NAPHTHALENESULFONIC ACID, 5,7-DINITRO-8-HYDROXY- see FBZ100
2-NAPHTHALENESULFONIC ACID, 8-HYDROXY-5,7-DINITRO-(8CI) see FBZ100
1-NAPHTHALENESULFONIC ACID, 3-HYDROXY-4-(2-HYDROXY-1-NAPHTHYLAZO)-, SODIUM SALT see HLI100
1-NAPHTHALENESULFONIC ACID, 2-((2-HYDROXY-1-NAPHTHALENYL)AZO)-, MONOSODIUM SALT see NAP100
1,4,5,8-NAPHTHALENETETRACARBOXYLIC ACID see NAP250
2-NAPHTHALENOL see NAX000
2-NAPHTHALENOXYACETIC ACID see NBJ100
(β-NAPHTHALENYLOXY)ACETIC ACID see NBJ100
NAPHTHANIL RED 3G BASE see KDA050
NAPHTHANTHRACENE see BBC250
NAPHTHOELAN RED RL BASE see MMF780
α-NAPHTHOFLAVONE see NBI100
α-NAPHTHOFLUORENE see NAU600
α-NAPHTHOIC ACID see NAV490
1-NAPHTHOIC ACID see NAV490
2-NAPHTHOIC ACID, 3-HYDROXY- see HMX520
β-NAPHTHOL see NAX000
2-NAPHTHOL see NAX000
NAPHTHOLACTAM see NAX100
1,8-NAPHTHOLACTAM see NAX100
2-NAPHTHOL, 5-AMINO- see ALJ500
NAPHTHOL B see NAX000
NAPHTHOL B.O.N. see HMX520
α-NAPHTHONITRILE see NAY090
1-NAPHTHONITRILE see NAY090
3H-NAPHTHO(2,1-b)PYRAN-2-CARBOXYLIC ACID, 3-OXO-, ETHYL ESTER see OOI200
4H-NAPHTHO(1,2-b)PYRAN-4-ONE, 2-PHENYL- see NBI100
1,4-NAPHTHOQUINONE, 2-HYDROXY- see HMX600
1,4-NAPHTHOQUINONE, 2-METHOXY- see MEY800
NAPHTHOSTYRIL see NAX100
2-NAPHTHOXYACETIC ACID see NBJ100
β-NAPHTHOXYACETIC ACID see NBJ100
β-NAPHTHYL ALCOHOL see NAX000
α-NAPHTHYLCARBOXYLIC ACID see NAV490
NAPHTHYLENEETHYLENE see AAF275
α-NAPHTHYLFLAVONE see NBI100
O-(2-NAPHTHYL)GLYCOLIC ACID see NBJ100
β-NAPHTHYL HYDROXIDE see NAX000
α-NAPHTHYLNITRILE see NAY090
1-NAPHTHYLNITRILE see NAY090
(2-NAPHTHYLOXY)ACETIC ACID see NBJ100
α-NAPHTHYL RED see PEJ600
NAPHTOELAN FAST RED 3GL BASE see KDA050
β-NAPHTOL (GERMAN) see NAX000
2-NAPHTOL (FRENCH) see NAX000
NARCOGEN see TIO750
NARKOSOID see TIO750
NATRIUMARSENIT (GERMAN) see ARJ500
NATRIUMFLUORACETAAT (DUTCH) see SHG500
NATRIUMFLUORACETAT (GERMAN) see SHG500
NATURAL MUSK AMBRETTE see OKU100
NAVINON BLUE BC see DFN425
NAVINON BLUE RSN see IBV050
NAVINON BLUE RSN REDDISH SPECIAL see IBV050
NAVINON BRILLIANT BLUE RCL see DFN425
NCC 45 see CAT775
NCI-C00442 see DUV600
NCI-C02211 see NMC100
NCI-C02551 see CAH500
NCI-C02711 see CAJ750
NCI-C02799 see FMV000
NCI-C03612 see BBO500
NCI-C04535 see DFF809
NCI-C04546 see TIO750
NCI-C04580 see PCF275
NCI-C04604 see HCI000
NCI-C06508 see BDX000
NCI-C07272 see TGK750
NCI-C50646 see CBF700
NCI-C54637 see PJQ100
NCI-C54728 see DTD800
NCI-C54944 see DEP600
NCI-C54955 see DEP800
NCI-C54999 see CPD750
NCI-C55367 see BPX000
NCI-C55481 see EGV400
NCI-C55856 see CCP850
NCI-C55947 see TDY250
NCI-C56326 see AAG250
NCI-C56542 see SGB550
NCI-C56633 see TKV000
NCI-C60162 see FCA100
NCI-C60311 see CNA250
NCI-C60355 see KDA050
NCI C61290 see CMO500
NCI-C61392 see BGB275
NCI-C61405 see HEO000
NDS see DKH250
NECATORINA see CBY000
NECATORINE see CBY000
NEMA see PCF275
NEOANTICID see CAT775
NEO-CARDIAMINE see GCE600
NEO-CEBICURE see SAA025
NEO-CULTOL see MQV750
NEOLAMIN see TES800
NEOLITE F see CAT775
NEOSPIRAN see GCE600
NEW YORK RED see NAP100
NIACIN BENZYL ESTER see NCR040
NIAGARA BLUE 4B see CMO500
NIAGARA SKY BLUE see CMO500
NIALK see TIO750
NICOTINALDEHYDE see NDM100
NICOTINAMIDE-ADENINE DINUCLEOTIDE see CNF390
NICOTINEALDEHYDE see NDM100
NICOTINEAMIDE ADENINE DINUCLEOTIDE see CNF390
NICOTINIC ACID, BENZYL ESTER see NCR040
NICOTINIC ACID NITRILE see NDW515
NICOTINIC ALCOHOL see NDW510
NICOTINIC ALDEHYDE see NDM100
NICOTINONITRILE see NDW515
NICOTINYL ALCOHOL see NDW510
NIGALAX see PPN100
NIGRIN see SMA000
NIHONTHRENE BLUE BC see DFN425
NIHONTHRENE BLUE RSN see IBV050
NIHONTHRENE BRILLIANT BLUE RCL see DFN425
NIHONTHRENE BRILLIANT BLUE RP see IBV050
NIHONTHRENE BRILLIANT VIOLET 4R see DFN450
NIHONTHRENE BRILLIANT VIOLET RR see DFN450
NIHONTHRENE OLIVE R see DUP100

NIKKOL S.C.S see HCP900
NIPAGALLIN A see EKM100
NIPAHEPTYL see HBO650
NIPA NO. 48 see EKM100
NIPPON DIRECT SKY BLUE see CMO500
NIPPON KAGAKU SAFRANINE GK see GJI400
NIPPON KAGAKU SAFRANINE T see GJI400
NITER see PLL500
NITRAN see DUV600
NITRE see PLL500
NITRIC ACID, CADMIUM SALT see CAH000
NITRIC ACID, CADMIUM SALT, TETRAHYDRATE see CAH250
NITRIC ACID, POTASSIUM SALT see PLL500
NITRILE ADIPICO (ITALIAN) see AER250
NITRILO-2,2′,2″-TRIETHANOL see TKP500
2,2′,2″-NITRILOTRIETHANOL see TKP500
NITRITO D'AMILE see ILW100
2-NITRO-p-ACETANISIDIDE see NEK100
5-NITROBARBITURIC ACID see NET550
o-NITROBENZACETONITRILE see NFN600
p-NITROBENZAMIDE see NEV525
4-NITROBENZENEACETIC ACID see NII510
2-NITROBENZENEACETONITRILE see NFN600
4-NITROBENZENEDIAZONIUM FLUOBORATE see NEZ200
p-NITROBENZENEDIAZONIUM TETRAFLUOROBORATE see NEZ200
4-NITROBENZENEDIAZONIUM TETRAFLUOROBORATE(1⁻) see NEZ200
p-NITROBENZONITRILE see NFI010
4-NITROBENZONITRILE see NFI010
4-NITROBENZYL ACETATE see NFM100
p-NITROBENZYLACETATE see NFM100
m-NITROBENZYL CHLORIDE see NFN010
3-NITROBENZYL CHLORIDE see NFN010
p-NITROBENZYLIDENEACETOPHENONE see NFS505
2-NITROBENZYL NITRILE see NFN600
1-(p-NITROBENZYL)PYRIDINE see NFO750
m-NITROBROMOBENZENE see NFQ090
o-NITROBROMOBENZENE see NFQ080
p-NITROBROMOBENZENE see NFQ100
2-NITROBROMOBENZENE see NFQ080
3-NITROBROMOBENZENE see NFQ090
4-NITROBROMOBENZENE see NFQ100
4-NITROCALONE see NFS505
4-NITROCHALCONE see NFS505
2-NITRO-4-CHLOROANILINE see KDA050
p-NITROCINNAMIC ACID see NFT440
4-NITROCINNAMIC ACID see NFT440
4-NITRO-o-CRESOL see NFV010
2-NITRO-N,N-DIMETHYLANILINE see NFW600
5-NITROFURALDEHYDE DIACETATE see NGB800
5-NITRO-2-FURAN METHANDIOL DIACETATE see NGB800
5-NITROFURFURAL DIACETATE see NGB800
5-NITROFURFURYLIDENE DIACETATE see NGB800
NITROGENOL see HCP800
3,3′-NITROIMINODIPROPIONIC ACID see NHI600
4′-NITROOXANILIC ACID see NHX100
o-NITROPHENETOLE see NIC990
(p-NITROPHENYL)ACETIC ACID see NII510
(4-NITROPHENYL)ACETIC ACID see NII510
2-(p-NITROPHENYL)ACETIC ACID see NII510
(o-NITROPHENYL)ACETONITRILE see NFN600
(2-NITROPHENYL)ACETONITRILE see NFN600
3-(4-NITROPHENYL)ACRYLIC ACID see NFT440
((4-NITROPHENYL)AMINO)OXOACETIC ACID see NHX100
p-NITROPHENYLDIAZONIUM FLUOBORATE see NEZ200
p-NITROPHENYL ETHER see NIM600
N-(p-NITROPHENYL)GLYCINE see NIN100
2-(p-NITROPHENYL)HYDRAZIDEACETIC ACID see NIQ550
N-(4-NITROPHENYL)OXAMIC ACID see NHX100
3-(4-NITROPHENYL)-1-PHENYL-2-PROPEN-1-ONE see NFS505
3-(4-NITROPHENYL)PROPENOIC ACID see NFT440
4-NITROPHTHALIMIDE see NIX100
NITROPORE OBSH see BBS300
5-NITRO-2,4,6(1H,3H,5H)-PYRIMIDINETRIONE see NET550
p-NITROSALICYLIC ACID see NJF100
4-NITROSALICYLIC ACID see NJF100
4-NITRO-SALICYLSAEURE see NJF100
p-NITROSODIETHYLANILINE see NJW600
β-NITROSTYRENE see NMC100
γ-NITROSTYRENE see NMC100
p-NITROSTYRYL PHENYL KETONE see NFS505
4-NITRO-o-TOLUENEDIAZONIUM TETRAFLUOROBORATE see NMO700
2-NITRO-α-α-α-TRIFLUORO-p-CRESOL see NMQ100

NITROXIDE, BIS(1,1-DIMETHYLETHYL) (9CI) see DEF090
NITROXIDE, DI-tert-BUTYL see DEF090
NITRO-p-XYLENE see NMS520
NITRYL KWASU NIKOTYNOWEGO see NDW515
NITTO DIRECT SKY BLUE 5B see CMO500
NK ESTER 3G see MDN510
NOCCELER BG see TGX550
NONANENITRILE see OES300
n-NONANENITRILE see OES300
NONANONITRILE see OES300
n-NONANONITRILE see OES300
NONANOYL PEROXIDE see NNA100
NON-FER-AL see CAT775
NON-FLOCCULATING GREEN G 25 see PJQ100
4-NONYLPHENOL see NNC510
NOPCAINE see PAQ200
5-NORBORNENE-2,3-DICARBOXYLIC ANHYDRIDE, 1,4,5,6,7,7-HEXACHLORO- see CDS050
NOR-psi-EPHEDRINE see NNM510
psi-NOREPHEDRINE see NNM510
d-NOR-psi-EPHEDRINE see NNM510
NORLEUCAMINE see PBV505
NORMO-LEVEL see NNR400
NORMOSTEROL see NNR400
2-NORPINENE-2-CARBOXALDEHYDE, 6,6-DIMETHYL- see FNK150
NORPSEUDOEPHEDRINE, (+)- see NNM510
d-NORPSEUDOEPHEDRINE see NNM510
NOVISMUTH see BKW100
NOXA see NBJ100
2-NOXA see NBJ100
NS (carbonate) see CAT775
NS 100 (carbonate) see CAT775
NS 200 (filler) see CAT775
NSC 3060 see PKV500
NSC-20527 see BOD600
NSC-24343 see OMC300
NSC 45383 see SMA000
NU-2121 see NDW510
NUBIAN YELLOW TB see PEJ600
NUJOL see MQV750
NUODEX V 1525 see DEJ100
NUTMEG OIL see NOG500
NUTMEG OIL, EAST INDIAN see NOG500
NYACOL see SCH000
NYACOL 830 see SCH000
NYACOL 1430 see SCH000
NYANTHRENE BLUE BFP see DFN425
NYANTHRENE BRILLIANT VIOLET 4R see DFN450
NYANTHRENE OLIVE R see DUP100
NYLOQUINONE PINK B see AKO350
NYLOQUINONE YELLOW 2R see DUW500
NZ see CAT775

OA-A 1102 see CAT775
OCTA see CPB120
OCTAACETYLSUCROSE see OAF100
OCTABENZONE see HND100
OCTADECAFLUORODECAHYDRONAPHTHALENE see PCG700
OCTADECAFLUOROOCTANE see OAP100
OCTADECANAMINE ACETATE see OAP300
1-OCTADECANAMINIUM, N,N-DIMETHYL-N-OCTADECYL-, CHLORIDE (9CI) see DXG625
OCTADECANOIC ACID, ALUMINUM SALT see AHH825
OCTADECANOIC ACID, AMMONIUM SALT see ANU200
OCTADECANOIC ACID, BARIUM CADMIUM SALT (4:1:1) (9CI) see BAI800
OCTADECANOIC ACID, CADMIUM SALT see OAT000
OCTADECANOIC ACID, 1-METHYLETHYL ESTER (9CI) see IRL100
(Z)-9-OCTADECENOIC ACID BUTYL ESTER see BSB000
OCTADECYLAMINE, ACETATE see OAP300
α-OCTADECYLETHER of GLYCEROL see GGA915
1-O-OCTADECYLGLYCEROL see GGA915
3-(OCTADECYLOXY)-1,2-PROPANEDIOL see GGA915
OCTADECYL SODIUM SULFATE see OBG100
1-OCTADEDECYLAMINE ACETATE see OAP300
2,6-OCTADIENAL, 3,7-DIMETHYL-, (E)- see GCU100
OCTAFLUORO-sec-BUTENE see OBM000
OCTAFLUOROISOBUTENE see OBM000
OCTAFLUOROISOBUTYLENE see OBM000
OCTAMETHYLCYCLOTETRASILOXANE see OCE100
OCTANAL, 7-HYDROXY-3,7-DIMETHYL-, DIMETHYL ACETAL see LBO050
2-OCTANAMINE see OEK010
OCTAN n-BUTYLU (POLISH) see BPU750

OCTANEDINITRILE (9CI)

OCTANEDINITRILE (9CI) see OCW050
OCTANE, HEPTADECAFLUORO-1-IODO- see PCH325
OCTANE, 1,1,1,2,2,3,3,4,4,5,5,6,6,7,7,8,8-HEPTADECAFLUORO-8-IODO-(9CI) see PCH325
OCTANE, 1-IODO- see OFA100
OCTANENITRILE see OCW100
OCTANE, OCTADECAFLUORO- see OAP100
OCTANOIC ACID, CADMIUM SALT (2:1) see CAD750
OCTANOIC ACID, HEXYL ESTER see HFS600
OCTANOIC ACID, METHYL ESTER see MHY800
OCTANOIC ACID, PENTADECAFLUORO- see PCH050
OCTANOIC ACID, SODIUM SALT see SIX600
2-OCTANOL see OCY090
2-OCTANOL, 8,8-DIMETHOXY-2,6-DIMETHYL-(9CI) see LBO050
3-OCTANOL, 3,7-DIMETHYL- see LFY510
OCTANONITRILE see OCW100
n-OCTANOYL PEROXIDE (DOT) see CBF705
OCTAN WINYLU (POLISH) see VLU250
1-OCTEN-3-OL ACETATE see ODW028
6-OCTEN-1-OL, 3,7-DIMETHYL- see DTF410
OCTENYL ACETATE see ODW028
OCTOWY ALDEHYD (POLISH) see AAG250
OCTOWY BEZWODNIK (POLISH) see AAX500
2-OCTYLAMINE see OEK010
GKg-n-OCTYL-γ-n-BUTYROLACTONE see OES100
OCTYL CYANIDE see OES300
n-OCTYL CYANIDE see OES300
1-OCTYL CYANIDE see OES300
OCTYL IODIDE see OFA100
n-OCTYL IODIDE see OFA100
1-OCTYL IODIDE see OFA100
1-n-OCTYL IODIDE see OFA100
OCTYL METHYL KETONE see OFE050
OCTYL SODIUM SULFATE see OFU200
4-OCTYN-3,6-DIOL, 3,6-DIMETHYL- see DTF850
1-OCTYNE see OGI025
1-OCTYN-3-OL see OGI050
ODB see DEP600
ODCB see DEP600
OHIO RED see NAP100
OIL of HEERABOL-MYRRH see MSB775
OIL MIST, MINERAL (OSHA, ACGIH) see MQV750
OIL of MYRISTICA see NOG500
OIL of NUTMEG see NOG500
OK PRE-GEL see SLJ500
OKTAMETHYLCYKLOTETRASILOXAN see OCE100
OLEIC ACID, BUTYL ESTER see BSB000
OLITREF see DUV600
OLIV OSTANTHRENOVY R see DUP100
OMADINE see MCQ700
OMI see FAB000
OMS-0971 see CMA100
OMYA see CAT775
OMYA BLH see CAT775
OMYACARB F see CAT775
OMYALENE G 200 see CAT775
OMYALITE 90 see CAT775
OPALINE GREEN G 1 see PJQ100
OPLOSSINGEN (DUTCH) see FMV000
OPTICAL BLEACH 13-61 see OOI200
OPTIVAL see PLY275
OPTOCHINIDIN see QFS100
ORALITH RED SR WATER SOLUBLE see NAP100
ORANGE SALT NRD see CEG800
ORGATRAX see HOR470
ORPIMENT see ARI000
ORTHANILIC ACID see SNO100
ORTHOARSENIC ACID see ARB250
ORTHOARSENIC ACID HEMIHYDRATE see ARC500
ORTHOCHROME YELLOW GGW see SIT850
ORTHODIAZINE see OJW200
ORTHODICHLOROBENZENE see DEP600
ORTHODICHLOROBENZOL see DEP600
OS-CAL see CAT775
OSTANTHREN BLUE BCL see DFN425
OSTANTHREN BLUE BCS see DFN425
OSTANTHREN BLUE RS see IBV050
OSTANTHREN BLUE RSN see IBV050
OSTANTHREN BLUE RSZ see IBV050
OSTANTHRENE BLUE RS see IBV050
OSTANTHREN OLIVE R see DUP100
OTAN see OMC300
7-OXABICYCLO(4.1.0)HEPTANE-3-CARBOXYLIC ACID, 7-OXABICYCLO(4.1.0)HEPT-3-YLMETHYL ESTER see EBO050
7-OXABICYCLO(2.2.1)HEPTANE, 1-ISOPROPYL-4-METHYL-(6CI) see IKC100
7-OXABICYCLO(2.2.1)HEPTANE, 1-METHYL-4-(1-METHYLETHYL)-(9CI) see IKC100
6-OXABICYCLO(3.1.0)HEXANE see EBO100
OXACYCLOHEPTADEC-8-EN-2-ONE, (Z)- see OKU100
OXALIC ACID, DIAMMONIUM SALT see ANO750
OXALIC ACID, DIPOTASSIUM SALT see PLN300
OXAMIDE, N,N'-BIS(2-HYDROXYETHYL)DITHIO- see BKD800
OXAMIDE, N,N'-DIBENZYLDITHIO- see DDF700
OXANAL FAST BLUE G see APG700
OXANAL FAST RED SW see SEH475
OXANTIN see OMC300
3-OXAPENTANEDIOIC ACID see ONQ100
OXATONE see OMC300
OXIDATION BASE see NAN505
2-OXIMINO-3-BUTANONE see OMY910
OXIRANE, 2,2'-(1,2-ETHANEDIYLBIS(OXYMETHYLENE))BIS-(9CI) see EEA600
OXIRANE, (((2-ETHYLHEXYL)OXY)METHYL)-(9CI) see GGY100
OXIRANE, (METHOXYMETHYL)-(9CI) see GGW600
OXIRANE, (4-METHYLPHENOXY)METHYL)-(9CI) see GGY175
OXIRANE, TRIFLUORO(TRIFLUOROMETHYL)- see HDF050
2-OXOBENZIMIDAZOLE see ONI100
3-OXO-N-(2-CHLOROPHENYLBUTANAMIDE) see AAY600
OXODIACETIC ACID see ONQ100
2-OXOHEXAMETHYLENIMINE see CBF700
OXOMETHANE see FMV000
3-OXO-N-(2,4-METHYLPHENYL)BUTANAMIDE see OOI100
3-OXO-3H-NAPHTHO(2,1-b)PYRAN-2-CARBOXYLIC ACID ETHYL ESTER see OOI200
OXONIUM, TRIETHYL-, TETRAFLUOROBORATE(1⁻) see TJL600
2-OXOPYRIDINE see OOO200
4-OXO-2,2,6,6-TETRAMETHYLPIPERIDINE see TDT770
p-OXYBENZOESAEUREHEPTYLESTER see HBO650
OXYBISACETIC ACID see ONQ100
2,2'-OXYBISACETIC ACID see ONQ100
OXYBIS((3-AMINOPROPYL)DIMETHYLSILANE) see OPC100
OXYBIS(4-NITROBENZENE) see NIM600
1,1'-OXYBIS(4-NITROBENZENE) see NIM600
OXYBULAN see PFR325
7-OXYCOUMARIN see HJY100
OXYDE de CARBONE (FRENCH) see CBW750
OXYDIACETIC ACID see ONQ100
2,2'-OXYDIACETIC ACID see ONQ100
OXYDIETHANOLIC ACID see ONQ100
OXYLAN see PFR325
OXYMETHYLENE see FMV000
OXYNEURINE see GHA050
OXYPHENIC ACID see CCP850
4-OXYPIPERIDOL see HOI300
OXYQUINOLINE BENZOATE see HOE100
OZP 9 see DXB450

PAG see NNR400
PAINT WHITE see BKW100
PALAFUGE see PFR325
PALANIL PINK RF see AKO350
PALANTHRENE BLUE BC see DFN425
PALANTHRENE BLUE BCA see DFN425
PALANTHRENE BLUE GPT see IBV050
PALANTHRENE BLUE GPZ see IBV050
PALANTHRENE BLUE RPT see IBV050
PALANTHRENE BLUE RPZ see IBV050
PALANTHRENE BLUE RSN see IBV050
PALANTHRENE BRILLIANT BLUE R see IBV050
PALANTHRENE OLIVE R see DUP100
PALANTHRENE PRINTING BLUE KRS see IBV050
PALMITIC ACID, HEXADECYL ESTER see HCP700
PALMITYL ACETATE see HCP100
PALMITYLDIMETHYLAMINE see HCP525
PALMITYL PALMITATE see HCP700
PAPER RED 4B see DXO850
PAPETHERINE HYDROCHLORIDE see PAH260
PARABENCIL see PFR325
PARABENCILFENOL see PFR325
PARABUTYRALDEHYDE see TNC100
PARACIDE see DEP800
PARA CRYSTALS see DEP800
PARADI see DEP800
PARADIAZINE see POL590
PARADICHLORBENZOL (GERMAN) see DEP800
PARADICHLOROBENZENE see DEP800

PARADICHLOROBENZOL see DEP800
PARADONE BLUE RC see DFN425
PARADONE BLUE RS see IBV050
PARADONE BRILLIANT BLUE R see IBV050
PARADONE OLIVE R see DUP100
PARADONE PRINTING BLUE FRS see IBV050
PARADOW see DEP800
PARAFFIN OIL see MQV750
PARAFORM see FMV000
PARALACTIC ACID see LAG010
PARAMOTH see DEP800
PARA NONYL PHENOL see NNC510
PARANUGGETS see DEP800
PARAZENE see DEP800
PARCLAY see KBB600
PAROL see MQV750
PAROLEINE see MQV750
PCA see TDV300
PCON see KDA050
PCONA see KDA050
PDB see DEP800
PDCB see DEP800
PEACH ALDEHYDE see HBN200
PEACH LACTONE see HBN200
PEERLESS see KBB600
PELAGOL GREY C see CCP850
PELARGONITRILE see OES300
PELARGONONITRILE see OES300
PELARGONOYL PEROXIDE see NNA100
PELARGONYL PEROXIDE see NNA100
PELARGONYL PEROXIDE, technically pure (DOT) see NNA100
PENETECK see MQV750
PENFORD GUM 380 see SLJ500
PENICILLIN G PROCAINE see PAQ200
PENITE see SEY500
PENRECO see MQV750
PENTADECAFLUOROOCTANOIC ACID see PCH050
PENTADECAFLUORO-n-OCTANOIC ACID see PCH050
PENTAERYTHRITOL, TETRAACETATE see NNR400
PENTAFLUOROBENZOIC ACID see PBD275
PENTAFLUOROCHLOROBENZENE see PBE100
PENTAFLUOROPHENYL CHLORIDE see PBE100
1,1,3,3,5-PENTAMETHYL-4,6-DINITROINDANE see MRU300
PENTAMETHYLENE DIIODIDE see PBH125
PENTAMETHYLENEDITHIOCARBAMATE see PIY500
N,N-PENTAMETHYLENEUREA see PIL525
1-PENTANAMINE see PBV505
PENTANE, 1-BROMO- see AOF800
PENTANE, 1,5-DIIODO- see PBH125
PENTANE, 1,2-EPOXY- see ECE525
PENTANE, 1-IODO- see IET500
PENTANENITRILE (9CI) see VAV300
PENTANOIC ACID, 4,4′-AZOBIS(4-CYANO-(9CI) see ASL500
PENTANOIC ACID, 4,4-BIS((1,1-DIMETHYLETHYL)DIOXY)-, BUTYL ESTER see BHL200
PENTANOIC ACID, METHYL ESTER (9CI) see VAQ100
PENTANOIC ACID, 4-OXO-, BUTYL ESTER (9CI) see BRR700
2-PENTANOL, 3-ETHYL- see EOB300
3-PENTANOL, 3-ETHYL- see TJP550
PENTASILOXANE, DODECAMETHYL- see DXS800
2-PENTENE, 2-METHYL- see MNK100
1-PENTYLALLYL ACETATE see ODW028
PENTYLAMINE see PBV505
n-PENTYLAMINE see PBV505
1-PENTYLAMINE see PBV505
PENTYL BENZOATE see PBV800
n-PENTYL BENZOATE see PBV800
PENTYL BROMIDE see AOF800
n-PENTYL BROMIDE see AOF800
1-PENTYL BROMIDE see AOF800
PENTYL IODIDE see IET500
n-PENTYL IODIDE see IET500
1-PENTYL IODIDE see IET500
o-PENTYLPHENOL see AOM325
2-PENTYLPHENOL see AOM325
PENZAL N 300 see PAQ200
PEPTAZIN BAFD see BDK800
PEPTISANT 1O see BDK800
PEPTON 22 see BDK800
PERAWIN see PCF275
PERCHLOORETHYLEEN, PER (DUTCH) see PCF275
PERCHLOR see PCF275

PERCHLORAETHYLEN, PER (GERMAN) see PCF275
PERCHLORETHYLENE see PCF275
PERCHLORETHYLENE, PER (FRENCH) see PCF275
PERCHLOROETHANE see HCI000
PERCHLOROETHYLENE see PCF275
PERCHLOROMETHANE see CBY000
PERCLENE see PCF275
PERCLOROETILENE (ITALIAN) see PCF275
PERCOSOLVE see PCF275
PERFECTA see MQV750
PERFLUOROBENZOIC ACID see PBD275
PERFLUOROCAPRYLIC ACID see PCH050
PERFLUOROCTANOIC ACID see PCH050
PERFLUORODECAHYDRONAPHTHALENE see PCG700
PERFLUORODECALIN see PCG700
PERFLUORODECYL IODIDE see IEU075
PERFLUOROHEPTANECARBOXYLIC ACID see PCH050
PERFLUORO-1-IODODECANE see IEU075
PERFLUORO-1-IODOOCTANE see PCH325
PERFLUOROISOBUTYLENE (ACGIH) see OBM000
PERFLUORO(METHYLOXIRANE) see HDF050
PERFLUOROOCTANE see OAP100
n-PERFLUOROOCTANE see OAP100
PERFLUOROOCTANOIC ACID see PCH050
PERFLUOROOCTYL IODIDE see PCH325
PERFLUOROPROPYLENE OXIDE see HDF050
2-PERHYDROAZEPINONE see CBF700
PERIETHYLENENAPHTHALENE see AAF275
PERILAX see PPN100
PERILITON BRILLIANT PINK R see AKO350
2(1H)-PERINAPHTHAZOLONE see NAX100
PERK see PCF275
PERKADOX SE 8 see CBF705
PERKADOX SE 10 see DAH475
PERKLONE see PCF275
PERLITON YELLOW RR see DUW500
PERM-A-CHLOR see TIO750
PERMANENT GREEN TONER GT-376 see PJQ100
PERMANENT YELLOW 2K see DUW500
PERNAMBUCO EXTRACT see LFT800
PERNITHRENE BLUE BC see DFN425
PERNITHRENE BLUE RS see IBV050
PERNITHRENE OLIVE R see DUP100
PEROXIDE, BIS(1-OXODECYL) see DAH475
PEROXIDE, BIS(1-OXONONYL) see NNA100
PEROXIDE, BIS(1-OXOOCTYL) (9CI) see CBF705
PEROXIDE, 1-HYDROPEROXYCYCLOHEXYL 1-HYDROXYCYCLOHEXYL see CPC300
PEROXIDE, OCTANOYL see CBF705
PEROXYBENZOIC ACID, 1,1,4,4-TETRAMETHYLTETRAMETHYLENE ESTER see DRJ100
PEROXYCARBONIC ACID, OO-tert-BUTYL O-ISOPROPYL ESTER see BSD150
PEROXYDICARBONIC ACID, DICYCLOHEXYL ESTER see DGV650
PEROXYDICARBONIC ACID, DISODIUM SALT see SJB400
1,1′-PEROXYDICYCLOHEXANOL see PCM100
PEROXYHEXANOIC ACID, 2-ETHYL-, tert-PENTYL ESTER see AOM200
PEROXYISOBUTYRIC ACID, tert-BUTYL ESTER see BSC600
PERPARINE HYDROCHLORIDE see PAH260
PERPARIN HYDROCHLORIDE see PAH260
PERPERINE HYDROCHLORIDE see PAH260
PERSEC see PCF275
PERSIA-PERAZOL see DEP800
PERSICOL see HBN200
PETRIN see PCR150
PETROGALAR see MQV750
PETROLATUM, liquid see MQV750
PETZINOL see TIO750
PFD see PCG700
PFIB see OBM000
PFOA see PCH050
PHENAMINE PURPURINE 4B see DXO850
PHENAMINE SKY BLUE A see CMO500
PHENANTHRENE, 2-METHYL- see MNN520
PHENAZINIUM, 3,7-DIAMINO-2,8-DIMETHYL-5-PHENYL-, CHLORIDE see GJI400
1-PHENAZINOL, 6-METHOXY-, 5,10-DIOXIDE see HLT100
PHENETHYL ALCOHOL, α-METHYL- see PGA600
PHENETHYL ALCOHOL, β-METHYL- see HGR600
β-PHENETHYL-o-AMINOBENZOATE see APJ500
PHENETHYL ANTHRANILATE see APJ500
m-PHENETIDINE see PDK800
o-PHENETIDINE see PDK819
2-PHENETIDINE see PDK819

PHENETOLE, p-NITRO- see NIC990
PHENOHEP see HCI000
PHENOL, 2-AMINO-4-ARSENOSO-, SODIUM SALT see ARJ900
PHENOL, 4,4'-(3H-2,1-BENZOXATHIOL-3-YLIDENE)BIS-, S,S-DIOXIDE (9CI) see PDO800
PHENOL,4,4'-(3H-2,1-BENZOXATHIOL-3-YLIDENE)BIS(5-METHYL-2-(1-METHYLETHYL)-, S,S-DIOXIDE see TFX850
PHENOL, 4,4'-(3H-2,1-BENZOXATHIOL-3-YLIDENE)DI-, S,S-DIOXIDE see PDO800
PHENOL, p-(BENZYLAMINO)- see BDY750
PHENOL, 4-BROMO-2,5-DICHLORO- see LEN050
PHENOL, o-(tert-BUTYL)- see BSE460
PHENOL, 2-tert-BUTYL-5-METHYL- see BQV600
PHENOL, 2-tert-BUTYL-6-METHYL- see BRU790
PHENOL, 4-tert-BUTYL-2-METHYL- see BRU800
PHENOL, 2-CHLORO-5-METHYL- see CFE500
PHENOL, 4-CHLORO-5-METHYL-2-(1-METHYLETHYL)-(9CI) see CLJ800
PHENOL, 2,4-DIBROMO- see DDR150
PHENOL, 2,6-DIBROMO-4-NITRO- see DDQ500
PHENOL, 2,6-DI-tert-BUTYL- see DEG100
PHENOL, 2,5-DICHLORO- see DFX850
PHENOL, 3,4-DICHLORO- see DFY425
PHENOL, 3,5-DICHLORO- see DFY450
PHENOL, 4,5-DICHLORO-2-METHOXY- see DFL720
PHENOL, 2,6-DIETHYL- see DJU700
PHENOL, 3-(DIETHYLAMINO)-(9CI) see DIO400
PHENOL, m-(DIETHYLAMINO)- see DIO400
PHENOL, m-(DIMETHYLAMINO)- see HNK560
PHENOL, 4-((DIMETHYLAMINO)METHYL)-2,6-BIS(1,1-DIMETHYLETHYL)-(9CI) see FAB000
PHENOL, p-(α-α-DIMETHYLBENZYL)- see COF400
PHENOL, 3-(1,1-DIMETHYLETHYL)-, METHYLCARBAMATE (9CI) see BSG300
PHENOL, p-(2,4-DINITROANILINO)- see DUW500
PHENOL, 4-ETHENYL-, ACETATE see ABW550
PHENOL, p-ETHYL- see EOE100
PHENOL, 3-(ETHYLAMINO)-4-METHYL- see EGF100
PHENOL, o-FLUORO- see FKT100
PHENOL GLUCOSIDE see PDO300
PHENOL, 4,4'-IMINOBIS-(9CI) see IBJ100
PHENOL, 4,4'-IMINODI- see IBJ100
PHENOL, m-IODO- see IEV010
PHENOL, 3-IODO- see IEV010
PHENOL, m-ISOPROPYL- see IQX090
PHENOL, o-ISOPROPYL- see IQX100
PHENOL, 2-METHOXY-3,4,5,6-TETRACHLORO- see TBQ290
PHENOL, p-(α-METHYLBENZYL)- see PFD400
PHENOL, 4-METHYL-2,6-BIS(1-PHENYLETHYL)-(9CI) see MHR050
PHENOL, 3-(1-METHYLETHYL)- see IQX090
PHENOL, 2-(1-METHYLETHYL)-(9CI) see IQX100
PHENOL, 4-(1-METHYL-1-PHENETHYL)- see COF400
PHENOL, 4-METHYL-, SODIUM SALT (9CI) see SIM100
PHENOL, p-(METHYLTHIO)- see MPV300
PHENOL, p-NITRO-, HYDROGEN PHOSPHATE see BLA600
PHENOL, p-NONYL- see NNC510
PHENOL, o-PENTYL- see AOM325
PHENOL, 2-PENTYL-(9CI) see AOM325
PHENOL, 4-(1-PHENYLETHYL)- see PFD400
PHENOL, 4-(PHENYLMETHYL)-, CARBAMATE see PFR325
PHENOLPHTHALEIN, 3',3''-DIMETHYL- see CNX400
PHENOLPHTHALEIN, 4,5,6,7-TETRABROMO-3',3''-DISULFO-, DISODIUM SALT see HAQ600
PHENOL, 4,4'-(2-PYRIDINYLMETHYLENE)BIS-, DIACETATE (ESTER) see PPN100
PHENOL, 4,4'-(2-PYRIDYLMETHYLENE)DI-, DIACETATE (ester) see PPN100
PHENOL RED see PDO800
PHENOLSULFONEPHTHALEIN see PDO800
PHENOLSULFONIC ACID see HJH100
1-PHENOL-4-SULFONIC ACID ZINC SALT see ZIJ300
PHENOLSULFONPHTHALEIN see PDO800
PHENOLSULPHONPHTHALEIN see PDO800
PHENOLTETRABROMOPHTHALEINSULFONATE see HAQ600
PHENOL, 2,4,6-TRIMETHYL-(9CI) see MDJ740
PHENOL, p-VINYL-, ACETATE (6CI,7CI,8CI) see ABW550
PHENOTHIAZINE, 2-CHLORO- see CJL100
10H-PHENOTHIAZINE, 2-CHLORO- see CJL100
PHENOTHIAZINE, 2-(TRIFLUOROMETHYL)- see TKE775
PHENOTHIAZIN-5-IUM, 3-AMINO-7-(DIMETHYLAMINO)-, CHLORIDE see DUG700
o-PHENOXYANILINE see PDR490
2-PHENOXYANILINE see PDR490
2-PHENOXYBENZENAMINE see PDR490
2-PHENOXYPROPIONIC ACID see PDV725
PHENOZIN see ZIJ300
PHENYLACETATE SODIUM SALT see SFA200
PHENYLACETIC ACID SODIUM SALT see SFA200
PHENYLACETONE see MHO100
α-PHENYLACETONE see MHO100
3-PHENYLACRYLAMIDE see CMP970
trans-3-PHENYLACRYLIC ACID see CMP980
dl-PHENYLALANINE, 4-FLUORO-(9CI) see FLD100
PHENYLALLYL CINNAMATE see CMQ850
2-PHENYLAMINOPROPIONITRILE see AOT100
4-(PHENYLAZO)DIPHENYLAMINE see PEI800
PHENYLAZO α-NAPHTHYLAMINE see PEJ600
4-PHENYLAZO-1-NAPHTHYLAMINE see PEJ600
2-PHENYL-4H-1-BENZOPYRAN-4-ONE see PER700
3-PHENYL-BUTIN-1-OL-(3) see EQN230
PHENYLCARBONYLAMINOACETIC ACID see HGB300
PHENYLCHROMONE see PER700
2-PHENYLCHROMONE see PER700
PHENYLDIAZENECARBOXYLIC ACID 2-PHENYLHYDRAZIDE see DVY950
p-PHENYLENEBIS(DIMETHYLSILANE) see PEW725
1,4-PHENYLENEBIS(DIMETHYLSILANE) see PEW725
m-PHENYLENEDIAMINE, 4-METHOXY-, DIHYDROCHLORIDE see DBO100
p-PHENYLENEDIAMINESULFONIC ACID see PEY800
1,3-PHENYLENE-DI-4-AMINOPHENYL ETHER see REF070
o-PHENYLENEDIOL see CCP850
4,4'-(m-PHENYLENEDIOXY)DIANILINE see REF070
PHENYLESSIGSAEURE NATRIUM-SALZ see SFA200
PHENYLETHANEDIOL see SMQ100
1-PHENYLETHANONE see ABH000
2-PHENYLETHYL-o-AMINOBENZOATE see APJ500
PHENYLETHYL ANTHRANILATE see APJ500
2-PHENYLETHYL ANTHRANILATE see APJ500
PHENYLETHYLENE GLYCOL see SMQ100
1-PHENYLETHYLENE GLYCOL see SMQ100
4-(PHENYLETHYLIDENE)PHENOL see PFD400
4-(1-PHENYLETHYL)PHENOL see PFD400
PHENYL β-D-GLUCOPYRANOSIDE see PDO300
PHENYL β-D-GLUCOSIDE see PDO300
PHENYL GLYCOL see SMQ100
threo-1-PHENYL-1-HYDROXY-2-AMINOPROPANE see NNM510
2-PHENYL-2-IMIDAZOLINE see PFJ300
PHENYLMERCURIC OLEATE see PFP100
PHENYLMERCURY OLEATE see PFP100
PHENYLMETHANE see TGK750
PHENYLMETHANESULFONYL FLUORIDE see TGO300
PHENYL METHYL KETONE see ABH000
PHENYLMETHYL METHYL KETONE see MHO100
4-(PHENYLMETHYL)PHENOL CARBAMATE see PFR325
N-(PHENYLMETHYL)-1H-PURIN-6-AMINE see BDX090
2-(PHENYLMETHYL)PYRIDINE see BFG600
PHENYLMETHYLSULFONYL FLUORIDE see TGO300
2-PHENYL-4H-NAPHTHO(1,2-b)PYRAN-4-ONE see NBI100
PHENYL PHOSPHATE ((PhO)$_2$(HO)PO) see PHE300
1-PHENYL-2-PROPANOL see PGA600
2-PHENYLPROPAN-1-OL see HGR600
1-PHENYL-2-PROPANONE see MHO100
3-PHENYLPROPENAMIDE see CMP970
3-PHENYL-2-PROPENAMIDE see CMP970
(E)-3-PHENYL-2-PROPENOIC ACID see CMP980
3-PHENYL-2-PROPEN-1-YL CINNAMATE see CMQ850
3-PHENYL-2-PROPENYL 3-PHENYL-2-PROPENOATE see CMQ850
β-PHENYLPROPYL ALCOHOL see HGR600
2-PHENYLPROPYL ALCOHOL see HGR600
PHENYL 2-PYRIDYL KETONE see PGE760
PHENYLSILANE see SDX250
PHENYLSULFOHYDRAZIDE see BBS300
PHENYLSULFONYL HYDRAZIDE see BBS300
PHENYLSULFONYLHYDRAZINE see BBS300
2-(PHENYLTHIO)BENZOIC ACID see PGL270
N-PHENYL-p-TOLUENESULFONAMIDE see TGN600
PHENYL(TRICHLOROMETHYL)CARBINOL see TIR800
PHLOROPROPIONONE see TKP100
PHLOROPROPIOPHENONE see TKP100
PHORTISOLONE see PLY275
PHOSPHINE, TRIBUTYL- see TIA300
2-PHOSPHOLENE, 3-METHYL-1-PHENYL-, 1-OXIDE see MNV800
PHOSPHONIC ACID, METHYLENEDI-, TETRAETHYL ESTER see MJT100
PHOSPHONIC ACID, (2-(TRIETHOXYSILYL)ETHYL)-, DIETHYL ESTER see DKD500
PHOSPHONIUM, BUTYLTRIPHENYL-, BROMIDE see BSR900
PHOSPHONOTHIOIC DICHLORIDE, ETHYL- see EOP000
PHOSPHORIC ACID, BIS(p-NITROPHENYL) ESTER see BLA600
PHOSPHORIC ACID, DIMETHYL ESTER see PHC800
PHOSPHORIC ACID, DIPHENYL ESTER see PHE300
PHOSPHORIC ACID, 2,2,2-TRICHLOROETHYL ESTER see TIO800

PHOSPHOROUS ACID, TRIS(2-ETHYLHEXYL) ESTER see TNI300
PHOSPHORSAEURE-BIS-(p-NITRO-PHENYLESTER) see BLA600
PHOSVEL PHENOL see LEN050
PHOSPHORUS TRIAMIDE, HEXAMETHYL- see HEK100
PHTHALALDEHYDIC ACID see FNK010
PHTHALAMIDE see BBO500
PHTHALAMIDE, N,N,N',N'-TETRAETHYL- see GCE600
PHTHALETHAMIDE see GCE600
o-PHTHALIC ACID BIS(DIETHYLAMIDE) see GCE600
o-PHTHALIC ACID DIAMIDE see BBO500
PHTHALIC ACID, DICYCLOHEXYL ESTER see DGV700
PHTHALIC ACID, DIDODECYL ESTER see PHW550
PHTHALIC ACID, DIISOPROPYL ESTER see PHW600
PHTHALIC ACID, TETRACHLORO- see TBT100
PHTHALIMIDE, N-HYDROXY- see HNS600
PHTHALIMIDE, 4-NITRO- see NIX500
PHTHALOCYANINE BRILLIANT GREEN see PJQ100
PHTHALOCYANINE GREEN see PJQ100
PHTHALOCYANINE GREEN LX see PJQ100
PHTHALOCYANINE GREEN V see PJQ100
PHTHALOCYANINE GREEN VFT 1080 see PJQ100
PHTHALOCYANINE GREEN WDG 47 see PJQ100
o-PHTHALYLBIS(DIETHYLAMIDE) see GCE600
PHYLLEMBLIN see EKM100
PIAZINE see POL590
3-PICOLINE, 6-AMINO- see AMA010
β-PICOLYL ALCOHOL see NDW510
γ-PICOLYL ALCOHOL see POR810
PICRYL CHLORIDE see PIE530
PIGMENT ANTHRAQUINONE DEEP BLUE see IBV050
PIGMENT BLUE 60 see IBV050
PIGMENT BLUE ANTHRAQUINONE see IBV050
PIGMENT BLUE ANTHRAQUINONE V see IBV050
PIGMENT DEEP BLUE ANTHRAQUINONE see IBV050
PIGMENT FAST GREEN G see PJQ100
PIGMENT FAST GREEN GN see PJQ100
PIGMENT GREEN 7 see PJQ100
PIGMENT GREEN PHTHALOCYANINE see PJQ100
PIGMENT GREEN PHTHALOCYANINE V see PJQ100
PIGMENT WHITE 18 see CAT775
PINACOLYL ALCOHOL (6CI) see BRU300
PINENE (DOT) see PIH245
2-PINEN-10-OL, ACETATE (6CI,7CI,8CI) see MSC050
PIPECOLATE see HDS300
PIPECOLIC ACID see HDS300
PIPECOLINIC ACID see HDS300
α-PIPECOLINIC ACID see HDS300
1-PIPERAZINECARBOXYLIC ACID, ETHYL ESTER see EHG050
PIPERAZINE, 1,4-DIMETHYL- see LIQ500
PIPERAZINE ETHYLCARBOXYLATE see EHG050
1-PIPERIDINECARBODITHIOIC ACID, compounded with PIPERIDINE see PIY500
PIPERIDINE-N-CARBONIC ACID AMIDE see PIL525
1-PIPERIDINECARBOXAMIDE see PIL525
2-PIPERIDINECARBOXYLIC ACID (9CI) see HDS300
4-PIPERIDINECARBOXYLIC ACID (9CI) see ILG100
PIPERIDINE, 2,6-DIMETHYL- see LIQ550
PIPERIDINIUM see PIY500
4-PIPERIDINONE, 2,2,6,6-TETRAMETHYL-(9CI) see TDT770
PIPERIDINOOXY, 4-HYDROXY-2,2,6,6-TETRAMETHYL- see HOI300
PIPERIDINOOXY, 2,2,6,6-TETRAMETHYL- see TDT800
1-PIPERIDINYLOXY, 4-HYDROXY-2,2,6,6-TETRAMETHYL-(9CI) see HOI300
1-PIPERIDINYLOXY, 2,2,6,6-TETRAMETHYL-(9CI) see TDT800
4-PIPERIDONE, 2,2,6,6-TETRAMETHYL- see TDT770
PIPEROLINIC ACID see HDS300
PIP-PIP see PIY500
PIRITON see CLX300
PLASTHALL 503 see BSB000
PLASTORESIN RED SR see NAP100
P.M. 346 see EIT100
PMO 10 see PFP100
PMSF see TGO300
POLARONIL see CLX300
POLCARB see CAT775
POLYCHLORO COPPER PHTHALOCYANINE see PJQ100
POLYESTER TGM 3 see MDN510
POLYMO GREEN FBH see PJQ100
POLYMO GREEN FGH see PJQ100
POLYMON BLUE 3R see IBV050
POLYMON GREEN G see PJQ100
POLYMON GREEN 6G see PJQ100
POLYMON GREEN GN see PJQ100
POLYOXYMETHYLENE GLYCOLS see FMV000
PONOLITH FAST VIOLET 4RN see DFN450
PONSOL BLUE BCS see DFN425
PONSOL BLUE BF see DFN425
PONSOL BLUE BFD see DFN425
PONSOL BLUE BFDP see DFN425
PONSOL BLUE BFN see DFN425
PONSOL BLUE BFND see DFN425
PONSOL BLUE BFP see DFN425
PONSOL BLUE GZ see IBV050
PONSOL BLUE RCL see IBV050
PONSOL BLUE RPC see IBV050
PONSOL BRILLIANT BLUE R see IBV050
PONSOL OLIVE AR see DUP100
PONSOL OLIVE ARD see DUP100
PONSOL RP see IBV050
PONTACHROME YELLOW GS see SIT850
PONTACYL SKY BLUE 4BX see CMO500
PONTAMINE SKY BLUE 5BX see CMO500
POROFOR BSH see BBS300
POROFOR-BSH-PULVER see BBS300
POROFOR ChKhZ 9 see BBS300
POTASSIUM ACID ARSENATE see ARD250
POTASSIUM ARSENATE see ARD250
POTASSIUM ARSENITE see PKV500
POTASSIUM DIHYDROGEN ARSENATE see ARD250
POTASSIUM GLYCEROPHOSPHATE see PLG810
POTASSIUM HYDROGEN ARSENATE see ARD250
POTASSIUM α-HYDROXYPROPIONATE see PLK650
POTASSIUM ISOPROPYL XANTHANATE see IRG050
POTASSIUM ISOPROPYLXANTHATE see IRG050
POTASSIUM ISOPROPYL XANTHOGENATE see IRG050
POTASSIUM LACTATE see PLK650
POTASSIUM METAARSENITE see PKV500
POTASSIUM NEUTRAL OXALATE see PLN300
POTASSIUM NITRATE see PLL500
POTASSIUM OXALATE see PLN300
POTASSIUM PEROXIDE see PLP250
PP 5 see PCG700
PRAGMOLINE see CMF260
PRECISION CLEANING AGENT see TJE100
PREDNESOL see PLY275
PREDNISOLONE DISODIUM PHOSPHATE see PLY275
PREDNISOLONE 21-PHOSPHATE DISODIUM see PLY275
PREDNISOLONE SODIUM PHOSPHATE see PLY275
PREDSOL see PLY275
PREDSOLAN see PLY275
PREEGLONE see DWX800
PREGNA-1,4-DIENE-3,20-DIONE, 11-β,17,21-TRIHYDROXY-, 21-(DIHYDROGEN PHOSPHATE), DISODIUM SALT see PLY275
PREPARED CHALK see CAT775
PRIMOL 335 see MQV750
PROCAINE BENZYLPENICILLINATE see PAQ200
PROCAINE PENICILLIN G see PAQ200
PRODALUMNOL see SEY500
PRODALUMNOL DOUBLE see SEY500
PRODOX 131 see IQX100
PROGALLIN A see EKM100
PROPANAL, 2-METHYL-2-(METHYLSULFINYL)-, O-((METHYLAMINO)CARBONYL)OXIME see MLW750
PROPANAL, 2-METHYL-2-(METHYLTHIO)-, OXIME see MLX800
PROPANAMIDE, N,N-DIETHYL- see DJX250
PROPANAMIDE, N-METHYL-(9CI) see MOS900
1-PROPANAMINIUM, 3-CARBOXY-2-HYDROXY-N,N,N-TRIMETHYL-, CHLORIDE, (+−)-(9CI) see CCK655
1-PROPANAMINIUM, 2-HYDROXY-N,N,N-TRIMETHYL-, CHLORIDE (9CI) see MIM300
PROPANE, 1,2-DIBROMO-1,1,2,3,3,3-HEXAFLUORO- see DDO450
PROPANE, 1,2-DIBROMO-2-METHYL- see DDQ150
PROPANE, 1,3-DIIODO- see TLR050
PROPANEDINITRILE, ((3-CHLOROPHENYL)HYDRAZONO)-(9CI) see CKA550
1,3-PROPANEDIOL, 2,2-BIS((ACETYLOXY)METHYL)-, DIACETATE (9CI) see NNR400
1,3-PROPANEDIOL, 2-ETHYL-2-(HYDROXYMETHYL)- see HDF300
1,2-PROPANEDIOL, 3-(OCTADECYLOXY)- see GGA915
1,3-PROPANEDIONE, 2-BROMO-1,3-DIPHENYL- see BNH100
1,2-PROPANEDITHIOL see PML300
PROPANE, 1,2-EPOXY-3-((2-ETHYLHEXYL)OXY)- see GGY100
PROPANE, 1,2-EPOXY-1,1,2,3,3,3-HEXAFLUORO- see HDF050
PROPANE, 1,2-EPOXY-3-METHOXY- see GGW600
PROPANE, 1,2-EPOXY-2-METHYL- see IIQ600
PROPANE, 1,2-EPOXY-3-(p-TOLYLOXY)- see GGY175
PROPANENITRILE, 3-ANILINO- see AOT100

PROPANENITRILE, 3-(PHENYLAMINO)- see AOT100
PROPANEPEROXOIC ACID, 2-METHYL-, 1,1-DIMETHYLETHYL ESTER see BSC600
1,2,3-PROPANETRICARBOXYLIC ACID, 2-HYDROXY-, TRILITHIUM SALT, HYDRATE see LHD150
PROPANOIC ACID, 3-BROMO-2-OXO-(9CI) see BOD550
PROPANOIC ACID, 2,3-DIBROMO-(9CI) see DDS500
PROPANOIC ACID, 2-HYDROXY-, (S)-(9CI) see LAG010
PROPANOIC ACID, 2-HYDROXY-2-METHYL-, ETHYL ESTER (9CI) see ELH700
PROPANOIC ACID, 2-HYDROXY-, MONOPOTASSIUM SALT (9CI) see PLK650
1-PROPANOL, 3-CHLORO- see CKP725
2-PROPANOL, 1-(CYCLOHEXYLAMINO)- see CPG700
2-PROPANONE, 1,3-DIHYDROXY- see OMC300
2-PROPANONE, 1,1-DIPHENYL- see MJH910
2-PROPANONE, 1-PHENYL- see MHO100
1-PROPANONE, 1-(2,4,6-TRIHYDROXYPHENYL)- see TKP100
2-PROPENAMIDE, N-(BUTOXYMETHYL)-(9CI) see BPM660
2-PROPENAMIDE, N-(1,1-DIMETHYLETHYL)-(9CI) see BPW050
2-PROPENAMIDE, 3-PHENYL-(9CI) see CMP970
1-PROPENE, 2-METHOXY- see MFL300
PROPENE, 3,3,3-TRIFLUORO- see TKH015
1-PROPENE, 3,3,3-TRIFLUORO-(9CI) see TKH015
PROPENE, 3,3,3-TRIFLUORO-2-(TRIFLUOROMETHYL)- see HDC450
2-PROPENOIC ACID, 2-CHLORO-, ETHYL ESTER see EHH200
2-PROPENOIC ACID, 2-(DIMETHYLAMINO)ETHYL ESTER (9CI) see DPB300
2-PROPENOIC ACID, HEPTYL ESTER see HBL150
2-PROPENOIC ACID, 3-(2-HYDROXYPHENYL)-, (E)-(9CI) see CNU850
2-PROPENOIC ACID, 3-(1H-IMIDAZOL-4-YL)-(9CI) see UVJ440
2-PROPENOIC ACID, 2-METHYL-, 2-(DIETHYLAMINO)ETHYL ESTER (9CI) see DIB300
2-PROPENOIC ACID, 2-METHYL-, 1,2-ETHANEDIYLBIS(OXY-2,1-ETHANEDIYL) ESTER(9CI) see MDN510
2-PROPENOIC ACID, 2-METHYL-, 2-(ETHENYLOXY)ETHYL ESTER (9CI) see VQA150
2-PROPENOIC ACID, 2-METHYL-, 2-HYDROXYPROPYL ESTER (9CI) see HNV500
2-PROPENOIC ACID, 2-METHYL-, OXYBIS(2,1-ETHANEDIYLOXY-2,1-ETHANEDIYL)ESTER see TCE400
2-PROPENOIC ACID, 3-(4-NITROPHENYL)- see NFT440
2-PROPENOIC ACID, 3-PHENYL-, (E)-(9CI) see CMP980
2-PROPENOIC ACID, 3-PHENYL-, 3-PHENYL-2-PROPENYL ESTER (9CI) see CMQ850
2-PROPENOIC ACID, 3-(3,4,5-TRIMETHOXYPHENYL)-(9CI) see CMQ100
2-PROPEN-1-ONE, 1-(2-HYDROXYPHENYL)-3-PHENYL- see HJS900
2-PROPEN-1-ONE, 3-(4-NITROPHENYL)-1-PHENYL-(9CI) see NFS505
2-PROPENYL DISULPHIDE see AGF300
PROPIONALDEHYDE, 2-METHYL-2-(METHYLSULFINYL)-, O-(METHYLCARBAMOYL)OXIME see MLW750
PROPIONALDEHYDE, 2-METHYL-2-(METHYLTHIO)-, OXIME see MLX800
PROPIONALDEHYDE, 3,3,3-TRIFLUORO- see TKH020
PROPIONAMIDE, N,N-DIBUTYL- see DEH300
PROPIONAMIDE, N,N-DIETHYL- see DJX250
PROPIONAMIDE, N-METHYL- see MOS900
PROPIONIC ACID, 2,3-DIBROMO- see DDS500
PROPIONIC ACID, 3,3'-NITROIMINODI- see NHI600
PROPIONIC ACID, 2-PHENOXY- see PDV725
PROPIONITRILE, 3-ANILINO- see AOT100
PROPIONYLPHLOROGLUCINOL see TKP100
PROPIOPHENONE, 2-BROMO- see BOB550
PROPIOPHENONE, 2',4',6'-TRIHYDROXY- see TKP100
PROPIOPHLOROGLUCINE see TKP100
PROPYLAMINE, 1,2-DIMETHYL- see AOE200
PROPYL BUTANOATE see PNF100
PROPYL BUTYRATE see PNF100
PROPYLENE GLYCOL MONOACETATE see PNL100
PROPYLENE GLYCOL MONOMETHYL ETHER ACETATE see PNL265
PROPYLENE OXIDE HEXAFLUORIDE see HDF050
PROPYLESTER KYSELINY MASELNE see PNF100
PROPYLOXIRANE see ECE525
PROPYLUREA see PNX550
1-PROPYLUREA see PNX550
N-PROPYLUREA see PNX550
PROTOCATECHUIC ACID see POE200
PROTOPET see MQV750
PROXEL PL see BCE475
PROZORIN see PLY275
PS 100 (carbonate) see CAT775
PSEUDOBUTYLENE see BOW255
PSEUDOIONONE see POH525
PSEUDO-α-ISOMETHYL IONONE see TMJ100
PSEUDONOREPHEDRINE see NNM510
PSEUDOUREA, 2-(p-CHLOROBENZYL)-2-THIO-, MONOHYDROCHLORIDE see CEQ800
PSEUDOUREA, 2-METHYL-2-THIO-, SULFATE see MPV790

PSP see PDO800
PSP (INDICATOR) see PDO800
PULEGON see POI615
PULEGONE see POI615
(+)-PULEGONE see POI615
d-PULEGONE see POI615
(+)-(R)-PULEGONE see POI615
(R)-(+)-PULEGONE see POI615
PURECAL see CAT775
PURECALO see CAT775
1H-PURIN-6-AMINE, N-(PHENYLMETHYL)-(9CI) see BDX090
1H-PURIN-6-AMINE, SULFATE see AEH500
PURPURIN 4B see DXO850
PURPURINE 4B see DXO850
PV-FAST GREEN G see PJQ100
PYCARIL see NCR040
PYKARYL see NCR040
2H-PYRAN-2-ONE, 4-HYDROXY-6-METHYL- see THL800
PYRAZINE see POL590
PYRAZOLE, 3,5-DIMETHYL- see DTU850
PYRAZOLE, 3-METHYL- see MOX010
PYRAZOLE, 5-METHYL- see MOX010
3-PYRAZOLIN-5-ONE, 4-AMINO-2,3-DIMETHYL-1-PHENYL- see AIB300
2-PYRAZOLIN-5-ONE, 3-METHYL- see MOX100
2-PYRAZOLIN-5-ONE, 3-METHYL-1-p-TOLYL- see THA300
3H-PYRAZOL-3-ONE, 4-AMINO-1,2-DIHYDRO-1,5-DIMETHYL-2-PHENYL- see AIB300
1,6-PYRENEDIONE see PON340
1,8-PYRENEDIONE see PON350
3,8-PYRENEDIONE see PON340
3,10-PYRENEDIONE see PON350
1,6-PYRENEQUINONE see PON340
1,8-PYRENEQUINONE see PON350
3,8-PYRENEQUINONE see PON340
3,10-PYRENEQUINONE see PON350
PYRIDAZINE see OJW200
3-PYRIDINALDEHYDE see NDM100
2-PYRIDINAMINE, 5-METHYL- see AMA010
PYRIDINE, 4-ACETYL- see ADA365
p-PYRIDINEALDEHYDE see IKZ100
3-PYRIDINEALDEHYDE see NDM100
4-PYRIDINEALDEHYDE see IKZ100
PYRIDINE, 2-BENZOYL- see PGE760
PYRIDINE, 2-BENZYL- see BFG600
PYRIDINE, 2-BROMO- see BOM600
PYRIDINE, 3-BUTYL- see BSJ550
PYRIDINE-3-CARBALDEHYDE see NDM100
PYRIDINE-4-CARBALDEHYDE see IKZ100
PYRIDINE-3-CARBINOL see NDW510
GKb-PYRIDINECARBONALDEHYDE see NDM100
3-PYRIDINECARBONITRILE (9CI) see NDW515
3-PYRIDINECARBOXALDEHYDE (9CI) see NDM100
4-PYRIDINECARBOXALDEHYDE (9CI) see IKZ100
γ-PYRIDINECARBOXAMIDE see ILC100
4-PYRIDINECARBOXAMIDE (9CI) see ILC100
3-PYRIDINECARBOXYLIC ACID, PHENYLMETHYL ESTER see NCR040
PYRIDINE, 2-(p-CHLORO-α-(2-(DIMETHYLAMINO)ETHYL)BENZYL)- see CLX300
PYRIDINE, 2,4-DIMETHYL-(9CI) see LIY990
2-PYRIDINEMETHANETHIOL see POR790
3-PYRIDINEMETHANOL see NDW510
4-PYRIDINEMETHANOL see POR810
PYRIDINE, 3-METHYL-, 1-OXIDE see MOY550
PYRIDINE, 4-METHYL-, 1-OXIDE see MOY790
3-PYRIDINENITRILE see NDW515
PYRIDINE, 4-(p-NITROBENZYL)- see NFO750
PYRIDINE, 2-(PHENYLMETHYL)-(9CI) see BFG600
2-PYRIDINEPROPANAMINE, γ-(4-CHLOROPHENYL)-N,N-DIMETHYL-(9CI) see CLX300
PYRIDINE-2-THIOCARBINOL see POR790
2-PYRIDINETHIOL see TFP300
2-PYRIDINETHIOL, 1-OXIDE see MCQ700
2(1H)-PYRIDINETHIONE see TFP300
PYRIDINE, 2,4,6-TRIMETHYL- see TME272
PYRIDINE, 4-VINYL- see VQK590
PYRIDINIUM, 3-CARBAMOYL-1-β-D-RIBOFURANOSYL-, HYDROXIDE, 5',5'-ESTER with ADENOSINE 2'-(DIHYDROGEN PHOSPHATE) 5'-(TRIHYDROGEN PYROPHOSPHATE), inner salt see CNF400
PYRIDINIUM, 3-CARBAMOYL-1-β-D-RIBOFURANOSYL-, HYDROXIDE, 5'-ESTER with ADENOSINE 5'-5'-(TRIHYDROGEN PYROPHOSPHATE), inner salt see CNF390
PYRIDINIUM, 1-DODECYL-, CHLORIDE see LBX050
PYRIDINIUM, 1,1'-ETHYLENEDI-, DIBROMIDE see EIT100

PYRIDINIUM, 1-HEXADECYL-, BROMIDE see HCP800
2-PYRIDINOL see OOO200
3-PYRIDINOL see PPH025
4-PYRIDINOL see HOB100
2-PYRIDINONE see OOO200
2(1H)-PYRIDINONE (9CI) see OOO200
2(1H)-PYRIDINONE, 1-METHYL-(9CI) see MPA075
2-PYRIDIYLAMINE see PPH300
9H-PYRIDO(3,4-b)INDOLE, 1,2,3,4-TETRAHYDRO-6-METHOXY-1-METHYL- see MLJ100
3-PYRIDOL see PPH025
α-PYRIDONE see OOO200
PYRIDONE-2 see OOO200
2-PYRIDONE see OOO200
2(1H)-PYRIDONE see OOO200
2(1H)-PYRIDONE, 1-METHYL- see MPA075
3-PYRIDYLALDEHYDE see NDM100
1-(3-PYRIDYL)BUTANE see BSJ550
3-PYRIDYLCARBINOL see NDW510
4-PYRIDYLCARBINOL see POR810
3-PYRIDYLCARBONITRILE see NDW515
3-PYRIDYLCARBOXALDEHYDE see NDM100
3-PYRIDYLMETHANOL see NDW510
4-PYRIDYLMETHANOL see POR810
4,4'-(2-PYRIDYLMETHYLENE)DIPHENOL DIACETATE see PPN100
2-PYRIDYLMETHYL MERCAPTAN see POR790
4-(4-PYRIDYL)PYRIDINE see BGO600
PYRILAX see PPN100
2-PYRIMIDINAMINE (9CI) see PPH300
PYRIMIDINE, 2-AMINO- see PPH300
4,6-PYRIMIDINEDIOL see PPP100
2,4(1H,3H)-PYRIMIDINEDIONE, 5-AMINO- see AMW100
2(1H)-PYRIMIDINETHIONE, 4-AMINO- see TFH800
2,4,6(1H,3H,5H)-PYRIMIDINETRIONE, 5-NITRO-(9CI) see NET550
2(1H)-PYRIMIDINONE, 4-AMINO- see CQM600
4(1H)-PYRIMIDINONE, 2,3-DIHYDRO-5-METHYL-2-THIOXO-(9CI) see TFQ650
4(1H)-PYRIMIDINONE, 6-HYDROXY- see PPP100
PYRINEX see CMA100
PYROCATECHIN see CCP850
PYROCATECHINIC ACID see CCP850
PYROCATECHOL see CCP850
PYROCATECHUIC ACID see CCP850
PYROMELLITIC ACID see PPQ630
PYROMELLITIC ACID ANHYDRIDE see PPQ635
PYROMELLITIC ACID DIANHYDRIDE see PPQ635
PYROMELLITIC ANHYDRIDE see PPQ635
PYROMELLITIC DIANHYDRIDE see PPQ635
PYRROLE, 1-FURFURYL- see FPX100
1-PYRROLIDINYLOXY, 3-CARBOXY-2,2,5,5-TETRAMETHYL- see TDV300
PYRUVIC ACID, BROMO- see BOD550
PZ see CAT775

Q-D 86P see DXG625
QUARTZ GLASS see SCK600
QUATERNARIO LPC see LBX050
QUATERNIUM 5 see DXG625
QUATERNIUM 13 see TCB200
QUEENSGATE WHITING see CAT775
QUESYL see DEL300
QUILONE see LGO100
QUINALDINIUM, 1-ETHYL-, IODIDE see EPD600
α-QUINIDINE see CMP910
QUINIDINE BISULFATE see QFS100
QUINIDINE SULFATE (1581) (salt) see QFS100
2-QUINOLINAMINE (9CI) see AMK700
3-QUINOLINEAMINE see AMK725
QUINOLINE, 2-AMINO- see AMK700
QUINOLINE, 3-AMINO- see AMK725
QUINOLINE, 4,7-DICHLORO- see DGJ250
QUINOLINE, 6-DODECYL-1,2-DIHYDRO-2,2,4-TRIMETHYL- see SAU480
2-QUINOLINETHIOL see QOJ100
2(1H)-QUINOLINETHIONE see QOJ100
8-QUINOLINOL, BENZOATE (ester) see HOE100
8-QUINOLINOL, 5-CHLORO- see CLD600
8-QUINOLINOL, 5,7-DICHLORO- see DEL300
8-QUINOLINOL, 2-METHYL- see HMQ600
QUINOLOR see DEL300
QUINOXALINE, 2,3-DICHLORO- see DGJ950
QUINOXALINE, 2,3-DIMETHYL- see DTY700
2-QUINOXALINOL see QSJ100
2(1H)-QUINOXALINONE see QSJ100
2-QUINUCLIDINEMETHANOL, α-4-QUINOLYL-5-VINYL- see CMP910

3-QUINUCLIDINOL see QUJ990
QUIXALIN see DEL300

R 10 see CBY000
R 143a see TJY900
RACEPHEDRINE see EAW100
RAMAPO see PJQ100
RASORITE 65 see DXG035
RATBANE 1080 see SHG500
R 114B2 see FOO525
RB-BL see IHC550
RCRA WASTE NUMBER P010 see ARB250
RCRA WASTE NUMBER P011 see ARH500
RCRA WASTE NUMBER P058 see SHG500
RCRA WASTE NUMBER P112 see TDY250
RCRA WASTE NUMBER U001 see AAG250
RCRA WASTE NUMBER U018 see BBC250
RCRA WASTE NUMBER U034 see CDN550
RCRA WASTE NUMBER U070 see DEP600, DEP800
RCRA WASTE NUMBER U071 see DEP800
RCRA WASTE NUMBER U072 see DEP800
RCRA WASTE NUMBER U074 see DEV000
RCRA WASTE NUMBER U076 see DFF809
RCRA WASTE NUMBER U092 see DOQ800
RCRA WASTE NUMBER U122 see FMV000
RCRA WASTE NUMBER U131 see HCI000
RCRA WASTE NUMBER U158 see MJM200
RCRA WASTE NUMBER U210 see PCF275
RCRA WASTE NUMBER U211 see CBY000
RCRA WASTE NUMBER U220 see TGK750
RCRA WASTE NUMBER U228 see TIO750
RE 5030 see BSG300
RECOLITE RED LYS see NAP100
11935 RED see NAP100
RED BALL see CAT775
RED BASE CIBA VI see KDA050
RED BASE CIBA X see MMF780
RED BASE 3 GL see KDA050
RED BASE IRGA VI see KDA050
RED BASE IRGA X see MMF780
RED BASE NRL see MMF780
RED 3G BASE see KDA050
RED No. 205 see NAP100
RED RL BASE see MMF780
REGLON see DWX800
REGLONE see DWX800
RELITON YELLOW R see DUW500
REMYLINE Ac see SLJ500
RESAMINE RED RB see NAP100
RESAMINE RED RC see NAP100
RESINATED INDO BLUE B 85 see DFN425
β-RESORCINALDEHYDE see REF100
RESORCINOL, 2-AMINO-, HYDROCHLORIDE see AML600
RESORCINOL, 4-BENZYL- see BFI400
RESORCINOL, DIBENZOATE see BHB100
RESORCINOL DIMETHYL ETHER see REF025
RESORCINOL OXYDIANILINE see REF070
β-RESORCYLALDEHYDE see REF100
α-RESORCYLIC ACID see REF200
β-RESORCYLIC ALDEHYDE see REF100
RETARDILLIN see PAQ200
RETINAL (9CI) see VSK985
all-trans-RETINAL see VSK985
all-E-RETINAL see VSK985
trans-RETINAL see VSK985
RETINAL, all-trans- see VSK985
RETINALDEHYDE see VSK985
RETINENE see VSK985
RETINENE 1 see VSK985
α-RETINENE see VSK985
RHODINOL see DTF410
RIBOFLAVIN-ADENINE DINUCLEOTIDE see RIF100
RIBOFLAVINE-ADENINE DINUCLEOTIDE see RIF100
RIBOFLAVIN 5'-(TRIHYDROGEN DIPHOSPHATE), 5'-5'-ESTER with ADENOSINE (9CI) see RIF100
RICE STARCH see SLJ500
RIKELATE CALCIUM see CAR780
RINEPTIL see HBM490
R JUTAN see CAT775
RO-1-5155 see NDW510
RODINOL see DTF410
ROFOB 3 see AHH825

ROMANTHRENE BLUE FRS see IBV050
ROMANTRENE BLUE FBC see DFN425
ROMANTRENE BLUE FRS see IBV050
ROMANTRENE BLUE GGSL see IBV050
ROMANTRENE BLUE RSZ see IBV050
ROMANTRENE BRILLIANT BLUE FR see IBV050
ROMANTRENE BRILLIANT BLUE R see IBV050
RONIACOL see NDW510
β-ROSORCALDEHYDE see REF100
ROWALIND see NDM100
ROYAL WHITE LIGHT see CAT775
5278 R.P. see SMA000
RP 13907 see TKP100
13907 R.P. see TKP100
"522" RUBBER ACCELERATOR see PIY500
RUBIDIUM CARBONATE see RPB200
RUBRIMENT see NCR040
RUBRINE C see GHA050
RUFOCHROMOMYCIN see SMA000
RUFOCROMOMYCINE see SMA000
RV 1 see SIT850
RX 2557 see CAT775

SA 97 see HGI100
SA 1500 see AHH825
SACCHAROSE ACETATE ISOBUTYRATE see SNH050
SACCHAROSONIC ACID see SAA025
SAFRANIN see GJI400
SAFRANINE see GJI400
SAFRANINE A see GJI400
SAFRANINE B see GJI400
SAFRANINE G see GJI400
SAFRANINE GF see GJI400
SAFRANINE J see GJI400
SAFRANINE O see GJI400
SAFRANINE OK see GJI400
SAFRANINE SUPERFINE G see GJI400
SAFRANINE T see GJI400
SAFRANINE TH see GJI400
SAFRANINE TN see GJI400
SAFRANINE Y see GJI400
SAFRANINE YN see GJI400
SAFRANINE ZH see GJI400
SAFRANIN T see GJI400
SAIB see SNH050
SALICOYL HYDRAZIDE see HJL100
SALICYCLOHYDRAZINE see HJL100
SALICYLALDEHYDE, 5-CHLORO- see CLD800
SALICYLANILIDE, 4′,5-DIBROMO- see BOD600
SALICYL HYDRAZIDE see HJL100
SALICYLIC ACID, 5-CHLORO- see CLD825
SALICYLIC ACID, 3,5-DINITRO- see HKE600
SALICYLIC ACID, HYDRAZIDE see HJL100
SALICYLIC ACID, MONOAMMONIUM SALT see ANT600
SALICYLIC ACID, 4-NITRO- see NJF100
SALICYLIC HYDRAZIDE see HJL100
SALICYL-VASOGEN see ANT600
SALICYL YELLOW see SIT850
SALTPETER see PLL500
SAMARON PINK RFL see AKO350
SANDOTHRENE BLUE NG see DFN425
SANDOTHRENE BLUE NGR see DFN425
SANDOTHRENE BLUE NGW see DFN425
SANDOTHRENE BLUE NRSC see IBV050
SANDOTHRENE BLUE NRSN see IBV050
SANDOTHRENE OLIVE N2R see DUP100
SANDOTHRENE VIOLET N 4R see DFN450
SANDOTHRENE VIOLET N 2RB see DFN450
SANDOTHRENE VIOLET 4R see DFN450
SANKUMIN see HGI100
SANTOCEL see SCH000
SANTOCHLOR see DEP800
SANTOFLEX DD see SAU480
SANYO CYANINE GREEN see PJQ100
SANYO FAST ORANGE SALT RD see CEG800
SANYO FAST RED RL BASE see MMF780
SANYO FAST RED SALT RL see MMF780
SANYO PHTHALOCYANINE GREEN FB PURE see PJQ100
SANYO PHTHALOCYANINE GREEN F6G see PJQ100
SANYO THRENE BLUE IRN see IBV050
SARCOLACTIC ACID see LAG010
SAXOL see MQV750

SCH 10159 see TIO800
SCHEELES GREEN see CNN500
SCHEELE'S MINERAL see CNN500
SCHULTZ No. 1228 see IBV050
SD 4901 see BDX090
SECONDARY AMMONIUM ARSENATE see DCG800
SEGNALE LIGHT GREEN G see PJQ100
SEGNALE RED R see NAP100
SEPRISAN see HCP800
SERISOL FAST YELLOW A see DUW500
SETACYL YELLOW P-BS see DUW500
SG-67 see SCH000
SHERWOOD GREEN A 4436 see PJQ100
SHIKISO DIRECT SKY BLUE 5B see CMO500
SHINNIPPON FAST RED 3GL BASE see KDA050
SHIPRON A see CAT775
SHOWA CHROME YELLOW GG see SIT850
SHS see HCP900
SIEGLE FAST GREEN G see PJQ100
SIFERRIT see IHG100
SIGNAL RED see NAP100
SILACYCLOBUTANE see SCF550
SILANE, BENZYLCHLORODIMETHYL- see BEE800
SILANE, CHLOROTRIPHENYL- see TMU800
SILANE, (3-CYANOPROPYL)TRICHLORO- see COR550
SILANE, DICHLOROMETHYL(3,3,3-TRIFLUOROPROPYL)- see DFS700
SILANE, DIHYDROXYDIPHENYL- see DMN450
SILANE, HYDROXYTRIPHENYL- see HOM300
SILANE, OXYBIS((3-AMINOPROPYL)DIMETHYL)- see OPC100
SILANE, PHENYL- see SDX250
SILANE, p-PHENYLENEBIS(DIMETHYL)- see PEW725
SILANE, 1,4-PHENYLENEBIS(DIMETHYL)-(9CI) see PEW725
SILANE, TRICHLORO(3-CYANOPROPYL)- see COR550
SILICA AEROGEL see SCH000
SILICA, AMORPHOUS see SCH000
SILICA, AMORPHOUS FUMED see SCH000
SILICA, AMORPHOUS FUSED see SCK600
SILICA, FUSED see SCK600
SILICA, VITREOUS see SCK600
SILICETANE see SCF550
SILICIC ANHYDRIDE see SCH000
SILICON DIOXIDE see SCK600
SILICON DIOXIDE (FCC) see SCH000
SILIKILL see SCH000
SILOTON RED R see NAP100
SILVER ACETATE see SDI800
SILVER(I) ACETATE see SDI800
SILVER(1+) ACETATE see SDI800
SILVER MATT POWDER see TGB250
SILVER MONOACETATE see SDI800
SILVER W see CAT775
SILYLBENZENE see SDX250
SINCALINE see CMF800
SINKALIN see CMF800
SINKALINE see CMF800
SIPEX OLS see OFU200
SK-BISACODYL see PPN100
SKIMMETIN see HJY100
SKIMMETINE see HJY100
SKY BLUE 4B see CMO500
SKY BLUE 5B see CMO500
SL 700 see CAT775
SLAB OIL (9CI) see MQV875
SMITHKO KALKARB WHITING see CAT775
SN see SMA000
SNOWCAL see CAT775
SNOWCAL 5SW see BKW100
SNOWFLAKE WHITE see CAT775
SNOW TEX see KBB600
SNOW TOP see CAT775
SOCAL see CAT775
SOCAL E 2 see CAT775
SODANIT see SEY500
SODIO, FLUOROACETATO di (ITALIAN) see SHG500
SODIUM ACID ARSENATE see ARC000
SODIUM ACID ARSENATE, HEPTAHYDRATE see ARC250
SODIUM ALIZARINESULFONATE see SEH475
SODIUM ALIZARIN-3-SULFONATE see SEH475
SODIUM ARSENATE see ARC000, ARD500, ARD600
SODIUM ARSENATE (DOT) see ARD750
SODIUM ARSENATE DIBASIC, anhydrous see ARC000
SODIUM ARSENATE, DIBASIC, HEPTAHYDRATE see ARC250

SODIUM ARSENATE HEPTAHYDRATE see ARC250
SODIUM ARSENITE see SEY500
SODIUM ARSENITE (liquid) see SEZ000
SODIUM ARSENITE, liquid (solution) (DOT) see SEY500
SODIUM ARSENITE, solid (DOT) see SEY500
SODIUM ATP see AEM100
SODIUM BENZENEACETATE see SFA200
SODIUM BENZENESULFONATE see SJH050
SODIUM BENZOSULFONATE see SJH050
SODIUM BIBORATE see DXG035
SODIUM BROMOSULFALEIN see HAQ600
SODIUM BROMOSULFOPHTHALEIN see HAQ600
SODIUM BROMSULPHALEIN see HAQ600
SODIUM BROMSULPHTHALEIN see HAQ600
SODIUM CALCIUM EDETATE see CAR780
SODIUM CAPRATE see SGB600
SODIUM CAPRINATE see SGB600
SODIUM CAPRYLATE see SIX600
SODIUM CAPRYL SULFATE see OFU200
SODIUM CARBONATE see SJB400
SODIUM CETYL SULFATE see HCP900
SODIUM p-CRESOLATE see SIM100
SODIUM p-CRESOXIDE see SIM100
SODIUM CYANURATE see SGB550
SODIUM DECANOATE see SGB600
SODIUM-n-DECANOATE see SGB600
SODIUM DECANOIC ACID see SGB600
SODIUM 2,6-DICHLOROINDOPHENOL see SGG650
SODIUM 2,6-DICHLOROINDOPHENOLATE see SGG650
SODIUM DIHYDROGEN ARSENATE see ARD600
SODIUM DIHYDROGEN ORTHOARSENATE see ARD600
SODIUM FLUOACETATE see SHG500
SODIUM FLUOACETIC ACID see SHG500
SODIUM FLUORACETATE de (FRENCH) see SHG500
SODIUM FLUOROACETATE see SHG500
SODIUM HEXADECYL SULFATE see HCP900
SODIUM HEXAFLUOROARSENATE see SHM000
SODIUM ISOCYANURATE see SGB550
SODIUM-α-KETOBUTRATE see SIY600
SODIUM LAURYL SULFOACETATE see SIB700
SODIUM LITHOL see NAP100
SODIUM LITHOL RED see NAP100
SODIUM LITHOL RED 20-4018 see NAP100
SODIUM METAARSENATE see ARD500
SODIUM METAARSENITE see SEY500
SODIUM p-METHYLPHENOLATE see SIM100
SODIUM p-METHYLPHENOXIDE see SIM100
SODIUM 4-METHYLPHENOLATE see SIM100
SODIUM MONOFLUOROACETATE see SHG500
SODIUM MONOHEXADECYL SULFATE see HCP900
SODIUM MONOHYDROGEN ARSENATE see ARD500
SODIUM MONOOCTADECYL SULFATE see OBG100
SODIUM MONOSTEARYL SULFATE see OBG100
SODIUM MYRISTATE see SIN900
SODIUM m-NITROBENZENEAZOSALICYLATE see SIT850
SODIUM OCTADECYL SULFATE see OBG100
SODIUM OCTANOATE see SIX600
SODIUM n-OCTANOATE see SIX600
SODIUM OCTYL SULFATE see OFU200
SODIUM OCTYL SULPHATE see OFU200
SODIUM ORTHOARSENATE see ARD750
SODIUM ORTHOARSENITE see ARJ500
SODIUM 2-OXOBUTANOATE see SIY600
SODIUM 2-OXOBUTYRATE see SIY600
SODIUM PALMITYL SULFATE see HCP900
SODIUM PERCARBONATE see SJB400
SODIUM PEROXYDICARBONATE see SJB400
SODIUM PHENOL TETRABROMOPHTHALEIN see HAQ600
SODIUM PHENYLACETATE see SFA200
SODIUM PHENYLSULFONATE see SJH050
SODIUM PREDNISOLONE PHOSPHATE see PLY275
SODIUM STEARYL SULFATE see OBG100
SODIUM SUCCINATE see SJW100
SODIUM SULFOBROMOPHTHALEIN see HAQ600
SODIUM SULPHOBROMOPHTHALEIN see HAQ600
SODIUM TETRABORATE see DXG035
SODIUM TETRABORATE (Na₂B₄O₇) see DXG035
SODIUM TETRADECANOATE see SIN900
SODUXIN see SJW100
SOFTON 1000 see CAT775
SOLANTHRENE BLUE B see DFN425
SOLANTHRENE BLUE F-SBA see DFN425

SOLANTHRENE BLUE RS see IBV050
SOLANTHRENE BLUE RSN see IBV050
SOLANTHRENE BLUE SB see DFN425
SOLANTHRENE BRILLIANT VIOLET F 2R see DFN450
SOLANTHRENE OLIVE R see DUP100
SOLANTHRENE R FOR SUGAR see IBV050
SOLEAL see OMC300
SOLFAST GREEN see PJQ100
SOLFAST GREEN 63102 see PJQ100
SOLNAPYRIN-A see AIB300
SOLOCHROME YELLOW WN see SIT850
SOLUCORT see PLY275
SOLVENT BLUE 78 see BKP500
SOLVENT BLUE 93 see BKP500
SOPANOX see TGX550
SORGHUM GUM see SLJ500
SORMETAL see CAR780
SOS see OFU200
SPANISH WHITE see BKW100
SPECIAL M see AHA275
SPECIAL TERMITE FLUID see DEP600
SPECTRA-SORB UV 24 see DMW250
SPECTRA-SORB UV 531 see HND100
SPECTROLENE RED RL see MMF780
SPENT IRON MASS (UN1376) (DOT) see IHG100
SPENT IRON SPONGE (UN1376) (DOT) see IHG100
SPINACEN see SLG800
SPINACENE see SLG800
SPRIDINE HYDROCHLORIDE see SLG600
SPRIDINE TRIHYDROCHLORIDE see SLG600
SQ 4609 see BDX090
SQUALANE see SLG700
SQUALEN see SLG800
SQUALENE see SLG800
(E,E,E,E)-SQUALENE see SLG800
trans-SQUALENE see SLG800
SR 209 see TCE400
SRA GOLDEN YELLOW VIII see DUW500
SS 30 (carbonate) see CAT775
SS 50 (carbonate) see CAT775
SSB₁ see TES800
SSB 100 see CAT775
STADALAX see PPN100
STANCLERET 157 see DEJ100
STANDAMUL 1616 see HCP700
STANNANE, DICHLORODIMETHYL- see DUG825
STANNANE, FLUOROTRIPHENYL- see TMV850
STANN RC 40F see DEJ100
STANWHITE 500 see CAT775
STARAMIC 747 see SLJ500
STARCH see SLJ500
α-STARCH see SLJ500
STARCH, CORN see SLJ500
STARCH DUST see SLJ500
STA-RX 1500 see SLJ500
STAVINOR SN 1300 see DEJ100
STAVINOR 1300SN see DEJ100
STAYPRO WS 7 see HBO650
STEARIC ACID, ALUMINUM SALT see AHH825
STEARIC ACID, AMMONIUM SALT see ANU200
STEARIC ACID, BARIUM CADMIUM SALT (4:1:1) see BAI800
STEARIC ACID, CADMIUM SALT see OAT000
STEARIC ACID, ETHYL ESTER see EPF700
STEARIC ACID, ISOBUTYL ESTER see IJN100
STEARIC ACID, ISOPROPYL ESTER see IRL100
STEPAN C40 see MLC800
STEROGENOL see HCP800
STEROLAMIDE see TKP500
STILBENE, (E)- see SLR100
(E)-STILBENE see SLR100
trans-STILBENE see SLR100
2,2′-STILBENEDISULFONIC ACID, 4,4′-BIS((4,6-DIANILINO-s-TRIAZIN-2-YL)AMINO)-, DISODIUM SALT see DXB450
2,2′-STILBENEDISULFONIC ACID, 4,4′-DIAMINO- see FCA100
STREPTONIGRAN see SMA000
STREPTONIGRIN see SMA000
STRONTIUM CHROMATE (1:1) see SMH000
STRONTIUM CHROMATE (VI) see SMH000
STRONTIUM CHROMATE 12170 see SMH000
STRONTIUM YELLOW see SMH000
STURCAL D see CAT775
STYRACIN see CMQ850

STYRENE GLYCOL

STYRENE GLYCOL see SMQ100
STYRENE, p-METHYL- see VQK700
STYRENE, β-NITRO- see NMC100
STYRENE, α-β,β-TRIFLUORO- see TKH310
STYROLFENOL see PFD400
STYROLYL ALCOHOL see SMQ100
SUBERONITRILE see OCW050
SUCCINIC ACID, CADMIUM SALT (1:1) see CAI750
SUCCINIC ACID, 2,3-DIMERCAPTO- see DNV610
SUCCINIC ACID, DIMETHYL ESTER see SNB100
SUCCINIC ACID, DISODIUM SALT see SJW100
SUCCINIC ACID, SULFO-, 1,4-DIOCTYL ESTER, SODIUM SALT see SOD300
SUCROSE ACETATE ISOBUTYRATE see SNH050
SUCROSE ACETOISOBUTYRATE see SNH050
SUCROSE, DIACETATE HEXAISOBUTYRATE see SNH050
SUCROSE, OCTAACETATE see OAF100
p-SULFAMOYLBENZOIC ACID see SNL840
4-SULFAMOYLBENZOIC ACID see SNL840
p-SULFAMYLBENZOIC ACID see SNL840
SULFANILIC ACID see SNN600
o-SULFANILIC ACID see SNO100
SULFANILSAEURE see SNN600
SULFOACETIC ACID 1-DODECYL ESTER, SODIUM SALT see SIB700
SULFOACETIC ACID DODECYL ESTER S-SODIUM SALT see SIB700
SULFOBROMOPHTHALEIN see HAQ600
SULFOBROMOPHTHALEIN SODIUM see HAQ600
SULFOBROMPHTHALEIN see HAQ600
SULFOCARBOLIC ACID see HJH100
p-SULFONAMIDOBENZOIC ACID see SNL840
SULFONE, BIS(4-FLUORO-3-NITROPHENYL) see DKH250
SULFONIUM, TRIMETHYL-, IODIDE, OXIDE see TMG800
SULFONPHTHAL see PDO800
SULFOPARABLUE see SOB600
SULFOSUCCINIC ACID 1,4-DIOCTYL ESTER SODIUM SALT see SOD300
SULFOXONIUM, TRIMETHYL-, IODIDE see TMG800
SULFURIC ACID, CADMIUM(2+) SALT see CAJ000
SULFURIC ACID, CADMIUM SALT, HYDRATE see CAJ250
SULFURIC ACID, CADMIUM SALT, TETRAHYDRATE see CAJ500
SULFURIC ACID, MONOOCTADECYL ESTER, SODIUM SALT see OBG100
SULFURIC ACID, MONOOCTYL ESTER, SODIUM SALT see OFU200
SULPHANILIC ACID see SNN600
SULPHENTAL see PDO800
SULPHOBROMOPHTHALEIN see HAQ600
SULPHOBROMOPHTHALEIN SODIUM see HAQ600
SULPHONTHAL see PDO800
SULPHURIC ACID, CADMIUM SALT (1:1) see CAJ000
SUMISORB 130 see HND100
SUNBURST RED see NAP100
SUNCHROMINE YELLOW GG see SIT850
SUNLIGHT 700 see CAT775
SUPER 3S see CAT775
SUPER 1500 see CAT775
SUPERBRESILINE see LFT800
SUPERCOAT see CAT775
SUPER COBALT see CNA250
SUPER CRAB-E-RAD-CALAR see CAM000
SUPER DAL-E-RAD see CAM000
SUPER DAL-E-RAD-CALAR see CAM000
SUPERLYSOFORM see FMV000
SUPERMITE see CAT775
SUPER MULTIFEX see CAT775
SUPER-PFLEX see CAT775
SUPER SSS see CAT775
SUPER-TREFLAN see DUV600
SUPRACET FAST PINK 2R see AKO350
SUPRACET FAST YELLOW 2R see DUW500
SUPRACET YELLOW RR see DUW500
SURFEX MM see CAT775
SURFIL S see CAT775
SU SEGURO CARPIDOR see DUV600
SUSPENSO see CAT775
SUXIMER see DNV610
SWEDISH GREEN see CNN500
SWEENEY'S ANT-GO see ARD750
SYLACAUGA 88B see CAT775
SYMULER FAST BLUE 6011 see IBV050
SYMULER FAST VIOLET R see DFN450
SYMULON RED RL BASE see MMF780
SYNFLORAN see DUV600
SYNTEN YELLOW P 2R see DUW500
SYNTHALINE GREEN see PJQ100
SYNTHETIC AMORPHOUS SILICA see SCH000
SYNTHETIC INDIGO see BGB275
SYNTHETIC INDIGO TS see BGB275
SYRINGIC ACID see SPE700
T 130-2500 see CAT775
TA see THS825
TA 12 see TAN750
TALOFLOC see DXG625
TAMA PEARL TP 121 see CAT775
TA-33MP see TAN750
TANAN see TDT800
TANANE see TDT800
TANCAL 100 see CAT775
TANOL see HOI300
T.A.P.E. see NNR400
TAPIOCA STARCH see SLJ500
TAPON see SLJ500
TBPMC see BSG300
TBT see BSP500
TCTNB see TJE200
TDS see TES800
TDS (NEUROTROPE) see TES800
TEBAC see BFL300
TECH PET F see MQV750
TEDMA see MDN510
TELEMIN see PPN100
TEMASEPT see BOD600
TEMIK OXIME see MLX800
TEMIK SULFOXIDE see MLW750
TEMPO see TDT800
TEMPOL see HOI300
TEOF see TJL600
TERBAM see BSG300
TEREPHTHALIC ACID see TAN750
TEREPHTHALIC ACID, DIBUTYL ESTER see DEH700
TEREPHTHALIC ACID, DIETHYL ESTER see DKB160
TERGITOL ANIONIC 7 see HCP900
TERMITKIL see DEP600
TERMOFLEKS A see BKO600
TERMOSOLIDO GREEN FG SUPRA see PJQ100
β-TERPINEOL see TBD775
TERTRACID LIGHT BLUE SE see APG700
TERTRODIRECT BLUE F see CMO500
TERTRODIRECT RED 4B see DXO850
TERTROPIGMENT RED LR see NAP100
TETACIN see CAR780
TETACIN-CALCIUM see CAR780
TETAZINE see CAR780
TETLEN see PCF275
TETRAACETIL PENTOETRIOL see NNR400
TETRAACETYL PENTESTRIOL see NNR400
TETRABROMOPHENOLSULFOPHTHALEIN see HAQ600
TETRABROMOSULFOPHTHALEIN see HAQ600
TETRABROMSULFTHALEIN see HAQ600
TETRACAP see PCF275
1,2,4,5-TETRACARBOXYBENZENE see PPQ630
TETRACHLOORETHEEN (DUTCH) see PCF275
TETRACHLOORKOOLSTOF (DUTCH) see CBY000
TETRACHLOORMETAAN see CBY000
TETRACHLORAETHEN (GERMAN) see PCF275
TETRACHLORKOHLENSTOFF, (GERMAN) see CBY000
TETRACHLORMETHAN (GERMAN) see CBY000
3,4,5,6-TETRACHLORO-1,2-BENZENEDICARBOXYLIC ACID see TBT100
TETRACHLOROCARBON see CBY000
TETRACHLOROETHENE see PCF275
TETRACHLOROETHYLENE (DOT) see PCF275
1,1,2,2-TETRACHLOROETHYLENE see PCF275
TETRACHLOROGUAIACOL see TBQ290
TETRACHLOROMETHANE see CBY000
TETRACHLOROPHTHALIC ACID see TBT100
TETRACHLORURE de CARBONE (FRENCH) see CBY000
TETRACLOROETENE (ITALIAN) see PCF275
TETRACLOROMETANO (ITALIAN) see CBY000
TETRACLORURO di CARBONIO (ITALIAN) see CBY000
2,6,10,14,18,22-TETRACOSAHEXAENE, 2,6,10,15,19,23-HEXAMETHYL-, (all-E)- see SLG800
TETRACOSANE, 2,6,10,15,19,23-HEXAMETHYL- see SLG700
1-TETRADECANAMINIUM, N,N,N-TRIMETHYL-, BROMIDE (9CI) see TCB200
TETRADECANOIC ACID, SODIUM SALT (9CI) see SIN900
4,8,12-TETRADECATRIENOIC ACID, 5,9,13-TRIMETHYL-, 3,7-DIMETHYL-2,6-OCTADIENYLESTER, (E,E,E)- see GDG200
TETRADECYLTRIMETHYLAMMONIUM BROMIDE see TCB200

TETRADONIUM BROMIDE see TCB200
TETRAETHYLAMMONIUM PERCHLORATE (dry) (DOT) see TCD002
N,N,N′,N′-TETRAETHYL-1,2-BENZENEDICARBOXAMIDE see GCE600
TETRA(2-ETHYLBUTOXY) SILANE see TCD250
TETRA(2-ETHYLBUTYL) ORTHOSILICATE see TCD250
N,N,N′,N′-TETRAETHYL-2-BUTYNYLENEDIAMINE see BJA250
TETRAETHYLENE GLYCOL DIMETHACRYLATE see TCE400
N,N,N′,N′-TETRAETHYLPHTHALAMIDE see GCE600
TETRAFINOL see CBY000
TETRAFORM see CBY000
TETRAHYDRO-3-FURANOL see HOI200
TETRAHYDROFURFURYL BROMIDE see TCS550
TETRAHYDROLINALOOL see LFY510
1,4,4a,8a-TETRAHYDRO-1,4-METHANONAPHTHALENE-5,8-DIONE see TCU650
TETRAHYDROPHTHALIC ACID IMIDE see TDB100
TETRAHYDROPHTHALIMIDE see TDB100
1,2,3,6-TETRAHYDROPHTHALIMIDE see TDB100
Δ⁴-TETRAHYDROPHTHALIMIDE see TDB100
1,2,3,4-TETRAHYDROSTYRENE see CPD750
TETRAHYDRO-2-THIOPHENONE see TDC800
1,4,9,10-TETRAHYDROXYANTHRACENE see LEX200
3,4,6a,10-TETRAHYDROXY-6a,7-DIHYDROBENZ(b)INDENO(1,2-d)PYRAN-9(6H)-ONE see HAO600
TETRAISOPROPOXIDE TITANIUM see IRN200
TETRAISOPROPOXYTITANIUM see IRN200
TETRAISOPROPYL ORTHOTITANATE see IRN200
TETRAISOPROPYL TITANATE see IRN200
TETRAKIS(ISOPROPOXY)TITANIUM see IRN200
TETRALENO see PCF275
TETRALEX see PCF275
N,N,N′,N′-TETRAMETHYLBENZIDINE see BJF600
N,N,N′,N′-TETRAMETHYL-p,p′-BENZIDINE see BJF600
TETRAMETHYLENE CYANIDE see AER250
TETRAMETHYLENEDIAMINE, N,N-DIETHYL-4-METHYL- see DJQ850
TETRAMETHYLENE DIIODIDE see TDQ400
TETRAMETHYLENE IODIDE see TDQ400
N,N,N′,N′-TETRAMETHYLHEXAMETHYLENE DIAMINE see TDR100
N,N,N′,N′-TETRAMETHYL-1,6-HEXANEDIAMINE see TDR100
TETRAMETHYLPIPERIDINE NITROXIDE see TDT800
2,2,6,6-TETRAMETHYLPIPERIDINE-1-OXYL see TDT800
TETRAMETHYLPIPERIDINOL N-OXYL see HOI300
2,2,6,6-TETRAMETHYLPIPERIDINONE see TDT770
2,2,6,6-TETRAMETHYL-4-PIPERIDINONE see TDT770
TETRAMETHYLPIPERIDINOOXY see TDT800
2,2,6,6-TETRAMETHYLPIPERIDINOOXY see TDT800
2,2,6,6-TETRAMETHYLPIPERIDINOOXYL see TDT800
2,2,6,6-TETRAMETHYLPIPERIDINYLOXY see TDT800
2,2,6,6-TETRAMETHYL-1-PIPERIDINYLOXY see TDT800
2,2,6,6-TETRAMETHYLPIPERIDONE see TDT770
2,2,6,6-TETRAMETHYL-4-PIPERIDONE see TDT770
2,2,6,6-TETRAMETHYLPIPERIDOXYL see TDT800
2,2,5,5-TETRAMETHYL-1-PYRROLIDINYLOXY-3-CARBOXYLIC ACID see TDV300
TETRANITROMETHANE see TDY250
TETRAPHENE see BBC250
TETRA PINK B see DNT300
TETRAPROPYLGERMANE see TED650
TETRAPROPYLGERMANIUM see TED650
TETRASOL see CBY000
TETRAVEC see PCF275
TETROGUER see PCF275
TETROPIL see PCF275
TF see TJE100
TGM 3 see MDN510
TGM 4 see TCE400
TGM 3PC see MDN510
TGM 3S see MDN510
TGT see GGY150
TH 564 see DTU850
THALLIUM OXIDE see TEL040
THALLOUS OXIDE see TEL040
THALO GREEN No.1 see PJQ100
2-THENOIC ACID see TFM600
THERALAX see PPN100
4-THIA-1-AZABICYCLO(3.2.0)HEPTANE-2-CARBOXYLIC ACID, 3,3-DIMETHYL-7-OXO-6-(2-PHENYLACETAMIDO)-, compd. with 2-(DIETHYLAMINO)ETHYL p-AMINOBENZOATE (1:1) see PAQ200
THIACYCLOPENTANONE-2 see TDC800
THIACYCLOPENTAN-2-ONE see TDC800
9-THIAFLUORENE see TES300
THIAMIDIN F see TES800
THIAMIN DISULFIDE see TES800
THIAMINE DISULFIDE see TES800

2-THIAZOLAMINE, 4,5-DIHYDRO-(9CI) see TEV600
THIAZOLE see TEU300
5-THIAZOLECARBOXYLIC ACID, 2-AMINO-4-(TRIFLUOROMETHYL)-, ETHYL ESTER see AMU550
THIAZOLE, 4,5-DIHYDRO-2-METHYL- see DLV900
THIAZOLE, 4-METHYL-5-VINYL- see MQM800
2-THIAZOLIDINIMINE see TEV600
2-THIAZOLINE, 2-AMINO- see TEV600
2-THIAZOLINE, 2-METHYL- see DLV900
p-THIOANISIDINE see AMS675
THIOBENZOIC ACID see TFC550
γ-THIOBUTYROLACTONE see TDC800
4-THIOBUTYROLACTONE see TDC800
THIOCARBOSTYRIL see QOJ100
m-THIOCRESOL see TGO800
3-THIOCRESOL see TGO800
THIOCYTOSINE see TFH800
2-THIOCYTOSINE see TFH800
THIOFACO T-35 see TKP500
THIOINDIGO see DNT300
THIOINDIGO RED B see DNT300
THIOINDIGO RED S see DNT300
THIOLAN-2-ONE see TDC800
p-THIOMETHOXYANILINE see AMS675
THIOPEROXYDICARBONIC ACID, DIBUTYL ESTER see BSS550
α-THIOPHENECARBOXYLIC ACID see TFM600
2-THIOPHENECARBOXYLIC ACID see TFM600
2-THIOPHENIC ACID see TFM600
2(3H)-THIOPHENONE, DIHYDRO-(8CI,9CI) see TDC800
2-THIOPHENONE, TETRAHYDRO- see TDC800
2-THIOPYRIDINE see TFP300
THIOPYRIDONE-2 see TFP300
THIOTHYMINE see TFQ650
2-THIOTHYMINE see TFQ650
THRETHYLENE see TIO750
THYMINE, 2-THIO- see TFQ650
THYMOL, 6,6′-(3H-2,1-BENZOXATHIOL-3-YLIDENE)DI-, S,S-DIOXIDE see TFX850
THYMOL BLUE see TFX850
THYMOL, 6-CHLORO- see CLJ800
THYMOL METHYL ETHER see MPW650
THYMOLSULFONEPHALEIN see TFX850
THYMOLSULFONEPHTHALEIN see TFX850
THYMOLSULFOPHTHALEIN see TFX850
THYMOLSULPHONPHTHALEIN see TFX850
THYMOSULFONPHTHALEIN see TFX850
THYMYL METHYL ETHER see MPW650
TIN see TGB250
TIN (α) see TGB250
TINA PINK B see DNT300
TIN, DIMETHYL-, DICHLORIDE see DUG825
TIN FLAKE see TGB250
TIN, FLUOROTRIPHENYL- see TMV850
TINON BLUE GF see DFN425
TINON BLUE GL see DFN425
TINON BLUE RS see IBV050
TINON BLUE RSN see IBV050
TINON OLIVE 2R see DUP100
TINON VIOLET B 4RP see DFN450
TINON VIOLET 4R see DFN450
TINON VIOLET 2RB see DFN450
TINOPAL BHS see FCA100
TIN POWDER see TGB250
TISULAC see LAG010
TITANIUM(4+) ISOPROPOXIDE see IRN200
TITANIUM ISOPROPYLATE see IRN200
TITANIUM TETRAISOPROPOXIDE see IRN200
TITANIUM TETRAISOPROPYLATE see IRN200
TITANIUM TETRA-n-PROPOXIDE see IRN200
TL 336 see BOB550
TL 869 see SHG500
TL 1026 see CAG000
TL 1070 see CAG500
TL 1182 see CAI000
TL 1473 see DGJ250
TM 1 (filler) see CAT775
TMA see TKU700, TKV000, TLD500
TMAN see TKV000
m-TMI see IKG800
TMP see HDF300
TMP (ALCOHOL) see HDF300
TMPN see HOI300
TN 3J see DEJ100

TNCB see PIE530
TNM see TDY250
TOLUEEN (DUTCH) see TGK750
TOLUEN (CZECH) see TGK750
TOLUENE see TGK750
TOLUENE, m-BROMO- see BOG300
TOLUENE, o-BROMO- see BOG260
TOLUENE, 4-CHLORO-3,5-DINITRO-α-α-α-TRIFLUORO- see CGM225
TOLUENE, α-CHLORO-m-NITRO- see NFN010
TOLUENE, 2-CHLORO-6-NITRO- see CJG825
TOLUENE-2,3-DIAMINE see TGY800
TOLUENE-3,5-DIAMINE, α-α-α-TRIFLUORO-(8CI) see TKB775
p-TOLUENEDIAZONIUM FLUOROBORATE see TGM450
o-TOLUENEDIAZONIUM, 4-NITRO-, TETRAFLUOROBORATE(1⁻) see NMO700
p-TOLUENEDIAZONIUM, TETRAFLUOROBORATE (1⁻) see TGM450
p-TOLUENEDIAZONIUM TETRAFLUOROBORATE (7CI) see TGM450
TOLUENE, p,α-DICHLORO- see CFB600
TOLUENE, 2,4-DICHLORO- see DGM700
TOLUENE, α-α-DICHLORO-α-FLUORO- see DFL100
TOLUENE, o-FLUORO- see FLZ100
p-TOLUENESULFANILIDE see TGN600
p-TOLUENESULFONANILIDE see TGN600
p-TOLUENESULFONYLANILIDE see TGN600
α-TOLUENESULFONYL FLUORIDE see TGO300
m-TOLUENETHIOL see TGO800
TOLUENE, 2,3,6-TRICHLORO- see TJD600
p-TOLUIDINE, 2-BROMO- see BOG500
o-TOLUIDINE, 6-CHLORO- see CLK227
m-TOLUIDINE, 6-CHLORO-α-α-α-TRIFLUORO- see CEG800
m-TOLUIDINE, N,N-DIMETHYL- see TLG100
p-TOLUIDINE, N,N-DIMETHYL- see TLG150
o-TOLUIDINE, 4-NITRO- see MMF780
TOLUOL (DOT) see TGK750
TOLUOLO (ITALIAN) see TGK750
TOLUSAFRANINE see GJI400
TOLU-SOL see TGK750
2,3-TOLUYLENEDIAMINE see TGY800
o-TOLYLBIGUANIDE see TGX550
1-o-TOLYLBIGUANIDE see TGX550
m-TOLYL BROMIDE see BOG300
2-TOLYL BROMIDE see BOG260
o-TOLYLBROMIDE see BOG260
o-TOLYLDIGUANIDE see TGX550
2,3-TOLYLENEDIAMINE see TGY800
1-p-TOLYLETHENE see VQK700
m-TOLYLMERCAPTAN see TGO800
1-(p-TOLYL)-3-METHYLPYRAZOLONE-5 see THA300
TONASO see CAT775
TONKALIDE see HDY600
TONOCHOLIN B see CMF260
N-TOSYLANILINE see TGN600
TOUKALIDE see HDY600
TOYOFINE TF-X see CAT775
TP-95 see DDT500
TP 222 see CAT775
TP 121 (filler) see CAT775
TPN see CNF400
β-TPN see CNF400
TPN (NUCLEOTIDE) see CNF400
TRAN-Q DIHYDROCHLORIDE see HOR470
TRANQUIZINE DIHYDROCHLORIDE see HOR470
TREFANOCIDE see DUV600
TREFICON see DUV600
TREFLAM see DUV600
TREFLAN see DUV600
TREFLANOCIDE ELANCOLAN see DUV600
TRI-4 see DUV600
TRIACETATE d'HYDROXYHYDROQUINONE see HLF600
TRIACETIC ACID LACTONE see THL800
TRIACETONAMIN see TDT770
TRIACETONAMINE see TDT770
TRIACETONE AMINE see TDT770
1,2,4-TRIACETOXYBENZENE see HLF600
TRIACRYLFORMAL see THM900
TRIACRYLOYLHEXAHYDROTRIAZINE see THM900
TRI(N-ACRYLOYL)HEXAHYDROTRIAZINE see THM900
TRIACRYLOYLHEXAHYDRO-s-TRIAZINE see THM900
1,3,5-TRIACRYLOYLHEXAHYDROTRIAZINE see THM900
TRIACRYLOYLPERHYDROTRIAZINE see THM900
TRIAD see TIO750
TRIAETHANOLAMIN-NG see TKP500
s-TRIAMINOBENZENE see THN775

1,3,5-TRIAMINOBENZENE see THN775
sym-TRIAMINOBENZENE see THN775
TRIASOL see TIO750
1,3,5-TRIAZIN-2-AMINE, 4-METHOXY-6-METHYL-(9CI) see MGH800
s-TRIAZINE, 2-AMINO-4-METHOXY-6-METHYL- see MGH800
s-TRIAZINE, HEXAHYDRO-1,3,5-TRIACRYLOYL- see THM900
s-TRIAZINE, HEXAHYDRO-1,3,5-TRIMETHYL- see HDW100
1,3,5-TRIAZINE, HEXAHYDRO-1,3,5-TRIMETHYL- see HDW100
1,3,5-TRIAZINE, HEXAHYDRO-1,3,5-TRIS(1-OXO-2-PROPENYL)-(9CI) see THM900
s-TRIAZINE-2,4,6(1H,3H,5H)-TRIONE, MONOSODIUM SALT see SGB550
1,3,5-TRIAZINE-2,4,6(1H,3H,5H)-TRIONE, MONOSODIUM SALT (9CI) see SGB550
s-TRIAZINE-2,4,6(1H,3H,5H)-TRIONE, TRIS(2,3-EPOXYPROPYL)- see GGY150
s-TRIAZINE-2,4,6(1H,3H,5H)-TRIONE, 1,3,5-TRIS(2,3-EPOXYPROPYL)- see GGY150
s-TRIAZINE-2,4,6(1H,3H,5H)-TRIONE, 1,3,5-TRIS(OXIRANYLMETHYL)-(9CI) see GGY150
s-TRIAZOLE see THS825
1H-1,2,4-TRIAZOLE (9CI) see THS825
1H-v-TRIAZOLO(4,5-d)PYRIMIDIN-7-AMINE see AIB340
TRIBASIC ALUMINUM STEARATE see AHH825
TRIBROMOARSINE see ARF250
TRIBROMOETHENE see THV100
TRIBROMOETHYLENE see THV100
1,1,2-TRIBROMOETHYLENE see THV100
2,4,5-TRIBROMOIMIDAZOLE CADMIUM SALT (2581) see THV500
TRIBUTYLFOSFIN see TIA300
TRIBUTYLPHOSPHINE see TIA300
TRI-n-BUTYLPHOSPHINE see TIA300
TRICADMIUM DINITRIDE see TIH000
TRICALCIUMARSENAT (GERMAN) see ARB750
TRICALCIUM ARSENATE see ARB750
1,2,4-TRICARBOXYBENZENE see TKU700
TRICHLOORETHEEN (DUTCH) see TIO750
TRICHLOORETHYLEEN, TRI (DUTCH) see TIO750
TRICHLORAETHEN (GERMAN) see TIO750
TRICHLORAETHYLEN, TRI (GERMAN) see TIO750
TRICHLORAN see TIO750
TRICHLORETHENE (FRENCH) see TIO750
TRICHLORETHYLENE, TRI (FRENCH) see TIO750
TRICHLOR-3-KYANPROPYLSILAN see COR550
TRICHLOROACETALDEHYDE see CDN550
2,2,2-TRICHLOROACETALDEHYDE see CDN550
TRICHLOROARSINE see ARF500
s-TRICHLOROBENZENE see TIK300
1,2,3-TRICHLOROBENZENE see TIK100
1,2,6-TRICHLOROBENZENE see TIK100
1,3,5-TRICHLOROBENZENE see TIK300
sym-TRICHLOROBENZENE see TIK300
vic-TRICHLOROBENZENE see TIK100
TRICHLOROETHANAL see CDN550
TRICHLOROETHENE see TIO750
TRICHLOROETHYLENE see TIO750
1,2,2-TRICHLOROETHYLENE see TIO750
TRICHLOROETHYL PHOSPHATE see TIO800
2,2,2-TRICHLOROETHYL PHOSPHATE see TIO800
α-(TRICHLOROMETHYL)BENZENEMETHANOL see TIR800
α-(TRICHLOROMETHYL)BENZYL ALCOHOL see TIR800
TRICHLOROMETHYLPHENYL CARBINOL see TIR800
3,5,6-TRICHLORO-2-PYRIDINOL-O-ESTER with O,O-DIETHYL PHOSPHOROTHIOATE see CMA100
2,3,6-TRICHLOROTOLUENE see TJD600
TRICHLOROTRIFLUOROETHANE see TJE100
1,1,1-TRICHLORO-2,2,2-TRIFLUOROETHANE see TJE100
TRICHLORO-1,3,5-TRINITROBENZENE see TJE200
sym-TRICHLOROTRINITROBENZENE see TJE200
1,3,5-TRICHLORO-2,4,6-TRINITROBENZENE see TJE200
TRICHLORURE d'ARSENIC (FRENCH) see ARF500
TRI-CLENE see TIO750
TRICLOFOS see TIO800
TRICLORETENE (ITALIAN) see TIO750
TRICLOROETILENE (ITALIAN) see TIO750
TRICLOS see TIO800
TRIDECANE, 1-BROMO- see BOI500
TRIELINA (ITALIAN) see TIO750
TRIETHANOLAMINE see TKP500
TRIETHANOLAMINE (ACGIH) see TKP500
TRIETHOXONIUM FLUOROBORATE see TJL600
TRIETHYLAMINE, HYDROCHLORIDE see TJO050
N,N,N-TRIETHYLBENZENEMETHANAMINIUM CHLORIDE see BFL300
TRIETHYLBENZYLAMMONIUM CHLORIDE see BFL300
TRIETHYLCARBINOL see TJP550
TRIETHYLENE GLYCOL DIMETHACRYLATE see MDN510
TRIETHYLMETHANOL see TJP550

TRIETHYLOLAMINE see TKP500
TRIETHYLOXONIUM BOROFLUORIDE see TJL600
TRIETHYLOXONIUM FLUOBORATE see TJL600
TRIETHYL OXONIUM FLUOROBORATE see TJL600
TRIETHYLOXONIUM TETRAFLUOROBORATE see TJL600
TRIFLORAN see DUV600
TRIFLUORALIN (USDA) see DUV600
TRIFLUOROARSINE see ARI250
α,α,α-TRIFLUORO-2,6-DINITRO-N,N-DIPROPYL-p-TOLUIDINE see DUV600
1,1,1-TRIFLUOROETHANE see TJY900
(TRIFLUOROETHENYL)BENZENE see TKH310
1,1,1-TRIFLUOROFORM see TJY900
5-(TRIFLUOROMETHYL)-1,3-BENZENEDIAMINE see TKB775
4-(TRIFLUOROMETHYL)-2,6-DINITRO-N,N-DIPROPYLANILINE see DUV600
TRIFLUOROMETHYLETHYLENE see TKH015
2-(TRIFLUOROMETHYL)PHENOTHIAZINE see TKE775
(TRIFLUOROMETHYL)TRIFLUOROOXIRANE see HDF050
TRIFLUOROMETHYL TRIFLUOROVINYL ETHER see TKG800
1,1,1-TRIFLUOROPROPENE see TKH015
3,3,3-TRIFLUOROPROPENE see TKH015
3,3,3-TRIFLUORO-1-PROPENE see TKH015
TRIFLUOROPROPIONALDEHYDE see TKH020
3,3,3-TRIFLUOROPROPIONALDEHYDE see TKH020
3,3,3-TRIFLUOROPROPYLENE see TKH015
TRIFLUORO SELENIUM HEXAFLUORO ARSENATE see TKH250
α-β,β-TRIFLUOROSTYRENE see TKH310
α-α-α-TRIFLUOROTOLUENE-3,5-DIAMINE see TKB775
TRIFLUORO(TRIFLUOROMETHYL)OXIRANE see HDF050
3,3,3-TRIFLUORO-2-(TRIFLUOROMETHYL)PROPENE see HDC450
TRIFLURALIN see DUV600
TRIFLURALINA 600 see DUV600
TRIFLURALINE see DUV600
TRIFUREX see DUV600
TRIGLYCIDYL ISOCYANURATE see GGY150
1,3,5-TRIGLYCIDYL ISOCYANURATE see GGY150
N,N',N''-TRIGLYCIDYL ISOCYANURATE see GGY150
1,3,5-TRIGLYCIDYLISOCYANURIC ACID see GGY150
TRIGONOX 17/40 see BHL200
2,3,4-TRIHYDROXYBENZALDEHYDE see TKN800
TRI(HYDROXYETHYL)AMINE see TKP500
α¹,α³,2-TRIHYDROXYMESITYLENE see HLV100
1,1,1-TRI(HYDROXYMETHYL)PROPANE see HDF300
2,4,6-TRIHYDROXYPROPIOPHENONE see TKP100
2',4',6'-TRIHYDROXYPROPIOPHENONE see TKP100
TRIHYDROXYTRIETHYLAMINE see TKP500
2,2',2''-TRIHYDROXYTRIETHYLAMINE see TKP500
TRIIODOARSINE see ARG750
TRIIRON TETRAOXIDE see IHC550
TRIISOPROPOXYALUMINUM see AHC600
TRIKEPIN see DUV600
TRILENE see TIO750
TRIM see DUV600
TRIMAR see TIO750
TRIMELLIC ACID ANHYDRIDE see TKV000
TRIMELLIC ACID-1,2-ANHYDRIDE see TKV000
TRIMELLITIC ACID see TKU700
TRIMELLITIC ACID CYCLIC-1,2-ANHYDRIDE see TKV000
TRIMELLITIC ANHYDRIDE see TKV000
3,4,5-TRIMETHOXYBENZOIC ACID, METHYL ESTER see MQF300
TRIMETHOXYBOROXIN see TKZ100
TRIMETHOXYBOROXINE see TKZ100
3,4,5-TRIMETHOXYCINNAMIC ACID see CMQ100
3,4,5-TRIMETHOXYPHENYLACRYLIC ACID see CMQ100
3-(3,4,5-TRIMETHOXYPHENYL)-2-PROPENOIC ACID see CMQ100
TRIMETHYLAMINE see TLD500
TRIMETHYLAMINE, anhydrous (DOT) see TLD500
TRIMETHYLAMINE, aqueous solution (DOT) see TLD500
TRIMETHYLAMINE, aqueous solutions containing not more than 30% of trimethylamine (DOT) see TLD500
m,N,N-TRIMETHYLANILINE see TLG100
N,N,3-TRIMETHYLANILINE see TLG100
N,N,4-TRIMETHYLANILINE see TLG150
p,N,N-TRIMETHYLANILINE see TLG150
2,4,6-TRIMETHYLBENZALDEHYDE see MDJ745
N,N,3-TRIMETHYLBENZENAMINE see TLG100
N,N,N-TRIMETHYLBENZENEMETHANAMINIUM METHOXIDE see TLM100
2,4,6-TRIMETHYLBENZOIC ACID see IKN300
TRIMETHYLCARBINOL see BPX000
α-α-4-TRIMETHYLCYCLOHEXANEMETHANOL see MCE100
cis-3,3,5-TRIMETHYLCYCLOHEXANOL see TLO510
TRIMETHYLENE CHLOROHYDRIN see CKP725
TRIMETHYLENE DIIODIDE see TLR050

N,N,N-TRIMETHYLETHANAMINIUM IODIDE see EQC600
TRIMETHYLGLYCINE see GHA050
TRIMETHYLGLYCOCOLL see GHA050
1,3,5-TRIMETHYLHEXAHYDRO-1,3,5-TRIAZINE see HDW100
1,3,5-TRIMETHYLHEXAHYDRO-sym-TRIAZINE see HDW100
1,3,3-TRIMETHYL-2-METHYLENEINDOLINE see TLU100
TRIMETHYLMYRISTYLAMMONIUM BROMIDE see TCB200
TRIMETHYLOLPROPANE see HDF300
1,1,1-TRIMETHYLOLPROPANE see HDF300
2,4,6-TRIMETHYLPHENOL see MDJ740
2,4,6-TRIMETHYLPYRIDINE see TME272
TRIMETHYL SULPHOXONIUM IODIDE see TMG800
N,N,N-TRIMETHYL-1-TETRADECANAMINIUM BROMIDE see TCB200
(E,E,E)-5,9,13-TRIMETHYL-4,8,12-TETRADECATRIENOIC ACID 3,7-DIMETHYL-2,6-OCTADIENYL ESTER see GDG200
TRIMETHYLTETRADECYLAMMONIUM BROMIDE see TCB200
2,6,9-TRIMETHYLUNDECA-2,6,8-TRIEN-10-ONE see TMJ100
3,6,10-TRIMETHYL-3,5,9-UNDECATRIEN-2-ONE see TMJ100
2,4,6-TRIMETYLOFENOL see MDJ740
2,4,6-TRINITROANISOLE see TMK300
2,4,6-TRINITROCHLOROBENZENE see PIE530
2,8,9-TRIOXA-5-AZA-1-SILABICYCLO(3.3.3)UNDECANE, 1-ETHENYL- see VQK600
2,8,9-TRIOXA-5-AZA-1-SILABICYCLO(3.3.3)UNDECANE, 1-METHYL- see MPI650
2,8,9-TRIOXA-5-AZA-1-SILABICYCLO(3.3.3)UNDECANE, 1-VINYL- see VQK600
s-TRIOXANE, 2,4,6-TRIPROPYL- see TNC100
1,3,5-TRIOXANE, 2,4,6-TRIPROPYL- see TNC100
TRIPHENYLSILYL CHLORIDE see TMU800
TRIPHENYLTIN FLUORIDE see TMV850
TRIPHOSPHOPYRIDINE NUCLEOTIDE see CNF400
TRI-PLUS see TIO750
2,4,6-TRIPROPYL-s-TRIOXANE see TNC100
TRIS(N-ACRYLOYL)HEXAHYDROTRIAZINE see THM900
TRIS(ACRYLOYL)HEXAHYDRO-s-TRIAZINE see THM900
TRIS(EPOXYPROPYL)ISOCYANURATE see GGY150
TRIS(2,3-EPOXYPROPYL)ISOCYANURATE see GGY150
TRIS(2-ETHYLHEXYL)PHOSPHITE see TNI300
TRIS(2-HYDROXYETHYL)AMINE see TKP500
TRIS(HYDROXYMETHYL)PROPANE see HDF300
1,1,1-TRIS(HYDROXYMETHYL)PROPANE see HDF300
TRISTAR see DUV600
TRITHIOCARBONIC ACID, CYCLIC ETHYLENE ESTER see EJQ100
TRITYL BROMIDE see TNP600
TRIULOSE see OMC300
TROGUM see SLJ500
TROJACETONOAMINY see TDT770
TROLAMINE see TKP500
TRONAMANG see MAP750
TsPB see HCP800
TUAMINE see HBM490
TUAMINOHEPTANE see HBM490
TULABASE FAST RED RL see MMF780
TVS-MA 300 see DEJ100
TVS-N 2000E see DEJ100
T-WD602 see TJE100
TYRIAN BLUE I-RSN see IBV050
TYRIAN BRILLIANT BLUE I-R see IBV050
TYRIAN OLIVE I-R see DUP100
TYRIAN RED A-5B see DNT300
TYZOR TPt see IRN200

U 6245 see DTU850
U 21221 see DSK950
UF 2 see DMW250
UF 4 see HND100
ULCOLAX see PPN100
ULTRA-PFLEX see CAT775
UMBELLIFERON see HJY100
UMBELLIFERONE see HJY100
γ-UNDECALACTONE see HBN200
UNDECANOIC ACID, 4-HYDROXY-, γ-LACTONE see HBN200
γ-UNDECANOLACTONE see HBN200
γ-UNDECANOLIDE see HBN200
4-UNDECANOLIDE see HBN200
1,4-UNDECANOLIDE see HBN200
3,5,9-UNDECATRIEN-2-ONE, 6,10-DIMETHYL- see POH525
3,5,9-UNDECATRIEN-2-ONE, 3,6,10-TRIMETHYL- see TMJ100
10-UNDECENOIC ACID, ETHYL ESTER see EQD200
γ-UNDEKALAKTON see HBN200
UNDEVELOPED LITHOL TONER see NAP100
UNIBUR 70 see CAT775
UNIFLEX BYO see BSB000
UNIMOLL 66 see DGV700

UNIPHAT A20 see MHY800
UNIPHAT A30 see MHY650
UNIPHAT A40 see MLC800
UNISPIRAN see GCE600
UNIVERM see CBY000
URACIL, 5-AMINO- see AMW100
URACIL, 6-AMINO-2-THIO- see AMS750
URACIL DESOXYURIDINE see DAZ050
UREA, sec-BUTYL- see BSS300
UREA, tert-BUTYL- see BSS310
UREA, N,N-DIETHYL-(9CI) see DKD650
UREA, 1,1-DIETHYL- see DKD650
UREA, 1,1-DIMETHYL- see DUM150
UREA, (1,1-DIMETHYLETHYL)-(9CI) see BSS310
UREA, N,N''-1,2-ETHANEDIYLBIS-(9CI) see EJC100
UREA, 1,1'-ETHYLENEDI- see EJC100
UREA, ISOPROPYL- see IRR100
UREA, (1-METHYLETHYL)-(9CI) see IRR100
UREA, PROPYL- see PNX550
URIDINE, 2'-DEOXY- see DAZ050
UROCANIC ACID see UVJ440
UROCANINIC ACID see UVJ440
USAF AM-2 see DNP700
USAF AN-11 see CAM000
USAF DO-10 see GHA100
USAF DO-19 see CPG700
USAF DO-40 see BMR100
USAF DO-49 see HDH200
USAF DO-53 see BHB300
USAF EK-496 see ABH000
USAF EK-2680 see TGO800
USAF EK-P-4382 see CIH100
USAF MA-9 see BLA800
USAF MA-13 see CEG800
USAF MK-1 see DDF700
USAF MK-5 see BKD800
USAF ND-60 see SOB600
USAF PD-57 see TEV600
USAF SN-31 see DVB850
USAF XR-20 see PPQ630
UULCOL SCARLET 2G see NAP100
UV 1 see HND100
UV 24 see DMW250
UV 531 see HND100
UVA 1 see HND100
UVINUL 408 see HND100
UVINUL D 49 see BKH200

VAC see VLU250
VALACIDIN see SMA000
VALERIC ACID, 4,4'-AZOBIS(4-CYANO)- see ASL500
VALERIC ACID, 4,4-BIS(tert-BUTYLPEROXY)-, BUTYL ESTER see BHL200
VALERIC ACID, METHYL ESTER see VAQ100
VALERONITRILE see VAV300
n-VALERONITRILE see VAV300
VARISOFT 100 see DXG625
VAT BLUE 1 see BGB275
VAT BLUE 4 see IBV050
VAT BLUE 6 see DFN425
VAT BLUE KD see DFN425
VAT BLUE O see IBV050
VAT BLUE OD see IBV050
VAT BRIGHT VIOLET K see DFN450
VAT BRILLIANT VIOLET K see DFN450
VAT BRILLIANT VIOLET KD see DFN450
VAT BRILLIANT VIOLET KP see DFN450
VAT FAST BLUE BCS see DFN425
VAT FAST BLUE R see IBV050
VAT GREEN B see DFN425
VAT RED 5B see DNT300
VAT SKY BLUE K see DFN425
VAT SKY BLUE KD see DFN425
VAT SKY BLUE KP 2F see DFN425
VERATRIC ACID see VHP600
VERMOESTRICID see CBY000
VERSAL BLUE GGSL see IBV050
VERSAL GREEN G see PJQ100
VERSENE CA see CAR780
VESTROL see TIO750
VETSPEN see PAQ200
VEVETONE see CAT775
VI-CAD see CAE250

VICALIN see BKW100
VICKNITE see PLL500
VICRON see CAT775
VICRON 31-6 see CAT775
VIDINE see CMF800
VIENNA WHITE see CAT775
VIGOT 15 see CAT775
VINCUBINA see TDT770
VINCUBINE see TDT770
VINILE (ACETATO di) (ITALIAN) see VLU250
VINYLACETAAT (DUTCH) see VLU250
VINYLACETAT (GERMAN) see VLU250
VINYL ACETATE see VLU250
VINYL A MONOMER see VLU250
VINYLCARBAZOLE see VNK100
N-VINYLCARBAZOLE see VNK100
9-VINYLCARBAZOLE see VNK100
1-VINYLCYCLOHEXENE see VNZ990
4-VINYLCYCLOHEXENE see CPD750
1-VINYLCYCLOHEXENE-3 see CPD750
1-VINYLCYCLOHEX-3-ENE see CPD750
4-VINYLCYCLOHEXENE-1 see CPD750
4-VINYL-1-CYCLOHEXENE see CPD750
VINYLE (ACETATE de) (FRENCH) see VLU250
N-VINYLKARBAZOL see VNK100
2-(VINYLOXY)ETHYL METHACRYLATE see VQA150
4-VINYLPHENYL ACETATE see ABW550
p-VINYLPHENOL ACETATE see ABW550
4-VINYLPYRIDINE see VQK590
VINYLSILATRAN see VQK600
VINYLSILATRANE see VQK600
1-VINYLSILATRANE see VQK600
VINYLSTYRENE see DXQ740
p-VINYLTOLUENE see VQK700
4-VINYLTOLUENE see VQK700
1-VINYL-2,8,9-TRIOXA-5-AZA-1-SILABICYCLO(3.3.3)UNDECANE see VQK600
VIOLET KYPOVA 1 see DFN450
VIOLET PIGMENT 31 see DFN450
VITABLEND see PAQ200
VITAMIN A ALDEHYDE see VSK985
VITAMIN A1 ALDEHYDE see VSK985
trans-VITAMIN A ALDEHYDE see VSK985
VITAMIN B_1 DISULFIDE see TES800
VITAMIN D see VSZ095
VITICOLOR see OMC300
VITRAN see TIO750
VITREOUS QUARTZ see SCK600
VONDACEL BLUE HH see CMO500
VONDACID BLUE SE see APG700
VULCAFIX BLUE R see BGB275
VULCAFIX FAST BLUE SD see IBV050
VULCAFIX RED R see NAP100
VULCAFOR BLUE A see BGB275
VULCAFOR FAST BLUE 3R see IBV050
VULCAL FAST GREEN F2G see PJQ100
VULCANOSINE DARK BLUE L see BGB275
VULCANOSINE FAST BLUE GG see IBV050
VULCANOSINE FAST GREEN G see PJQ100
VULCOL FAST BLUE S see IBV050
VULCOL FAST GREEN F2G see PJQ100
VULKACIT 1000 see TGX550
VULKASIL see SCH000
VULNOC AB see ANB100
VYAC see VLU250
VYNAMON BLUE A see BGB275
VYNAMON BLUE 3R see IBV050
VYNAMON GREEN BE see PJQ100
VYNAMON GREEN BES see PJQ100
VYNAMON GREEN GNA see PJQ100

WAREFLEX see DDT500
WEEDTRINE-D see DWX800
WEGLA TLENEK (POLISH) see CBW750
WESTROSOL see TIO750
W-GUM see SLJ500
WHICA BA see CAT775
WHITCARB W see CAT775
WHITE MINERAL OIL see MQV750, MQV875
WHITE-POWDER see CAT775
WHITING see CAT775
WHITON 450 see CAT775
WICKENOL 127 see IRL100

WICRON see HGI100
WINNOFIL S see CAT775
WINZER SOLUTION see BEL900
WITCARB see CAT775
WITCARB P see CAT775
WITCARB REGULAR see CAT775
W-13 STABILIZER see SLJ500

XANTHATE see EQG600
XANTHIC ACID, ISOPROPYL-, POTASSIUM SALT see IRG050
XANTHOGENIC ACID see EQG600
p-XYLENE, 2-NITRO- see NMS520
2,6-XYLENOL, 4-BROMO- see BOL303
XYLOSE, D- see XQJ300
(D)-XYLOSE see XQJ300

YAMADA FAST RED RL BASE see MMF780
YASOKNOCK see SHG500

YORK WHITE see CAT775
YUDOCHROME YELLOW GGN see SIT850

ZESET T see VLU250
ZG 301 see CAT775
ZINC-m-ARSENITE see ZDS000
ZINC ARSENITE, solid (DOT) see ZDS000
ZINC p-HYDROXYBENZENESULFONATE see ZIJ300
ZINC METAARSENITE see ZDS000
ZINC METHARSENITE see ZDS000
ZINC PHENOLSULFONATE see ZIJ300
ZINC p-PHENOL SULFONATE see ZIJ300
ZINC SULFOCARBOLATE see ZIJ300
ZINC SULFOPHENATE see ZIJ300
ZINN (GERMAN) see TGB250
ZMA see ZDS000
ZOTOX see ARB250, ARH500
ZOTOX CRAB GRASS KILLER see ARB250

References

ABANAE Antibiotics Annual. (New York, NY) 1953–60. For publisher information, see AMACCQ

ABBIA4 Archives of Biochemistry and Biophysics. (Academic Press, 111 Fifth Ave., New York, NY 10003) V.31–, 1951–

ABCHA6 Agricultural and Biological Chemistry. (Maruzen Co. Ltd., P.O.Box 5050 Tokyo International, Tokyo 100–31, Japan) V.25–, 1961–

ABMGAJ Acta Biologica et Medica Germanica. (Berlin, Germany) V.1–41, 1958–82. For publisher information, see BBIADT

ACIEAY Angewandte Chemie, International Edition in English. (Verlag Chemie GmbH, Postfach 1260/1280, D6940, Weinheim, Germany) V.1–, 1962–

ACNSAX Activitas Nervosa Superior. (Avicenum, Malostranske nam. 28, Prague 1, Czech Republic) V.1–, 1959–

ADREDL Archives of Dermatological Research. (Springer-Verlag New York, Inc., Service Center, 44 Hartz Way, Secaucus, NJ 07094) V.253–, 1975–

AECTCV Archives of Environmental Contamination and Toxicology. (Springer-Verlag New York, Inc., Service Center, 44 Hartz Way, Secaucus, NJ 07094) V.1–, 1973–

AEHLAU Archives of Environmental Health. (Heldreff Publications, 4000 Albemarle St., N.W., Washington, D.C. 20016) V.1–, 1960–

AEMIDF Applied and Environmental Microbiology. (American Society for Microbiology, 1913 I St., N.W., Washington, DC 20006) V.31–, 1976–

AEPPAE Naunyn-Schmiedeberg's Archiv fuer Experimentelle Pathologie und Pharmakologie. (Berlin, Germany) V.110–253, 1925–66. For publisher information, see NSAPCC

AEXPBL Archiv fuer Experimentelle Pathologie und Pharmakologie. (Leipzig, Germany) V.1–109, 1873–1925. For publisher information, see NSAPCC

AFDOAQ Association of Food and Drug Officials of the United States, Quarterly Bulletin. (Editorial Committee of the Association, P.O. Box 20306, Denver, CO 80220) V.1–, 1937–

AFREAW Advances in Food Research. (Academic Press, 111 Fifth Ave., New York, NY 10003) V.1–, 1948–

AGACBH Agents and Actions, A Swiss Journal of Pharmacology. (Birkhaeuser Verlag, P.O. Box 34, Elisabethenststrasse 19, CH–4010, Basel, Switzerland) V.1–, 1969/70–

AGGHAR Archiv fuer Gewerbepathologie und Gewerbehygiene. (Berlin, Germany) V.1–18, 1930–61. For publisher information, see IAEHDW

AHBAAM Archiv fuer Hygiene und Bakteriologie. (Munich, Germany) V.101–154, 1929–71. For publisher information, see ZHPMAT

AICCA6 Acta Unio Internationalis Contra Cancrum. (Louvain, Belgium) V.1–20, 1936–64. For publisher information, see IJCNAW

AIHAAP American Industrial Hygiene Association Journal. (AIHA, 475 Wolf Ledges Pkwy., Akron, OH 44311) V.19–, 1958–

AIMDAP Archives of Internal Medicine. (American Medical Association, 535 N. Dearborn St., Chicago, IL 60610) V.1–, 1908–

AIMEAS Annals of Internal Medicine. (American College of Physicians, 4200 Pine St., Philadelphia, PA 19104) V.1–, 1927–

AIPAAV Annales de l'Institut Pasteur. (Paris, France) V.1–123, 1887–1972. For publisher information, see ANMBCM

AIPTAK Archives Internationales de Pharmacodynamie et de Therapie. (Editeurs, Institut Heymans de Pharmacologie, De Pintelaan 135, B–9000 Ghent, Belgium) V.4–, 1898–

AISFAR Archivio Italiano di Scienze Farmacologiche. (Modena, Italy) 1932–65. Discontinued.

AISSAW Annali dell'Istituto Superiore di Sanita. (Istituto Poligrafico dello Stato, Libreria dello Stato, Piazza Verdi, 10 Rome, Italy) V.1–, 1965–

AJEBAK Australian Journal of Experimental Biology and Medical Science. (Univ. of Adelaide Registrar, Adelaide, S.A. 5000, Australia) V.1–, 1924–

AJHYA2 American Journal of Hygiene. (Baltimore, MD) V.1–80, 1921–64. For publisher information, see AJEPAS

AJIMD8 American Journal of Industrial Medicine. (Alan R. Liss, Inc., 150 Fifth Ave., New York, NY 10011) V.1–, 1980–

AJMEAZ American Journal of Medicine. (Yorke Medical Group, 666 Fifth Ave., New York, NY 10103) V.1–, 1946–

AJOPAA American Journal of Ophthalmology. (Ophthalmic Publishing Co., 435 N. Michigan Ave., Chicago, Il 60611) V.1–, 1918–

AJPEAG American Journal of Public Health and the Nation's Health. (New York, NY) V.18–60, 1928–60. For publisher information, see AJHEAA

AJSNAO American Journal of Syphilis and Neurology. (St. Louis, MO) V.1–19, 1917–35. For publisher information, see AJSGA3

ALBRW* Albright and Wilson Inc., P.O. Box 26229, Richmond, VA 23260–6229

AMIHAB AMA Archives of Industrial Health. (Chicago, IL) V.11–21, 1955–60. For publisher information, see AEHLAU

AMIHBC AMA Archives of Industrial Hygiene and Occupational Medicine. (Chicago, IL) V.2–10, 1950–54. For publisher information, see AEHLAU

AMPMAR Archives des Maladies professionnelles de Medecine du Travail et de Securite sociale. (Masson et Cie, Editeurs, 120 Blvd. Saint-Germain, P–75280, Paris 06, France) V.7–, 1946–

AMRL** Aerospace Medical Research Laboratory Report. (Aerospace Technical Div., Air Force Systems Command, Wright-Patterson Air Force Base, OH 45433)

ANASAB Anaesthesia. (Blackwell Scientific, Osney Mead, Oxford OX2 OEL, England) V.1–, 1946–

ANREAK Anatomical Record. (Alan R. Liss, Inc., 150 Fifth Ave., New York, NY 10011) V.1–, 1906/08–

ANSUA5 Annals of Surgery. (J.B. Lippincott Co., Keystone Industrial Park, Scranton, PA 18512) V.1–, 1885–

ANTBAL Antibiotiki. (Moscow, USSR) V.1–29, 1956–84. For publisher information, see AMBIEH

ANTCAO Antibiotics and Chemotherapy. (Washington, DC) V.1–12, 1951–62. For publisher information, see CLMEA3

ANYAA9 Annals of the New York Academy of Sciences. (The Academy, Exec. Director, 2 E. 63rd St., New York, NY 10021) V.1–, 1877–

AOHYA3 Annals of Occupational Hygiene. (Pergamon Press, Headington Hill Hall, Oxford OX3 OBW, England) V.1–, 1958–

APFRAD Annales Pharmaceutiques Francaises. (Masson et Cie, Editeurs, 120 Blvd. Saint-Germain, P–75280, Paris 06, France) V.1–, 1943–

APPHAX Acta Poloniae Pharmaceutica. (Ars Polona-RUCH, P.O. Box 1001, P-00 068 Warsaw, 1, Poland) V.1–, 1937–

APTOA6 Acta Pharmacologica et Toxicologica. (Munksgaard, 35 Norre Sogade, DK-1370, Copenhagen K, Denmark) V.1–, 1945–

APTOD9 Abstracts of Papers, Society of Toxicology. Annual Meetings. (Academic Press, 111 Fifth Ave., New York, NY 10003)

ARGEAR Archiv fuer Geschwulstforschung. (VEB Verlag Volk und Gesundheit Neue Gruenstr. 18, DDR-102 Berlin, German Democratic Republic) V.1–, 1949–

ARPAAQ Archives of Pathology. (American Medical Assn., 535 N. Dearborn St., Chicago, IL 60610) V.5, no.3–, V.50, no.3, 1928–50; V.70–99, 1960–75.

ARSIM* Agricultural Research Service, USDA Information Memorandum. (Beltsville, MD 20705)

ARTODN Archives of Toxicology. (Springer-Verlag, Heidelberger, Pl. 3, D–1 Berlin 33, Germany) V.32–, 1974–

ARZNAD Arzneimittel-Forschung. (Edition Cantor Verlag fuer Medizin und Naturwissenschaften KG, D-7960 Aulendorf, Germany) V.1-, 1951-

ASBIAL Archivio di Scienze Biologiche. (Cappelli Editore, Via Marsili 9, I-40124 Bologna, Italy) V.1-, 1919-

ATXKA8 Archiv fuer Toxikologie. (Berlin, Germany) V.15-31, 1954-74. For publisher information, see ARTODN

AXVMAW Archiv fuer Experimentelle Veterinaermedizin. (S. Hirzel Verlag, Postfach 506, DDR-701 Leipzig, Germany) V.6-, 1952-

BBMS** "Medicaments du Systeme nerveux vegetalif" Bovet, D., and F. Bovet-Nitti, New York, S. Karger, 1948

BBRCA9 Biochemical and Biophysical Research Communications. (Academic Press Inc., 111 Fifth Ave., New York, NY 10003) V.1-, 1959-

BCFAAI Bollettino Chimico Farmaceutico. (Societa Editoriale Farmaceutica, Via Ausonio 12, 20123 Milan, Italy) V.33-, 1894-

BCPCA6 Biochemical Pharmacology. (Pergamon Press Inc., Maxwell House, Fairview Park, Elmsford, NY 10523) V.1-, 1974-

BCSTB5 Biochemical Society Transactions. (Biochemical Society, P.O. Box 32, Commerce Way, Whitehall Rd., Industrial Estate, Colchester CO2 8HP, Essex, England) V.1-, 1973-

BCTKAG Bromatologia i Chemia Toksykologiczna. (Ars Polona-RUCH, P.O. Box 1001, P-00 068 Warsaw, 1, Poland) V.4-, 1971-

BDCGAS Berichte der Deutschen Chemischen Gesellschaft. (Leipzig/Berlin) V.1-61, 1868-1928. For publisher information, see CHBEAM

BECTA6 Bulletin of Environmental Contamination and Toxicology. (Springer-Verlag New York, Inc., Service Center, 44 Hartz Way, Secaucus, NJ 07094) V.1-, 1966-

BEXBAN Bulletin of Experimental Biology and Medicine. Translation of BEBMAE. (Plenum Publishing Corp., 233 Spring St., New York, NY 10013) V.41-, 1956-

BICMBE Biochimie. (Masson et Cie, Editeurs, 120 Blvd. Saint-Germain, P-75280, Paris 06, France) V.53-, 1971-

BIMADU Biomaterials. (Quadrant Subscription Services Ltd., Oakfield House, Perrymount Rd., Haywards Heath, W. Sussex, RH16 3DH, England) V.1-, 1980-

BIMEA5 Biologie Medicale. (Paris) V.1-60, 1903-71; new series 1972- V.4, 1975

BIOFX* BIOFAX Industrial Bio-Test Laboratories, Inc., Data Sheets. (1810 Frontage Rd., Northbrook, IL 60062)

BIPMAA Biopolymers. (John Wiley and Sons, 605 Third Ave., New York, NY 10158) V.1-, 1963-

BIREBV Biology of Reproduction. (Society for the Study of Reproduction, 309 West Clark Street, Champaign, IL 61820) V.1-, 1969-

BJCAAI British Journal of Cancer. (H.K. Lewis and Co., 136 Gower St., London WC1E 6BS, England) V.1-, 1947-

BJEPA5 British Journal of Experimental Pathology. (H.K. Lewis and Co., 136 Gower St., London WC1E 6BS, England) V.1-, 1920-

BJIMAG British Journal of Industrial Medicine. (British Medical Journal, 1172 Commonwealth Ave., Boston, MA 02134) V.1-, 1944-

BJPCAL British Journal of Pharmacology and Chemotherapy. (London, England) V.1-33, 1946-68. For publisher information, see BJPCBM

BMJOAE British Medical Journal. (British Medical Association, BMA House, Tavistock Square, London WC1H 9JR, England) V.1-, 1857-

BSIBAC Bolletino della Societa Italiana di Biologia Sperimentale. (Casa Editrice Libraria V. Idelson, Via Alcide De Gasperi, 55, 80133 Naples, Italy) V.2-, 1927-

BSPII* SPI Bulletin. (Society of the Plastics Industry, 250 Park Ave., New York, NY 10017)

CALEDQ Cancer Letters. (Elsevier Publishing, P.O. Box 211, Amsterdam C, Netherlands) V.1-, 1975-

CANCAR Cancer. (J. B. Lippincott Co., E. Washington Sq., Philadelphia, PA 19105) V.1-, 1948-

CBCCT* "Summary Tables of Biological Tests" National Research Council Chemical-Biological Coordination Center. (National Academy of Science Library, 2101 Constitution Ave., N.W., Washington, DC 20418)

CCCCAK Collection of Czechoslovak Chemical Communications. (Academic Press, 24-28 Oval Rd., London NW1 7DX, England) V.1-, 1929-

CCPTAY Contraception. (Geron-X, Publishers, P.O. Box 1108, Los Altos, CA 94022) V.1-, 1970-

CCROBU Cancer Chemotherapy Reports, Part 1. (Washington, DC) V.52, no.6-, V.59, 1968-75. For publisher information, see CTRRDO

CFRGBR Code of Federal Regulations. (U.S. Government Printing Office, Supt. of Doc., Washington, DC 20402)

CGCGBR Cytogenetics and Cell Genetics. (S. Karger AG, Arnold-Boecklin Str. 25, CH-4011 Basel, Switzerland) V.12-, 1973-

CHDDAT Comptes Rendus Hebdomadaires des Seances de l'Academie des Sciences, Serie D. (Centrale des Revues Dunod-Gauthier-Villars, 24-26 Blvd. de l'Hopital, 75005 Paris, France) V.262-, 1966-

CHTPBA Chimica Therapeutica. (Editions DIMEO, Arcueil, France) V.1-8, 1965-73.

CIIT** Chemical Industry Institute of Toxicology, Docket Reports. (POB 12137, Research Triangle Park, NC 27709)

CJMIAZ Canadian Journal of Microbiology. (National Research Council of Canada, Administration, Ottawa, Canada K1A OR6) V.1-, 1954-

CKFRAY Ceskoslovenska Farmacie. (PNS-Ustredni Expedice Tisku, Jindriska 14, Prague 1, Czech Republic) V.1-, 1952-

CLDND* Compilation of LD50 Values of New Drugs. (J.R. MacDougal, Dept. of National Health and Welfare, Food and Drug Divisions, 35 John St., Ottawa, Ontario, Canada)

CMAJAX Canadian Medical Association Journal. (CMA House, Box 8650, Ottawa K1G OG8, Ontario, Canada) V.1-, 1911-

CMMUAO Chemical Mutagens. Principles and Methods for Their Detection (Plenum Publishing Corp., 233 Spring St., New York, NY 10013) V.1-, 1971-

CMTRAG Chemotherapia. (Basel, Switzerland) V.1-12, 1960-67. For publisher information, see CHTHBK

CNCRA6 Cancer Chemotherapy Reports. (Bethesda, MD) V.1-52, 1959-68. For publisher information, see CCROBU

CNJGA8 Canadian Journal of Genetics and Cytology. (Genetics Society of Canada, 151 Slater St., Suite 907, Ottawa, Ont. K1P 5H4, Canada) V.1-, 1959-

CNREA8 Cancer Research. (Waverly Press, Inc., 428 E. Preston St., Baltimore, MD 21202) V.1-, 1941-

CODEDG Contact Dermatitis. Environmental and Occupational Dermatitis. (Munksgaard, 35 Norre Sogade, DK 1370 Copenhagen K, Denmark) V.1-, 1975-

COREAF Comptes Rendus Hebdomadaires des Seances de l'Academie des Sciences. (Paris, France) V.1-261, 1835-1965. For publisher information, see CHDDAT

CPBTAL Chemical and Pharmaceutical Bulletin. (Pharmaceutical Society of Japan, 12-15-501, Shibuya 2-chome, Shibuya-ku, Tokyo, 150, Japan) V.6-, 1958-

CRNGDP Carcinogenesis. (Information Retrieval, 1911 Jefferson Davis Highway, Arlington, VA 22202) V.1-, 1980-

CRSBAW Comptes Rendus des Seances de la Societe de Biologie et de Ses Filiales. (Masson et Cie, Editeurs, 120 Blvd. Saint-Germain, P-75280, Paris 06, France) V.1-, 1849-

CSLNX* U. S. Army Armament Research and Development Command, Chemical Systems Laboratory, NIOSH Exchange Chemicals. (Aberdeen Proving Ground, MD 21010)

CTOIDG Cosmetics and Toiletries. (Allured Publishing Corp., P.O. Box 318, Wheaton, IL 60187) V.91-, 1976-

CTOXAO Clinical Toxicology. (New York, NY) V.1-18, 1968-81. For publisher information, see JTCTDW

CYGEDX Cytology and Genetics. English Translation of Tsitologiya i Genetika. (Allerton Press, Inc., 150 Fifth Ave., New York, NY 10011) V.8-, 1974-

CYTBAI Cytobios. (The Faculty Press, 88 Regent St., Cambridge, England) V.1-, 1969-

CYTOAN Cytologia. (Maruzen Co. Ltd., P.O. Box 5050, Tokyo International, Tokyo 100-31, Japan) V.1-, 1929-

DABBBA Dissertation Abstracts International, B: The Sciences and Engineering. (University Microfilms, A Xerox Co., 300 N. Zeeb Rd., Ann Arbor, MI 48106) V.30–, 1969–

DADEDV Drug and Alcohol Dependence. (Elsevier Sequoia SA, POB 851, CH-1001 Lausanne, Switzerland) V.1–, 1975–

DANKAS Doklady Akademii Nauk S.S.S.R. (v/o Mezhdunarodnaya Kniga, Kuznetskii Most 18, Moscow G-200, Russia.) V.1–, 1933–

DCTODJ Drug and Chemical Toxicology. (Marcel Dekker, POB 11305, Church St. Station, New York, NY 10249) V.1–, 1977/78–

DECRDP Drugs under Experimental and Clinical Research. (J.R. Prous Pub., Apartado de Correos 1179, Barcelona, Spain) V.1–, 1977–

DEMAEP Dental Materials. (Munksgaard International Pub., POB 2148, DK-1016 Copenhagen K, Denmark) V.1–, 1985–

DFSCDX Developments in Food Science. (Elsevier Science Pub. Co., Inc. 52 Vanderbilt Ave., New York, NY 10017) V.1–, 1978–

DIPHAH Dissertationes Pharmaceuticae. (Warsaw, Poland) V.1–17, 1949–65. For publisher information, see PJPPAA

DOWCC* Dow Chemical Company Reports. (Dow Chemical U.S.A., Health and Environment Research, Toxicology Research Lab., Midland, MI 48640)

DRUGAY Drugs. International Journal of Current Therapeutics and Applied Pharmacology Reviews. (ADIS Press Ltd., 18/F., Tung Sun Commercial Centre, 194–200 Lockhart Road, Wanchai, Hong Kong)

DTESD7 Developments in Toxicology and Environmental Science. (Elsevier, Scientific Publishing Co., POB 211, 1000 AE Amsterdam, Netherlands) V.1–, 1977–

DTLVS* "Documentation of Threshold Limit Values for Substances in Workroom Air." For publisher information, see 85INA8

DUPON* E. I. Dupont de Nemours and Company, Technical Sheet. (1007 Market St., Wilmington, DE 19898)

ECREAL Experimental Cell Research. (Academic Press, 111 Fifth Ave., New York, NY 10003) V.1–, 1950–

EESADV Ecotoxicology and Environmental Safety. (Academic Press, 111 Fifth Ave., New York, NY 10003) V.1–, 1977–

EJCAAH European Journal of Cancer. (Pergamon Press, Headington Hill Hall, Oxford OX3 OEW, England) V.1–, 1965–

EJMBA2 Egyptian Journal of Microbiology. (National Information and Documentation Centre, A1-Tahrir St., Awqaf P.O. Dokki, Cairo, Egypt) V.7–, 1972–

EJMCA5 European Journal of Medicinal Chemistry. Chimie Therapeutique. (Centre National de la Recherche Scientifique, 3 rue J.B. Clement, F-92290 Chatenay-Malabry, France) V.9–, 1974–

EJTXAZ European Journal of Toxicology and Environmental Hygiene. (Paris, France) v.7–9, 1974–76. For publisher information, see TOERD9

EMMUEG Environmental and Molecular Mutagenesis. (Alan R. Liss, Inc., 4 E. 11th St., New York, NY 10003) V.10–, 1987–

ENDOAO Endocrinology. (Williams and Wilkins Co., Dept. 260, P.O. Box 1496, Baltimore, MD 21203) V.1–, 1917–

ENMUDM Environmental Mutagenesis. (Alan. R. Liss, Inc., 150 Fifth Ave., New York, NY 10011) V.1–, 1979–

ENVRAL Environmental Research. (Academic Press, 111 Fifth Ave., New York, NY 10003) V.1–, 1967–

EPASR* United States Environmental Protection Agency, Office of Pesticides and Toxic Substances. (U.S. Environmental Protection Agency, 401 M St., S.W., Washington, DC 20460) History Unknown

EQSSDX Environmental Quality and Safety, Supplement. (Academic Press, 111 Fifth Ave., New York, NY 10003) V.1–, 1975–

ESKGA2 Eisei Kagaku. (Nippon Yakugakkai, 2–12–15 Shibuya, Shibuya-Ku, Tokyo 150, Japan) V.1–, 1953–

ESKHA5 Eisei Shikenjo Hokoku. Bulletin of the National Hygiene Sciences. (Kokuritsu Eisei Shikenjo, 18–1 Kamiyoga 1 chome, Setagaya-ku, Tokyo, Japan) V.1–, 1886–

EVETBX Environmental Entomology. (Entomological Society of America, 4603 Calvert Rd., College Park, MD 20740) V.1–, 1972–

EVHPAZ EHP, Environmental Health Perspectives. Subseries of DHEW Publications. (U.S. Government Printing Office, Superintendent of Documents, Washington, DC 20402) No.1– 1972–

EVSRBT Environmental Science Research. (Plenum Publishing Corp., 233 Spring St., New York, NY 10013) V.1–, 1972–

EXMDA4 International Congress Series – Excerpta Medica. (Elsevier North Holland, Inc., 52 Vanderbilt Ave., New York, NY 10017) No.1– 1952–

EXPEAM Experientia. (Birkhaeuser Verlag, P.O. Box 34, Elisabethenst 19, CH-4010, Basel, Switzerland) V.1–, 1945–

FAATDF Fundamental and Applied Toxicology (Official Journal of the Society of Toxicology, 475 Wolf Ledges Parkway, Akron, OH 44311) V.1–, 1981–

FAONAU Food and Agriculture Organization of United Nations, Report Series. (FAO-United Nations, Room 101, 1776 F Street, NW, Washington, DC 20437)

FATOAO Farmakologiya i Toksikologiya (Moscow). (v/o Mezhdunarodnaya Kniga, Kuznetskii Most 18, Moscow G–200, Russia) V.2–, 1939– For English translation, see PHTXA6 and RPTOAN.

FAVUAI Fiziologicheski Aktivnye Veshchestva. Physiologically Active Substances. (Akademiya Nauk Ukrainskoi S.S.R., Kiev, Russia) No.1– 1966–

FCTOD7 Food and Chemical Toxicology. (Pergamon Press, Headington Hill Hall, Oxford OX3 OBW, England) V.20–, 1982–

FCTXAV Food and Cosmetics Toxicology. (Pergamon Press Ltd., Maxwell House, Fairview Park Elmsford, NY 10523) V.1–19, 1963–81.

FDWU** Uber die Wirkung Verschiedener Gifte Auf Vogel, Ludwig Forchheimer Dissertation. (Pharmakologischen Institut der Universitaet Wuerzburg, Germany, 1931)

FEPRA7 Federation Proceedings, Federation of American Societies for Experimental Biology. (9650 Rockville Pike, Bethesda, MD 20014) V.1–, 1942–

FESTAS Fertility and Sterility. (American Fertility Society, 1608 13th Ave. S., Birmingham, AL 35205) V.1–, 1950–

FMCHA2 Farm Chemicals Handbook. (Meister Publishing, 37841 Euclid Ave., Willoughy, OH 44094)

FPNJAG Folia-Psychiatrica et Neurologica Japonica. (Folia Publishing Society, Todai YMCA Bldg., 1–20–6 Mukogaoka, Bunkyo-Ku, Tokyo 113, Japan) 1947–

FRPSAX Farmaco, Edizione Scientifica. (Casella Postale 114, 27100 Pavia, Italy) V.8–, 1953–

FRXXBL French Demande Patent Document. (Commissioner of Patents and Trademarks, Washington, DC 20231)

GAFCC* GAF Material Safety Data Sheet. (GAF Chemicals Corporation, 1361 Alps Road, Wayne, NJ 07470)

GANNA2 Gann. Japanese Journal of Cancer Research. (Tokyo, Japan) V.1–75, 1907–84. For publisher information, see JJCREP

GASTAB Gastroenterology. (Elsevier North Holland, Inc., 52 Vanderbilt Avenue, New York, NY 10017) V.1–, 1943–

GDIKAN Gifu Daigaku Igakubu Kiyo. Papers of the Gifu University School of Medicine. (Gifu Daigaku Igakubu, 40 Tsukasa–cho, Gifu, Japan) V.15–, 1967–

GENEA3 Genetica (The Hague). (Dr. W. Junk bv Publishers, POB 13713, 2501 ES The Hague, Netherlands) V.1–, 1919–

GENRA8 Genetical Research. (Cambridge University Press, P.O. Box 92, Bentley House, 200 Euston Rd., London NW1 2DB, England) V.1–, 1960–

GEPHDP General Pharmacology. (Pergamon Press Inc., Maxwell House Fairview Park, Elmsford, NY 10523) V.1–, 1970–

GISAAA Gigiena i Sanitariya. (English Translation is HYSAAV). (v/o Mezhdunarodnaya Kniga, Kuznetskii Most 18, Moscow G–200, Russia) V.1–, 1936–

GNAMAP Gigiena Naselennykh Mest. Hygiene in Populated Places. (Kievskii Nauchno - Issledovatel'skii Institut Obshchei i Kommunol'noi Gigieny, Kiev, Russia) V.7–, 1967–

GTPPAF Gigiena Truda i Professional'naya Patologiya v Estonskoi SSR. Labor Hygiene and Occupational Pathology in the Estonian SSR. (Institut Eksperimental'noi i Klinicheskoi Meditsiny Ministerstva Zdravookhraneniya Estonskoi SSR, Tallinn, USSR) V.8–, 1972–

GTPZAB Gigiena Truda i Professional'nye Zabolevaniia. Labor Hygiene and Occupational Diseases. (v/o Mezhdunarodnaya Kniga, Kuznetskii Most 18, Moscow G–200, Russia) V.1–, 1957–

GUCHAZ "Guide to the Chemicals Used in Crop Protection" Information Canada, 171 Slater St., Ottawa, Ontario, Canada (Note that although the DPIM cites data from the 1968 edition, this reference has been superseded by the 1973 edition.)

GWXXAW German Patent Document. (U.S. Patent Office, Science Library, 2021 Jefferson Davis Highway, Arlington, VA 22202)

HBAMAK "Abdernalden's Handbuch der Biologischen Arbeitsmethoden." (Leipzig, Germany)

HBTXAC "Handbook of Toxicology, Volumes I–V" Philadelphia, W.B. Saunders, 1956–1959

HDTU** Pharmakologische Prufung von Analgetika, Gunter Herrlen Dissertation. (Pharmakologischen Institut der Universitat Tubingen, Germany, 1933)

HEREAY Hereditas. (J.L. Toernqvist Book Dealers, S–26122 Landskrona, Sweden) V.1–, 1947–

HSZPAZ Hoppe-Seyler's Zeitschrift fuer Physiologische Chemie. (Walter de Gruyter and Co., Genthiner Street 13, D–1000, Berlin 30, Federal Republic of Germany) V.21–, 1895/96–

HUTODJ Human Toxicology. (Macmillan Press Ltd., Houndmills, Bassingstoke, Hants., RG21 2XS, UK) V.1–, 1981–

HYSAAV Hygiene and Sanitation: (English Translation of Gigiena Sanitariya). (Springfield, VA) V.29–36, 1964–71. Discontinued.

IAAAAM International Archives of Allergy and Applied Immunology. (S. Karger, Postfach CH–4009, Basel Switzerland) V.1–, 1950–

IAPWAR International Journal of Air and Water Pollution. (London, England) V.4, No.1–4, 1961. For publisher information, see ATENBP

IARC** IARC Monographs on the Evaluation of Carcinogenic Risk of Chemicals to Man. (World Health Organization, International Agency for Research on Cancer, Lyon, France) (Single copies can be ordered from WHO Publications Centre U.S.A., 49 Sheridan Avenue, Albany, NY 12210)

IGSBAL Igaku to Seibutsugaku. Medicine and Biology. (1–11–4 Higashi-Kanda, Chiyoda, Tokyo 101, Japan) V.1–, 1942–

IJANDP International Journal of Andrology. (Scriptor Publisher ApS, 15 Gasvaerksvej, DK–1656 Copenhagen V, Denmark) V.1–, 1978–

IJBBBQ Indian Journal of Biochemistry and Biophysics. (Council of Scientific and Industrial Research, Publication and Information Director, Hillside Rd., New Delhi 110012, India) V.8–, 1971–

IJCNAW International Journal of Cancer. (International Union Against Cancer, 3 rue du Conseil-General, 1205 Geneva, Switzerland) V.1–, 1966–

IJNEAQ International Journal of Neuropharmacology. (Oxford, England/New York, NY) V.1–8, 1962–69. For publisher information, see NEPHBW

IJOCAP Indian Journal of Chemistry. (Council of Scientific and Industrial Research, Publication and Information Director, Hillside Rd., New Delhi 110012, India) V.1–13, 1963–75.

IJRBA3 International Journal of Radiation Biology and Related Studies in Physics, Chemistry and Medicine. (Taylor and Francis Ltd., 4 John Street, London WC1N 2ET, UK) V.1–, 1959–

IJSIDW Indian Journal of Pharmaceutical Sciences. (Indian Journal of Pharmaceutical Sciences, Kalina, Santa Cruz (East), Bombay 400 029, India) V.40, no.2–, 1978–

IMMUAM Immunology. (Blackwell Scientific Publications, Osney Mead, Oxford OX2 OEL, England) V.1–, 1958–

IMSUAI Industrial Medicine and Surgery. (Chicago, IL/Miami, FL) V.18–42, 1949–73. For publisher information see IOHSA5

INHEAO Industrial Health. (2051 Kizukisumiyoshi-cho, Nakahara-ku, Kawasaki, Japan) V.1–, 1963–

INVRAV Investigative Radiology. (J.B. Lippincott Co., E. Washington Sq., Philadelphia, PA 19105) V.1– 1966–

IPSTB3 International Polymer Science and Technology. (Rapra Technology Ltd., Shawbury, Shrewsbury, Shropshire SY4 4NR, UK)

IRGGAJ Internationales Archiv fuer Gewerbepathologie und Gewerbehygiene. (Heidelberg, Germany) V.19–25, 1962–69. For publisher information, see IAEHDW

ITCSAF In Vitro. (Tissue Culture Association, 12111 Parklawn Dr., Rockville, MD 20852) V.1–, 1965–

JACSAT Journal of the American Chemical Society. (American Chemical Society Publications, 1155 16th St., N.W., Washington, DC 20036) V.1–, 1879–

JACTDZ Journal of the American College of Toxicology. (Mary Ann Liebert, Inc., 500 East 85th St., New York, NY 10028) V.1–, 1982–

JADSAY Journal of the American Dental Association. (American Dental Assoc., 211 E. Chicago Ave., Chicago, IL 60611) V.9–, 1922–

JAFCAU Journal of Agricultural and Food Chemistry. (American Chemical Society Publications, 1155 16th St., N.W., Washington, DC 20036) V.1–, 1953–

JAMAAP JAMA, Journal of the American Medical Association. (American Medical Association, 535 N. Dearborn St., Chicago, IL 60610) V.1–, 1883–

JANSAG Journal of Animal Science. (American Society of Animal Science, 309 West Clark Street, Champaign, IL 61820) V.1–, 1942–

JANTAJ Journal of Antibiotics. (Japan Antibiotics Research Association, 2-20-8 Kamiosaki, Shinagawa-ku, Tokyo, Japan) V.2–5, 1948–52; V.21–, 1968–

JAPMA8 Journal of the American Pharmaceutical Association, Scientific Edition. (Washington, DC) V.29–49, 1940–60. For publisher information, see JPMSAE

JBCHA3 Journal of Biological Chemistry. (American Society of Biological Chemists, Inc., 428 E. Preston St., Baltimore, MD 21202) V.1–, 1905–

JCINAO Journal of Clinical Investigation. (Rockefeller University Press, 1230 York Avenue, New York, NY 10021) V.1–, 1924–

JEEMAF Journal of Embryology and Experimental Morphology. (P.O. Box 32, Commerce Way, Colchester CO2 8HP. Essex, England) V.1–, 1953–

JEPTDQ Journal of Environmental Pathology and Toxicology. (Park Forest South, IL) V.1–5, 1977–81.

JESEDU Journal of Environmental Science and Health, Part A: Environmental Science and Engineering. (Marcel Dekker, POB 11305, Church St. Station, New York, NY 10249) V.A11– 1976–

JETOAS Journal europeen de Toxicologie. (Paris, France) V.1–6, 1968–72. For publisher information, see TOERD9

JGMIAN Journal of General Microbiology (P.O. Box 32, Commerce Way, Colchester CO2 8HP, Essex, England) V.1–, 1947–

JHEMA2 Journal of Hygiene, Epidemiology, Microbiology and Immunology. (Avicenum, Zdravotnicke Nakladatelstvi, Malostranske namesti 28, Prague 1, Czech Republic) V.1–, 1957–

JHTCAO Journal of Heterocyclic Chemistry. (Hetero Corporation, POB 16000 MH, Tampa, FL 33687) Vol.1–, 1964–

JIDEAE Journal of Investigative Dermatology. (Williams and Wilkins Co., 428 E. Preston St., Baltimore, MD 21202) V.1–, 1938–

JIDHAN Journal of Industrial Hygiene. (Baltimore, MD/New York, NY) V.1–17, 1919–35. For publisher information, see AEHLAU

JIHTAB Journal of Industrial Hygiene and Toxicology. (Baltimore, MD/New York, NY) V.18–31, 1936–49. For publisher information, see AEHLAU

JIMMBG Journal of Immunological Methods. (Elsevier North Holland Inc., 52 Vanderbilt Ave., New York, NY 10017) V.1–, 1971–

JJATDK JAT, Journal of Applied Toxicology. (Heyden and Son, Inc., 247 S. 41st St., Philadelphia, PA 19104) V.1–, 1981–

JJCREP Japanese Journal of Cancer Research (Gann). (Elsevier Science Publishers B.V., POB 211, 1000 AE Amsterdam, Netherlands) V.76–, 1985–

JJIND8 JNCI, Journal of the National Cancer Institute. (U.S. Government Printing Office, Superintendent of Documents, Washington, DC 20402) V.61–, 1978–

JJPAAZ Japanese Journal of Pharmacology. (Nippon Yakuri Gakkai, c/o Kyoto Daigaku Igakubu Yakurigaku Kyoshitu Sakyo-ku, Kyoto 606, Japan) V.1–, 1951–

JLCMAK Journal of Laboratory and Clinical Medicine. (C.V. Mosby Co., 11830 Westline Industrial Dr., St. Louis, MO 63141) V.1–, 1915–

JMCMAR Journal of Medicinal Chemistry. (American Chemical Society Pub., 1155 16th St., N.W., Washington, DC 20036) V.6–, 1963–

JMPCAS Journal of Medicinal and Pharmaceutical Chemistry. (Washington, DC) V.1–,5, 1959–62. For publisher information, see JMCMAR

JNCIAM Journal of the National Cancer Institute. (Washington, DC) V.1–, 60, No.6, 1940–78. For publisher information, see JJIND8

JNSVA5 Journal of Nutritional Science and Vitaminology. (Business Center Acad. Soc., Japan) V.19–, 1973–

JOANAY Journal of Anatomy. (Cambridge University Press, The Pitt Building, Trumpington Street, Cambridge CB2 1RP, UK) V.51–, 1916–

JOBAAY Journal of Bacteriology. (American Society for Microbiology, 1913 I St., N.W., Washington, DC 20006) V.1–, 1916–

JOCMA7 Journal of Occupational Medicine. (American Occupational Medical Association, 150 N. Wacker Dr., Chicago, IL 60606) V.1–, 1959–

JOENAK Journal of Endocrinology. (Biochemical Society Publications, P.O. Box 32, Commerce Way, Whitehall Industrial Estate, Colchester CO2 8HP, Essex, England) V.1–, 1939–

JOGBAS Journal of Obstetrics and Gynaecology of the British Commonwealth. (London, England) V.68–81, 1961–74. For publisher information, see BJOGAS

JOIMA3 Journal of Immunology. (Williams and Wilkins Co., 428 E. Preston St., Baltimore, MD 21202) V.1–, 1916–

JONUAI Journal of Nutrition. (Journal of Nutrition, Subscription Dept., 9650 Rockville Pike, Bethesda, MD 20014) V.1–, 1928–

JOPHDQ Journal of Pharmacobio-Dynamics. (Pharmaceutical Society of Japan, 12–15–501, Shibuya 2-chome, Shibuya-Ku, Tokyo 150, Japan) V.1–, 1978–

JPBAA7 Journal of Pathology and Bacteriology. (London, England) V.1–96, 1892–1968. For publisher information, see JPTLAS

JPCEAO Journal fuer Praktische Chemie. (Johann Ambrosius Barth Verlag, Postfach 109, DDR–701, Leipzig, Germany) V.1–, 1834– (Several new series, but continuous vol. nos. also used)

JPETAB Journal of Pharmacology and Experimental Therapeutics. (Williams and Wilkins Co., 428 E. Preston St., Baltimore, MD 21202) V.1–, 1909/10–

JPHYA7 Journal of Physiology. (Cambridge University Press, P.O. Box 92, Bentley House, 200 Euston Rd., London NW1 2DB, England) V.1–, 1878–

JPIFAN Japan Pesticide Information. (Japan Plant Protection Assoc., 1–43–11, Komagome, Toshima-ku, Tokyo 170, Japan) No.1–, 1969–

JPMSAE Journal of Pharmaceutical Sciences. (American Pharmaceutical Assoc., 2215 Constitution Ave., N.W., Washington, DC 20037) V.50–, 1961–

JPPMAB Journal of Pharmacy and Pharmacology. (Pharmaceutical Society of Great Britain, 1 Lambeth High Street, London SEI 5JN, England) V.1–, 1949–

JRBED2 Journal of Reproductive Biology and Comparative Endocrinology. (P.G. Institute of Basic Medical Sciences, Dept. of Endocrinology, Taramani, 600 113, India) V.1–, 1981

JRPFA4 Journal of Reproduction and Fertility. (Journal of Reproduction and Fertility Ltd., 22 New Market Rd., Cambridge CB5 8D7, England) V.1–, 1960–

JSICAZ Journal of Scientific and Industrial Research, Section C: Biological Sciences. (New Delhi, India) V.14–21, 1955–62. For publisher information, see IJEBA6

JTEHD6 Journal of Toxicology and Environmental Health. (Hemisphere Publ., 1025 Vermont Ave., N.W., Washington, DC 20005) V.1–, 1975/76–

JTSCDR Journal of Toxicological Sciences. (Editorial Office, Higashi Nippon Gakuen Univ., 7F Fuji Bldg., Kita 3, Nishi 3, Sapporo 060, Japan) V.1–, 1976–

KCRZAE Kauchuk i Rezina. (v/o Mezhdunarodnaya Kniga, Kuznetskii Most 18, Moscow G-200, USSR) V.11–, 1937–

KHFZAN Khimiko-Farmatsevticheskii Zhurnal. Chemical Pharmaceutical Journal. (v/o Mezhdunarodnaya Kniga, Kuznetskii Most 18, Moscow G-200, Russia) V.1–, 1967–

KODAK* Kodak Company Reports. (343 State St., Rochester, NY 14650)

KRMJAC Kurume Medical Journal. (Kurume Igakkai, c/o Kurume Daigaku Igakubu, 67, Asahi-machi, Kurume, Japan) V.1–, 1954–

KSRNAM Kiso to Rinsho. Clinical Report. (Yubunsha Co., Ltd., 1–5, Kanda Suda-Cho, Chiyoda-ku, KS Bldg., Tokyo 101, Japan) V.1–, 1960–

LANCAO Lancet. (7 Adam St., London WC2N 6AD, England) V.1–, 1823–

LIFSAK Life Sciences. (Pergamon Press, Maxwell House, Fairview Park, Elmsford, NY 10523) V.1–8, 1962–69; V.14–, 1974–

LONZA# Personal Communication from LONZA Ltd., CH-4002, Basel, Switzerland, to NIOSH, Cincinnati, OH 45226

MADCAJ Medical Annals of the District of Columbia. (Washington, DC) V.1–43, 1932–74. Discontinued.

MDCHAG Medicinal Chemistry: A Series of Monographs. (Academic Press, 111 Fifth Ave., New York, NY 10003) V.1–, 1963–

MEIEDD Merck Index. (Merck and Co., Inc., Rahway, NJ 07065) 10th ed. 1983– Previous eds. had individual CODENS

MELAAD Medicina del Lavoro. Industrial Medicine. (Via S. Barnaba, 8 Milan, Italy) V.16–, 1925–

MEPAAX Medycyna Pracy. Industrial Medicine. (ARs-Polona-RUSH, POB 1001, 00-068 Warsaw 1, Poland) V.1–, 1950–

MEWEAC Medizinische Welt. (F.K. Schattauer Verlag, Postfach 2945, D-7000 Stuttgart 1, Fed. Rep. Ger.) V.1–18, 1927–1944; V.1–, 1950–

MEXPAG Medicina Experimentalis. (Basel, Switzerland) V.1–11, 1959–64; V.18–19, 1968–69. For publisher information, see JNMDBO

MIKBA5 Mikrobiologiya. (v/o Mezhdunarodnaya Kniga, Kuznetskii Most 18, Moscow G-200, USSR) V.1–, 1932–

MIVRA6 Microvascular Research. (Academic Press, 111 Fifth Ave., New York, NY 10003) V.1–, 1968–

MLDCAS Medecine legale et Dommage corporel. (Paris, France) V.1–7, 1968–74. Discontinued.

MOPMA3 Molecular Pharmacology. (The American Society for Pharmacology and Experimental Therapeutics, 9650 Rockville Pike, Bethesda, MD 20014) V.1–, 1965–

MPHEAE Medicina et Pharmacologia Experimentalis. (Basel, Switzerland) V.12–17, 1965–67. For publisher information, see PHMGBN

MUREAV Mutation Research. (Elsevier/North Holland Biomedical Press, P.O. Box 211, 1000 AE Amsterdam, Netherlands) V.1–, 1964–

MUTAEX Mutagenesis. (IRL Press Ltd. 1911 Jefferson Davis Highway, Suite 907, Arlington, VA 22202) V.1–, 1986–

NATUAS Nature. (Macmillan Journals Ltd., Brunel Rd., Basingstoke RG21 2XS, UK) V.1–, 1869–

NATWAY Naturwissenschaften. (Springer-Verlag, Heidelberger Platz 3, D-1000 Berlin 33, Federal Republic of Germany) V.1–, 1913–

NCITR* National Cancer Institute Carcinogenesis Technical Report Series. (Bethesda, MD 20014) No. 0–205. For publisher information, see NTPTR*

NCIUS* Progress Report for Contract NO. PH-43-64-886, Submitted to the National Cancer Institute by the Institute of Chemical Biology, University of San Francisco. (San Francisco, CA 94117)

NCNSA6 National Academy of Sciences, National Research Council, Chemical-Biological Coordination Center, Review. (Washington, DC)

NCPBBY National Clearinghouse for Poison Control Centers, Bulletin. U.S. Department of Health, Education, and Welfare (Washington, DC)

NDRC** National Defense Research Committee, Office of Scientific Research and Development, Progress Report.

NETOD7 Neurobehavioral Toxicology. (ANKHO International, Inc., P.O. Box 426, Fayetteville, NY 13066) V.1–2, 1979–80, For publisher information, See NTOTDY

NEZAAQ Nippon Eiseigaku Zasshi. Japanese Journal of Hygiene. (Nippon Eisei Gakkai, c/o Kyoto Daigaku Igakubu Koshu Eiseigaku Kyoshita, Yoshida Konoe-cho, Sakyo-ku, Kyoto, Japan) V.1–, 1946–

NIHBAZ National Institutes of Health, Bulletin. (Bethesda, MD)

NNGADV Nippon Noyaku Gakkaishi. (Pesticide Science Society of Japan, 43–11, 1-Chome, Komagome, Toshima-ku, Tokyo 170, Japan) V.1–, 1976–

NPIRI* Raw Material Data Handbook, V.1 Organic Solvents, 1974. (National Association of Printing Ink Research Institute, Francis McDonald Sinclair Memorial Laboratory, Lehigh University, Bethlehem, PA 18015)

NTIS** National Technical Information Service. (Springfield, VA 22161) (formerly U. S. Clearinghouse for Scientific and Technical Information)

NTPTR* National Toxicology Program Technical Report Series. (Research Triangle Park, NC 27709) No.206–

NUCADQ Nutrition and Cancer. (Franklin Institute Press, POB 2266, Phildelphia, PA 19103) V.1–, 1978–

NURIBL Nutrition Reports International. (Geron-X, Inc., POB 1108. Los Altos, CA 94022) V.1–, 1970–

NYKZAU Nippon Yakurigaku Zasshi. Japanese Journal of Pharmacology. (Nippon Yakuri Gakkai, 2-4-16, Yayoi, Bunkyo-Ku, Tokyo 113, Japan) V.40–, 1944–

NZMJAX New Zealand Medical Journal. (Otago Daily Times and Witness Newspapers, P.O. Box 181, Dunedin C1, New Zealand) V.1–, 1900–

OIGZDE Osaka-shi Igakkai Zasshi. Journal of Osaka City Medical Association. (Osaka-shi Igakkai, c/o Osaka-shiritsu Daigaku Igakubu, 1–4–54 Asahi-cho, Abeno-ku, Osaka, 545, Japan) V.24–, 1975–

ONCOBS Oncology. (S. Karger AG, Postfach CH-4009 Basel, Switzerland) V.21–, 1967–

OYYAA2 Oyo Yakuri. Pharmacometrics. (Oyo Yakuri Kenkyukai, Tohoku Daigaku, Kitayobancho, Sendai 980, Japan) V.1–, 1967–

PAACA3 Proceedings of the American Association for Cancer Research. (Waverly Press, 428 E. Preston St., Baltimore, MD 21202) V.1–, 1954–

PABIAQ Pathologie et Biologie. (Paris, France) V.1–16, 1953–68. For publisher information, see PTBIAN

PAREAQ Pharmacological Reviews. (Williams and Wilkins, 428 E. Preston St., Baltimore, MD 21202) V.1–, 1949–

PATHAB Pathologica. (Via Alessandro Volta, 8 Casella Postale 894, 16128 Genoa, Italy) V.1–, 1908–

PBPHAW Progress in Biochemical Pharmacology. (S. Karger AG, Postfach CH-4009 Basel, Switzerland) V.1–, 1965–

PCBPBS Pesticide Biochemistry and Physiology. (Academic Press, 111 Fifth Ave., New York, NY 10003) V.1–, 1971–

PCJOAU Pharmaceutical Chemistry Journal. English Translation of KHFZAN. (Plenum Publishing Corp., 233 Spring St., New York, NY 10013) No.1– 1967–

PCOC** Pesticide Chemicals Official Compendium, Association of the American Pesticide Control Officials, Inc. (Topeka, Kansas, 1966)

PEMNDP Pesticide Manual. (British Crop Protection Council, 20 Bridgport Rd., Thornton Heath CR4 7QG, UK) V.1–, 1968–

PESTC* Pesticide and Toxic Chemical News. (Food Chemical News, Inc., 400 Wyatt Bldg., 777 14th St., N.W. Washington, DC 20005) V.1–, 1972–

PEXTAR Progress in Experimental Tumor Research. (S. Karger AG, Postfach CH–4009 Basel, Switzerland) V.1–, 1960–

PGPKA8 Problemy Gematologii i Perelivaniia Krovi. Problems of Hematology and Blood Transfusion. (v/o Mezhdunarodnaya Kniga, Kuznetskii Most 18, Moscow G–200, Russia) V.1–, 1956–

PHARAT Pharmazie. (VEB Verlag Volk und Gesundheit, Neue Gruenstr 18, 102 Berlin, Germany) V.1–, 1946–

PHBUA9 Pharmaceutical Bulletin. (Tokyo, Japan) V.1–5, 1953–57. For publisher information, see CPBTAL

PHJOAV Pharmaceutical Journal. (Pharmaceutical Press, 17 Bloomsbury Sq., London WC1A 2NN, England) V.131–, 1933–

PHMCAA Pharmacologist. (American Society for Pharmacology and Experimental Therapeutics, 9650 Rockville Pike, Bethesda, MD 20014) V.1–, 1959–

PHMGBN Pharmacology: International Journal of Experimental and Clinical Pharmacology. (S. Karger AG, Postfach CH-4009 Basel, Switzerland) V.1–, 1968–

PHTXA6 Pharmacology and Toxicology. Translation of FATOAO. (New York, NY) V.20–22, 1957–59. Discontinued.

PJACAW Proceedings of the Japan Academy. (Tokyo, Japan) V.21–53, 1945–77. For publisher information, see PJABDW

PJPPAA Polish Journal of Pharmacology and Pharmacy. (ARS-Polona-Rush, POB 1001, 00–068 Warsaw 1, Poland) V.25–, 1973–

PLRCAT Pharmacological Research Communications. (Academic Press, 111 Fifth Ave., New York, NY 10003) V.1–, 1969–

PMDCAY Progress in Medical Chemistry. (American Elsevier Publishing Co., 52 Vanderbilt Ave., New York, NY 10017) V.1–, 1961–

PMRSDJ Progress in Mutation Research. (Elsevier North Holland, Inc., 52 Vanderbilt Ave., New York, NY 10017) V.1–, 1981–

PNASA6 Proceedings of the National Academy of Sciences of the United States of America. (The Academy, Printing and Publishing Office, 2101 Constitution Ave., Washington, DC 20418) V.1–, 1915–

PRKHDK Problemi na Khigienata. Problems in Hygiene. (Durzhavno Izdatel'stvo Meditsina i Zizkultura, Pl. Slaveikov 11, Sofia, Bulgaria) V.1–, 1975–

PROTA* "Problemes de Toxicologie alimentaire," Truhaut, R., Paris, France, L'Evolution pharmaceutique, (1955?)

PSEBAA Proceedings of the Society for Experimental Biology and Medicine. (Academic Press, 111 Fifth Ave., New York, NY 10003) V.1–, 1903/04–

QJMEA7 Quarterly Journal of Medicine. (Oxford University Press, Press Road, Neasden, London NW 10 0DD, England) V.1–, 1932–

QJPPAL Quarterly Journal of Pharmacy and Pharmacology. (London, England) V.2–21, 1929–48. For publisher information, see JPPMAB

RADLAX Radiology. (Radiological Society of North America, 20th and Northampton Sts., Easton, PA 18042) V.1–, 1923–

RAREAE Radiation Research. (Academic Press, 111 Fifth Ave., New York, NY 10003) V.1–, 1954–

RBPMAZ Revue Belge de Pathologie et de Medecine Experimentale. (Brussels, Belgium) V.18–31, 1947–65. For publisher information, see PTEUA6

RCOCB8 Research Communications in Chemical Pathology and Pharmacology. (PJD Publications, P.O. Box 966, Westbury, NY 11590) V.1–, 1970–

RCRVAB Russian Chemical Reviews. (Chemical Society, Publications Sales Office, Burlington House, London W1V 0BN, England) V.29–, 1960–

RCTEA4 Rubber Chemistry and Technology. (Div. of Rubber Chemistry, American Chemical Society, University of Akron, Akron, OH 44325) V.1–, 1928–

REPMBN Revue d'Epidemiologie, Medecine sociale et Sante publique. (Masson et Cie, Editeurs, 120 Blvd. Saint-Germain, P-75280, Paris 06, France) V.1–, 1953–

RESJAS Journal of the Reticuloendothelial Society. (1964) (Res: Journal of the Reticuloendothelial Society, New York) V.1(1)–14(6), 1964–73.

RMCHAW Revista Medica de Chile. (Sociedad Medica de Santiago, Esmeralda 678, Casilla 23-d, Santiago, Chile) V.1–, 1872–

RMNIBN Revista Medico-Chirurgicala. (Societatea de Medici si Naturalisti, Bulevardul Independentei, Iasi, Romania) V.35–, 1924–

RPTOAN Russian Pharmacology and Toxicology. Translation of FATOAO. (Euromed Publications, 97 Moore Park Rd., London SW6 2DA, England) V.30–, 1967–

SAIGBL Sangyo Igaku. Japanese Journal of Industrial Health. (Japan Association of Industrial Health, c/o Public Health Building, 78 Shinjuku 1–29-8, Shinjuku-Ku, Tokyo, Japan) V.1–, 1959–

SAMJAF South African Medical Journal. (Medical Association of South Africa, Secy., P.O. Box 643, Cape Town, S. Africa) V.6–, 1932–

SAPHAO Skandinavisches Archiv fuer Physiologic. (Karolinska Institutet,

Editorial Office, Stockholm, Sweden) V.1–83, 1899–1940. Superseded by APSCAX

SCCUR* Shell Chemical Company. Unpublished Report. (2401 Crow Canyon Rd., San Ramon, CA 94583)

SchF## Personal communication from F.W. Schaller, Inco Ltd., Park 80 West Plaza Two, Saddle Brook, NJ 07662, May 16, 1986

SCIEAS Science. (American Assoc. for the Advancement of Science, 1515 Massachusetts Ave., NW, Washington, DC 20005) V.1–, 1895–

SEIJBO Senten Ijo. Congenital Anomalies. (Nihon Senten Ijo Gakkai, Kyoto 606, Japan) V.1–, 1960–

SIGAAE Shika Igaku. Odontology. (Osaka Shika Gakkai, 1 Kyobashi, Higashi-Ku, Osaka, Japan) V.1–, 1930–

SIZSAR Sapporo Igaku Zasshi. Sapporo Medical Journal. (Sapporo Igaku Daigaku, Nishi-17-Chome, Minami-1-jo, Chuo-ku Sapporo 060, Japan) V.3–, 1952–

SKEZAP Shokuhin Eiseigaku Zasshi. Journal of the Food Hygiene Society of Japan. (Nippon Shokuhin Eisei Gakkai, c/o Kokuritsu Eisei Shikenjo, 18–1, Kamiyoga 1-chome, Setagaya-Ku, Tokyo, Japan) V.1–, 1960–

SMBUA9 Stanford Medical Bulletin. (Palo Alto, CA) V.1–20, 1942–62. Discontinued.

SOGEBZ Soviet Genetics. Translation of GNKAA5. (Plenum Publishing Corp., 233 Spring St., New York, NY 10013). V.2–, 1966–

SOVEA7 Southwestern Veterinarian. (College of Veterinary Medicine, Texas A and M University, College Station, TX 77843) V.1–, 1948–

SPEADM Special Publication of the Entomological Society of America. (4603 Calvert Rd., College Park, MD 20740) (Note that although the DPIM cites data from the 1974 edition, this reference has been superseded by the 1978 edition.)

STEVA8 Science of the Total Environment. (Elsevier Scientific Publishing Co., P.O. Box 211, Amsterdam C, Netherlands) V.1–, 1972–

SWEHDO Scandinavian Journal of Work, Environment and Health. (Haartmaninkatu 1, FIN-00290 Helsinki 29, Finland) V.1–, 1975–

TABIA2 Tabulae Biologicae. (The Hague, Netherlands) V.1–22, 1925–63. Discontinued.

TCMUD8 Teratogenesis, Carcinogenesis, and Mutagenesis. (Alan R. Liss, Inc., 150 Fifth Ave., New York, NY 10011) V.1–, 1980–

TECSDY Toxicological and Environmental Chemistry. (Gordon and Breach Science Pub. Inc., 1 Park Ave., New York, NY 10016) V.3(3/4)–, 1981–

TGNCDL "Handbook of Organic Industrial Solvents" 2nd ed., Chicago, IL, National Association of Mutual Casualty Companies, 1961

THERAP Therapie. (Doin, Editeurs, 8 Place de l'Odeon, Paris 6, France) V.1–, 1946–

TJADAB Teratology, A Journal of Abnormal Development. (Wistar Institute Press, 3631 Spruce St., Philadelphia, PA 19104) V.1–, 1968–

TOANDB Toxicology Annual. (Marcel Dekker, POB 11305, Church St. Station, New York, NY 10249) 1974/75–

TOFOD5 Tokishikoroji Foramu. Toxicology Forum. (Saiensu Foramu, c/o Kida Bldg., 1–2–13 Yushima, Bunkyo-ku, Tokyo, 113, Japan) V.6–, 1983–

TOIZAG Toho Igakkai Zasshi. Journal of Medical Society of Toho University. (Toho Daigaku Igakubu Igakkai, 5–21–16, Omori, Otasku, Tokyo, Japan) V.1–, 1954–

TOLED5 Toxicology Letters. (Elsevier Scientific Publishing Co., P.O. Box 211, Amsterdam, Netherlands) V.1–, 1977–

TOXIA6 Toxicon. (Pergamon Press, Headington Hill Hall, Oxford OX3 OBW, England) V.1–, 1962–

TOXID9 Toxicologist. (Society of Toxicology, Inc., 475 Wolf Ledge Parkway, Akron, OH 44311) V.1–, 1981–

TPKVAL Toksikologiya Novykh Promyshlennykh Khimicheskikh Veshchestv. Toxicology of New Industrial Chemical Sciences. (Akademiya Meditsinskikh Nauk S.S.R., Moscow, Russia) No.1– 1961–

TRENAF Kenkyu Nenpo - Tokyo-toritsu Eisei Kenkyusho. Annual Report of Tokyo Metropolitan Research Laboratory of Public Health. (24–1, 3 Chome, Hyakunin-cho, Shin-Juku-Ku, Tokyo, Japan) V.1–, 1949/50–

TSCAT* Office of Toxic Substances Report. (U. S. Environmental Protection Agency, Office of Toxics Substances, 401 M Street SW, Washington, DC 20460)

TSPMA6 Travaux de la Societe de Pharmacie de Montpellier. (Soc. de Pharmacie de Montpellier, Faculte de Pharmacie de Montpellier, Ave. Ch.-Flahault, 34060 Montpellier, France) V.1–, 1942–

TXAPA9 Toxicology and Applied Pharmacology. (Academic Press, 111 Fifth Ave., New York, NY 10003) V.1–, 1959–

TXCYAC Toxicology. (Elsevier/North-Holland Scientific Publishers Ltd., 52 Vanderbilt Ave., New York, NY 10017) V.1–, 1973–

TXMDAX Texas Medicine. (Texas Medical Assoc., 1905 N. Lamar Blvd., Austin, TX 78705) V.60–, 1964–

UCDS** Union Carbide Data Sheet. (Industrial Medicine and Toxicology Dept., Union Carbide Corp., 270 Park Ave., New York, NY 10017)

VDGPAN Verhandlungen der Deutschen Gesellschaft fuer Pathologie. (Gustav Fischer Verlag, Postfach 53, Wollgrasweg 49, 7000 Stuttgart-Hohenheim, Germany) V.1–, 1898–

VHTODE Veterinary and Human Toxicology. (American College of Veterinary Toxicologists, Office of the Secretary-Treasurer, Comparative Toxicology Laboratory, Kansas State University, Manhattan, Kansas 66506) V.19–, 1977–

VINIT* Vsesoyuznyi Institut Nauchnoi i Tekhnicheskoi Informatsii (VINITI). All-Union Institute of Scientific and Technical Information. (Moscow, USSR)

WEHRBJ Work, Environment, Health. (Helsinki, Finland) V.1–11, 1962–74. For publisher information, see SWEHDO

WRPCA2 World Review of Pest Control. (London, England) V.1–10, 1962–71. Discontinued.

XENOBH Xenobiotica. (Taylor and Francis Ltd., 4 John St., London WC1N 2ET, England) V.1–, 1971–

XEURAQ U. S. Atomic Energy Commission, University of Rochester, Research and Development Reports. (Rochester, NY)

YAKUD5 Gekkan Yakuji. Pharmaceuticals Monthly. (Yakugyo Jihosha, Inaoka Bldg., 2-36 Jinbo-cho, Kandu, Chiyoda-ku, Tokyo 101, Japan) V.1–, 1959–

YHHPAL Yaoxue Xuebao. Acta Pharmaceutica Sinica. Pharmaceutical Journal. (China International Book Trading Corp., POB 2820, Beijing, Peop. Rep. China) V.1–, 1953– (Suspended 1966–78)

YKKZAJ Yakugaku Zasshi. Journal of Pharmacy. (Nippon Yakugakkai, 12–15–501, Shibuya 2-chome, Shibuya-ku, Tokyo 150, Japan) No.1–, 1881–

YKYUA6 Yakkyoku. Pharmacy. (Nanzando, 4–1–11, Yushima, Bunkyo-ku, Tokyo, Japan) V.1– 1950–

ZAARAM Zentralblatt fuer Arbeitsmedizin und Arbeitsschutz. (Dr. Dietrich Steinkopff Verlag, Saalbaustr 12,6100 Darmstadt, Germany) V.1–, 1951–

ZAPOAK Zeitschrift fuer Allgemeine Mikrobiologie. Morphologie, Physiologie, Genetik und Oekologie der Mikroorganismen. (Akademie-Verlag GmbH, Liepziger Str. 3–4, DDR-108 Berlin, German Democratic Republic) V.1–, 1960–

ZDKAA8 Zdravookhranenie Kazakhstana. Public Health of Kazakhstan. (v/o Mezhdunarodnaya Kniga, Kuznetskii Most 18, Moscow G–200, Russia) V.1–, 1941–

ZEKBAI Zeitschrift fuer Krebsforschung. (Berlin, Germany) V.1–75, 1903–71. For publisher information, see JCROD7

ZGEMAZ Zeitschrift fuer die Gesamte Experimentelle Medizin. (Berlin, Germany) V.1–139, 1913–65. For publisher information, see REXMAS

ZHINAV Zeitschrift fuer Hygiene und Infektionskrankheiten. (Berlin, Germany) V.11–151, 1892–1965. For publisher information, see MMIYAO

ZHYGAM Zeitschrift fuer die Gesamte Hygiene und Ihre Grenzgebiete. (VEB Georg Thieme, Hainst 17/19, Postfach 946, 701 Leipzig, Germany) V.1–, 1955–

ZKMAAX Zhurnal Eksperimental'noi i Klinicheskoi Meditsiny. (v/o Mezhdunarodnaya Kniga, 121200 Moscow, USSR) V.2–, 1962–

ZNTFA2 Zeitschrift fuer Naturforschung. (Wiesbaden, Federal Republic of Germany) V.1, No.1–12 1946. For publisher information, See ZENBAX

14CYAT "Industrial Hygiene and Toxicology, 2nd rev. ed.," F. A. Patty, ed., (New York, Interscience Publishers), 1958–63

14KTAK "Boron, Metallo-Boron Compounds and Boranes" R.M. Adams, (New York, Wiley), 1964

26UZAB "Pesticides Symposia" collection of papers presented at the Sixth and Seventh Inter-American Conferences on Toxicology and Occupational Medicine, Miami, Florida, Univ. of Miami, School of Medicine, 1968–70

27ZWAY "Heffter's Handbuch der Experimentelle Pharmakologie."

28ZEAL "Pesticide Index"E.H. Frear, ed., (State College, PA, College Science Publications, 1969) (Note that although the DPIM cites data from the 1969 edition, this reference has been superseded by the 5th edition, 1976.)

28ZPAK "Sbornik Vysledku Toxixologickeho Vysetreni Latek A Pripravku" J.V. Marhold, (Institut Pro Vychovu Vedoucicn Pracovniku Chemickeho Prumyclu Praha, Czech Republic), 1972

28ZRAQ "Toxicology and Biochemistry of Aromatic Hydrocarbons" H. Gerarde, (New York, Elsevier), 1960

29ZWAE "Practical Toxicology of Plastics" R. Lefaux, (Cleveland, Ohio, Chemical Rubber Company), 1968

34ZIAG "Toxicology of Drugs and Chemicals" W.B. Deichmann, (New York, Academic Press), 1969

35WYAM "In Vitro Metabolic Activation in Mutagenic Testing" Proceedings of the Symposium on the Role of Metabolic Activation in Producing Mutagenic and Carcinogenic Environmental Chemicals, Research Triangle Park, N.C., Feb. 9–11, 1976, F.J. De Serres et al., eds., (New York, Elsevier North Holland), 1976

38MKAJ "Patty's Industrial Hygiene and Toxicology" 3rd rev. ed., G.D. Clayton, and F.E. Clayton, eds., (New York, John Wiley & Sons, Inc.), 1978–82. Vol. 3 originally pub. in 1979; pub. as 2nd rev. ed. in 1985.

41HTAH "Aktual'nye Problemy Gigieny Truda. Current Problems of Labor Hygiene" N.Y. Tarasenko, ed., (Moscow, Pervyi Moskovskii Meditsinskii Inst.), 1978

48RKAL "Industrial and Environmental Xenobiotics: Metabolism and Pharmacokinetics of Organic Chemicals and Metals, Proceedings of an International Conference," (Prague, 1980, Gut, I., et al., (Berlin, Germany, Springer-Verlag), 1981

50EXAK "Formaldehyde Toxicity" Conference, 1980, J.E. Gibson, ed., (Washington, DC, Hemisphere Publishing Corp.), 1983

50NNAZ "Polynuclear Aromatic Hydrocarbons: Mechanisms, Methods and Metabolism, Papers of the 8th International Symposium, Columbus, OH, 1983," M. Cooke, and A.J. Dennis, eds., (Columbus, OH, Battelle Press), 1985

54DEAI "Occupational Health in the Chemical Industry, Proceedings of the International Congress, 11th, Alberta, Canada, 1983" R. R. Oxford, et al, eds., (Calgary, Alberta, Canada, Univ. of Calgary), 1984.

85ARAE "Agricultural Chemicals, Books I, II, III, and IV" W.T. Thomson, (Fresno, CA, Thomson Publications), 1976/77 revision

85ARAE "Agricultural Chemicals, Books I, II, III, and IV" W.T. Thomson, (Fresno, CA, Thomson Publications), 1976/77 revision

85CYAB "Chemistry of Industrial Toxicology" H.B. Elkins, 2nd Ed., (New York, J. Wiley), 1959

85DCAI "Poisoning; Toxicology, Symptoms, Treatments" J.M. Arena, 2nd Ed., (Springfield, Illinois, C. C. Thomas), 1970

85DJA5 "Malformations Congenitales des Mammiferes" H. Tuchmann-Duplessis, (Paris, Masson et Cie), 1971

85DKA8 "Cutaneous Toxicity"V.A. Drill and P. Lazar, eds., (New York, Academic Press), 1977

85DPAN "Wirksubstanzen der Pflanzenschutz und Schadlingsbekampfungsmittel" Werner Perkow, (Berlin, Verlag Paul Parey), 1971–1976

85ECAN "Metal Toxicity in Mammals, Vol. 2: Chemical Toxicity of Metals and Metalloids," B. Venugopal and T.D. Luckey, (New York, Plenum Press), 1978

85ERAY "Antibiotics: Origin, Nature, and Properties" T. Korzyoski, Z. Kowszyk-Gindifer, and W. Kurylowicz, eds., (Washington DC, American Society for Microbiology), 1978

85ESA3 "Merck Index; an Encyclopedia of Chemicals, Drugs, and Biologicals", 11th ed., (Rahway, NJ 07065, Merck & Co., Inc.) 1989

85GDA2 "CRC Handbook of Antibiotic Compounds, Volumes 1–9" Berdy, Janos. (Boca Raton, Florida, CRC Press), 1980

85GMAT "Toxicometric Parameters of Industrial Toxic Chemicals Under Single Exposure" N.F. Izmerov, et al. (Moscow, Centre of International Projects, GKNT), 1982

85INA8 "Documentation of the Threshold Limit Values and Biological Exposure Indices" 5th ed., (Cincinnati, Ohio, American Conference of Governmental Industrial Hygienists, Inc.), 1986

85IXA4 "Structure et Activite Pharmacodyanmique des Medicaments du Systeme Nerveux Vegetatif" D. Bovet, and F. Bovet-Nitti, (New York, S. Karger), 1948

85JCAE "Prehled Prumyslove Toxikologie; Organicke Latky" Marhold, J., (Prague, Czech Republic, Avicenum), 1986

A Guide to Using This Book

SAX Number – Entries are indexed in order by this alphanumeric code.
See Introduction: paragraph 1, p. xix

CAS: – The Americal Chemical Society's Chemical Abstracts Service number. A complete CAS number cross-index is located in Section 4, Vol. I.
See Introduction: paragraph 4, p. xix

Entry Name – A complete entry name and synonym cross-index is located in Section 5, Vol. I.
See Introduction: paragraph 2, p. xix

DOT: – The four digit hazard code assigned by the U.S. DOT.
See Introduction: paragraph 6, p. xix

mf: – the molecular formula
mw: – the molecular weight
See Introduction: paragraphs 7 and 8, p. xx

PROP: – Physical properties including solubility and flammability data. May contain a definition of the entry.
See Introduction: paragraph 10, p. xx

SYNS: – Synonyms for the entry. A complete synonym cross-index appears in Section 5, Vol. I.
See Introduction: paragraph 11, p. xx

TOXICITY DATA: – Data for skin and eye irritation, mutation, teratogenic, reproductive, carcinogenic, human, and acute lethal effects.
See Introduction: paragraphs 12-16, pp. xx–xxix

Toxic and Hazard Reviews – These are text summaries of the toxicity, fire, reactivity, incompatibilities, and other dangerous properties of the material.
See Introduction: paragraph 20, p. xxxi